MTI and Pulsed Doppler Radar with MATLAB®

Second Edition

For a listing of recent titles in the *Artech House Radar Library*,
turn to the back of this book.

MTI and Pulsed Doppler Radar with MATLAB®

Second Edition

D. Curtis Schleher

ARTECH HOUSE
BOSTON | LONDON
artechhouse.com

Library of Congress Cataloging-in-Publication Data
A catalog record for this book is available from the U.S. Library of Congress.

British Library Cataloguing in Publication Data
A catalogue record for this book is available from the British Library.

ISBN-13: 978-1-59693-414-6

Cover design by Vicki Kane

MTI and Pulsed Doppler Radar, First Edition, was originally published by Artech House, ©1991 D. Curt Schleher

685 Canton Street
Norwood, MA 02062

10 9 8 7 6 5 4 3 2 1

Contents

Preface

The ability of radar to detect targets in a noise background is well documented. However, most real-world radars must also contend with competing extraneous returns (called clutter) which typically result from ground, sea, rain, and other hydrometers, chaff, birds, insects, and aurora reflections. In most applications the returns from clutter exceed those from real targets by several orders of magnitude. The only practical way to extract real targets from clutter is by using *MTI and Pulsed Doppler Radars,* which exploit the differential Doppler shifts between the real targets and the clutter. The ability of MTI and PD radars to detect moving targets provides a critical function for the operation of most modern radars. However, losses experienced by Doppler processing radars can significantly exceed those of noise-limited radars, resulting in reduced detection ranges than those that might be predicted for noise-limited only radar.

MTI and Pulsed Doppler Radar with MATLAB® provides an in-depth treatment of both theoretical and practical considerations involved with the system design and performance of such radars. The book emphasizes a basic understanding of radar clutter properties and how MTI or PD radar functions extract moving radar targets from a clutter background. A detailed theoretical basis is established for the optimum detection of moving radar targets in clutter which is related to MTI and PD implementations. Detailed derivations are given for the various theoretical relationships used in these systems, followed by practical examples to illustrate their use.

The characteristics of a number of modern Doppler radar systems are described throughout the book. These applications include civil and military air surveillance radars, AEW and AI radars, missile air defense radar systems, SAR surveillance systems, and GMTI and ISAR systems. Performance estimates are provided for these systems, which include a MATLAB design tool for surveillance of airborne targets. Also discussed are major advancements to the technology used in these systems such as MIMIC solid state transmitters, phased array antennas, airborne electronically steered array antennas (AESA), Interferometric GMTI, Space Time Adaptive Processing (STAP), Hot Clutter, Interferometric SAR, transversal filters, Doppler beam sharpening, ISAR, and MTI combined with circular polarized antennas. In addition, a major thrust of this book is the inclusion of MATLAB software to allow the design and analysis of these techniques.

MATLAB software is provided to allow readers to both design and analyze MTI and PD radar systems without going through the tedious process of trying to apply the 730 detailed technical equations in the book. There are over 100 MATLAB programs in the book whose application is illustrated in 32 design examples interspersed throughout the book. Many of the programs exploit the excellent GUI capability of MATLAB by using dialog boxes followed by menu bars that provide high resolution graphics. With this technique the MATLAB program code is transparent to the user who overwrites the default values to suit his or her application and selects the environment through the menu bars. Of course, users who are skilled in MATLAB programming can modify the MATLAB code using the excellent MATLAB Editor.

MATLAB, which derives its name from *Matrix Laboratory,* is ideally suited to complement the matrix and vector approach utilized throughout the book to analyze the processing performed in Doppler radars. This approach reflects my 15-year experience in teaching a course to graduate students on this subject at the Naval Postgraduate School using the original text upon which this volume is based. One of the pedagogical techniques employed in the course was to require students to generate MATLAB programs that provided graphical outputs of specific results required in the design of Doppler radars. This resulted in a deep understanding of the underlying theory and also taught the students how to program equations that are written in standard publishing book format into MATLAB code.

This book serves as a sequel to *MTI and Pulse Doppler Radar,* published in 1991, which at that time brought together much of the significant material on this subject. The present book expands on much of the material in *MTI and Pulse Doppler Radar,* gives detailed derivations of the theoretical results and examples, and adds extensive MATLAB examples on MTI and PD radars. In addition, a chapter is provided to cover specialized applications of pulsed Doppler radars, including such diverse topics as Interferometric GMTI and SAR, STAP, Doppler beam sharpening, inverse synthetic aperture radars (ISAR), and an MTI design tool.

Chapter 1 covers the fundamental principles of Doppler processing and gives an overview of significant results relating to MTI and PD radars. The characteristics of a number of modern Doppler radars are described and their design criteria and performance compared. The overall radar detection problem in clutter and a receiver noise background is discussed from the viewpoint of a radar resolution cell. Optimal MTI and PD radar solutions are described. The properties of various clutters are described through the use of the environmental diagram. This leads to a discussion of both Gaussian and non-Gaussian clutter models, which are useful for Doppler radar analysis. The chapter ends with an in-depth description of the various moving target detector (MTD) configurations, which is one of the more powerful implementations of modern Doppler processing in air surveillance radars. The MTD is an example of a low-PRF (pulse repetition frequency) PD radar processor, which provides almost optimal detection performance against moving targets while incorporating various false alarm control mechanisms such as Doppler filter CFAR (constant false alarm rate) and digital clutter maps.

Chapter 2 describes waveform design and Doppler filtering principles as they apply to MTI and PD radars. In general, the waveform for a PD radar can be classified as a low-PRF (unambiguous range), high-PRF (unambiguous Doppler), or medium-PRF (ambiguous in both range and Doppler) design. Each type of waveform interacts with the clutter in a different way and requires a different form of signal processing. The choice of the waveform is strongly influenced by the application and performance requirements. Doppler filter requirements are generally dependent on the type of waveform utilized. The transversal type of Doppler filter is optimum, but also hardware-intensive. Comb-type filters (i.e., delay line cancellers) are used in MTI applications or in front of Doppler filter banks as a clutter notch filter. Contiguous banks of Doppler filters are often implemented digitally using the FFT algorithm. Modern radars use digital signal processors, which are discussed in detail at the end of that chapter.

Chapter 3 discusses the detection performance that can be achieved by using PD and MTI radars. Performance is related to the usual probability of detection (Pd) and probability of false alarm (Pfa) criteria generally used in radar analysis. These are related to the MTI improvement factor, which establishes a standard measure of performance for Doppler radars. The improvement factor is related to two distinct measures. The first is the action of an ideal Doppler processor on the clutter spectrum. The second is due to system instabilities such as transmitter and stable local oscillator (STALO) noise. In general, system instabilities form a performance floor, which is achieved if appropriate radar and Doppler filter parameters are designed into the system.

Chapter 4 discusses the subject of radar clutter. We describe the average back-scattering coefficient and spectra for land, sea, weather, and chaff clutter, as determined from various experimental programs. Parametric approximations are given for the data. Various clutter models such as the Gaussian, Rician, log-normal, Weibull, and K-distribution are described in detail. Several environmental models that can be used for computer simulations of clutter are also described. The performance of an MTI processor preceded by a circular polarized antenna is developed.

Chapter 5 discusses the theoretical bases of MTI and pulsed Doppler radars. This treatment starts from the general problem of detecting n Doppler-shifted coherent signals with random initial phase in a background of Gaussian clutter and receiver noise. The resulting linear transversal filter with complex weights followed by an envelope detector is shown to be the optimal procesor. The MTI solution is shown to apply when all target velocities are equally likely, while the PD solution applies when the target Doppler signature is known a priori. The performance of the optimal processor is then compared with that of a quasioptimal processor, which consists of an MTI followed by a coherent integrator (e.g., fast Fourier transform, FFT), and further with an MTI followed by an incoherent integrator.

Chapter 6 treats various aspects of MTI system design. We describe different types of coherent and noncoherent MTIs. Performance drivers for surface-based and airborne types of systems are quantified. We include a section on the MTI performance limitations of a nonlinear receiver, using receiver distortion theory. Recursive and nonrecursive MTI filter design theory is given with design examples. The generally ambiguous velocity response of MTI systems results in blind speeds about the radar's PRF, which can be eliminated by staggering the PRF. Staggering the radar's PRF, in turn, results in a modified MTI velocity response, which is a function of the stagger pattern and the order of the canceller employed. Design curves are provided that show that high detection losses result when a staggered PRF is not used.

Chapter 7 discusses various aspects of pulsed Doppler system design. General PD configurations and waveforms are described with emphasis on performance limitations. The problem of resolving range ambiguities in high-PRF designs through the transmission of multiple PRFs or FM ranging waveforms is discussed in detail. Modern designs of AESA airborne intercept (AI) radars are discussed in detail. The chapter concludes with a discussion of main-lobe and altitude-line clutter performance requirements. This becomes the critical design parameter in many high-PRF PD radars due to the folding of clutter from multiple ranges into a single range cell, which hence competes with the target response.

Chapter 8 treats a number of specialized uses of the Doppler effect in radar systems. The discussions are self-contained and provide sufficient detail to allow a clear understanding of the principles involved. The topics covered are Doppler beam sharpening, synthetic aperture radar (SAR), Interferometric SAR, space time adaptive processing (STAP), hot clutter, ISAR, and an interactive design tool using MATLAB for air surveillance radars.

The enclosed CD-ROM contains the MATLAB programs, color figures, and a short tutorial on MATLAB, including instructions on how to use the disc.

As is appropriate, the author acknowledges a number of people who provided indispensable help in preparing the manuscript for publication. My wife, Carol, typed the entire manuscript and its many revisions, including the detailed equations that appear throughout the book. David K. Barton constructively reviewed the entire manuscript and suggested many useful revisions, which were incorporated into the text. Also, over the years, the author has received useful comments from the many students who utilized the original text and developed MATLAB code as part of their studies.

D. Curtis Schleher
Camano Island, Washington
2010

CHAPTER 1

Fundamental Principles and Overview

Moving target indication and pulsed Doppler radars provide the capability to detect target reflections having differential radial motion with respect to an interfering background (called clutter) that might typically consist of reflections from terrain, sea, weather, or chaff. MTI and PD radars can be operated from either a fixed based or a moving platform such as a ship, aircraft, or satellite [1–4].

A Doppler radar is defined by the IEEE standard radar definitions as one that uses the Doppler effect to determine the radial component of radar-target velocity or to select targets having particular radial velocities. When a Doppler radar uses pulse transmissions it is called pulsed Doppler (PD) radar. Further, moving target indication (MTI) is defined as a technique that enhances the detection and display of moving radar targets, for which Doppler processing is one method of implementation. These formal definitions point to the central purpose of MTI and PD radars, which is to use the Doppler effect to separate small moving targets from competing extraneous *clutter* returns, which typically result from reflections obtained from the ground, sea, rain and other hydrometers, chaff, birds, insects, and auroras.

The ability of MTI and PD radars to detect moving targets provides a critical function for the operation of most modern radars. A typical civil application is in air surveillance, where small nontransponder-equipped aircraft must be detected over land in heavy ground and weather clutter for air traffic control purposes. Military applications range from the detection of low-flying aircraft and cruise missiles from a surface-based radar to airborne applications such as AWACS, airborne early warning (AEW), and airborne intercept (AI) radars, which involves looking down into a heavy surface clutter environment.

Although similar in nature, there is a distinct theoretical difference between MTI and PD radars. The MTI radar ideally functions to eliminate clutter through the use of a comb filter, the stopband of which is placed in regions of heavy clutter concentrations. Moving targets are passed through those velocity regions not occupied by the clutter. PD radars function to resolve and enhance targets within a particular velocity band while rejecting clutter and other echoes outside the velocity band of interest. PD radars generally use a bank of contiguous Doppler filters matched to the target response through the velocity range of interest. The bank of Doppler filters serves to coherently integrate the target's response with respect to noise and, hence, the PD radar provides the potential for enhanced detection performance as compared to the MTI radar.

1.1 MTI and PD Radar Applications

A typical application of Doppler radars is surveillance to detect aircraft or missiles moving over terrain through possible weather disturbances. The function of the MTI or PD radar in this situation is to reject the returns from terrain and weather while retaining the returns from the aircraft, thereby allowing target detection.

When the radar is surface based, the terrain reflections, called *ground clutter,* usually are stronger than expected target returns. The ground clutter extends out to a range where terrain features that cause clutter are masked due to the earth's curvature. This can be for ranges on the order 10 to 20 nautical miles (nmi) over relatively flat terrain to more than 50 nmi when cities and mountainous terrain are within the radars' field of view, and more than 100 nmi when ducted propagation is present. When the radar is airborne, the ground clutter generally extends to the full operating range of the radar.

The intensity of weather clutter is in proportion to the fourth power of frequency from low frequencies up to the millimeter-wave region. This type of clutter becomes significant at L-band frequencies and above, and it differs from ground and sea clutter in that a volumetric distribution of scatterers is involved, rather than a surface distribution. Weather clutter generally extends to the maximum operating range of the radar.

The ASR-9, depicted in Figure 1.1, is a two-dimensional (2-D) air traffic control radar operating at S-band (2.7–2.9 GHz) that provides a detection range of 60 nmi on a small 5 m^2 aircraft. The array antenna located above the radar antenna provides data for the complementary air traffic control radar beacon system (ATCRBS), which allows the identity and altitude of detected targets to be determined for cooperating targets.

Parameters for the ASR-9 are given in Table 1.1. Most important is that this radar was the first to employ a digital moving target detector (MTD) processor that is described in detail later in this chapter. The primary function of the MTD processor is to virtually eliminate false alarms that had plagued previous versions of this type of radar. The MTI stability factor of 50 dB is primarily determined by the capability of the Klystron transmitter. The major features employed to

Figure 1.1 ASR-9/ASR-12 air traffic control radar. (*From:* [76].)

Table 1.1 Characteristics of Modern Doppler Surveillance Radar

	ASR-9	*ASR-12*	*ARSR-4*	*TPS-59*
Type	2D	2D	3D	3D
Targets	A,W	A,W	A,W	A,TBM
P_t (kW)	1200	22	60	54
P_A (kW)	1.575	1.45	1.94	9.7
f (GHz)	2.7–2.9	2.7–2.9	1.215–1.4	1.215–1.4
PRF (pps)	1,200	1,200	216/72	277–833
PW (μs)	1.05	55,1	150	100–800
G_T (dB)	33.5	33.5	35	38.9
G_R (dB)	33.5/32.5	33.5/32.5	40	38.3
θ_{az} (°)	1.3	1.3	1.41	3.4
ϕ_E (°)	4.8	4.8	2-Stacked	1.7
Polarization	Linear or CP	Linear or CP	Linear or CP	Linear
Scan rate (°/s)	75	60/72/90	30	36/72
Noise Temp (K)	627	627	438	540
Doppler BW (Hz)	150	150	PRF	PRF
Signal Proc.	MTD	MTD	8-pulse MTI	MTI
PC PW (μs)	NA	1.4 NLFM	1.3 NLFM	0.4, 1.6

combat false alarms are: (1) a wide dynamic range quasi-optimum linear Doppler filter, (2) a clutter map that linearly suppresses the clutter level in each range-azimuth resolution cell to a value that can be accommodated by the Doppler filter, (3) adaptive constant false alarm rate (CFAR) thresholding in each Doppler cell of interest to suppress weather clutter returns, (4) censoring techniques that eliminate pulse-type interference that might be received from other radars, and (5) a block pulse repetition frequency (PRF) stagger to eliminate blind speeds and aliasing of targets in Doppler cells suppressed by the weather clutter CFAR [5]. The ASR-9 includes a dedicated processing channel to measure and display precipitation reflectivity. At selected sites it also measures Doppler wind velocity to support detection of low-angle wind shear.

The dual-beam antenna feed shown in Figure 1.1 provides both a low and a high beam. The radar transmits on the lower beam, but receives on both beams. The lower beam is activated when receiving long-range targets and the higher beam is used to detect short-range targets. The net effect is that ground clutter and surface targets are decoupled in the higher receiving beam, thereby enhancing the target-to-clutter ratio by 15 to 20 dB. In addition, low-altitude targets, such as birds and insects, which are problems at short ranges, are also attenuated in the higher beam.

The ASR-12 is an evolution of the ASR-9 that builds on MTD technology while employing a solid-state transmitter using bipolar transistor technology [6]. The solid-state transmitter provides higher reliability and maintainability than its tube counterpart while increasing the potential for a higher MTI stability factor. An individual S-band bipolar transistor is limited to a peak power of about 100 W [7, 8], so that the outputs of a large number of transistors must be combined to produce the 22 kW peak power transmitted pulse.

A long transmitted pulse length of 55 μs provides equivalent pulse energy (1.375J) to that provided by the ASR-9 [9], but its use is not without consequences. Pulse compression (PC) is required to maintain the resolution (700 ft) similar to that of the ASR-9, however, the range sidelobes created by the PC must be minimized to prevent large weather returns from spilling over into adjacent cells necessitating the use of nonlinear FM coding. A 1 μs short pulse is transmitted directly following the long pulse at a different carrier frequency to enable the detection of short-range targets.

In the dual-beam antenna the timing of the transition between the low and high beams is determined on the basis of the clutter situation for each site. The low beam, pointed just above the horizon, uses the long pulse to detect far-out targets, while the high beam uses both the short and long pulses to detect close-in targets.

The ARSR-4, depicted in Figure 1.2, is a solid-state, long-range (250-nmi), three-dimensional (3-D), stacked-beam surveillance radar operating in L-band (1.215 to 1.4 GHz) that supports defense of the national airspace and en-route civil ATC. The parameters of the ARSR-4 are given in Table 1.1 [10]. Requirements for ATC use include detection of a 2.2 m^2 Swerling 1 target over mountainous terrain in 4 mm/h rain at ranges of 5 to 180 nmi. Military requirements include over water detection of a 0.1 m^2 cruise missile target from a 6,500-ft-high site under sea state 5 conditions and detection of a 1 m^2 air target over mountainous terrain at ranges between 5 and 92 nmi [11].

The phased-array feed shown in Figure 1.2 is offset from the reflector to avoid blockage and consists of 600 horizontally and vertically polarized radiators in 23 rows that are mounted on a cylindrical surface. This enables the radar to radiate and receive left-or right-handed circular polarization or linear vertical polarization in each 11.25° azimuth sector depending on weather conditions. The ARSR-4 uses the same sense circular polarization for transmitting and receiving in the target channel to discriminate against weather clutter, which is sufficient in L-band, and a weather reception channel is included that uses receiving polarization opposite to that transmitted [11].

The elevation coverage is divided among nine stacked receiving beams each with about 2° beamwidth that are floodlighted on transmission. At any instant of time, either a stack of five lower beams or five upper beams processes the returns. The stacks overlap so that the fifth beam is common to both stacks. The lower

(a) (b)

Figure 1.2 (a, b) ARSR-4 long-range surveillance radar. (*From:* [77], [75, p. 65].)

beams that process the far-out returns are narrow compared to the upper beams that process the shorter range returns, which provide uniform height accuracy of 3,000 ft throughout the radar's coverage area.

The ARSR-4 employs an 8-pulse MTI canceller with a nine-period PRF stagger to detect targets in surface clutter. The 8-pulse canceller provides a wide rejection spectral notch [11] sufficient to eliminate heavy mountainous clutter whose spectrum is spread by antenna scanning modulation. This corrects a defect of the ARSR-3 that used a 3-pulse MTI to reject mountain clutter, which proved inadequate. The nine-period stagger eliminates blind speeds over the velocity response that extends from 20 to 3,000 knots. When the radar coverage extends over water, it measures the mean sea velocity in each 5.6° by 8 nmi zone and motion corrects the MTI's clutter notch to 0.5 knot accuracy.

The L-band solid-state transmitter provides a peak power of 60 kW that is diplexed in frequency with 83 MHz separation. The total pulsewidth is 150 μs (90 μs + 60 μs), whereas the PRF is 216 Hz in the lower stack and 72 Hz in the high stack to reduce the susceptibility to second-time-around mountain or urban clutter. As with the ASR-12, nonlinear-FM coding is used in the PC system to prevent time sidelobes from weather returns spilling over into adjacent range cells.

The contrast in Doppler processing system design between the ARSR-4 and ASR-9/12 is apparent. The requirement for both systems involves the detection of airborne targets in both surface and weather clutter. The terminal ASR-9/12 dictates the use of S-band to provide the required azimuth resolution from a modest size antenna that can be located at an airfield without a radome. This raises the intensity of weather clutter to where it is not practical to eliminate it entirely using CP, resulting in the need for MTD-type processing. The en-route ARSR-4 operates at L-band where the reduced intensity weather clutter can be further attenuated by 21 dB using CP processing. The surface clutter is then processed using an MTI processor where an 8-pulse design is required to eliminate high-reflectivity mountain clutter. To smooth the MTI's velocity response, a 9-period stagger is needed. This further complicates the design since MTI radars with staggered PRFs do not cancel second-time-around clutter expected from mountain or urban clutter. This requires a special mode using a double MTI canceller and a low PRF to extend the unambiguous range. The military application of the radar requires protection against jamming. The ARSR-4 uses burst frequency-agility transmission for this purpose [11]. However, the MTI must maintain a stable frequency over the 8-pulse MTI processing time that makes the radar vulnerable to responsive jamming techniques.

The solid-state TPS-59 depicted in Figure 1.3 is an L-band (1.125 to 1.4 GHz), long-range, 3-D tactical air surveillance planar phased-array radar [12]. The antenna rotates mechanically in azimuth and electronically scans a pencil beam in elevation from 0-to-19° against aircraft targets and 0-to-60° against theater ballistic missiles (TBMs) using phase steering in an active aperture configuration. The TPS-59(V)3 is designed to detect 0.1 m^2 targets flying with speeds up to Mach 7 at a range of about 400 nmi.

The 30-by-15-ft planar array antenna is composed of 54 identical feed row networks, each with 24 dipole elements. The row transmit-receive (T/R) assembly consisting of the transmitter, preamplifier, phase shifter, and duplexer is located directly behind the row feed. The vertical row-to-row distribution is accomplished

Figure 1.3 TPS-59 air surveillance tactical radar. (*From:* [78].)

by three column feeds. The row transmitter consists of 20 silicon bipolar transmitters, each providing 55 W whose outputs after combining losses provides a power of about 1,000 W.

A short-range waveform is used to cover 3 to 100 nmi, whereas a long-range waveform covers 100 nmi to the elevation envelope limit of about 300 nmi. The short-range waveform, consisting of a simple pulse followed in 252 μs by a linear 2.5 MHz FM pulse, is repeated three or six times to synthesize a family of MTI waveforms. Three- pulse MTI waveforms are employed in beams containing ground clutter, whereas 6-pulse MTI waveforms are used when ground and weather clutter are simultaneously present. The long-range waveform contains up to four contiguous pulses, offset 3.75 MHz pulse to pulse, with identical 625 kHz linear FM codes. The number of pulses used is a function of the maximum range and the elevation beam position.

Eight pencil beams are used to search the long-range interval (0° to 9° elevation) and 11 to cover the short ranges (0° to 20° elevation). The three lowest elevation beams in the short-range interval are assumed to contain ground clutter and use 3-pulse MTI waveforms. Height finding accuracy is 1,000 ft below the 100-nmi range linearly increasing to 3,000 ft at 300 nmi. Ground clutter is suppressed by 51.4 dB and weather clutter by 32.8 dB.

The TPS-59 and ARSR-4 are examples of L-band solid-state 3-D surveillance radars using different design approaches to accomplish measurement of the target's elevation position. Both radars use multipulse MTI processing to create wide rejection notches that discriminate against ground clutter. The ARSR-4 uses CP to discriminate against weather clutter [13], whereas the TPS-59 uses 6-pulse MTI processing in the short-range (under 100 nmi) beam positions. A limitation of the elevation electronic scan used in the TPS-59 is that the total number of pulses available for MTI processing in an azimuth target dwell time must be distributed among the 19 beam positions. To provide unambiguous range, the long-range beams employ a PRF of 277 Hz that at most allows two hits in each beam position. This is inadequate for MTI operation and, hence, weather clutter and chaff can not be rejected beyond the 100 nmi range. The PRF is increased to 833 Hz in the short-range beams, allowing up to six pulses to be utilized to suppress both ground and weather clutter. An advantage of the scanning pencil beam approach is that the full aperture is effective in both transmitting and receiving the elevation beams, thereby concentrating more energy on the target and somewhat reducing weather clutter intake.

The Erieye S-band (3.1 to 3.3 GHz), solid-state AEW radar is depicted in Figure 1.4. The phased-array radar is contained in a dorsal fin radome that is mounted on top of the aircraft's fuselage. It consists of back-to-back planar arrays consisting of 192 ram-air-cooled elements that are driven by T/R modules located between the planar arrays. Each 8m by 0.7m array can phase scan a 0.8° by 9° beam through coverage of ±60°. This leaves a low-sensitivity blind gap of 60° directly ahead and behind the aircraft that can be covered by the aircraft flying a racetrack pattern. Scanning can be accomplished simultaneously (with reduced power) or sequentially in the two arrays [14–16].

AEW radars take advantage of the increased line of sight (LOS) to low-altitude targets afforded by a medium- to high-altitude platform. This solves the terrain-shielding problem associated with surface-based early warning radars, but introduces another significant problem. The AEW radar must now "look down" into the surface (ground or water) at high grazing angles, which causes a radar clutter return of high magnitude that competes with the normal target return. Moving targets can be separated from the clutter using Doppler processing techniques. Operation over water has been feasible for many years, but overland operation requires subclutter visibilities on the order of 50 to 80 dB, which have only recently become practical.

The Wedgetail AEW&C system, also shown in Figure 1.4, employs an L-band multirole electronically scanned antenna (MESA) radar mounted in a dorsal structure on the top of the Boeing 737 fuselage. The MESA radar provides 360° coverage by using three antennas consisting of two back-to-back sideway-looking phase arrays and a third endfire array antenna—called the "top hat"—positioned high in the dorsal fin that allows both front and back looks. The two 18 by 6 ft sidelooking antennas each provide ±60° coverage while the 25 by 5 ft endfire array

Figure 1.4 Dorsal fin AEW radars: Erieye; and Wedgetail. (*From:* [17, 79].)

provides ±30° fore and aft coverage. The use of an L-band frequency allows incorporation of an integral IFF capability but has the potential for interference with the JTIDS data link and Global Positioning System (GPS).

The detection performance of the Wedgetail is not generally available, but can be estimated by comparison with the TPS-59, which operates in the same frequency range and utilizes a similar architecture. First the MESA utilizes 288 transmitting modules compared to 540 in the TPS-59, which reduces the power by a factor of 0.533. The sidelooking aperture is reduced by a factor of $(18 \cdot 6)/(30 \cdot 15) = 0.24$. The detection performance of the TPS-59 is quoted as 200 nmi with a $P_d = 0.9$ on a 1 m^2 fluctuating target [18, p. 704]. The resulting estimate for the MESA under the same target conditions would be $200 \cdot (0.533 \cdot 0.24)^{1/4} = 119.6$ nmi. An estimate for the endfire array performance can be made by assuming that the azimuth aperture length is proportional to the square root of the actual length [18, p. 619]. Hence the aperture factor would be reduced to $(\sqrt{25} \cdot 5)/(30 \cdot 15) = 0.0556$, and the range estimate in the forward and aft sectors to 68.7 nmi. In addition, the azimuth resolution in the endfire region would be reduced to approximately 9° from the 2.5° in the sidelooking region.

Three general types of Doppler radar processing systems are employed in AEW systems. The pulsed Doppler approach uses a high radar PRF, which is at least twice the Doppler frequency corresponding to the sum of the aircraft's velocity and the target's closing velocity. This creates a clear Doppler region without competing clutter returns, where narrowband filters can be used to detect moving targets. The system is highly range ambiguous, requiring special techniques, usually employing a computer, to resolve the ambiguities. The high PRF has the advantage of increasing the radiated energy for the same peak power as compared to low- and medium-PRF designs.

The signal processing that occurs in PD radar can be visualized by a range–velocity matrix. Each cell of the matrix has a range dimension, which corresponds to the radar's range resolution (usually, the range is equivalent to the transmitted pulsewidth), and a velocity or Doppler frequency dimension, which corresponds to the reciprocal of the time the radar beam dwells on the target. The detection process consists of examining the energy in each cell and comparing it with a threshold formed from surrounding cells to determine the presence of a target. This process is too complex to be performed by the radar operator; therefore, automatic detection techniques are generally employed to accomplish this function. After detection and resolution of range ambiguities, a synthetic display output is generated for operator viewing.

Examples of PD AEW radars are the U.S. and NATO AN/APY-1 and -2, which are used in the E-3A/B/C/D/E/F AWACS aircraft (described in Chapter 7). This system operates in S-band, which allows good angular resolution in both azimuth and elevation with ultralow antenna sidelobes [19], but must cope with weather clutter returns whose intensity is proportional to the fourth power of frequency. The AWACS employs a 30 ft diameter rotodome, which contains a 7.3 m by 1.5 m slotted-waveguide array antenna whose outstanding feature is ultralow sidelobes down 50 dB and lower from the peak of the antenna's main lobe. Both systems employ elevation scanning to allow determination of target height, standard IFF secondary radars (1.03 to 1.09 GHz), and elaborate communications equipment including data links to perform C^3 functions [20].

A second type of Doppler radar processing that is used in AEW radar systems is the airborne moving target indicator (AMTI). This type of processing removes the effect of aircraft motion by antenna design and signal processing (described in Chapter 6). Once platform motion effects are removed, the same signal processing techniques used in surface MTIs are applicable. Because an unambiguous range is desired, this type of AEW radar usually has low PRFs on the order of 300 PPS (270 nmi). The low PRF results in many target "blind speeds," which are compensated by using a staggered PRF. If, for example, a 4:5 stagger ratio is used, the first blind speed for a 300 PPS would be at 6,000 Hz. Operating at UHF frequencies (420 to 450 MHz), this would represent a closing velocity between the AEW aircraft and the target that varies from 1,855 to 4,161 knots. This consideration causes a frequency selection in the lower radar bands (UHF). However, it is difficult to achieve low antenna sidelobes at UHF frequencies due to excitation of the AEW's aircraft structure, which tends toward a frequency selection in the radar L-band as a reasonable compromise [21].

An example of AMTI AEW radar is the APS-145, which is contained in the E-2C aircraft. The performance with regard to subclutter visibility is somewhat less than that provided by PD AEW radars, but is still adequate for overland operation. The primary disadvantages of this type of radar are the larger main antenna lobe beamwidths and the higher antenna sidelobes, both of which result from low-frequency radar operation. The wide antenna beamwidth reduces the overall radar coverage in the presence of a main beam jammer. The higher antenna sidelobes make the radar more susceptible to sidelobe jamming and interception, increase vulnerability to antiradiation missiles (ARMs) [5], and also reduce radar performance in clutter.

A third type of AEW radar utilizes a medium-PRF design that is ambiguous in both the range and Doppler domains [22]. The primary reason for operating above the low-PRF region is to improve the radar's ability to contend with main lobe clutter and ground moving targets. The primary reason for operating below the high-PRF region is to allow the radar to detect targets with low or negative closing rates in sidelobe clutter. The Erieye is an example of medium-PRF AEW radar. The PRFs utilized in this radar are not available in the open literature, but it is possible to make an intelligent estimate based on the 80-dB improvement factor claimed for this radar. From equation (3.21), the standard deviation (σ_{pm}) of the main beam clutter can be calculated as 20 Hz based on an aircraft speed of 600 knots and 8 m antenna dimension. Assuming the use of a double canceller MTI, Figure 3.5 reveals that $\sigma_{pm}/\text{PRF} = 0.002$, resulting in a medium-PRF estimate of 10 kHz.

The primary advantage of the Erieye AEW system is that it can be installed on small commuter-sized jet or turboprop aircraft, which results in a considerable savings over the larger AWACS and E-2C aircraft that employ 30 ft rotodomes to house the rotating antennas. The estimated cost for operating the Erieye AEW is \$500/hr as compared to \$2,700/hr for the E-2C and \$8,380/hr for the AWACS.

Comparison of the detection ranges of the three types of AEW systems is difficult because significant parameters are not published by their manufacturers.

However, we can make some educated guesses, as shown in Table 1.2. The ASR-9, whose parameters are well known, is used as a reference. Its detection range on a 5 m^2 Swerling 1 fighter sized target is 53 nmi. We can then estimate the detection ranges of the three AEW radars by comparing their respective power-aperture products with that of the ASR-9 using the relationship $R = k[P_{ave} A_R t_s]^{1/4}$ where A_R is the effective antenna receiving aperture and t_s is the time to search the search volume that is assumed constant for all of the radars [5].

Examination of the Erieye's T/R module [23] reveals the use of a single transistor, which is probably of the bipolar silicon type providing a peak power of 50 W with a 15% duty factor from each of the 192 modules resulting in a total of 1,440 W average power [8]. Note that the Erieye's detection range is determined broadside to the array and will be diminished by the factor $[\cos(\theta_s)]^{1/4}$ as the phased array scans away from the broadside. A uniform illumination is used on transmitting to maximize the power output of the active aperture array, whereas a 50 dB Chebyshev taper is used for receiving that, including losses, provides an average 43 dB sidelobe level. An adaptive sidelobe canceller is used to protect against sidelobe jamming [24]. The estimated parameters of the APS-145 radar used by the E-2C Hawkeye can be found in [25]. The peak power provided by the AWACS Klystron amplifier is estimated at 1.2 MW by comparison to a similar transmitter in the ASR-9. The high-PRF of the AWACS must be unambiguous in Doppler so that its value can be estimated from the relationship PRF = $2(v_{AWACS} + v_{TAR})/\lambda$. The operating speed of the AWACS is about 300 knots, whereas those of typical targets such as the Fulcrum or Flanker aircraft have a maximum speed of 1,300 knots. The resulting estimated PRF is about 35 kHz, and the estimated pulsewidth is 1 μs through similarity with the ASR-9 transmitter. The resulting average power estimate is 42 kW.

The comparison given in Table 1.2 indicates that the AWACS and E-2C are in a different class than the Erieye. However, on an economical basis, the Erieye might compare favorably because more aircraft can be used to provide operational surveillance at a lower cost. Note that the range comparisons are on a first-order basis. In the high-PRF system, range ambiguities must be resolved by transmitting two or three different PRFs on different scans. Medium-PRF designs generally require seven or eight different PRFs to resolve range ambiguities, whereas low-PRF designs utilize PRF staggers to prevent blind velocity regions All of these effects cause detection losses that must be evaluated on an individual basis.

The SA-15 GAUNTLET missile system, depicted in Figure 1.5, is a low- to medium-altitude SAM system capable of engaging not only aircraft and helicopters

Table 1.2 Estimated Detection Ranges on Fighter-Sized Aircraft

System	*Frequency*	*PRF*	P_{ave} *(kW)*	A_R (m^2)	*Scan Time (sec)*	*Range (nmi)*[a]
ASR-9	S-Band	Low	1.575	2.044	4.8	53
Erieye	S-Band	Medium	1.440[b]	3.147	10	69
E-2C	UHF	Low	18.75[b]	7.03	10	161
E-3D	S-Band	High	42[b]	3.505	10	166

[a]From a 5-m^2 Swerling 1 target.
[b]Estimated.

Figure 1.5 SA-15 GAUNTLET short-range missile system. (*From:* [29].)

but also UAVs, precision-guided weapons, and various types of guided missiles. The SA-15 is designed to be a completely autonomous air defense system, capable of surveillance, command-and-control functions, missile launch, and guidance functions from a single vehicle. It carries eight ready missiles stored in two containers holding four missiles each. The SA-15 has the capability to automatically track and destroy two targets simultaneously in any weather and at any time of the day. The SA-15 is designed to hit targets flying at speeds from 36 to 2,500 km/hr. It can engage both slowly moving hovering helicopters and those on the ground with turning rotors [26–28].

The SA-15 is mounted on a tracked chassis as shown in Figure 1.5. Its C-band frequency-scanned MTI target acquisition radar and associated IFF antenna are located at the rear. The K_u-band limited-scan, phased-array radar is located in the front. Mounted on either side of the phased-array radar are the electro-optical television system and a small beacon tracking monopulse radar that initially acquires the missile [29].

The target acquisition (TAR) and engagement (TER) radars provide a number of unique features. Coherent in-phase (I) and quadrature (Q) MTI processing is provided in each of the multibeams of the 3-D TAR. The MTI can be tuned either manually or automatically using a Doppler clutter map. This allows target acquisition in heavy ground clutter in the lower beams and performance in chaff and weather clutter in the upper beams. The TER functions to track both the targets and the beacon in the missile while providing guidance commands to the missile. It uses an FFT Doppler processor to reject ground clutter and chaff while enhancing target returns from aircraft and helicopters. Both target acquisition and engagement radars use pulse compression to enhance resolution and provide increased performance in clutter and noise jamming.

The 1.5 kW average power TAR is a 3-D, C-band frequency-scanned acquisition radar that continuously mechanically scans in azimuth at a 60 rpm rate. It features pulse compression combined with a coherent MTI processor. In the elevation plane, the beam is electronically frequency stepped to provide coverage of 32° with eight 4° beam positions. Any three beam positions can be transmitted on each revolution, requiring several revolutions to cover all of the beam positions. The lowest beam position is transmitted twice as often as the other beam positions to provide better low-altitude coverage, which is advantageous when engaging helicopters or low-flying cruise missiles.

In normal operation, the TAR transmits a long pulse consisting of several subpulses. Each subpulse is transmitted on a different frequency using LFM pulse compression. Because the antenna is frequency scanned, each subpulse corresponds to a specific beam based on its frequency. It changes the set of beams on every scan in a preset pattern and repeats the pattern every frame. The TAR also employs other scan modes used under varying conditions. The other scans are a three-beam mode and a single-beam mode. In the single-beam mode, the TAR transmits a single-frequency long pulse and increases the frame time thereby concentrating more energy on the target. This mode is advantageous in a jamming environment.

The TAR transmitter is a coherent system that apparently utilizes a Klystron final amplifier with approximately 1.5 kW average power and about 6% bandwidth. The typical TAR waveform uses a multiperiod stagger, but reverts to a constant PRF to synchronize with the repetition rate of the TER when targets are in track.

The TAR antenna uses a parabolic cylinder with a serpentine line source array to illuminate the reflector. The frequency steps allow it to move through the various beam positions in elevation. The total electronic antenna coverage is about 30° through all beam positions, but the antenna can be further tilted mechanically by another 30°. The polarization is linear horizontal. The blockage caused by the serpentine feed results in relatively high sidelobes, which increases the vulnerability of the TAR to stand-off noise jamming. More advanced versions of the SA-15 use a planar array to correct this deficiency.

The SA-15 uses digital adaptive MTI processing. The adaptive MTI provides an adaptive notch that can be tuned in frequency to cancel returns from chaff and weather clutter. The notch is steered in each beam using a clutter velocity map that is generated on a multiple scan and range cell basis. Stationary clutter is generally prevalent in the lower beam position and decoupled from the higher beam positions by the antenna's elevation sidelobes. The MTI uses a multiperiod stagger to eliminate blind speeds and mitigate interference with adjacent SA-15 units. For the SA-15 TAR scanning clutter is significant due to the rapid antenna rotation. Staggering the PRF also reduces the improvement factor because it distorts the filter response relative to an MTI that is not staggered and generally a loss of about 5 to 10 dB results in the velocity pattern around the first blind speed. All target detections are processed through the MTI units and hence the detection performance of the SA-15 critically depends on the performance of the MTI units.

The TER functions to track a maximum of two engaged targets and missile beacon returns. It also sends guidance commands to the missiles directing them toward the engaged targets. The targets and missiles must all be generally located within a narrow 3° azimuth by 7° elevation angular sector to be within the TER's field of view (FOV). However, the phased-array antenna can be defocused to provide a larger FOV to capture the missile beacon signal during initial acquisition. The TER limited-scan, phased-array antenna is mounted on a turret (see Figure 1.5) that can be mechanically rotated through 360° in azimuth and tilted in elevation. The TER apparently uses a monopulse receiver and low-PRF fast Fourier transform (FFT) Doppler processor.

The central feature of the TER is a limited-scan, phased-array antenna with three-wavelength element spacing to control grating lobes that is mounted on the front of the SA-15 turret. The array aperture is apparently about 5 ft in diameter. Acquisition is accomplished by raster scanning about the selected target position determined from the TAR. The phased array probably uses ferrite phase shifters typical for this type of design. Vertical polarization is generally used in this type of design to reduce multipath and prevent cross coupling from the horizontally polarized TAR.

The TER transmitter provides approximately 600 W average power with multiple operating frequencies in K_u-band. The TER transmits several different waveforms depending on whether it is tracking the target or missile beacon or providing guidance commands to the missile. The EO/TV monitor allows the operator to monitor target tracking and to switch to optical auto tracking if necessary.

The detection performance of the TER is determined by the ability of the FFT to both reject ground, weather, and chaff clutter while coherently integrating the target returns. Critical to clutter rejection is the adaptive tuning compensation of the FFT so that the target spectrum is centered between PRF lines and windowing that provides low filter sidelobes in the spectral regions of heavy clutter concentrations.

At launch the TER phased-array antenna tracks the target and the vertically launched missile is out of its FOV. The function of the missile capture system is to passively track the missile beacon signal immediately after the missile's launch. It consists of a small adjunct tracking antenna located in a hemispherical radome attached to the upper right side of the TER antenna (see Figure 1.5).

The SA-15 system also contains an integral IFF system that is part of the target acquisition radar. After the TAR has detected a potential target, the IFF system functions to determine whether the target is friendly. To accomplish this function an IFF reflector antenna composed of a vertical grating with two feed systems is attached to, and mounted above, the TAR antenna. The reflector is skewed behind, and rotates with, the primary TAR antenna to provide for immediate interrogation of a potential target.

The SA-15 illustrates how modern Doppler radar can be integrated into an air defense weapons system to meet an operational need that involves the engagement of precision guided weapons (such as cruise missiles, ARMs, JDAMS) in addition to high-performance aircraft and helicopters. The MTI and associated clutter notch in the TAR allows nap-of-the-earth and low RCS threats to be engaged in a matter of seconds under all environmental conditions including heavy ground clutter, weather, and chaff. The phased array in the TER and associated FFT processor allows simultaneous engagement of two targets while tracking the attacking missile beacons and issuing guidance commands to the missiles.

1.2 The Doppler Effect

The physical mechanism that allows detection of the aircraft in the surveillance radar application is a unique transformation that shifts the frequency of the reflected return from each scatterer within the radar's FOV in direct proportion to its relative radial motion with respect to the radar. The transformation is due to the compression or extension of the radar wave caused by the relative radial motion of the scatterer with respect to the radar, and is generally known as the *Doppler effect*. The magnitude of the Doppler shift frequency can be computed by noting that the phase of the return from a scatterer, when received at the radar is given by

$$\theta = \frac{4\pi r}{\lambda} \tag{1.1}$$

where r is the range to the scatterer and λ is the radar's apparent wavelength. The instantaneous frequency from a scatterer moving radially with respect to the radar is then given by

$$f_d = \frac{1}{2\pi}\frac{d\theta}{dt} = \frac{2v_r}{\lambda} \tag{1.2}$$

where v_r is the scatterer's radial velocity vector. The sign of v_r is positive for a target that is closing on the radar and negative for a receding target. The apparent wavelength for a closing target is given by

$$\lambda = \frac{c + v_r}{f_t + f_d} \tag{1.3}$$

where f_t is the radar's transmitted frequency. Substituting this into (1.2) results in

$$f_d = f_t\frac{2v_r}{c}\left[1 + \frac{v_r}{c} + \ldots\right] \tag{1.4}$$

where the error term from using (1.2) is equal to v_r/c. For example, the error in using (1.2) for a 3,000 m/s radial velocity target is a negligible 10^{-5}. Doppler signatures from aircraft will generally be different from those of ground, sea, or weather clutter, thus allowing detection through use of appropriate Doppler filters.

► Example 1 Target Trajectory from Multiple Doppler Measurements

Doppler measurements from continuous-wave (CW) radar, such as police radar or projectile velocity radar, can be used to determine the trajectory of the target. The basic Doppler equation can be written as

$$f_d = \frac{2v_{\max}\cos(\alpha)}{\lambda}$$

where α is the angle between the velocity vector and the target's LOS. Assuming the target is moving in a straight line at a constant velocity, the geometry of the target with respect to the radar can be described as illustrated in Figure EX-1.1 [30].

Doppler Position Measurement Using MATHCAD (Dopplerpos.mcd)

Doppler Measurements

$f_0 := -17{,}320.50807$ Hz $f_1 := -19{,}702.4210$ Hz $f_2 := -19{,}895.24645$ Hz

System Parameters

$$\tau := 0.1 \text{ sec} \qquad \lambda := 0.02\text{m}$$

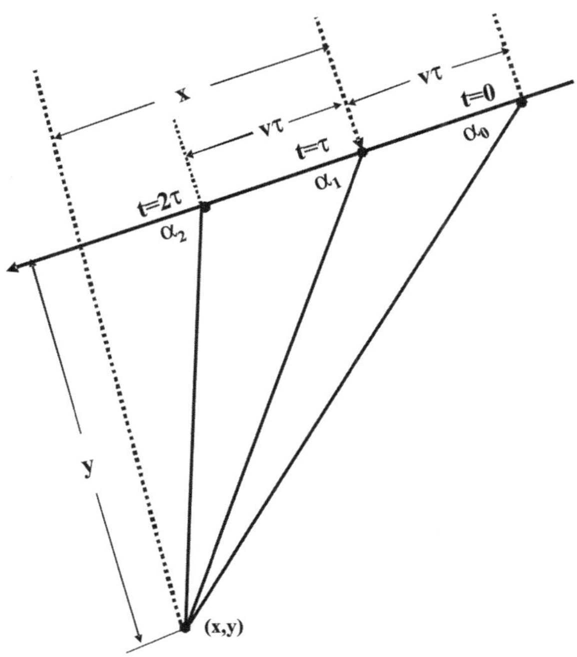

Figure EX-1.1 Target motion relative to sensor at (x, y).

Initial Guess

$$v := -300\,\frac{\text{m}}{\text{s}} \qquad x := 50\text{m} \qquad y := 50\text{m}$$

Determine Position and Velocity
Given

$$f_0 = \frac{v}{\lambda}\cdot\frac{2}{\sqrt{1+\left(\frac{y}{x+v\cdot\tau}\right)^2}} \qquad f_1 = \frac{v}{\lambda}\cdot\frac{2}{\sqrt{1+\left(\frac{y}{x}\right)^2}} \qquad f_2 = \frac{v}{\lambda}\cdot\frac{2}{\sqrt{1+\left(\frac{y}{x-v\cdot\tau}\right)^2}}$$

find

$$(v, x, y) = \begin{pmatrix} -200 \\ 28.6603 \\ -5 \end{pmatrix} \qquad v := -200\,\frac{\text{m}}{\text{s}} \qquad x := 28.6603\text{m} \qquad y := 5\text{m}$$

▶

1.3 Doppler Processing Fundamentals

Figure 1.6 illustrates effective operation of MTI radar. The normal video display shows ground clutter out to a range of 40 nautical miles. The MTI display shows elimination of the ground clutter while retaining aircraft target reports. The moving targets are recorded for three antenna revolutions, resulting in the triple blips shown [31].

The function of most radars is to supply position location data on targets of interest. In MTI and PD radars, this is a two-stage process whereby the radar first uses Doppler filtering to separate the target signal from the surroundings, which consists of reflections from terrain, weather, and other extraneous targets, and then performs detection and location measurement on the extracted target. The focus in MTI and PD radars is generally on the Doppler filtering process. However, the overall performance of the radar involves a consideration of the overall detection and location process.

1.3.1 Resolution Considerations [32, 33]

The performance of MTI or PD radar is intimately related to its resolution properties. Resolution is defined as the ability of the radar to distinguish between two closely spaced targets in at least one of the state variables used to describe the target vector. State variables associated with radar targets generally form a four-dimensional space consisting of an azimuth angle, elevation angle, range, and radial velocity (Doppler). The angle states are generally decoupled from the range-velocity states, thereby allowing the resolution properties of the radar to be described by

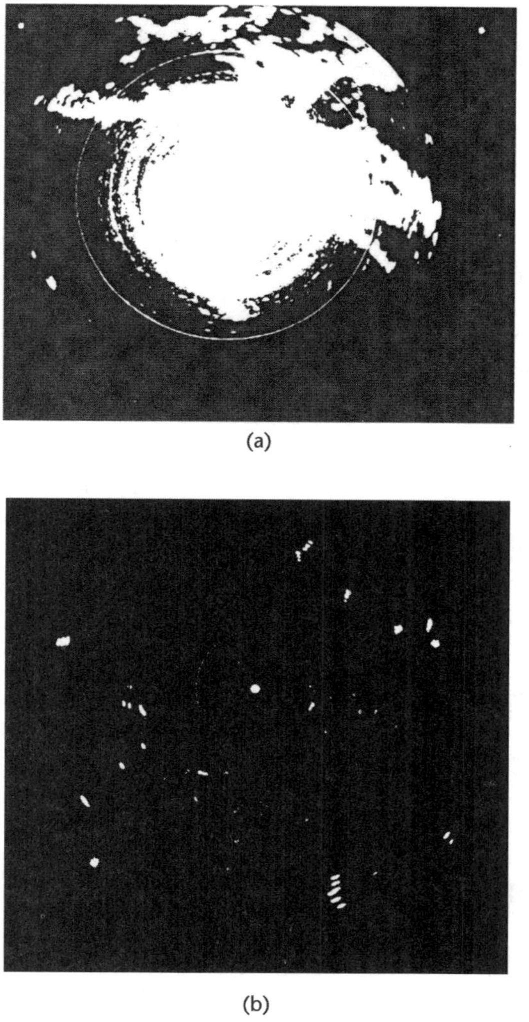

Figure 1.6 Effective MTI radar performance: (a) normal display; and (b) MTI display. (*From:* [31, p. 15.2].)

two vectors: one consisting of the angle dimensions (azimuth and elevation angles) and the other the temporal-spectral (range-Doppler) dimensions.

Radar waveform design is concerned with resolution in the range-Doppler dimensions. The radar ambiguity function, which displays a complex envelope of the response of the matched filter receiver to the radar transmitter waveform reflected from a point scatterer, is useful for visualizing the resolution properties of the radar waveform. In 3-D spaces, this diagram displays the response to time delay (range) in one dimension, Doppler shift (frequency shift) in an orthogonal direction, and the magnitude of the receiver response in the third, orthogonal dimension. The desirable thumbtack type of response produced by a noise-like waveform, as shown in Figure 1.7, depicts the features of the ambiguity function response.

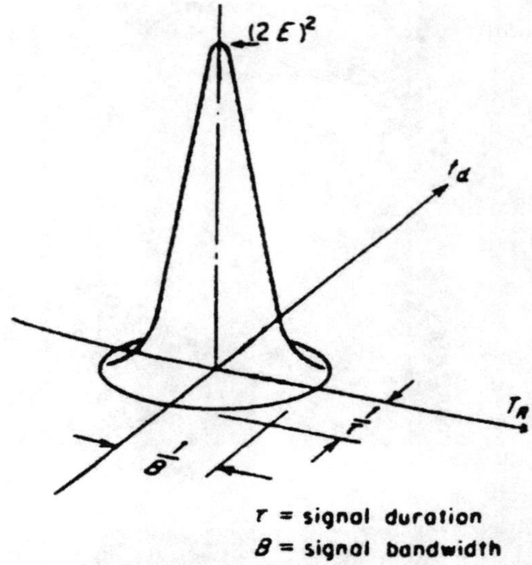

Figure 1.7 Thumbtack radar ambiguity diagram. (*From:* [5, p. 30].)

Resolution in the range-Doppler domain can be visualized by superimposing two ambiguity functions displaced by the range-Doppler spacing of the scatterers. With the idealized diagram of Figure 1.7, which is characterized by an isolated central spike, we can see that the range resolution is enhanced by transmitting a greater bandwidth (B) because it compresses the extent in the range plane. Doppler resolution is increased by using a longer transmitted waveform (T) because it compresses the extent in the Doppler plane. However, the volume under the ambiguity diagram is fixed so that the range and Doppler resolution must be balanced against each other (i.e., the radar waveform uncertainty principle).

MTI and PD radars represent two alternative waveform strategies for detecting targets based on Doppler filtering principles. The distinction between MTI and pulse Doppler radar is not the use of low, medium, or high PRFs, but the use of a passband-stopband filter for MTI as compared to a bank of coherently integrating filters for pulsed Doppler radars. Hence, there are medium-PRF MTI systems and low-PRF pulsed Doppler systems (e.g., the Moving Target Detector). These distinctions are recognized in Chapter 2.

1.3.2 MTI Doppler Filters in a Surface Clutter Environment

MTI radars use a low-PRF waveform design that is generally unambiguous in range, but highly ambiguous in Doppler response. For surface clutter (sea or ground clutter), this means that only the clutter returned from scatterers within the azimuth beamwidth and effective pulsewidth dimensions competes with the target. The total energy returned from the clutter is minimized by matching the azimuth beamwidth and effective pulsewidth as closely as possible to the corresponding target dimensions. This is usually impossible in the azimuth dimension (a 45 ft aperture at L-band provides a 1° beamwidth), but can be approached in the range dimension using pulse compression to balance resolution and energy. The resulting resolution cell for low-PRF radar is usually much larger than the target dimensions, which allows for an appreciable amount of ground or sea clutter energy into the receiver.

Figure 1.8 depicts the resolution cell associated with surface-based, low-PRF radars superimposed on a field of randomly oriented, elemental, stationary point scatterers. When illuminated by the radar transmitter, reflections from those scatterers contained within the resolution cell will form the surface clutter return. Under conditions where the scatterers are numerous and no single scatterer dominates, we can show that the ground clutter signal is a Gaussian distributed random process. The power spectral density of the process is determined by the internal radial motion of the clutter scatterer (due to wind-driven vegetation) and antenna scanning motion, which induces spectral spreading into the scatterer's return [1].

A simple target can be represented as a point reflector moving with respect to the elemental clutter scatterers. The target's radial motion causes this return to be shifted in frequency with respect to the clutter scatterers, the returned signal spectra that are centered around the transmitter's carrier frequency. The target's spectral width, and hence its power spectral density, is determined by the antenna's scanning motion, which induces spectral spreading onto the target.

The target and surface clutter frequency spectra are depicted in Figure 1.9, where both the underlying (equivalent CW) and sample data versions are shown. Based on the somewhat idealized target and clutter model just described, synthesis of an optimum MTI detection processor is possible by using statistical detection theory (see Chapter 5).

1.3.2.1 Optimum MTI Filter Theory

The fundamental Doppler filtering problem to be solved is the detection of n coherently Doppler-shifted target pulses embedded in a background of Gaussian distributed interference consisting of clutter and receiver noise. The initial radio-

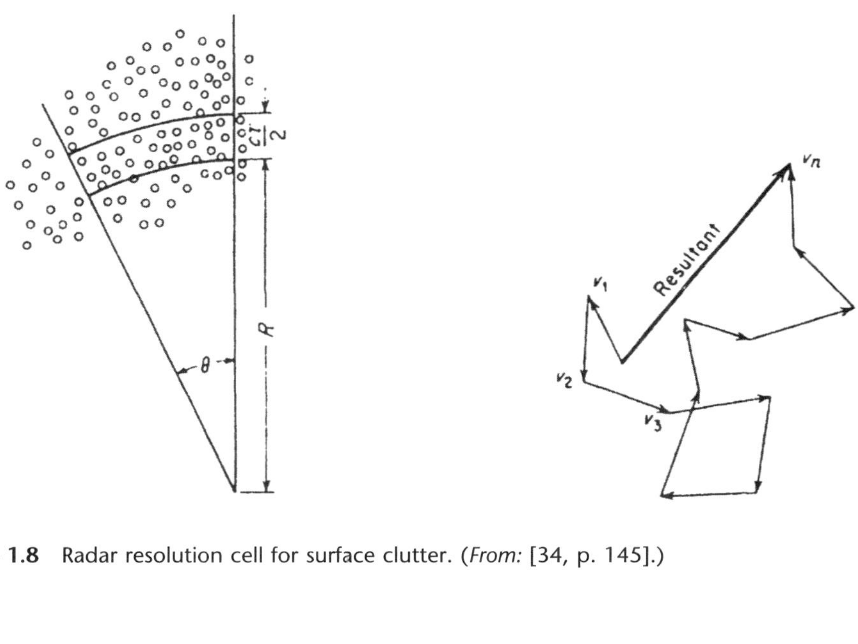

Figure 1.8 Radar resolution cell for surface clutter. (*From:* [34, p. 145].)

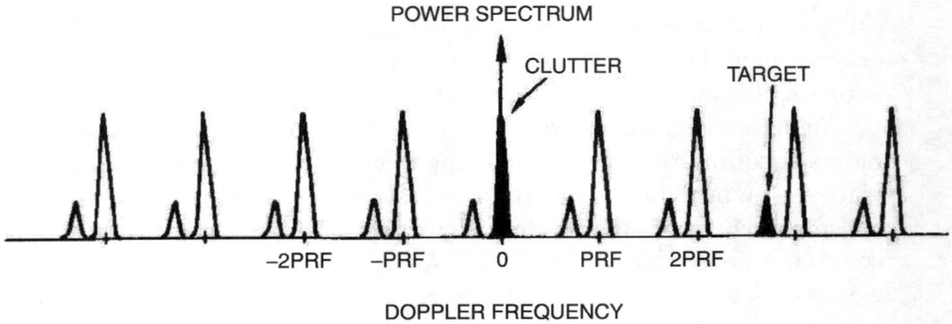

Figure 1.9 Target and clutter frequency spectra. (*From:* [35, p. 37].)

frequency (RF) phase of a coherent pulse train returned from the target is uncertain and, hence, will be randomly distributed with respect to the phase of the coherently transmitted waveform.

The detailed solution to this problem is discussed in Chapter 5. However, at this point, we will discuss some general features of the solution. First, the general architecture of the optimum Doppler processor consists of a linear transversal filter that uses complex weights (amplitude and phase) followed by an envelope detector (see Section 5.1). The assumed Gaussian statistics of the interference results in a linear filter structure, whereas the envelope detector results from the initial random phase of the coherent pulse train [36]. The complex weights of the transversal filter are determined by the interference (clutter and receiver noise) covariance matrix, and the assumed complex distribution of the signal's Doppler distribution.

The MTI solution to the optimum Doppler processor is obtained when there is no a priori knowledge of the Doppler shift of the target return. In other words, all possible Doppler shifts are equally likely. As shown in Figure 1.10, the resulting MTI processor consists of two identical quadrature channel transversal filters that use real weights, of which one channel processes the in-phase (I) component and the other processes the quadrature (Q) channel. The signals from the two channels

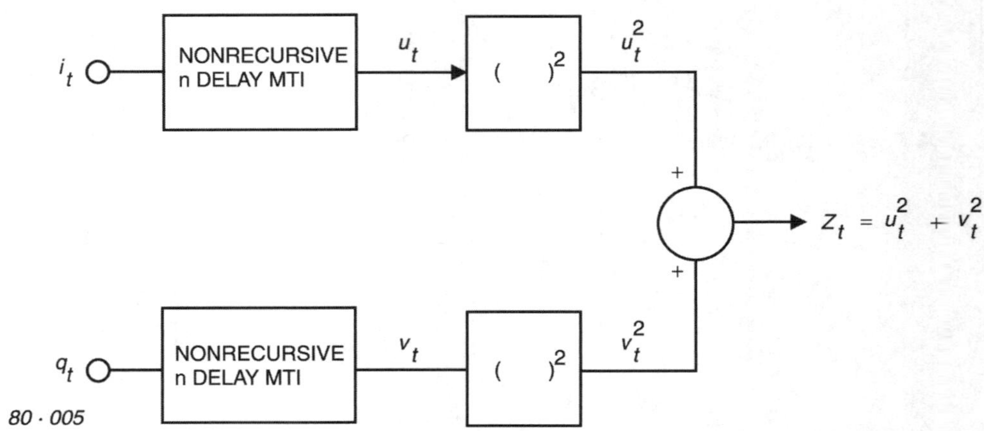

Figure 1.10 Block diagram of quadrature channel MTI. (*From:* [37, p. 300].)

are then envelope detected, squared, and summed to form the MTI processor output. Note that the MTI processor provides no detection improvement against receiver noise and in fact introduces correlation across the output pulses that reduce the available integration gain (see Example 18).

The resulting MTI Doppler filter response has a comb-like shape that attenuates regions (stopband) of any clutter concentration and provides minimal attenuation (passband) in regions with little or no clutter energy. A typical single-channel MTI filter response is shown superimposed on the target and ground clutter spectra in Figure 1.11.

A feature of the low-PRF MTI design is that the range measurements are unambiguous, whereas Doppler responses are generally ambiguous, creating so-called blind speeds that occur in small regions about the radial target velocities corresponding to multiples of the radar's PRF. To allow detection of targets in the blind regions, staggering or jittering the MTI radar's PRF is generally necessary, which has the effect of "filling in" the MTI's velocity response to cover the region of interest.

1.3.3 MTI Doppler Filters in a Surface and Volume Clutter Environment

The resolution states that apply to weather or chaff returns (sky or volumetric clutter) are the radar's azimuth and elevation angles and effective pulsewidth. Figure 1.12 depicts how these states all formed into a 3-D resolution cell, where orthogonal dimensions consist of the arc subtended by the radar's azimuth and elevation beamwidths and the range equivalent to the radar's effective pulsewidth.

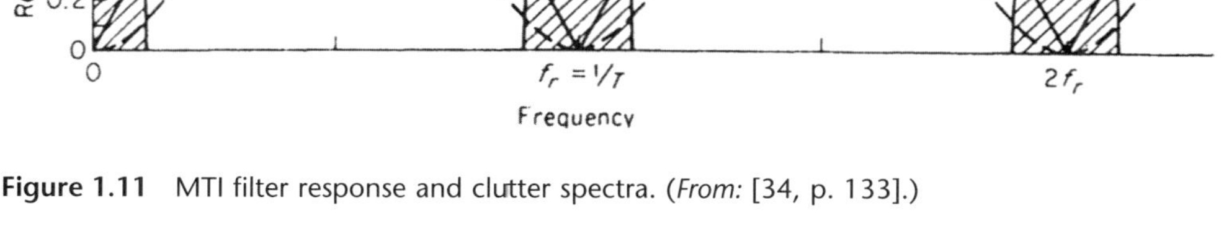

Figure 1.11 MTI filter response and clutter spectra. (*From:* [34, p. 133].)

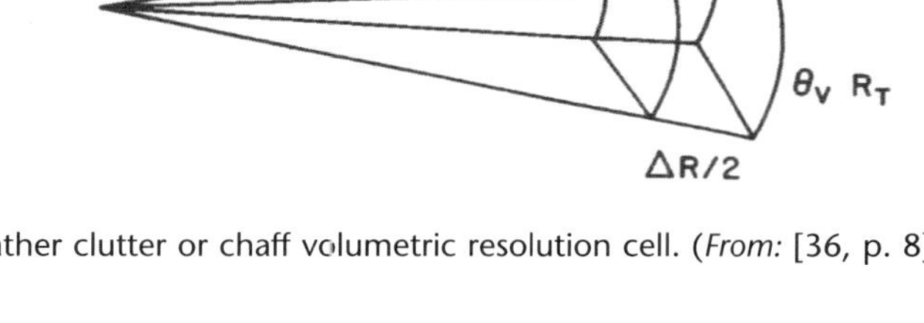

Figure 1.12 Weather clutter or chaff volumetric resolution cell. (*From:* [36, p. 8].)

The reflections from all weather disturbances (rain, snow, fog, and clouds) and chaff particles contained within the 3-D volumetric resolution cell combine to form the weather clutter or chaff returns. Thus, the magnitude of sky clutter is weighted by the effect of both the azimuth and elevation antenna patterns. In addition to this difference, sky clutter has several other distinctive features that distinguish it from ground clutter. Notable among these are the variable mean velocity (Doppler frequency), which is proportional to wind speed, and the considerable spectral spreading due to the effects of wind shear and turbulence.

Design of an optimum MTI filter that provides detection of a moving target in the presence of sky clutter follows the same general principles as those described for ground clutter. However, the variability of sky clutter (weather or chaff) spectra requires some form of adaptive control of the MTI's frequency response. This can be accomplished by monitoring the clutter returns in real time and adjusting the MTI's complex filter weights to minimize its clutter output. The maximum entropy algorithm is sometimes used to estimate the characteristics of the clutter spectra for this purpose [38].

A difficult MTI design problem results when both surface and sky clutter are present simultaneously. Figure 1.13 depicts an approach that uses dual-offset MTI filters to simultaneously process both types of clutter. Adaptive control of the MTI's filter weights is necessary to provide reasonable performance in this situation.

The mean Doppler frequency associated with sky clutter often exceeds the radar's PRF when an MTI design is used. This causes an *aliasing* (spectral foldover)

TYPE OF MTI SYSTEM	NO. PULSES NEEDED IN FILTER	CLUTTER REJECTION		CONCLUSION
		GROUND	SKY	
ZERO REF CANCELLER 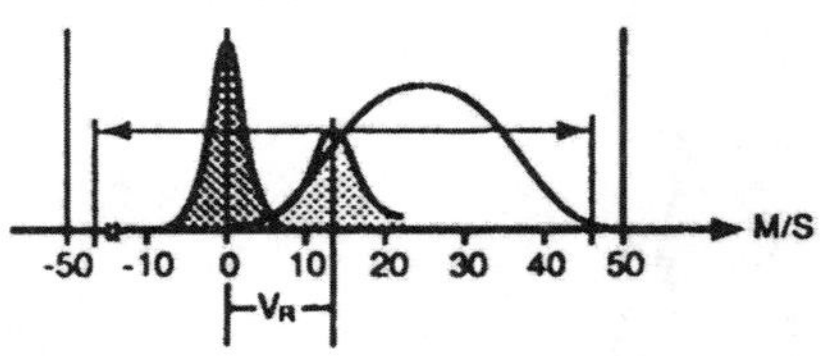	3	VERY GOOD	POOR	APPLY WHEN ONLY GROUND CLUTTER EXISTS
OFFSET CANCELLER 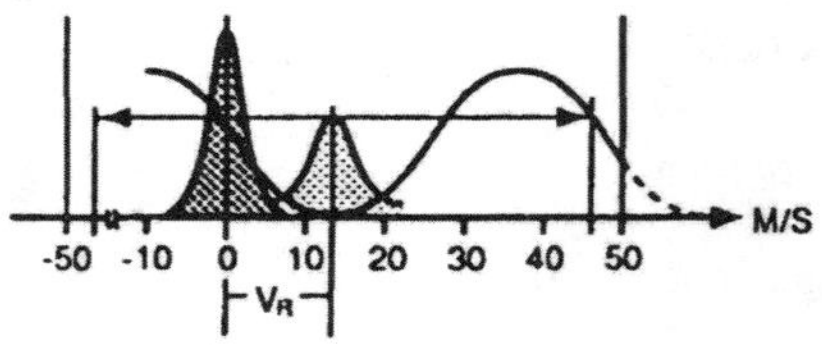	3	POOR	FAIR	APPLY WHEN ONLY MILD SKY CLUTTER EXISTS
DUAL OFFSET CANCELLERS 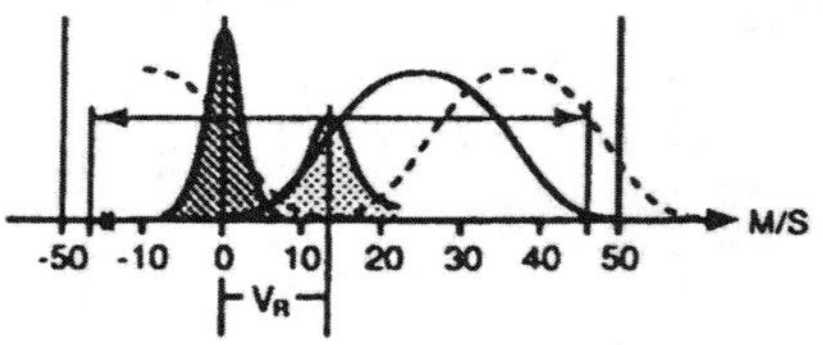	5	VERY GOOD	FAIR	APPLY FOR GROUND AND MILD SKY CLUTTER

Figure 1.13 MTI processing of weather clutter or chaff with ground clutter. (*From:* [36, p. 9].)

effect, whereby the Doppler frequency spectrum of the clutter is folded into the region between zero and the first PRF line. The same effect occurs for targets, which often are also aliased. This results in a situation where the aliased target spectrum may overlap the sky clutter spectra in the MTI, although the target has a different true Doppler than the clutter. The general solution to this problem is to change the radar's PRF, thereby causing the aliased spectra to change their relative positions within the first PRF spectral region.

This general approach is employed in the Moving Target Detector [1]. An early version of the MTD used an MTI to suppress ground clutter, followed by an FFT Doppler filter bank, the function of which was both to suppress sky clutter and to provide coherent integration. The FFT-generated filter bank, as depicted in Figure 1.14, divided the MTI passband into eight contiguous filters. The filters containing weather clutter were suppressed by sensing the heavy clutter content in these filters using a cell-averaging CFAR to control their detection thresholds. A second PRF was then employed to shift the aliased targets into different Doppler filters, thereby allowing them to be detected. The MTI-FFT Doppler filter bank combination is replaced in later designs by a more efficient 8-pulse transversal filter design that provides the same function.

1.3.4 Pulse Doppler Fundamentals [2]

The pulse Doppler radar operates on the same general principle as MTI radar, but the objectives and implementation are somewhat different. One basic difference is that the high-PRF PD radar operates at a PRF that provides unambiguous Doppler frequency returns from the target. This involves using a PRF that is at least twice the Doppler frequency, corresponding to the maximum closing speed between the radar and target. High-PRF PD radars are usually highly ambiguous in the range dimension. Medium-PRF radars are generally ambiguous in both range and Doppler.

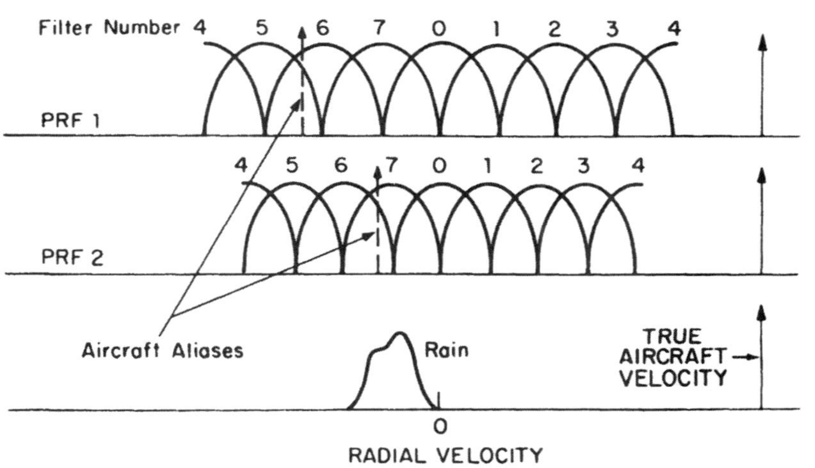

Figure 1.14 MTD processing of weather clutter. (*From:* [39, p. 40].)

In high-PRF PD the unambiguous Doppler allows implementation of a signal processing filter that provides a high degree of clutter rejection. The ambiguous range causes clutter returns from many radar resolution cells to fold into the target cell, thereby increasing the clutter power that competes with the target. Also, the resolution of the many possible target ranges is not a trivial problem and usually involves transmitting multiple PRFs with associated computer processing to determine the correct target range [3, 40].

Pulsed Doppler radars (both high- and medium-PRF) are used in airborne applications (e.g., AWACS and AI radars) where low-PRF MTI radars are impractical, except at very low transmitter frequencies, due to spectrum spreading effects from platform motion. Both high- and medium-PRF PD radars are also used in ground-based applications against low-flying, penetrating targets, where modest ranges are required. High-PRF pulsed Doppler radars are particularly effective against chaff and weather clutter.

1.3.4.1 Pulsed Doppler Waveform and Clutter Considerations [2, 4, 40]

A block diagram of a simple PD radar and its transmitted waveform are depicted in Figure 1.15. The waveform consists of a coherent RF pulse train (the starting RF phase of successive pulses is separated by an integral number of wavelengths) resulting in a line spectra whose adjacent spectral lines are separated by the PRF (f_r).

Target and clutter spectra for a single-spectral-line transmission from a moving platform are shown in Figure 1.16. Because of the sidelobes, the frequencies of the fixed clutter returns can vary $\pm 2\ v_r/\lambda$ from the transmitter frequency. The clutter possesses a high peak, resulting from the clutter return in the main beam. For closing targets ($v_t > v_r$), the returns will be shifted in frequency by $2\ v_r/\lambda$, and therefore appear in the clutter-free portion of the frequency spectrum.

When the return is mixed with a coherent "sidestep" oscillator, as shown in the block diagram, the output video spectrum is folded over about one-half of the PRF. When processed through a bank of contiguous Doppler filters, the resulting target output, which appears in the appropriate Doppler filter, is simply the Doppler return associated with the central line of the transmitter spectrum. In practice, some form of single-sideband detection process would be employed instead of the simple mixing system shown in Figure 1.15. With this system, the minimum PRF is given by

$$f_r \geq 2\frac{v_f + v_r}{\lambda} \tag{1.5}$$

where $v_r = v_f + v_t$ is the maximum relative target closing velocity.

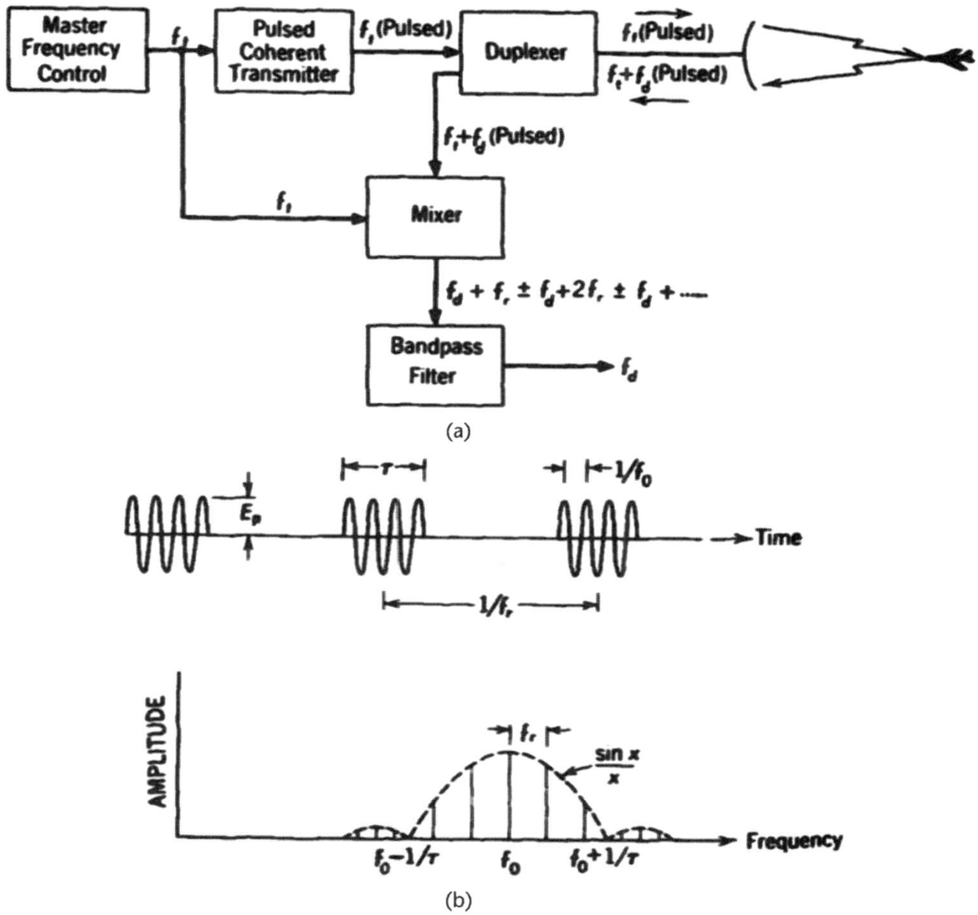

Figure 1.15 Elemental pulsed Doppler system. (*From:* [2, p. 25].)

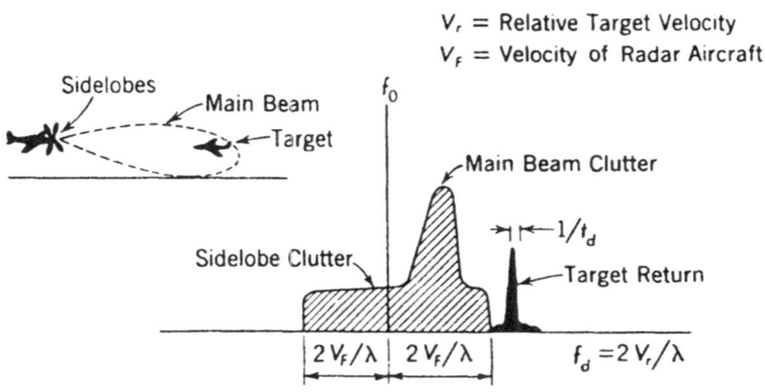

Figure 1.16 Target and clutter spectra for a single-spectral-line transmission. (*From:* [2, p. 322].)

The bandwidth (B_D) of the Doppler filter generally determines the Doppler resolution of the PD radar. In scanning radar, the bandwidth is often selected so that all of the target returns received during the time that the radar dwells on the target are collected within a single Doppler filter. For this situation, the appropriate relationship

$$B_D = \frac{1}{T_d} \tag{1.6}$$

is generally applicable, where T_d is the dwell time of the main beam on the target. This results in target returns being coherently integrated in the filter:

$$n = \frac{f_r}{B_D} \tag{1.7}$$

The PD radar thus approximates a matched filter that provides optimum detection in a noisy environment.

A variation of the pulsed Doppler waveform is the uniform pulse burst (see Figure 1.17), which consists of the finite train of n pulses transmitted at an interpulse period T, where generally $nT \leq T_d$. The period nT is the coherent processing interval (CPI). Dwell time T_d is filled with several CPIs. The use of mixed predetection (coherent) and postdetection integration (matched to T_d) provides for good overall detection performances for the PD radar when changing the transmitter frequency or PRF on a block to block (i.e., CPI) basis. The block-to-block frequency changes are useful in applying frequency agility to the PD radar, and PRF changes are useful in resolving range ambiguities.

Calculating the ground clutter spectrum return to airborne PD radar is a particularly complex procedure due to the multiple range ambiguities and the spread of the clutter spectrum that results from platform motion. Insight into the ground clutter spectrum can be gained from Figure 1.18, which shows the lines of constant Doppler returns, called isodops, obtained from a horizontally moving platform on which concentric rings are superimposed, representing the ground scatterer returns, collected within a particular range gate. The isodops represent the intersection between the ground and the surface of a cone about the radar's velocity vector. Because the angle between this vector and every point on the cone

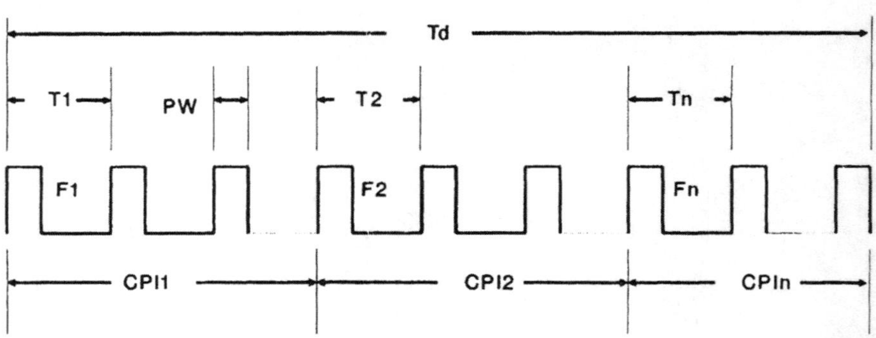

Figure 1.17 Uniform pulse burst waveform. (*From:* [36, p. 13].)

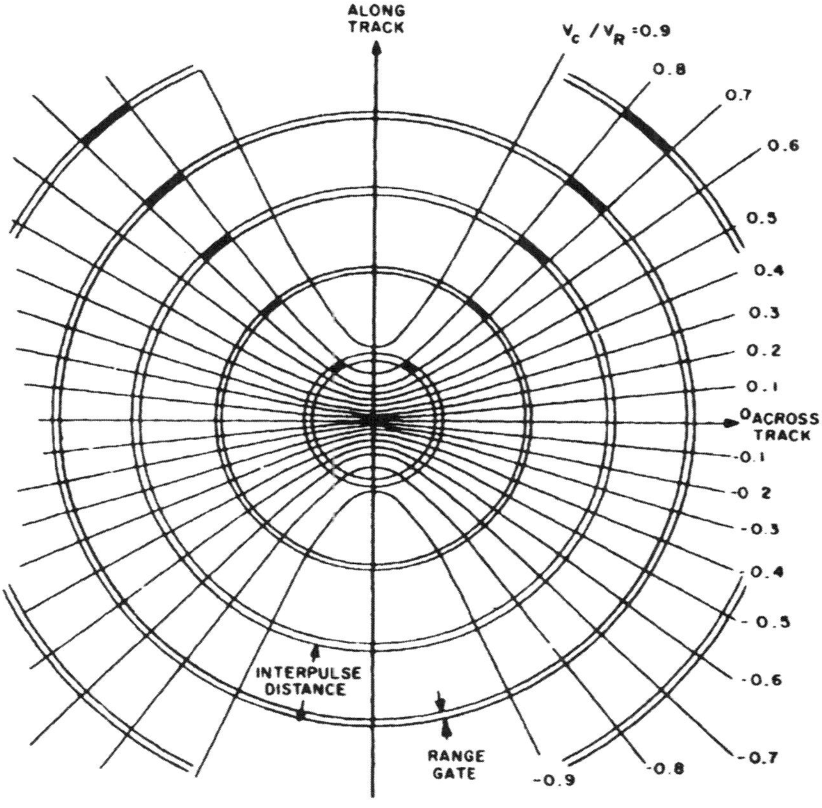

Figure 1.18 Ground clutter isodops from airborne pulsed Doppler radar. (*From:* [31, p. 19-8].)

is the same, the return from every point along the hyperbolic line has the same Doppler frequency.

The ground clutter spectrum associated with a particular Doppler frequency increment is obtained by summing those returns from the incremental shaded areas (shown in Figure 1.18) between two isodops separated by the incremental frequency difference. The ground clutter power spectral density (PSD) for a particular frequency increment can then be obtained through the following procedure:

1. Compute the incremental surface area represented by each shaded area.
2. Multiply each incremental surface area by the backscattering coefficient (σ^0 with dimensions m^2/m^2) for the appropriate terrain and grazing angle to obtain the equivalent radar cross-sectional area.
3. Substitute this cross-sectional area into the radar equation with appropriate weighting for the two-way antenna pattern and variable slant range to obtain the incremental power returned from each elemental area.
4. Sum the returns from all range-ambiguous incremental areas to obtain the overall incremental PSD.
5. Repeat the process for all incremental frequencies over $f_d(\text{max}) = \pm 2v_f/\lambda$ to obtain the complete spectrum.

The resulting shape of the PSD for ground clutter associated with airborne PD radar is generally like that shown in Figure 1.19. Note that for a stationary PD radar the ground clutter spectrum becomes concentrated about each spectral line, which is similar to that of an MTI radar (recall Figure 1.9), except that the power must be summed over the various range ambiguities. The width of the spectrum for the stationary case is due to internal motion and scanning-induced spectral spreading.

1.3.4.2 Optimum Pulsed Doppler Filter Theory [36, 41]

The general structure of an optimum PD filter is that of a linear complex-weight transversal filter followed by an envelope detector, which is identical to that of the optimum MTI filter previously discussed. However, the weights of the linear transversal filter are optimized under the assumption that the Doppler frequency of the target is precisely known at a particular point in the Doppler domain. The range of possible target Dopplers is then filled with a suitable number of point-optimized, complex-weighted transversal Doppler filters. The Doppler filters are usually separated by a frequency equal to PRF/n, where n is the number of pulses in a CPI.

A quasi-optimum PD processor used in a number of applications is the cascaded MTI coherent integrator, in which the coherent integrator is implemented by using the FFT algorithm. This processor generally provides fewer detection performances than the optimum transversal filter, but utilizes considerably less hardware.

1.4 The Environmental Diagram [1, 35]

The principal means of discerning target echoes from clutter returns in a PD or MTI radar is through Doppler processing. Consequently, the way in which the clutter is distributed across the bands of possible frequencies, the clutter's Doppler spectrum, and the relationship of the spectrum to the Doppler frequencies of anticipated targets critically influence the radar's design. One way to examine the clutter and target's frequency distribution is by using the environmental diagram, which provides a pictorial representation of the clutter model.

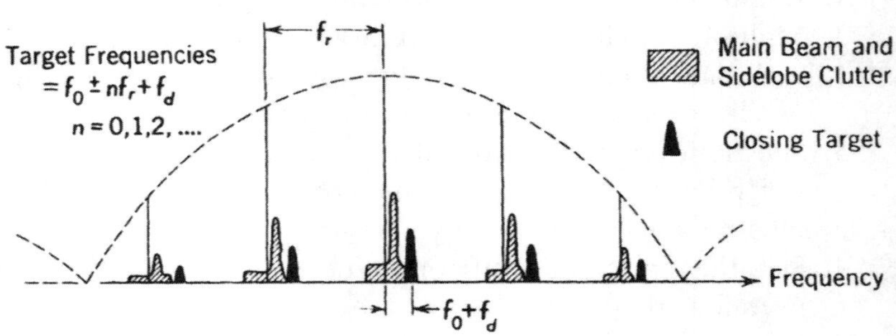

Figure 1.19 Power spectral density for an airborne pulsed Doppler radar. (*From:* [2, p. 323].)

An environmental diagram for a low-PRF, L-band (1.3-GHz) air traffic control (ATC) radar is depicted in Figure 1.20 [1]. The limits on the radial velocity of various types of clutter are given on the ordinate, and the range extent is indicated along the abscissa. In the diagram, land clutter is shown to extend from the origin to 20 nmi, which represents the radar horizon for this particular location. The vertical extent of the bar is the standard deviation of the velocity spectrum for ground clutter returns from undeveloped wooded land with a 30 knot wind. Ground clutter is depicted by a solid bar to indicate that it has the highest amplitude return of the clutter shown.

Sea clutter is modeled for a sea state 4 condition, where a mean velocity of 3.4 m/s and a standard deviation of 1 m/s are representative. The mean velocity considers the radar to be looking directly into the wind or waves. The cross-hatching in Figure 1.20 is denser near the radar, indicating that the reflectivity is higher near the radar than at lower grazing angles near the horizon. As with ground clutter, sea clutter extends to the radar horizon.

Rain clutter is modeled with an initial mean velocity of 28 m/s, which is appropriate for a moderate altitude with a wind driving the rain directly toward the radar. Like sea clutter, the mean velocity of the precipitation clutter will drop

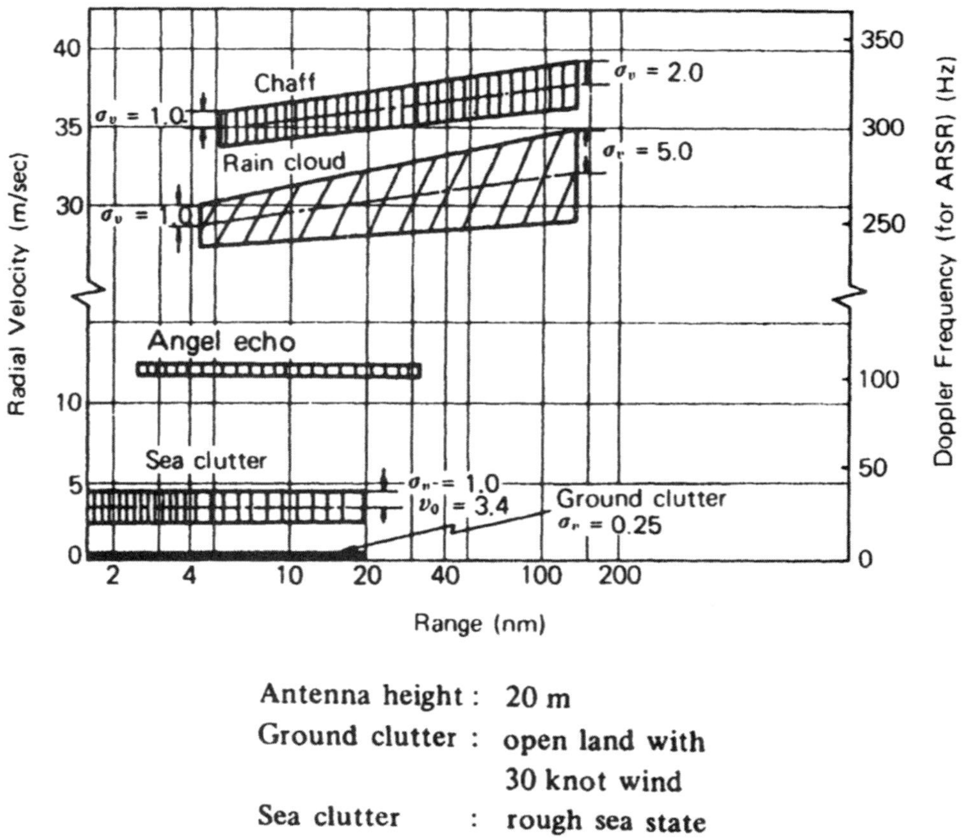

Figure 1.20 Environmental diagram for ATC radar. (*From:* [1, p. 425].)

to zero under crosswind conditions. A linear gradient in the wind velocity with altitude (called wind shear) is postulated [42]. This causes the mean velocity of the rain clutter to increase (a linear model is assumed) with range due to the increased height intercepted by the center of the radar beam. The wind shear effects also cause the standard deviation of the rain clutter to increase with range due to the increasing vertical extent subtended by the radar beam at longer ranges. The rain clutter is shown extending to the full operating range of the radar. The standard deviation of the rain clutter spectrum varies from 1 to 5 m/s at the radar's maximum range.

A high-altitude chaff corridor is modeled with characteristics analogous to those of rain clutter. The amplitude is assumed to be greater than the rain clutter, but with reduced vertical extent. The standard deviation of the chaff spectrum varies from 1 to 2 m/s at the radar's maximum range.

Angel clutter is generally due to reflections returned from flocks of migrating birds. The return is primarily from the bird's body, which for large birds is resonant at L-band [36, p. 231]. The spectral spread is thus due to the differential radial velocity among the various birds forming the flock, which is assumed to be relatively narrow. The mean velocity is modeled at 15 m/s, which assumes that the birds are generally moving in a radial direction with respect to the radar. The angel return is shown to a radar horizon of about 30 nmi, which corresponds to a target height of 300 ft for a 20 m antenna height.

The environmental diagram shows the various radial velocity parameters expressed in meters per second, which can be converted to Doppler frequency (as in Figure 1.20) through multiplying by the factor $2/\lambda$, where λ is the radar's wavelength in meters. The cross-hatching indicates the normalized reflectivity per unit frequency, rather than the received PSD. Thus the cross-hatching for the rain and chaff shown in Figure 1.20 has a low density in the shear region, indicating that the PSD of these echoes is lower for given reflectivity than for signals with narrow spectral width, such as ground clutter. This may be significant for pulsed Doppler radars having a Doppler bandwidth that is less than the standard deviation of the clutter spectral spread.

Figure 1.21 depicts typical received power as a function of range for the various clutters shown in the environmental diagram [1]. Also shown are the signal strength of typical aircraft targets at altitudes of 10,000 and 30,000 ft as received by the radar's dual-beam antenna system. Note that the returns from the various clutters generally would prevent detection of the targets without the use of some form of Doppler processing.

The ground clutter return is the strongest shown in the diagram. The return exceeds the receiver noise level by more than 100 dB and the target returns by 20 to 50 dB. Precipitation returns are comparable to target returns, but will increase as the fourth power of frequency if the radar's L-band transmitting frequency is increased. Sea clutter returns generally exceed target returns, but require less suppression than ground clutter to allow target detection. At short ranges, angel clutter returns are comparable to target returns and generally exceed receiver noise levels.

An alternative representation of the environmental diagram is shown in Figure 1.22 [35]. In this three-dimensional presentation, the power level of the return is

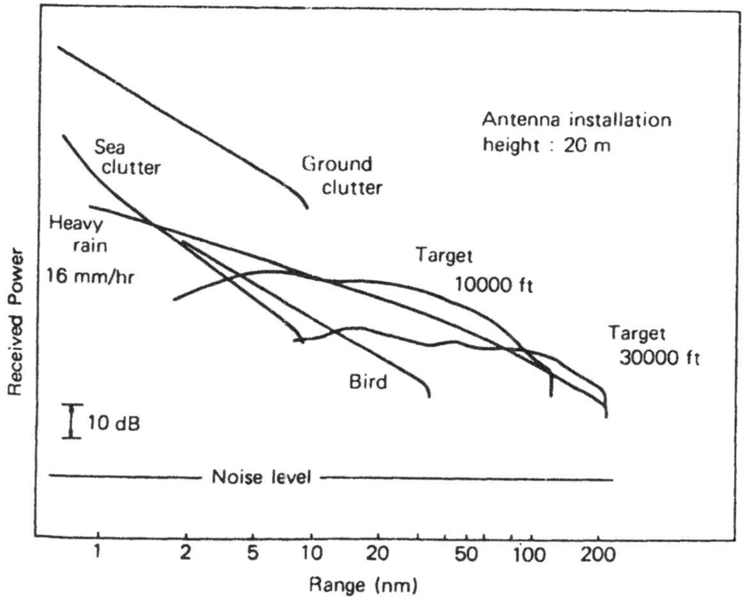

Figure 1.21 Typical received power frcm clutter and targets. (*From:* [1, p. 425].)

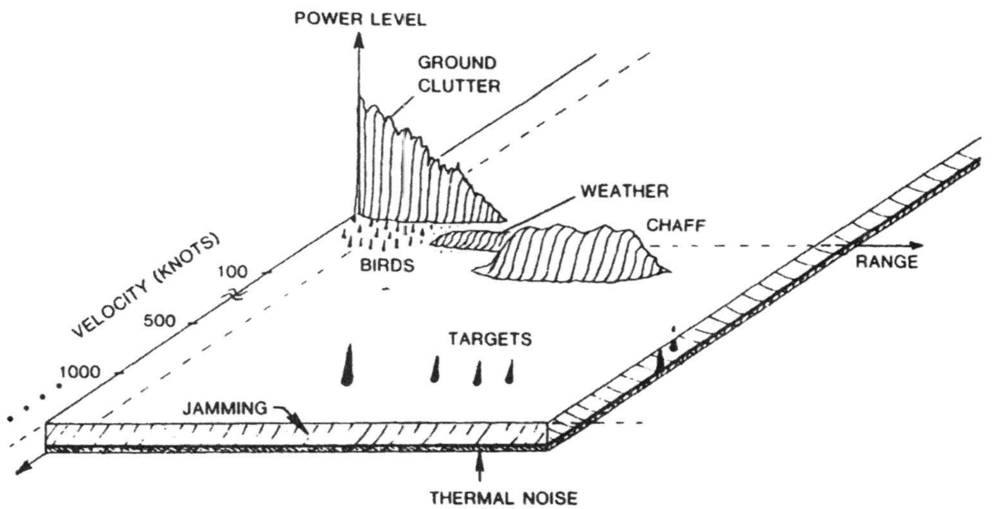

Figure 1.22 Alternative form of environmental diagram. (*From:* [35, p. 29].)

shown as the ordinate and the range-velocity matrix is shown in the orthogonal horizontal plane. Thus, this form of the environmental diagram presents the information of Figures 1.15 and 1.16 in a single diagram.

The coordinate system used by both the alternative environmental diagram and the radar ambiguity diagram (see Figure 1.7) are the same. By drawing both diagrams using the same scale, we are able to gain insight into the clutter response of different waveform designs. Preferably, with this method, the ambiguity diagram is drawn as a transparent overlay so that its origin can be superimposed on the positions of the various clutter scatterers in the environmental diagram [42]. This presentation would then reveal an intersection of a high sidelobe of the radar waveform with, for example, a low-intensity rain clutter return, or, alternatively, the intersection with a high-intensity ground clutter return, which would reveal a problem.

The environmental diagram reveals the wide variability of clutter in the range, velocity, and amplitude dimensions. One characteristic is the wide dynamic range of clutter, which when coupled with the desirability of linear processing, presents a formidable problem to achieving the optimum Doppler filter processing described in the previous sections. This has led to a number of clutter processing procedures to supplement the Doppler filtering processing. The general objectives of these techniques are: (1) to reduce the clutter intake of the radar (e.g., pulse-compression, dual-beam receiver antenna, circular polarization), (2) to normalize the clutter return (e.g., sensitivity time control, instantaneous automatic gain control), and (3) to suppress or censor particular range cells that have heavy clutter returns (e.g., clutter maps, cell-averaging CFAR).

Use of the last of these techniques is a tacit admission that building a practical Doppler processor that suppresses all clutter returns may be impossible. The philosophy of dealing with this clutter "poke-through" is then to desensitize the radar in those resolution cells where processing the clutter is impossible, so that an inordinately large number of false alarms do not occur.

1.5 Clutter Models [1]

A clutter model assumption is implicit in the design or analysis of MTI or PD radar. The function of the clutter model is to represent the random clutter process in such a manner as to determine system performance. Ideally, the model accurately reflects the clutter process in a realistic matter. Practically, the clutter database is rarely adequate to achieve this objective, and the art of modeling the clutter involves a compromise between accurate representation and analytic convenience.

1.5.1 The Gaussian or Rayleigh Clutter Model [1, 43]

The simplest clutter model is the Gaussian model, whereby the random clutter process is represented as a multidimensional Gaussian process. This model has been employed almost exclusively in MTI and PD radar analysis. An attractive feature of this model is that the clutter process is completely specified by the mean value and covariance function. The Gaussian model assumption leads to an optimum Doppler processor that is linear.

The Gaussian or Rayleigh envelope clutter model is applicable to distributed clutter return from a spatially continuous distribution of scatterers where no subset of scatterers predominates. This type of clutter is usually associated with weather clutter, chaff, sea clutter observed with low-resolution radar (pulsewidth $> 0.5\ \mu s$) or with a high-resolution radar at high grazing angles ($\psi > 5°$), and land clutter observed from high grazing angles ($\psi > 5°$) over undeveloped terrain.

In the Gaussian model, the clutter signals in the predetection portions of the radar receiver can be represented as a random process as follows:

$$c_t = x_t \cos \omega_c t - y_t \sin \omega_c t \tag{1.8}$$

The term *random process* connotes that the value of c_t does not depend in a completely definite way on the independent variable t as will be the case for a causal process. The clutter signal information is carried by the jointly normal, zero-mean, identically spectrally distributed, lowpass, independent quadrature processes x_t and y_t. A significant property of the c_t process is that any linear operation (such as Doppler filtering) on the c_t signal does not disturb the Gaussian nature of the distribution, but rather modifies its covariance matrix. Quadrature detection is often used to extract the slowly varying lowpass (the time variation is consistent with the radar's processing bandwidth) quadrature processes x_t and y_t, which are then jointly processed. When Doppler filtering is performed in an MTI and PD radar on the quadrature x_t and y_t processes, the act is generally termed *coherent* (predetection) processing.

The voltage envelope of c_t is given by the random process

$$v_t = \sqrt{x_t^2 + y_t^2} \tag{1.9}$$

which represents the signal obtained after linear envelope detection. The first probability density function (pdf) process of v_t describes the probability of finding the envelope-detected voltage v_t in the range $(v, v + dv)$ at any specific time. The envelope-detected voltage v_t has a Rayleigh distributed first pdf given by

$$p(v) = \frac{v}{\sigma^2} \exp\left(-\frac{v^2}{2\sigma^2}\right) \tag{1.10}$$

where σ is related to the clutter power by the relation

$$P_c = 2\sigma^2 \tag{1.11}$$

For this reason the clutter model is often termed the *Rayleigh clutter model*. However, note that the equivalent radar cross-section (RCS) of the clutter does not follow a Rayleigh pdf, but rather an exponential pdf, as described later.

1.5.2 The RCS of Clutter

The radar cross section, σ, of an object is defined as the area intercepting that amount of power which, when scattered isotropically, produces an echo equal to that received from the object [44]. For example, a sphere with radius a that is large with respect to a wavelength $(a > \lambda)$ scatters isotropically and its mean RCS is given by

$$\overline{\sigma} = \pi a^2 \tag{1.12}$$

The power (P_R) received from a target at range R can be computed using the standard radar equation given by

$$P_R = \frac{P_t G_t^2 \lambda^2 \overline{\sigma}}{(4\pi)^3 R^4} \tag{1.13}$$

where P_t is the transmitted power, G_t is the antenna's gain, and λ is the radar's wavelength. Note from this relationship that the mean echo power received by the radar is directly proportional to the mean RCS ($\overline{\sigma}$).

The RCS of extended targets is generally a sensitive function of orientation and wavelength and hence exhibits a range of values. For this situation, an appropriate way to describe the target's RCS is by using its pdf $p(\sigma)$, the average value of which is given by $\overline{\sigma}$. Through examination of the standard radar equation, one can deduce that the power received by the radar from an extended target exhibits a probability distribution identical to that of its RCS and, hence,

$$p(P_R) = p(\sigma) \tag{1.14}$$

Because power is proportional to the voltage squared, the relationship

$$p(v^2) = p(\sigma) \tag{1.15}$$

is established. Hence, by applying the transformation $\sigma = v^2$ to the Rayleigh voltage distribution for clutter, we can determine that the first probability distribution for the RCS of the Gaussian clutter model is given by the exponential distribution

$$p(\sigma) = \frac{1}{\overline{\sigma}} \exp\left(\frac{\sigma}{\overline{\sigma}}\right), \qquad \sigma \geq 0 \tag{1.16}$$

where $\overline{\sigma}$ is the clutter's mean RCS. Note that this distribution is completely determined by its mean value $\overline{\sigma}$ and also that the clutter's RCS (σ) is a statistical quantity that varies over a wide dynamic range.

1.5.2.1 Normalized Clutter Backscattering Coefficient

The return from surface or volume clutter is caused by reflections from all scatterers contained within the radar's resolution cell (see Sections 1.3.2 and 1.3.3). Conventional use of the concept of the equivalent RCS replaces a large amount of elemental clutter scatterers contained within a radar resolution cell with a single equivalent scatterer, which returns the same amount of power to the radar and exhibits the same statistical characteristics.

To calculate the RCS of this equivalent scatterer, a normalized parameter called the *backscattering coefficient per unit area* σ^0 (m^2/m^2) is used for surface clutter, and the *backscattering coefficient per unit volume* η (m^2/m^3) applies to volumetric clutter. The equivalent clutter RCS is then obtained by multiplying the appropriate backscattering coefficient by either the surface area or volume contained within the radar's resolution cell. The effective RCS for surface clutter is then given by

$$\sigma_{cx} = \sigma^0 A_c \tag{1.17}$$

where A_c is the surface area of the clutter's resolution cell (see Figure 1.3) . Note that the clutter's RCS (σ_{cx}) is proportional to σ^0 so that those variates have identical statistical properties. In general, the conventional way to specify the backscattering coefficient is by its mean value σ^0, which is extensively tabulated. For low grazing angles ($\psi < 10°$) and narrow azimuthal beamwidths, the resulting mean clutter power received at the radar can then be found by substituting the clutter's RCS value

$$\overline{\sigma_{cx}} = \overline{\sigma^0} R \theta_{az} \left(\frac{c\tau_e}{2}\right) \sec(\psi) \tag{1.18}$$

into the standard radar equation, where σ^0 is the tabulated clutter's mean reflectivity per unit surface area, R is the slant range to the clutter patch, θ_{az} is the one-way radar's 3 dB azimuth beamwidth in radians, τ_e is the radar's effective pulsewidth, and c is the velocity of light.

The use of the one-way azimuth beamwidth θ_{az} is in accordance with Goldstein's original formulation [43, p. 484] that defined σ^0. This is significant because σ^0 is derived experimentally by measuring the power returned from a clutter patch and then adjusting this using Goldstein's formulation, which includes the use of the antenna's one-way azimuth beamwidth. Thus, any beam shape loss or pattern propagation factor is embedded in the value of σ^0 and, hence, it would be redundant to include it by including loss terms modifying (1.18) [44, p. 36]. Another viewpoint that supports the use of the one-way azimuth beamwidth is expressed in [36, p. 177; 40, p. 15.10]. These references assert that the large variation of experimentally determined σ^0 (see Figure 4.48) makes it insignificant to include beam shape factors on the order of 1.5 dB. Several references [18, p. 406; 42, p. 274] advocate the use of the two-way azimuth beamwidth, which they claim provides higher accuracy than the one-way azimuth beamwidth. This would be correct if σ^0 were derived using the two-way azimuth beamwidth. However, this evidence is not generally available and the two-way azimuth beamwidth is never included in a radar specification.

Figure 1.23 shows the projection of the clutter patch's surface area onto a plane perpendicular to the radar line of sight. This projected incident area (A_i)

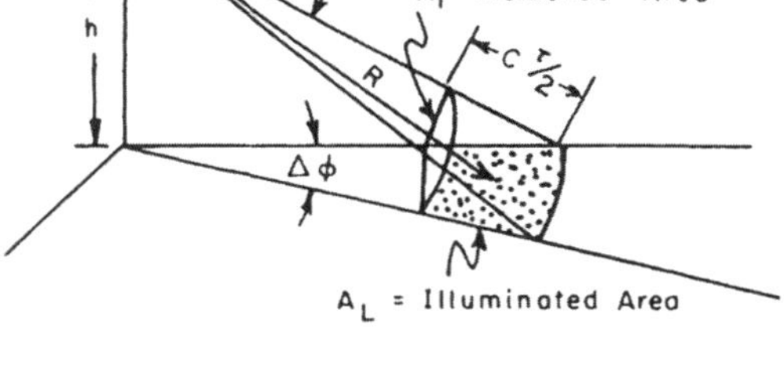

Figure 1.23 Surface clutter area geometry. (*From:* [46, p. 171].)

Table 1.3 Typical Backscattering Coefficients

Terrain	σ^0 *(dB)*	γ *(dB)*	
Smooth water	–53	–45.4	
Desert	–20	–12.4	$\psi = 10°$
Wooded area	–15	–7.4	$f = 10$ GHz
Cities	–7	0.6	

determines how much transmitted energy will be intercepted by the surface area (A_L) and is given by the relation [45]

$$A_i = \frac{c\tau_e}{2} \tan(\psi) R\theta_{az} \tag{1.19}$$

Note that A_i generally conforms to the usual concept of the capture area of a radar target and, through multiplication by a proportionality factor γ, can be transformed into the effective clutter RCS. The resulting RCS is given by [45]

$$\sigma_{cx} = \gamma \frac{c\tau_e}{2} \tan(\psi) R\theta_{az} \tag{1.20}$$

Comparing this to the alternative representation provides the relationship

$$\sigma^0 = \gamma \sin(\psi) \tag{1.21}$$

This representation is sometimes called the *constant Gamma model* for terrain backscatter [47], and its advantage over using the σ^0 representation is that the parameter is independent of the grazing angle. Typical surface backscattering coefficients for both σ^0 and γ are given in Table 1.3 for a 10° grazing angle and an X-band radar [22, p. 296].

► Example 2 Backscatter Coefficient (σ^0) from Constant Gamma Model

The parameters for the constant gamma sea clutter model are given in Table EX-2.1; those for the land clutter model are given in Table EX-2.2 [48]. Note that the sea clutter model is a function of frequency, whereas the land clutter model is independent of frequency.

The constant gamma model is related to the backscatter coefficient (σ^0) by the relation

$$\sigma^0 = \gamma \sin(\psi)$$

where ψ is the grazing angle. Gamma is usually expressed in decibels and can be found from the relation

$$\gamma_{\text{dB}} = \text{product} - 10 \log(\lambda)$$

Table EX-2.1 Sea Clutter Parameters

SS	*Product* $\lambda\ \gamma$ *(dB)*
0	–58
1	–52
2	–46
3	–40
4	–34
5	–28
6	–22
7	–16
8	–10
9	–4

Table EX-2.2 Land Clutter Parameters

Terrain Type	γ *(dB)*
Mountains	–5
Urban	–5
Wooded Hills	–10
Rolling Hills	–12
Farmland, Desert	–15
Flatland	–20
Smooth Surface, Roads	–25

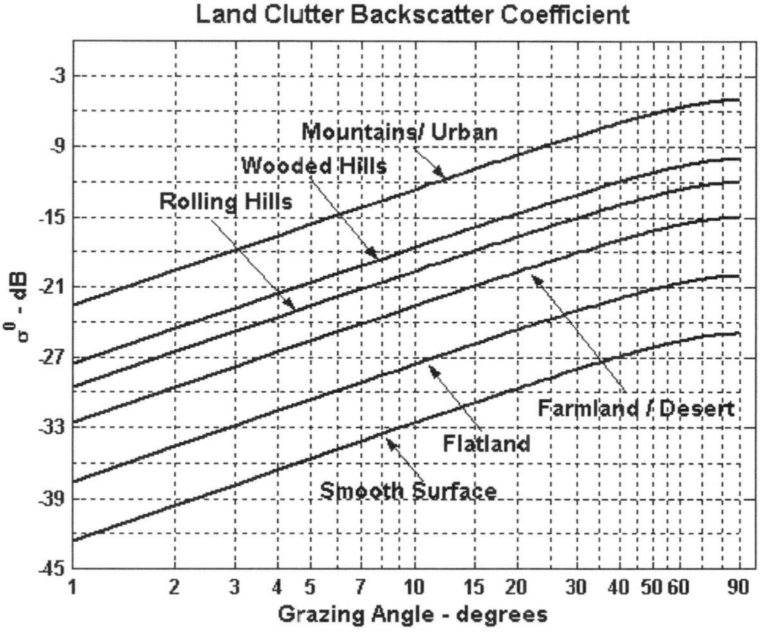

Figure EX-2.1 Land clutter backscatter coefficient.

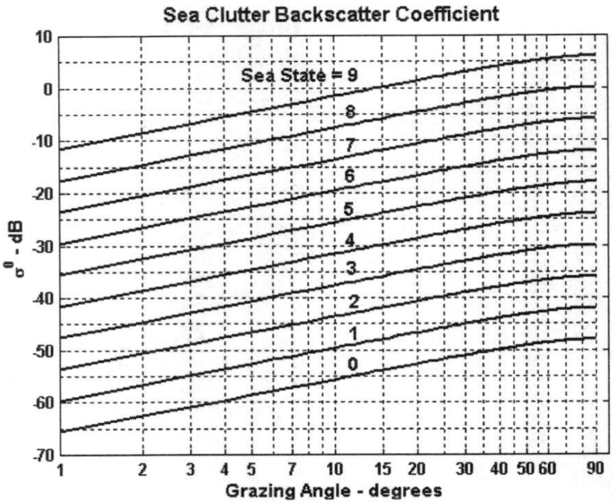

Figure EX-2.2 Sea clutter backscatter coefficient.

The MATLAB Program (land_clt_bartongl.m) computes the backscatter coefficient (σ^0 in decibels) as a function of the grazing angle ψ for land clutter. The backscatter coefficient (σ^0 in decibels) is plotted in Figure EX-2.1 against the grazing angle ψ for all terrain types using the MATLAB function semilogx.

A second MATLAB program, sea_clt_bartongl.m, computes the backscatter coefficient σ^0 in decibels as a function of the grazing angle ψ for sea clutter. The user input to this program is frequency. The backscatter coefficient (σ^0 in decibels) is plotted in Figure EX-2.2 against the grazing angle ψ for all sea states using the MATLAB function semilogx. A plot for $f = 3$ is shown in Figure EX-2.2.

▶

The effective RCS of weather or chaff clutter is found through multiplying the radar's volumetric resolution cell (recall Figure 1.12) by the volumetric backscatter coefficient η (m^2/m^3). The volumetric backscatter coefficient can be approximated over the microwave region for weather clutter in m^2/m^3 as

$$\eta = 7f^4 r^{1.6} 10^{-12} \tag{1.22}$$

where f is the radar frequency in gigahertz, and r is the rainfall rate in millimeters per hour [18]. The fourth-power variation with frequency generally makes this type of clutter significant at frequencies above UHF for most radar. The radar's volume resolution cell can be found by multiplying the areas subtended by the radar's circular antenna beam pattern (possibly elliptical for fan beams) at a particular range by the range equivalent to the radar's effective transmitted pulsewidth. The effective RCS of weather clutter is given by

$$\sigma_{cx} = \frac{\pi \eta \theta_{az} \phi_{el} R^2 \left(\frac{c\tau_e}{2}\right)}{4} \tag{1.23}$$

where η is the backscattering coefficient in m^2/m^3, θ_{az} is the one-way azimuth beamwidth in radians, ϕ_{el} is the one-way elevation beamwidth in radians, R is the range in meters to the resolution cell, c is the velocity of light in meters per second, and τ_e is the radar's effective pulsewidth in seconds [18]. Note that the one-way antenna beam patterns are used in the relationship for the effective RCS [40, p. 15.10], which assumes η includes the effect of beam shape factors ($L_p = 2.468$ dB for two coordinates).

The ratio of the clutter power to target power in a resolution cell is given by

$$\frac{P_c}{P_t} = \frac{\sigma_{cx}}{\sigma_t} \tag{1.24}$$

where σ_{cx} is the mean clutter effective RCS, and σ_t is the target's mean RCS. The above relation assumes that beam shape losses for the target and rain clutter are equal [42, p. 234]. Generally the clutter power exceeds the target power and MTI or PD processing is required to allow target detection. The *Improvement Factor* (I) defines the average increase in the signal-to-clutter power ratio due to Doppler filtering and is given by [1]

$$I = \frac{\left(\frac{P_S}{P_C}\right)_O}{\left(\frac{P_S}{P_C}\right)_{IN}} \tag{1.25}$$

The signal-to-clutter power at the output of the Doppler filter can then be found as

$$\left(\frac{P_S}{P_C}\right)_O = \frac{\sigma_t I}{\sigma_{cx}} \tag{1.26}$$

Solving this relationship for the Doppler filter improvement factor results in

$$I = \frac{\left(\frac{P_S}{P_C}\right)_O}{\frac{\sigma_t}{\sigma_{cx}}} \tag{1.27}$$

where the detection performance desired from the MTI or PD radar (e.g., $P_d = 0.9$, $P_{fa} = 10^{-6}$) determines the required output signal-to-clutter performance (e.g., $P_s/P_c = 13.2$ dB).

1.5.3 The Clutter Spectrum and Autocorrelation Function

A primary feature of clutter, which makes it distinguishable from receiver noise, is that clutter is highly correlated from one radar pulse to the next, whereas receiver noise is usually independent on a pulse-to-pulse basis. This property is reflected in the clutter's PSD or equivalent autocorrelation function.

The first pdf (see Section 1.3.1), which gives the distribution of the dynamic range or amplitude fluctuation of the clutter, and the second pdf, $p(v_1, v_2; \tau)$, which gives the differential probability of measuring v_t in the range $(v_1, v_1 + dv_1)$ at time t_1 and v_t in the range $(v_2, v_2 + dv_2)$ at time t_2, describes the rate of clutter fluctuations. Because we are assuming a stationary process (statistics are invariant with time), time enters into consideration only as the difference $\tau = t_1 - t_2$. As τ approaches zero, we would expect the values of v_1 and v_2 to coalesce and, as τ becomes infinite, the values would become independent of one another, with each being determined by its respective first pdf. This action is described by a parameter called the *correlation coefficient* (ρ), which is one if $\tau = 0$, indicating perfect correlation and zero as τ increases to infinity, indicating no correlation.

The autocorrelation function $R(\tau)$ is obtained by finding the expected value or average of the product of values of the clutter process v_t at two times $(t, t + \tau)$:

$$R(\tau) = E(v_t v_{t+\tau}) \tag{1.28}$$

When $\tau = 0$, this second-order moment function assumes its maximum value equal to the mean squared value (power) of v_t. For a stationary random process, the correlation coefficient ρ can be expressed in terms of the autocorrelation function as

$$\rho = \frac{R(\tau) - m^2}{\sigma^2} \tag{1.29}$$

where m is the mean, and σ^2 is the variance of the process. The function

$$M(\tau) = R(\tau) - m^2 \tag{1.30}$$

called the *covariance function*, is often used by statisticians to describe the random process. In pulsed radar systems, the covariance function is defined only at pairs of instants (e.g., $t_1, t_2; t_1, t_3$) for which the signals are defined. The values of the covariance function can then be arranged in covariance matrix form:

$$\mathbf{M}_x = \sigma_x^2 \begin{bmatrix} 1 & \rho_{12} & \rho_{13} & \cdots & \rho_{1n} \\ \rho_{21} & 1 & \rho_{23} & \cdots & \rho_{2n} \\ \rho_{31} & \rho_{32} & 1 & \cdots & \rho_{3n} \\ \vdots & \vdots & \vdots & \cdots & \vdots \\ \rho_{n1} & \rho_{n2} & \rho_{n3} & \cdots & 1 \end{bmatrix} \tag{1.31}$$

The covariance matrix is extensively used in MTI analyses to determine the effect of Doppler filtering on the random clutter process.

For random processes another useful concept is to decompose signal power into elemental frequency components, this decomposition being described by the PSD function. Frequency generally implies a rate at which the signal changes. When the signal has high-frequency components, it is expected to exhibit variations in the time domain, which will be reflected in the autocorrelation function. Because the autocorrelation function is a time-domain signal, it can be Fourier transformed to obtain a frequency-domain signal. In accordance with the Wiener-Khintchine theorem, the Fourier transform of the autocorrelation function of a random process results is the PSD of that process. The PSD of various clutters has been measured and is a fundamental parameter used in the analysis of MTI and PD radars.

The PSD associated with the randomly fluctuating portion of ground clutter RCS (discrete clutter targets are excluded) is generally given by the Gaussian-shaped spectrum

$$S_p(f) = \frac{P_c}{\sqrt{2\pi\sigma_f^2}} \exp\left(\frac{-f^2}{2\sigma_f^2}\right) \tag{1.32}$$

where P_c is the clutter power and σ_f is the standard deviation of the clutter spectrum, which is related to the rms velocity clutter spread (σ_v) by [1]

$$\sigma_f = \frac{2\sigma_v}{\lambda} \tag{1.33}$$

The autocorrelation function, which is found by the Fourier transform of the PSD, is given by

$$R(\tau) = P_c \exp\left(\frac{-\tau^2}{2\sigma_\tau^2}\right) \tag{1.34}$$

where

$$\sigma_\tau = \frac{1}{2\pi\sigma_f} \tag{1.35}$$

The normalized autocorrelation coefficient is given by

$$\rho_p = \frac{R_p(\tau)}{P_c} = \exp\left(\frac{-\tau^2}{2\sigma_\tau^2}\right) \tag{1.36}$$

Typical power spectral densities for various types of the clutter at 1 GHz are shown in Figure 1.24 (also see Table 1.4).

In coherent MTI processing, the PSD and autocorrelation function of the quadrature random processes x_t and y_t are of interest. We can show that

$$\rho_p = \rho_x^2 = \rho_y^2 \tag{1.37}$$

relates the normalized autocorrelation function of the clutter power envelope (proportional to clutter RCS) and the clutter x_t and y_t processes [43]. Thus the PSD

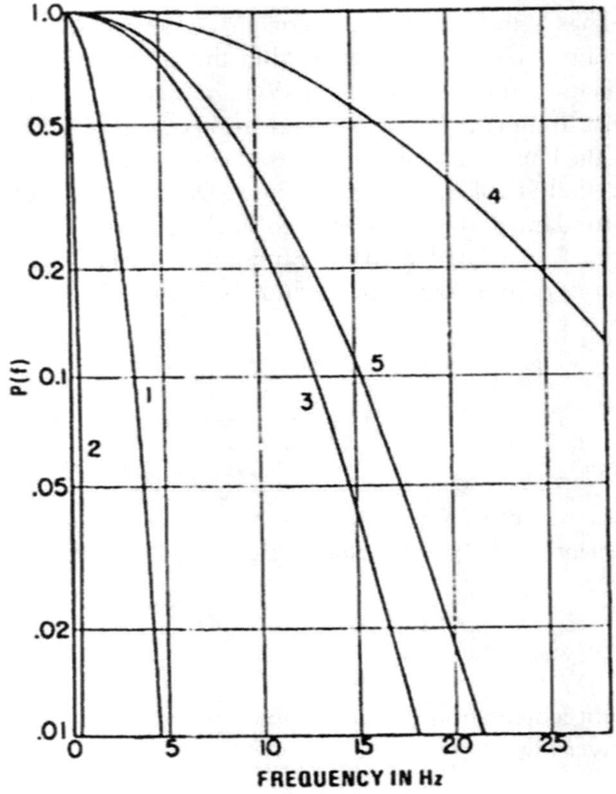

Figure 1.24 Power spectral densities of various types of clutter. (*From:* [44, pp. 159–160].)

Table 1.4 Clutter Types and Coefficient for $P(f) = \exp[-a(f/f_0)^2]$

Clutter Types	*a*	*Curve in Figure 1.24*
Heavily wooded hills, 20-mph wind	$2.3 \cdot 10^{17}$	1
Sparsely wooded hills, calm day	$3.9 \cdot 10^{19}$	2
Sea echo, windy day	$1.4 \cdot 10^{16}$	3
Rain clouds	$2.8 \cdot 10^{15}$	4
Chaff	$1.0 \cdot 10^{16}$	5

of the power envelope can be found by convolving the power spectral density of the coherent quadrature components, resulting in

$$S_p(f) = S_x(f) * S_x(f) \tag{1.38}$$

Hence, for Gaussian-shaped spectra, all PSDs and autocorrelation functions are Gaussian shaped, and the standard deviation of the power envelope can be found from those of the coherent quadrature components by using

$$\sigma_p^2 = \sigma_x^2 = \sigma_y^2 \tag{1.39}$$

When working with experimental and published data, care must be exercised to ascertain whether the standard deviation cited refer to the coherent (σ_x) or noncoherent (σ_p) standard deviations.

The environmental diagram (refer to Figure 1.20) gives some idea of the spectral spread and mean frequency of various types of clutter. In applications involving sea clutter, weather clutter, chaff, and angel clutter, we need to model clutter that has relative radial motion with respect to the radar. This is accomplished by modifying the Gaussian-shaped ground clutter PSD by the mean Doppler shift frequency (f_d), resulting in

$$S_p(f) = P_c \exp\left[\frac{-(f - f_d)^2}{2\sigma_f^2}\right] \tag{1.40}$$

Experimental evidence [44, 49] suggests that the clutter spectrum may have significantly longer tails than indicated by a Gaussian-shaped spectrum. A model suggested for ground clutter [1, 49], measured with an X-band radar, postulates a power law spectral density shape given by

$$S(f) = \frac{k}{\left(1 + \left|\frac{f}{f_c}\right|^n\right)} \tag{1.41}$$

where n varies from 3.3 to 2.2 and f_c from 0.8 to 1.9 depending on radar frequency and wind speed [18]. The power law dependence results in a significantly greater magnitude at high frequencies than the Gaussian-shaped spectrum.

More recent measurements employing a wide dynamic range receiver indicate that the clutter spectrum from windblown trees can be closely approximated by using an exponentially shaped spectrum down to the –60-dB power levels [50, 51]. Typical measured spectra on both calm and windy conditions are depicted in Figure 1.25. The implication of these measurements is that the previously measured spectral data may have been distorted by receiver nonlinearities that caused wider spectral spreads at low power levels (see Section 3.2.5 and Figure 3.20).

The power spectral density, valid from UHF to X-band, is given by the two-sided exponential

$$S(f) = \frac{r}{r+1}\delta(f) + \frac{1}{r+1}\frac{\beta\lambda}{4}\exp\left(-\frac{\beta\lambda}{2}|f|\right) \tag{1.42}$$

where β is the shape parameter whose typical values are given in Table 1.5, and r is the ratio of the steady component (reflection from nonmoving objects) to that of the fluctuating windblown components.

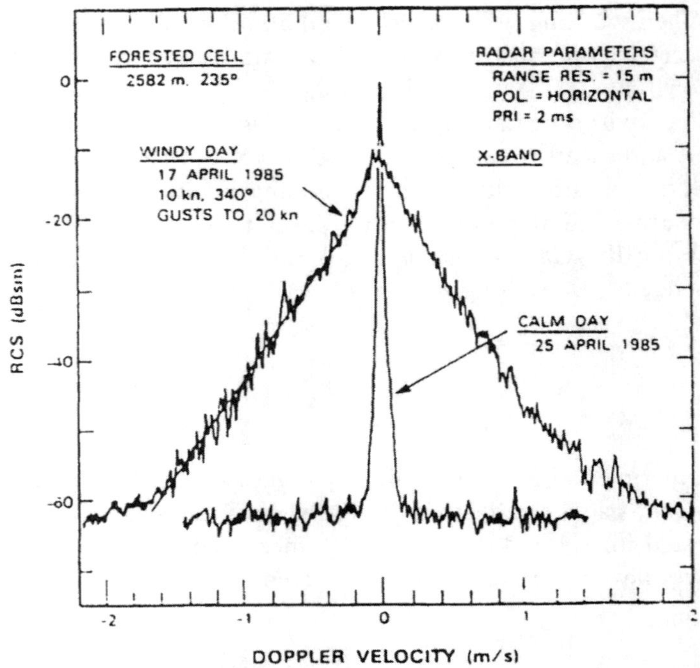

Figure 1.25 Power spectra of X-band radar from windblown trees. (*From:* [50, p. 14].)

Table 1.5 Exponential Shape Parameter as a Function of Wind Speed

		β (s/m)	
Wind Condition	*Wind Speed (knots)*	*Typical*	*Worst Case*
Light air	1–6	12	—
Breezy	6–12	8	—
Windy	12–25	5.7	5.2
Gale force wind	25–45	4.3	3.8

▶ Example 3 MTI Improvement Factor for Exponential Ground Clutter

The exponential ground clutter model describes the PSD for windblown trees and other vegetation over a frequency range from UHF to X-band. In this example we develop a MATLAB program that calculates the MTI improvement factor using binomial cancellers on the order of 1 through 7.

The covariance function of the clutter can be derived by applying the Fourier transform to the expression for the PSD [see (1.42)] resulting in

$$R(\tau) = \frac{r}{r+1} + \frac{1}{r+1}\frac{(\beta\lambda)^2}{(\beta\lambda)^2 + (4\pi\tau)^2}$$

In addition, we need two empirical relationships given by [51]

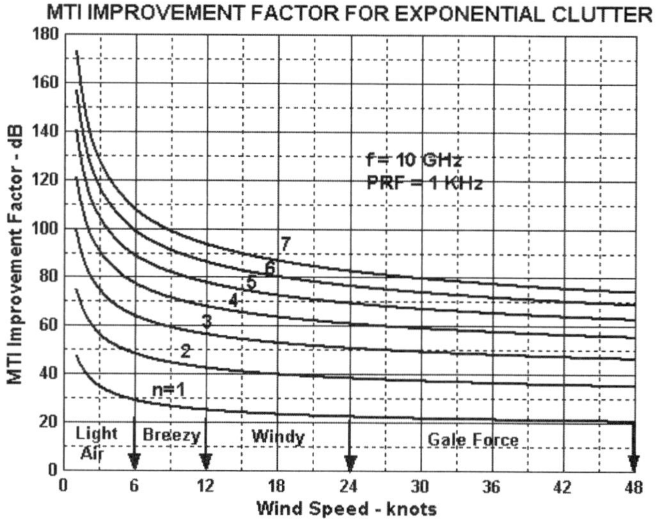

Figure EX-3.1 MTI improvement factor for windblown trees with binomial cancellers on the order of 1 through 7.

$$r = \frac{394}{w^{1.55} f_{\mathrm{GHz}}^{1.21}} \qquad \beta = \frac{1}{0.105\left(\log(w) + 0.476\right)}$$

where w is the wind speed in knots, and f_{GHz} is the frequency in gigahertz.

MTI binomial weights are given by:

$$w_k = (-1)^k \frac{n!}{k!(n-k)!}$$

where n is the number of MTI delay sections and k is the weight number 0, 1, 2, . . . , n. The MTI improvement factor is then given by:

$$I = \frac{\mathbf{w}^T \mathbf{M}_S \mathbf{w}^*}{\mathbf{w}^T \mathbf{R}_N \mathbf{w}^*} \qquad \mathbf{w}^T = [w_0, w_1, \ldots, w_n]$$

$$\mathbf{M}_S = \begin{pmatrix} 1 & 0 & 0 \\ 0 & 1 & 0 \\ 0 & 0 & 1 \end{pmatrix} \qquad \mathbf{R}_N = \begin{pmatrix} 1 & \rho_1 & \rho_2 \\ \rho_1 & 1 & \rho_1 \\ \rho_2 & \rho_1 & 1 \end{pmatrix}$$

$$\rho(iT) = R(\tau = iT)$$

where w^T is the transposed weight vector, $\mathbf{M}_S$ is the signal covariance matrix, and $\mathbf{R}_N$ is the clutter's covariance matrix. MATLAB program impfac_expgl.m gives the resulting plot for f = 10 GHz and PRF = 1,000 Hz is shown in Figure EX-3.1.

▶

1.5.4 Non-Gaussian Clutter Models

Experimental measurements indicate that when land or sea clutter is viewed with high-resolution radar (pulsewidth < 0.5 μs) or at low grazing angles ($\psi < 5°$), a larger dynamic range is observed than will be predicted by a Gaussian (Rayleigh envelope) clutter model. Figure 1.26 shows a series of photographs taken with a medium-resolution ($\theta_{az} = 1.3°$, $\tau = 2$ μs), ground surveillance radar viewing mountainous terrain clutter, where the clutter dynamic range exceeds 70 dB [31]. These photos were taken by inserting a variable attenuator into the receiver channel to reduce the amplitude of the clutter with respect to receiver noise.

The wide dynamic range of clutter presents several problems for MTI and PD radar. First, any nonlinearity inserted into the receiver to compress the dynamic range is known to spread the spectrum of the clutter. This significantly reduces the detection performance of linear Doppler filter processors, which are designed on the basis of the optimum Doppler filter theory discussed in Section 1.3.4.2. Second, the optimum Doppler filter theory, which leads to a linear processor and is based on a Gaussian clutter model, may no longer be applicable. These questions lead to the need for a non-Gaussian clutter model, which allows prediction of the detection performance of MTI and PD radar in clutter of wide dynamic range. Regardless of the specific model selected, the inherent detection performance of MTI or PD radar apparently will be degraded in clutter of wide dynamic range, compared to that which can be achieved with a Gaussian clutter model.

The determination of a non-Gaussian clutter model is a difficult problem that is the subject of much current research. From the theoretical viewpoint, we need to specify the complete multidimensional pdf given by

$$p(x_1, x_2, \ldots, x_n; t_1 + \tau, t_2 + \tau, \ldots, t_n + \tau) \tag{1.43}$$

which completely defines a stationary random process. For a stationary Gaussian random process, this information is embodied in its mean (amplitude) and covariance processes (or equivalent PSD) functions, which form the basis of most experimental measurements. The specification of non-Gaussian processes requires high-order moments (beyond the second-order covariance function), which are not generally available from experimental data. This situation has resulted in a number of approaches for specifying non-Gaussian clutter models, which results in an incomplete, although useful, specification of the non-Gaussian clutter process.

The first approach to modeling a non-Gaussian clutter process involves a specification of the first pdf $p(\sigma^0)$, which describes the variation of the clutter's backscattering coefficient per unit area σ^0 (m^2/m^2). Recall, that σ^0 is proportional to the effective clutter RCS ($A = \sigma^0 A_c$) as well as the power ($P_c = k\sigma_{cx}$) returned from the clutter cell so that all variates have the same statistical properties.

The variation of σ^0 may be measured on a spatial basis from cell to cell with the understanding that the clutter is homogeneous in the sense that the terrain over which the statistics are measured is the same for each cell used to form the statistics. The probability distribution formed for σ^0 on a ensemble basis can be used as the probability distribution for independent samples from a single cell, provided that the random clutter process is *ergodic* (i.e., ensemble and time statistics are identical) [52].

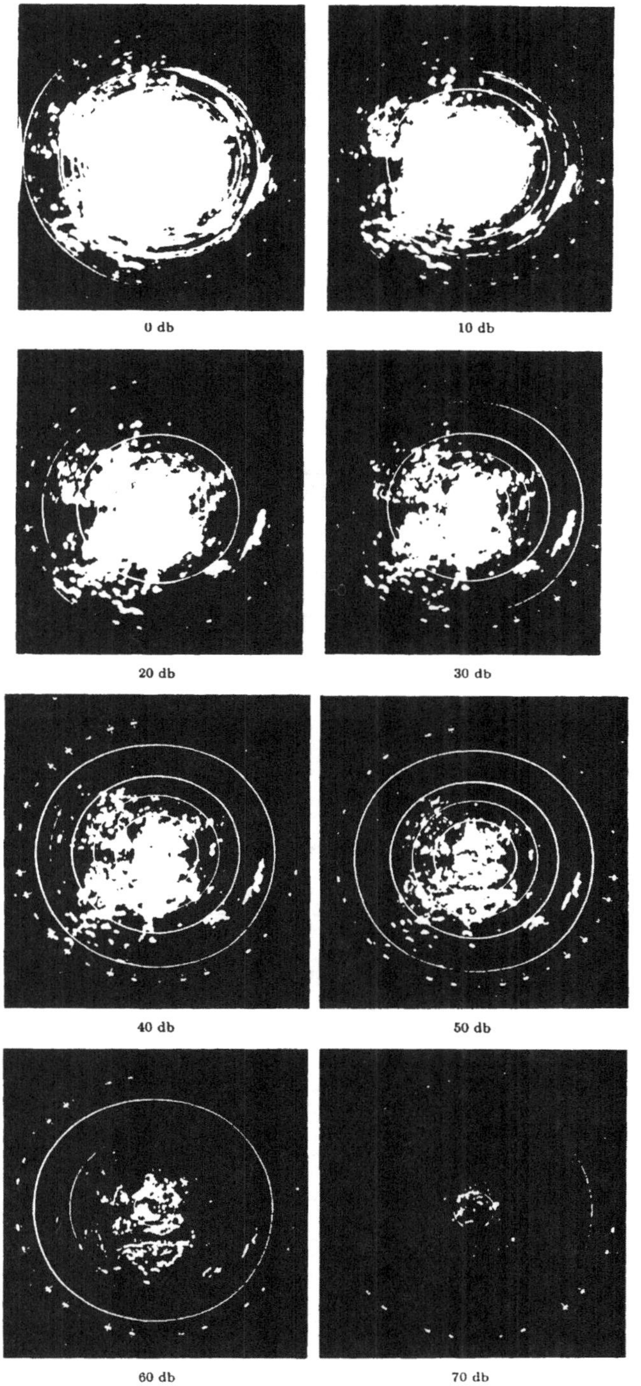

Figure 1.26 Wide dynamic range clutter returns over mountainous terrain. (*From:* [31, pp. 17-12, 17-13].)

Figure 1.27 depicts the first pdf for Rocky Mountains ground clutter taken with S-band radar having a resolution cell defined by a 2 μs pulsewidth and a 1.5° beamwidth [31, 45]. The amplitude probability distribution for the backscatter coefficient σ^0 is plotted on log-Rayleigh coordinates and has the property that a Weibull probability distribution plots as a straight line. The close fit of the straight line to data points in Figure 1.27 indicates that the Weibull probability distribution provides a close match to the experimental data and may represent a good model for this type of clutter.

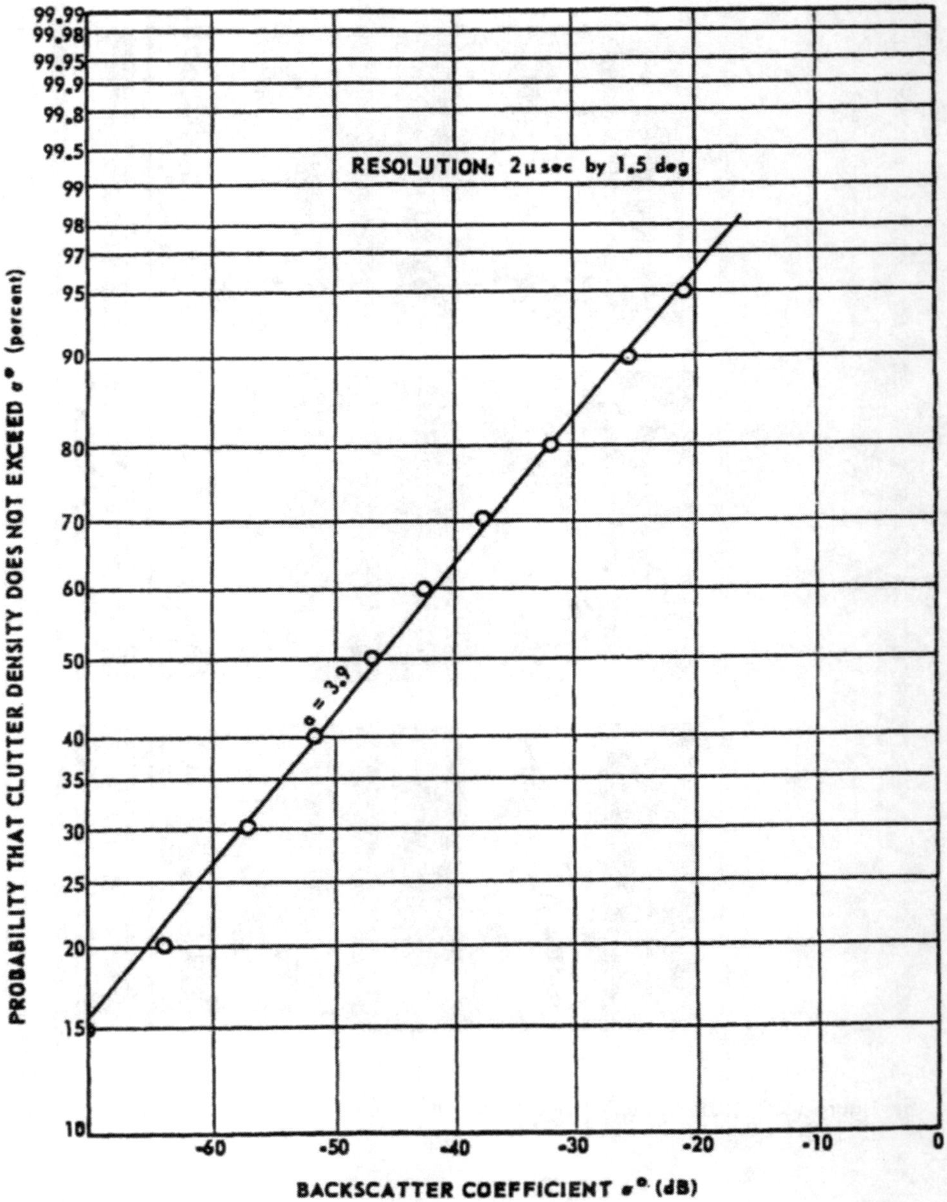

Figure 1.27 Spatial distribution for S-band Rocky Mountain ground clutter. (*From:* [53, p. 8].)

Note that in Figure 1.27 the data only extend to the 0.99 probability point, whereas detection calculations involving false alarm probabilities usually require values several orders of magnitude beyond this point. Thus illustrated is a typical problem in clutter measurements in that rarely are enough independent data points available to define the shape of the probability distribution in the tail region.

Consider estimating tail probability clutter distribution given by

$$Q(y) = 1 - F(y) \tag{1.44}$$

where $F(y)$ is the clutter's cumulative probability distribution function. Usually the observed data points are formed into a histogram by counting the number of observations that occurs above a preset threshold point, Y. The estimate for the cumulative tail pdf at level Y is then

$$\hat{Q}(y) = \frac{1}{N}\sum_{i=1}^{N} D_Y(y_i) \tag{1.45}$$

where N is the total number of samples, and $D_Y(y)$ has a value of one if the data point exceeds Y and zero otherwise. The expected value of the estimates for $Q(y)$ is

$$E[\hat{Q}(y)] = Q(y) \tag{1.46}$$

and the standard deviation is

$$SD[\hat{Q}(y)] = \left[\frac{Q(y)}{N}\right]^{1/2} \tag{1.47}$$

The first relationship indicates that $\hat{Q}(y)$ is an unbiased estimate for $Q(y)$, and the second specifies how many data points are necessary to estimate $Q(y)$ with high precision. A precise estimate is achieved when its standard deviation is much smaller than its mean, resulting in $NQ(y) \gg 0$ [54]. Thus, to make a precise estimate of the tail distribution of the clutter's probability distribution for Q equals 10^{-4} would require on the order of 10^5 independent data points, illustrating the difficulty of estimating such a distribution.

Several probability distributions have been empirically fitted to clutter data. The most popular are the log-normal, Weibull, and K-distributions. Each of these non-Gaussian models is intended to represent the effective RCS distribution (first probability distribution of σ^0) of the long-term average clutter for measured clutter data in a single clutter resolution cell. These distributions are useful for calculating detection probabilities of fast-scanning radars, where multiple independent clutter samples are obtained on a scan-to-scan basis [55–59]. However, care must be exercised when using these distributions within a single radar scan for a conventional search radar. This relates to the nonstationary behavior of non-Gaussian clutter.

One method that has been used to relate these non-Gaussian probability distributions to conventional radar detection calculations is as follows. From physical reasoning, there are usually a sufficient number of clutter scatterers (the minimum number is on the order of five scatterers) within the radar resolution cell to result in a clutter return having a Gaussian (Rayleigh envelope) distribution. The Gaussian distribution has an associated PSD that determines the second-order statistics, which are useful for Doppler filtering detection calculations. Further, the Gaussian distribution is assumed to have a variable mean σ^0, which is a random variable selected from the non-Gaussian first probability spatial distribution of the measured clutter data. The temporal situation is assumed to be represented by two correlation functions: a short-term fast correlation function (orders of 10 ms), applicable over a single radar scan associated with the Gaussian model's PSD, and a longer term correlation function associated with the variable mean σ^0 for which there is complete correlation within the radar's dwell on the clutter resolution cell and significant decorrelation over the period within the radar scan (usually on the order of 5 to 10 sec).

The detection probability can be found by using this model. First, compute the probability of detection for the Gaussian clutter model as a function of the variable backscatter coefficient $P_d(s \mid \sigma^0)$ and then average over the specified by the non-Gaussian first pdf, $p(\sigma^0)$, which results in [37]

$$\overline{P_d}(s) = \int_0^\infty P_d(s \mid \sigma^0) p(\sigma^0)\, d\sigma^0 \tag{1.48}$$

Note that this is the same procedure as that used to compute the probability of detection for the Swerling target models, except that in the preceding relationship s represents the signal-to-interference (receiver noise plus clutter power residuals after Doppler filtering) power ratio and σ^0 represents the clutter's RCS probability density rather than that of the actual target [5, 60, 61]. Extension to Swerling fluctuating targets is easily made by averaging $\overline{P_d}(s)$ over the expected target distributions (Rayleigh for Swerling 1 and 2 targets) [60].

The cumulative probability distribution functions representing the effective RCS of the clutter for Rayleigh, Weibull, log-normal, and K-distributed clutter models are given in Table 1.6. Note that the Weibull and log-normal distributed RCSs (σ) are normalized to their median values (σ_m), whereas the Rayleigh and K-distributed RCSs are normalized to their mean values ($\overline{\sigma}$). The Rayleigh distribution model is part of the Weibull family of distributions and can be obtained from the Weibull distribution by setting $\beta = 1$.

The presentation of the probability distributions for the effective clutter RCS given in Table 1.6 is facilitated by plotting the distribution under consideration on special probability graph paper, which results in a straight-line plot on a decibel scale. The Weibull distribution can be expressed as

$$\sigma_{\text{dB}} = \sigma_{\text{mdB}} + \frac{1.592}{\beta} + \frac{10}{\beta}\log\left[\ln\left(\frac{1}{1 - P(\sigma)}\right)\right] \tag{1.49}$$

Table 1.6 Probability Distributions for Effective Clutter RCS (σ) [36]

Clutter Model	*Probability Density* $p(\sigma)$	*Probability Distribution Clutter RCS,* $P(\sigma)$
Rayleigh	$\dfrac{\exp\left(\dfrac{-\sigma}{\overline{\sigma}}\right)}{\overline{\sigma}}$	$1-\exp\left(\dfrac{-\sigma}{\overline{\sigma}}\right)$
Weibull	$\dfrac{\beta \ln 2}{\sigma_m}\left(\dfrac{\sigma}{\sigma_m}\right)^{\beta-1}\exp\left[-\ln 2\left(\dfrac{\sigma}{\sigma_m}\right)^{\beta}\right]$	$1-\exp\left[-\ln 2\left(\dfrac{\sigma}{\sigma_m}\right)^{\beta}\right]$
Log-Normal	$\dfrac{\exp\left[-\ln^2(\sigma/\sigma_m)/2\sigma_p^2\right]}{\sqrt{2\pi}\sigma_p\,\sigma}$	$\dfrac{1}{2}\left[1+\mathbf{erf}\left(\dfrac{\ln\left(\dfrac{\sigma}{\sigma_m}\right)}{\sqrt{2}\sigma_p}\right)\right]$
K-Distribution	$\dfrac{2\nu\left(\dfrac{\nu\sigma}{\overline{\sigma}}\right)^{\nu-1/2}\mathbf{K}_{\nu-1}\left[2\left(\dfrac{\nu\sigma}{\overline{\sigma}}\right)^{1/2}\right]}{\mathbf{\Gamma}(\nu)}$	$1-\dfrac{2\left(\dfrac{\nu\sigma}{\overline{\sigma}}\right)^{\nu}\mathbf{K}_{\nu}\left[2\left(\dfrac{\nu\sigma}{\overline{\sigma}}\right)^{1/2}\right]}{\Gamma(\nu)}$
	$\overline{\sigma}$ = mean clutter RCS	σ_m = median clutter RCS

which plots on log-Rayleigh probability graph paper as a straight line. The straight line can be determined by connecting the 0.5 probability point $\sigma_{dB} = \sigma_{mdB}$ with the 0.9999 probability point given by

$$\sigma_{dB} = \sigma_{mdB} + \frac{11.235}{\beta} \tag{1.50}$$

Values of the parameter β for the Weibull distribution range from about $\beta = 0.25$ for mountainous land clutter to $\beta = 1$ (Rayleigh clutter) for low-resolution, high-grazing angle land and sea clutter [1].

For log-normal clutter, the RCS distribution plots as a straight line on normal probability graph paper using a decibel scale. The straight line is determined by connecting the 0.5 probability $\sigma_{dB} = \sigma_{mdB}$ with the 0.9999 probability point given by

$$\sigma_{dB} = \sigma_{mdB} + 16.153\sigma_p \tag{1.51}$$

Values of the parameter σ_p range from approximately $\sigma_p = 0.7$ for rain clutter to over $\sigma_p = 2$ for certain types of land clutter [1].

A parameter used to characterize non-Gaussian clutter is the mean-to-median ratio. The higher this ratio, the longer the tails of the probability distribution. The mean-to-median ratio ($\rho = \overline{\sigma}/\sigma_m$) is given by

$$\rho = \frac{\overline{\sigma}}{\sigma_m} = \frac{\Gamma\left(1+\dfrac{1}{\beta}\right)}{(\ln 2)^{1/\beta}} \tag{1.52}$$

for the Weibull distribution and

$$\rho = \exp\left(\frac{\sigma_p^2}{2}\right) \tag{1.53}$$

for the log-normal distribution. No simple analytical expression appears to be available for the K-distribution. Figure 1.28 depicts the mean-to-median ratio for the Weibull, log-normal, and K-distributions. Note that the mean-to-median ratio for the Rayleigh ($\beta = 1$) distribution is given by

$$\frac{\overline{\sigma}}{\sigma_m} = \frac{1}{\ln 2} = 1.442 \tag{1.54}$$

Figure 1.29 depicts the Rayleigh, Weibull ($\beta = 0.5$), log-normal ($\sigma_p = 1.689$), and K-distribution ($\nu = 0.5$) plotted on log-Rayleigh probability graph paper for a mean-to-median ratio of $\rho = 4.163$, all normalized to their median values. The Rayleigh and Weibull distributions plot as straight lines. The K-distribution in the tail region is very similar to the Weibull distribution, and the log-normal distribution has the largest tail values.

The K-distribution was developed to represent non-Gaussian sea clutter data, and is claimed to model both the spatial and temporal correlation characteristics of the clutter [62]. The compound form of the K-distributed sea clutter model identifies two components of clutter fluctuations. The first component is a spatially varying mean level associated with the sea swell structure. This component has a long correlation time (on the order of seconds) and is unaffected by frequency agility. The second component (called the *speckle* component) is due to the return

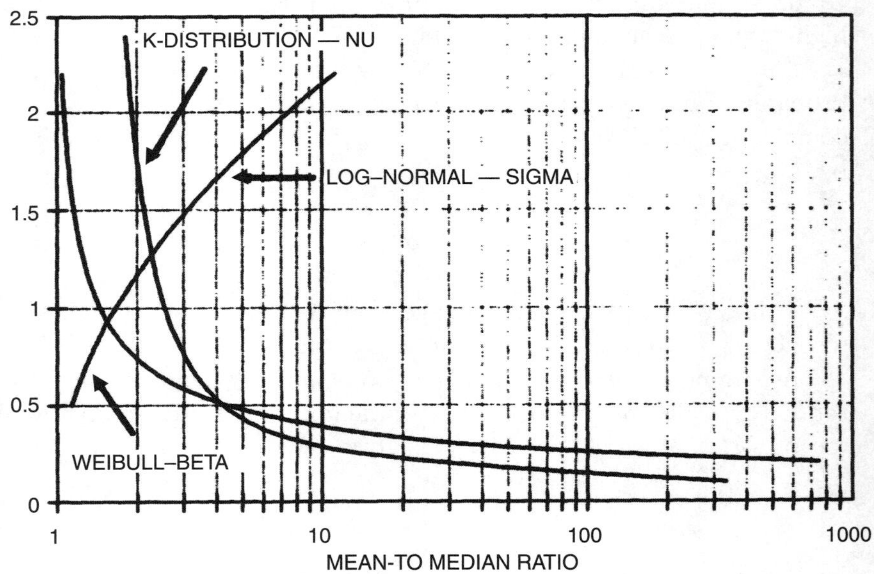

Figure 1.28 Mean-to-median ratios for Weibull, log-normal, and K-distributions. (*From:* [36, p. 35].)

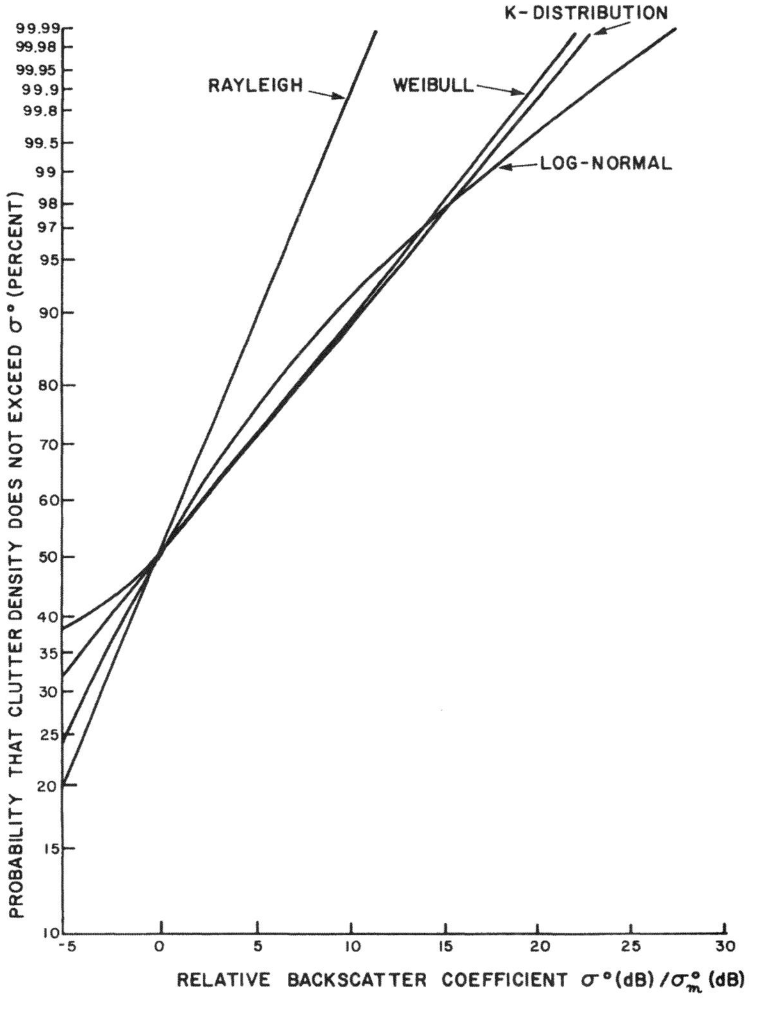

Figure 1.29 Rayleigh, Weibull, log-normal, and K-distributed probability distributions. (*From:* [36, p. 36].)

from multiple scatterers within the radar's resolution cell, and exhibits the general characteristics of Gaussian (Rayleigh envelope) clutter. This component has a relatively short correlation time (order of 10 ms) and can be decorrelated by using frequency utility.

The K-distribution is derived by assuming that the fast component consists of a Rayleigh distributed variate, conditioned on its mean value, which is chi-distributed. The chi-distribution is an empirical fit to measured sea clutter data [63]. Thus, the K-distributed model, like the Weibull and log-normal models, is based on an empirical fit to measured data, but has the virtue of some physical justification. However, the analytical properties of the K-distribution are not as well developed as those of the Weibull or log-normal distributions.

1.6 The Moving Target Detector

The moving target detector (MTD) is a digital signal processor specifically designed to allow low-PRF two-dimensional (2-D) Doppler air surveillance radar to function in a multifaceted clutter environment. The processor is designed to provide highly reliable detection of small, low-flying aircraft in the simultaneous presence of ground clutter, precipitation clutter, and angel clutter (e.g., refer to the environmental diagram in Figure 1.20).

The original MTD-I processor was developed for the ASR-9 radar, which is a medium-range (60 nmi), 2-D air surveillance terminal radar that operates at most major airports. ASR-9 operates at S-band (2.7 to 2.9 GHz) with a pulsewidth of 1.05 μs, a 1.3° azimuth beam width, an antenna rotation rate of 12.5 rpm, a PRF of 1,200 pulse/s, and a peak power of 1,200 kW. Typical performance in ground clutter is depicted by Figure 1.30, where the left photograph shows the normal video return and the right MTD photograph shows a controlled aircraft track over moderate-intensity ground clutter. Figure 1.31 indicates analogous performance of the MTD amid strong weather clutter. The controlled target shown in both MTD photographs was a small, single-engine, general aviation aircraft (RCS ~ 1 m^2). The tracks depicted in the MTD photograph show that no detections are lost due to clutter effects and false alarms are negligible. Further, the tracks were maintained while the aircraft flew tangentially to the radar, which illustrates the *superclutter visibility* capability of the MTD processor.

A block diagram of an early version of the MTD processor (MTD-I) is depicted in Figure 1.32 [64]. The basic elements of the processor are: (1) a coherent MTI filter for rejecting ground clutter; (2) an FFT Doppler filter bank, which provides coherent integration, used in the suppression of weather clutter and chaff; (3) a digital clutter map, which allows tangential targets to be detected over ground clutter, also used to suppress large clutter false alarms; (4) CFAR processing, which suppresses false alarms; and (5) block PRF staggering, which eliminates MTI blind speeds, second-time-around clutter, and stationary targets.

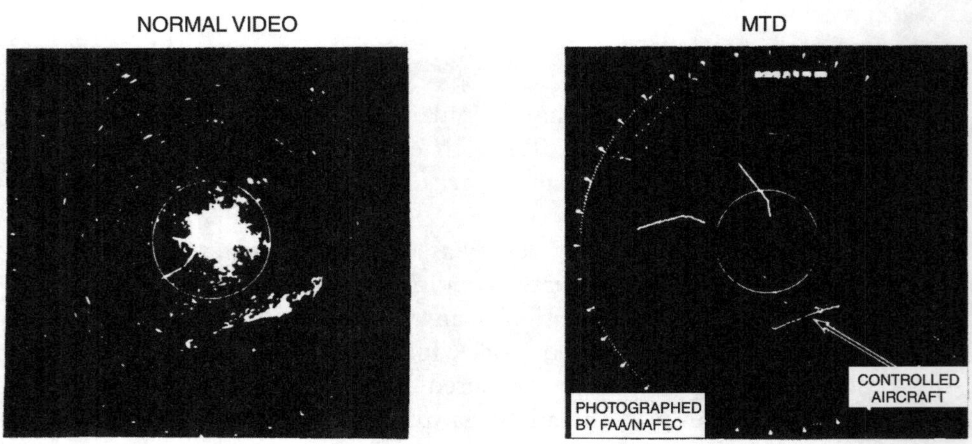

Figure 1.30 Performance of the MTD in ground clutter. (*From:* [39, p. 66].)

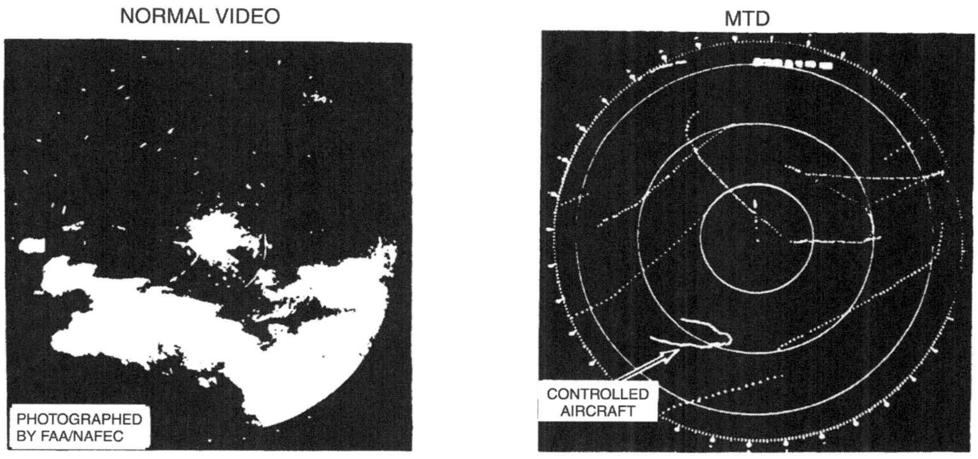

Figure 1.31 Performance of the MTD in precipitation clutter. (*From:* [39, p. 7].)

The basic idea of the MTD processor is to break the radar coverage into a large number of elemental range-azimuth-velocity (Doppler) cells and to select those containing aircraft targets on the basis of the instantaneous radial velocity, motion, and extent of the real targets as compared to the characteristic of extraneous targets including clutter. The ASR-9 MTD processor contains 365,000 range-azimuth cells (1/16 nmi by 3/4 degree) each containing eight velocity cells (150-Hz Doppler bandwidth) for a total of 2,920,000 cells.

The MTD processor shown in Figure 1.32 separates the input signal into two channels, depending on the return's radial velocity. The upper channel contains an MTI that suppresses ground clutter by 45 to 50 dB, and a Doppler filter bank that accepts targets and weather clutter, the radial velocity of which is within the MTI passband. The lower channel accepts targets and ground clutter returns having radial velocities that fall within the lowest Doppler filter's passband (±75 Hz). The output of this lower channel is used to generate a ground clutter map by averaging eight consecutive radar scans of each range-azimuth cell.

The function of the upper channel is to suppress both ground and weather clutter while passing all moving targets, except those that fall into the low-radial-velocity class associated with the lower channel. The ground clutter is suppressed by the MTI action, whereas the weather clutter, which tends to fall within the first three Doppler filters, is suppressed by a mean-level CFAR circuit. The mean-level CFAR sets the threshold in each Doppler cell by averaging the returns from the same Doppler cell in 16 range cells, 8 on either side of the cell of interest. Weather returns, which are generally distributed throughout this 1-mile-plus range extent, fill the Doppler cells of interest, thereby creating a high threshold that suppresses the return in this combination of range-Doppler cells.

Several complications arise in low-PRF ambiguous-velocity designs of this type. First, all targets having radial Doppler velocity that fall in a region about the multiples of the first PRF line (1,200 Hz, which corresponds to 125 knots, 250 knots, and so on) are suppressed by the MTI. This is the *blind-speed effect* of the

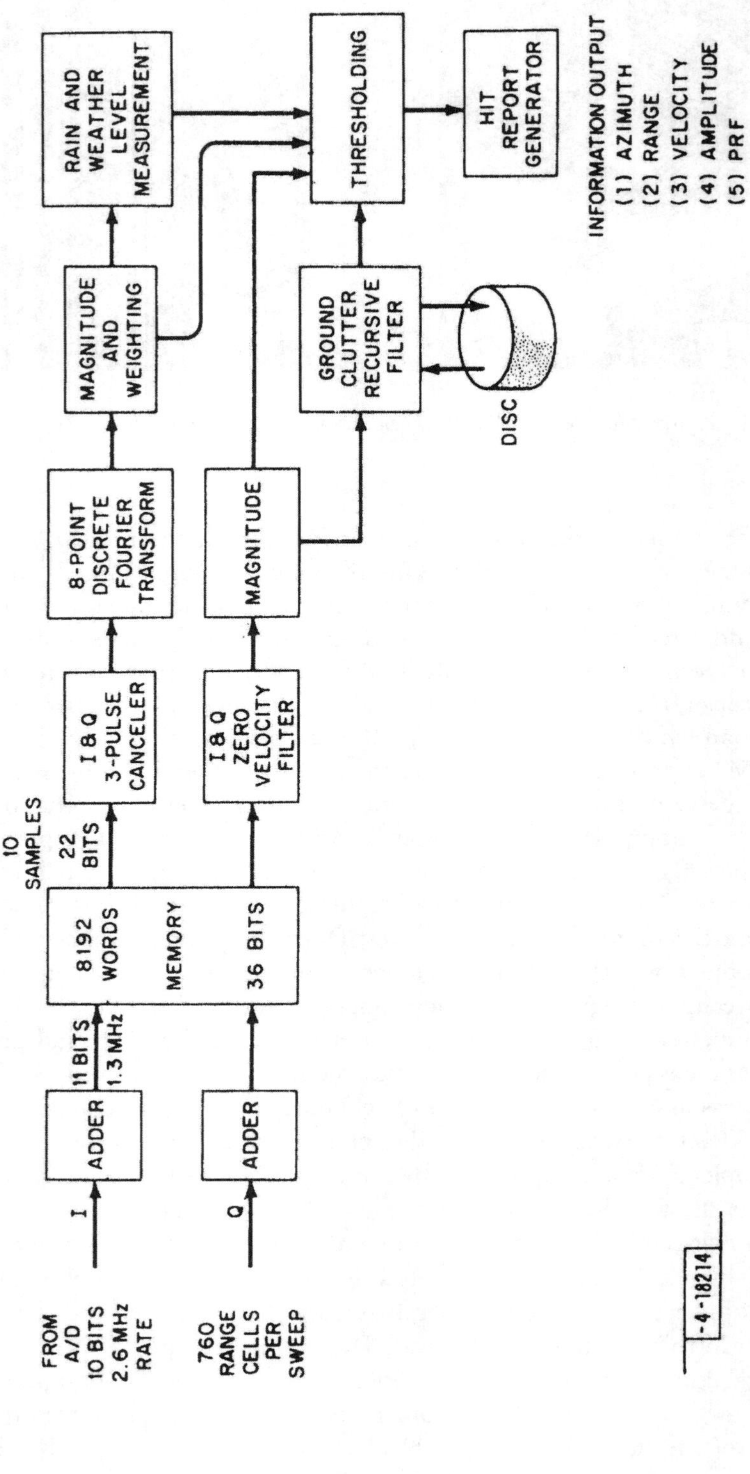

Figure 1.32 MTD block diagram. (*From:* [64, p. 5].)

MTI. Second, all targets having radial velocities that are ambiguous with respect to Doppler filter cells occupied by the weather clutter are suppressed. This target suppression is due to the folding effect of low-PRF radar, whereby the Doppler returns of all targets are folded within the region between zero frequency and the first PRF line.

The blind-speed problem and suppression of targets by weather clutter (recall Figure 1.14) is alleviated by using a block-to-block PRF stagger, in which a block in the MTD processor of Figure 1.32 corresponds to 10 consecutive pulses. The need to maintain the same PRF (and transmitter frequency) for 10 pulses can be understood by considering that the 3-pulse MTI canceller must be filled before its output can be applied to the eight-point FFT Doppler filter bank. The ASR-9 MTD uses a block stagger of about 20% between returns obscured by weather clutter or by the MTI blind-speed effect to translate returns into adjacent Doppler filters so that they can be detected. The significant effect of this procedure is that 20 pulses must be processed before target detection can be made.

A particular feature of the MTD processor that distinguishes it from earlier designs is its management of false alarms, which is accomplished by using adaptive CFAR thresholding of each range-Doppler cell. Two types of mean levels are formed in the MTD processor. First, the clutter map is formed using a fading-memory recursive filter that operates on the output of the Doppler filter centered at zero frequency. The recursive filter has a bandwidth that corresponds to approximately eight radar scans (32 seconds). The second type of mean level is formed from the output of each nonzero velocity Doppler filter averaged in a moving window detector whose width is 16 range cells centered about the Doppler cell of interest.

The mean-level clutter map is used to form the thresholds for the zero velocity Doppler filter shown in the lower channel of Figure 1.32 and the eight Doppler filters of the upper channel that contain residual clutter components folded about the PRF lines. This would provide superclutter visibility whereby large cross-sectional tangential targets (zero radial velocity) could be detected over clutter that otherwise might have fallen into the MTI's blind-speed zone and thus been eliminated. The moving window mean-level outputs for Doppler filters 2 through 7 are used to form the threshold for their respective Doppler filters. This threshold is used to provide a CFAR for weather clutter, which can have a mean radial velocity of up to 60 knots and a spectral spread of 25 to 30 knots. In addition, a combination of both mean-level filters is used to form the threshold of the Doppler filter adjacent to the zero velocity filter to allow for ground clutter spillover.

In addition to the MTI, Doppler filtering, and CFAR processing, a number of other censoring techniques are used in the MTD processor. A pulse type of interference might be received from other radars, which would not be eliminated due to the averaging effects of the CFAR processor. To guard against this condition, the magnitude of any one return is checked against the average of the return over the coherent processing interval (8 to 10 pulses). If any pulse exceeds the average by five times, all information from the range-azimuth cell is discarded. Also, saturation of the analog-to-digital (A/D) converter on any pulse causes the information over the whole coherent processing interval to be deleted.

The multiplicity of range-Doppler-azimuth cells in the MTD processor increases the possibility of a large target causing multiple range, Doppler, and azimuth cell reports. Reports in adjacent cells that appear to originate from a single target can be grouped to ease the problem of track initiation in the radar data processing subsystem. Also, the target amplitude and Doppler reports on the target can be used to eliminate small cross-section and low-speed radar angels (birds and insects) returns before these targets' reports enter the radar data processing tracking system.

The classic Doppler radar design problem addressed by the MTD is the detection of a moving target return in the presence of both ground and precipitation clutter (see Section 1.3.3). The MTD-I handles ground clutter through a combination of a coherent double canceller MTI followed by an 8-point FFT Doppler filter bank where the sidelobes of the Doppler filter bank further attenuate the residual clutter from the MTI canceller. This cascade can be shown to approximate the linear optimum transversal filter for this situation, but is inefficient in the sense that the same improvement factor performance can be achieved by using a lower order transversal filter (the optimum transversal filter would use 8 versus 10 pulses for the MTD-I) [65].

The realization of quasi-optimum performance for the Doppler processor requires linear operation to mitigate the notorious effects of limiting that spreads the clutter spectrum, thereby causing a reduction in the improvement factor (see Section 3.2.5 and Figure 3.23). The linear dynamic range of the MTD-I is approximately 50 to 60 dB due to the use of a 10-bit A/D converter in the digital processor. When clutter returns exceed this limit (due primarily to large discrete ground clutter returns), one possibility is to use the clutter map to control a linear attenuator in these limit cells to maintain linear operation.

The suppression of precipitation clutter in the MTD-I is accomplished by generating a Doppler filter bank in each range-azimuth angle cell that divides the spectrum between PRF lines into n Doppler cells, each with bandwidth $f_B = f_r/n$, where n is the number of applied pulses used to form the filter (i.e., eight pulses) and f_r is the radar's PRF. This situation is depicted in Figure 1.33, which shows how the MTD resolution cells are spread through the radar's range, azimuth angle, and Doppler coverage. When precipitation clutter (or chaff returns) occurs, it generally will be confined to one or two adjacent Doppler cells within each Doppler filter bank of those located within the range-angle extent of the clutter cloud. The return in each Doppler cell containing the clutter can then be attenuated in accordance with a normalizing CFAR signal formed by examining similar Doppler cells contained in contiguous range cells about the test cell's range (i.e., ±1 nmi) at the same azimuth angle. This leaves the other Doppler cells, which are free of precipitation clutter returns, available to detect moving targets, thus providing subweather clutter visibility. Also, superclutter visibility is provided in those Doppler cells containing weather clutter that have target returns sufficiently stronger than the precipitation clutter return.

Although precipitation returns are generally unambiguous in Doppler (i.e., $|f_d| \leq 55$ knots), aircraft target returns are not. This causes target returns to be lost due to spectrum folding into Doppler cells suppressed by precipitation clutter return (recall Section 1.3.3 and Figure 1.19). The solution to this problem employed by the MTD is to use a block stagger where two coherent processing intervals (CPIs) are consecutively transmitted at PRFs that differ by about 20%. Thus, if the targets are suppressed on one PRF, they will be folded into another nonsuppressed

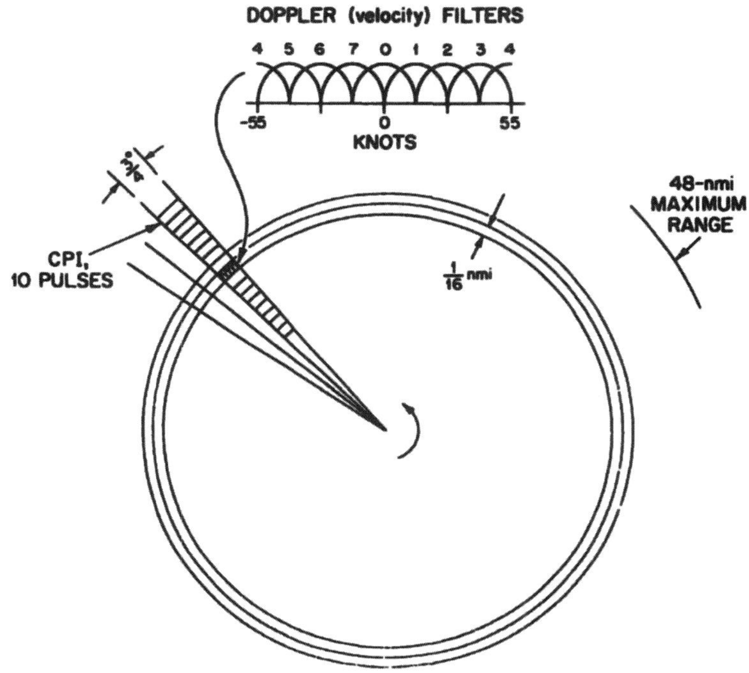

Figure 1.33 MTD resolution cells. (*From:* [64, p. 35].)

Doppler filter on the second PRF and thus detected. The block stagger used in the MTD also suppresses blind speeds. These are associated with the low-PRF employed and multiple-time-around clutter returns that occur when the radar is operated under conditions of anomalous propagation or in regions containing reflections from mountainous terrain and urban areas, so that the radar waves intercept ground targets at distances greater than that associated with the first interpulse period.

A critical issue involved with the MTD type of processor that uses Doppler filter banks to suppress precipitation clutter concerns the sidelobes of the Doppler filters. If the sidelobes are too high, strong precipitation clutter spectrum components located outside the passband of any particular Doppler filter will cause the filter to suppress targets due to CFAR action that follows the filter. In the MTD-I, the Doppler filters consist of a cascade of three separate filters: a 3-pulse canceller for ground clutter suppression, an 8-point FFT for precipitation clutter suppression, and post-FFT frequency-domain weighting to reduce the filter's sidelobe level. Further, the double canceller used 10 bits of signal information and coefficient precision (to provide approximately 60 dB of ground and discrete clutter rejection), whereas the FFT carried a lesser number of bits (approximately 3 or 4 bits) because this was all that would be required to form the filters. The resulting Doppler filter response is depicted in Figure 1.34 for two of the filters. The filter closest to the ground clutter response exhibits poor sidelobes, which are approximately –5 dB below the principal response. This is due in part to the use of a double canceller in front of the FFT, which attenuates the main responses. Also, the small number of bits employed in the FFT further limits the achievable sidelobe level.

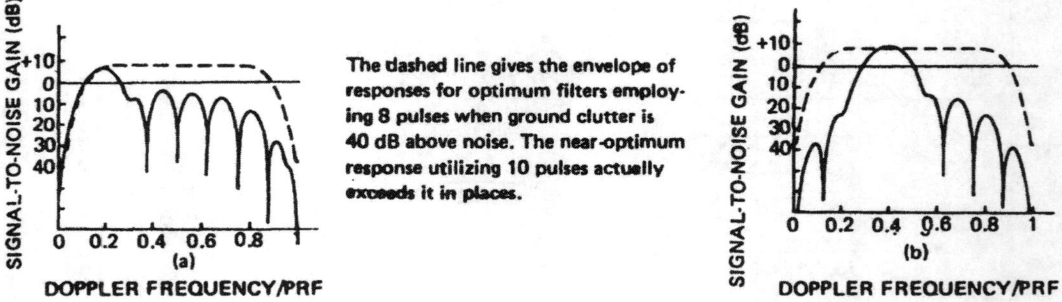

Figure 1.34 Doppler filter response of the MTD-I processor. (*From:* [66, p. 42].)

In the MTD-II processor shown in Figure 1.35, the Doppler filtering is made more efficient, and the sidelobe levels of the filters are reduced by 10 dB [67]. The filters are synthesized by the cascade of a single canceller and eight 7-pulse complex transversal filters. The single canceller reduces the zero-velocity ground clutter and carries the full dynamic range of the signal information (10 bits). The point optimized transversal filters are equally spaced to cover the PRF interval with sufficient overlap to minimize filter straddling losses. The filters are formed using 5-bit weighting coefficients. The improved sidelobe response of the MTD-II is depicted in Figure 1.36.

Note that the efficiency of the Doppler portion of the MTD processor is improved in the MTD-II, where 8-pulse transversal filters were used to provide essentially the same results as with the 10-pulse filters used in the MTD-I. This reduces the number of pulses required to implement an MTD-type processor from 20 to 16 pulses (i.e., for two CPI bursts). This improvement results because, although the first two pulses into the MTI are useful in suppressing ground clutter, they are essentially wasted in the formation of the Doppler filter bank.

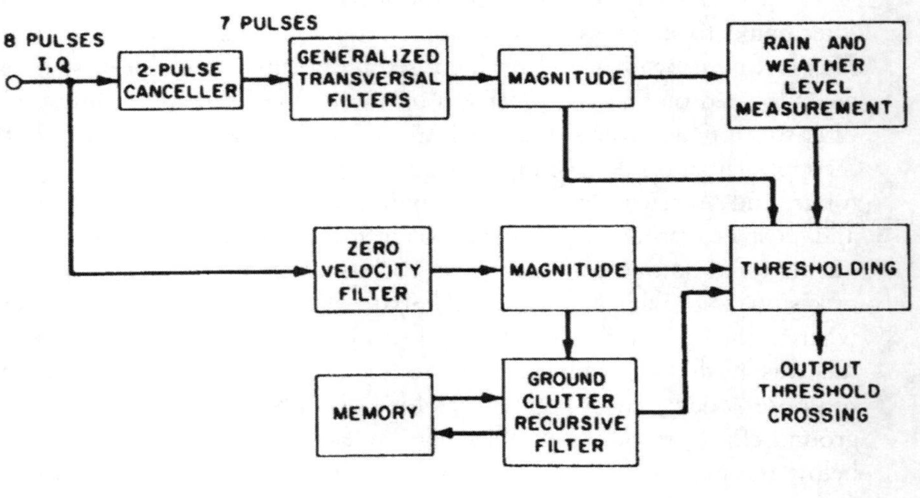

Figure 1.35 The MTD-II processor. (*From:* [67, p. 512].)

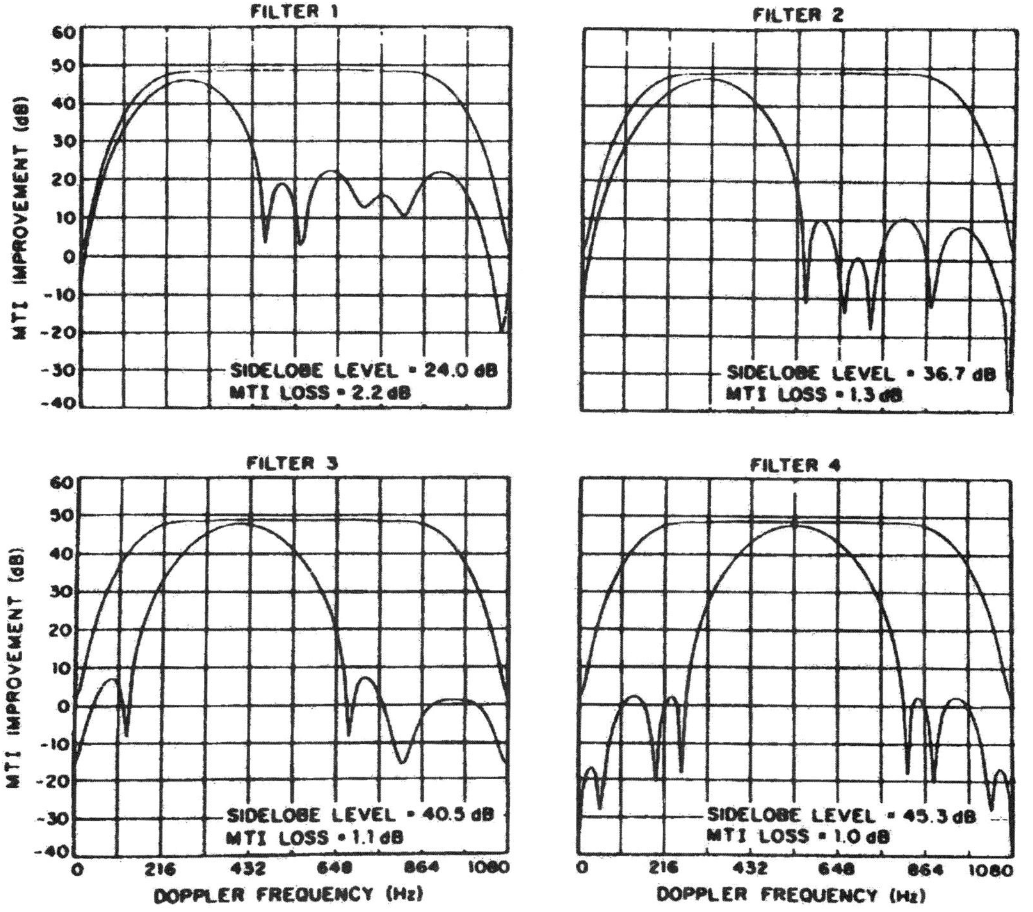

Figure 1.36 MTI improvement factor for the MTD-II processor. (*From:* [67, p. 512].)

Additional improvements included in the MTD-II are: (1) transversal filter design of the zero-velocity filter using Chebyshev criteria to provide uniform response with low sidelobes in this region; (2) clutter map with one cell for each range-azimuth cell rather than one cell for each range-CPI; (3) nonzero velocity Doppler filters with lower sidelobes to provide better performance in weather clutter; (4) block stagger CPIs of 900 and 1,100 μs providing PRFs of 1111 and 900 Hz, respectively; (5) correlation and interpolation algorithms to optimize single-target reports of aircraft, automobiles, or birds; (6) area CFAR to eliminate single CPI false alarms of birds, interference, and weather clutter; and (7) scan-to-scan correlation to eliminate returns not consistent with aircraft targets [67].

The number of bits required for the coefficient of an MTD-type processor is open to some controversy. If a pure transversal filter configuration (no MTI) is used, each filter in the bank (e.g., eight filters) must be designed to suppress zero-velocity clutter (large discrete and ground terrain clutter) and may require up to 12-bit precision for coefficient weightings [66]. If an MTI prefilter is used, then the transversal filters may require 8- to 9-bit precision in the filter coefficients to achieve minimum sidelobes, which are of the order of –40 dB below the main response [34]. Sidelobes are also affected by imbalances in the amplitude response of the in-phase (I) and quadrature (Q) channels. For example, a 4% imbalance would raise the sidelobe level on the order of 6 dB [68].

Another factor that reduces the sidelobes of transversal-type Doppler filters occurs when multiple-time-around clutter is processed. For example, when processing an n-pulse burst with low sidelobe weighting, the second-time-around clutter echoes will arrive with an extra time delay so that the first pulse will receive the second weighting and the last pulse will be lost. The erroneous weightings raise the sidelobe level as shown in Figure 1.37 for an 8-pulse processor with a Dolph-Chebyschev type of weighting [69].

A significant disadvantage of the MTD processor is the relatively large number of pulses that must be transmitted to achieve the desired performance level. This restricts its use to 2-D surveillance radars or 3-D radars that use the stacked beam approach. Further, the need for a relatively long coherent burst transmission at a stable PRF (i.e., 8 to 10 pulses) makes the MTD susceptible to electronic attack (EA) [5]. A responsive jammer could measure the frequency of the first transmitted pulse and then center the jammer to spot jam the following pulses. Also, the requirement for a stable PRF precludes the use of pulse-to-pulse jitter, which is one of the most effective techniques against deceptive and camouflage jammers that rely on anticipating the radar transmitted pulse.

The MTD-III design incorporated into the ASR-9 is based on the MTD-II design with a number of improvements. The transversal filter weighting coefficients are implemented using 12 bits rather than the 5 bits in the MTD-II, thus providing a 6 to 17 dB increase in rain rejection sidelobes and an optimally shaped ground clutter rejection notch effective in eliminating A/D converter bias. Doppler filter sidelobes are –40 dB for rain clutter and –44 dB for ground clutter. Sensitivity time control (STC) was added to reduce the large ground clutter returns that exceeded the available dynamic range of the system. In mountain clutter an alternate filter design that provides 52 dB of attenuation at the expense of less rain clutter rejection can be selected. An improved CFAR algorithm provided better resolution and cluster editing, allowing two closely spaced targets to be resolved. Further, an algorithm was added that allowed two closely separated targets in azimuth with unequal amplitudes to be resolved [68, 71, 72].

1.6.1 Clutter Maps

One innovation employed in the MTD type of processor is the use of a clutter map for false alarm control against a spatially nonstationary variable clutter background (see Section 1.5.4). This is contrasted with a conventional cell-averaging CFAR that provides good performance in a spatially stationary background but poor performance when the central cell surrounding the target cell has nonuniform statistics [73]. The conventional cell-averaging CFAR can be modified by using a split gate or by employing nonparametric detection techniques to improve its performance in nonstationary interference. However, when conditions are appropriate for using a clutter map CFAR, this type of CFAR control will generally

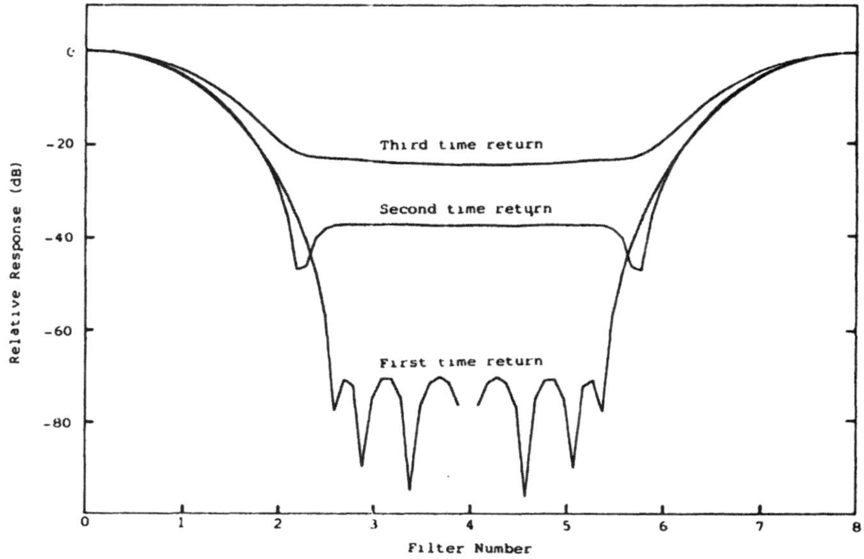

(a) Response of 8-pulse processor.

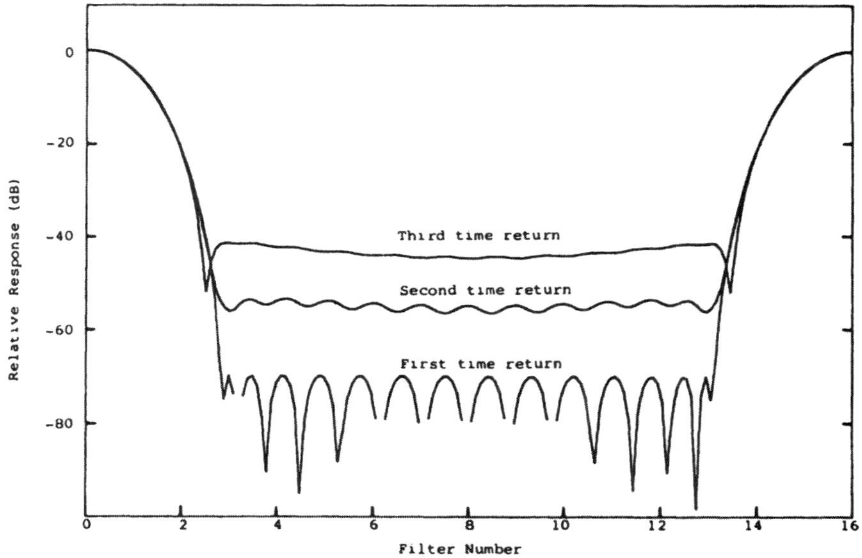

(b) Response of 16-pulse processor.

Figure 1.37 Response of a transversal filter with Dolph-Chebyshev weighting to ambiguous clutter: (a) response of 8-pulse processor; and (b) response of a 16-pulse processor. (*From:* [70, pp. 2089–2090c].)

outperform (i.e., have less CFAR loss than) the conventional and derivative forms of cell time-averaging CFAR processors.

The basic difference between the conventional cell averaging and the clutter map type of CFAR concerns the manner in which samples are obtained to estimate the background interference level and the method used to censor the target return from the background estimate. In the conventional CFAR, the background estimate is made with a single radar scan by examining statistically independent range (and possible Doppler) cells contiguous to the test cell, which is excluded from the estimate. In the clutter map approach, the background in the test cell is examined on a scan-to-scan basis where clutter returns are excluded through Doppler discrimination (the MTD uses the zero-velocity Doppler filter to estimate the mean clutter level) or through a target detection test [74]. Implicit in the clutter map CFAR method is that independent clutter samples are available on a scan-to-scan basis. This occurs naturally in most conventional surveillance radars whose scan frame time is on the order of 5 to 10 sec, but the clutter correlation time due to scan modulation may be on the order of 100 to 500 ms.

Clutter-map type CFAR processors generally use a first-order recursive filter, as shown in Figure 1.38, to estimate the mean clutter level. This type of filter requires only one memory storage cell for each clutter map resolution cell. Because there can be many clutter map resolution cells (e.g., three 65,000 range-azimuth cells in the ASR-9 MTD), the total memory could be excessive if a moving-window type of averager (e.g., with 16 samples) were used as in the conventional CFAR processor. In the recursive filter type of estimator, a fraction α of the present sample (e.g., $\alpha = 1/8$ in the MTD) is added to $1 - \alpha$ of the previous estimate to form the new estimate. This produces an exponential type of smoothing action, where the effect of each sample fades at an exponential rate.

The state equation for the first-order recursive filter used in the clutter map CFAR is given by

$$\hat{y}_j = (1 - \alpha)\hat{y}_{j-1} + \alpha x_j \tag{1.55}$$

where $\hat{y}_j$ is the jth output of the power estimate, x_j is the jth sample of the interference (clutter and noise) power, and α is the filter's gain coefficient. The expected value of the power estimate is given by

$$E(\hat{y}_j) = P_c[1 - (1 - \alpha)^j] \tag{1.56}$$

which approaches $E(y) = P_c$ for large values of j. Note that the value of the clutter power in each cell is proportional to the backscatter coefficient σ^0, which is assumed to be a random variable in the spatially nonstationary clutter model described in

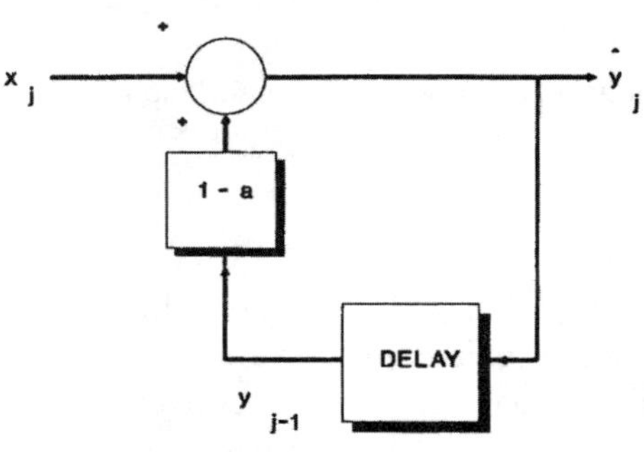

Figure 1.38 Clutter map recursive filter. (*From:* [36, p. 47].)

Section 1.5.4. Thus, the clutter map CFAR estimates the backscattering coefficient σ^0 on a cell-to-cell basis and hence accommodates both spatially homogeneous and nonhomogeneous clutter distributions.

The variance of the recursive filters estimate of the clutter power is given by

$$\operatorname{var}(\hat{y}_j) = \alpha^2 \sigma_x^2 \frac{[1 - (1 - \alpha)^{2j}]}{[1 - (1 - \alpha)^2]} \tag{1.57}$$

where σ_x^2 is the variance of the clutter power random variable. For large values of j this approaches

$$\operatorname{var}(\hat{y}) = \frac{\alpha \sigma_x^2}{2 - \alpha} \tag{1.58}$$

which illustrates the smoothing effect of the filter because the variance is reduced by the factor

$$n' = \frac{2 - \alpha}{\alpha} \tag{1.59}$$

which is equivalent to the use of n' cells in a conventional cell-averaging CFAR [73].

One use of the clutter map type of CFAR is to provide superclutter visibility in those clutter cells where the target return significantly exceeds the clutter return. In the MTD processor, this allows strong tangential targets (those with small radial velocities) to be detected over weak ground clutter or tangentially moving precipitation returns. This is particularly advantageous in the formation and maintenance of ground tracks that occur in subsequent data processing.

A simplified analysis of the CFAR loss associated with the clutter map CFAR processor shown in Figure 1.35 can be made for the situation where a Swerling 1 type fluctuating target is to be detected in an unknown power Rayleigh clutter background, where the clutter is completely correlated within a single radar scan [73]. The probability detection for a square-law detector and a fixed threshold (y_b) is given by

$$P_{di} = \exp\left[\frac{-y_b}{1 - \left(\frac{P_s}{P_c}\right)}\right] \tag{1.60}$$

where P_s/P_c is the signal-to-clutter power ratio. The threshold is estimated by the clutter map such that

$$y_b = \frac{C\hat{y}_j}{P_c} \tag{1.61}$$

where C is a fixed threshold offset, $\hat{y}_j$ is the clutter power estimate from the recursive filter, and P_c is the clutter power. The value for P_d can then be written as

$$P_d = \exp\left[\frac{-C\hat{y}_j}{P_c\left(1 + \frac{P_s}{P_c}\right)}\right] \tag{1.62}$$

which is a function of the estimated clutter power (y_j). If we average over the clutter power estimate, this results in

$$\overline{P}_{di} = \int_0^\infty \exp\left[\frac{-C\hat{y}_j}{P_c\left(1 + \frac{P_s}{P_c}\right)}\right] P(\hat{y})\, d\hat{y} \tag{1.63}$$

which can be recognized as the characteristic function

$$M\left[S = \frac{C}{P_c + P_s}\right] \tag{1.64}$$

of y evaluated at the argument of the exponential. The false alarm probability can then be evaluated from this relationship by setting $P_s/P_c = 0$:

$$P_{fa} = M\left(S = \frac{C}{P_c}\right) \tag{1.65}$$

The recursive filter of Figure 1.35 provides an estimate of the clutter, formed from the current and previous samples as

$$\hat{y}_j = \alpha \sum_{l=0}^{\infty} (1 - \alpha)^l x_{j-l} \tag{1.66}$$

The probability density of each x clutter sample is given by

$$p(x) = \frac{\exp\left(\frac{-x}{P_c}\right)}{P_c} \tag{1.67}$$

so that the characteristic function of $\hat{y}$ equals

$$M(S) = \prod_{l=0}^{\infty} \frac{1}{\alpha(1 + \alpha)^l P_c S} \tag{1.68}$$

which, on substitution into the probability detection relationship, becomes

$$\overline{P_{d1}} = \prod_{l=0}^{\infty} \left[1 + \frac{C\alpha(1-\alpha)^l}{1 + \frac{P_s}{P_c}} \right]^{-1} \tag{1.69}$$

This relationship evaluated at $P_s/P_c = 0$ results in

$$P_{fa} = \prod_{l=0}^{\infty} \frac{1}{1 + C\alpha(1-\alpha)^l} \tag{1.70}$$

The CFAR loss can be evaluated by first solving for the P_{fa} relationship for C and then substituting this and an appropriate value of P_s/P_c into the P_d relationship to provide a particular probability of detection. The known power (non-CFAR) value of γ_k can be calculated from

$$\gamma_k = \frac{\ln P_{fa}}{\ln \overline{P_{d1}}} - 1 \tag{1.71}$$

The CFAR loss in decibels is then found from

$$L = 10 \log\left(\frac{P_s}{P_c} - \gamma_k\right) \tag{1.72}$$

This is plotted in Figure 1.39 as a function of the recursive filter gain α evaluated for $P_d = 0.9$ and $P_{fa} = 10^{-6}$.

Figure 1.39 shows that small CFAR loss results from using a small filter gain coefficient (α). This occurs because increasing α corresponds to using a longer data window to estimate the clutter power. Conversely, map settling time is reduced as α is increased as is shown in Figure 1.40 [73]. Thus, a compromise between low CFAR loss and adequate settling time is required. A filter gain $\alpha = 0.125$ is used in the MTD, which represents a compromise between these two factors.

Settling time becomes important for clutter transients caused by either changing or moving clutter. One example of this situation occurs in the MTD processor, where relatively large zero-velocity returns have been observed to occur from weather fronts moving rapidly through the radar's coverage. The clutter map then reacts to the changing clutter power by adapting its threshold to the new clutter condition. However, a false alarm will be reported until the map reaches its steady-state condition. In the MTD processor, this situation was handled in the data processing by not allowing tracks to be initiated on zero-velocity targets.

The digital clutter map obtained with the MTD processor is depicted in Figure 1.41 [39]. The MTD uses 10-bit storage words to preserve the large dynamic range of the clutter signal. The ground clutter patch indicated in the right-hand corner was used for subclutter visibility tests using the MTD. The expected subclutter visibility of the MTD (in the Doppler filter channels) is approximately 41 dB, provided the clutter return is within the linear dynamic range of the receiver-processor, which is on the order of 45 dB measured at the A/D converter. To

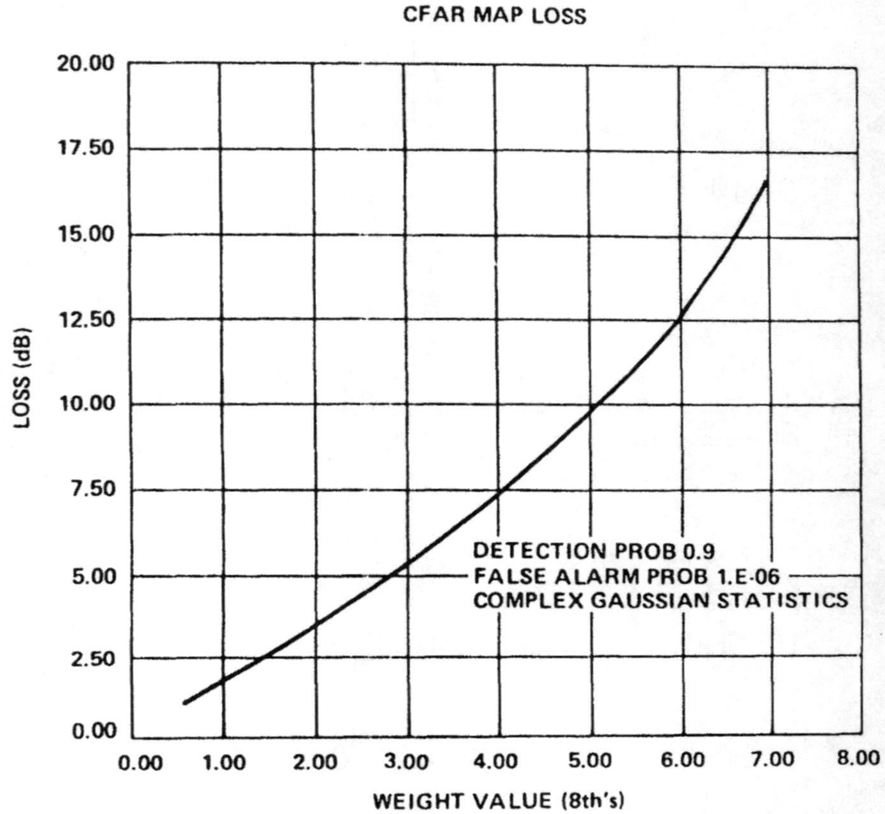

Figure 1.39 Clutter map CFAR loss. (*From:* [74, p. 421].)

accommodate ground clutter returns that exceed the linear dynamic range, the MTD raises the detection thresholds of the Doppler filter channels in direct proportion to the measured clutter level. This in general has the effect of censoring those cells that contain clutter returns beyond the subclutter visibility capability of the Doppler processor. Figure 1.42 depicts a situation that might occur in spotty ground clutter where the clutter cells containing severe clutter returns are censored from the output. The probability of the radar data processor tracker losing track is generally low in the situation. This is an example of interclutter visibility that generally improves as the radar's resolution is increased.

The minimum dimension of a clutter cell used in a clutter map is the radar's surface resolution cell (the radar range corresponding to the effective pulsewidth and the azimuth beamwidth). If a stationary target exists within this cell, then it will be suppressed by the CFAR action as its return becomes part of the estimated clutter return. A slowly moving target also will be partially suppressed if its return is contained within the clutter cell for several data samples. The minimum map-free velocity is a measure of this effect and is defined as the minimum target velocity that will cause a target to be suppressed by the clutter map's CFAR action. In the range dimension, this velocity is given by

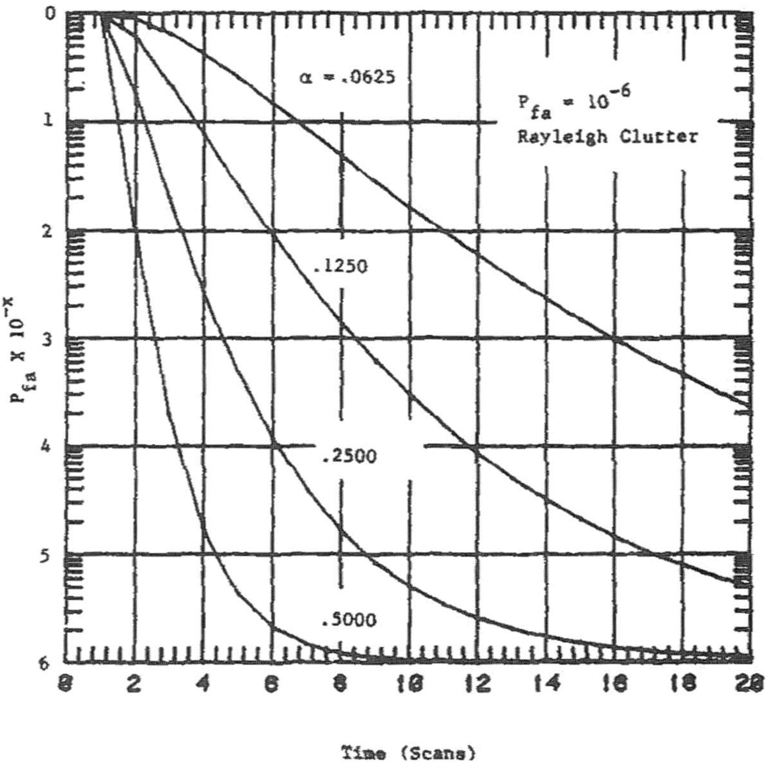

Figure 1.40 Clutter map settling time as a function of clutter map gain. (*From:* [73, p. 6].)

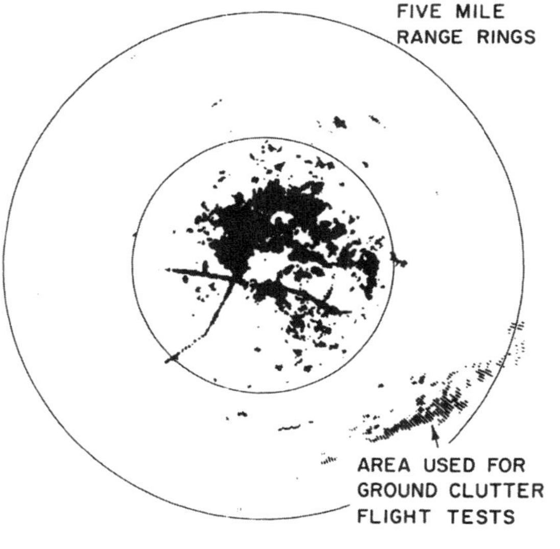

Figure 1.41 Digital clutter map of ground clutter. (*From:* [64, p. 126].)

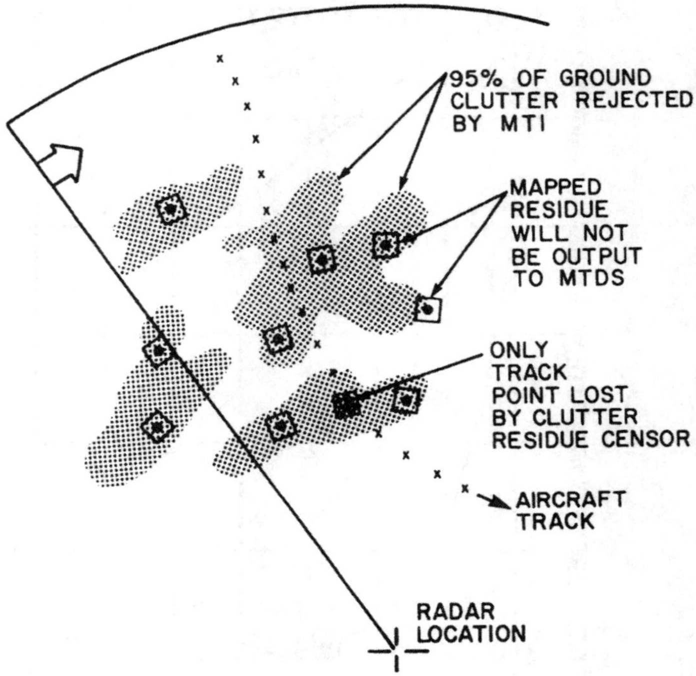

Figure 1.42 Aircraft tracking using MTD processing. (*From:* [36, p. 52].)

$$v_{mfr} = \frac{\left(\frac{c\tau_e}{2}\right)}{T_s} \tag{1.73}$$

where τ_e is the radar's effective pulsewidth and T_s is the interscan period. For the ASR-9 ($\tau_e = 1.05$ μs, $\theta_{az} = 1.3°$, $T_s = 4.8$ sec), the minimum map-free velocity in the range dimension is then a radial velocity of about 65 knots. In the azimuth dimension, the relationship equals

$$v_{mft} = \frac{\theta R}{T_s} \tag{1.74}$$

where θ is the azimuth beamwidth, R is the range to the clutter cell, and T_s is the interscan period. For the ASR-9 at 10 nmi, the minimum map-free tangential velocity becomes 170 knots.

References

[1] Schleher, D. C., (ed.), *MTI Radar*, Dedham, MA: Artech House, 1978.

[2] Povejsil, D., R. Raven, and P. Waterman, *Airborne Radar*, New York: D. Van Nostrand, 1961.

[3] Barton, D. K., (ed.), *CW and Doppler Radar*, Vol. 7 of *Radars*, Dedham, MA: Artech House, 1978.

[4] Stimson, G., *Introduction to Airborne Radar*, Hughes Aircraft Company, El Segundo, CA, 1983.

[5] Schleher, D. C., *Introduction to Electronic Warfare*, Dedham, MA: Artech House, 1986.

[6] Cole, E., et al., "ASR-12: A Next Generation Solid State Air Traffic Control Radar," *Proc. 1998 IEEE Radar Conference*, Dallas, TX, May 11–14, 1998, pp. 9–14.

[7] Olsen, F., "Microwave Solid State Power Amplifier Performance: Present and Future," *Microwave Journal*, February 1995, pp. 24–46.

[8] Brookner, E., *Aspects of Modern Radar*, Norwood, MA: Artech House, 1988.

[9] Martineau, M., and E. Cole, "Weather Detection Design Issues for a Solid State ASR," *Proc. IEEE Int. Radar Conference*, Washington, D.C., May 1995, pp. 536–541.

[10] Weber, M., "FAA Surveillance Radar Data as a Complement to the WSR-880 Network," *Proc. National Astronomy and Ionosphere Center 9th Conference on Aviation, Range and Aerospace Meteorology*, Orlando, FL, September 2000, Paper J1.10, American Meteorological Society.

[11] Lay, R., "Mission Determines Equality Between Radars," *Proc. IEEE Int. Radar Conference*, Washington, D.C., May 1995, pp. 5–8.

[12] Lain, G., and E. Gersten, "AN/TPS-59 Overview," *Proc. IEEE Int. Radar Conference*, Washington, D.C., May 1975, pp. 527–532.

[13] Barton, D., "Radar Today," *Microwave Journal*, Vol. 49, No. 1, January 2005, pp. 24–36.

[14] Van Nederveen, G., "EMB-145SA and RS: Brazil's New Eye in the Sky," *Aerospace Power Journal*, Winter 2000, pp. 87–89.

[15] Nordwell, B., "Sweden Develops New Radar Computer for Gripen and Airborne Early Warning," *Aviation Week*, July 2, 1990.

[16] Kronhamn, T., "AEW Performance Improvements with the Erieye Phased Array Radar," *Proc. IEEE National Radar Conference*, Boston, MA, May 1993, pp. 34–39.

[17] Hendrix, R., *Aerospace System Improvements Enabled by Modern Phased Array Radar*, Northrop Grumman, October 2002, pp. 1–17.

[18] Skolnik, M., *Introduction to Radar Systems*, 3rd ed., New York: McGraw-Hill, 2001.

[19] Schrank, H., "Low Sidelobe Phased Array Antennas," *IEEE Antenna and Propagation Society Newsletter*, April 1983, pp. 5–9.

[20] *E-3 Sentry AWACS*, http://globalsecurity.org/military/systems/aircraft/e-3.htm.

[21] ap Rhys, T., and G. Andrews, "AEW Radar Antennas," AGARD Conference Preprint No. 139, *Antennas for Avionics*, 1974; also in D. C. Schleher, (ed.), *MTI Radar*, Dedham, MA: Artech House, 1978.

[22] Stimson, G., *Introduction to Airborne Radar*, 2nd ed., Mendham, NJ: SciTech, 1998.

[23] Dernery, A., "ERIEYE: An Airborne Active Phased Array Antenna," *Cost 260 Management Committee and Working Group Meeting*, Gothenburg, Sweden, May 2–5, 2001.

[24] Heed, M., "The Erieye Phased Array Antenna—From a Systems Viewpoint," *Proc. 2000 Conference on Phased Array Systems and Technology*, Dana Point, CA, May 2000, pp. 391–395.

[25] Lt. Col. Chiou, "Vulnerability of E-2C to Standoff Jamming," Master's Thesis, Naval Postgraduate School, Monterey, CA, September 1995.

[26] Fiszer, M., and J. Gruszczynski, "Russia's Roving SAMs," *Journal of Electronic Defense*, July 2002, pp. 47–56.

[27] http://enemyforces.com/missiles/tor.htm.

[28] http://globalsecurity.org/military/worlds/russia/sa-15.htm.

[29] http://greekmilitary.net/airdefence.htm.

[30] Webster, R., "An Exact Trajectory Solution from Doppler Shift Measurements," *IEEE Trans. on Aerospace and Electronic Systems*, Vol. AES-14, No. 5, March 1982, pp. 249–252.

[31] Skolnik, M., (ed.), "MTI Radar" and "Airborne MTI," Chapters 17 and 18 in *Radar Handbook*, 1st ed., New York: McGraw-Hill, 1970.

[32] Rihaczek, A., *Principles of High Resolution Radar*, Norwood, MA: Artech House, 1996.

[33] Barton, D. K., (ed.), *Radar Resolution and Multipath Effects*, Vol. 4 in *Radars*, Dedham, MA: Artech House, 1975.

[34] Skolnik, M. I., *Introduction to Radar Systems*, 1st ed., New York: McGraw-Hill, 1962.

[35] Hansen, V. G., "Signal Processing Applications to Radar," *Electronic Progress*, Vol. 28, No. 1, 1987.

[36] Schleher, D. C., *MTI and Pulsed Doppler Radar*, Norwood, MA: Artech House, 1991.

[37] Schleher, D. C., "MTI Detection Performance in Rayleigh and Log-Normal Clutter," *Proc. IEEE Int. Radar Conference*, Washington, D.C., April 1980, pp. 299–304.

[38] Haykin, S., "Radar Signal Processing," *IEEE ASSP Magazine*, April 1985.

[39] Cartledge, L., and R. O'Donnell, *Description and Performance Evaluation of the Moving Target Detector*, Report FAA-RD-76-191, MIT Lincoln Laboratory, Lexington, MA, April 1977.

[40] Skolnik, M., (ed.), "Pulsed Doppler Radar," Chapter 19 in *Radar Handbook*, 2nd ed., New York: McGraw-Hill, 1990.

[41] Andrews, G., *Optimal Radar Doppler Processors*, Report 7727, Naval Research Laboratory, May 1974.

[42] Nathanson, F., *Radar Design Principles*, 2nd ed., Mendham, NJ: SciTech, 1999.

[43] Kerr, D., *Propagation of Short Radio Waves*, New York: McGraw-Hill, 1951.

[44] Long, M. W., *Radar Reflectivity of Land and Sea*, Dedham, MA: Artech House, 1983.

[45] D. K. Barton, (ed.), *Radar Clutter*, Vol. 5 of *Radars*, Dedham, MA: Artech House, 1975.

[46] Cosgriff, R., W. Peake, and R. Taylor, *Terrain Scattering Properties for Sensor Design, Terrain Handbook*, Ohio State University, 1960; also in D. Barton, (ed.), *Radars*, Vol. 5, *Radar Clutter*, Dedham, MA: Artech House, 1977.

[47] Barton, D. K., "Land Clutter Models for Radar Design and Analysis," *Proc. IEEE*, Vol. 73, No. 2, February 1985, pp. 198–204.

[48] Barton, D., and W. Barton, *Modern Radar System Analysis Software, Version 2.0*, Norwood, MA: Artech House, 1993.

[49] Fishbein, W., S. Graveline, and O. Rittenback, *Clutter Attenuation Analysis*, Technical Report ECOM 2808, U.S. Army Electronics Command, Ft. Monmouth, NJ, March 1967; also in D. C. Schleher, (ed.), *MTI Radar*, Dedham, MA: Artech House, 1978.

[50] Billingsley, J., and J. Larrabee, *Measured Spectral Extent of L- and X-Band Radar Reflections from Wind-Blown Trees*, Report CMT-57, MIT Lincoln Laboratory, Lexington, MA, February 1987.

[51] Billingsley, J., *Exponential Decay in Wind Blown Radar Ground Clutter Spectra: Multifunction Measurements and Model*, Technical Report 997, MIT Lincoln Laboratory, Lexington, MA, July 29, 1996.

[52] Papoulis, A., *Probabilities, Random Variables and Stochastic Processes*, New York: McGraw-Hill, 1965.

[53] Boothe, R., *The Weibull Distribution Applied to the Backscatter Coefficient*, Report RE-TR-69-15, U.S. Army Missile Command, Redstone Arsenal, AL, June 1969; also in D. C. Schleher, (ed.), *Automatic Detection and Radar Data Processing*, Dedham, MA: Artech House, 1980.

[54] Mitchell, R., "Importance Sampling Applied to Simulation of False Alarm Statistics," *IEEE Trans. on Aerospace and Electronic Systems*, Vol. AES-17, No. 1, January 1981, pp. 13–24.

[55] Schleher, D. C., (ed.), *Automatic Detection and Radar Data Processing*, Dedham, MA: Artech House, 1980.

[56] Trunk, G., *Non-Rayleigh Sea Clutter: Properties and Detection of Targets*, Rep. 7986, Naval Research Laboratory, June 1976.

[57] Schleher, D. C., "Radar Detection in Log-Normal Clutter," *Proc. IEEE Int. Radar Conference*, Washington, D.C., April 1975, pp. 262–264; also in D. C. Schleher, (ed.), *Automatic Detection and Radar Data Processing*, Dedham, MA: Artech House, Dedham, MA, 1980.

[58] Schleher, D. C., "Radar Detection in Weibull Clutter," *IEEE Trans. on Aerospace and Electronic Systems*, Vol. AES-12, No. 6, November 1976, pp. 736–743; also in D. C. Schleher, (ed.), *Automatic Detection and Radar Data Processing*, Dedham, MA: Artech House, 1980.

[59] Schleher, D. C., "Harbor Surveillance Radar Detection Performance," *IEEE J. of Oceanic Engineering*, Vol. OE-2, No. 4, October 1977, pp. 318–321; also in D. C. Schleher, (ed.), *Automatic Detection and Radar Data Processing*, Dedham, MA: Artech House, 1980.

[60] Swerling, P., "Probability of Detection for Fluctuating Targets," *IRE Trans. on Information Theory*, Vol. IT-6, April 1960, pp. 269–308.

[61] DiFranco, J., and W. Rubin, *Radar Detection*, Dedham, MA: Artech House, 1980.

[62] Ward, K., and S. Watts, "Radar Sea Clutter," *Microwave J.*, Vol. 28, No. 6, June 1985, pp. 109–121.

[63] Watts, S., "Radar Detection Prediction in Sea Clutter using the Compound K-Distribution Model," *Proc. IEE*, Vol. 132, Part F, No. 7, December 1985, pp. 613–620.

[64] Cartledge, L., and R. O'Donnell, *Description and Performance Evaluation of the Moving Target Detector*, Report FAA-RD-76-190, MIT Lincoln Laboratory, Lexington, MA, March 1977.

[65] Schleher, D. C., "Performance Comparison of MTI and Coherent Doppler Processors," *Proc. IEE Int. Radar Conference*, London, November, 1982, pp. 154–158.

[66] Taylor, J., "Sacrifices in Radar Clutter Suppression Due to Compromises in Implementation of Digital Doppler Filters," *Proc. IEE Int. Radar Conference*, London, October 1982, pp. 46–50.

[67] O'Donnell, R., and C. Muehe, "Automated Tracking for Aircraft Surveillance Radar Systems," *IEEE Trans. on Aerospace and Electronic Systems*, Vol. AES-15, No. 4, July 1979, pp. 508–517.

[68] Taylor, J., E. D'Addio, and G. Galati, "Effects of Finite Word Length on FIR Filters for MTD Processing," *Proc. IEE*, Vol. 130, Part F, No. 6, October 1983.

[69] Barton, D. K., *Radar System Analysis*, Norwood, MA: Artech House, 1988.

[70] Ward, H., "The Effect of Bandpass Limiting on Noise with a Gaussian Spectrum," *Proc. IEEE*, Vol. 57, No. 11, November 1969, pp. 2089–2090c.

[71] Taylor, J., and G. Druning, "Design of New Airport Surveillance Radar (ASR-9)," *Proc. IEEE*, Vol. 73, February 1985, pp. 284–289.

[72] Cole, E., et al., "Novel Accuracy Algorithms for the Third Generation MTD," *Proc. IEEE Radar Conference*, 1986, pp. 44–47.

[73] Khoury, E., and J. Hoyle, "Clutter Maps Design and Performance," *Proc. National Radar Conference*, Atlanta, March 1984, pp. 1–7.

[74] Nitzberg, R., "Clutter Map CFAR Analysis," *IEEE Trans. on Aerospace and Electronic Systems*, Vol. AES-22, No. 4, July 1986, pp. 419–421.

[75] Blake, B., ed., "Janes Radar and Electronic Warfare 1989–1990," Janes Information Group, Surrey, U.K., p. 65.

[76] http://www.radartutorial.eu/19karteil/pic/img2031.jpg.

[77] http://www.radomes.org.

[78] http://www.defenseindustrydaily.com/images/ELEC.AN-TPS-59_Mobile_TRAD_Radar_USMC_jg.jpg.

[79] http://www.ausairpower.net.

CHAPTER 2

Doppler Radar Waveform Design and Filtering

In Chapter 1, we discussed the theory of optimum Doppler filter design for detection of a Doppler frequency-shifted coherent pulse train. The MTI solution was identified as appropriate when there was no a priori knowledge about the Doppler shift of the target return, and the pulsed Doppler design resulted when it was assumed that the Doppler shift of the target return was precisely known. In practice, the PD design is implemented by filling the expected region of target Dopplers with a suitable number of point-optimized complex weighted linear transversal Doppler filters. Although the structure of the Doppler filters depends on the assumption associated with the velocity (Doppler) characteristics of the expected targets, the detailed filter parameters depend on the power spectral density characteristics of the clutter, which, in turn, depend on the radar's waveform design.

The primary consideration in Doppler radar waveform design is concerned with the range and Doppler ambiguities associated with the particular waveform under consideration. One way to visualize this situation is through the use of the radar ambiguity function. An idealized form of the ambiguity surface for a coherent pulse train with a large number of pulses is shown in Figure 2.1. Note that range ambiguities occur at multiples of the pulse repetition intervals given by $T = 1/f_r$, where f_r is the radar's pulse repetition frequency. Doppler (velocity) ambiguities are also related to the radar's PRF and are given by $v_1 = \lambda f_r/2$, where λ is the radar's wavelength. Thus, the radar's PRF (f_r) determines both range and Doppler ambiguities and is the critical parameter in a Doppler radar's waveform design.

By combining the relations for unambiguous range and velocity the result is

$$R_u v_u = \frac{2.3565 \cdot 10^4}{f_{\text{GHz}}} \tag{2.1}$$

where R_u is the unambiguous range in nautical miles, v_u is in knots, and the radar's frequency is in gigahertz. This equation indicates that a low radar transmitter frequency is required when unambiguous range and velocity are design requirements.

The radar ambiguity function provides the response of a matched filter receiver to a point scatterer whose range and Doppler are centered about the target. Consider first fixed-base radar where ground clutter occurs at ranges randomly distributed throughout the range delay space and at zero Doppler. By superimposing the radar

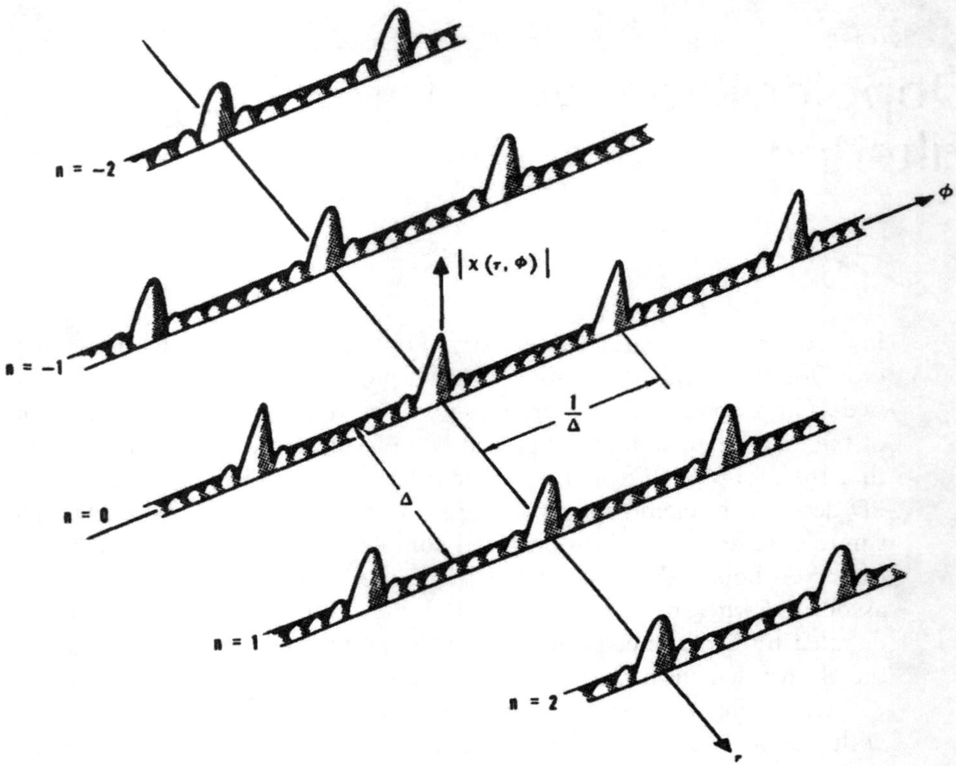

Figure 2.1 Idealized radar ambiguity surface for a coherent pulse train with a large number of pulses. (*From:* [1, p. 3-24].)

ambiguity response onto the target and clutter, we can see that the clutter response will be distributed throughout the range dimension. If the target possesses a differential Doppler with respect to the clutter, however, it can be separated in the Doppler domain by appropriate filtering. If the target range is to be measured unambiguously, then the multiple responses in the range domain should be separated by greater than the range delay associated with the maximum expected target range. This results in the low-PRF radar waveform design appropriate for a fixed-base MTI radar operating against ground clutter. Note that if the Doppler associated with the target exceeds the radar's PRF, then the Doppler generally can be separated by Doppler filtering but the radar's Doppler response will be ambiguous. When the target's Doppler responses occur at multiples of the radar's PRF, then the targets cannot be separated by Doppler filtering, resulting in the so-called *blind speeds* associated with low-PRF Doppler radar design. In this case, we must jitter or stagger the radar's PRF to achieve target detection.

Next, consider the case of airborne radar operating against ground clutter. The aircraft motion will cause the return from the clutter scatterers to have a Doppler response that depends on the relative closing velocity between the aircraft and a particular scatterer. The magnitude of the response will depend on the two-way antenna pattern in the direction of the scatterer. Hence, in effect, the aircraft motion spreads the Doppler return of the clutter throughout the Doppler region corresponding to the aircraft's velocity ($f_d = \pm 2v_r/\lambda$). If a low-PRF design is used with ambiguous Doppler response, then sidelobe clutter will generally appear within the same Doppler region as the target. Therefore, the low-PRF airborne radar clutter improvement factor generally cannot exceed the magnitude of the radar antenna's sidelobe two-way radiation pattern when compared to the magnitude of the antenna's peak response in its main lobe.

To circumvent this difficulty, the radar's PRF can be increased to greater than $f_d = 2v_{ac}/\lambda$ plus the maximum Doppler frequency of the target

$$\mathrm{PRF} > \frac{2(v_{ac} + v_{tar})}{\lambda} \tag{2.2}$$

Then the target's response is unambiguous in Doppler, and closing targets can be separated from the clutter by using appropriate Doppler filtering. This type of radar waveform design results in a high-PRF radar appropriate for an airborne PD radar. Note that this type of radar generally is highly ambiguous in range, which requires PRF coding to resolve the range ambiguities.

However, another significant difference between the high- and low-PRF airborne radar designs affects the strength of the clutter return. In the low-PRF radar design, only clutter returns occurring within the same range resolution cell as the target compete with the target. In a high-PRF radar design, clutter returns from shorter ambiguous ranges compete with the target return in a range resolution cell. Because the shorter range clutter returns are much stronger than those at the target, the high-PRF radar must provide a much greater clutter improvement factor than the low-PRF design for equal detection performance. Another consequence of this effect is that detection performance in high-PRF radars is limited against tail-aspect targets because these targets produce Doppler returns that must compete against range ambiguous sidelobe clutter Doppler returns.

► Example 4 Magnification of Main Beam Clutter in High-PRF Radar

In a high-PRF radar the magnitude of main beam clutter is increased over that from low-PRF radar due to the reflections from ambiguous clutter returns in proximity to the radar. This situation is depicted in the diagram shown in Figure EX-4.1.

The RCS of clutter for low-PRF radar is given by

$$\sigma_{LPRF} = \sigma^0 \theta_{az} R_{tar} R_\tau \sec \Psi$$

where σ^0 is the clutter backscatter coefficient, θ_{az} is the azimuth beamwidth in radians, R_{tar} is the target range, R_τ is the equivalent pulsewidth range, and Ψ is the grazing angle with respect to the target. The radar horizon in meters for clutter can then be calculated as

$$R_{hor} = 4124\sqrt{h_{ant}}, \quad [\mathrm{m}]$$

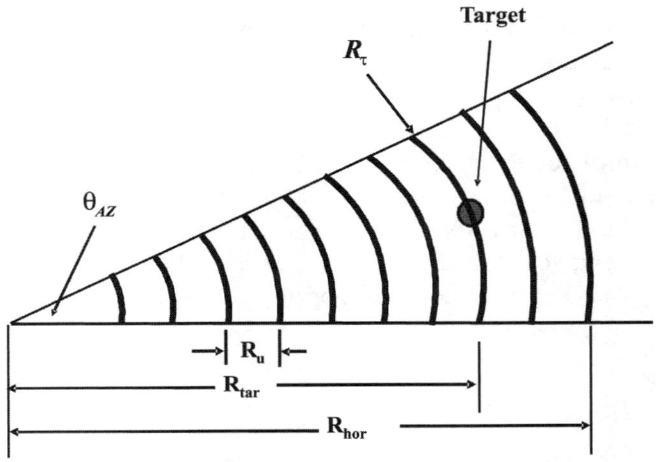

Figure EX-4.1 Ambiguous clutter in high-PRF radar.

where h_{ant} is the antenna height in meters. The integer number of ambiguous clutter patches within the radar horizon is then given by

$$n = \text{floor}\left(\frac{R_{hor}}{R_u}\right)$$

where $R_u = c/2$ PRF is the unambiguous range. In addition, the integer number of clutter patches between the target and the radar is given in compatible units by

$$n_x = \text{floor}\left(\frac{R_{tar} - h_{ant}}{R_u}\right)$$

The clutter RCS can then be determined by adjusting the relative strength of each ambiguous clutter patch with a range equal to or less than the target by $[(R_{tar}/(R_{tar} - iR_u)]^4$ for $i = 0, 1, 2, \ldots, n_x$ and between the horizon and the target by $[R_{tar}/(R_{tar} + iR_u)]^4$ for $i = 1, 2, \ldots, n - n_x$ resulting in

$$\sigma_{HPRF} = \sigma_{LPRF}\left[\left(\frac{R_{tar}}{R_{tar} + (n - n_x)R_u}\right)^3 + \ldots + 1 + \left(\frac{R_{tar}}{R_{tar} - R_u}\right)^3 + \ldots + \left(\frac{R_{tar}}{R_{tar} - n_x R_u}\right)^3\right]$$

This equation can generally be approximated as

$$\sigma_{HPRF} = \sigma_{LPRF}\left[\left(\frac{R_{tar}}{R_{tar} - n_x R_u}\right)^3\right]$$

where the term in brackets is the *magnification factor* that relates the high-PRF clutter RCS to that of the low-PRF RCS.

The Gaussian elevation antenna pattern illuminates the various ambiguous clutter patches with a two-way power gain given by

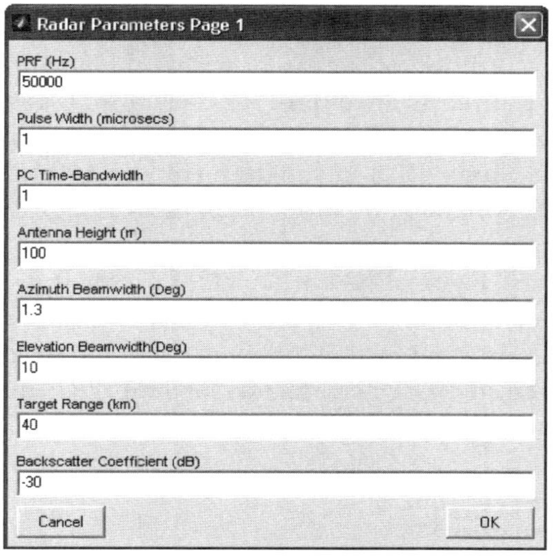

Figure EX-4.2 Dialog box for high-PRF radar.

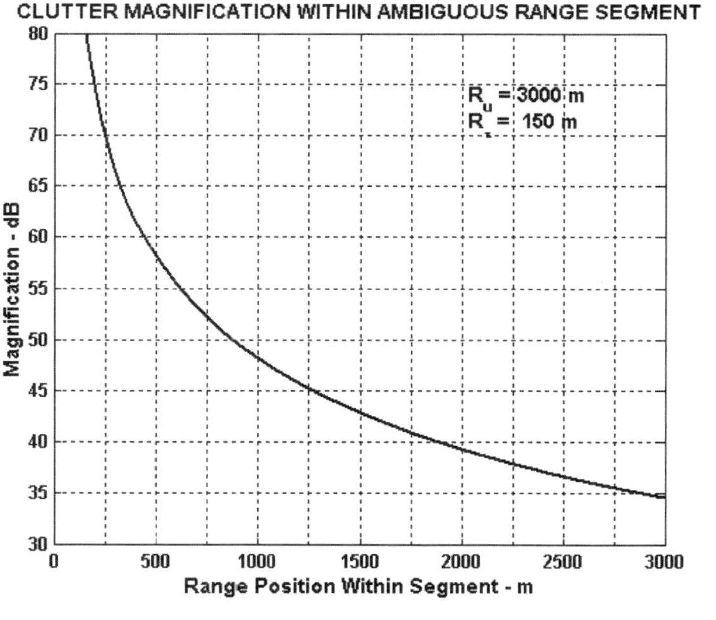

Figure EX-4.3 Clutter magnification within PRF range segment.

$$F(\phi) = \exp\left[-5.55\left(\frac{\phi_{el} - \phi_{clut}}{\phi_{elbw}}\right)^2\right]$$

where ϕ_{el} is the elevation angle to the target, ϕ_{clut} is the elevation angle to the clutter patch, and ϕ_{elbw} is the elevation one-way, 3 dB beamwidth.

A MATLAB program called high_prf_clutc.m calculates the clutter RCS magnification factor both with (CMag_dB) and without (BMag_dB) elevation antenna weighting. In addition, the program plots the high-PRF magnification factor as a function of the position of the target within the unambiguous range segment (R_u). This effect is utilized in medium-PRF radars that transmit multiple PRFs to find quiet clutter zones to compete with the target return. Also the program calculates the equivalent clutter RCS in square meters and dBsm.

The MATLAB dialog box for a hypothetical high-PRF radar is shown in Figure EX-4.2. Different situations can be calculated and plotted by overwriting the dialog box parameters. Figure EX-4.3 depicts the variation of the magnification factor for various positions of the target within the unambiguous range segment (R_u). The MATLAB printout is shown below. Note that the large *clutter magnification* requires a large improvement factor for high-PRF radar, placing a premium on transmitter and STALO stability.

```
BMag_dB = 48.1506
CMag_dB = 41.0334
RCS_clut_sm = 1.7271e+006
RCS_clut_dBsm = 62.3731
```

▶

A compromise airborne radar waveform design uses a medium-PRF waveform that is ambiguous in both the range and Doppler domains. This type of radar waveform can provide performance against closing targets that is somewhat less than that provided by a high-PRF design and performance against tail-aspect targets that is somewhat less than that provided by a low-PRF design. The medium-PRF airborne radar design, therefore, is useful when all-aspect coverage is required and the tail-aspect limitations of high-PRF designs or the closing aspect limitations of low-PRF designs cannot be tolerated. As described in Chapter 7, this leads to multimode designs in most modern AI radars.

The medium-PRF airborne radar waveform finds application in situations where the radar must look down in heavy ground clutter against tail-aspect targets. These targets appear in the sidelobe clutter region of the range-Doppler clutter surface shown in Figure 2.2 for the high-PRF waveform. Hence, any Doppler filter placed in this region collects both target and clutter returns in contrast to the noise-limited region that offers good sensitivity for high-radial-speed closing targets. The heavy sidelobe clutter shown in Figure 2.2 for the high-PRF design is emphasized by the collection of close-in clutter due to the highly range-ambiguous nature of this type of waveform. The medium-PRF range-Doppler clutter surface also depicted in Figure 2.2 shows a reduction of the sidelobe clutter's magnitude due to spreading of the range ambiguities. However, the targets now must generally compete with sidelobe clutter, which occurs at all ranges and radial velocities, thereby placing a premium on the low-sidelobe antenna design.

Medium- and high-PRF waveform designs are also useful in surface-based radars for detecting low-altitude targets in heavy ground clutter or rejecting sky clutter (weather or chaff). In low-PRF radars weather or chaff returns often are aliased into the first Doppler region, where they overlap fundamental and aliased target Doppler returns.

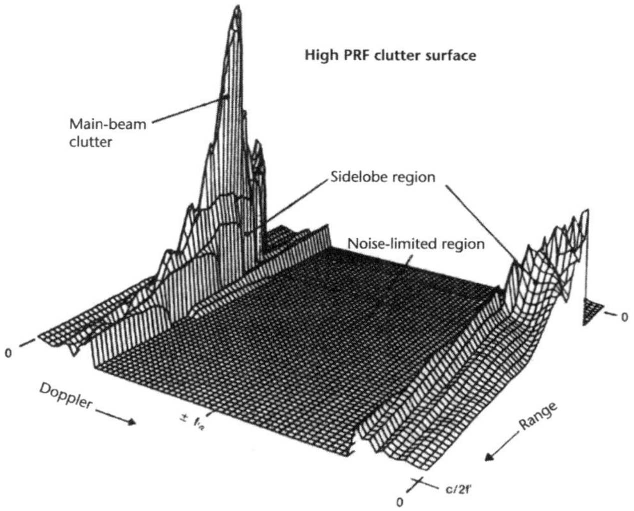

High PRF radar signals offer good sensitivity to approaching targets moving at high speed. This range-Doppler diagram shows noise-limited performance in which ground echo does not compete with such targets.

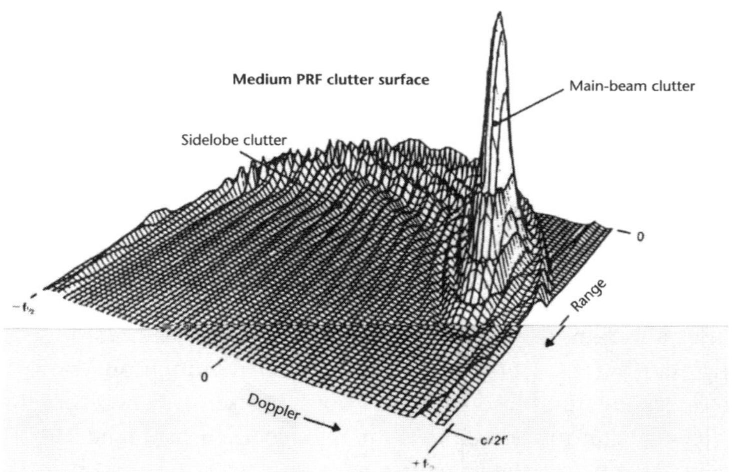

Medium PRF radar signals are subject to sidelobe clutter at all ranges, but careful antenna design can reduce this limitation to satisfactory levels. Medium PRF can be a good compromise between high and low PRF signals.

Figure 2.2 High- and medium-PRF airborne radar range-Doppler clutter surface. (*From:* [2, p. 132].)

2.1 Definition of PRF Classes [3]

The preceding discussion points to the desirability of classifying Doppler radar modes on the basis of their PRFs. What has become common practice, particularly in the airborne radar field, is to classify radar modes as low-, medium-, and high-PRF radars. The generally accepted definitions are as follows [3]:

- In low-PRF radar operation, the maximum range the radar is designed to handle falls into the first range zone given approximately in nautical miles by $R_u = 80/f_r$, where f_r is the radar's PRF in kilohertz. In the absence of a return from beyond this zone, range is unambiguous.
- In a high-PRF operation radar, the observed Doppler frequencies of all significant targets are unambiguous, but the range generally is highly ambiguous.
- In medium-PRF radar operation, both range and Doppler frequency are ambiguous.

The blind speeds ($f_{d\,blind}$) for the radar waveform occur when the return Doppler corresponds to multiples ($n = 1, 2, \ldots$) of the radar's PRF ($f_{d\,blind} = nf_r$). The corresponding blind speeds are given by

$$v_{d\,blind} = \frac{n\lambda f_r}{2} \tag{2.3}$$

The unambiguous range is given by

$$R_u = \frac{c}{2f_r} \tag{2.4}$$

where f_r is the radar's PRF. Using these equations, unambiguous velocity (first blind speed) can be plotted against unambiguous range and PRF, which is depicted in Figure 2.3.

As an example, consider the situation for fixed-base S-band (3-GHz) air surveillance radar with a maximum range of 50 nmi that searches for targets with radial velocities in the ±300-knot region. Reference to Figure 2.3 indicates that setting the first blind speed at 330 knots results in a PRF of approximately 3,100 Hz, which provides a maximum unambiguous range of approximately 26 nmi. Alternatively, setting the PRF at 1,600 Hz results in an unambiguous range of about 50 nmi, but reduces the first blind speed to about 155 knots. Therefore, unambiguous range and velocity cannot be achieved at the specified frequency for the example given. However, reducing the transmitter frequency to L-band (1 GHz) would allow unambiguous operation in both range and Doppler.

In the air surveillance radar example, the S-band version operating at a PRF of 1,600 Hz would be classified as low-PRF radar. The same radar operating at a PRF of 3,100 Hz would be classified as medium-PRF radar. The L-band version operating at a PRF of 1,600 Hz would generally be classified as a low-PRF radar even though it provides unambiguous velocity information. In general, the PRF class does not depend on the numerical value of the radar's PRF in a precise way, but rather is strongly influenced by the radar's operating frequency and the maximum expected range of both target and clutter returns and their relative velocities with respect to the radar.

Airborne radars used on high-performance aircraft generally are required to perform a number of functions that require the use of several PRF classes. X-band multimode radar might use a low-PRF mode, typically from 250 to 4,000 Hz;

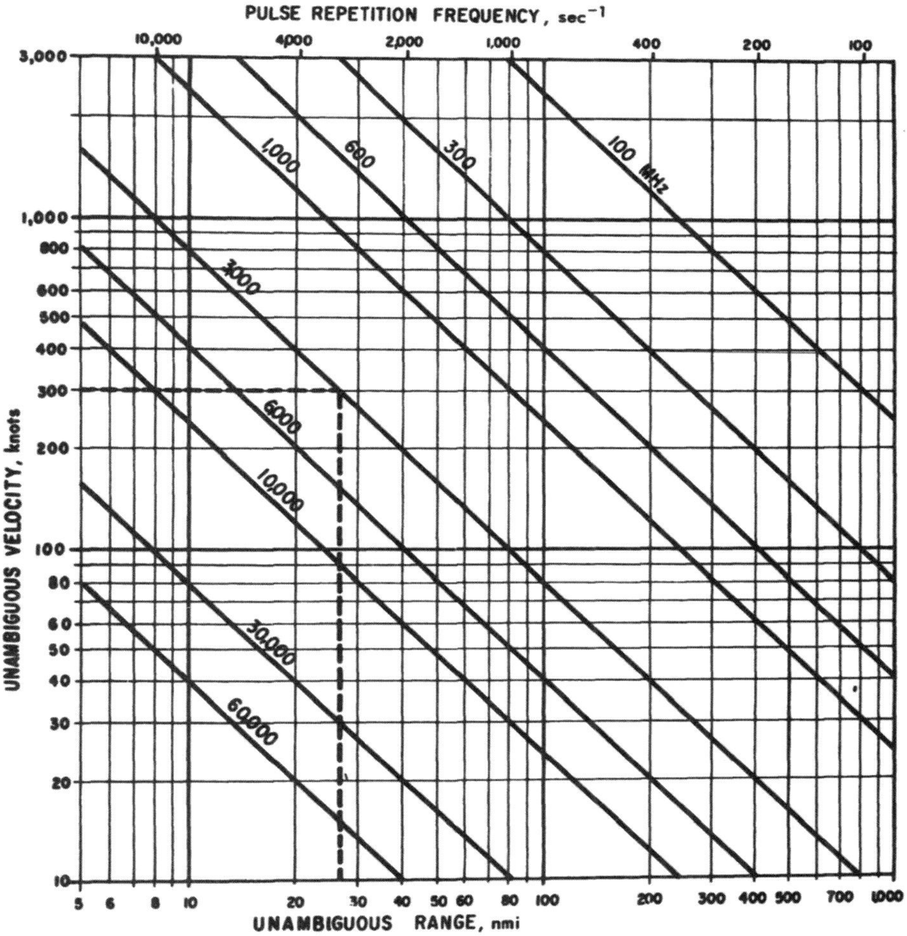

Figure 2.3 Unambiguous velocities versus unambiguous range for carrier frequencies from 100 to 60,000 MHz. (*From:* [4, p. 473].)

PRFs in the medium category, from 10 to 20 kHz; and PRFs in the high category, from 100 to 300 kHz [3]. The three categories of PRF are typically interlaced as the radar scans through its coverage volume (see Section 7.3).

2.2 Signal Processing and System Comparisons

Each PRF choice has advantages and limitations that relate to a number of complex factors. The range and Doppler ambiguities of the radar waveform generally determine the radar's response to target and clutter returns. This determines the radar's signal processing configuration and, ultimately, the radar's overall performance capability.

.2.1 Low-PRF Radar

Consider the low-PRF airborne radar whose ground clutter Doppler PSD for the antenna pointing along the ground track is depicted in Figure 2.4 [5, p. 190]. The generally ambiguous Doppler responses of this type of design cause target and clutter returns to be folded into the region between the reference Doppler (Doppler return from the ground corresponding to the spatial direction of the antenna's main lobe peak radiation pattern) and the first blind speed. The objective of the Doppler comb filter is to eliminate all of the clutter returns while retaining the entire target returns in this region.

This becomes increasingly difficult as the number of clutter and target returns that are aliased into this region increases, causing possible overlapping clutter and target returns. Note that clutter sidelobe returns are translated into the filter's passband due to aircraft motion and, hence, cannot generally be eliminated by Doppler filtering. The width of the central clutter lobe along the ground track generally is set by antenna scanning modulation induced onto each clutter scatterer illuminated by the antenna's main lobe radiation pattern.

In Figure 2.5, the ground clutter Doppler PSD is depicted for the antenna pointing away from the aircraft's ground track. The width of the central clutter lobe is now set by the differential Doppler responses caused by the different closing velocities for the velocity vectors to each clutter scatterer and that to the Doppler reference vector. In particular, the effective width of the main lobe can be shown to be given by $\sigma_p = kv_{ac} \sin \theta / D_{az}$ [5, p. 21] where v_{ac} is the aircraft's ground track velocity in meters per second, θ is the antenna's azimuth pointing angle relative to the aircraft's ground track, and D_{az} in meters is the antenna's aperture width in the azimuth dimension. Note that for high-performance aircraft the Doppler region available for target detection is severely restricted.

At low PRFs the target's Doppler frequency may be many times the PRF, so the target is about as likely to appear at any one point, within a span of frequencies equal to the PRF, as at any other. Therefore, the probability of a target being in

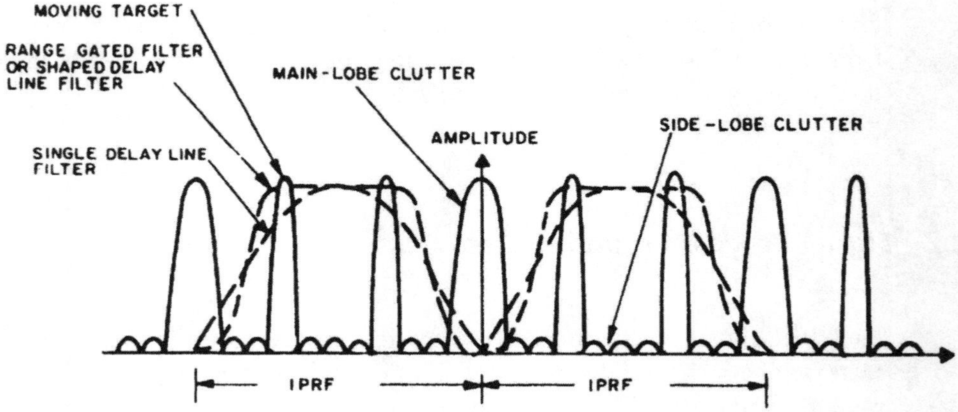

Figure 2.4 MTI filter and clutter spectrum (antenna pointing along the ground track). (*From:* [5, p. 190].)

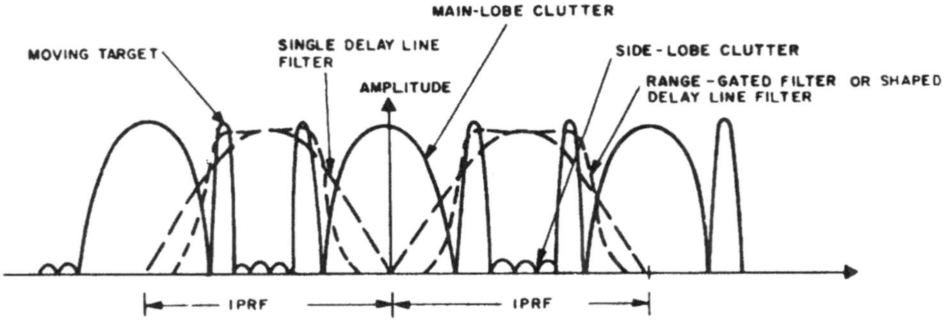

Figure 2.5 MTI filter and clutter spectrum (antenna pointing away from the ground track). (*From:* [5, p. 190].)

the blind zones at any one time is roughly equal to the ratio of the width of the rejection notches to the PRF.

Figure 2.6 shows two applications of low-PRF airborne radar that are designed to have equal ratios of Doppler-clear to blind regions. The first is typical of AEW radar (e.g., the E-2C radar), where the speed of the radar-bearing aircraft is modest and a large antenna can be mounted in a rotodome above the aircraft. For the example depicted (v_{ac} = 300 knots, θ = 90°, and D_{az} = 20 feet), the Doppler-clear region will be at least as wide as the blind region at a PRF as low as 200 Hz. Typical transmitter frequencies for low-PRF AEW radars are UHF (420 to 450 MHz) or L-band (1,215 to 1,400 MHz). The first blind velocity for a 200-Hz PRF would be about 135 knots at UHF and 45 knots for L-band.

The second application, shown in Figure 2.6, is that of an airborne interceptor radar in which the antenna size is limited and radar speeds may be high. For

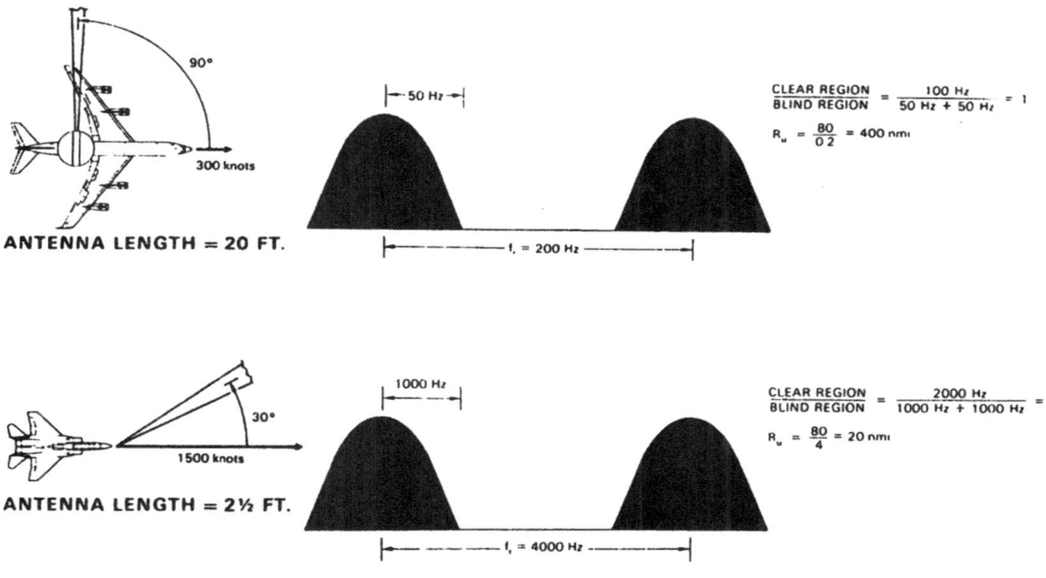

Figure 2.6 Ground clutter spectra for two applications of low-PRF airborne radar. (*From:* [3, p. 138].)

low-PRF designs, the two recourses open are to limit the look angle (±30° in the example) or to use the highest possible PRF. Typical transmitter frequencies for this type of radar are X-band (8.5 to 10.7 GHz) and K_u-band (15.7 to 17.7 GHz). The first blind velocity for a 4,000-Hz PRF would be about 117 knots at X-band and 71 knots at K_u-band.

Low-PRF airborne radar operation has several advantages. The unambiguous range capability allows: (1) direct measurement of the target's range; (2) sensitivity time control (STC) in the receiver to equalize returns as a function of range, thereby reducing the overall dynamic range; and (3) minimizing clutter intake by range gating to just those in the target's range resolution cell.

Limitations on low-PRF airborne radars involve: (1) poor look-down capability where excessive target returns are rejected along with main lobe clutter; (2) Doppler ambiguities generally too severe to be resolved; and (3) ground moving targets indistinguishable from airborne moving targets. These limitations are generally too severe for airborne interceptor radar, forcing use of other PRF classes. Low-PRF waveforms are most useful in AEW radars and for surface-based radars that do not have the severe clutter intake of look-down airborne interceptor radars.

A block diagram of a typical coherent low-PRF radar digital signal processor is depicted in Figure 2.7. The receiver output is fed to a synchronous detector that converts it to in-phase and quadrature video signals. The frequency of the reference signal supplied to the detector is such as to place the central line of main lobe clutter at zero frequency. An A/D converter samples the video signals at intervals matching the width of the transmitted pulses. The output of the converter, therefore, is a stream of numbers representing I and Q components of the returns from successive range increments. The numbers are sorted by range increment into separate range bins.

An MTI comb filter clutter canceller in each I and Q channel is used to reduce the main lobe clutter component. To reduce the main lobe clutter residue in the output of the canceller as well as minimize the amount of noise and simultaneously received sidelobe clutter with which a target must compete, the output of each clutter canceller is generally integrated in a bank of Doppler filters. At the end of every filter integration time, the magnitude of each desired filter's output is detected. When the integration time is less than the length of the time-on-target for the radar beam, postdetection integration (PDI) may be provided. A threshold detector is set at a predetermined amount, higher than the average of the output of the corresponding filters for several range bins on either side. This controls the probability of false alarms while providing adequate sensitivity for detecting targets. A frequency offset is required to adjust the synchronous detector's reference frequency to the changing center frequency (due to changing antenna look angles) of the ground clutter spectrum. This usually is accomplished to a first order using aircraft navigation data coupled with the antenna's known look angle. Fine tracking uses a clutter spectral tracking servo that adjusts the reference frequency so that main lobe clutter is centered within the canceller's rejection notch.

Figure 2.8 depicts a single-channel, low-PRF signal processor employed in older or less sophisticated modern radars. This type of processor uses only the I channel from the synchronous detector, resulting in an average 3-dB loss in signal energy and the introduction of blind phases (Doppler phases causing zero output). Elimina-

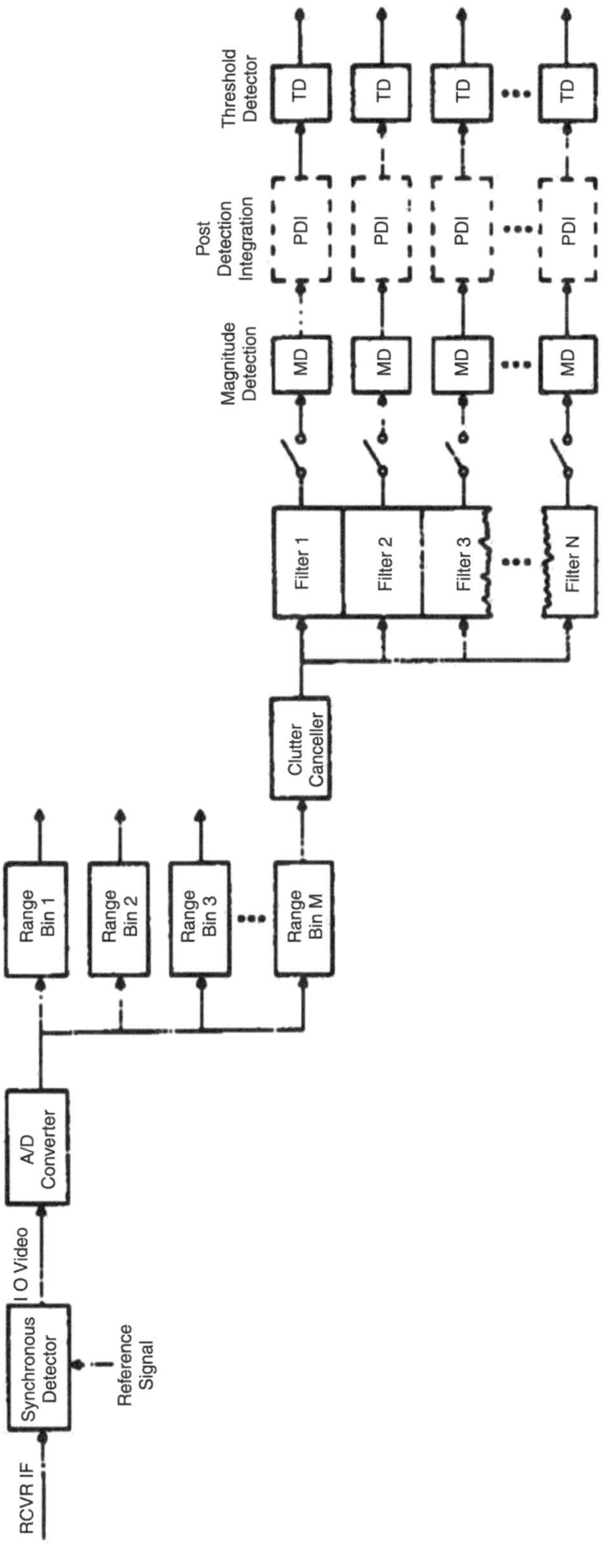

Figure 2.7 Signal processor for low-PRF radar with Doppler filters. (*From:* [3, p. 340].)

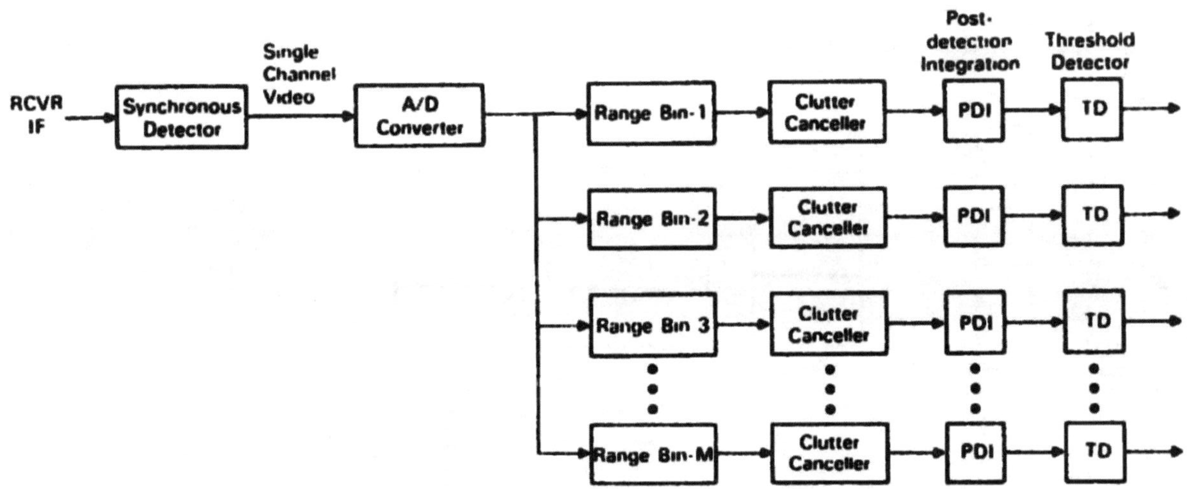

Figure 2.8 Single-channel, low-PRF signal processor. (*From:* [3, p. 346].)

tion of the Q-channel Doppler filters and magnitude detectors considerably simplifies the processing hardware.

2.2.2 High-PRF Radar

A typical PSD for airborne high-PRF radar is depicted in Figure 2.9. All main lobe and sidelobe Doppler returns from ground clutter are located in the region

$$f_d = \pm \frac{2v_r}{\lambda} \tag{2.5}$$

where the velocity of the aircraft carrying the radar is v_r. The absence of clutter returns in the region between the band of central-line sidelobe clutter frequencies and the first repetition of this band opens up a region that is clutter free. In this region, which corresponds to targets closing on the airborne radar at a speed greater than the radar's ground speed, detection can be made against a noisy background using a matched-filter signal processor. From an operational viewpoint, long detection ranges are desired for these approaching targets.

The high-PRF radar allows the average transmitter power to be maximized by increasing the PRF. This further enhances the detection capability against closing targets. At X-band typical PRFs for airborne radars are on the order of 100 to 300 kHz with duty factors of up to 50%. Achieving the same amount of average power with low-PRF radar, subject to a transmitter peak power limitation, requires increasing the radar's pulsewidth and employing large amounts of pulse compression to provide the degree of range resolution essential to low-PRF radar operation.

The principal limitation of high-PRF airborne radar operation is that detection performance may be degraded by sidelobe and backlobe clutter when operating

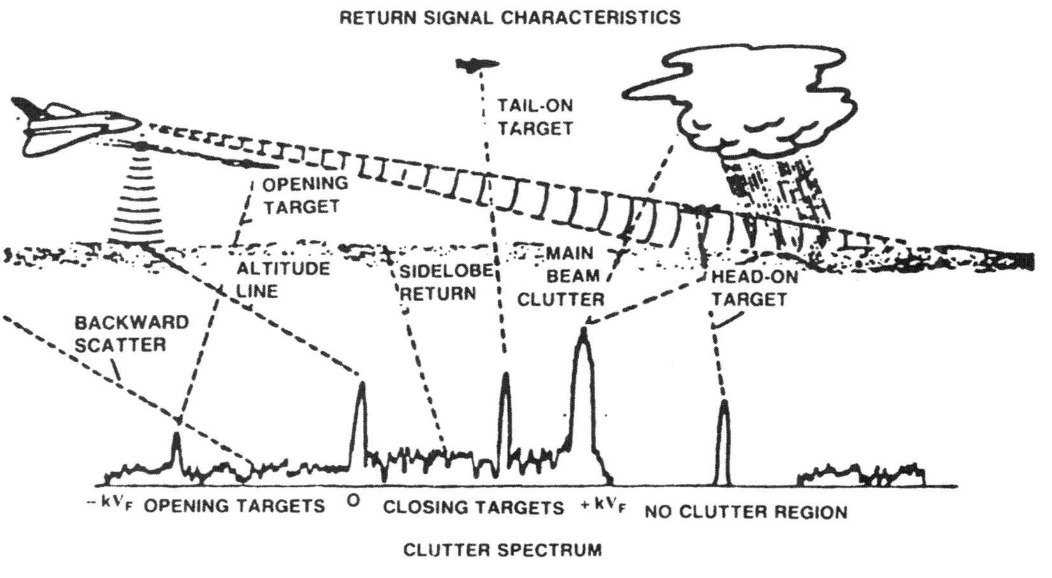

Figure 2.9 High-PRF radar clutter and target spectra. (*From:* [6, p. 283].)

against low closing-rate and opening-rate tail-aspect targets. As we can see from Figure 2.9, these targets must compete with sidelobe and backlobe clutter with spectral responses that overlap the target's spectral response. At the high PRFs typically employed in modern high-performance aircraft, the highly ambiguous range response causes sidelobe clutter from virtually all ranges to fold within the range interval corresponding to the interpulse period between successive radar transmissions. Consequently, the sidelobe clutter can be rejected only by resolving the return in Doppler frequency. Much of the sidelobe clutter falling within the same resolvable frequency increment as a target's return will have been reflected from much shorter ranges and so will be quite strong. Thus, unless the target is at a short range or has a very large RCS, its return may be lost in the clutter.

Another disadvantage of the high-PRF radar waveform is the necessity to resolve the target's range ambiguities to obtain the target's range. This can be accomplished by either switching the radar PRFs or by using a frequency-modulated (FM) transmitter. In either case, the losses incurred in resolving ambiguities usually cause a significant reduction in overall detection range compared to that calculated on a power aperture basis. Because range measurement reduces detection range, where maximum detection range is desired, a special velocity search mode may be provided in which range is not measured.

High-PRF radars, when operating at high duty factors, experience significant eclipsing losses that occur due to returns being received while the radar is transmitting and the receiver blanked. This may be minimized by switching PRFs or providing multiple-range gates. However, switching PRFs reduces the detection range in the Doppler-clear region.

A typical high-PRF airborne radar analog signal processor is depicted in Figure 2.10. The receiver output is first applied to a bandpass filter that passes only the central spectral band, thereby converting its output into a continuous signal, and all subsequent processing is handled as in CW radar. The output of the central band filter is passed through a filter that removes the return signals having zero Doppler frequency, consisting of any transmitter spillover and altitude return, composed of specular returns from the terrain directly under the aircraft. Along with them, the echoes of any zero-closing-rate targets will unavoidably be rejected.

The entire Doppler spectrum is then shifted in frequency to center the main lobe clutter in the rejection notch of the main lobe clutter (MLC) filter. The Doppler spectrum output of the MLC filter is then multiplexed into two subbands. The first subband spans the Doppler clutter clear region, and the second covers the sidelobe and backlobe clutter regions. By applying automatic gain control separately in each subband, a strong return in one band is prevented from desensitizing the other bands.

Signal levels equalized, the subbands are recombined and applied to a bank of Doppler filters. The output of each Doppler filter is then detected and compared against a detection threshold. The Doppler filter output builds up for a time $T_i = 1/B_D$, where B_D is the Doppler filter's bandwidth. When T_i is less than T_D (the time on target for the radar antenna beam), then the detector outputs are added up in postdetection integrators before comparison to separate threshold detectors.

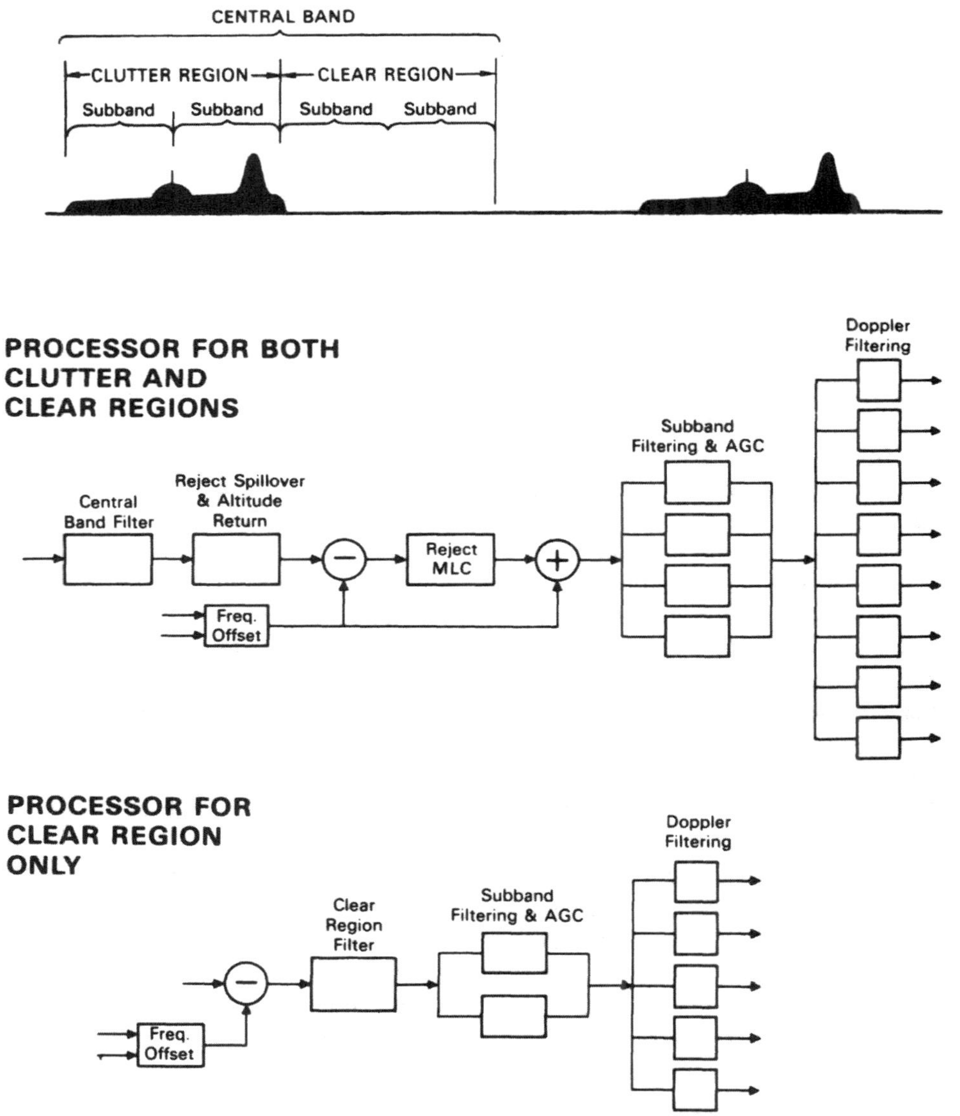

Figure 2.10 Signal processor for interleaved high- and medium-PRFs and for only high-PRF modes. (*From:* [3, p. 379].)

2.2.3 Medium-PRF Radar [2, 7, 8]

Medium-PRF airborne radar was conceived as a solution to the problems of detecting tail-aspect targets in the presence of both main lobe and strong sidelobe clutter, thereby providing good overall coverage from all aspects. For modest operating ranges, the PRF can be set high enough to provide adequate separation between the periodic repetitions of the main lobe clutter spectrum without incurring particularly severe range ambiguities. Main lobe clutter can then be rejected through Doppler filtering, and individual targets can be extracted from sidelobe clutter using a combination of Doppler filtering and range resolution discrimination. Also ground moving targets, being similar in characteristic to main lobe clutter, can be rejected using a medium-PRF design.

The key feature of medium-PRF airborne radar operation is its ability to detect tail-aspect moving targets in a sidelobe clutter background. Increasing the PRF over low-PRF operation opens the clear region between main lobe clutter responses that occur at multiples of the radar's PRF (e.g., recall Figure 2.6). This makes it easier to construct practical Doppler filters to both reject main lobe clutter and bracket the target's response, thereby rejecting that sidelobe clutter not in the target's Doppler resolution cell (e.g., $B_D = 1/T_d$, where T_d is the time on target).

However, in the process of increasing the PRF, range ambiguities are created that allow close-in main lobe and sidelobe clutter to enter the range resolution cell containing the target. Main lobe clutter can be rejected through the use of a comb filter (e.g., MTI canceller) just as for a low-PRF design except that the rejection capability must be increased due to the ambiguous clutter. Sidelobe clutter must be rejected through increased Doppler resolution (e.g., bandpass filtering), which serves the dual function of attenuating the clutter outside the target's Doppler resolution cell while minimizing the unavoidable sidelobe clutter intake within the Doppler filter containing the target. A prime consideration in the design of medium-PRF radar involves the minimization of the unavoidable sidelobe clutter intake that ultimately limits the radar's performance.

The clutter intake into medium-PRF airborne radar is regulated through sorting by both range and Doppler frequency. Sorting by range is accomplished using range gates that isolate the returns received from relatively narrow strips of ground at constant range. Because of the range ambiguities, the return passed by each gate will come from several strips of ground. In Figure 2.11, the concentric circles represent the projection of the radar's ambiguous ranges onto the ground through which the clutter return entering a single range gate must pass.

Sorting of sidelobe clutter by Doppler frequency is accomplished by applying the output of each range gate to a bank of Doppler filters. They will isolate the returns received from strips of ground lying between lines of constant angle relative to the radar's velocity, which are called *isodops* (see Section 1.3.4.1). Because of Doppler ambiguities, any one filter will pass the returns from several strips. In Figure 2.11, the narrow strips between the hyperbolic lines represent the strips of ground providing clutter returns to a single Doppler filter.

The shaded areas in Figure 2.11 depict those ground patches contributing returns to the output of a single range-gated Doppler filter. Note that each Doppler filter receives only a portion of the total sidelobe return passed by a single range gate. Increasing range and Doppler resolution reduces the shaded areas in Figure 2.11, thereby minimizing the unavoidable clutter intake that competes with the target returns in each Doppler filter.

The level of sidelobe clutter in a Doppler filter for low-, medium-, and high-PRF designs of X-band radar is depicted in Figure 2.12 [8]. In the medium-PRF design, because most of the sidelobe clutter comes from relatively short ranges, the background of clutter against which targets must be detected is generally stronger than the background noise falling in the passband of a Doppler filter. Therefore, detection performance is limited by the target-to-sidelobe clutter ratio rather than the signal-to-noise (S/N) ratio. As the target's range is continuously increased, a point

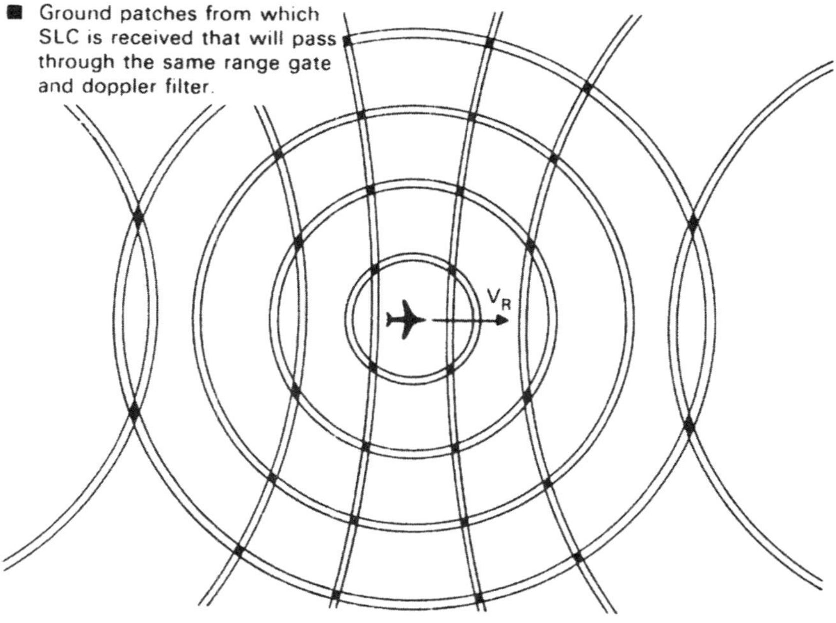

Figure 2.11 Ground patches providing sidelobe clutter returns in medium-PRF airborne radar. (*From:* [3, p. 358].)

eventually is reached where targets become lost in the clutter. However, before this range is reached, there are alternating regions when the target return exceeds the clutter return and those where the target is obscured by the clutter. These regions are called *range blind zones* (e.g., blind speeds in low-PRF radar). The positions of the radar blind zones shift with changes in the PRF. Fortunately, the shift is such that the same PRF switching used to reduce Doppler blind zones is also effective in reducing range blind zones. Therefore, by applying multiple PRFs in the medium-PRF design, smoothing the detection performance for range blind zones is possible in the same manner as Doppler blind zones are reduced in a low-PRF radar design.

From the preceding discussion, the key design issue for medium-PRF airborne radars clearly is the minimization of sidelobe clutter. Reducing the antenna's sidelobes is the most fundamental action that can be taken in this regard. Recent developments in this area have produced ultralow-sidelobe designs (i.e., relative one-way sidelobe responses –40 dB or more below the main beam response), but these are difficult to achieve in typical high-performance airborne installations due to installation and radome considerations.

Another action that can reduce sidelobe clutter is to increase the radar's range resolution by narrowing the radar's effective pulsewidth and correspondingly narrowing the radar's range gates. Narrowing the radar's effective pulsewidth reduces the sidelobe clutter level in the same proportion by which the pulsewidth is narrowed. Additional range gates are required to provide full range coverage and more Doppler filters must be formed, because a separate bank of Doppler filters is required for each range gate.

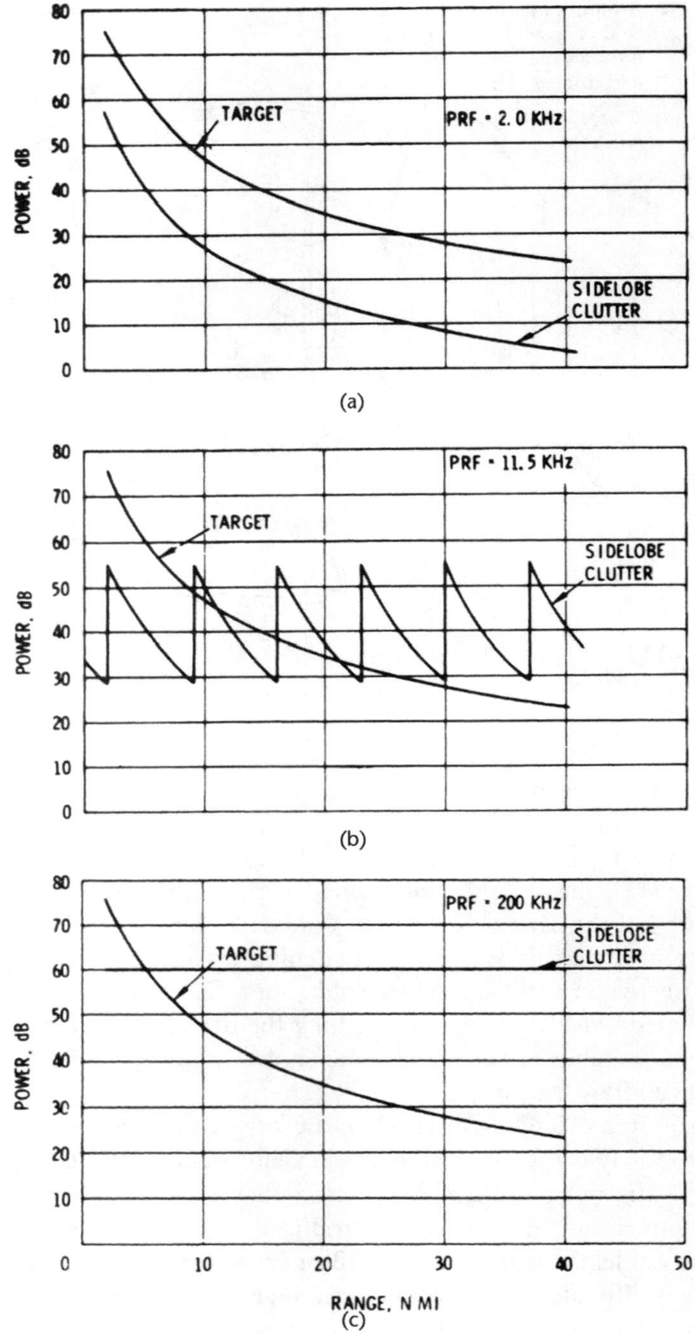

Figure 2.12 (a–c) Sidelobe clutter and target responses for low-, medium-, and high-PRF airborne radar. (*From:* [8, p. 272].)

One way to narrow the effective radar pulsewidth without reducing the average transmitted power is pulse compression. A common practice is to maximize the average transmitter power without incurring an unacceptable loss due to eclipsing and then use enough pulse compression to achieve the desired range resolution. An alternative approach that minimizes eclipsing is to transmit very narrow pulses of higher peak power.

The sidelobe clutter with which a target must compete can be further reduced by narrowing the passbands of the Doppler filters. To retain all of the power in the target return, the filters must be wide enough to pass the target's spectrum. The target spectrum generally is inversely proportional to the radar's time on target (T_d) plus an allowance for internal target motion and acceleration effects.

An important form of sidelobe clutter in medium-PRF airborne radars results from large RCS targets on the ground (e.g., buildings and water towers, etc.), which are particularly prevalent in urban or mountainous areas. Also included in this class of clutter are unwanted moving ground vehicular traffic that, because of the range and Doppler ambiguities, conflicts with the detection of airborne targets.

Although airborne radar is vulnerable to such targets when operating at low PRF, it is more vulnerable at medium PRFs due to the severe range ambiguities in this type of operation. One way to reduce these unwanted targets is to provide the radar with a guard channel that consists of a separate receiver processor whose input is supplied by a small horn antenna mounted on the main radar antenna. The width of the horn's main lobe is sufficient to encompass the entire region illuminated by the radar antenna's principal sidelobes, and the gain of the horn's main lobe is greater than that of any of the main antenna's sidelobes. Any detectable target in the radar antenna's sidelobes will produce a stronger output from the guard receiver than from the main receiver. Conversely, any detectable target in the main lobe will produce a stronger signal than in the guard receiver. Consequently, by comparing the output of the two receivers and inhibiting the output of the main receiver when that of the guard receiver is stronger, we can eliminate the unwanted targets in the radar's sidelobe region.

At medium PRFs, signal processing is handled very much like it is at low PRFs with three main exceptions. First, because range ambiguities preclude the use of sensitivity time control, additional automatic gain control is needed to avoid saturation of the A/D converter. Second, the passband of the Doppler filters employed must be made considerably narrower to exclude the excess sidelobe clutter that builds up due to Doppler ambiguities. Third, additional processing is required to resolve range and Doppler ambiguities.

The block diagram of a typical medium-PRF airborne radar digital signal processor is depicted in Figure 2.13 [3]. The first step in processing the receiver output is to translate the Doppler spectrum to in-phase and quadrature channel baseband signals for digital sampling. The shift must be dynamically controlled to account for changes in radar velocity and antenna look angle. The *I* and *Q* outputs are sampled at intervals on the order of the transmitted pulsewidth and digitized.

However, to reduce the dynamic range required of the A/D converter, automatic gain control is provided ahead of the converter. To accomplish this, the converter's output is monitored and a continuously updated profile of the output over the course of the interpulse period is stored. On the basis of this profile, a gain control

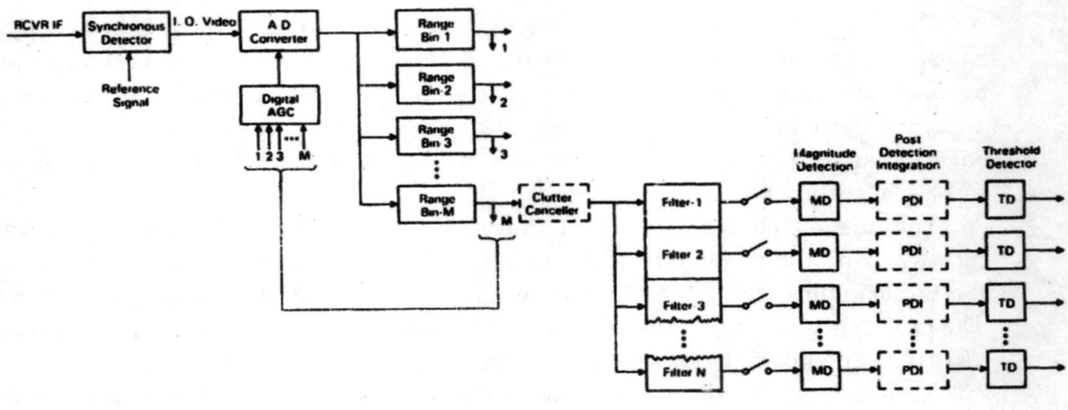

Figure 2.13 Signal processor for medium-PRF radar. (*From:* [3, p. 359].)

signal is produced and applied to the amplifiers ahead of the A/D converter. By reducing the gain when the main lobe clutter and strong close-in sidelobe clutter are being received, the control signal keeps the converter from being saturated while maintaining the input to the converter well above the local noise level when a weaker return is being received. Because the control signal is derived after the return has been digitized, this is called digital automatic gain control (DAGC). The use of DAGC in medium-PRF radar causes target suppression, in the same way that STC would, for targets with apparent ranges that place them in strong clutter regions.

The output of the A/D converter is then sorted into range bins, after which main lobe clutter is largely eliminated by using a clutter canceller. The clutter canceller serves the dual purpose of reducing both the main lobe clutter and the dynamic range into the Doppler filter bank. The return in each clutter cancelled range bin is then applied to a Doppler filter bank as depicted in Figure 2.14. Note that the first and last few Doppler filter outputs are discarded to eliminate ground moving targets.

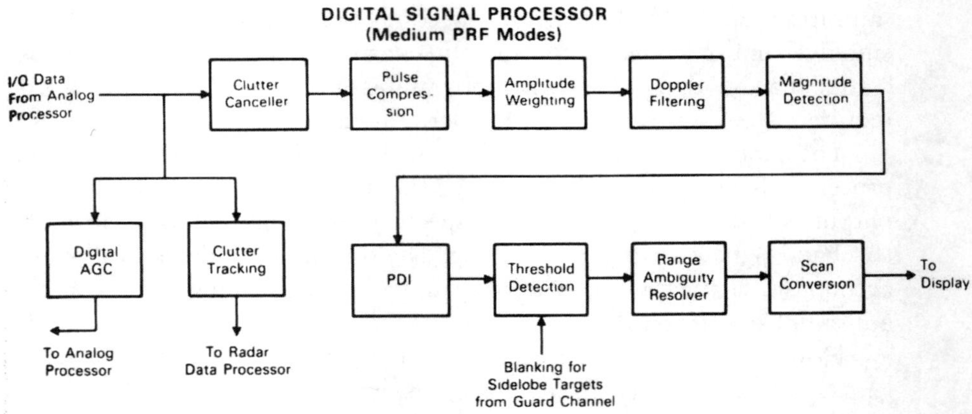

Figure 2.14 Medium-PRF radar clutter canceller and Doppler filter bank. (*From:* [9, p. 509].)

The output (I and Q) of each Doppler filter is then magnitude detecte end of each processing integration period. The integration period (T_i) is i proportional to the Doppler filter's bandwidth ($T_i = 1/B_D$). If T_i is less than the time on target (T_d), then each magnitude-detected Doppler filter output is subsequently integrated in a postdetection integrator. The integrated output is then applied to a threshold detector, which is adaptively set to keep the probability of clutter producing false alarms acceptably low. The adaptive threshold is generally set by averaging range and Doppler bins on either side of the one in question.

The apparent target range and Doppler is determined by observing which range-gated Doppler bin provides the output. Because both range and Doppler are ambiguous, PRF switching is required to resolve these ambiguities.

2.3 Doppler Filtering

The heart of any MTI or PD radar is the Doppler filter that extracts the desired Doppler-shifted target from the interfering clutter background. The basic filter associated with an MTI radar functions to attenuate those Doppler frequency regions associated with any clutter return while passing all other Doppler spectral regions. Because the spectral region associated with ground clutter tends to be concentrated about the PRF spectral lines (e.g., recall Figure 2.4), the periodic MTI filter response assumes a comb-like shape and hence is called a *comb filter*.

The pulsed Doppler filter response, on the other hand, is generally matched to the target's Doppler-shifted spectral response. Because the magnitude of the target's Doppler frequency shift relative to the radar can appear anywhere in the region bounded by the Doppler frequency corresponding to the sum of the maximum radar platform speed (v_r) and target speed (e.g., $v_m = \pm|v_r + v_t|$), the target spectral region must be filled with a bank of Doppler filters covering all possible expected target Doppler shifts. Hence, the PD filter response generally consists of a bank of Doppler filters that fill most of the region between the transmitted waveform's PRF spectral lines, as shown in Figure 2.15.

The MTI and PD filtering responses are not mutually exclusive and many systems employ a cascade of the two types of responses. In a cascade, the MTI filter is generally used first to eliminate much of the clutter before it reaches the bandpass Doppler filters. This reduces the requirements on the skirt selectivity of the many bandpass Doppler filters, which otherwise would have to both eliminate the clutter and match the target response. In a digital design, which might typically use the FFT algorithm, this simplifies the filter arithmetic that would carry the full dynamic range of the clutter (e.g., 10 to 12 bits) to where only the target's dynamic range need be carried (e.g., 3 to 4 bits).

2.3.1 Coherent and Noncoherent MTI

The ground clutter return from surface-based low-PRF radar was described in Section 1.3.2 and depicted in Figure 1.8. This return consists of the vector sum from the reflections of many elemental scatterers that combine to form a waveform at the carrier frequency of the radar, whose amplitude slowly fluctuates due to

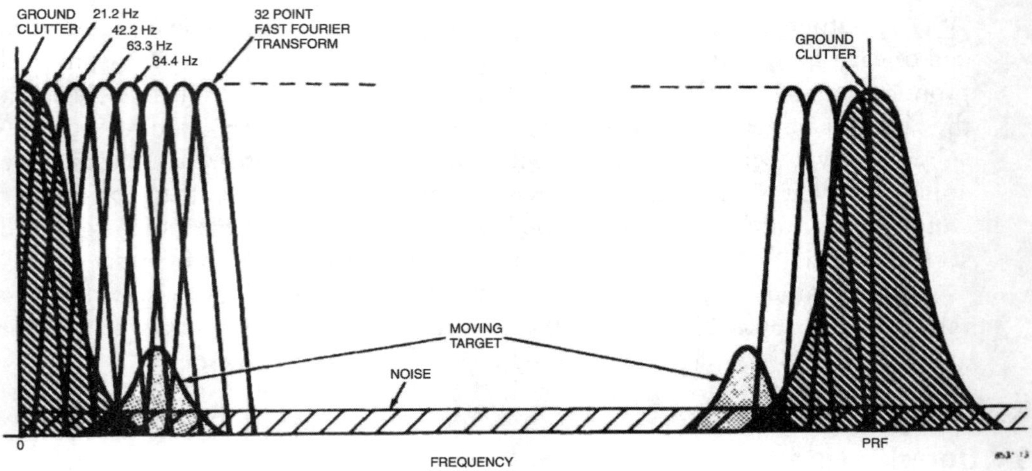

Figure 2.15 Pulse Doppler filter response. (*From:* [10, p. 81].)

both the internal scatterer motion and induced radar antenna scanning motion. The waveform associated with this return can be visualized through the use of a phasor diagram as shown in Figure 2.16.

The phasor diagram shows the instantaneous composite signal (r_t) due to ground clutter, target, and receiver noise components with respect to an in-phase reference I_t (x-axis of phasor diagram) and a quadrature reference Q_t (y axis of phasor diagram), where both references are synchronized to the carrier frequency of the radar (ω_c) but displaced in phase by 90°. In this diagram, the clutter signal (c_t) slowly fluctuates in both phase (ϕ_c) and amplitude at a rate corresponding to its spectral spread. The target vector rotates about the clutter vector at a rate

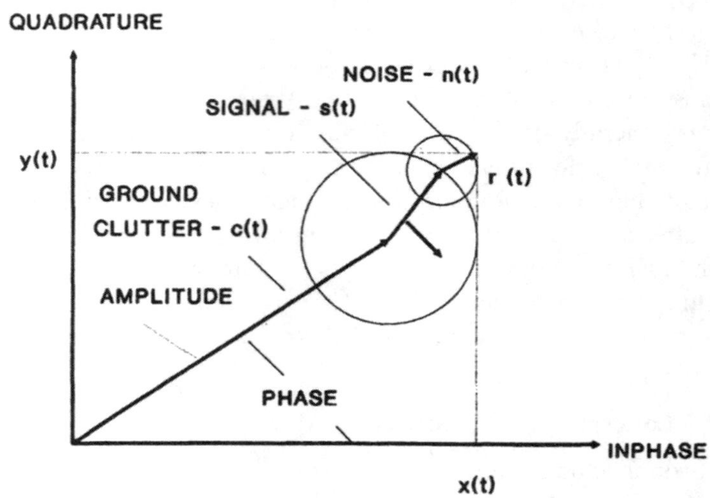

Figure 2.16 Phasor diagram of Doppler shifted target in ground clutter and receiver noise. (*From:* [11, p. 75].)

that corresponds to its Doppler frequency shift (ω_d) with respect to the clutter. The noise component fluctuates rapidly both in amplitude and phase about its mean value such that its position varies randomly each time the phasor diagram is sampled.

In the phasor diagram, the magnitude of the vector r_t represents the envelope of the return signal. This envelope is processed in a noncoherent MTI (see Figure 2.17) to provide detection of the Doppler-shifted target. When no target is present, the envelope fluctuates slowly due to: (1) internal clutter motion, (2) antenna scanning modulation, and (3) platform motion; the envelope fluctuates rapidly due to the receiver noise component. However, when the target is present, the envelope varies at a rate corresponding to the target's Doppler shift. Targets can be detected by using a clutter canceller. The simplest form of this is the single delay line canceller depicted in Figure 2.18, which functions by subtracting the envelope signal returned from one previous interpulse period ($T = 1/f_r$) from the present signal. The output of the single delay line canceller thus responds to the change in the envelope signal from successive radar transmitted pulses. When a Doppler-shifted target is present, this fluctuation is emphasized; when no target is present, the fluctuation is minimized. The fluctuation will generally be maximum when the phase shift of the target in an interpulse period ($\Delta\phi = 2\pi v_d T$) equals π radians. This occurs when the Doppler frequency shift of the target is at one-half of the radar PRF ($f_d = f_r/2$), which is sometimes called an *optimum speed target*.

Note that if the target vector in the phasor diagram of Figure 2.16 is parallel to the clutter vector, there is no output of the MTI filters for an optimum speed

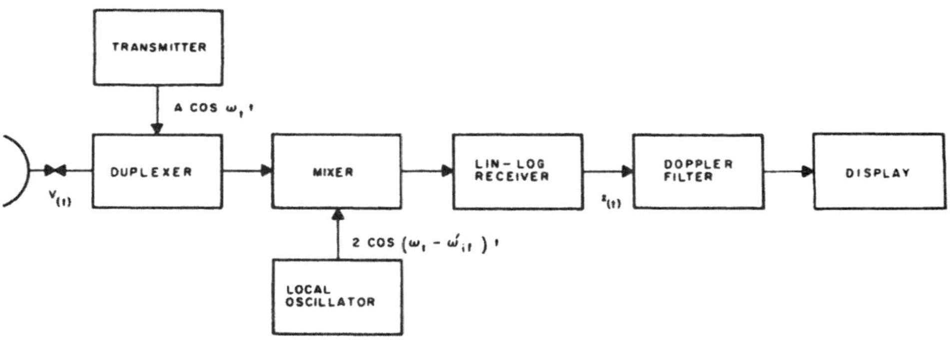

Figure 2.17 Block diagram of noncoherent MTI. (*From:* [5, p. 198].)

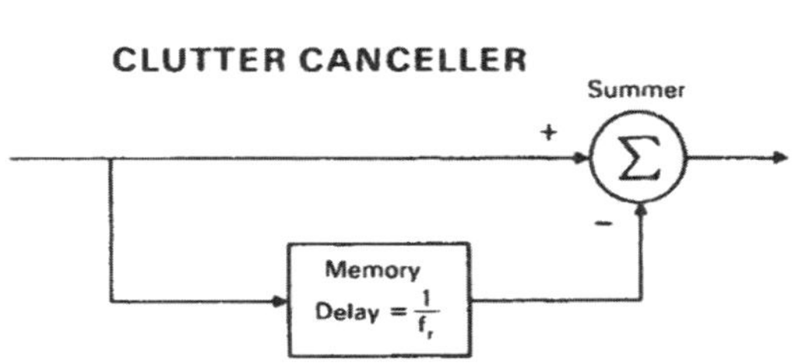

Figure 2.18 Single delay line canceller. (*From:* [3, p. 341].)

target. This is known as a *blind phase*, which is inherent in single-channel MTI designs. Also note that the MTI filter's power gain response, when averaged over all target velocities, provides equal response to both signal and noise components and hence does not alter the S/N ratio.

The noncoherent MTI just described is also called a *clutter referenced MTI*, because clutter must be present for this type of MTI to function. To circumvent this problem, the output of the noncoherent MTI can be switched between MTI and normal video, with the MTI selected only when there is sufficient clutter for it to operate. The performance of the noncoherent MTI will also be limited by the use of a relatively noisy reference (the clutter itself) compared to the stable internal reference used in coherent MTI.

A coherent MTI utilizes the phase information (ϕ_r) illustrated in the phasor diagram of Figure 2.16. When only clutter is present, the phase of the vector r_t with respect to the in-phase reference I_t is primarily pertubated by the receiver noise vector. When a target is present, the phase varies at a rate determined primarily by the Doppler frequency of the target. The bipolar video signal produced by phase detecting the return signal against the coherent in-phase reference signal (I_t) is depicted in Figure 2.19 for a single interpulse period and also for several interpulse periods superimposed on the same sweep. When a moving target is present, a butterfly type of pattern is generated; otherwise, the trace is relatively stable except for the receiver noise output component.

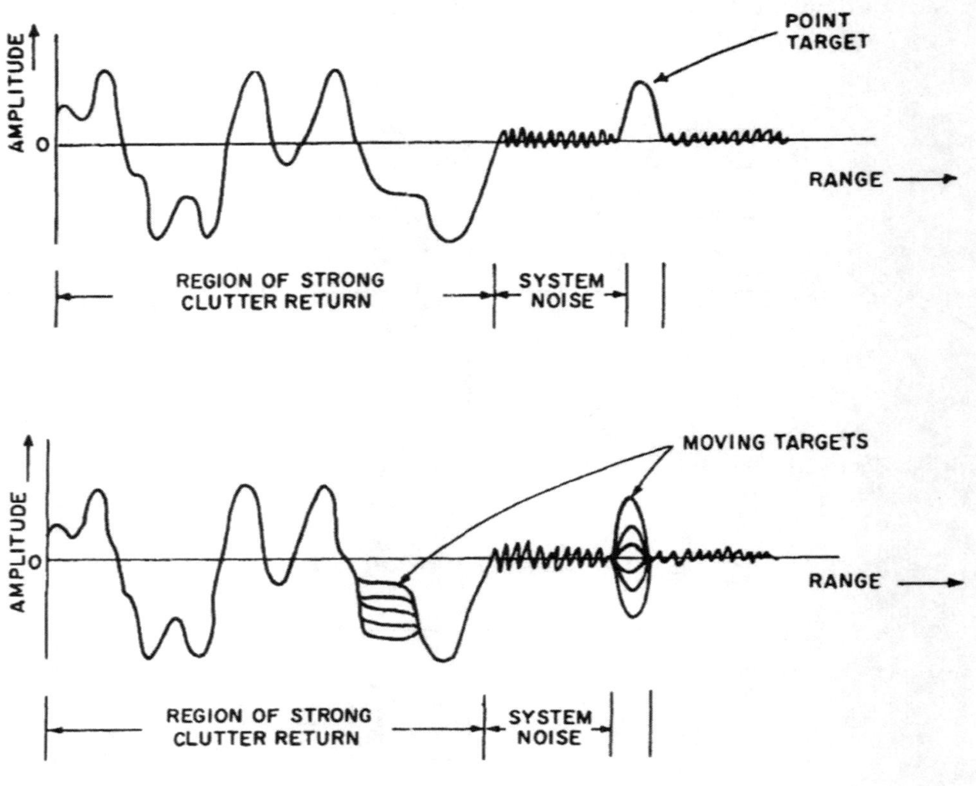

Figure 2.19 Bipolar video from a coherent MTI system. (*From:* [12, p. 15.5].)

Examination of Figure 2.19 indicates that the same type of delay line canceller used in noncoherent MTI systems will also be effective in processing the phase-detected bipolar video output of a coherent MTI system. The basic action of the delay line canceller in this case is to suppress those phase-detected signals that result from clutter that remains stable from one pulse period to the next and to accentuate those target signals that fluctuate from one interpulse period to the next. This action is equally effective whether processing envelope-detected amplitude data or phase-detected bipolar video.

Note that the coherent MTI does not require clutter to operate, which is a distinct advantage relative to noncoherent MTI. Nonetheless, the blind phase problem will remain when a single reference channel is used. The solution is to add a second quadrature channel using reference vector Q_t, the phase of which is shifted 90° with respect to the first reference vector. From Figure 2.16, we can see that the phase of the Doppler- shifted target produces a blind phase in the first channel while its maximum output is in the second quadrature channel.

The use of a single-channel coherent MTI, as depicted earlier in Figure 2.8, results in a sensitivity loss when compared to fully coherent MTI systems that process both in-phase and quadrature channels. The most severe loss occurs for nonfluctuating targets with high detection probabilities when no postdetection integration is used. For example, the relative detection loss has been evaluated as 13.2 dB for a $P_d = 0.9$ with a $P_{fa} = 10^{-6}$. The loss is reduced to 2.75 dB for $P_d = 0.5$ with $P_{fa} = 10^{-6}$ and overall the relative loss is generally insensitive to the false alarm probability. The relative loss (compared to that for a nonfluctuating target) is reduced for Swerling 1 fluctuating targets with high detection probabilities ($L = 10.66$ dB, $P_d = 0.9$, $P_{fa} = 10^{-6}$), but reduced for moderate detection probabilities ($L = 4.35$ dB, $P_d = 0.5$, $P_{fa} = 10^{-6}$) [13].

The cause of the sensitivity loss in a single-channel coherent MTI is due to a modification of the detection statistics. The relative loss is reduced in a noisy background by using postdetection integration [14, 15]. The relative loss with postdetection integration can be related to a reduction in the number of effective pulses integrated by a factor of 2, which causes a loss in the integrated S/N ratio of 1.5 to 2.5 dB [14]. However, the analyses apparently have not considered the results of residual clutter components, which can result in degraded performance with postdetection integration [16].

The single-channel coherent MTI was used extensively in analog processors due to its simplicity. Most modern MTIs utilize digital processing, which allows use of quadrature-channel MTI systems as depicted earlier in Figure 2.7.

2.3.1.1 Coherence

Generally, coherence implies a consistency, or continuity, in the phase of a signal's carrier frequency from one pulse to the next. Coherent MTI operation requires that the phase of the receiving reference signal into the MTI's phase detector be consistent with the radar's transmitted signal. A common way of achieving this coherence in coherent MTI or PD radar is depicted in the block diagram shown in Figure 2.20.

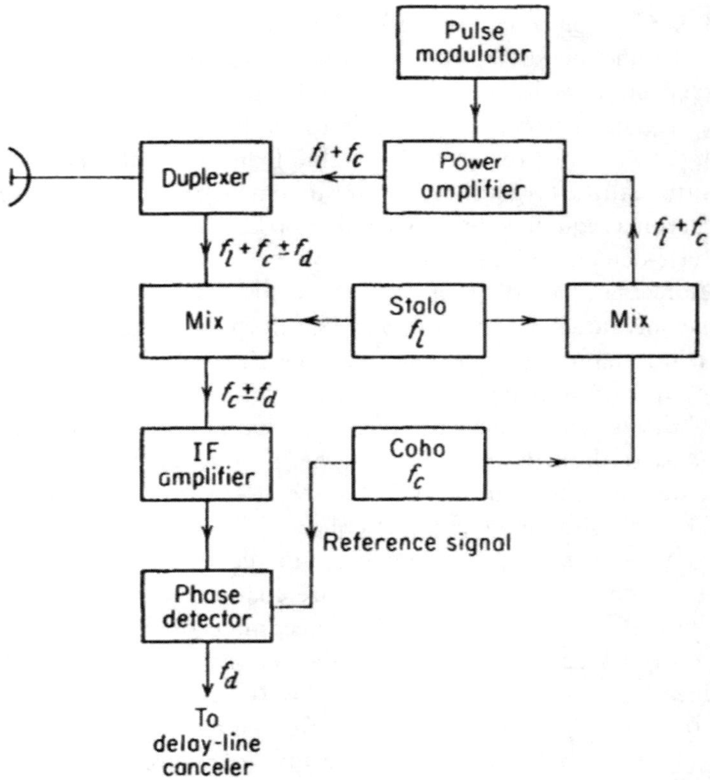

Figure 2.20 Block diagram of MTI radar with master oscillator power amplifier. (*From:* [17, p. 111].)

The system shown in Figure 2.20 is called the master oscillator power amplifier (MOPA) approach, because the basic coherent reference signal is developed as a CW signal within the radar. This signal is then amplified in a pulsed power amplifier such as a Klystron, traveling-wave tube, or solid-state amplifier and radiated through the radar antenna.

The MOPA system generally uses two stable oscillators to develop the coherent signals needed in the radar receiver and transmitter. A stable coherent oscillator (COHO) develops a signal at the radar's intermediate frequency (e.g., 60 MHz) that is used as the coherent reference signal in the MTI's phase detector. Another stable local oscillator (STALO) signal is used as the receiver's local oscillator to heterodyne the received RF signal to a lower intermediate frequency signal, which is compared against the COHO in the MTI's phase detector.

The transmitter signal is formed by mixing the COHO (f_c) and STALO (f_s) signals resulting in a transmitter signal at $f_t = f_s + f_c$. The Doppler-shifted return signal ($f_r = f_s \pm f_d$) is then mixed with the STALO resulting in a signal at $f_c \pm f_d$, which is compared with the COHO in the phase detector. The net effect of this process is an addition and subtraction of the STALO signal in the signal path so that the absolute value of the phase (or frequency) of the STALO is cancelled provided it does not change over the processing time of the MTI.

The MOPA approach can be viewed as providing coherence by cutting a stream of pulses out of a CW signal. An alternate method of achieving coherence, called the *coherent-on-receive approach*, is depicted in Figure 2.21. This approach is useful in situations that employ a power oscillator (e.g., magnetron) instead of a power amplifier as the radar transmitter. In a pulsed power oscillator, the phase of the RF from one pulse to the next is not coherent. However, by locking the phase of the COHO onto each transmitted pulse, we can achieve quasicoherence during each phase immediately following an interpulse period. Note that multiple-time-around range-ambiguous clutter and targets are not coherent when using this basic scheme as they are with the MOPA approach. This can result in large range-ambiguous clutter returns (e.g., mountains, cities) that are not cancelled by this type of MTI system.

In the coherent-on-receive block diagram shown in Figure 2.21, a portion of the transmitted signal is mixed with the STALO output to produce a sample pulse at IF, that is phase coherent with the transmitted output. The IF pulse is then applied to the COHO and locks the phase of the COHO onto that of the transmitter pulse. Operation of the system is then functionally equivalent to that of the MOPA system shown in Figure 2.20. Performance in this type of system is generally lower than that of the MOPA system due to the difficulty in achieving a high degree of coherence with a locked COHO and the previously described multiple-time-around clutter cancellation problem.

2.3.2 Doppler Filter Characteristics in the Frequency Domain

The single delay line canceller depicted earlier in Figure 2.18 is an example of an elemental Doppler filter, whose basic action can best be visualized by considering

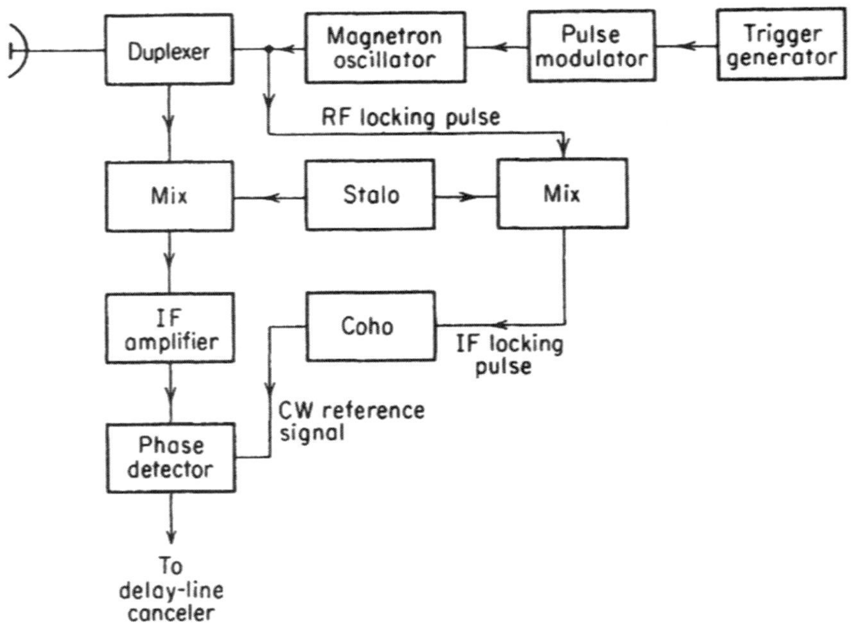

Figure 2.21 Block diagram of coherent-on-receive MTI radar. (*From:* [18, p. 106].)

its operation in the time domain. Assume that the target vector signal depicted in Figure 2.16 is applied to the single delay line canceller. This vector rotates at the Doppler frequency-shifted rate of the target and, when its phase is sampled at the radar's PRF, provides an unchanged output for rotations of $2n\pi$, where $n = 0, 1, 2, 3, \ldots$. When this unchanged vector is applied to the delay line canceller, a null output results (due to the differencing of successive samples). This nulled output is denoted as the *blind speeds* of the MTI Doppler filter. Examination of the single delay line canceller's characteristics shows that these occur where the target's Doppler frequency shift is either zero or a multiple of the radar's PRF ($f_d = nf_r$, $n = 0, 1, 2, 3, \ldots$). When an optimum-speed target is applied to the single delay line canceller (i.e., $f_d = nf_r/2$, $n = 1, 3, 5, \ldots$), then the output is maximum or twice that of the input target vector. Further consideration of various Doppler-shifted target vectors applied to the single delay line canceller results in the well-known sinusoidal-shaped voltage amplitude characteristic ($v_c = |2\sin(\pi f_d/f_r)|$) depicted in Figure 2.22.

Thus, the single delay line canceller's amplitude characteristic has a periodic comb filter structure that repeats at multiples of the radar's PRF and is a function of the target's Doppler shift with respect to the radar. When viewed as a Doppler filter, the rejection notches are concentrated at Doppler frequency shifts, which occur at frequencies corresponding to multiples of the radar's PRF (i.e., $f_d = nf_r$, $n = 0, 1, 2, 3, \ldots$), whereas the acceptance passband generally occurs at frequencies between the rejection notches with a maximum response for optimum speed targets (i.e., $f_d = nf_r/2$, $n = 1, 3, 5, \ldots$).

The single delay line canceller is a linear system and can be analyzed using classical Fourier transform theory [19]. The impulse response of the single canceller is given by

$$h(t) = \delta(t) - \delta(t - T) \tag{2.6}$$

The frequency response is found by Fourier transforming the impulse response as

$$H(\omega) = 1 - \exp(-j\omega T) = 2j\exp\left(\frac{-j\omega T}{2}\right)\sin\left(\frac{\omega T}{2}\right) \tag{2.7}$$

The magnitude of the frequency response is then

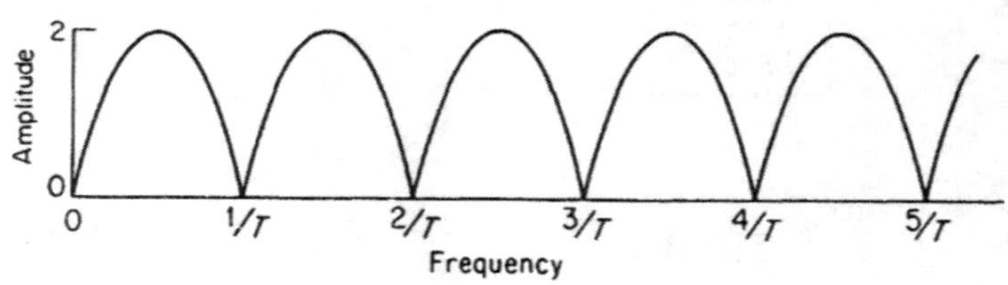

Figure 2.22 Single delay line canceller frequency responses. (*From:* [18, p. 107].)

$$|H(\omega)| = \left|2\sin\left(\frac{\omega T}{2}\right)\right| \tag{2.8}$$

The rejection notch of the single delay line canceller is not always wide enough to provide adequate clutter cancellation. The notch width can be increased by using a cascade of two single cancellers, which is called a *double canceller*. The frequency response of this configuration is the product of the response of the two single cancellers, which is given by

$$|H(\omega)| = \left|4\sin^2\left(\frac{\omega T}{2}\right)\right| \tag{2.9}$$

The increased notch width is depicted in Figure 2.23, which shows both the frequency and impulse response of both single and double delay line cancellers.

Another way of characterizing the double delay line canceller is through its impulse response, which is given by

$$h(t) = \delta(t) - 2\delta(t - T) + \delta(t - 2T) \tag{2.10}$$

In this form, it can be seen that the double delay line canceller processes three pulses using a tapped delay line (i.e., taps at nT for $n = 0, 1, 2$) with weights

$$\mathbf{w} = (1, -2, 1) \tag{2.11}$$

This general type of filter, depicted in Figure 2.24, is called a *transversal* or *nonrecursive Doppler filter*. In its most general form, complex weights (phase and amplitude) are employed (see Chapter 5); however, in a basic MTI canceller, only real weights are used. For MTI use, one way of achieving the effect of complex weights is to use transversal filters with real weights in both the I and Q channels. Further, as a note of interest, the basic cascade of two single cancellers, rather than a transversal filter, was usually preferred in analog MTI applications, because any amplitude imbalance in the first canceller was compensated for in the second canceller. This is no longer a problem because digital implementations are generally used in most modern MTIs.

Cascades of single delay line cancellers produce an amplitude frequency response given by

$$|H(\omega)| = \left|2\sin\left(\frac{\omega T}{2}\right)\right|^n \tag{2.12}$$

where n is the number of single delay line cancellers. We can show by convolving the impulse response of a single canceller n times [5, p. 23] that the impulse response of such a cascade uses weights with binomial coefficients given by

$$w_k = (-1)^k\binom{n}{k} = \frac{(-1)^k n!}{k!(n-k)!} \tag{2.13}$$

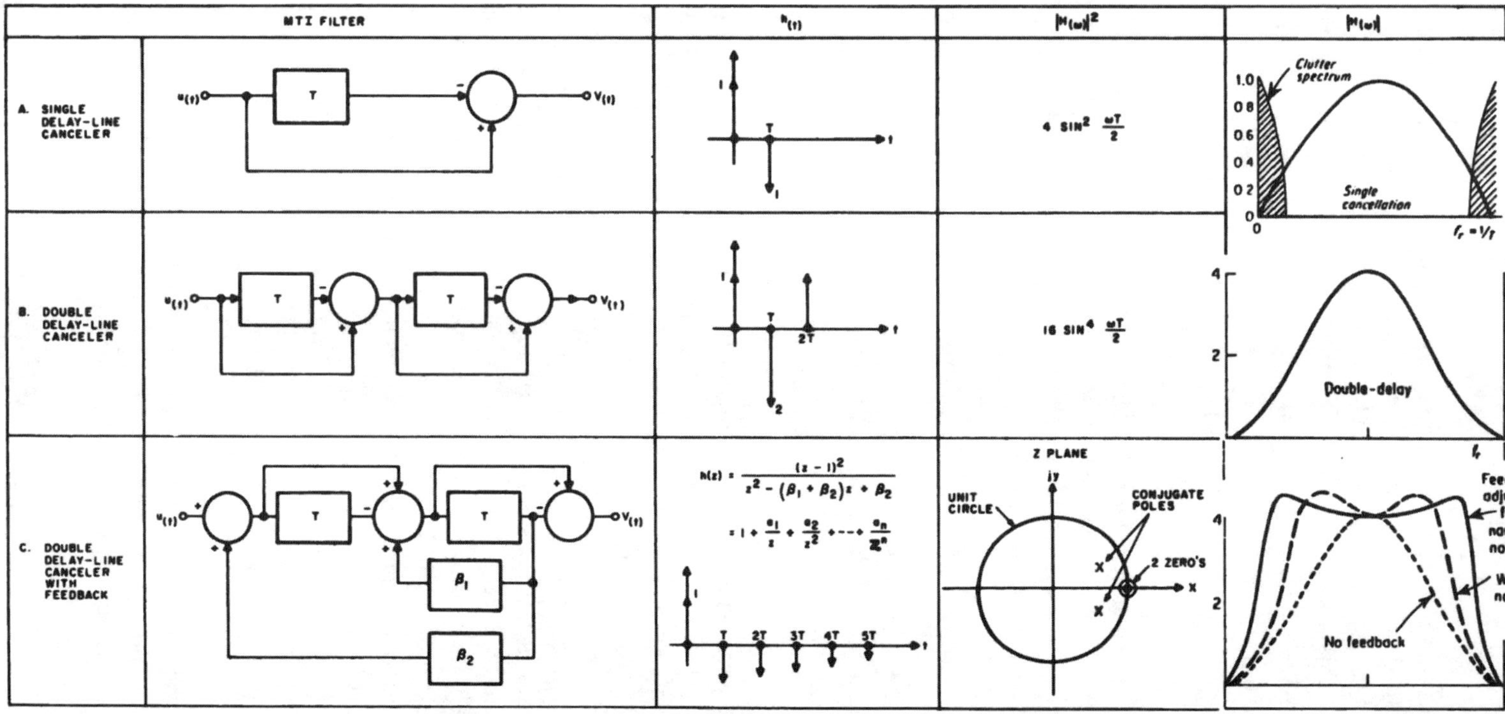

Figure 2.23 Frequency and impulse response of delay line cancellers. (*From:* [11, p. 83].)

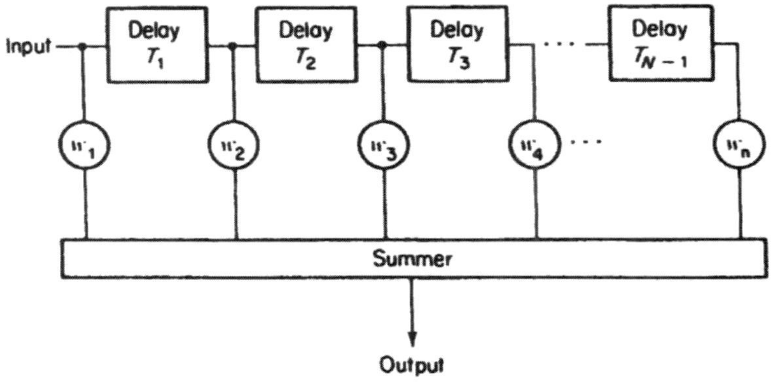

Figure 2.24 General transversal filter. (*From:* [17, p. 119].)

where k is the weight of the kth tap and n is the number of delay sections. The normalized power transfer function ($|H(\omega)|^2$ of an MTI having the indicated number of cancellers is shown in Figure 2.25. MTIs with binomial coefficients are known to produce close to optimum MTI improvement factors. Note that the MTI's acceptance response becomes progressively narrower as the number of delay line cancellers is increased. In practice, generally no more than three delay sections are used because the excessive apparent clutter improvement factor (not supported by MTI stability considerations) lowers the target response due to the narrowing of the MTI filter's acceptance band.

An alternative approach to the transversal filters is the recursive MTI filter whose canonical form is depicted in Figure 2.26. This type of Doppler filter employs both feedforward and feedback paths and is sometimes called a *shaped delay line canceller* [5, p. 224]. The type of frequency response produced by this type of MTI canceller is depicted in Figure 2.23. The ability to shape the passband is an attribute of this MTI filter, but its poor transient response is a definite disadvantage. The poor transient response results from the feedback paths in this type of filter, which allow interfering pulses to recirculate through the filter for many interpulse periods (e.g., recall Figure 2.23).

Recursive MTI Doppler delay line filters can be synthesized from classical filter design theory by using the z-transform [21]. The z-transform is given by

$$z = \exp(pT) \tag{2.14}$$

where $p = j\omega$ and T is the radar's interpulse interval. The z-transform of a unit impulse delayed in time is

$$\delta(t - nT) \Leftrightarrow z^{-n} \tag{2.15}$$

Hence, the z-transform for the impulse response of a single delay line canceller would be given by

$$H(z) = 1 - \frac{1}{z} = \frac{z - 1}{z} \tag{2.16}$$

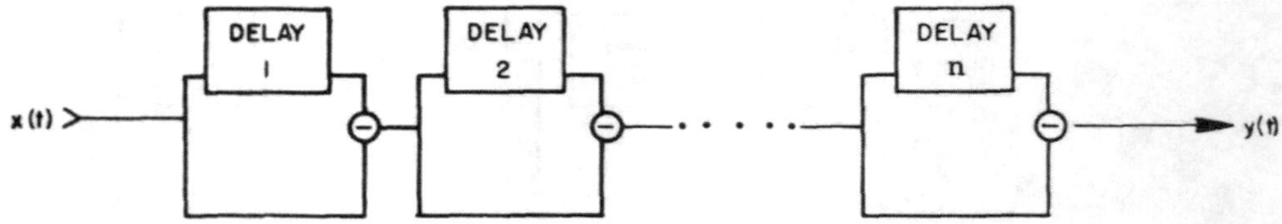

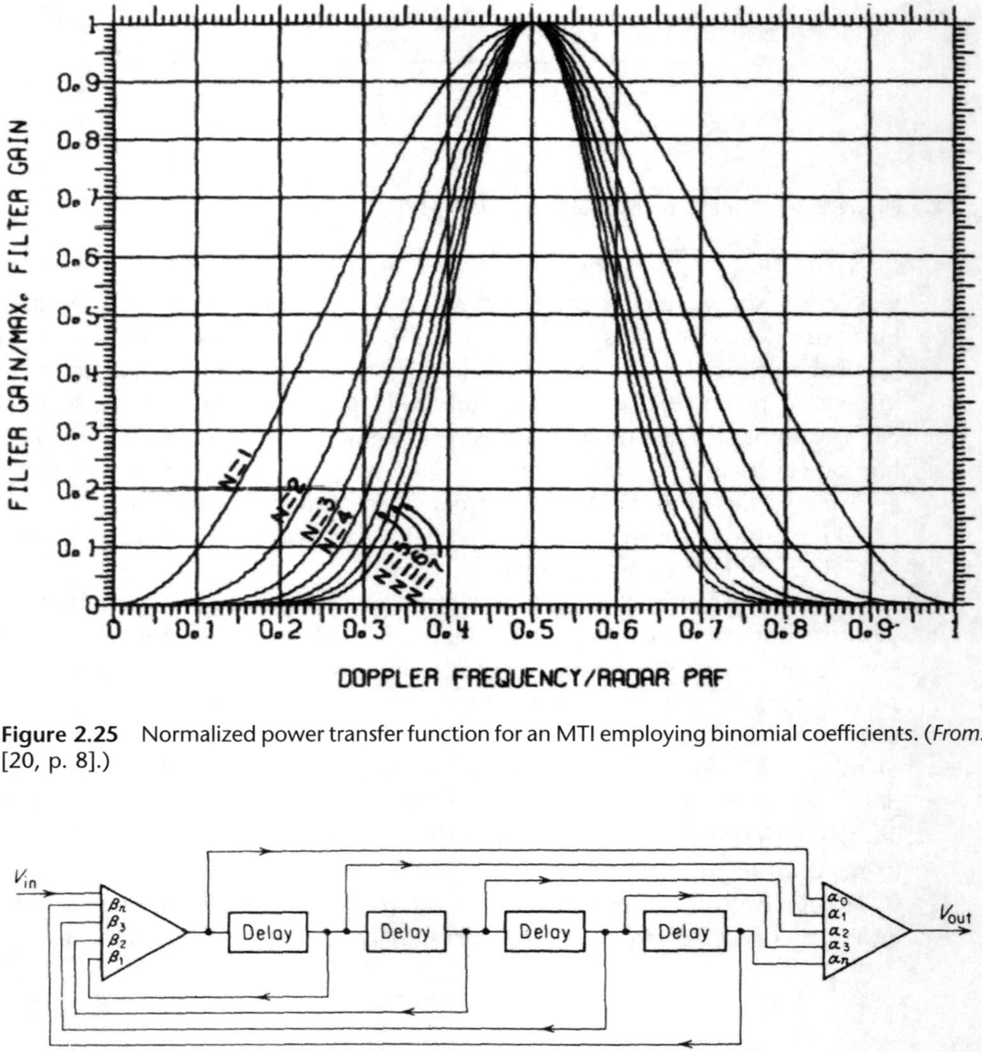

Figure 2.25 Normalized power transfer function for an MTI employing binomial coefficients. (*From:* [20, p. 8].)

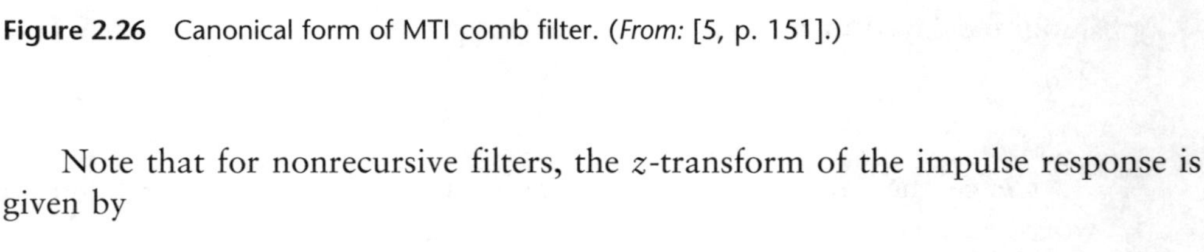

Figure 2.26 Canonical form of MTI comb filter. (*From:* [5, p. 151].)

Note that for nonrecursive filters, the z-transform of the impulse response is given by

$$H(z) = \sum_{k=0}^{n} w_k z^{-k} \tag{2.17}$$

where k is the tap number, n is the number of delay lines, and w_k represents the tap weights.

The amplitude frequency response of an MTI filter can be characterized by its pole-zero patterns. In conventional filter design, the pole-zero pattern is plotted in the p-plane, where $p = j\omega$ is the variable utilized in the Laplace transform of the filter's impulse response. The transformation utilized in the z-transform [$z = \exp(pT)$] maps the left-hand p-plane into the unit circle in the z-plane as depicted in Figure 2.27. The transformation is effective because it allows one period of the filter's transfer characteristic to be examined by a single rotation about the unit circle in the z-plane.

The single delay line canceller has a zero at $z = 1$ and a pole at $z = 0$, as depicted in Figure 2.27. The amplitude response for any point on the unit circle, located at an angular position of Ω radians,

$$f_d = \frac{\Omega f_r}{2\pi} \tag{2.18}$$

can be found using the following procedure [11, pp. 621–622]. First, form vectors from the point on the unit circle to all zeros and poles. Then, the amplitude function is given by the product of all zero vectors divided by the product of all pole vectors. For the single canceller pole-zero patterns shown in Figure 2.27, the amplitude response is given by the length of the vector from the zero to the point on the unit circle. Simple geometry shows this to be equal to

$$|H(\omega)| = \sqrt{2(1 - \cos \omega T)} = 2 \sin\left(\frac{\omega T}{2}\right) \tag{2.19}$$

However, the advantage of the z-transform method stems from the availability of a simple transformation that allows poles and zeros in the p-plane (obtained from classical network theory) to be transformed into the z-plane [21]. Thus standard Butterworth, Chebyshev, Bessel, or elliptical filter characteristics can be easily translated into shaped delay line filter designs. When a standard lowpass filter design is to be translated into a highpass MTI canceller design, the appropriate

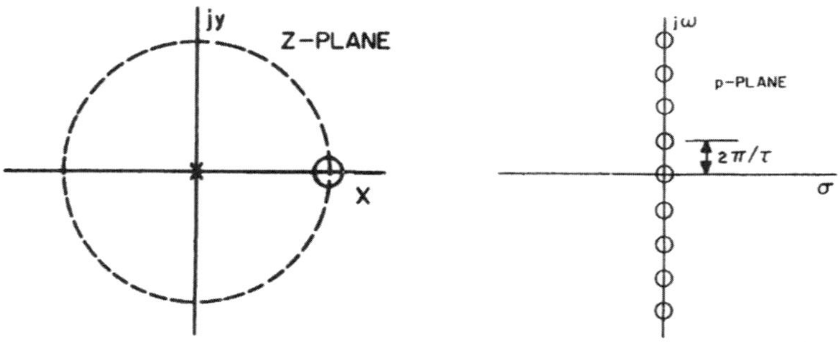

Figure 2.27 Pole-zero pattern in *p*- and *z*-planes for single delay line canceller. (*From:* [5, pp. 164–166].)

bilinear transformation that transforms the poles and zero in the p-plane (lowpass filters p-plane) to the MTI filter's z-plane is given by

$$z = \frac{p + \Omega}{p - \Omega} \tag{2.20}$$

where Ω is a scale adjusting factor related to the filter's acceptance bandwidth given by [11, pp. 630–631]

$$\Omega = \omega_{co} \tan\left(\frac{\omega_{co} T}{2}\right) \tag{2.21}$$

An example of a double shaped canceller design is given in Figure 2.23. The transfer function in the z-plane for this type of design is given by

$$H(z) = \frac{(z - 1)^2}{z^2 - (\beta_1 + \beta_2)z + \beta_2} \tag{2.22}$$

and its response is shown in Figure 2.23. This type of design can be used to provide the varying filter shapes that are useful in situations with variable clutter spectral spreads such as might be found in radars employing variable antenna scan rates or airborne radars that scan away from the ground track.

► Example 5 Design of Shaped MTI Cancellers

The MTI improvement factor for surface-based search radar using binomially weighted MTI cancellers is generally limited by spectral spreading of the ground clutter spectrum due to antenna scanning. Assuming a Gaussian-shaped clutter power spectrum, its standard deviation is given by

$$\sigma_S = \frac{\mathrm{PRF}}{3.77 n_B}$$

where n_B is the number of hits per 3-dB antenna beamwidth. Substituting this into the relation for the MTI improvement factor results in

$$I_n = \frac{n_B^{2n}}{n!(1.388)^n}$$

where n is the canceller order or number of delay sections in the processor.

Using the parameters of the ASR-9 radar given in Table 1.1 results in $n_B = 20.8$ hits and MTI improvement factors of 24.9, 46.9, and 67 dB, respectively, for single, double and triple binomial MTI cancellers. Because the specification for the ASR-9 requires a 50-dB MTI improvement factor, that would indicate the use of a triple canceller. Unfortunately, the MTI improvement factor is limited by stability considerations to the order of 50 dB and, hence, the full capability of the triple canceller is not achieved. Also the passband of binomial cancellers is diminished as the order increases (see Figure 2.25), resulting in a loss of 60% of the target Dopplers for a triple canceller. The solution to this problem is the use of a shaped MTI canceller that allows us to trade excess improvement factor for increased passband.

The design of shaped cancellers is best approached using z-transforms, but fortunately utilizing MATLAB's Signal Processing Toolbox [34] circumvents the need for understanding the intricate details of this method. MATLAB provides Butterworth and Chebyshev lowpass prototype filters that are converted to highpass digital filters with equivalent characteristics. The z-transform transfer function is given by

$$H(z) = \frac{b(1) + b(2)z^{-1} + \ldots + b(n+1)z^{-n}}{1 + a(2)z^{-1} + \ldots + a(n+1)z^{-n}}$$

A Butterworth-shaped MTI filter is designed by making this assignment [34]:

$$[\mathbf{b}, \mathbf{a}] = \text{butter}(n, w_n, \text{'high'})$$

where the vectors **a** and **b** are the coefficients of $H(z)$, n is the order, $w_n = f_{co}/(\text{PRF}/2)$, and f_{co} is the frequency where the normalized transfer function has a magnitude of 0.707. The Chebyshev-shaped MTI filter is designed by making this assignment [34]:

$$[\mathbf{b}, \mathbf{a}] = \text{cheby1}(n, R_p, \mathbf{w}_n, \text{'high'})$$

where R_p is the passband ripple in decibels. The frequency response is found using [34]

$$[\mathbf{h}, \mathbf{w}] = \text{freqz}(\mathbf{b}, \mathbf{a}, n)$$

where **h** is the complex frequency response, and $\mathbf{w}_n$ is a vector containing the frequency points.

The shaped MTI canceller design is accomplished in the MATLAB program called shaped_design.m. Figure EX-5.1 shows Butterworth and Chebyshev filters for f_{co}/PRF = 0.1, PRF = 1,200 Hz, $R_p = 1$ dB, and $n = 3$. Also shown for reference is the response of a triple canceller. The response of Butterworth-shaped cancellers for various values of f_{co}/PRF are depicted in Figure EX-5.2.

All of the shaped MTI filters shown have passbands greater than that of a triple canceller. The MATLAB program shaped_impfac1.m provided computes the MTI improvement factor for Butterworth- and Chebyshev-shaped MTI cancellers. The result using ASR-9 parameters is depicted in Figure EX-5.3. Examination of Figure EX-5.3 indicates that an f_{co}/PRF of 0.15 provides a 50-dB MTI improvement factor. Selecting this filter provides a loss of 32% of the target Dopplers, which is about half the loss associated with a triple canceller.

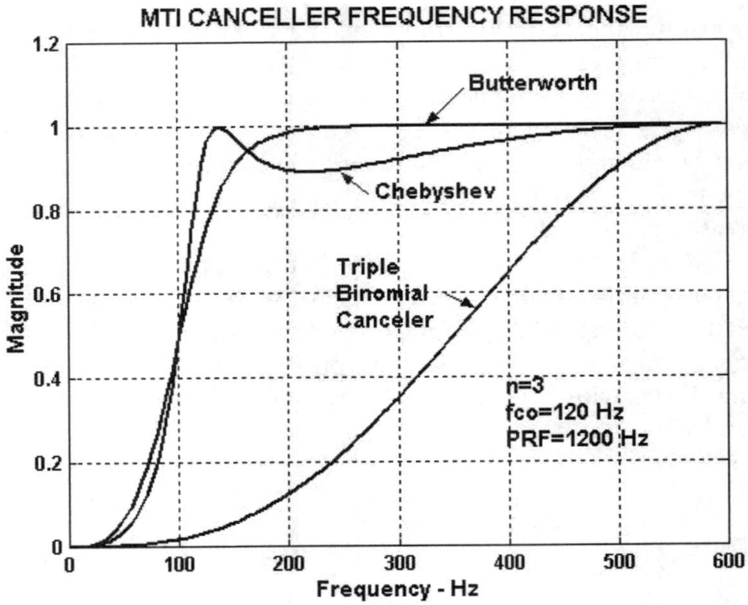

Figure EX-5.1 Frequency response of shaped MTI canceller.

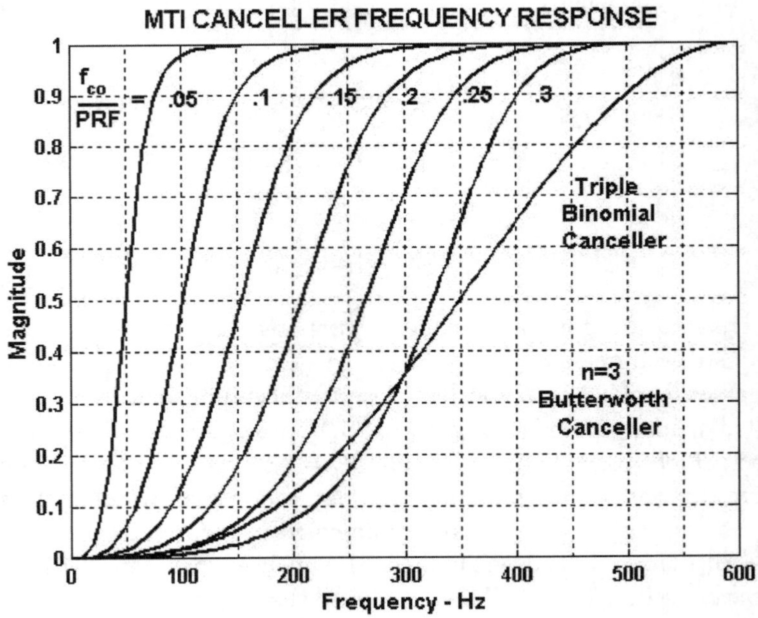

Figure EX-5.2 Shaped Butterworth MTI canceller with various cutoffs.

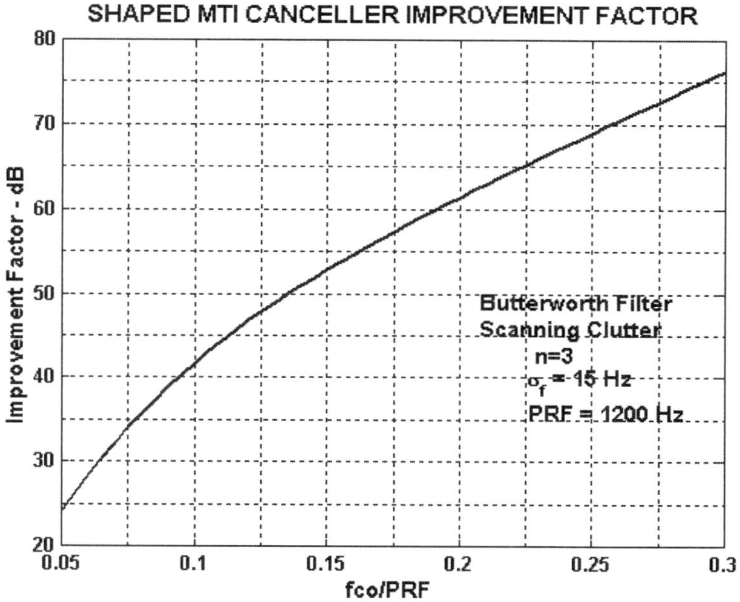

Figure EX-5.3 MTI improvement factor for shaped MTI canceller.

Shaped MTI cancellers with adjustable cutoff frequencies can be used in combination with a variable scan rate antenna to provide a constant MTI improvement factor. This is useful in concentrating more energy on targets by slowing the antenna scan rate to provide a higher probability of detection.

▶

2.3.3 Digital Doppler Filters

In modern radars, signal processing is accomplished digitally, reflecting both the capability and the economy associated with modern digital components. All of the Doppler filtering functions described in the previous sections can be accomplished digitally. Although the theoretical capabilities of analog and digital signal processing are equivalent, the detailed implementation is significantly different. Basically, in digital signal processing, the input waveform is converted to a set of numbers that is then manipulated in accordance with a set of algorithms to accomplish the desired target detection function.

A block diagram of the basic conversion from the input analog waveform into digital numbers is shown in Figure 2.28. The first part of this process is a conversion of the radar's input waveform into lowpass I and Q video signals. This conversion is accomplished by using synchronous detectors (i.e., phase detectors) that multiply the input IF signal with quadrature reference signals coherently related to the radar transmitting signal. The purpose of this conversion is to translate the signal to baseband where a minimal sampling rate preserves the information content of the input waveform.

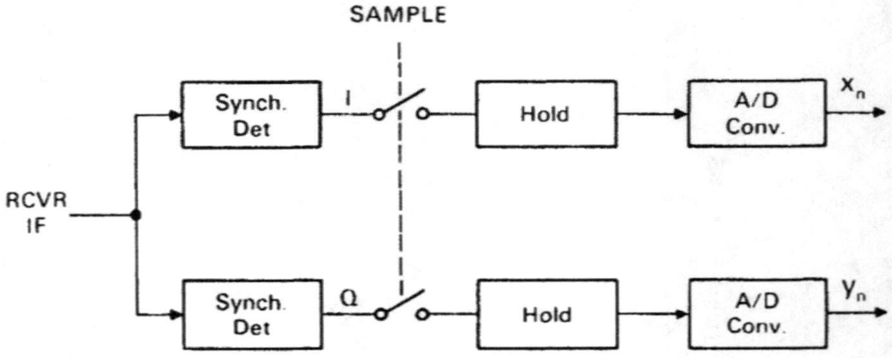

Figure 2.28 Block diagram of digital conversion in digital signal processing. (*From:* [3, p. 246].)

The next step in the digital conversion process is to sample I and Q video waveforms. Successive samples are taken at intervals corresponding to the radar's range resolution cell (i.e., the radar range corresponding to the effective transmitter pulsewidth) throughout the range coverage of the radar. This range gating process results in a sampling rate equal to the PRF times the number of range gates. Each sample thus represents the return of a single pulse for a given range increment—or, if range is ambiguous, from a number of range increments separated by the ambiguous range (R_u).

The sampling is generally followed by a holding circuit. Separate circuits are provided from the outputs of both the I and Q phase detectors. The overall sample-and-hold circuits sense the instantaneous values of the waveform at precisely the right time and hold the sample long enough for them to be converted to digital numbers, then dump them, and the process repeats.

The A/D converter is the device that translates the sample-and-hold voltage output into a digital number. A/D converters generally work by comparing the voltage of each sample with a succession of progressively higher voltages of precisely known values. When the closest of these is found, the converter produces a binary number equal to the known voltage. Separate A/D converters are required for both the I and Q samples.

The output of the A/D converters due to one transmitter pulse is then a set of numbers in signal space $s = (x_1, x_2, \ldots, x_m; y_1, y_2, \ldots, y_m)$, where m is the range gate number and the pair (x_n, y_n), $n = 1, 2, \ldots, m$, represents the return in a particular range gate. The function of the signal processor is to take the sets of numbers from successive radar transmitter pulse returns $(s_1, s_2, \ldots, s_t)$ and manipulate them to perform the desired target detection. The general procedure is to form a decision function $D(z)$ where $z = f(\mathbf{s}_1, \mathbf{s}_2, \ldots, \mathbf{s}_t)$, which is compared against a criterion (e.g., threshold) to decide on the presence or absence of a target.

A key function performed within the signal processor is Doppler filtering. It is generally performed by examining the set of returns within a particular range gate so that the number set into the Doppler filter is given by $\mathbf{s} = (x_{n1}, x_{n2}, \ldots, x_{nl}; y_{n1}, y_{n2}, \ldots, y_{nl})$, where n is the range gate number and l is the number of pulses processed. This range-gated Doppler filter then has the capacity to produce l complex (I and Q) Doppler filters for each range gate, resulting in an overall number of nl complex Doppler filters. When MTI Doppler filters are involved, only one complex filter is produced for each range gate, resulting in the formation of n Doppler filters.

Doppler filters are generally formed in a particular range gate by taking the weighted sum of I quadrature returns ($x_{n1}, x_{n2}, \ldots, x_{nl}; y_{n1}, y_{n2}, \ldots, y_{nl}$). In simple signal processors, the weighted sums are stored in shift registers as depicted in Figure 2.29. The shift register has one storage position for each range increment to be processed. Each time a new product is formed by multiplying the input sample by an appropriate weight, the stored sums are shifted one position to the right. The new product is thus added to the stored sum of the previous products, resulting in a vector output equal to $\Sigma_{i=1}^{l} \mathbf{w}_i \mathbf{x}_{nl}$ for the situation depicted in Figure 2.29. When all return pulses have been processed, the weighted sum output of the shift register is then used to find the magnitude of the vectors in the I and Q channels.

A block diagram of a single delay line canceller is depicted in Figure 2.30. The vector outputs of the shift register using the weights (1, –1) are equal to $(\mathbf{I}, \mathbf{Q}) = x_{n1} - x_{n2}, y_{n1} - y_{n2}$. Note that the weights in both I and Q channels are equal, and no cross coupling between I and Q channels is involved (i.e., the complex weights are real). The output of the single canceller is formed by finding the magnitude of the **I** and **Q** vectors, which is given by $z = \sqrt{|I|^2 + |Q|^2}$. Because this is difficult to accomplish digitally, an approximation algorithm is frequently employed.

One algorithm [3, 22] used to find the magnitude of the vector sum of I and Q channels is as follows. First, subtract the magnitude of the **I** vector from the magnitude of the **Q** vector to determine which is smaller. For example, assume |**Q**| is the smaller vector. Then divide |**Q**| by 2, which is simply accomplished in binary arithmetic by shifting the number to the right by one binary place. Then add the result to the larger quantity, resulting in $z = |I| + |Q|/2$ for the example. The error of the approximation in terms of the S/N ratio to achieve a specific probability of detection varies with the value of the phase angle between the vector and in-phase component, but generally is less than a fraction of a decibel [22].

2.3.3.1 Digital Doppler Filter Banks

The most general form of a digital Doppler filter of the transversal type is depicted in Figure 2.31. This uses two channels: a real channel and an imaginary channel, each employing a shift register. For each target return from the same range increment, the filter receives an in-phase (x_n) and a quadrature (y_n) input. These inputs are multiplied by both real (w^R) and imaginary (w^I) weights.

The analysis of transversal filters is best handled by expressing the pertinent quantities in complex form. In this notation, a signal return from a range increment is represented by using its in-phase (x_n) and quadrature (y_n) parts as $s_n = x_n + jy_n$. The transversal filter weights applied to this signal element are given by

$$\mathbf{w}_n = \mathbf{w}_n^R + j\mathbf{w}_n^I \tag{2.23}$$

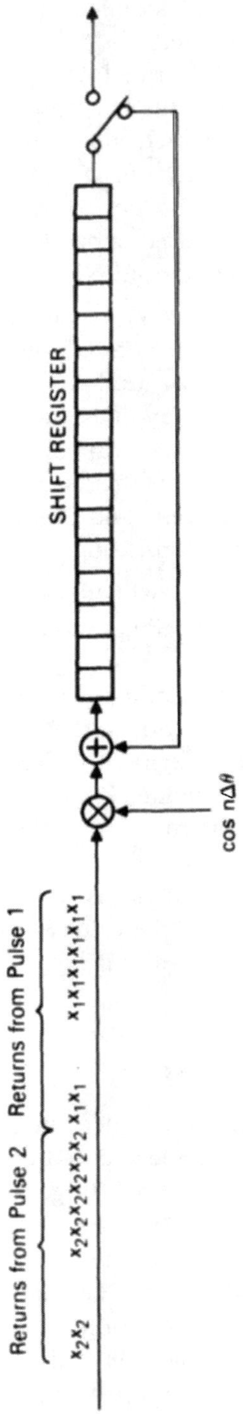

Figure 2.29 Formation of digital filters using shift registers. (*From:* [3, p. 246].)

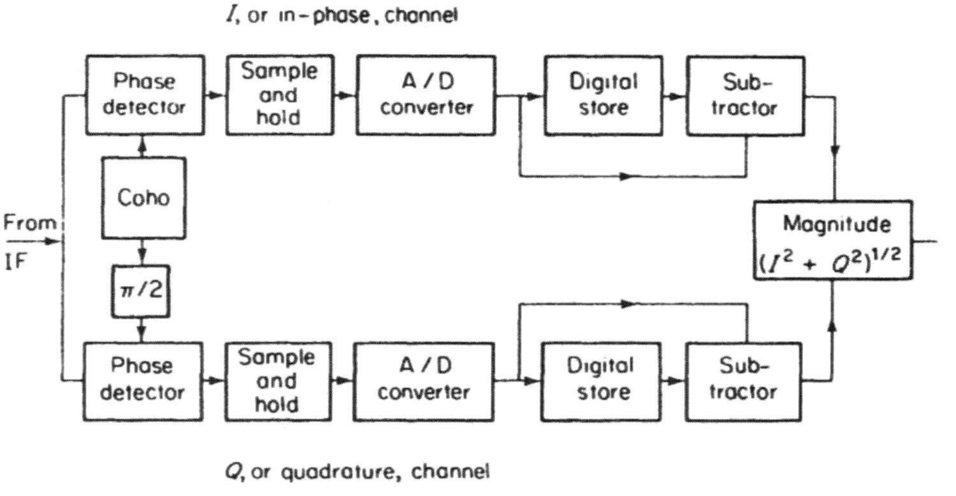

Figure 2.30 Block diagram of a digital single canceller MTI. (*From:* [17, p. 138].)

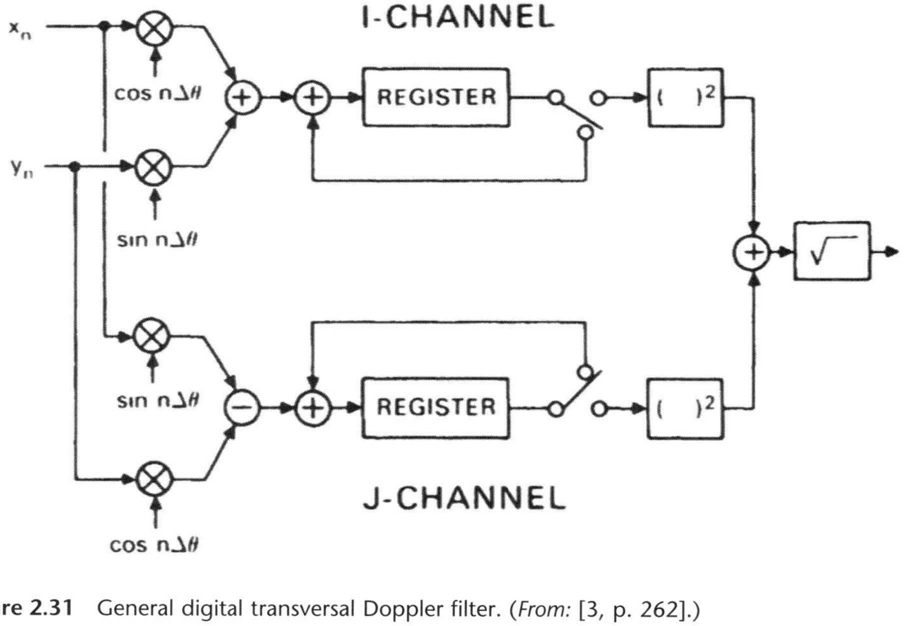

Figure 2.31 General digital transversal Doppler filter. (*From:* [3, p. 262].)

The output of the transversal filter is then given by

$$\mathbf{c}_n = \mathbf{r}_n + j\mathbf{i}_n = \left(\mathbf{w}_n^R + j\mathbf{w}_n^I\right)(\mathbf{x}_n + j\mathbf{y}_n) \tag{2.24}$$

The real part of this output is therefore given by

$$r_n = \mathbf{w}_n^R \mathbf{x}_n - \mathbf{w}_n^I \mathbf{y}_n \tag{2.25}$$

and the imaginary part equals

$$i_n = \mathbf{w}_n^I \mathbf{x}_n + \mathbf{w}_n^R \mathbf{y}_n \tag{2.26}$$

The basic action of the Doppler filter can be understood from the phasor diagram depicted in Figure 2.32. This diagram shows the target signal return that rotates at the Doppler shift rate (f_d) superimposed on two sets of coordinates displaced by an angle (θ_m). The first set of coordinates represents the in-phase and quadrature reference signals (recall Figure 2.16). The target signal return has an in-phase component of x_n and a quadrature component of y_n in this coordinate system. The filter formation is accomplished through the projection of target inputs onto the second coordinate system, which represents the filter's incremental output. If the filter is tuned to the target's Doppler, then the second coordinate system rotates in synchrony with the target vector. Otherwise, if tuned to a different Doppler frequency, the second set of coordinates rotates at a rate different than the target vector. The overall filter output is the sum of all projections that result from the l returns processed by the filter. This sum is maximized when the filter coordinate system rotates at the same rate as the target Doppler. When this occurs, the net effect is a coherent integration of the target return (this is sometimes called *Coherent Integration of Doppler echoes*).

The algorithm that defines the computations required to form a bank of digital Doppler filters can be deduced from Figure 2.32. The projection of the in-phase and quadrature components of the successive samples of the input waveform (x_n and y_n) onto the rotating coordinates (i and j) is given by

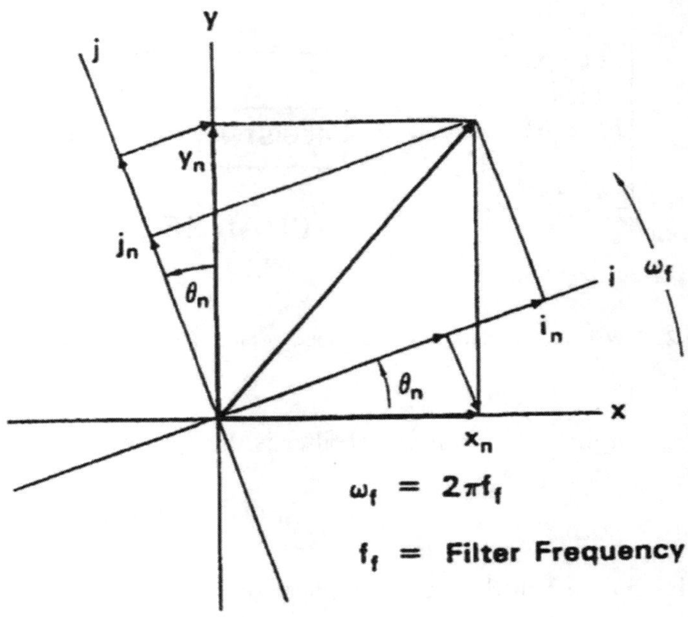

Figure 2.32 Phasor diagram of a coherent integrating digital Doppler filter. (*From:* [3, p. 256].)

$$r_n = x_n \cos \theta_n + y_n \sin \theta_n, \quad i_n = y_n \cos \theta_n - x_n \sin \theta_n \tag{2.27}$$

Comparison of these projections to the complex representation of a transversal filter allows the evaluation of the real weight as

$$w_n^R = \cos \theta_n \tag{2.28}$$

and the imaginary weight as

$$w_n^I = -\sin \theta_n \tag{2.29}$$

Alternatively, the complex weight can be expressed as

$$w_n = \exp[-j\theta_n] \tag{2.30}$$

This indicates that the Doppler filter bank's characteristics are a function of only the rotational position of the parameter θ_n.

Next, the initial position and rotational rate of the filter coordinate system are determined by setting $\theta_n = (p-1)\Delta\theta_n$, where $p = 1, 2, 3, \ldots, l$ is the pulse number and $\Delta\theta_n$ is the incremental angle through which the coordinate system rotates when the filter is tuned to Doppler frequency f_f. By causing the filter and signal coordinates to coincide for $n = 1$, the filter's weight is normalized so that $w_{n1} = 1(w_{n1}^R = 1, w_{n1}^I = 0)$.

Next, the initial position and rotational rate of the filter coordinate system are determined by setting $\theta_n = (p-1)\Delta\theta_n$, where $p = 1, 2, 3, \ldots, l$ is the pulse number and $\Delta\theta_n$ is the incremental angle through which the coordinate system rotates when the filter is tuned to Doppler frequency f_f. By causing the filter and signal coordinates to coincide for $n = 1$, the filter's weight is normalized so that $w_{n1} = 1\left(w_{n1}^R = 1, w_{n1}^I = 0\right)$.

The Doppler filter bank for an 8-pulse processor is depicted in Figure 2.33. This filter bank is formed by finding the magnitude of the weighted sums of the real and imaginary incremental projections onto the filter's coordinate system. Note the periodic nature of the Doppler filter bank whereby the filters recur periodically with a period equal to the radar's PRF. This requires that the filter's coordinate system rotate through an angle of 2π radians in the time required to process l pulses (e.g., $l = 8$). In addition, filter tuning requires that

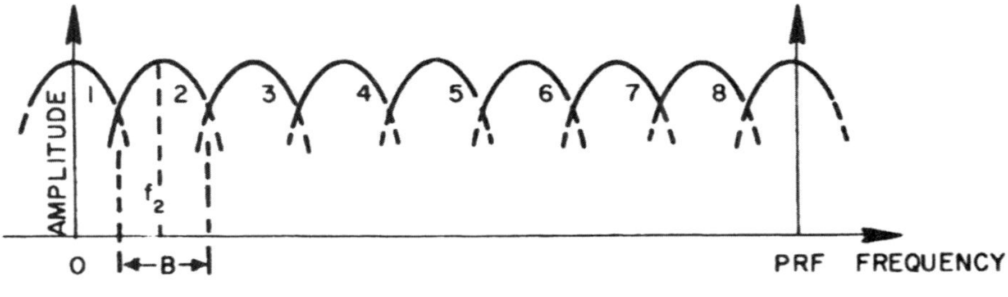

Figure 2.33 Coherent filter bank. (*From:* [23, p. 5].)

$$\Delta\theta_n = \frac{2\pi f_f}{\text{PRF}} \tag{2.31}$$

where f_f is the Doppler frequency to which the filter is tuned. The relationship for the incremental angle rotation of the filter's coordinate system can then be expressed as $\Delta\theta_n = 2\pi(k-1)/l$, where $k = 1, 2, \ldots, l$ is the filter number. This result is

$$\theta_n = \frac{2\pi(p-1)(k-1)}{l} \tag{2.32}$$

where p is the pulse number, k is the filter number, and l is the total number of pulses processed. Substituting this into the weight relationship results in

$$\mathbf{w}_s = \exp\left[\frac{-j2\pi(p-1)(k-1)}{l}\right] \tag{2.33}$$

as the complex weight factor that must be applied to input sample p to form the kth filter of the Doppler filter bank.

For example, when two samples ($l = 2$) are processed, the first filter ($k = 1$) weights become (1, 1), which implies coherent integration of fixed target echoes, and the second filter ($k = 2$) has weights of (1, –1), which provides for the coherent integration of an optimum speed target. For a 4-pulse sample, the weights for the corresponding four filters become (1, 1, 1, 1), (1, $-j$, –1, j), (1, –1, 1, –1), and (1, j, –1, $-j$). Note that for more than two pulse samples, the weights generally become complex, which is necessary to coherently integrate Doppler-shifted target returns.

2.3.3.2 The Discrete Fourier Transform

In the previous sections, Doppler filtering has been described as the algorithmic process of projecting the radar's input return signal (x_n, y_n) onto a rotating filter coordinate system (i, j) and adding up the projections. The rate at which the coordinates rotate relative to the x and y axes is the frequency to which the filter is tuned.

Alternatively, Doppler filtering can be viewed from a spectral analysis viewpoint, which involves performing a discrete Fourier transform (DFT) on the signal. The algorithms for both procedures are identical. The Fourier transform approach has the advantage that the computational procedures using Fourier analysis are well developed with many fast algorithms, the most notable of which is the fast Fourier transform. The FFT provides a dramatic reduction in the immense volume of computing otherwise required to form a large filter bank with the DFT.

The discrete Fourier transform is given by

$$F(p, k) = \sum_{p=1}^{l} s_n(p) w^{(p-1)(k-1)} \tag{2.34}$$

where $s_n(p) = x_n + jy_n$ is the complex signal at sample time $p = 1, 2, \ldots, l$, where l is the number of sample points, $k = 1, 2, \ldots, l$ is the output transform point, and $w = \exp(-j2\pi/l)$ is the complex weight kernel (incremental angle rotation of filter coordinates). The equivalent normalized transfer function of the DFT can be shown [5] equal to

$$|H(\omega)| = \left| \frac{\sin\left[\pi\left(\frac{lf}{f_r} - k + 1\right)\right]}{l \sin\left[\frac{\pi}{l}\left(\frac{lf}{f_r} - k + 1\right)\right]} \right| \tag{2.35}$$

The transfer function is depicted in Figure 2.34 for both a rectangular window (uniform input weighting) and an amplitude taper that provides substantially reduced sidelobes.

A method for providing a reduced filter sidelobe structure is depicted in Figure 2.35. This process, called *amplitude weighting*, is applied to the digitized video (after A/D conversion) before it is applied to the Doppler filters. Following each transmission, the I and Q receiving signals (x_n, y_n) returned from each range increment are multiplied by a weighting coefficient. The coefficient is changed from one pulse to the next according to a prescribed pattern [e.g., Chebyshev, Hamming, or Hann (i.e., Hanning) weighting], which is repeated for each train of pulses to be integrated. Amplitude tapers can be selected to reduce sidelobes to very low levels (e.g., –35 to –60 dB); however, in this process, the filter's main lobe response is significantly widened (e.g., Hamming weighting; –43-dB sidelobes, 1.30 3 dB bandwidth expansion). Alternatively, post-DFT weighting can reduce filter sidelobes [5].

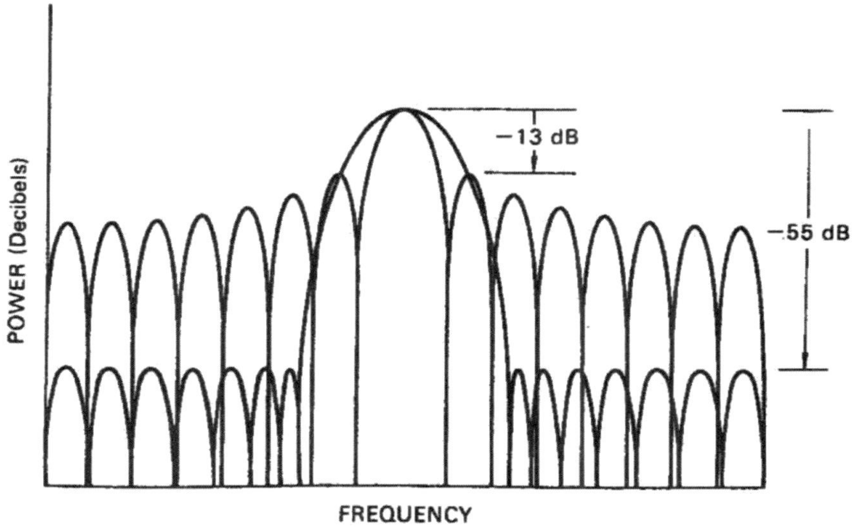

Figure 2.34 Transfer function of a DFT with uniform and weighted inputs. (*From:* [3, p. 264].)

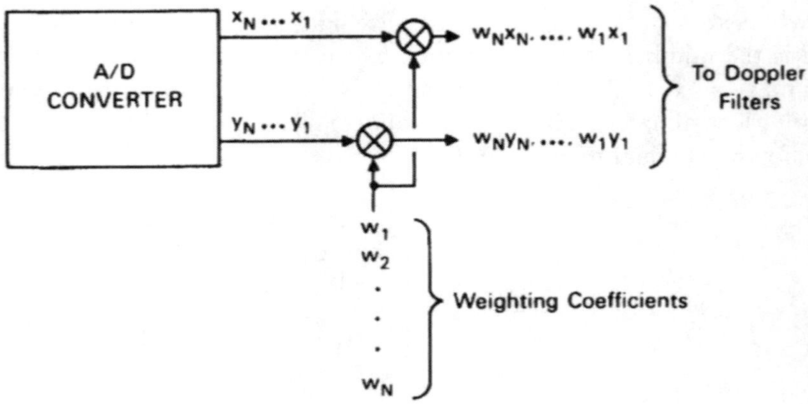

Figure 2.35 Reducing Doppler filter sidelobes through amplitude weighting. (*From:* [3, p. 263].)

► Example 6 Cascaded MTI-FFT Doppler Filters

Cascaded MTI-FFT Doppler filters are useful in surface-based search radars that must operate in an environment consisting of both ground and weather clutter. Due to the fourth-power dependence of weather clutter reflectivity on frequency, these types of processors are appropriate for radars that operate at frequencies above L-band. Basically the MTI is used to eliminate the ground clutter while the FFT processes the clutter due to weather or chaff. The original ASR-9 MTD processor described in Chapter 1 used this technique.

To analyze this situation, first model the FFT as a transversal filter with weights given by

$$\mathbf{w}_{\mathrm{FFT}} = \exp\left[\frac{-2\pi j(p-1)(K-1)}{N}\right]$$

where $p = 1, 2, \ldots, N$; K is the filter number; and N is the number of FFT points. The power transfer function can be determined as

$$|\mathbf{H}(\omega)|^2 = \mathbf{w}_{\mathrm{FFT}}^T \mathbf{M}_S \mathbf{w}_{\mathrm{FFT}}^*$$

where the signal Hermitian covariance matrix is given by

$$\mathbf{M}_S = \begin{pmatrix} 1 & \cdots & e^{-jN\omega_d T} \\ \vdots & \ddots & \vdots \\ e^{+jN\omega_d T} & \cdots & 1 \end{pmatrix}$$

The power transfer function for an 8-point FFT with uniform weighting and 1,200-Hz PRF is computed using the MATLAB program fft_impfac.m. The transfer function is depicted in Figure EX-6.1. Note that the resulting eight filters divide

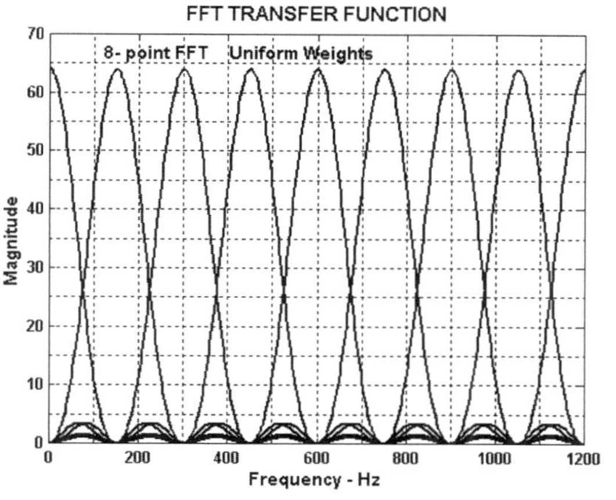

Figure EX-6.1 FFT transfer function with uniform weighting.

the frequency coverage between PRF lines into eight equal segments, each with 3 dB bandwidths of 134 Hz, power gains of $N^2 = 64$, and sidelobes of –13.2 dB. Ground clutter can be removed by discarding the filters that straddle the PRF lines. However, the ground clutter can still leak into adjacent filters through the high 13.2-dB sidelobes. The solution to the sidelobe problem is to use a window function that reduces the filter sidelobes. However, the window function increases the 3 dB bandwidth while reducing the power gain. For example, a Hamming window that reduces the sidelobes to –43 dB increases the 3 dB bandwidth to 195 Hz, while reducing the power gain to 35. A more flexible window is the Chebyshev window, which allows a particular sidelobe level to be selected while providing the minimum 3 dB bandwidth consistent with this level. Figure EX-6.2 shows the 8-point FFT filters designed using a –25-dB Chebyshev window, which reduces the FFT sidelobes

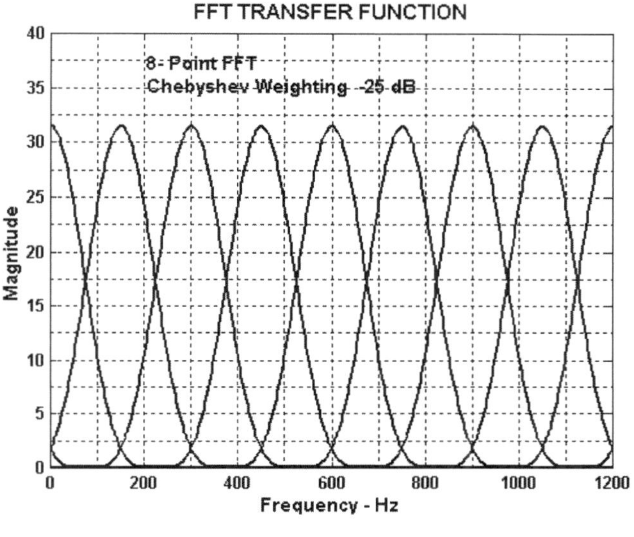

Figure EX-6.2 FFT transfer function with Chebyshev weighting.

to an indiscernible level. Examination of this figure shows that the 3 dB bandwidth is increased to 166 Hz, while the power gain is reduced by 3 dB.

Figure EX-6.3 shows the improvement factor when the FFT with Chebyshev –25 dB windowing is used against ground clutter with $\sigma_S = 15.3$ Hz and a PRF of 1,200 Hz (see Example 5 for ASR-9 parameters). The 35 dB performance is somewhere between the 24.9 dB provided by a single canceller and 46.9 dB by a double canceller. However, when the filters that straddle either side of the PRF are eliminated, the FFT provides about a 35 dB improvement factor with coverage from 200 to 1,000 Hz.

The general approach utilized in the MTD processor is to use a cascaded MTI-FFT combination. To analyze this situation, we can derive an equivalent weight that applies to the combined filter with length $N + m - 1$ where m is the number of pulses processed by the MTI. For example, if a single canceller is used with weights (1, –1), then its impulse response is given by

$$h(t) = \delta(t) - \delta(t + T)$$

where $\delta(t)$ is a unit impulse and T is the radar period. Applying these impulses to the FFT results in

$$\mathbf{w}_{\text{MTI-FFT}} = [\mathbf{w}_{\text{FFT}}, 0] - [0, \mathbf{w}_{\text{FFT}}]$$

For a double canceller followed by an FFT, the result is

$$\mathbf{w}_{\text{MTI-FFT}} = [\mathbf{w}_{\text{FFT}}, 0, 0] - 2[0, \mathbf{w}_{\text{FFT}}, 0] + [0, 0, \mathbf{w}_{\text{FFT}}]$$

As an example consider a 4-point FFT with $k = 2$ preceded by a single canceller. In such a case, the combined weight is given by

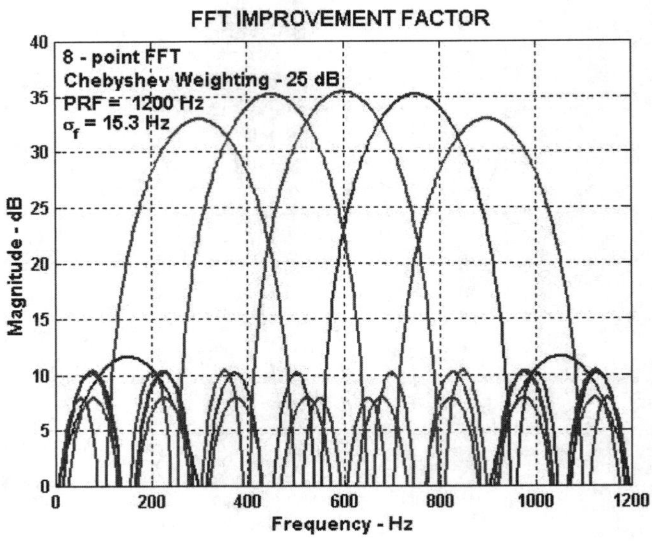

Figure EX-6.3 FFT improvement factor.

$$\mathbf{w}_{\text{MTI-FFT}} = [1, -1 - j, -1 + j, 1 + j, -j]$$

The weights can then be used to determine the improvement factor given by

$$I = \frac{\mathbf{w}^T \mathbf{M}_S \mathbf{w}^*}{\mathbf{w}^T \mathbf{R}_N \mathbf{w}^*}$$

where the clutter covariance Toeplitz matrix is given by

$$\mathbf{R}_N = \begin{pmatrix} 1 & \cdots & \rho(N+m-1) \\ \vdots & \ddots & \vdots \\ \rho(N+m-1) & \cdots & 1 \end{pmatrix}$$

The transfer function of the FFT-MTI can be found for the configuration used in the MTD (double MTI canceller followed by an 8-point FFT) by applying MATLAB program fft_impfac_9.m. The result is depicted in Figure EX-6.4. The FFT filter that straddles the PRF lines is suppressed by the MTI canceller, whereas the FFT response is generally modulated by the transfer function of the MTI.

The improvement factor for this configuration is shown in Figure EX-6.5. The maximum improvement factor exceeds 67 dB in the middle five filters. Hence, ground clutter is completely suppressed. Weather clutter is suppressed using an AGC in each of the five middle filters. The only problem is due to sidelobe cross-coupling of the weather clutter that, by design, is suppressed 30 dB. This is usually sufficient since the magnitude of weather clutter is generally moderate at S-band frequencies (see Example 7 in Chapter 3).

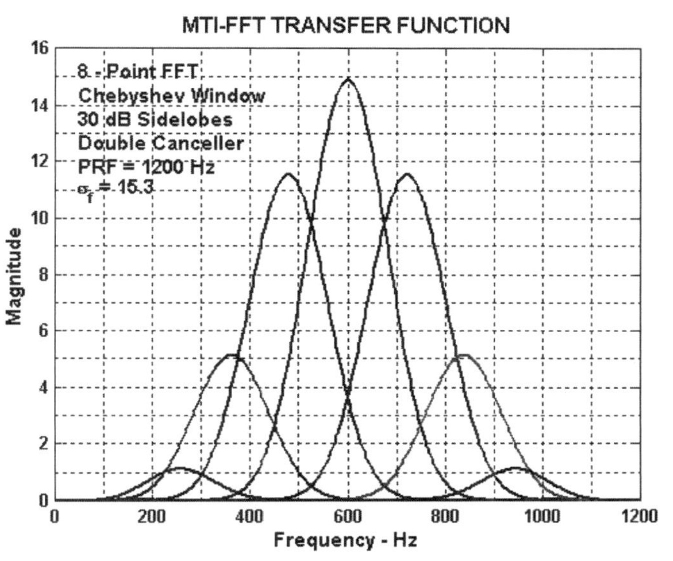

Figure EX-6.4 MTI-FFT transfer function.

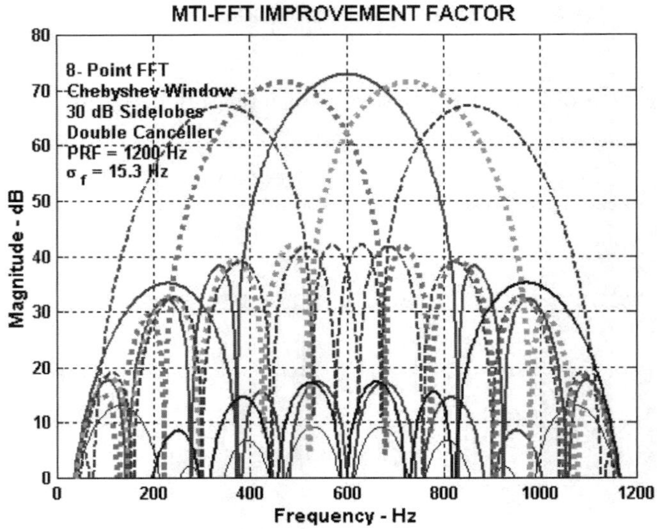

Figure EX-6.5 MTI-FFT improvement factor (ground clutter).

▶

2.3.3.3 The Fast Fourier Transform

To obtain an idea of the computations required to form a Doppler filter bank using DFT, consider the block diagram of the transversal filter implementation depicted earlier in Figure 2.31. First, each in-phase and quadrature sample must be multiplied by real weights and then summed, resulting in eight mathematical operations per sample. To form l filters, $8l$ operations per complex input sample are required. Because l pulses are required to completely form each filter, the overall requirement for each range cell is then $8l^2$ operations. The number of filter banks (N_B) that must be formed is equal to the number of range cells (i.e., one range-gated Doppler filter bank per range resolution cell). Because the samples occur at the radar's PRF rate, the total computation rate is given by

$$R_c = 8\ l\ N_B\ \text{PRF} \tag{2.36}$$

operations per second. For example, if we assume for airborne radar that a 16-sample DFT filter bank is to be formed using 200 range gates with a radar PRF of 8 kHz, the resulting computation rate is approximately 205 MOPS. This rate represents a formidable processing requirement that would severely stress the current digital array signal processors [7].

This example points to a requirement for substantially reducing the computational requirements involved in forming digital Doppler filter banks. This requirement is satisfied by the fast Fourier transform, an algorithm that evaluates the DFT in a computationally efficient manner. Basically, the FFT exploits the periodic structure of the DFT weighting coefficients to significantly reduce the number of arithmetic operations required to form the DFT.

The first step in forming a computationally efficient Doppler filter bank is to require the filter bank to span a range of frequencies equal to the sampling frequency (i.e., the PRF). The next step is to require that the number of filters in the bank be equal to an integer raised to some power, where the most popular choice is the radix-2 (i.e., $l = 2, 4, 8, \ldots, 2^m$); both steps result in a highly symmetrical distribution of the filter coefficients about the x and y axes (recall Figure 2.32), which, in turn, minimizes the number of multiplications required in the algorithm. The final step, which is critical in achieving high efficiency, is to add or subtract all inputs having a particular coefficient by the coefficients before multiplication instead of afterward. For this, all samples (x_n, y_n) that must be integrated to form the entire filter bank must be received before the multiplication operation can be accomplished. In other words, the algorithm cannot be performed online but must be accomplished in block fashion after all l samples have been received. The FFT reduces the number of computations from the order of l^2 required for the DFT to the order of $l \log_2(l)/2$. The dramatic reduction in the number of computations achieved by using the FFT as opposed to direct evaluation of the DFT is depicted in Figure 2.36.

The DFT can be conveniently written in matrix form as

$$[\mathbf{F}(k)] = [\mathbf{w}^{(p-1)(k-1)}][\mathbf{s}_{n(p)}] \tag{2.37}$$

where $\mathbf{w} = \exp(-j2\pi/l)$ and $k = 1, 2, \ldots, l$ is the output transform point and $p = 1, 2, \ldots, l$ is the input sample point. In this representation, $[w^{(p-1)(k-1)}]$ is an l by l square matrix.

Using the Cooley-Tukey algorithm, this matrix can be factored into $m = \log_2 l$ square matrices of dimension l by l [24]. The unique property of this factored form is that it allows formation of the FFT by using a cascade of repetitive operations where each operation is known as a *butterfly*. A flow graph of the conventional method for computing the DFT by using on the order of l^2 operations and the radix-2 FFT method by using $l \log_2(l)/2$ butterfly operations is depicted in Figure 2.37.

The butterfly operation in the radix-2 FFT is performed on a pair of samples in the mth stage to produce a corresponding pair of samples in the $m + 1$ stage. For an l-point FFT with $R_m = 2^m$, the butterfly operation can be expressed as the pair:

$$u_i(m+1) = u_i(m) + w^r v_{i+l/Rm}(m) \tag{2.38}$$

and

$$v_{i+l/Rm}(m+1) = u_i(m) - w^r v_{i+l/Rm}(m) \tag{2.39}$$

where u_i and $v_{i+l/Rm}$ are complex numbers representing the ith FFT data sample pair of the mth stage and r is an integer that is a function of the index I and stage m [26].

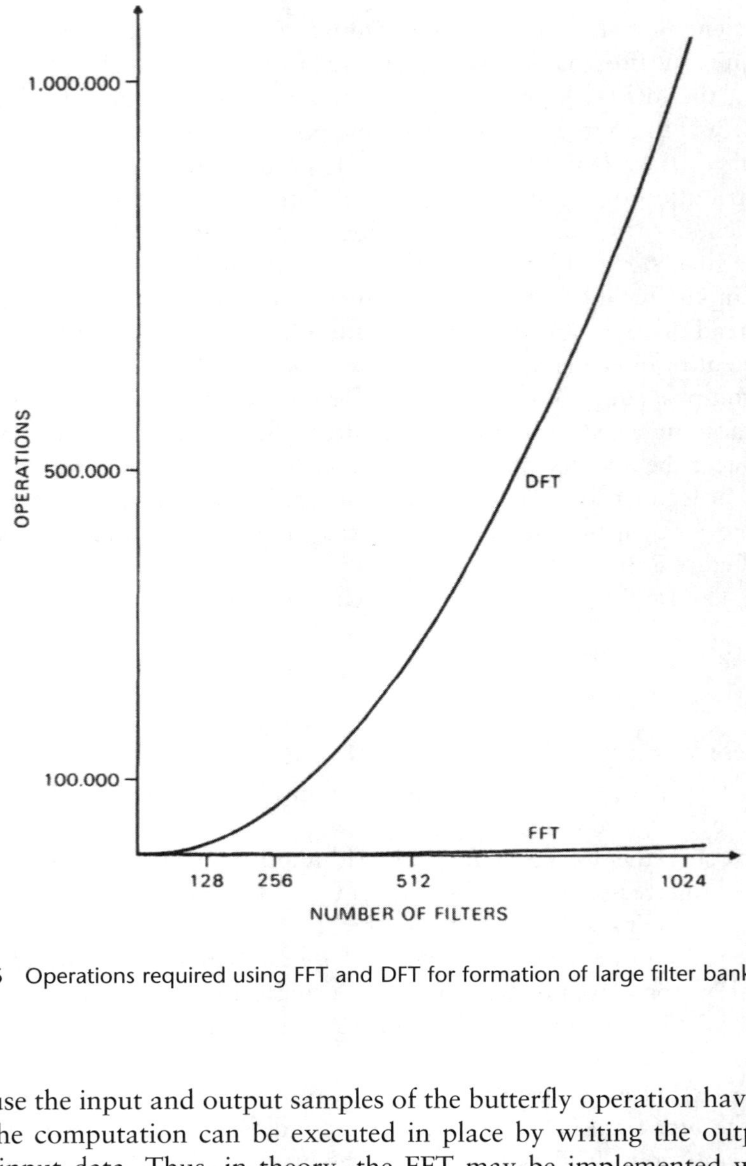

Figure 2.36 Operations required using FFT and DFT for formation of large filter banks. (*From:* [9, p. 353].)

Because the input and output samples of the butterfly operation have the same indices, the computation can be executed in place by writing the output results over the input data. Thus, in theory, the FFT may be implemented with only l complex storage locations, plus auxiliary storage registers to support the butterfly computation.

In the butterfly operation, the complex weight value w^r can be determined using the bit-reversal method [24]. This is done by writing $I - 1$ as a t-bit binary number, scaling this number by $t - m$ bits to the right, filling the newly opened bit positions of the left with zeros, and reversing the order of the bits to obtain the value for r. For example, in the 4-point FFT depicted in Figure 2.37, consider the second stage and third data sample (i.e., $m = 2$, $t = 2$). The binary number index is then given by 10, which, when scaled $t - m = 0$ bit to the right, results in the number 10. When bit-reversed, this results in 01, which yields $r = 1$ (i.e., w^1). The remaining weights for the 4-point FFT can be found as w^0 ($i = 1, 2$; $m = 1$ and $i = 1$; $m = 2$) and w^1 ($i = 3$; $m = 2$), which allows the complete transform to

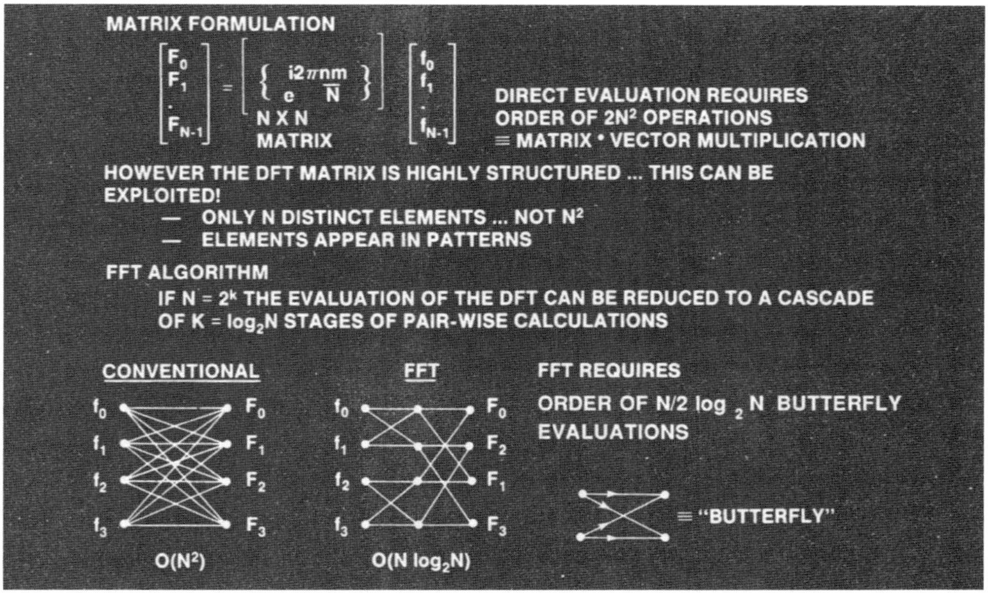

Figure 2.37 Flow graph of FFT using butterfly stages. (*From:* [25, p. 15].)

be found. Note that in the representation given in Figure 2.37, the weights were determined from a tree-graph construction (rather than a butterfly representation), but due to symmetry $w^2 = -w^0$ and $w^3 = -w^1$, which reconciles the two representations [24].

The bit-reversal process necessary to determine the butterfly weights is simply implemented in hardware form by using a counter that counts a bit-reversed notation [26]. For an 8-point FFT, a conventional 3-bit counter yields the successive integers 0, 1, 2, 3, 4, 5, 6, and 7. If the counter bit positions are reversed, then the resulting counter output is 0, 4, 2, 6, 1, 5, 3, and 7, which gives a one-to-one correspondence between the natural order sequence and the bit-reversed order sequence.

An analysis of the butterfly relationships shows that each butterfly can be accomplished with 4 real multiplications and 6 real additions (a total of 10 real operations). The butterfly operation can be considered a fundamental kernel associated with the FFT algorithm. The complexity of a signal processor that performs an FFT operation can be specified in terms of an equivalent number of butterfly operations (i.e., $l \log_2(l)/2$), and the computation time to perform a butterfly operation becomes a basic figure of merit for a radar signal processor [3, p, 277].

Many of the FFT algorithms were developed to calculate the DFT by using a high-level language on a general-purpose mainframe or a minicomputer. The emphasis in these algorithms was generally on minimizing the execution time and, hence, the cost of performing the DFT. The approach taken was to minimize the total number of operations performed by the arithmetic unit with particular emphasis on the total number of multiply operations, because this function represented the major limitation on processing speed. Consequently, most FFT algorithms are designed simply to minimize the total number of multiplications required to calculate the DFT, often by increasing the number of additions, memory access, and control complexity. Increased control complexity is not usually a major penalty in software-based systems because, in simplistic terms, the control part of an FFT algorithm implemented on a high-level machine comes for free; however, in hardware-based processors, this is not the case.

The hardware designer must consider control complexity and must also try to optimize the processor arithmetic unit, because the control structure usually represents the major portion of the hardware system cost. As a result, DFTs implemented on general-purpose computers are usually calculated by using highly composite data block sizes, where the arithmetic complexity can be reduced by decomposing the DFT into a large number of nested small-radix transforms, whereas hardware-based DFTs often use minimum block partitioning to simplify their control structure and microcode generation [27].

Emphasis in FFT hardware used for radar Doppler filtering applications is on throughput rate (rate at which data can be absorbed) rather than algorithm processing time. This leads to pipeline structures that allow increased amounts of data to be processed in a given algorithm processing time. The pipeline structure is analogous to a long pipe through which water is pumped, where water enters continuously rather than waiting for 1 gallon to exit the pipe before another enters. The pipeline structure thus effectively trades the ability to absorb data at a higher rate for processing time delay.

For example, consider the flow of data and computations through the 8-point radix-2 FFT depicted in Figure 2.38. The radix-2 configuration breaks the FFT into $l \log_2(l)/2 = 12$ nested butterfly operations, which are grouped into three stages with each providing four butterfly operations. Data pairs separated by a distance $l/2 = 4$ are fetched from the input storage buffer memory and operated on by the butterfly operations with the results returned to the original memory location. The process is repeated $l/2 = 4$ times until all l data points have been operated on. The first pass, therefore, consists of $l/2$ butterfly operations performed

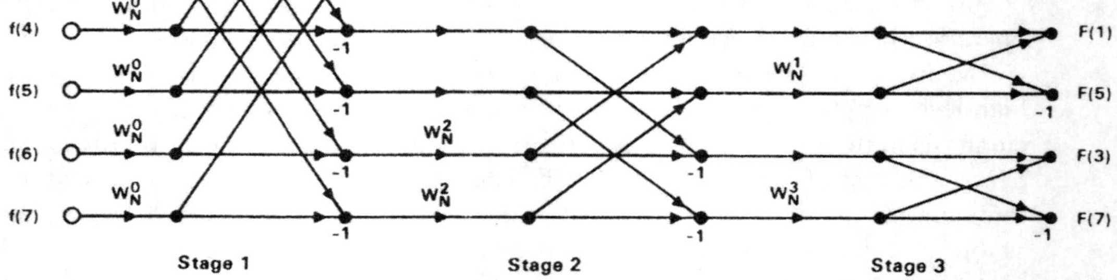

Figure 2.38 Flow of data through an 8-point radix-2 FFT. (*From:* [28, p. 70].)

on the data with a spacing of $l/2$. On the second pass, the process is repeated with a spacing of $l/4$. Successive passes use spacings that decrease by one-half until the spacing is such that adjacent data points are used. At this point, the desired spectrum has been computed and is resident in memory.

A pipeline FFT processor can be formed from the configuration depicted in Figure 2.38 by inserting storage registers between computational elements; this allows data for the second transform to be entered as soon as the immediate results of the first transform have been moved to the second state. A double buffer memory would seem to be required; one would accommodate new data entering the computational element, and the second would hold the data block being worked on. However, by using a special "shuffle" memory [29], we can process the data blocks through the pipelined FFT using only a single buffer memory equal to the block length of the FFT being processed (e.g., an 8-point FFT would require only eight words of memory). Note also that the rotational vector weight (w^r) is a function of the data displacement number and, hence, varies in a predictable way from one computational stage to the next.

The structure of the FFT algorithm is such that each of the $\log_2 l = M$ computational stages can be realized with a basic computational module, the general form of which is shown in Figure 2.39. The basic computational module is assembled to form a pipeline-FFT processor by combining M modules as depicted in Figure 2.40. Each module contains an arithmetic unit that computes the butterfly operation

$$[u(m+1) = u(m) + w^r v(m);\ v(m+1) = u(m) - w^r v(m)] \tag{2.40}$$

and a shift register (delay line) whose storage capacity is 2^{M-m}, where M is the total number of stages and m is the stage number. For example, the first stage ($m = 1$) of an 8-point FFT ($M = 3$) would use a shift register whose capacity is 2^2 = four words; the second stage, two words; and the third stage, one word.

Each pipeline FFT module alternately transfers flocks of data samples into the delay line and the arithmetic unit [30]. Initially, $l/2$ data samples are loaded into the delay line of the first-stage FFT module. Then, the first-stage arithmetic unit obtains a rotational weight vector (from a read-only memory) and begins its opera-

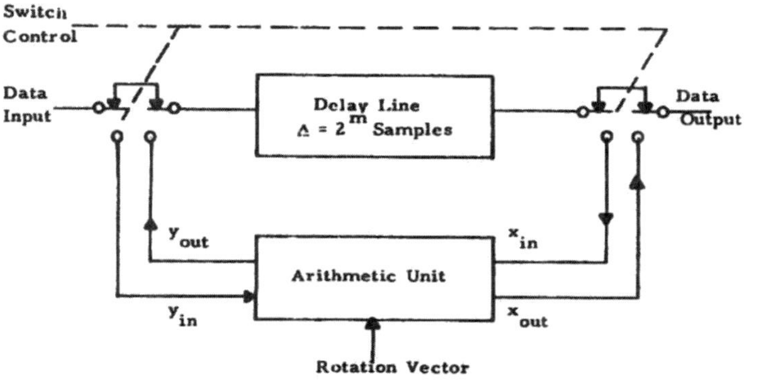

Figure 2.39 Pipeline FFT modules. (*From:* [29, p. 27].)

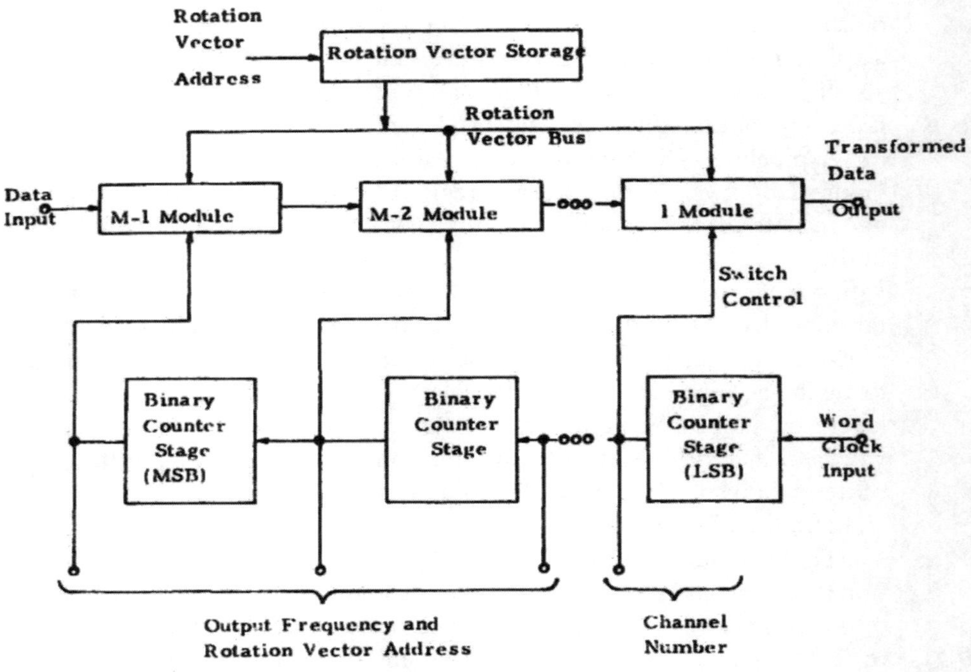

Figure 2.40 Pipeline FFT processor. (*From:* [29, p. 28].)

tion. The next $l/2$ data samples in conjunction with the first block stored in the delay line are loaded into the first FFT module to produce two successive complex butterfly outputs. One output is immediately transferred to the next stage, and the other output is sent to the delay line. Thus, in the interim period when the delay line is filled with fresh input data, the contents of the line containing the results of processing the earlier blocks are transferred to the next stage.

In the pipeline FFT processor shown in Figure 2.40, the rotational weight vector storage consists of a table of roots of unity (sines and cosines) shared by all M modules. In general, $l/2$ different rotational weight vectors must be read to process one block of l samples. If they are produced in the order required for the earlier stages, they may be obtained by strobing the memory storage in the proper time sequence. Thus, the sine and cosine weights can be obtained by providing a register for each arithmetic unit, all driven from a common bus. Note that $M = \log_2 l$ arithmetic units and $l - 1$ complex data points of storage are needed to implement the pipeline FFT processor depicted in Figure 2.40. Also, the first transform output is obtained immediately after the last data sample in the block of l is received.

The control mechanism in the pipeline FFT processor uses a binary counter driven in synchrony with the data stream. This counter generates a signal that designates the processing interval and switch position for each stage (see Figure 2.39). In general, when the counter output is supplying the output for the $k + 1$ stage, then the lower k bits of the counter, in normal order, give the channel number of the data currently being produced, whereas the upper $M - k$ bits, taken in bit-reversed order, give the frequency currently being produced. In addition, the counter supplies the address of the complex weights (sine and cosine) needed in the current computation of the $k + 1$ stage. Thus, the control mechanism contains all of the information needed to carry out the spectral analysis as well as to descramble the output data. In many applications, descrambling the output data is not necessary because the precise order of the output frequency samples is always known [29].

The pipeline FFT processor just described represents only one of several types of FFT processing hardware architectures that can be used. The types available, which range from the least hardware–lowest throughput to the most complex hardware–highest throughput, are called the (1) sequential, (2) cascade (pipeline FFT), (3) parallel-iterative, and (4) array processors [31].

The sequential FFT processor performs a single butterfly operation on $l \log_2 l/2$ pairs of data in a serial manner. The same memory can be used to store the input data, the intermediate results, and the resulting Fourier coefficients. Because the processor operates on batches of data, a real-time environment would dictate that a buffer memory precede or be incorporated into the data memory. The sequential FFT processor can be characterized as having: (1) one arithmetic unit, (2) $l \log_2 l/2$ butterfly operations performed sequentially, and (3) an execution time of $Bl \log_2 l/2$ where B is the time to perform the butterfly operation.

The cascade FFT processor introduces parallelism by using a separate arithmetic unit for each iteration stage, thereby increasing the throughput over the sequential processor by the factor $\log_2 l$. In the flow diagram of Figure 2.38, the arithmetic unit associated with each stage of the algorithm processes $l/2 = 4$ pairs of data. The buffering required for processing a continuous stream of data is incorporated directly into the processor organization in the form of a delay line (shift register). The cascade FFT processor can be characterized as having: (1) $M = \log_2 l$ arithmetic units and iterations performed in parallel, (2) $l/2$ operations performed sequentially, (3) an execution time of $Bl/2$ seconds, and (4) buffering incorporated within the processor in the form of time delays.

The parallel iterative FFT processor increases throughput by introducing parallelism into each processing stage. The processor performs the $l/2$ butterfly operations within each stage in parallel, and the $\log_2 l$ iterative operations are performed sequentially. The parallel iterative processor can be characterized as having: (1) $l/2$ arithmetic units and operations performed in parallel, (2) $M = \log_2 l$ iterations performed sequentially, and (3) an execution time of $B \log_2 l$ sec. In practice, due to cost constraints, this architecture is often combined with that of the sequential processor so that only the degree of parallelism actually needed to produce the required throughput is implemented.

The array processor is configured such that all $l \log_2 l/2$ butterfly operations are performed in parallel. This is accomplished by simultaneously pipelining $\log_2 l$ sets of data through the processor, which results in an execution time equal to that required to perform a basic butterfly operation. The FFT-array processor can be characterized as having: (1) $l/2 \log_2 l$ arithmetic units and operations performed in parallel, and (2) an execution time of B sec.

As an example, a 64-point FFT is to be performed using a butterfly unit that has an execution time of 0.5 μs ($l = 64$, $n = 6$, $B = 0.5$ μs). Table 2.1 compares

Table 2.1 FFT Hardware Architecture Data Rate to Compute 64-Point FFT

Architecture	*Butterfly Units*	*Execution Time* (μs)	*Data Rate (MHz)*
Sequential	1	96	0.67
Cascade	6	16	4
Parallel iterative	32	3	21.3
Array	192	0.5	128

B = 0.5 sec.

the various processing architectures, indicating the increasing data rate associated with an increased degree of parallel operation.

At present, FFT hardware emphasizes optimum use of standard digital integrated circuit chips to generate butterfly modules. Complete FFT processors can then be formed by using the butterfly modules in combination with appropriate digital control circuitry. One such hybrid unit, which provides a 4 to 16,384-point FFT, utilizes a radix-4 butterfly module that performs a basic butterfly operation in 100 ns [28].

Further increases in FFT hardware capacity involve utilizing very large scale integration (VLSI) circuit techniques to form a number of butterflies on a single chip. For example, one hardware survey projected that 16 butterflies could be implemented on a single 50 mm^2 chip [33]. A resulting problem with this approach is that the FFT butterfly interconnections lead to long wires between butterflies [33]. Not only do these wires consume a lot of area, but they require high-power semiconductors to drive this capacitive load at useful speeds. Thus, VLSI designs are not necessarily well matched to the implementation of the conventional FFT butterfly algorithm.

References

[1] Skolnik, M., (ed.), *Waveform Design*, Chap. 3 in *Radar Handbook*, 1st ed., New York: McGraw-Hill, 1970.

[2] Roulston, J., "Design Decisions Guide Airborne Radar," *Microwaves and RF*, March 1985.

[3] Stimson, G., *Introduction to Airborne Radar*, 2nd ed., Mendham, NJ: SciTech, 1998.

[4] Nathanson, F., *Radar Design Principles*, 2nd ed., New York: McGraw-Hill, 1991.

[5] Schleher, D. C., (ed.), *MTI Radar*, Dedham, MA: Artech House, 1978.

[6] Schleher, D. C., *Introduction to Electronic Warfare*, Dedham, MA: Artech House, 1986.

[7] Williams, F., and M. Radant, "Airborne Radar and the Three PRFs," *Microwave J.*, July 1983; also in M. Skolnik, (ed.), *Radar Applications*, New York: IEEE Press, 1989, pp. 272–276.

[8] Aronoff, E., and N. Greenblatt, "Medium PRF Radar Design and Performance," *20th Tri-Service Radar Symposium*, 1974; also in D. Barton, (ed.), *CW and Doppler Radars*, Vol. 7 of *Radars*, Dedham, MA: Artech House, 1978, pp. 261–275.

[9] Stimson, G., *Introduction to Airborne Radar*, 1st ed., Mendham, NJ: SciTech, 1983.

[10] Kennedy, P., "FFT Signal Processing for Non-Coherent Airborne Radars," *Proc. IEEE National Radar Conference*, 1984, pp. 79–83.

[11] Schleher, D. C., *MTI and Pulsed Doppler Radar*, Norwood, MA: Artech House, 1991.

[12] Shrader, W., and V. Gregers-Hansen, "MTI Radar," Chap. 15 in *Radar Handbook*, 2nd ed., M. Skolnik, (ed.), New York: McGraw-Hill, 1990.

[13] Nathanson, F., and P. Luke, "Loss from Approximations to Square-Law Detectors in Quadrature Systems with Postdetection Integration," *IEEE Trans. on Aerospace and Electronic Systems*, Vol. AES-8, No. 1, January 1972, pp. 75–77.

[14] Vannicola, V., "Fluctuation Loss and Diversity Gain for In-Phase Systems with Post-Detection Integration," *IEEE Trans.*, Vol. AES-9, No. 2, March 1973, pp. 290–295; also in D. Barton, (ed.), *The Radar Equation*, Vol. 2 of *Radars*, Dedham, MA: Artech House, 1974, pp.137–142.

[15] Weiss, M., and I. Gertner, "Loss in Single-Channel MTI with Post-Detection Integration," *IEEE Trans. on Aerospace and Electronic Systems*, Vol. AES-18, No. 2, March 1982, pp. 205–208.

[16] Schleher, D. C., "Performance Comparison of MTI and Coherent Doppler Processors," *Proc. IEEE Int. Radar Conference*, London, November 1982, pp. 154–158.

[17] Skolnik, M., *Introduction to Radar Systems*, 3rd ed., New York: McGraw-Hill, 2001.

[18] Skolnik, M., *Introduction to Radar Systems*, 2nd ed., New York: McGraw-Hill, 1980.

[19] Papoulis, A., *The Fourier Integral and Its Applications*, New York: McGraw-Hill, 1962.

[20] Andrews, G., *Optimal Radar Doppler Processors*, Report 7727, Naval Research Laboratory, Washington, D.C., May 1974.

[21] White, W., and A. Ruvin, *Recent Advances in the Synthesis of Comb Filters*, IRE National Convention Record, 1957, pp. 186–200; also in D. C. Schleher, (ed.), *MTI Radar*, Artech House, Dedham, MA, 1978.

[22] Filip, A., "A Baker's Dozen Magnitude Approximations and Their Detection Statistics," *IEEE Trans. on Aerospace and Electronic Systems*, Vol. AES-12, No. 1, January 1976, pp. 86–89.

[23] Andrews, G., *Performance of Cascaded MTI and Coherent Integration Filters in a Clutter Environment*, Report 7533, Naval Research Laboratory, Washington, D.C., March 1973.

[24] Brigham, E., and R. Morrow, "The Fast Fourier Transform," *IEEE Spectrum*, December 1967.

[25] Raytheon, *Electronic Progress*, Vol. XXVIII, No. 1, 1987, p. 15.

[26] Nussbaumer, H., *Fast Fourier Transform and Convolution Algorithms*, New York: Springer-Verlag, 1981.

[27] Curtis, T., and J. Wickenden, "Hardware-Based Fourier Transforms: Algorithms and Architectures," *IEE Proc.*, Vol. 130, Pt. F, No. 5, August 1983.

[28] Swartzlander, E., "Spectrum Analysis and the Unconventional Butterfly," *Defense Science and Engineering*, October 1985.

[29] Groginsky, H., and G. Works, "A Pipeline Fast Fourier Transform," *IEEE Eascon-69*, Washington, D.C., October 27–29, 1969, pp. 22–29; also in *IEEE Trans. on Computers*, Vol. C-19, November 1970.

[30] Brookner, E., (ed.), *Radar Technology*, Dedham, MA: Artech House, 1977.

[31] Bergland, G., "Fast Fourier Transform Hardware Implementations—An Overview," *IEEE Trans. on Audio and Electroacoustics*, Vol. 15, June 1969.

[32] Mavor, J., and P. Grant, "Operating Principles and Recent Developments in Analogue and Digital Signal Processing Hardware," *IEE Proc.*, Vol. 134, Pt. F, No. 4, July 1987.

[33] Ward, J., J. McCanmy, and J. McWhurter, "Bit-Level Systolic Array Implementation of the Winograd Fourier Transform Algorithm," *IEE Proc.*, Vol. 132, Pt. F., No. 6, October 1985.

[34] MATLAB Signal Processing Toolbox Users Guide, Version 4, The Math Works, Natick, MA, 1988.

CHAPTER 3

Doppler Radar Performance Measures

A basic consideration for any radar is the range at which the radar can detect the target. This obviously depends on the parameters of the radar (e.g., transmitter power, antenna gain, *et cetera*) and on the reflection characteristics of the target (e.g., radar cross section). However, a fundamental limitation is that the target must be detected in an interference background that, as a minimum, consists of the ever present receiver thermal noise. Because receiver noise is a random process, it must be described in terms of its statistical properties. Therefore, radar detection is a statistical problem, and when specified, the radar range must be stated in a statistical manner to be meaningful.

Marcum [1] and Swerling [2] were among the first to develop a consistent statistical theory of radar detection against a noisy receiving background. The two statistical parameters they used to qualify the statement of a radar's detection range were the *probability of detection* (e.g., $P_d = 0.9$) and the *probability of false alarm* (e.g., $P_{fa} = 10^{-6}$). The Marcum and Swerling (M-S) theory is documented by extensive sets of curves and computer-generated tabulations (see Example 32) that list the required S/N ratio to provide a specified P_d and P_{fa} for a given number of integrated pulses and Swerling target types (i.e., fluctuating and fixed) [3].

Calculation of the detection performance of a practical radar in a noisy background can be determined from the M-S theory by introducing the concept of detection loss. Detection loss refers to the additional S/N required to achieve a specified performance level relative to the M-S curves under conditions that deviate from the idealized ones embodied in the derivation of the standard curves. Many types of detection losses have been identified, and when these are considered, very accurate predictions of radar performance can be obtained that provide good agreement with actual test results [4].

The probability of detection analysis against a receiver noise background developed by Marcum and Swerling is not directly applicable to Doppler-type radars, which must operate in an interference background that contains clutter as well as receiver noise. The basic reason for this is that, because receiver noise (i.e., white noise) is uncorrelated from pulse to pulse, the corresponding analysis can be based on well-known statistical techniques for integrating independent statistics. This is not the case for Doppler radars because they must operate against clutter, which is generally highly correlated from pulse to pulse (the cancellation of fixed echoes depends on this). Even if the clutter were not so correlated, independence would be destroyed by the pulse-combining action of the Doppler filters. Consequently, the corresponding analysis for Doppler radars is more involved and therefore less highly developed than the M-S analysis.

The present state of detection analysis for Doppler radars in a clutter and noise background follows. First, a general theory for determining the probability of detection for a Doppler radar employing an optimum Doppler processor (i.e., transversal filter), as shown in Figure 3.1, is available and is discussed in Chapter 5 (see Section 5.1) [5]. This theory is useful because it establishes an upper bound on the detection probability of a Doppler radar. The theory also allows the detection probabilities to be determined for Doppler radars employing nonoptimum Doppler processors, which follow the general structure of a coherent linear Doppler filter followed by an envelope detector (see Section 5.1). For example, the analysis allows the determination of the detection probabilities of a coherent MTI followed by an envelope detector and an MTI-coherent integrator (FFT coherent filter bank) cascade, followed by an envelope detector, as depicted in Figure 3.2.

The problem of computing the probability of detection for a coherent MTI (or MTI-coherent integrator cascade), followed by a postdetection noncoherent integrator, as depicted in Figure 3.3, is difficult but susceptible to an analysis that decorrelates the correlated MTI output by using an orthogonal transformation to determine detection thresholds (see Section 5.3) [6]. To determine the probability of detection, we must expand the signal plus interference probability density into a Gram-Charlier series [5]. The tedious nature of this theory makes it more adaptable to the calculation of special situations rather than the generation of generalized results, such as those available from the M-S analysis.

This theory has been applied to two special situations. First, it has been used to determine loss when a target in receiver noise is passed through an MTI and

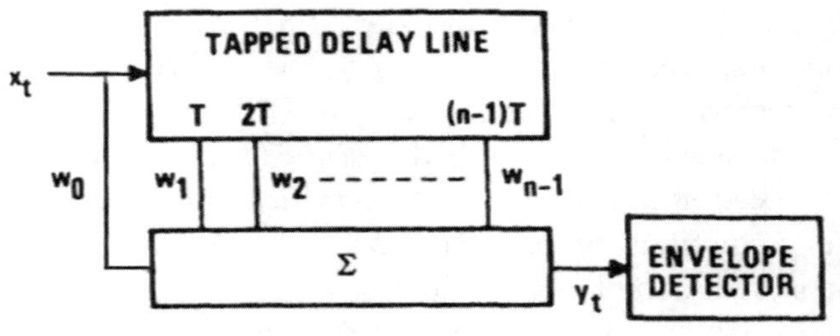

Figure 3.1 Optimum transversal Doppler filter processor. (*From:* [5, p. 157].)

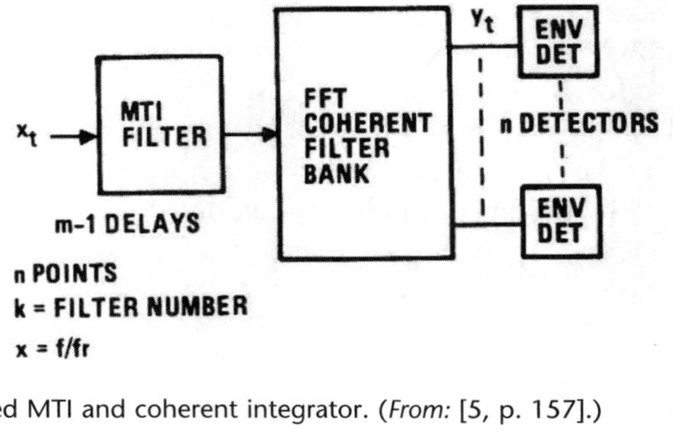

Figure 3.2 Cascaded MTI and coherent integrator. (*From:* [5, p. 157].)

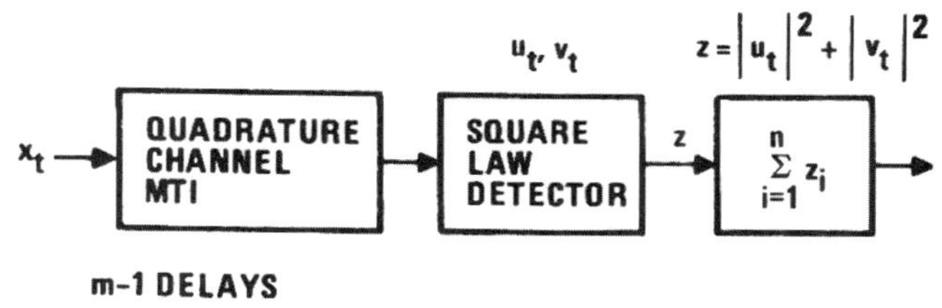

Figure 3.3 Cascaded MTI and noncoherent integrator. (*From:* [5, p. 157].)

then noncoherently integrated (see Section 5.3) [7]. Second, the theory has been applied to determine the detection loss when a target in clutter and receiver noise is passed through an MTI and then noncoherently integrated [5]. Note that the first case is a subset of the second case and also that the detection loss associated with the clutter plus noise situation is a function of the MTI's capability to suppress the clutter. A typical curve depicting the detection loss for a coherent single canceller MTI followed by a video integrator is given in Figure 3.4.

The next level of complexity in Doppler radar probability of detection analysis involves a noncoherent MTI where the MTI filtering occurs after envelope detection. Analytical techniques are available that would allow probability of detection calculations for this type of MTI by using a square-law envelope detector [8]. However, the complexities of this technique are such that it has been generally applied to second-moment (power) analyses rather than to determine detection probabilities [9, 10].

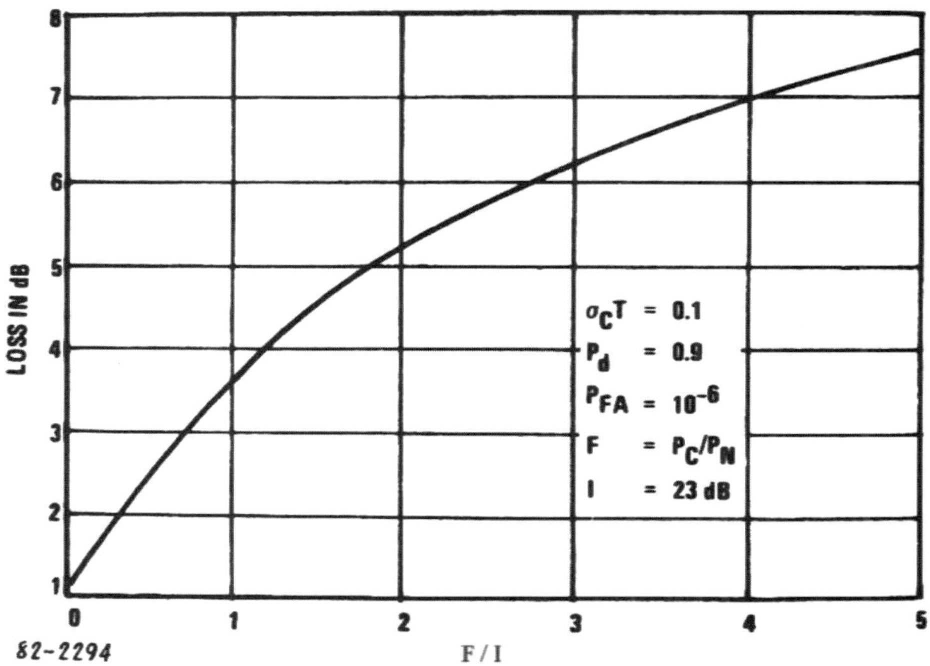

Figure 3.4 Video integration losses with a single canceller. (*From:* [5, p. 158].)

In practice, most Doppler radar processing techniques are compared on the basis of their ability to enhance the ratio of the target signal power to the interfering clutter power background. To facilitate this comparison, several figures of merit have been defined by the IEEE standards and in other references. The primary figure of merit employed is called the improvement factor, which generally encompasses the other criteria. The improvement factor criterion was generally developed by Emerson (who called it the *reference gain*) and accounts for both the attenuation of the clutter and enhancement of the target response by the Doppler filter [10]. Because the improvement factor criterion was initially developed for use with MTI systems, it assumes that the target's velocity is uniformly distributed over all possible values and ignores the effect of receiver noise on system performance. We can modify this criterion to apply to both receiver noise and clutter and, in certain situations, to relate to probability of detection analysis.

3.1 The MTI Improvement Factor (I_c)

The MTI improvement factor is defined as the signal-to-clutter (S/C) ratio at the output of the MTI system divided by the signal-to-clutter ratio at the input of the system, averaged uniformly over all target radial velocities of interest, which can be expressed as follows:

$$I_c = \frac{\left(\overline{S/C}\right)_o}{\left(\overline{S/C}\right)_i} \tag{3.1}$$

The improvement factor can be rewritten as the product

$$I_c = \left(\overline{S_o/S_i}\right)\frac{C_i}{C_o} \tag{3.2}$$

where the first term

$$\overline{G} = \left(\overline{S_o/S_i}\right) \tag{3.3}$$

is called the MTI *gain* (the average MTI response to targets) and the second

$$CA = \left(\frac{C_i}{C_o}\right) \tag{3.4}$$

is called the *clutter attenuation*. In this form, we can see that the improvement factor embodies both clutter attenuation and target enhancement effects in its definition. As defined, the improvement factor is independent of the target velocity and depends only on the weights of the MTI, the power spectrum of the clutter, and the number of pulses processed.

Another related MTI performance criterion is called the subclutter visibility (SCV), which is defined as the ratio by which the target echo power may be weaker than the coincident clutter echo power and still be detected with specified detection and false alarm probabilities. Target and clutter powers are measured on a single pulse return and all target radial velocities are assumed equally likely. The SCV is a measure of an MTI system's ability to detect a moving-target signal superimposed on clutter signals.

A subclutter visibility of 40 dB implies that a moving target can be detected in the presence of clutter, although the clutter echo power is 10,000 times the target echo. Two radars with the same SCV might not have the same ability to detect targets in clutter if the clutter resolution cell of one is greater than the other, causing it to accept a greater clutter power. Thus, a radar with a 10 μs pulse length and 10° beamwidth would generally need 20 dB more SCV than a radar with a 1 μs pulsewidth and 1° beamwidth for equal performance in a specified cluttered environment.

The subclutter visibility of a radar is generally less than the improvement factor associated with that radar. For example, assume that a 1 m^2 target was detected with $P_d = 0.9$ and $P_{fa} = 10^{-6}$ in a clutter-free environment. The radar would then require a subclutter visibility of 20 dB to detect the same target in an environment where the clutter background had an equivalent RCS of 100 m^2. However, an improvement factor of 20 dB in the radar's Doppler processor would produce an S/C ratio of only one at its output, which is insufficient to produce the specified P_d. This situation is further complicated if the radar employs postdetection integration, because the integration gain will be greater against noise than against clutter, which generally is highly correlated.

In the case of a Doppler processor followed by an envelope detector, the subclutter visibility can be directly related to the processor's improvement factor. A generalized relation (see Section 5.1.1) for the probability of detection of a Doppler processor is given by

$$P_d = Q\left\{\left[\frac{2P_s I_f}{P_n(1+F)}\right]^{1/2}, \left[2\ln\left(\frac{1}{P_{fa}}\right)\right]^{1/2}\right\} \tag{3.5}$$

where $F = P_c/P_n$ is the clutter-to-noise power ratio, P_s and P_n are the signal and noise powers, I_f is the interference improvement factor, P_{fa} is the probability of false alarm, and $Q(\cdot, \cdot)$ is Marcum's Q function [1]. In the preceding relationship, the interference improvement factor is given by

$$I_f = \frac{\mathbf{w}^T \mathbf{M}_s \mathbf{w}^*}{\mathbf{w}^T \mathbf{M}_x \mathbf{w}^*} \tag{3.6}$$

where $\mathbf{M}_s$ and $\mathbf{M}_x$ are the signal and interference covariance matrices, and $\mathbf{w}^T$ (transpose) and $\mathbf{w}^*$ (complex conjugate) form the complex weight vectors. Note that the Doppler processor functions to improve both the S/C and S/N ratios. This accounts for the use of I_f (the interference improvement factor) rather than I_c (the MTI improvement factor), which operates just on the clutter alone.

The interference improvement factor (I_f) can be expressed in terms of the MTI clutter improvement factor (I_c) as

$$I_f = \frac{1+F}{F} \cdot \left(\frac{I_c}{1 + \frac{I_c}{F}} \right) \tag{3.7}$$

Thus, as F (the clutter-to-noise power ratio) becomes large, I_f tends to approach I_c. Also, for a very large I_c, the maximum improvement against interference approaches $F + 1$, indicating a maximum improvement relating to the clutter-to-noise power ratio.

The general improvement factor relationship (I_f) can be specialized to the MTI improvement factor (I_c) case by setting $\mathbf{M}_s = \mathbf{I}$ (all target velocities are equally likely) and $\mathbf{M}_x = \mathbf{M}_c$, where $\mathbf{I}$ is the identity matrix and $\mathbf{M}_c$ is the clutter's covariance matrix. The expression for the probability of detection then becomes

$$P_d = Q\left\{ \left[\frac{2P_s}{P_n\left(1 + \frac{F}{I_c}\right)} \right]^{1/2}, \left[2 \ln\left(\frac{1}{P_{fa}}\right) \right]^{1/2} \right\} \tag{3.8}$$

which demonstrates that as the MTI improvement factor becomes large, the detection performance approaches that of the noise-limited case (i.e., the MTI provides no improvement in the S/N ratio).

The MTI improvement factor for the coherent MTI case can be evaluated as

$$I_c = \frac{\sum_{j=0}^{n-1} \mathbf{w}_j^2}{\sum_{j=0}^{n-1} \sum_{k=0}^{n-1} \mathbf{w}_j \mathbf{w}_k \, \rho_c(j-k)} \tag{3.9}$$

where

$$\overline{G} = \sum_{j=0}^{n-1} \mathbf{w}_j^2 \tag{3.10}$$

is the MTI gain, $\mathbf{w}_j$, are the MTI's real weights, $\rho_c(\cdot)$ is the clutter's correlation coefficient, and n is the number of pulses processed by the MTI. For a Gaussian-shaped spectrum, the correlation coefficient is given by

$$\rho_c(i-j) = \exp\left[\frac{-(i-j)^2 \Omega^2}{2} \right] = \rho_c(1)^{(i-j)\,2} \tag{3.11}$$

where $\Omega = \sigma_c T$ while σ_c is the standard deviation of the clutter spectrum in rad/s, and T is the radar's interpulse period.

Explicit evaluation of the MTI improvement factor requires a specification of the weights used in the MTI. Optimum weights can be determined by using Emerson's procedure (see Section 5.1.2) [10]. However, in practice, binomial weights (the binomial coefficients with alternating signs) are generally employed, because they provide performance approaching the optimum (for $n = 2$ pulses the binomial weights are optimum), and the values for the optimum and binomial weights are very nearly equal for narrow clutter spectra (i.e., $\sigma_c T \to 0$).

Table 3.1 lists the MTI gain and binomial weights for values of $n = 2$ to 8 pulses processed by the MTI. The binomial weight improvement factor for N delay sections ($N = n - 1$) is given by [9, p. 23]

$$I_c = \frac{1}{1 - 2\frac{N}{N+1}\rho_c(1) + 2\frac{N}{N+1}\frac{N-1}{N+2}\rho_c(2) + \cdots} \tag{3.12}$$

which can be approximated as

$$I_c = \frac{2^{n-1}}{(n-1)!(\sigma_c T)^{2(n-1)}}, \qquad \sigma_c T << 1 \tag{3.13}$$

The MTI improvement factor for binomial weights is depicted in Figure 3.5, and a comparison of the improvement factor for optimum and binomial weights is given in Figure 3.6. Note that the optimized weights provide the most advantage when: (1) more than two pulses are processed, (2) the clutter spectrum is narrow, and (3) the filter weights are accurate. (This follows from the small numerical difference for low $\sigma_c T$ between the optimum and binomial weights and relatively large difference in improvement factor.)

An alternative method for computing the MTI improvement factor in the spectral domain is given by

$$I_c = \frac{\overline{|H(\omega)|^2} \int_0^{2\pi f_r} S_i(\omega)\, d\omega}{\int_0^{2\pi f_r} |H(\omega)|^2 S_i(\omega)\, d\omega} \tag{3.14}$$

where $S_i(\omega)$ (the normalized clutter power spectral density) is

Table 3.1 MTI Power Gain and Binomial Weights

n	$G = \Sigma w_j^2$	*Binomial Weights*
2	2	1, −1
3	6	1, −2, 1
4	20	1, −3, 3, −1
5	70	1, −4, 6, −4, 1
6	252	1, −5, 10, −10, 5, −1
7	923	1, −6, 15, −20, 15, −6, 1
8	3432	1, −7, 21, −35, 35, −21, 7, −1

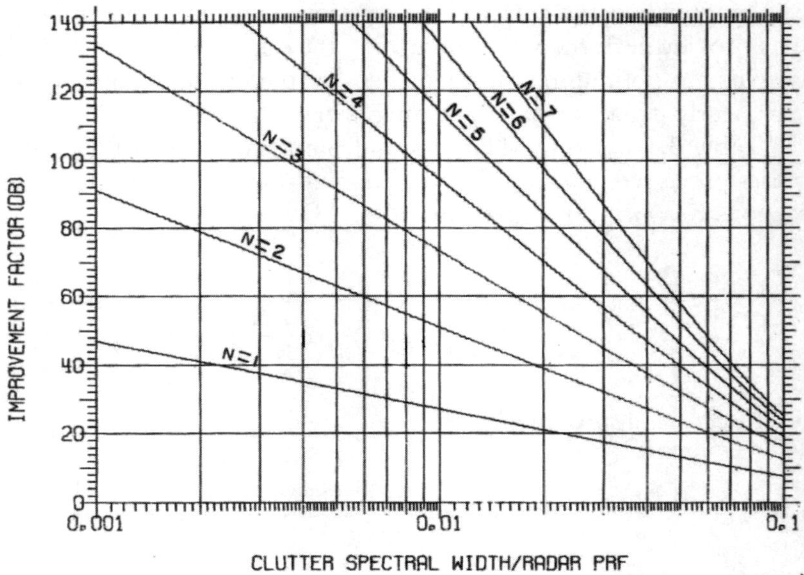

Figure 3.5 MTI improvement factor for binomial weights (n = 2 to 8). (*From:* [11, p. 10].)

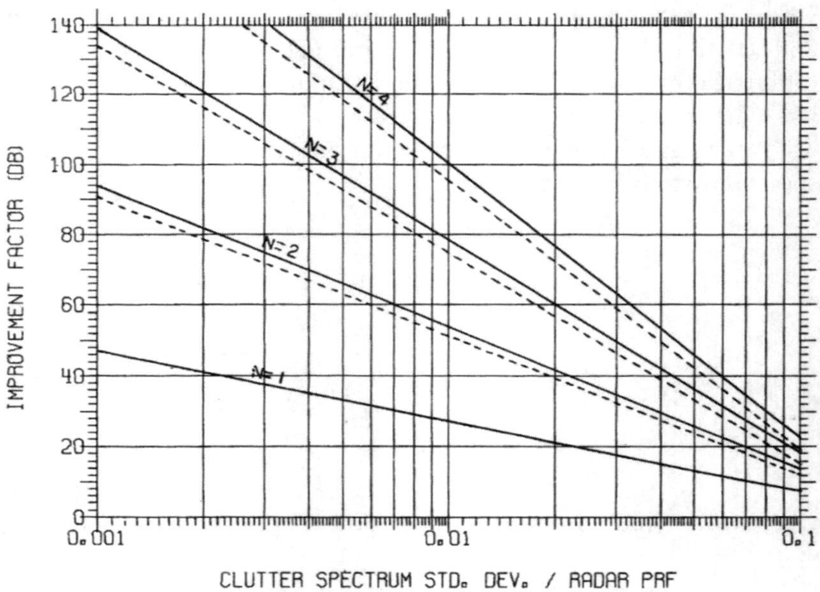

Figure 3.6 MTI improvement factor for optimum and binomial weights (N = 1 to 4). (*From:* [11, p. 10].)

$$S_i(\omega) = \frac{\exp\left(-\omega^2/2\sigma_c^2\right)}{\sqrt{2\pi\sigma_c^2}} \tag{3.15}$$

f_r is the radar's PRF, and $|H(\omega)|^2$ (the MTI filter's power transfer function for binomial weights) is

$$|H(\omega)|^2 = \left[2 \sin\left(\frac{\omega T}{2}\right)\right]^{2(n-1)} \tag{3.16}$$

Relationship (3.9) can be extended to compute the interference (clutter-plus-noise) improvement factor (I_f) by redefining $S_i(\omega)$ to be the clutter-plus-noise power spectral density and evaluating $|H(\omega)|^2$ at the frequencies of interest (see Section 5.2 for evaluation of the cascaded MTI and coherent integrator).

The MTI improvement factor of conventional surface-based search radars often is limited by the antenna scanning modulation induced onto the clutter return. The antenna scanning modulation causes a spreading of the clutter's spectrum, which has a maximum frequency that corresponds to the Doppler shift associated with the tip speed of the antenna's aperture. For analytic convenience, the antenna scanning clutter spectrum is customarily approximated by a Gaussian-shaped function, although this implies clutter frequency components greater than those associated with the antenna's tip speed. The MTI improvement factor due to antenna scanning modulation is best calculated using the spectral analysis method, as illustrated in the following text for a single-canceller type of MTI.

The one-way antenna pattern is assumed to be given by

$$G(\theta) = G_0 \exp\left(\frac{-2.773\theta^2}{\theta_B^2}\right) \tag{3.17}$$

where θ_B is the antenna's one-way 3 dB beamwidth. The scanning action of the antenna induces a Gaussian-shaped voltage modulation onto each clutter scatterer, which is given by $v(t) = v_o \exp(-2.773t^2/t_o^2)$, where the time on target is given by $t_o = n_B T$, n_B is the number of hits per 3 dB antenna beamwidth, and T is the radar's interpulse period. Next, the voltage modulation is Fourier transformed to find the voltage spectrum and then squared to find the clutter's power spectrum, resulting in

$$S(\omega) = P_o \exp\left[\frac{-(\omega t_o)^2}{2 \cdot 2.773}\right] \tag{3.18}$$

This can then be substituted into the relationship for the MTI improvement factor by using

$$|H(\omega)|^2 = \left[2 \sin\left(\frac{\omega T}{2}\right)\right]^2 \tag{3.19}$$

for the single-canceller MTI. However, because the clutter spectrum is Gaussian shaped, its standard deviation can be determined as $\sigma_c = 1.665/t_o$. Substitution of

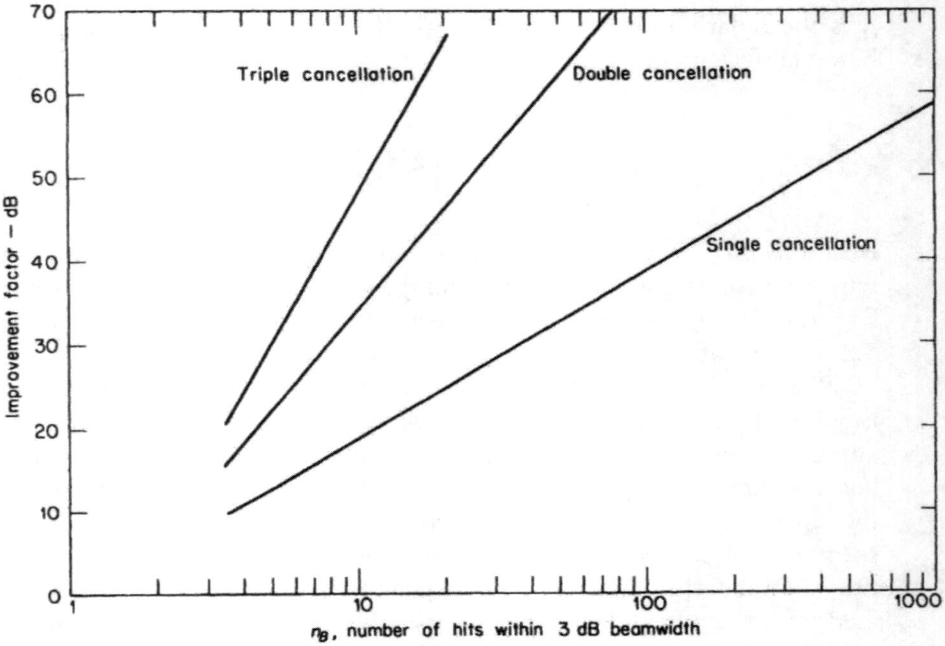

Figure 3.7 MTI improvement factor limitations due to antenna scanning. (*From:* [12, p. 135].)

$t_o = n_B T$ results in $\sigma_c T = 1.665/n_B$, which can be used to find the limitation to the MTI improvement factor caused by antenna scanning as

$$I_c = \frac{2}{(\sigma_c T)^2} = \frac{n_B^2}{1.386} \tag{3.20}$$

In a similar manner, the MTI improvement factor under an antenna scanning limitation for a double canceller can be determined as $I_c = n_B^4/3.844$ and a triple canceller as $I_c = n_B^6/15.985$. These values are plotted in Figure 3.7.

▶ Example 7 MTI Improvement Factor Requirements Analysis

The MTI improvement factor is defined as

$$I_c = \frac{\left(\frac{S}{C}\right)_{out}}{\left(\frac{S}{C}\right)_{in}}$$

where S is the signal power, C is the clutter power, and the equation is averaged over all target radial velocities of interest. This factor plays a central role in determining the detection range of MTI radar.

Examination of the radar equation reveals that the returned signal power is proportional to the radar cross section (σ_t) of the target. As described in Section 1.3.2, the power returned from the distributed clutter can similarly be described in terms of its equivalent radar cross section (σ_{cx}). This is logical because clutter is simply a conglomeration of a large number of target scatterers. Thus, we can form the ratio

$$\left(\frac{S}{C}\right)_{in} = \frac{\sigma_t}{\sigma_{cx}}$$

Furthermore, the amplitude probability distribution of the equivalent radar cross section of the clutter exhibits similar statistics to that of receiver noise (see Section 1.3.2). However, it is highly correlated from one sample to the next, which differentiates it from thermal noise. Because of this, property integration is not generally successful in significantly improving the signal-to-clutter ratio in contrast to its effect on the signal-to-noise ratio (see complete analysis in Example 18). The MTI detection situation is depicted in Figure EX-7.1. Note that the signal-to-clutter output ratio of the MTI has been conservatively replaced by the equivalent signal-to-noise ratio for a single pulse to allow utilization of detection statistics (P_d, P_{fa}, and Swerling target models) using the well-documented Marcum-Swerling theory. The result of this substitution is

$$\left(\frac{S}{C}\right)_{out} = \left[\frac{S}{N}\right]_1$$

where $[S/N]_1$ is the signal-to-noise ratio that applies for an appropriate single hit Swerling 0, 1, or 3 target model.

Swerling target model cases 0 (fixed target), 1, and 3 are generally appropriate for MTI-type radars because the transmitter frequency must be fixed for a time that corresponds to the number of pulses processed in the MTI processor.

The signal-to-noise ratio required to provide a specified P_d and P_{fa} for n integrated pulses on a steady target can be approximated by Albersheim's relationship. The relationship for a single pulse is given by

$$\left(\frac{S}{N}\right)_1 = A + 0.12AB + 1.7B$$

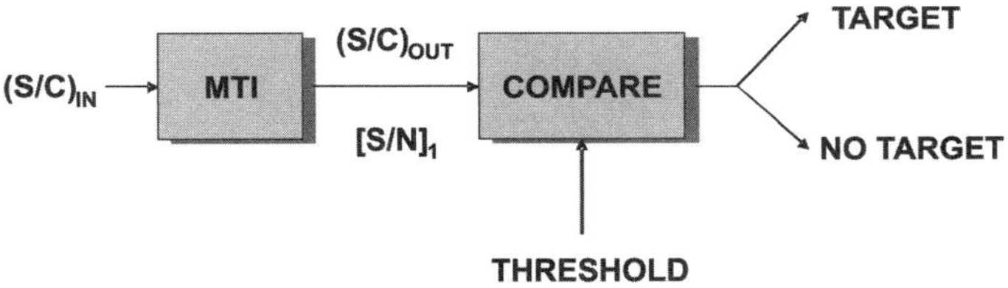

Figure EX-7.1 MTI detection block diagram.

where

$$A = \ln\left(\frac{0.62}{P_{fa}}\right) \qquad B = \ln\left(\frac{P_d}{1 - P_d}\right)$$

Further, the signal-to-noise ratio in decibels for n independent pulses can be approximated by

$$\left[\left(\frac{S}{N}\right)_n\right]_{dB} = -5\log(n) + \left(6.2 + \frac{4.54}{\sqrt{n + 0.44}}\right)\log\left(\frac{S}{N}\right)_1$$

This approximation generally has an error of less than 0.6 dB over the range of $P_d = 0.1$ to 0.9 with $P_{fa} = 10^{-3}$ to 10^{-6} and $n = 1$ to 8,096 (see Example 32).

To find the signal-to-noise ratio for Swerling cases 1 to 4 (fluctuating targets), a method from Barton is appropriate. The first step is to find the single-pulse ($n = 1$) signal-to-noise ratio for a Swerling case 1 target. This is given by

$$\left[\left(\frac{S}{N}\right)_1\right]_{dB}\Bigg|_{one} = 10\log\left(\frac{\ln(P_{fa})}{\ln(P_d)} - 1\right)$$

The fluctuation loss is defined as

$$FL = \left(\frac{S}{N}\right)_{dB}\Bigg|_{one} - 10\log\left(\frac{S}{N}\right)_1$$

The signal-to-noise ratio for the various Swerling-type targets are then found as

$$\left[\left(\frac{S}{N}\right)_n\right]_{dB}\Bigg|_{one} = \left[\left(\frac{S}{N}\right)_n\right]_{dB} + FL$$

$$\left[\left(\frac{S}{N}\right)_n\right]_{dB}\Bigg|_{two} = \left[\left(\frac{S}{N}\right)_n\right]_{dB} + \frac{FL}{n}$$

$$\left[\left(\frac{S}{N}\right)_n\right]_{dB}\Bigg|_{three} = \left[\left(\frac{S}{N}\right)_n\right]_{dB} + \frac{FL}{2}$$

$$\left[\left(\frac{S}{N}\right)_n\right]_{dB}\Bigg|_{four} = \left[\left(\frac{S}{N}\right)_n\right]_{dB} + \frac{FL}{2n}$$

The MATLAB program det_factor.m using function det_factors (P_d, P_fa, hits) provides the signal-to-noise ratio for both steady and Swerling-type targets. The inputs are the desired P_d, P_{fa}, and the specified number of independent pulses integrated (hits). The outputs include the signal-to-noise ratio for each of the target types in matrix form (det_facts(i,:) for $i = 1, 2, 3, 4, 5$).

The required MTI improvement factor can then be written as

$$I_c = \frac{\left[\frac{S}{N}\right]_1}{\left(\frac{S}{C}\right)_{in}}$$

where $[S/N]_1$ is the single-hit signal-to-noise ratio based on a specified P_d, P_{fa}, and Swerling type target [1]. Integration effects are covered in Example 18, but generally the integration gain is negligible for land and sea clutter except for systems with large improvement factors that reduce the clutter residue significantly below the noise level. In addition, the filtering action of the MTI attenuates the signal causing a loss in target power. When this loss (L_{MTI}) is included, the required MTI improvement factor can be expressed as

$$I_c = \frac{\left[\frac{S}{N}\right]_1 L_{\text{MTI}}}{\left(\frac{S}{C}\right)_{in}}$$

This equation can be used to synthesize the improvement factor necessary to provide a specified P_d and P_{fa}. When expressed in decibels this equation takes the following form:

$$I_{\text{dB}} = \left[\frac{S}{N}\right]_1\Bigg|_{\text{dB}} + L_{\text{MTI}}\Bigg|_{\text{dB}} - \left(\frac{\sigma_t}{\sigma_{cx}}\right)\Bigg|_{\text{dB}}$$

The next step is to determine the equivalent radar cross section of the clutter. For surface clutter using the pulse-limited case (see Section 4.1), which is applicable for low and medium grazing angles (ψ), the equivalent radar cross section of the clutter is given by

$$\sigma_{cx} = \sigma^0 R\,\theta_{az}\left(\frac{c\tau_e}{2}\right)\sec(\psi)$$

where σ^0 is the dimensionless backscatter coefficient that represents the radar cross section per unit area of surface, $(c\tau_e/2)$ is the range equivalent to a radar pulsewidth, τ_e is the radar pulsewidth, θ_{az} is the azimuth beamwidth in radians, ψ is the grazing angle, and R is the range to the surface clutter resolution cell containing the target. The equivalent radar cross section of volume clutter is given by

$$\sigma_{cx} = \eta R_\tau \theta_{az}\phi_{el} R_\tau^2$$

Table EX-7.1 ASR-9 Radar Parameters

Parameter	*Value*
Pulsewidth	1 μs
Probability of detection	0.9
Probability of false alarm	10^{-6}
Antenna height	20m
Azimuth beamwidth	1.3°
Elevation beamwidth	4.8°
Target RCS	1 m^2
Rain rate	4 mm/h
Frequency	3 GHz

where η is the volume clutter reflectivity (m^2/m^3), ϕ_{el} is the elevation beamwidth in radians, and the rest of the parameters are the same as for surface clutter. Note that for pulse compression radars, the compressed pulsewidth should be used to determine R_{τ}.

Using the MATLAB program mti_impfac.m, the improvement factor requirement for the ASR-9 radar is determined below. The pertinent parameters of the ASR-9 radar are shown in Table EX-7.1. Albersheim's approximation is used to determine the S/N ratio required to provide a specified P_d and P_{fa}. Both surface and volume clutter are considered. The radar horizon is computed for surface clutter since this limits the extent of the clutter.

The User Inputs to the Program:

- Swerling-type target (0, 1, 3)
- Range (km)
- Rain Rate (mm/h)

The User Outputs:

- Required improvement factor (surface)
- Required improvement factor (volume)

Example of MATLAB Printout:

```
swerl_type(0,1,3) = 1
R_hor(km) = 18.4524
Range(km) = 18.4524
sigma_0(dB) = -30
impfac_sur(dB) = 39.1233
rain rate(mm/hr) = 4
impfac_vol(dB) = 16.1426
```

▶

3.1.1 Noncoherent MTI Improvement Factor

The available MTI improvement factor for a noncoherent MTI (recall Figure 2.17) can be found by using a procedure similar to that used for a coherent MTI. The basic difference between coherent and noncoherent MTI is that the former employs synchronous (or phase) detection prior to MTI filtering, whereas the latter uses envelope detection. The synchronous detection used in the coherent MTI is a linear heterodyning operation that preserves the RF spectral shapes of the various target, clutter, and noise spectra. However, the envelope detection used in the noncoherent MTI widens the video target and clutter spectra, which generally results in lower performance than is available from a coherent MTI.

The noise-suppressed video spectra for coherent MTI and noncoherent MTI employing a square-law envelope detector are depicted in Figure 3.8. The essential action of the square-law envelope detector is to cause the coherent MTI spectrum to be convolved with itself [9, 13]. For Gaussian-shaped spectra, the resulting noncoherent MTI clutter spectrum ($C \times C$) has a variance given by $\sigma_{nc}^2 = 2\sigma_c^2$, whereas the target spectrum ($C \times T$) is $\sigma_{tnc}^2 = \sigma_t^2 + \sigma_c^2$ where σ_c^2 is the variance of the coherent spectrum and σ_t^2 is the variance of the coherent target spectrum. Note that the convolution action also causes the noncoherent MTI to have an additional target sideband, which is not present in a coherent MTI.

The MTI improvement factor for a noncoherent MTI with binomial weights can be deduced from the expression for the MTI improvement factor of a coherent MTI by multiplying by a factor of 2 (due to the additional sideband) and using the noncoherent MTI spectral variance $\left(\sigma_{nc}^2 = 2\sigma_c^2\right)$ as

$$I_{nc} = \frac{2 \cdot 2^{n-1}}{\left(\sqrt{2}\sigma_c T\right)^{2(n-1)}(n-1)!} = \frac{2}{(\sigma_c T)^{2(n-1)}(n-1)!} \tag{3.21}$$

for $\sigma_c T \ll 1$. The resulting MTI improvement factor for both coherent and noncoherent MTIs with binomial weights is tabulated in Table 3.2 where σ_c^2 is the

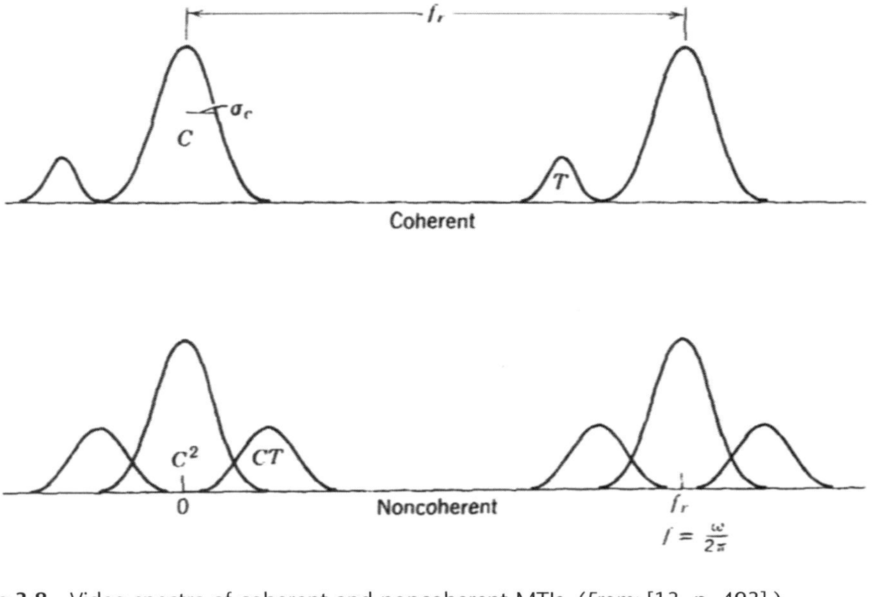

Figure 3.8 Video spectra of coherent and noncoherent MTIs. (*From:* [13, p. 493].)

Table 3.2 Coherent and Noncoherent MTI Improvement Factors for Binomial Weights ($\sigma_c T \ll 1$)

n	*Coherent*	*Noncoherent*
2	$2/(\sigma_c T)^2$	$2/(\sigma_c T)^2$
3	$2/(\sigma_c T)^4$	$1/(\sigma_c T)^4$
4	$4/3(\sigma_c T)^6$	$1/3(\sigma_c T)^6$

clutter spectrum variance and T is the radar's interpulse period. The resulting scanning limitations (n_B = number of pulses per 3 dB beamwidth) are listed in Table 3.3. Note that the improvement factors and scanning limitation are equal for a single MTI canceller ($n = 2$), whereas the coherent MTI provides progressively greater performance as the order (number of delay lines employed) of the MTI increases. Alternatively, a general expression for the noncoherent MTI improvement factor can be shown to be given by

$$I_{nc} = \frac{2\sum_{j=0}^{n-1} \mathbf{w}_j^2}{\sum_{j=0}^{n-1}\sum_{k=0}^{n-1} \mathbf{w}_j \mathbf{w}_k \rho_c^2(j-k)} \tag{3.22}$$

where $\mathbf{w}$ denotes the MTI weights and $\rho_c(\cdot)$ is the clutter's correlation coefficient (the ρ_c^2 term results from spectral convolution in the frequency domain that translates to multiplication in the time domain) [10].

The identical improvement factor provided by the single canceller ($n = 2$) for both the coherent and noncoherent MTI might suggest that the detection performance of the two types is equivalent. However, the detection performance of the coherent MTI is known to be optimum (see Section 5.1.2); hence, the performance of the suboptimal noncoherent MTI must be lower than that of the coherent MTI. The explanation for this apparent inconsistency is that the improvement factor criterion is based on a second moment analysis that compares only signal-to-clutter power, ignoring the shape of the pdfs involved under the alternate hypotheses. Hence, the translation of an MTI's improvement factor into a probability of detection is different for noncoherent and coherent MTIs, precluding a direct comparison of the two types on this basis. Unfortunately, the non-Gaussian nature of the

Table 3.3 Improvement Factor Scanning Limitation for Coherent and Noncoherent MTIs with Binomial Weights ($\sigma_c T \ll 1$)

n	*Coherent*	*Noncoherent*
2	$n_B^2/1.386$	$n_B^2/1.386$
3	$n_B^4/3.844$	$n_B^4/7.688$
4	$n_B^6/15.985$	$n_B^6/63.940$

probability distributions in the noncoherent MTI does not allow an analytical solution that represents the probability of detection in terms of the improvement factor.

These results for the MTI improvement factor of a noncoherent MTI primarily apply when a square-law envelope detector (the square of the envelope) is employed. An analysis of a noncoherent MTI employing a linear envelope detector indicates that the MTI improvement factor is reduced from that provided by a square-law envelope detector [14]. We can show that the clutter power residue from a single–canceller, noncoherent MTI using a linear envelope detector is given by

$$C_o = 4\sigma^2[1 - F(\rho)] \tag{3.23}$$

where $2\sigma^2$ is the input clutter power before envelope detection;

$$F(\rho) = \left(\frac{\pi}{4}\right) {}_2F_1\left(-\frac{1}{2}, -\frac{1}{2}, 1, \rho^2\right) \tag{3.24}$$

where ${}_2F_1(\cdot, \cdot, \cdot, \cdot)$ is the Gaussian hypergeometric function, and ρ is the clutter correlation coefficient [14]. This generalizes for higher order MTIs so that the clutter power residue is given by $C_o = \mathbf{w}^T\mathbf{M}'_c\mathbf{w}^*$ where $\mathbf{M}'_c$ is the coherent MTI's covariance matrix ($\mathbf{M}_c$) with ρ replaced by $F(\rho)$. Use of this relationship in combination with a Monte Carlo simulation to determine the MTI's target responses provides the curves depicted in Figure 3.9, which show the noncoherent MTIs' improvement factors for square-law and linear envelope detectors [14].

The curves depicted in Figure 3.9 show a deterioration of the MTI improvement factor for noncoherent MTIs that use a linear envelope detector instead of a square-law envelope detector. The decrease in performance varies with the clutter's spectral spread and the order (number of delay lines) of the MTI filter employed. This is significant for noncoherent MTIs because many practical detectors operate in the square-law region for small signals and the linear region for large applied signals. Because clutter signals are usually much larger than receiver noise, most practical detectors employed in noncoherent MTIs will probably work in the linear region.

3.1.2 The Cancellation Ratio

The cancellation ratio (CR) is a commonly used yardstick for measuring the instability noise generated within a Doppler radar and its effect on overall system performance. The CR is defined as that amount of attenuation creditable to the Doppler processor for truly fixed target returns when the latent amplifications (e.g., MTI gain) or losses have been normalized. The CR sets an upper bound on the improvement factor achievable by a particular Doppler radar design.

The CR is sometimes determined by presenting a completely correlated signal to the Doppler processor and measuring the power residue in the output. The input power is then divided by this residue, and the ratio, after accounting for the latent gain or loss, is the cancellation ratio. This measurement gives the CR for the Doppler processor, but does not account for any noise introduced by the transmitter

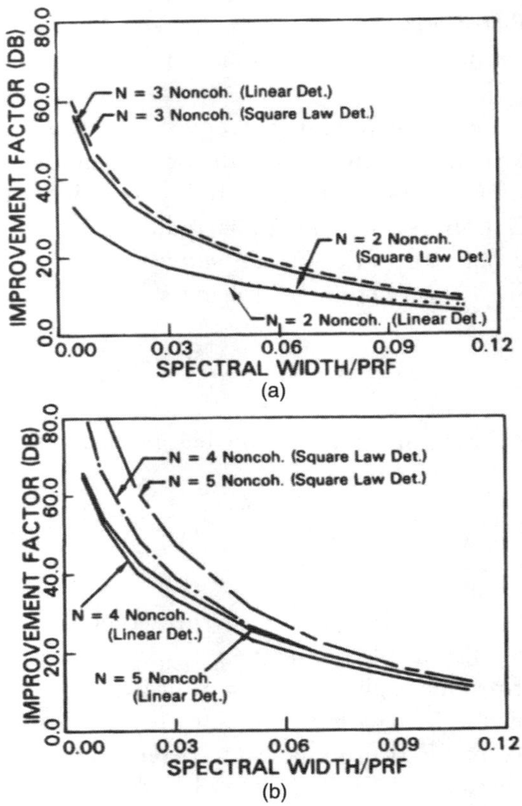

Figure 3.9 MTI improvement factors for noncoherent MTIs using linear and square-law envelope detectors. (*From:* [14, p. 402].)

or receiver. A more complete measurement involves searchlighting the radar onto a truly fixed target, but such a target is hard to find and great care must be exercised in performing the experiment so that the results will be meaningful. Many times, by judiciously combining both the direct and indirect measurements, we can determine the CR with reasonable accuracy.

There are many possible causes of system instability in both the transmitter and receiver. These instabilities are due to the inevitable imperfections in installed and operating equipment that give rise to system instability noise. The most prominent sources of system instability noise are:

- Pulse-time jitter;
- Pulsewidth jitter;
- Pulse amplitude jitter;
- Pulse distortion;
- Receiver phase and gain variations;
- Frequency modulation within the transmitted pulse;
- Transmitter, STALO, or COHO frequency drift;
- COHO locking instability;

- Gating noise;
- Quantization noise present in digital systems.

Note that pulse-time jitter does not generally apply to Doppler radars using digital processors where the system clock is started from the detected RF envelope of the transmitted pulse. Also, frequency modulation within the pulse and COHO locking instability generally apply to coherent-on-receive MTI systems (see Figure 2.21) and not MOPA systems (see Figure 2.20). In pulsed oscillator systems, the transmitter often exhibits frequency modulation (i.e., phase runout) within the pulse, causing the COHO to lock to a frequency which differs from that of the leading edge of the transmitted pulse. Generally, system instabilities are related to any mechanism that induces pulse-to-pulse (i.e., relative to the radar's interpulse interval) variations onto fixed target signals presented to the Doppler processor.

The CR limitation due to various peak pulse-to-pulse instabilities is listed in Table 3.4 [15]. In a pulse compression system, the compressed pulsewidth ($\tau_e = 1/B$) is reduced by the factor $\beta = B \cdot \tau$ from the uncompressed pulsewidth (τ), where B is the chirp bandwidth. The power residue in a pulse compression system for pulse timing and pulsewidth jitter is thus increased by this factor (β), which results in the CR being reduced by the time bandwidth factor $B\tau$. For simple pulsed systems, the time-bandwidth factor equals unity ($B\tau = 1$), which should be substituted into the expressions for the cancellation ratios for these types of systems. For phase-coded pulse compression systems (e.g., Barker code), the effective pulsewidth is the uncompressed pulsewidth divided by the number of elements in the code (e.g., $\tau_e = \tau/13$ for a 13-element Barker code) and, hence, the number of elements of the code is substituted for $B\tau$ (e.g., $B\tau = 13$) in these types of systems. Other factors used in Table 3.4 are Δf = interpulse frequency change, T_r = transmission time to and from the target, $\Delta\phi$ = interpulse phase change, and n = number of bits utilized in the A/D converter. For random fluctuations, the peak values

Table 3.4 Cancellation Ratio Limitation Due to System Instabilities

Pulse-to-Pulse Instability	*Cancellation Ratio*
Transmitter frequency	$(\pi \Delta f \tau)^{-2}$
STALO or COHO frequency	$(2\pi \Delta f T_R)^{-2}$
Transmitter phase shift	$(\Delta\phi)^{-2}$
COHO locking	$(\Delta\phi)^{-2}$
Pulse timing	$\dfrac{\tau^2}{[(\Delta t)^2 2B\tau]}$
Pulsewidth	$\dfrac{\tau^2}{[(\Delta t)^2 B\tau]}$
Pulse amplitude	$\left(\dfrac{A}{\Delta A}\right)^2$
Quantization noise	$\frac{3}{4}(2^n - 1)^2$

utilized in Table 3.4 can be replaced by their rms value to compute the average CR limitation (e.g., $\sigma_\phi = \sqrt{2}\,\Delta\phi$, $\sigma_\tau = \sqrt{2}\,\Delta\tau$, and so on) [10].

Doppler radars using digital signal processing are susceptible to quantization errors that occur in the system's A/D converter [36]. The A/D converter converts each sample to a binary number by comparing its voltage with a succession of progressively higher voltages of precisely known value. When the closest of these is found, the converter displays a binary number equal to the known voltage. Figure 3.10 illustrates the effect of applying a linearly changing voltage to the A/D converter whereby the output consists of the sum of two signals: a quantized replica of the original analog signal and a triangular error wave having peak amplitude equal to half the value of the least significant digit (LSD) or quantization step. For the uniformly distributed error depicted in Figure 3.10, the rms value of the quantization error voltage is given by LSD/$\sqrt{12}$.

In a coherent MTI, the A/D converter samples the bipolar video output of the MTI's synchronous detector (recall Figure 2.28). For an n-bit A/D converter, the LSD is given by $2v_p/(2^n - 1)$, where v_p is the peak signal level available from the MTI's synchronous detector. The rms value of the quantization noise is then given by

$$\sigma_q = \frac{2v_p}{\sqrt{12}\,(2^n - 1)} \tag{3.25}$$

If we assume that quantization noise behaves like thermal noise, the pulse-to-pulse power variation is given by $E(n_1 - n_2)^2 = 2\sigma_q^2$, resulting in an rms variation equal to $\sqrt{2}\sigma_q$. The fixed target signal out of the bipolar phase detector (synchronous detector) can be represented as $v_p \cos \phi$, where ϕ is the uniformly distributed phase difference between the target return signal and the MTI's coherent reference. The average fixed signal power from the synchronous detector is given by

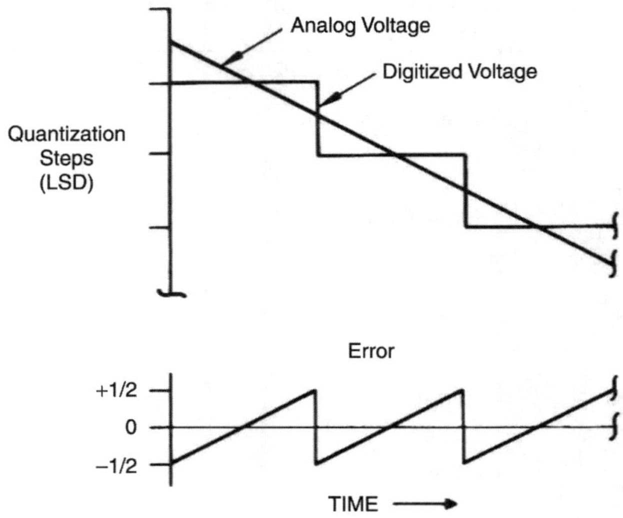

Figure 3.10 Quantization error for A/D converter. (*From:* [16, p. 312].)

$E(v_p \cos \phi)^2 = v_p^2/2$ resulting in an rms value of $v_p/\sqrt{2}$. The resulting cancellation ratio can then be derived [15] as the fixed signal power divided by the residue quantization noise power, which provides

$$\text{CR} = \frac{\left(\frac{v_p}{\sqrt{2}}\right)^2}{\left[\frac{2\sqrt{2}v_p}{(2^n - 1)\sqrt{12}}\right]^2} = 0.75\,(2^n - 1)^2 \tag{3.26}$$

The quantization noise CR limitation is tabulated in Table 3.5 as a function of the number of bits (n) employed in the A/D converter.

From an analytical viewpoint, a convenient approach is to break the fixed target returns into two parts: an unperturbed part (s_t) that is not affected by system instabilities and a perturbed part (n_t) caused by the system instabilities. The cancellation ratio is then given by

$$\text{CR} = \frac{\int S_s(\omega)\, d\omega}{\int \sum_{n=1}^{N} S_n(\omega)\, d\omega} \tag{3.27}$$

where $S_s(\omega)$ is the spectrum of the unperturbed part, $S_n(\omega)$ is the spectrum of each noiselike additive independent perturbation element, and N is the number of perturbation sources. Interchanging integration and summation results in

$$\text{CR} = \frac{1}{\frac{1}{\text{CR}_1} + \frac{1}{\text{CR}_2} + \ldots + \frac{1}{\text{CR}_n}} \tag{3.28}$$

where CR_n is the cancellation ratio for each individual effect. This expression allows the overall CR to be determined from the CRs of each individual effect [9].

Table 3.5 Cancellation Ratio Limitation for Quantization Noise

Number of Bits n	*Cancellation Ratio Limitation (dB)*
4	22.3
5	28.6
6	34.7
7	40.8
8	46.9
9	52.9
10	59.0
11	65.0
12	71.0

The effect of the CR on the overall improvement factor can be derived by assuming that the system instabilities (n_t) modulate the return signal. Because the instabilities are independent of the clutter returns, the autocorrelation function (ACF) of the received signal can be written as the product $R_{s+n}(\tau)\rho_c(\tau)$, where $R_{s+n}(\tau)$ is the ACF of the sum of the perturbed and unperturbed signal components and $\rho_c(\tau)$ is the normalized ACF of the clutter (see Section 3.1). The actual improvement factor is then given by

$$(I_{\text{actual}})^{-1} = \frac{\sum_{j=0}^{n-1} \sum_{k=0}^{n-1} \mathbf{w}_j \mathbf{w}_k \rho_c(j-k) \mathbf{R}_{s+n}(j-k)}{\mathbf{R}_{s+n}(0) \sum_{j=0}^{n-1} \mathbf{w}_j^2} \tag{3.29}$$

Substituting $R_{s+n}(j-k) = R_s(j-k) + R_n(j-k)$ and recognizing that $R_s(j-k)$ has a steady value $R_s(0)$ and $R_n(j-k)$ is a delta function with value $R_n(0)$ only at $j-k$ results in

$$(I_{\text{actual}})^{-1} = \frac{R_s(0)}{R_s(0) + R_n(0)} \left[I_{\text{ideal}}^{-1} + \frac{R_N(0)}{R_x(0)} \right] \tag{3.30}$$

Further, recognize that the CR = $R_s(0)/R_n(0)$ results in

$$(I_{\text{actual}})^{-1} = \left[\frac{\text{CR}}{1 + \text{CR}} \right] \left(I_{\text{ideal}}^{-1} + \text{CR}^{-1} \right) \tag{3.31}$$

which simplifies to

$$I_{\text{actual}}^{-1} = I_{\text{ideal}}^{-1} + \text{CR}^{-1} \tag{3.32}$$

for sufficiently large CR (15 dB or more). This relationship, which is plotted in Figure 3.11, indicates the effect of the CR on the ideal improvement factor (I_{ideal}), which is computed on the assumption that the instabilities have no effect on the clutter returns (i.e., CR $\rightarrow \infty$) [9].

As an example, consider a 3-GHz radar with a 100 nmi range that transmits a 2 μs pulse. We want to impose a 50 dB (316:1 voltage ratio) CR limitation on each instability component [15]. If a pulse oscillator is used, its peak pulse-to-pulse frequency change must be less than $\Delta f = 1/316\pi\tau = 504$ Hz, which is a stability of 2 parts in 10^7, and the COHO locking must be within $\Delta\phi = 1/316 = 3/16 \times 10^{-3} = 0.18°$. If a MOPA-type transmitter is used, the power amplifier must have a pulse-to-pulse phase change of less than $\Delta\phi = 1/316 = 0.18°$. The STALO or COHO pulse-to-pulse frequency change to provide performance at a 100 nmi range ($T_r = 12.4 \times 100 \times 10^{-6}$ s) must be less than $\Delta f = 1/632\pi T_r = 0.4$ Hz. This is short-term stability for the STALO of 1 part in 10^{10} and 1 part in 10^8 for the COHO assuming a 30 MHz COHO frequency. The pulse timing jitter must be less than

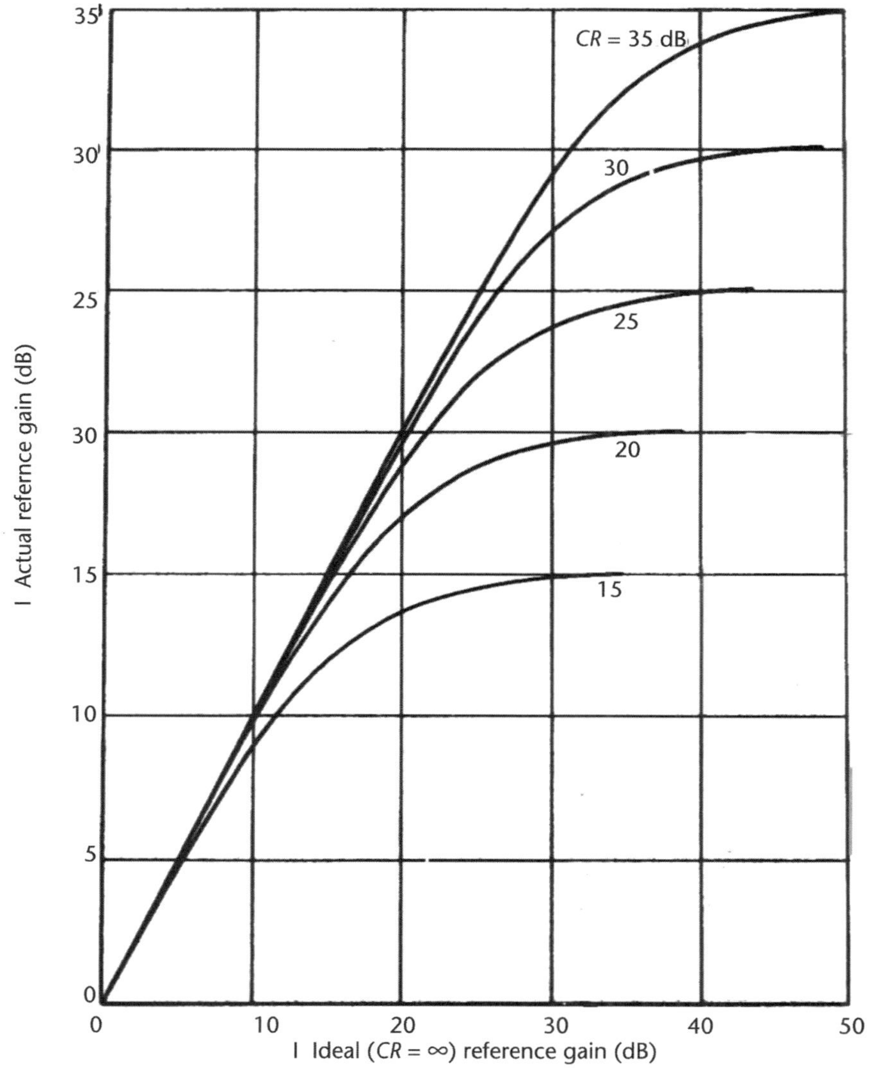

Figure 3.11 Improvement factor degradation due to system instability noise. (*From:* [10, p. 94].)

$\Delta t = \tau/316\sqrt{2} = 4.5 \times 10^{-9}$ s, the pulsewidth jitter less than $\Delta\tau = \tau/316 = 6 \times 10^{-9}$ s, and the pulse-to-pulse amplitude jitter less than $\Delta A/A = 1/316$ or 0.3%. For a digital processor, the A/D converter must use more than 9 bits.

3.2 Clutter Effects on Doppler Radar Performance

In the previous section, we discussed the general performance measures associated with Doppler radars and how they would affect system performance. The improvement factor was identified as the primary performance measure utilized to evaluate Doppler radars. For fully coherent systems followed by an envelope detector, the improvement factor is related directly to the probability of detection; otherwise, the improvement factor can be used as a measure to compare various alternative approaches. The improvement factor was shown to be composed of two components: one due to system instabilities, characterized as the cancellation ratio, and the other due to the effect of the Doppler filtering on the clutter signals, termed the ideal improvement factor (I_{ideal}). This section deals primarily with the determination of I_{ideal}, which is concerned with spreading the clutter spectrum due to factors induced by the radar as well as those intrinsic to the clutter.

The characteristic fluctuation of any particular patch of clutter return arises from the combined influence of several statistically independent factors. First among these is the relative motion of the reflecting scatterers constituting the clutter mass. This "internal motion"—the extent of which generally depends on wind velocity and such diverse factors as sea state, turbulence, and individual scatterer mobility—accounts for the basic randomness of the return.

In addition to the internal motion fluctuations, the echo intensity induced by the scanning motion of the antenna with respect to the individual clutter scatterers also will vary. The severity of this scanning noise depends on the ratio of the scanning rate to the antenna's beamwidth, which determines the time that the antenna illuminates a particular scatterer. As this dwell time is reduced, the scanning noise becomes more severe. The scanning modulation clutter spectrum is truncated at a Doppler frequency that corresponds to the tip speed of a mechanically scanning radar antenna. For low-PRF ground-based surveillance radar, the ground clutter's spectral width is determined primarily by the scanning modulation component. The scanning modulation component can be significantly reduced by "step scanning" the antenna, which is accomplished by freezing the antenna scanning motion (e.g., generally using an electronic scan) over the processing time of the Doppler processor.

A third factor affecting the clutter statistics is the so-called platform motion noise. This applies to airborne or shipborne installations, where the radar platform undergoes a translational motion with respect to the clutter mass. Motion noise is caused by the differential radial velocities (referenced to the radial velocity along the antenna's boresight) induced onto various clutter scatterers illuminated by the radar's antenna beam. This results in an increased spread of the clutter power spectrum, the extent of which depends directly on the planar translation velocity of the antenna and inversely on the antenna size. Thus, for large platform velocities and small antennas pointed away from their ground track, the motion noise can be severe, and it is the most important single factor affecting MTI performance in high-speed AI radars. As with scanning modulation noise, platform motion noise can be reduced by freezing the antenna (e.g., displaced phase center antenna), causing the antenna to appear stationary in space.

Another factor that tends to spread the clutter spectrum is the transmitter frequency instability. The first effect, which occurs during pulse-to-pulse transmitter frequency changes, is a partial decorrelation of the clutter returns similar to that which occurs by using frequency agility. This causes a pulse-to-pulse variation in the clutter signal at the output of the MTI's phase detector not cancelled in the Doppler processor. The second effect is from the noise sidebands (called *phase* or *frequency noise*) in every practical transmitter. These sidebands, which occur in the vicinity of the transmitter carrier frequency, are reflected by each clutter scatterer. The net result is that the clutter return of other spreading effects (i.e., internal motion, scanning, platform motion) is convolved with a noisy carrier rather than line spectra in forming the total clutter spectrum. This effect is most noticeable in high-PRF Doppler radars due to the folding of the many ambiguous clutter cells that compete with the target, thereby increasing the required Doppler processor improvement factor.

A general objective in Doppler radar processor design is to maintain a linear response over the full dynamic range of the clutter and design the processor with an improvement factor sufficient to reduce the clutter residue to the receiver noise level. In many cases, the dynamic range of the clutter is very large (e.g., see Figure 1.26 where the clutter dynamic range exceeds 60 dB) and requires large improvement factors. One solution to this problem employed in older MTIs is to use a limiter in the IF portion of the receiver. The nonlinear action of the limiter generally spreads the clutter spectrum.

The additional clutter spectral components generated by the limiter are caused in part by the sharp discontinuities in the clutter envelope as the clutter reaches the limit level. This is depicted in Figure 3.12, which shows the voltage response of a single-point scatterer illuminated by scanning radar for limiting and nonlimiting situations. The response for the nonlimiting situation provides a relatively smooth signal, which can be successively cancelled (the figure shows single, double, and triple cancellers) in the MTI processor to provide a high degree of improvement factor. When a signal having a peak that exceeds the limit level by 20 dB is applied to the MTI canceller, the resulting residues through the cancellers are much higher than in the nonlimited case, resulting in much lower improvement factors. In fact, in this case, additional stages of cancellation beyond a single canceller provide marginal additional improvement factor.

In low-PRF MTI radar designs, the Doppler frequency response is generally highly ambiguous (see Figure 2.3), resulting in blind speeds at Doppler frequencies corresponding to multiples of the radar's pulse repetition frequency. One way to resolve the blind speeds is to stagger the radar's PRF on a pulse-to-pulse basis. For example, if the periods of the staggered waveform have the relationship $n_1/T_1 = n_2/T_2 = \ldots n_N/T_N$, where $n_1, n_2, \ldots, n_N$ are integers, and if v_B is equal to the first blind speed of a nonstaggered waveform with a constant period equal to $T_{av} = (T_1 + T_2 + \ldots + T_N)$, the first blind speed is $v_1 = v_B(n_1 + n_2 + \ldots + n_N)/N$. However, when a staggered PRF is used, the improvement factor for a scanning radar is limited in proportion to the square of the number of hits per beamwidth (n_B). The scanning limitation is given by

$$I = \left[\frac{2.4n_B}{(\gamma - 1)}\right]^2 \tag{3.33}$$

where γ is the PRI stagger ratio [15].

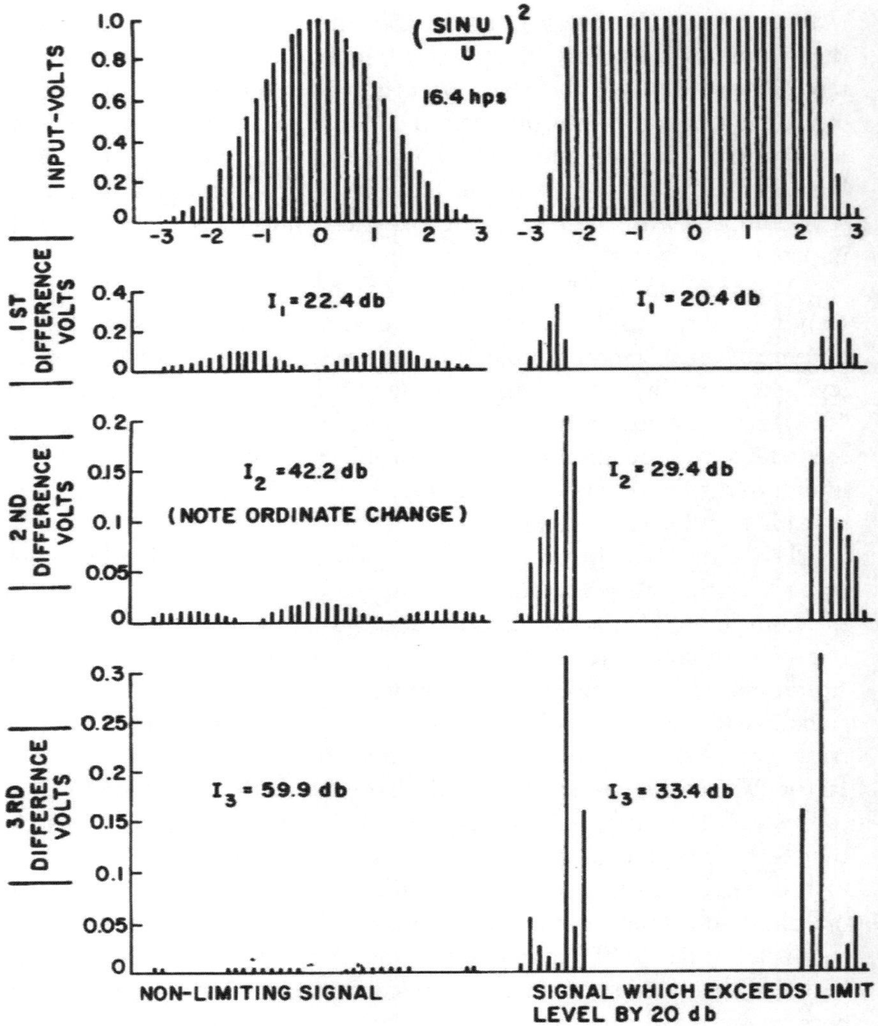

Figure 3.12 Improvement factor limitation caused by a limiter. (*From:* [15, p. 17-20].)

3.2.1 Internal Motion Clutter Spectrum

The internal motion clutter spectrum is generally assumed to be Gaussian shaped:

$$S(f) = P_c \frac{\exp\left(-f^2/2\sigma_f^2\right)}{\sqrt{2\pi}\sigma_f} \tag{3.34}$$

where σ_f is the standard deviation of the clutter spectrum in hertz and P_c is the clutter power. The corresponding ACF is given by

$$\rho(\tau) = P_c \exp\left(\frac{-\tau^2}{2\sigma_\tau^2}\right) \tag{3.35}$$

where $\sigma_\tau = 1/2\pi\sigma_f$. This can be related to the wind velocity through the relation $\sigma_f = 2\sigma_v/\lambda$, where σ_v is the standard deviation of the particle velocities about the mean velocity of the clutter mass. The improvement factor for a binary canceller is then given by

$$I_c = \frac{(\lambda f_r/\pi\sigma_v)^{2(n-1)}}{2^{3(n-1)} \cdot (n-1)!} \tag{3.36}$$

where λ is the radar wavelength (m/s), f_r is the pulse repetition frequency (Hz), σ_v is the rms velocity spread (m/s), and n is the number of pulses processed. Table 3.6 lists typical values measured for various clutter sources [9], and the improvement factor is plotted in Figure 3.13 for a double canceller.

An empirical relationship for the coherent spectral width (before envelope detection) of sea clutter based on a curve fit to experimental data is given by

Table 3.6 Apparent rms Velocity Spreads of Clutter Internal Motion

Description of Clutter	σ_v *(m/s)*
Sparsely wooded hills, calm day	0.005
Heavily wooded hills, 20-knot wind	0.3
Sea clutter, windy day	0.9
Chaff	1
Rain clouds	2

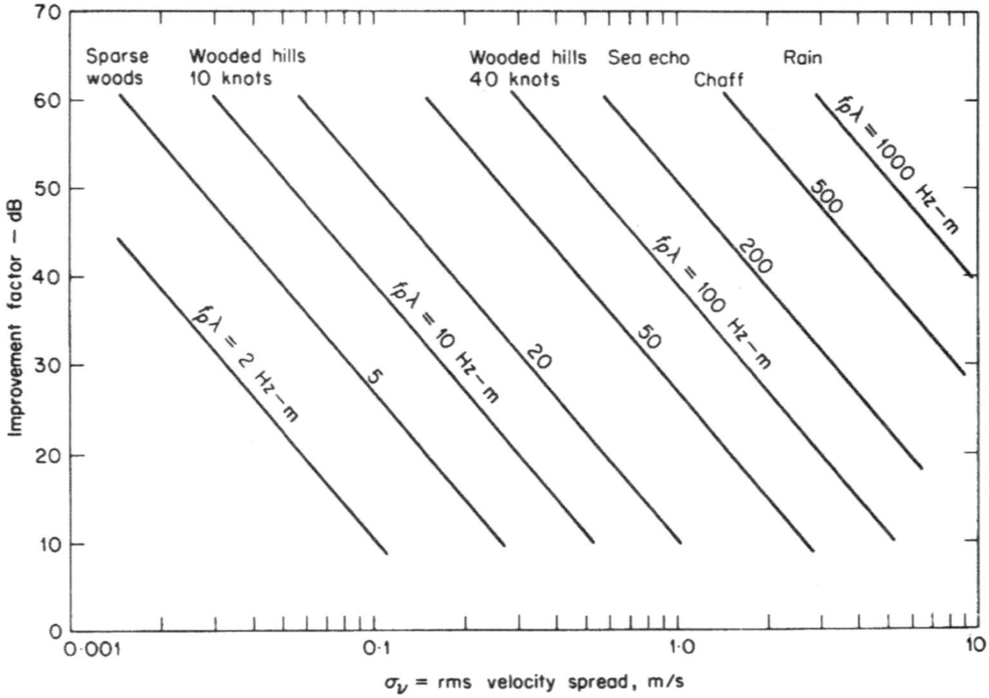

Figure 3.13 Improvement factor for internal motion limitation. (*From:* [12, p. 133].)

$\sigma_v = 0.051 v_w$, where v_w is the wind velocity in knots and σ_v (m/s) is the rms velocity spread [17]. This relation must be multiplied by $\sqrt{2}$ for the noncoherent spectral width (after envelope detection). A similar empirical relationship for wooded terrain type of land clutter is given by $\sigma_v = 0.0033 v_w^{1.26}$, where v_w is the wind speed in knots and σ_v (m/s) is the rms velocity spread [9, 17, 37].

MTI systems that search for slowly moving ground targets (e.g., combat surveillance radars) employ slowly scanning or staring antennas. In this type of system, internal motion clutter usually determines the overall clutter spectrum shape. Experiments have indicated that the spectrum measured with this type of radar would have a longer tail than would be predicted with a Gaussian-shaped spectrum.

Experimental evidence [9, 18–20] suggests that the clutter spectrum may have significantly longer tails than indicated by a Gaussian-shaped spectrum. A model suggested for ground clutter [19, pp. 153–154], measured with an X-band radar, postulates a power law spectral density shape given by

$$S(f) = \frac{k}{\left(1 + \left|\frac{f}{f_c}\right|^n\right)} \tag{3.37}$$

where n varies from 3.3 to 2.2 and f_c from 0.8 to 1.9 depending on radar frequency and wind speed. The power law dependence results in a significantly greater magnitude at high frequencies than the Gaussian-shaped spectrum.

Recent measurements employing a wide dynamic range receiver indicate that the clutter spectrum from windblown trees can be closely approximated by using an exponentially shaped spectrum down to the –60 dB power levels [20]. Typical measured spectra on both calm and windy conditions are depicted in Figure 1.25. The implication of these measurements is that the previously measured spectral data may have been distorted by receiver nonlinearities that caused wider spectral spreads at low power levels.

The power spectral density, valid from UHF to X-band, is given by the following two-sided exponential:

$$S(f) = \frac{r}{r+1}\delta(f) + \frac{1}{r+1}\frac{\beta\lambda}{4}\exp\left(-\frac{\beta\lambda}{2}|f|\right) \tag{3.38}$$

where β is the shape parameter whose typical values are given in Table 1.5 and r is the ratio of the steady component (reelection from nonmoving objects) to that of the fluctuating wind-driven components. The corresponding covariance function is given by

$$R(\tau) = \frac{r}{r+1} + \frac{1}{r+1}\frac{(\beta\tau)^2}{(\beta\tau)^2 + (4\pi r)^2} \tag{3.39}$$

The internal motion spectrum with power law dependence has significantly greater magnitude at high frequencies than the normally assumed Gaussian-shaped spectrum. When this effect dominates in determining the clutter shape, the MTI improvement factor will be poorer than that predicted by using a Gaussian-shaped spectrum. An example of a range-gated filter combat surveillance MTI used in a 20-knot wind resulted in a 7.5 dB degradation (i.e., from 28.5 to 21 dB) of the improvement factor for a cubic spectrum as compared to the Gaussian-shaped spectrum [9, 18].

The decorrelation effect of wind-driven scatterer motion discussed earlier describes the effect of scatterer motion relative to the center of gravity of the clutter mass. For ground clutter, the center of gravity of the clutter mass will be fixed, regardless of the wind velocity. However, for airborne clutter, such as rain clouds and chaff, the center of gravity will drift at the wind velocity (v_w) with respect to the ground. If the velocity is steady, this drift will produce a Doppler shift of the entire clutter spectrum away from the central ground-clutter frequency by an amount $f_w = 2v_w/\lambda$. The spectrum for Gaussian-shaped clutter is then given by

$$S(f) = P_c \frac{\exp\left[-(f - f_w)^2/2\sigma_f^2\right]}{\sqrt{2\pi\sigma_f^2}} \tag{3.40}$$

This mean velocity for rain and chaff, as well as for sea clutter, must be taken into account when designing the MTI processor.

3.2.2 Antenna Scanning Clutter Spectrum

As the radar antenna scans by a point target, it induces an amplitude modulation onto the target return that has a shape corresponding to the antenna's two-way pattern, as depicted in Figure 3.14. If the antenna has a Gaussian-shaped pattern with standard deviation σ_θ, the voltage spectrum will be Gaussian shaped with standard deviation $\sigma_w = \alpha/\sigma_\theta$, where the angular scan rate is α. If the clutter patch is assumed to be composed of many individual point scatterers, the resulting spectrum will be the sum of the spectra of the individual scatterers and have the same shape as that of the target.

A Gaussian one-way power antenna pattern is given by

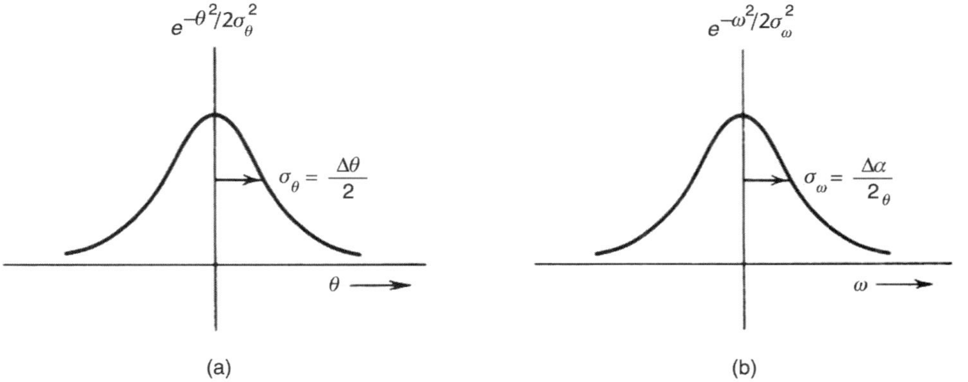

Figure 3.14 (a, b) Gaussian antenna pattern and resultant voltage spectrum. (*From:* [13].)

$$G(\theta) = G_o \exp\left(\frac{-2.773\theta^2}{\theta_B^2}\right) \tag{3.41}$$

where θ_B is the antenna beamwidth (in degrees). The standard deviation of this pattern is given by $\sigma_\theta = \theta_B/2.356$, which translates to a voltage spectrum standard deviation of $\sigma_w = 2.356\alpha/\theta_B$. The power spectrum (voltage squared) has a standard deviation that is reduced by the factor $\sqrt{2}$ resulting in $\sigma_c = 1.665/t_o$, where $t_o = \theta_B/\alpha$ is the time-on-target and α (degrees per second) is the scan rate. Alternately, the time-on-target can be expressed as $t_o = n_B T(\sigma_c T = 1.665/n_B)$, where n_B is the number of hits per 3-dB antenna beamwidth and T is the interpulse period of the radar. The net result is that the Gaussian-shaped scanning modulation clutter spectrum is given by

$$S(f) = P_c \frac{\exp\left(-f^2/2\sigma_f^2\right)}{\sqrt{2\pi}\sigma_f} \tag{3.42}$$

where P_c is the clutter power and

$$\sigma_f = \frac{1.665\alpha}{2\pi\theta_B} = \frac{0.833\alpha}{\pi\theta_B} \tag{3.43}$$

is the standard deviation of the power spectrum. The improvement factor for a binomial canceller is given by

$$I_c = \frac{2^{n-1}\left(\dfrac{f_r\theta_B}{1.665\alpha}\right)^{2(n-1)}}{(n-1)!} \tag{3.44}$$

or alternatively as

$$I_c = \frac{2^{n-1}\left(\dfrac{n_B}{1.665}\right)^{2(n-1)}}{(n-1)!} = \frac{n_B^{2(n-1)}}{(n-1)!(1.386)^{n-1}} \tag{3.45}$$

where f_r is the pulse repetition frequency, θ_B is the antenna's beamwidth, α is the antenna's scan rate, n_B is the number of hits per 3-dB beamwidth, and n is the number of pulses processed by the MTI canceller. The improvement factor limitation for scanning clutter is thus $I_1 = n_B^2/1.386$, $I_2 = n_B^4/3.844$, and $I_3 = n_B^6/15.983$ for binomial cancellers of order 1, 2, and 3, respectively.

The assumed Gaussian-shaped antenna pattern generally produces greater high-frequency components in the power spectral density than those associated with a real antenna pattern. This may be significant in analyzing high-performance MTIs, the improvement factor of which is limited by scanning modulation. To understand this effect, consider a uniformly illuminated antenna aperture scanning at constant speed that induces an amplitude modulation onto the target given by

$$v_t = V_m \frac{\sin^2 (a_t/2)}{(a_t/2)^2} \tag{3.46}$$

where $a = 5.6\alpha/\theta_B$, α is the scan rate (rad/s), and θ_B is the antenna's beamwidth [9]. Fourier transforming and squaring to find the scanning modulation power spectrum results in

$$S(\omega) = V_m^2 \left(1 - \frac{\omega}{\alpha}\right)^2 \tag{3.47}$$

for $\omega \leq a$ and zero for $\omega > a$. Thus, the spectrum is truncated at a frequency given by $f = 0.88\alpha/\theta_B$. Now, the 3-dB beamwidth of a rectangular aperture is given by $\theta_B = 0.88\lambda/D$, where D is the antenna's aperture width [16], resulting in $f = aD/\lambda$, which is precisely the Doppler frequency corresponding to the tip velocity of the antenna's aperture.

This analysis can be simplified by initially considering the antenna's illumination function rather than the antenna's beam pattern [9]. The PSD is then obtained by convolving and squaring the illumination function and scaling the frequency so that the truncation occurs at the tip velocity, given by $f = \alpha D/\lambda$. The voltage spectrum of the scanning clutter spectrum is obtained by using several standard illumination functions, as depicted in Figure 3.15.

3.2.3 Platform Motion Clutter Spectrum

If the radar platform is in motion, such as when the radar is installed on an aircraft or naval vessel, the radar motion induces a different velocity onto each clutter scatterer located within its field of view. In the low-PRF MTI case, the general clutter geometry is as depicted in Figure 3.16. The mean clutter velocity to the center of the ground clutter patch can be found by resolving the aircraft's Doppler frequency $f_{ac} = 2v_{ac}/\lambda$ into a radial translational component given by

$$f_c = f_{ac} \cos \phi_0 \cos \theta_0 \tag{3.48}$$

where v_{ac} is the platform velocity, λ is the radar wavelength, θ_0 is the antenna's azimuth angle, and ϕ_0 is the antenna's depression angle. This frequency (f_c) represents the center or mean frequency of the ground clutter patch spectrum due to platform motion. The MTI must generally adjust its rejection notch to correspond to this mean frequency. One method for accomplishing this adjustment is a clutter-locking system called TACCAR (time-averaged clutter coherent airborne radar), which adapts the MTI frequency null of the MTI to the mean clutter frequency using a voltage-controlled oscillator (VCO) in a phase-locked loop (PLL). However, because the adjustment frequency is a function of both the azimuth angle and range to the clutter patch, compensation is never perfect; and this must be taken into account when analyzing this type of MTI (see Section 3.2.7). Note also that in a noncoherent MTI, the average component is eliminated by the envelope detector, which is an advantage for this type of MTI on a moving platform.

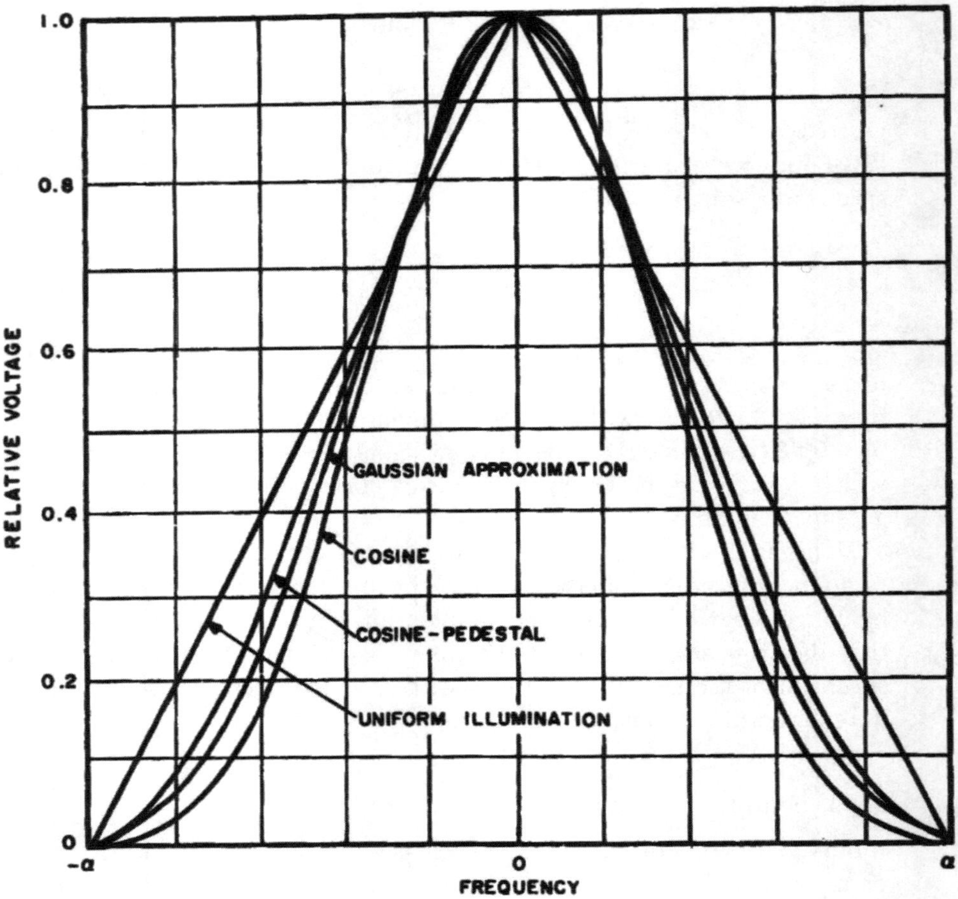

Figure 3.15 Voltage spectrum due to scanning antenna for various illumination functions. (*From:* [9, p. 210].)

Examination of Figure 3.16 reveals that the platform motion spectrum will have a spectral width determined by the variations in both azimuth and depression angles. For small depression angles, the main lobe rms width is generally determined by the radar's azimuth beamwidth (θ_B), which, in turn, is proportional to the antenna's aperture width (e.g., $\theta_B = 0.88\lambda/D$). The spectral width due to the antenna's azimuth beamwidth is found by subtracting the radial Doppler frequency components, which occur at the half-power edges of the azimuth beam as

$$\Delta f_0 = f_{ac} \cos \phi_0 \left[\cos\left(\theta_0 + \frac{\theta_B}{2} \right) - \cos\left(\theta_0 - \frac{\theta_B}{2} \right) \right] \quad (3.49)$$

for small azimuth beamwidths, this results in

$$\Delta f_\theta = f_{ac}\, \theta_B \cos \phi_0 \sin \theta_0 \quad (3.50)$$

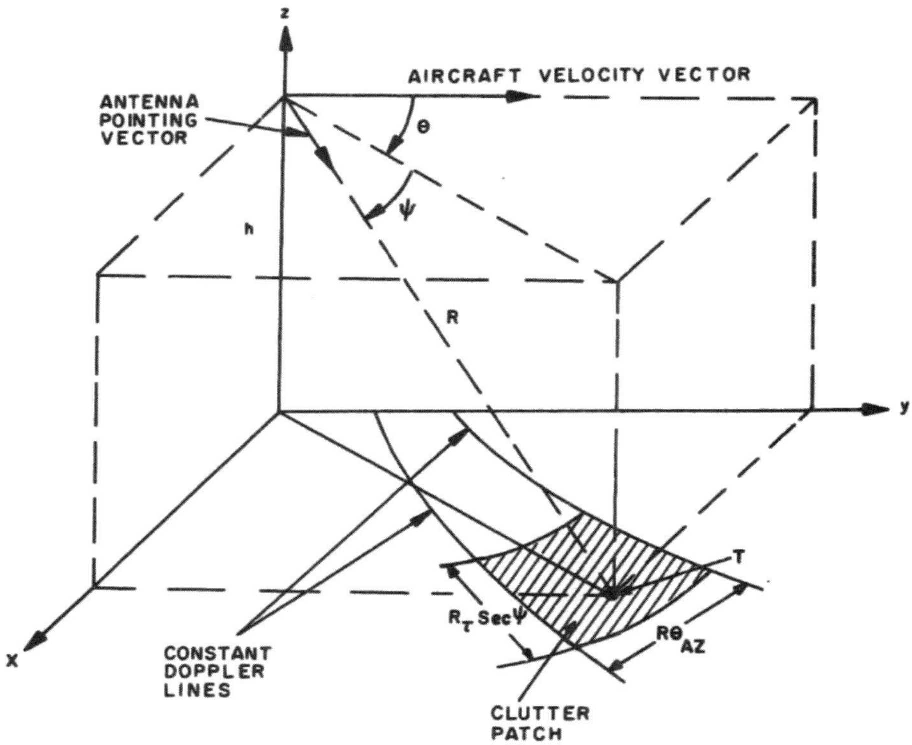

Figure 3.16 Airborne clutter geometry for MTI radar. (*From:* [9, p. 206].)

where f_{ac} is the platform's Doppler frequency, θ_B is the antenna's azimuth beamwidth (rad), ϕ_0 is the depression angle (assumed small), and θ_0 is the azimuth angle. For example, an X-band airborne radar ($f = 10$ GHz) moving at 300 knots with a 3° azimuth beamwidth would have a spectral width due to platform motion equal to 500 Hz at an azimuth angle of 90° and 250 Hz for an azimuth angle of 30°.

The preceding method for computing the Doppler spectral width due to the finite azimuth beamwidth breaks down at $\theta_0 = 0$ where it predicts a zero spectral width. The spread at $\theta_0 = 0$ can be found by considering the difference between the mean Doppler frequency (f_c) and the edge of the azimuth beam ($\theta_B/2$) as

$$\Delta f_\theta = f_{ac} \cos \phi_0 \left[1 - \cos\left(\frac{\theta_B}{2}\right)\right] = f_{ac} \frac{\cos \phi_0 \, \theta_B^2}{8} \tag{3.51}$$

for small azimuth beamwidths. For the previous example ($v = 300$ knots, $\theta_B = 3°$, $f = 10$ GHz), the spectral width at $\theta_0 = 0$ becomes a negligible 3 Hz.

A larger effect is due to the change of depression angle over the clutter patch. The magnitude of this effect can be derived in a way analogous to the method used in deriving the azimuth effect as

$$\Delta f_\phi = f_{ac} \Delta\phi \cos \theta_0 \sin \phi_0 \tag{3.52}$$

where $\Delta\phi$ is the depression angle spread over the clutter patch caused by the finite radar pulsewidth (see Figure 3.15). The value for $\Delta\phi = x/R$ where $x = R_\tau \tan\phi_0$ and $h = R\sin\phi_0$ can be found as $\Delta\phi = R_\tau \sin^3\phi_0 / h\cos\phi_0$ (rad), where R_τ is the range equivalent to the radar's pulsewidth (e.g., 1 μs equals 500 ft), ϕ_0 is the depression angle, and h is the platform height. The resulting Doppler spectral width due to the depression angle effect then becomes

$$\Delta f_\theta = \frac{f_{ac}\cos\theta_0 R_\tau \sin^3\phi_0}{h\cos\phi_0} \tag{3.53}$$

At $\theta_0 = 0$ for the previous example (v = 300 knots, f = 10 GHz) with a pulsewidth of 1 μs and an altitude of 10,000 ft, the Doppler spread becomes 72 Hz at a depression angle of 30° and only 2.6 Hz at a depression angle of 10°.

The azimuth and depression angle effects can be combined to form an overall spectral width (corresponding to antenna half-power points) relationship as follows:

$$\Delta f = f_{ac}\left\{\theta_B \cos\phi_0 \sin\theta_0 + \left[\frac{\theta_B^2\cos\phi_0\cos\theta_0}{8}\right] + \left[\frac{R_\tau \sin^3\phi_0\cos\theta_0}{h\cos\phi_0}\right]\right\} \tag{3.54}$$

where θ_B is the azimuth beamwidth, ϕ_0 is the depression angle, R_τ is the range equivalent of the radar's pulsewidth, and h is the platform altitude. For small depression angles, the relationship

$$\Delta f = f_{ac}\theta_B \sin\theta_0 = \frac{1.76 v_{ac}\sin\theta_0}{D} \tag{3.55}$$

is often appropriate, indicating that the spectrum spread is independent of frequency, whereas aircraft speed and antenna aperture size dictate the maximum clutter spread due to aircraft motion.

Another way of looking at platform motion noise is that the aircraft motion induces a form of scanning motion onto the clutter scatterers. This is illustrated in Figure 3.17, where to a moving observer (located on the aircraft) the clutter mass will appear to rotate about point o at an angular rate of $\omega = v_{ac}\sin\theta_0/R$. Consequently, the clutter scatterer at p will exhibit a relative radial velocity for small beamwidths approximately equal to $v_p = \omega R = v_{ac}\theta\sin\theta_0$.

Each clutter scatterer within the range resolution cell defining the clutter patch will be illuminated by the one-way voltage antenna pattern. The return voltage will then be weighted by the antenna's two-way voltage pattern or by the one-way antenna power pattern $G(\theta)$. The platform motion induces a Doppler spectral distribution that is a function of the angular displacement from the antenna sighting direction (i.e., $f = 2v\theta\sin\theta_0/\lambda$). The resulting power spectral density is given by $G^2(\theta = \lambda f/2v\sin\theta_0)$ and has a shape corresponding to the two-way power antenna pattern. This relationship indicates that the two-way power antenna sidelobes are important because they weight frequency components that appear in the passband

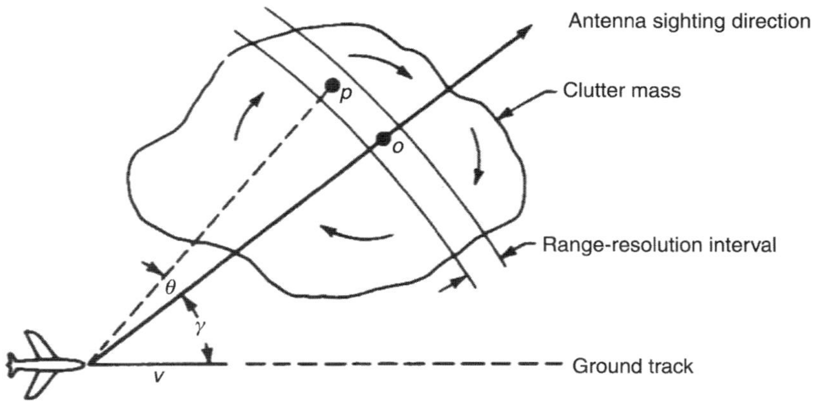

Figure 3.17 Relative rotation of clutter mass due to platform noise. (*From:* [10, p. 123].)

of the MTI (see Figure 2.5), which sets an upper limit on the performance of an airborne MTI.

For a Gaussian-shaped, one-way power antenna pattern given by

$$G(\theta) = G_0 \exp\left(\frac{-2.773\theta^2}{\theta_B^2}\right) \tag{3.56}$$

the power spectral density is given by

$$S(f) = P_0 \exp\left[\frac{-5.546f^2}{(f_{ac}\theta_B \sin\theta_0 \cos\phi_0)^2}\right] \tag{3.57}$$

The platform motion power spectral density can then be written as

$$S(f) = P_c \frac{\exp\left(-f^2/2\sigma_{pm}^2\right)}{\sqrt{2\pi}\sigma_{pm}} \tag{3.58}$$

where

$$\sigma_{pm} = 0.3f_{ac}\theta_B \sin\theta_0 \cos\phi_0 = \frac{0.528v_{ac} \sin\theta_0 \cos\phi_0}{D} \tag{3.59}$$

The improvement factor for an n-pulse coherent MTI binomial canceller with platform motion is limited by

$$I_c = \frac{2^{n-1}}{1.885(f_{ac}T\theta_B \sin\theta_0 \cos\phi_0)^{2(n-1)}(n-1)!} \tag{3.60}$$

which for a single canceller equals

$$I_c = \frac{1}{1.777(f_{ac} T \theta_B \sin\theta_0 \cos\phi_0)^2} \tag{3.61}$$

and for a double canceller equals

$$I_c = \frac{1}{6.313(f_{ac} T \theta_B \sin\theta_0 \cos\phi_0)^4} \tag{3.62}$$

For the previous example (f = 10 GHz, v_{ac} = 300 knots, $\theta_B = 3°$, $\theta_0 = 30°$, $\phi_0 = 0$) at a PRF of 5,000 Hz, the maximum improvement factor for a single canceller would be 23.5 dB and 44 dB for a double canceller.

Because the antenna sidelobes essentially cover 360°, the Doppler shifts received through the sidelobes range from $-2v_r/\lambda$ to $2v_r/\lambda$, where v_r is the platform velocity. In low-PRF airborne radar, these sidelobe returns are aliased through the entire MTI ambiguous frequency band. As a result, these sidelobe returns are not cancelled, and the MTI improvement factor is limited by the integrated main lobe-to-sidelobe ratio. This is the ratio of the integrated energy entering the system through the main lobe to that entering through the sidelobes for the two-way antenna pattern. Thus, the integrated main lobe-to-sidelobe ratio is a quality factor for the antenna used in this type of application [9].

Antennas are frequently specified on the basis of the amount of power radiated in their sidelobes compared to that radiated in their main lobe (e.g., –10 dB implies 10% power in the sidelobes). Ordinary antennas have sidelobes with 0 to –5 dBi rms sidelobes, low-sidelobe antennas have –5 to –20 dBi rms sidelobes, and ultralow-sidelobe antennas have rms sidelobes below –20 dBi. For example, a low-sidelobe antenna would provide an improvement factor limit from 10 to 40 dB in a low-PRF airborne MTI. Ultralow-sidelobe antennas are required to provide an improvement factor limit due to this effect on the order of 60 dB, which is required in high-performance AEW applications [9].

3.2.4 Transmitter Instability Spectrum Spreading

Pulse-to-pulse transmitter frequency instability causes two primary effects in a low-PRF coherent MTI system. First, for a fixed target, fluctuations occur at the output of the MTI's synchronous detector and are not cancelled by the MTI processor. This limitation is included in the CR determination for the MTI (see Table 3.4). Note also that this effect is not a factor for a noncoherent MTI. The second effect is to cause spectral spreading of the clutter spectrum due to the decorrelating effect of the pulse-to-pulse frequency changes on the clutter return, which is equivalent to the action of frequency agility on a distributed target.

To examine this effect, the clutter patch of Figure 3.18 is sliced in range at intervals separated by a radar wavelength (a 1-μs pulse at 10 GHz contains 10,000 slices). When the scatterers located in each slice are illuminated by the radar, they return in-phase signals. If the radar's transmitter frequency changes by an amount sufficient to cause a change of one in the number of wavelengths within one radar pulsewidth, the slices return a signal that contains all possible phases. Hence, the signal returns at the two frequencies are completely decorrelated.

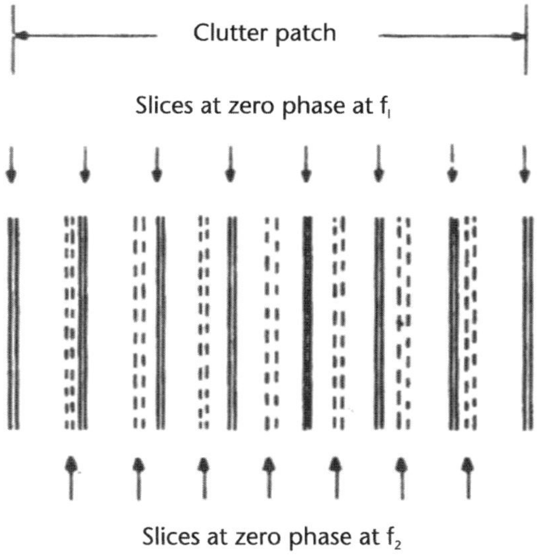

Figure 3.18 Phase of a clutter slice for frequencies f_1 and f_2. (*From:* [9, p. 215].)

We can show that the normalized autocorrelation function of the clutter return power for a rectangular pulse, which describes the correlation between return power P_1 at frequency f and return power P_2 for the same set of scatterers at frequency $f + \Delta f$, is given by

$$\rho = \left(\frac{\sin \pi\tau\Delta f}{\pi\tau\Delta f}\right)^2 \tag{3.63}$$

where τ is the radar's pulsewidth and Δf is the pulse-to-pulse frequency change [9]. The power spectral density can be found as

$$S(f) = P_0\left(1 - \frac{f}{\Delta f}\right), \qquad |f| \le \Delta f \tag{3.64}$$

Matching a Gaussian spectrum at the half-power points results in

$$S(f) = P_c \frac{\exp\left(f^2/2\sigma_i^2\right)}{\sqrt{2\pi}\sigma_i} \tag{3.65}$$

where $\sigma_i = 0.424\Delta f$.

3.2.5 Clutter Limiting Spectral Spreading

Modern radars generally attempt to maintain sufficient dynamic range to pass clutter peaks without saturation. When this is not possible (e.g., in a digital MTI processor because of limited A/D converter range), an intermediate-frequency (IF) limiter may be employed to keep the clutter's dynamic range within that of the MTI's receiver-processor. When an IF limiter is employed (as was prevalent in older MTI radars), the nonlinear action of the limiter spreads the clutter spectrum, causing a reduction in the MTI's improvement factor. Note that distortion of the clutter statistics by the limiter causes the optimum linear MTI processor dictated by the clutter's Gaussian statistics to be nonoptimum, and theoretically a nonlinear processor provides greater improvement in limited clutter than the generally employed linear MTI processor.

A convenient way to analyze the spectral spreading effect of clutter limiting is to employ a soft limiter (which has an error function characteristic) because this provides an analytic expression for the limiter's autocorrelation function (and power spectral density) [9, 21]. A block diagram of the IF limiter configuration is depicted in Figure 3.19, where the function of the bandpass filter is to eliminate harmonics of the IF caused by the nonlinear action of the limiter.

The normalized autocorrelation function of the limiter can be expressed as

$$\rho_l(\tau) = \frac{\sin^{-1}\left[\dfrac{\rho(\tau)}{(1+\alpha)}\right]}{\sin^{-1}\left[\dfrac{1}{(1+\alpha)}\right]} \tag{3.66}$$

where $\alpha = 0.637L/C$, L is the clutter's power saturation point, C is the clutter power, and $\rho(\tau)$ is the normalized ACF of the clutter [9, 21, 22]. For hard limiting, $\alpha = 0$ and the clutter's autocorrelation function reduces to

$$\rho_l(\tau) = \left(\frac{2}{\pi}\right)\sin^{-1}[\rho(\tau)] \tag{3.67}$$

The IF autocorrelation function of clutter with a Gaussian-shaped spectrum can be expressed as

$$\rho_l(\tau) = \exp\left[\frac{-(\sigma_c\tau)^2}{2}\right]\cos\omega_0\tau \tag{3.68}$$

where σ_c (rad/s) is the standard deviation of the clutter spread, τ is the time separation between samples, and ω_0 is the IF radian frequency.

Expanding the arcsine function into a power series:

$$\sin^{-1}(x) = \sum_{p=0}^{\infty} a_p x^{2p+1} \tag{3.69}$$

Figure 3.19 Block diagram of limiter and coherent MTI. (*From:* [21, p. 172].)

$$\rho_l(\tau) = k \sum_{p=0}^{\infty} C_p \exp\left[\frac{-(2p+1)(\sigma_c \tau)^2}{2}\right] \tag{3.70}$$

where

$$C_p = \frac{a_p b_p}{(1+\alpha)^{2p+1}}, \quad k = \frac{1}{\sum_{p=0}^{\infty} C_p} \tag{3.71}$$

and

$$a_p = \frac{1 \cdot 3 \cdot \ldots (2p-1)}{2 \cdot 4 \cdot \ldots (2p)(2p+1)} \tag{3.72}$$

where

$$b_p = \frac{\binom{2p+1}{p}}{2^{2p}} \tag{3.73}$$

The factor b_p results from expanding $\cos^n \omega_0 \tau$ and retaining only the $\cos \omega_0 \tau$ terms that pass through the bandpass filter; for example, $\cos^3 x = (\cos 3x + 3 \cos x)/4 \rightarrow 3 \cos x/4$. The corresponding spectrum is a sum of Gaussian-shaped spectral densities given by

$$S(\omega) = P_0 \sum C_p \exp\left[\frac{-\omega^2}{(2p+1)2\sigma_c^2}\right] \tag{3.74}$$

where each term has a progressively larger frequency spread with standard deviation given by $\sqrt{2p+1}\sigma_c$, but is diminished in amplitude by the coefficient of the power series (C_p).

The spectrum for limited Gaussian-shaped clutter is depicted in Figure 3.20 for various ratios of clutter power to limit power (C/L). The curve is plotted in terms of the prelimited 3 dB video bandwidth of the clutter spectrum, which is given by

$$B_3 = \sqrt{\left(\frac{2 \ln 2}{\pi^2}\right)}\sigma_c \tag{3.75}$$

Note that, as the ratio of clutter power to limit level increases, the clutter spectrum is increasingly spread until hard limiting is achieved. The increased clutter spectral spread will cause additional clutter components in the MTI's passband, which, in turn, will cause a reduction in the MTI's improvement factor.

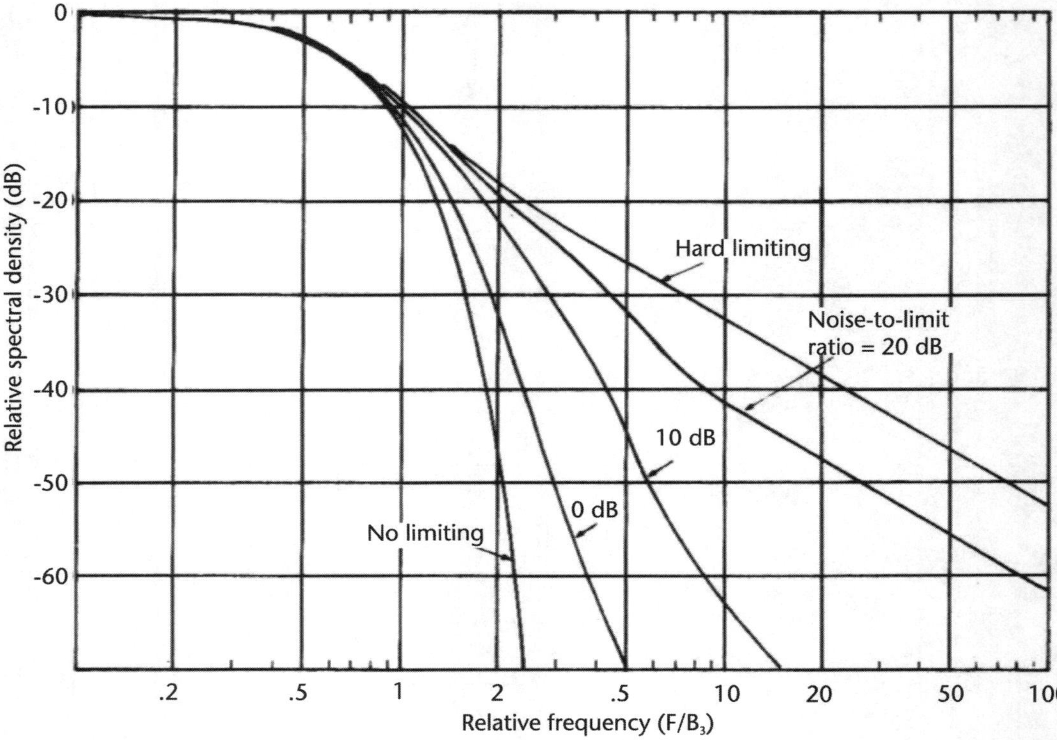

Figure 3.20 Power spectral density for limited Gaussian-shaped clutter. (*From:* [23, p. 247].)

The improvement factor for limited Gaussian-shaped clutter can be found by substituting the limited normalized ACF into the improvement factor expressions for a binomial canceller given in Section 3.1; that is,

$$I_1 = \frac{1}{1 - \rho_l(T)} \tag{3.76}$$

$$I_2 = \frac{3}{3 - 4\rho_l(T) + \rho_l(2T)} \tag{3.77}$$

$$I_3 = \frac{10}{10 - 15\rho_l(T) + 6\rho_l(2T) - \rho_l(3T)} \tag{3.78}$$

For example, the single canceller improvement factor is given by

$$I_1 = \frac{\sum_{p=0}^{\infty} C_p}{\sum_{p=0}^{\infty} C_p \left\{1 - \exp\left[\frac{-(2p+1)(\sigma_c T)^2}{2}\right]\right\}} \tag{3.79}$$

and

$$I_2 = \frac{\sum_{p=0}^{\infty} 3C_p}{\sum_{p=0}^{\infty} C_p \left\{3 - 4\exp\left[\frac{-(2p+1)(\sigma_c T)^2}{2}\right] + \exp[-2(2p+1)(\sigma_c T)^2]\right\}} \tag{3.80}$$

is the double canceller improvement factor.

As an example, consider a single canceller with hard limiting. The improvement factor is given by

$$I_1 = \frac{\sum_{p=0}^{\infty} a_p b_p}{\left\{1 - \exp\left[\frac{-(2p+1)(\sigma_c T)^2}{2}\right]\right\}} \tag{3.81}$$

which for sufficiently small $\sigma_c T$ can be written as

$$I_1 = \frac{2}{(\sigma_c T)^2}\left[\frac{\sum_{p=0}^{\infty} a_p b_p}{\sum_{p=0}^{\infty} a_p b_p (2p+1)}\right] \tag{3.82}$$

The first term can be recognized as the improvement factor without limiting, and the second factor can be identified as an asymptotic loss associated with the hard limiter. In this case, $\Sigma_{p=0}^{\infty} a_p b_p = 1.273$ and $\Sigma_{p=0}^{\infty} a_p b_p (2p + 1) = 6.722$, resulting in a loss of 7.2 dB. The hard limited improvement factor for $n = 100$ pulses would then be the nonlimited improvement factor $I_1 = 10 \log(100^2/1.388)$ $= 38.6$ dB diminished by the limiter loss (7.2 dB), resulting in $I_{l1} = 31.4$ dB.

Figure 3.21 depicts the improvement factor achieved with various levels of limiting for single, double, and triple cancellers [21]. In practice, a limiter will be set at a level that equalizes the clutter residuals to the receiver noise level. A typical situation is shown in Figure 3.22, which compares improvement factors for single, double, and triple cancellers for clutter 20 dB above the limit level (a situation common to most clutter areas) and for hard limiting. Note that very little advantage in improvement factor is gained by using higher order cancellers (i.e., triple and greater) as hard limiting is approached. In this situation, for typical MTI search radars employing 10 to 20 hits per scan, practical improvement factors of 20 to 25 dB with hard limiting appear achievable using a double canceller. The result is that targets are suppressed in those resolution cells containing strong clutter, which require heavy limiting. In other resolution cells, having weaker clutter (which is less than 20 dB above the limit level), targets will frequently be detected. This property, called *interclutter visibility*, points to the desirability of using moderately small spatial resolution cells in the radar to take advantage of the spotty nature of land clutter. This mode of operation allows MTI radars with limiting to achieve reasonable functional performance in many practical situations.

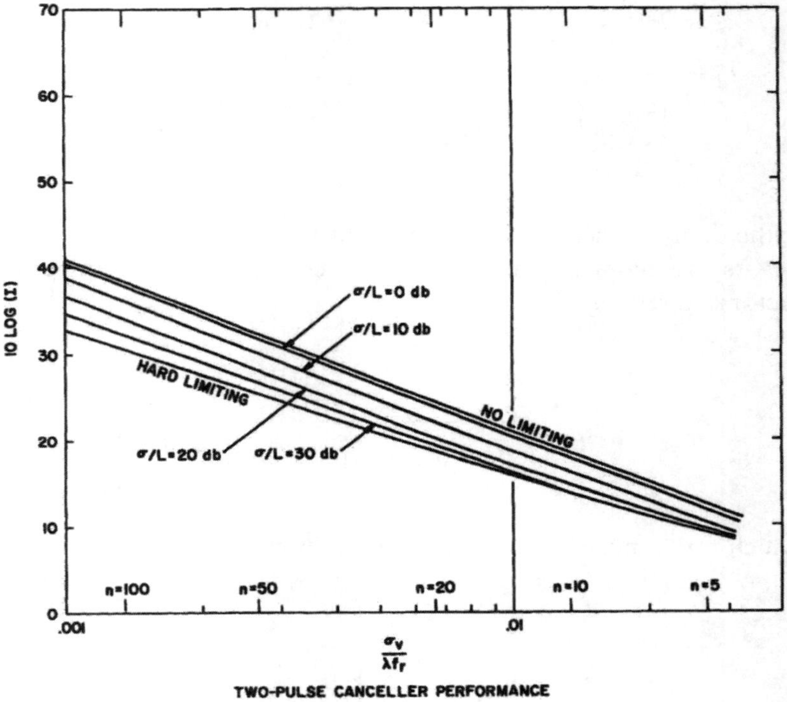

(a)

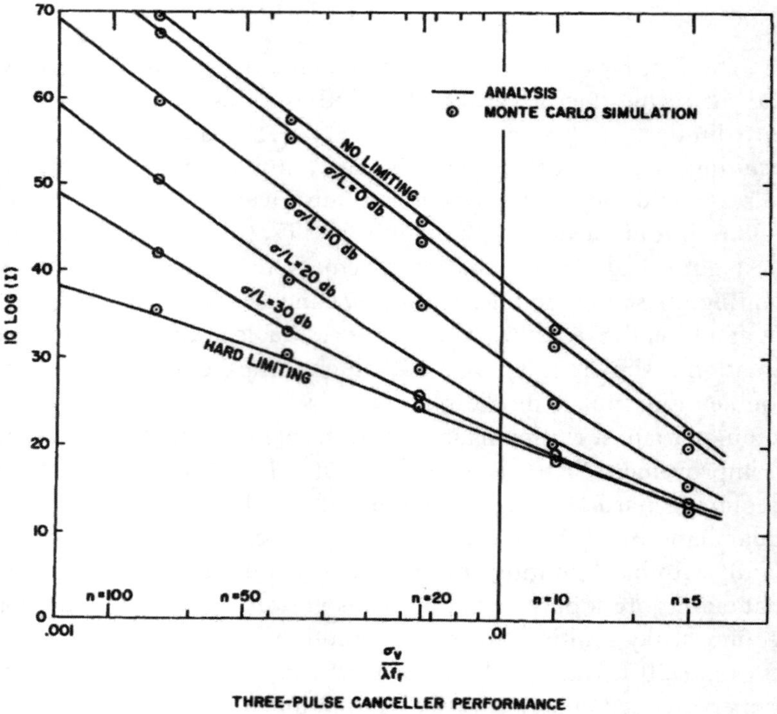

(b)

Figure 3.21 MTI improvement factor for (a) single, (b) double, and (c) triple cancellers for various clutter power to limit levels. (*From:* [24, p. 17-21].)

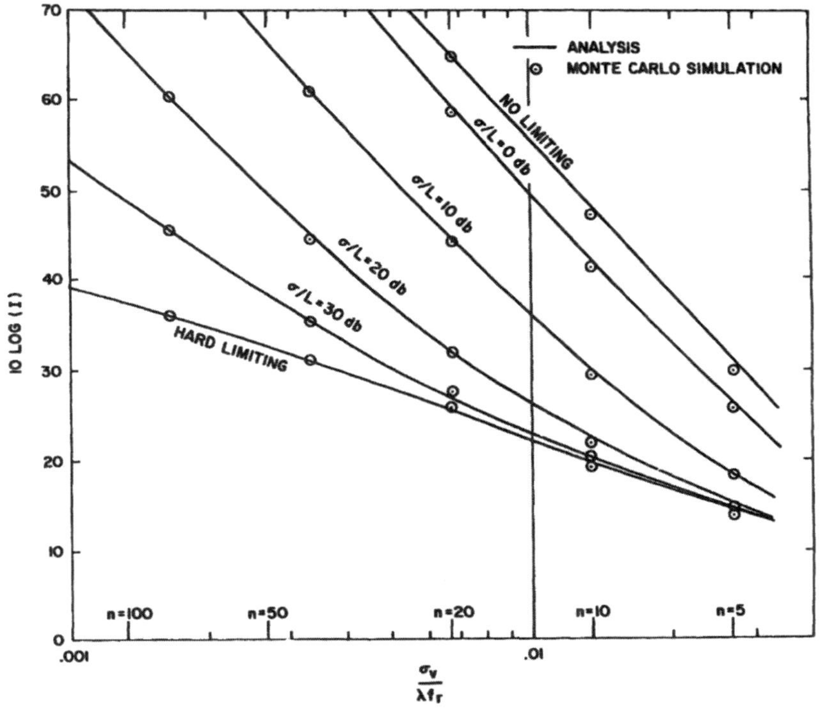

(c)

Figure 3.21 (continued).

Figure 3.23 compares the improvement factor available from an ASR-7 radar for single and double cancellers with and without limiting [9, 25]. An improvement factor of 46 dB is required to detect a 1 m^2 target (small target) at 15 nmi in moderately heavy ground clutter. As is obvious from the curve, this degree of performance is not available from an MTI that uses limiting to control the clutter's dynamic range.

The solution applied in the ASR-9 radar is to use a ground clutter map that measures the average clutter strength in each radar resolution cell. The measured clutter strength is then used to set a threshold that controls the false alarm rate to the specified value while allowing the MTI processor to operate in a linear mode. This provides the maximum MTI capability available from the radar and, in addition, provides *superclutter visibility* in those cells containing targets stronger than the clutter background.

The ground clutter map is a form of CFAR that provides operation in ground clutter (rather than receiver noise) and allows for the spotty nature of ground clutter where appreciable variations occur from one resolution cell to the next. The ground clutter map threshold approach tends to censor those resolution cells containing ground clutter with a dynamic range that exceeds the improvement factor capability of the MTI radar, but the applicability of this approach becomes questionable against moving clutter (e.g., weather or chaff) or from a moving platform (e.g., airborne radar).

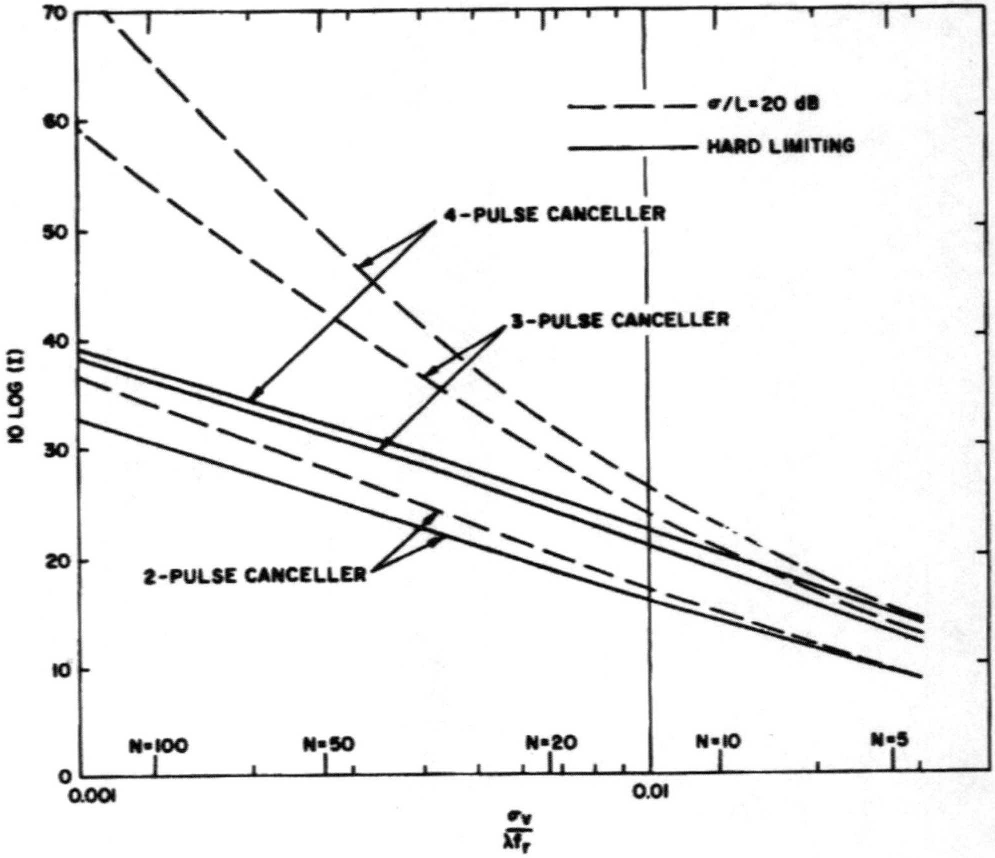

Figure 3.22 Comparison of limiting clutter in single, double, and triple cancellers. (*From:* [21, p. 172].)

3.2.6 Improvement Factor Restriction Caused by PRF Staggering

PRF staggering is used in low-PRF MTI radars to detect targets at blind speeds associated with targets having Doppler velocities ($f_d = 2v/\lambda$) that fall into the stopband of the MTI filter (recall Figures 2.4 and 2.5). The periodic nature of the MTI comb filter creates blind speeds when f_d equals or is a harmonic of the PRF. Staggering the interval between pulses makes possible the alteration of the MTI filter's passband shape and elimination of blind speeds below some very high speed.

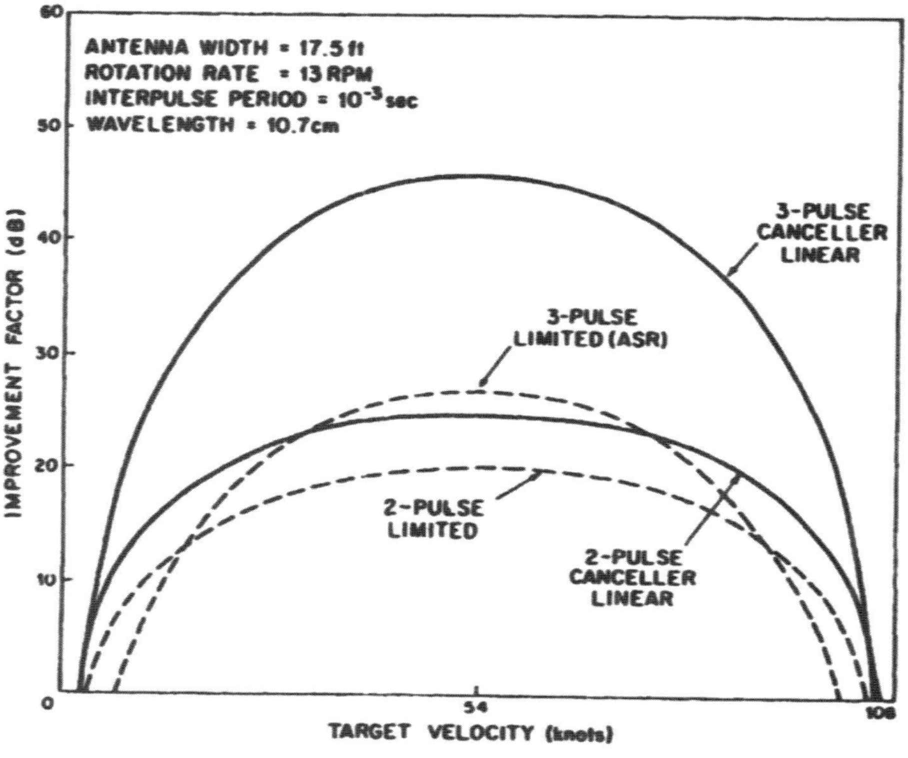

Figure 3.23 Improvement factors available with limiting and linear operation from airport surveillance radar. (*From:* [25, p. 717].)

► Example 8 Velocity Loss in MTI Systems

The velocity loss for MTI radar can be severe for a design that does not use staggered PRFs. The probability of detection for a single-sample Swerling case 1 target using a square-law envelope detector is given by [1, 2]

$$P_d(s) = \exp\left(-\frac{y_b}{1+\bar{s}}\right)$$

where y_b is the detection threshold and $\bar{s}$ is the average signal-to-noise ratio. The detection threshold is given by

$$y_b = \ln\left(\frac{1}{P_{fa}}\right)$$

where P_{fa} is the probability of false alarm. However the S/N ratio at the output of an MTI filter using binomial weights is a function of the target velocity and is given by

$$\bar{s}(\omega) = \frac{\bar{x}}{\bar{G}}\left(2\,\sin\left(\frac{\omega T}{2}\right)\right)^{2n}$$

where T is the MTI period, ω is the radian frequency, n is the number of cascaded single MTI binomial cancellers, $\bar{x}$ is the average S/N ratio into the MTI filter, and $\bar{G}$ is the average power gain that increases the noise power at the output of the MTI (see Section 3.1). The target radial velocity is assumed to be uniformly distributed over the target velocity response region. The frequencies of interest are folded into the first lobe of the comb filter that extends between zero frequency and the first blind speed given by $\omega_r = 2\pi/T$. The probability of detection (P_d) can then be found by averaging over this region, resulting in

$$P_d = \frac{1}{\omega_r}\int_0^{\omega_r} \exp\left(-\frac{y_b}{1+\frac{\bar{x}}{\bar{G}}\left(2\,\sin\left(\frac{\omega T}{2}\right)\right)^{2n}}\right) d\omega$$

Letting $\theta = \omega T/2$ results in the following simplified form, which can be solved for $\bar{x}$:

$$P_d = \frac{1}{\pi}\int_0^{\pi} \exp\left(-\frac{y_b}{1+\frac{\bar{x}}{\bar{G}}(2\,\sin\,\theta)^{2n}}\right) d\theta$$

using either MATLAB or MATHCAD.

The MATLAB program mti_velstagg.m using function fzero ('mti', x, options, yb, Pd, n) is used for both single and double cancellers to solve this equation for $\bar{x}$, where $\bar{G} = 2$ for a single canceller and 6 for a double canceller, and the function mti is given below:

```
function y = mti(x,yb,Pd,n)
N = 1024;
w = 0:pi/N:pi;
z = exp(-yb./(1+x*(2*sin(w)).^(2*n)));
y = (trapz(w,z)./pi)-Pd;
```

Figure EX-8.1 plots this loss ($\bar{s} - \bar{x}$) in decibels for $P_d = 0.5$ to 0.9 for both a single and double MTI canceller. Also shown is the loss associated with a 5:4 stagger followed by a single and double canceller. This figure explains the need for staggering the PRF of MTI radar to minimize the MTI velocity loss.

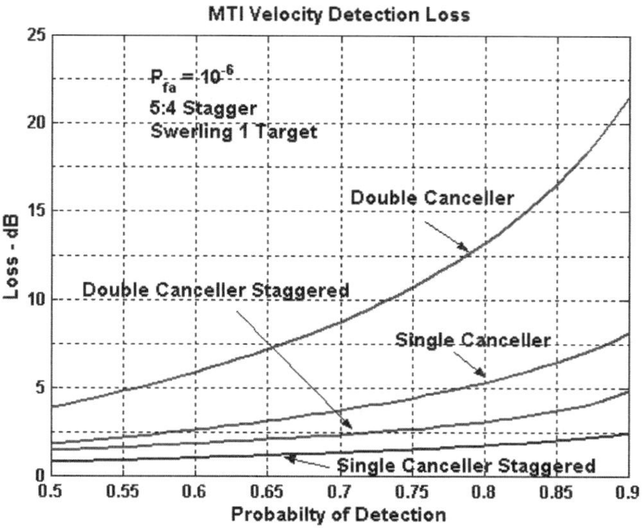

Figure EX-8.1 MTI velocity detection losses for single and double cancellers.

Figure 3.24 depicts a common method used to generate a staggered PRF: A short (relative to the average interpulse period) delay line is switched in and out of the signal path on alternate pulses. The transmitter trigger is also put through the same delay line on alternate pulses, causing the transmitter to be pulsed at the staggered rate. The stagger ratio is given by $\gamma = T_c + T_s/T_c - T_s$, where T_c is the average interpulse period and T_s is the short differential period. Note that the destaggered signal comes out at a constant rate and thus can be applied to a conventional canceller (e.g., single, double, or triple) with no modification to the canceller.

The PRF staggering modifies the passband shape of the MTI filter (relative to the nonstaggered shape), which causes a reduction in the MTI's improvement factor. To consider this effect, the two interpulse periods (see Figure 3.24) are written as $T_1 = (1 + e)T$ and $T_2 = (1 - e)T$. Applying this to a double canceller MTI results in an impulse response given by

$$h(t) = \delta[t + T(1 - e)] - 2\delta(t) + \delta[t - T(1 + e)] \tag{3.83}$$

This can be Fourier transformed to obtain the MTI's voltage transfer function as

$$H(\omega) = \exp[j\omega T(1 - e)] - 2 + \exp[-j\omega T(1 + e)] \tag{3.84}$$

The staggered power transfer function is then given by

$$|H(\omega)|^2 = 4[1 - 2\cos(\omega Te)\cos(\omega T) + \cos^2(\omega T)] \tag{3.85}$$

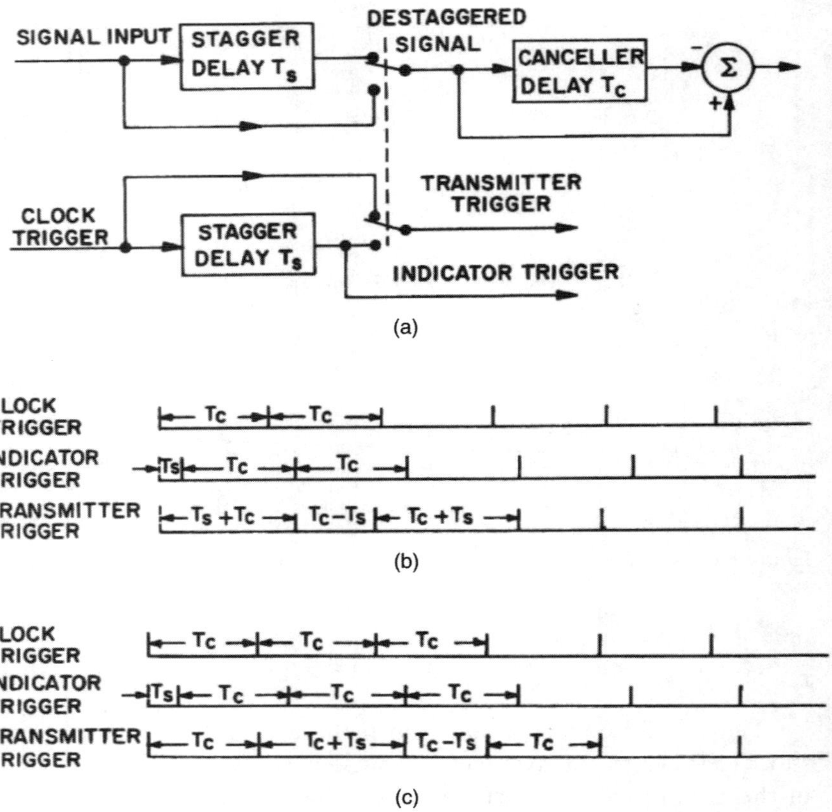

Figure 3.24 (a–c) Block diagram and waveforms for two- and three-period staggered PRFs. (*From:* [24, p. 17-39].)

which, for $e = 0$, reduces to the standard double canceller response:

$$|H(\omega)|^2 = 16 \sin^4\left(\frac{\omega T}{2}\right) \tag{3.86}$$

which is periodic $\omega = 2\pi/T$.

The staggered power transfer function is periodic for that value of ω for which $\cos(\omega T)$ and $\cos(\omega Te)$ are both +1 and −1. For $e << 1$, the approximate depth of the first null relative to the peak is then given by

$$\frac{\left|H\left(\frac{2\pi}{T}\right)\right|^2}{|H_{\max}|^2} = \left|\frac{1 - \cos(2\pi e)}{2}\right|^2 = (\pi e)^2 \tag{3.87}$$

or, equivalently, $[\pi(\gamma - 1)]^2/4$, where γ is the stagger ratio.

The double canceller improvement factor with PRF staggering can be found by first substituting the approximation $\cos(\omega\tau e) = 1 - (\omega\tau e)^2/2$ for $e << 1$ into the power transfer function, resulting in

$$|H(\omega)|^2 = \frac{16 \sin^4 (\omega T)}{2} + 4(\omega Te)^2 \tag{3.88}$$

Next, the residual clutter power after MTI filtering for the second term is found from $\int_{-\infty}^{\infty} |H(\omega)|^2 S(\omega)\, d\omega$ as $4(\sigma_c Te)^2$, where $S(\omega)$ is the Gaussian-shaped clutter distribution. The improvement factor is given by (see Section 3.1)

$$I = \frac{|H(\omega)|^2 \int_{-\infty}^{\infty} S(\omega)\, d\omega}{\int_{-\infty}^{\infty} |H(\omega)|^2 S(\omega)\, d\omega} \tag{3.89}$$

This results in $I_s = I_2/[1 + I_2(\sigma_c Te)^2/4]$, which for $I_2 >> 1$ results in $I_s = 4/(\sigma_c Te)^2$. Substituting $e = (\gamma - 1)/2$ and $n_B = 1.666/\sigma_c T$ yields

$$I_s = \left(\frac{2.4 n_B}{\gamma - 1}\right)^2 \tag{3.90}$$

as the MTI improvement factor limitation for PRF staggering in scanning clutter, where n_B is the number of hits per 3-dB beamwidth and γ is the stagger ratio. This relationship is depicted in Figure 3.25. For internal motion clutter, $\sigma_c T = 4\pi\sigma_v/\lambda$ (see Section 3.2.1), and the PRF staggering limitation becomes

$$I_s = \left(\frac{0.32}{\gamma - 1}\right)^2 \left(\frac{\lambda f_r}{\sigma_v}\right)^2 \tag{3.91}$$

where f_r is the pulse repetition frequency and σ_v is the rms velocity spread of the clutter (see Section 3.2.1). This relationship is depicted in Figure 3.26. The MTI improvement factor limitations due to PRF staggering tend to apply independently of the type of canceller employed following the staggering network.

Figure 3.27 illustrates the frequency response of an MTI used in long-range air traffic control radar with a four-period stagger having the ratio 25:30:27:31 [15]. The first blind speed is $(25 + 30 + 27 + 31)/4 = 28.25$ times that of a constant PRF waveform with the same average period. A disadvantage of the staggered PRF is its inability to cancel second-time-around and higher clutter echoes. Such clutter does not appear at the same range from pulse to pulse and thus produces uncancelled residues.

The problem of uncancelled multiple-time-around echoes in PRF staggering can be avoided by using a coherent PRF-diversity MTI, where the pulse repetition interval (PRI) is held constant until echoes from the longest range clutter are received and processed through the coherent MTI processor. The PRI is then changed on a burst-to-burst basis. After each constant-PRF burst is batch processed

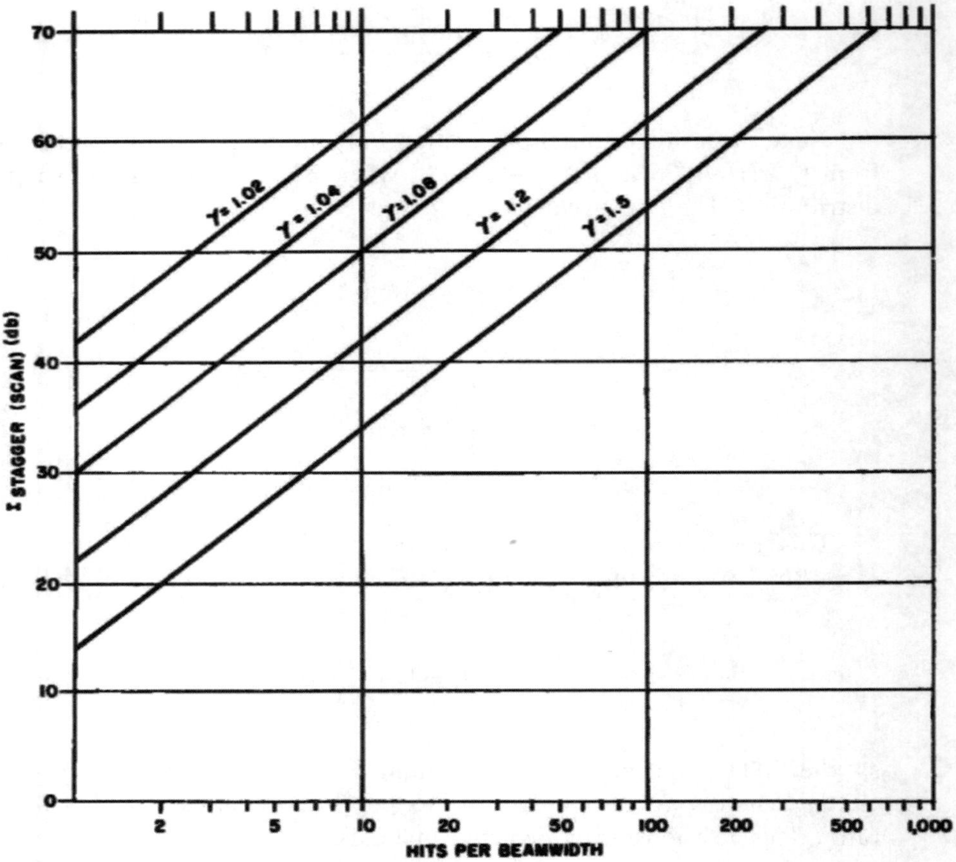

Figure 3.25 MTI improvement factor restrictions due to PRF staggering and scanning. (*From:* [24, p. 17-18].)

in the MTI canceller, the output must be blanked until the first of the longest range echoes from the next burst has passed the cancellers. Generally, each burst is processed coherently, followed by envelope detection, and then noncoherently integrated.

3.2.7 Improvement Factor Restriction Due to Motion Compensation Losses: TACCAR and DPCA

When a coherent MTI or PD radar is operated from a moving platform (e.g., airborne or naval radar), a Doppler shift is induced onto the clutter by the radar motion. This situation is depicted in Figure 3.16 for airborne radar in which the center of the ground clutter patch has an induced Doppler shift equal to $f_c = f_{ac} \cos \phi_0 \cos \theta_0$, where $f_{ac} = 2v_{ac}/\lambda$ is the Doppler due to the platform velocity, ϕ_0 is the antenna's depression angle, and θ_0 is the antenna's azimuth angle. In general, the mean Doppler shift to each clutter patch will be different (because either θ_0 or ϕ_0 will be different). Therefore, the clutter rejection notch needed to cancel clutter for airborne or naval radar cannot be fixed but must vary.

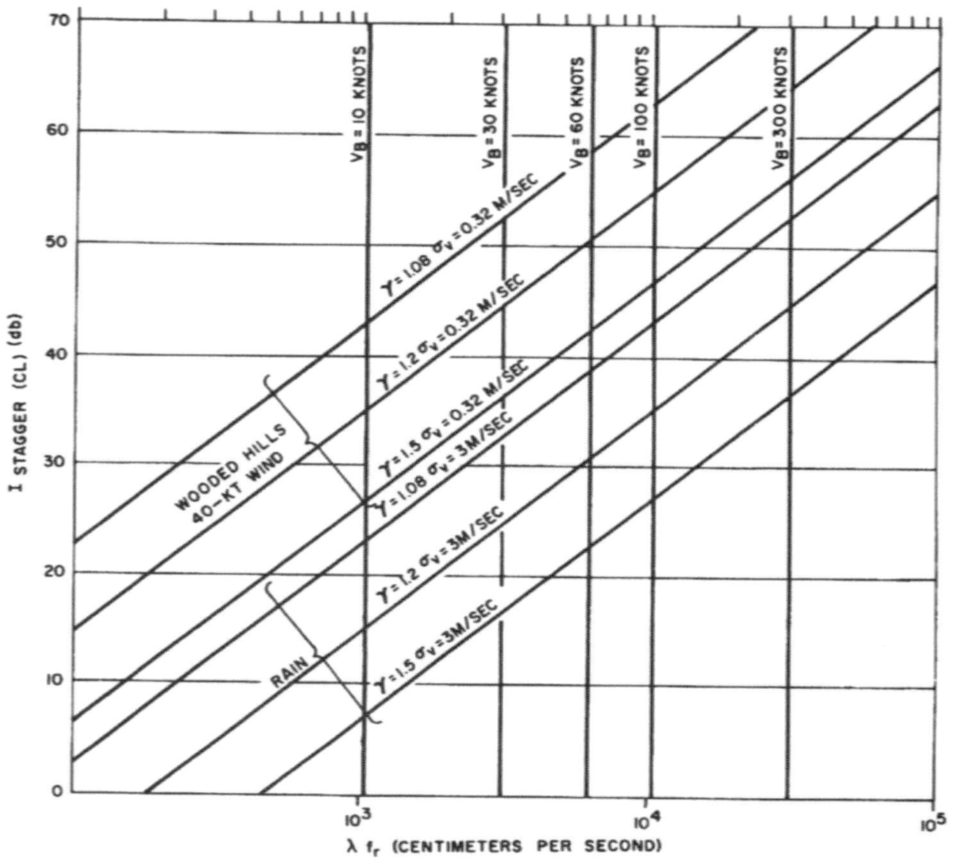

Figure 3.26 MTI improvement factor restrictions due to PRF staggering and internal clutter motion. (*From:* [24, p. 17-19].)

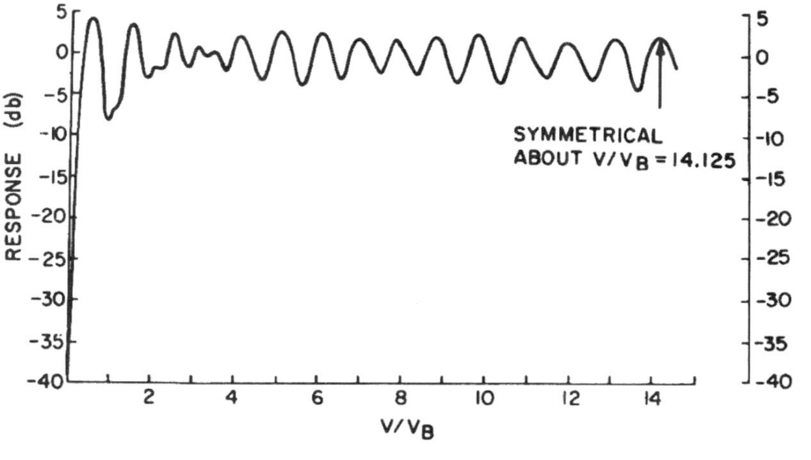

Figure 3.27 Frequency response of a four-period PRF stagger (25:30:27:31). (*From:* [24, p. 17-40].)

This problem is more difficult for airborne radar than shipborne radar because of the higher speeds and the greater range of depression angles, which results in a greater variation of the clutter spectrum.

Several methods are available to compensate the coherent Doppler processor for the mean Doppler shift induced by platform motion. One method, which is more appropriate against ground clutter for a high-PRF Doppler radar, uses the platform's inertial navigation system, which in conjunction with knowledge of the antenna's pointing direction provides sufficient information to appropriately shift the Doppler reference in the radar's processor (recall Figure 2.10). However, most Doppler processors employ an adaptive type of platform motion compensation that adjusts the rejection null of the frequency response of the Doppler processor to the mean Doppler frequency of the clutter spectrum. One implementation of this method adjusts the frequency of the coherent oscillator to compensate for the shift in the clutter Doppler frequency. This may be accomplished by mixing the output of the COHO with a signal from a tunable VCO to compensate for the shift in the clutter Doppler frequency [9, 26]. Another implementation inserts a phase shifter in one branch of an IF vector MTI canceller (I and Q channels) and adjusts its phase to shift the null of the frequency response (a phase shift in one branch corresponds to a frequency shift $2\pi f_d = \theta/T$, where T is the PRI) [12]. However, regardless of the motion compensation employed, there is generally an error in the adjustment of the Doppler processor null that restricts the improvement factor for these types of systems.

▶ Example 9 Improvement Factor for Motion-Compensated MTI

Motion compensation is used to adjust the mean frequency of the input clutter into an MTI to zero frequency. However, it is never perfect and results in a reduced improvement factor that depends on the clutter's frequency offset. To analyze this situation the improvement factor can be expressed as

$$I = G \cdot \mathrm{CA}$$

where G is the MTI's power gain and CA is the clutter attenuation. For Gaussian-shaped clutter the input clutter power is given by

$$P_c = \frac{1}{\sqrt{2\pi\sigma_f^2}} \int_0^{f_r} \left\{ \exp\left[-\frac{(f - fe_i)^2}{2\sigma_f^2}\right] + \exp\left[-\frac{(f - f_r - fe_i)^2}{2\sigma_f^2}\right] \right\} df$$

where f_r = PRF, fe_i is the offset frequency, and σ_f is the standard deviation of the clutter spectrum. The output power from the MTI canceller is given by

$$P_{co} = \frac{1}{\sqrt{2\pi\sigma_f^2}} \int_0^{f_r} 2\sin^{2n}\left(\frac{\pi f}{f_r}\right) \cdot \left\{ \exp\left[-\frac{(f - fe_i)^2}{2\sigma_f^2}\right] + \exp\left[-\frac{(f - f_r - fe_i)^2}{2\sigma_f^2}\right] \right\} df$$

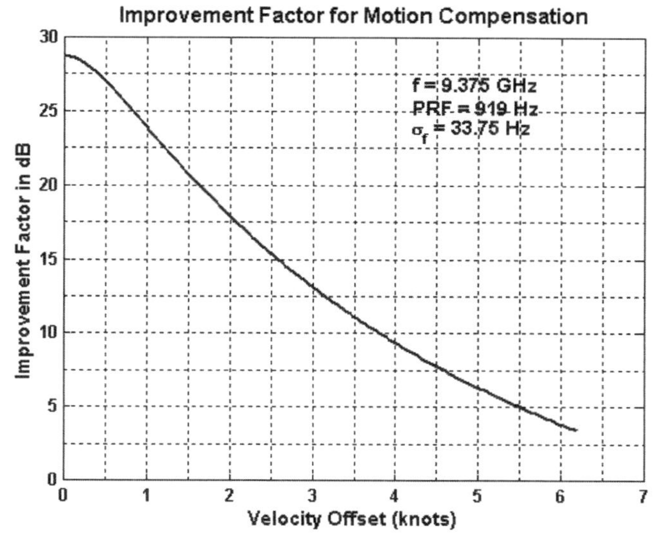

Figure EX-9.1 Improvement factor with motion compensation.

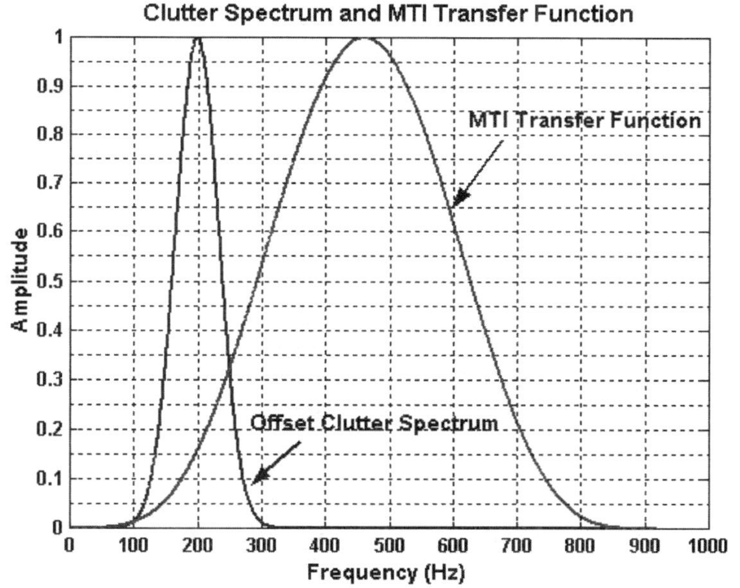

Figure EX-9.2 Offset clutter spectrum and double canceller transfer function.

where n is the number of cancellers. The power gain is given by

$$G = \frac{\displaystyle\int_0^{f_r} 2 \sin^{2n}\left(\frac{\pi f}{f_r}\right) df}{f_r}$$

MATLAB program motcoma.m is used to plot the improvement factor in decibels as a function of the velocity offset in knots for a double canceller, PRF = 919 Hz, $\sigma_f = 33.75$ Hz, and $f = 9{,}375$ MHz. This is shown in Figure EX-9.1. Also plotted in Figure EX-9.2 is the clutter spectrum (mean offset 200 Hz) and the MTI power transfer function.

▶

Figure 3.28 depicts the situation where the null of the Doppler filter is adjusted to a frequency that differs from the mean of the clutter spectrum. Figure 3.29 shows the resulting improvement factor for a single and double canceller as a function of the error frequency-to-PRF ratio (f_e/f_r) for a Gaussian-shaped clutter spectrum with various clutter spectrum standard deviations (σ_c). The improvement factor degrades as the clutter spectrum extends into more of the filter's passband,

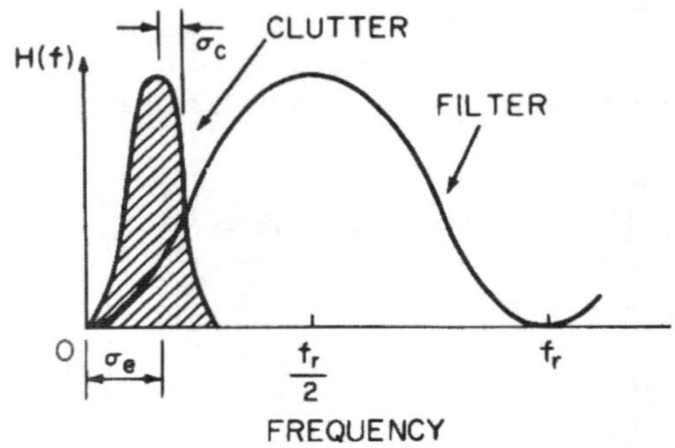

Figure 3.28 Effect of Doppler offset error on an MTI clutter filter. (*From:* [24, p. 18-5].)

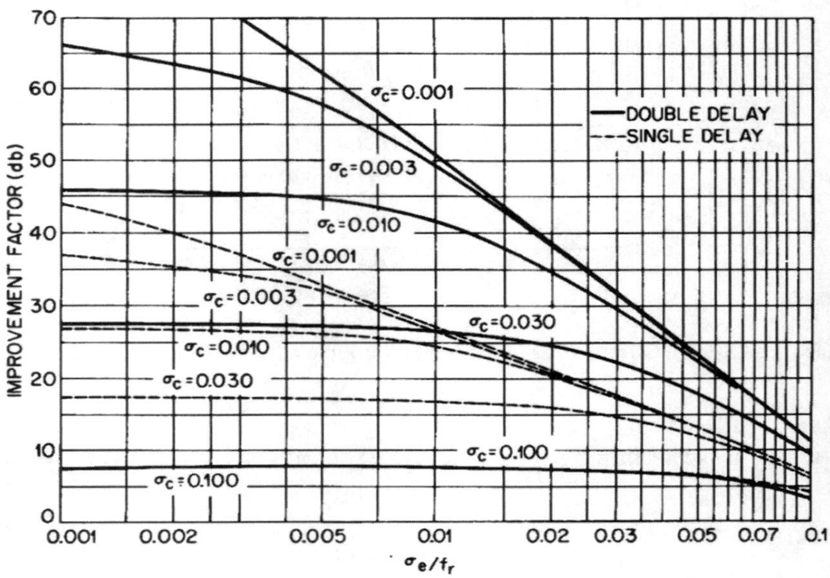

Figure 3.29 Improvement factor versus normalized Doppler offset as a function of clutter spectrum width. (*From:* [24, p. 18-5].)

which occurs as the frequency offset (f_e) becomes a larger fraction of the PRF (i.e., f_e/f_r increases) and also as the clutter spectral width becomes a larger fraction of the PRF (i.e., σ_c/f_r increases). In an airborne radar, the clutter spread (σ_c) due to platform motion is minimal when the radar's antenna points along the aircraft's flight path (see Section 3.2.3), whereas the offset error (f_e) is generally maximal under this condition, resulting in partially counterbalancing effects.

Time Average Clutter Coherent Airborne Radar (TACCAR) is a motion compensation technique used widely in both airborne and shipborne radars. A block diagram of one implementation of this technique is shown in Figure 3.30 [26]. The processing estimates the average Doppler shift of the returns from all objects illuminated by the radar's antenna and within some range sampling interval. This average Doppler is removed by shifting the relative difference frequency between the transmitter and the local oscillator.

The basic mechanism employed in TACCAR is a gated phase-locked loop, which is used to correct for the average Doppler shift during a selected interval of time as shown in Figure 3.30. For TACCAR, the phase-locked loop is gated on at a time interval that corresponds to the selected range interval, which contains the clutter to be used as the reference. The integrator and sample-and-hold network provides an estimate of the average Doppler within this selected range interval. This estimate is repeated during each PRI. The time constant of the TACCAR phase-locked loop must be large compared to the interpulse period to maintain the pulse-to-pulse coherency required for successful Doppler processing.

When TACCAR samples the clutter return from a particular range interval, the Doppler processor is compensated for the average Doppler frequency within this interval. Because the depression angle generally varies over the sample interval, this average Doppler frequency estimate is generally in error for every clutter patch. A performance evaluation of TACCAR must therefore consider the variation of the frequency error (f_e) of this estimate with range.

In the design of the usual TACCAR loop, a single sample range interval is selected, over which the average Doppler frequency (f_d) is developed. Because this estimate does not provide an optimum correction at all ranges, the problem is to select a range interval that will minimize the peak error over this interval. Defining the range extremes that bracket the estimation interval as $R_{\min}$ and $R_{\max}$ which have corresponding depression angles of $\phi_{\min}$ and $\phi_{\max}$, results in a clutter Doppler frequency estimate of

$$\hat{f}_d = f_{ac}\frac{\cos\,\theta_a\,(\cos\,\phi_{\max} + \cos\,\phi_{\min})}{2} \tag{3.92}$$

where f_{ac} is the radar's Doppler frequency and θ_a is the antenna's azimuth pointing angle. The TACCAR range error is then given by $f_e = \hat{f}_d - \overline{f_d}$, which is equal to

$$f_e = f_{ac}\cos\,\theta_a\left[\frac{\cos\,\phi_R - (\cos\,\phi_{\min})}{2}\right] \tag{3.93}$$

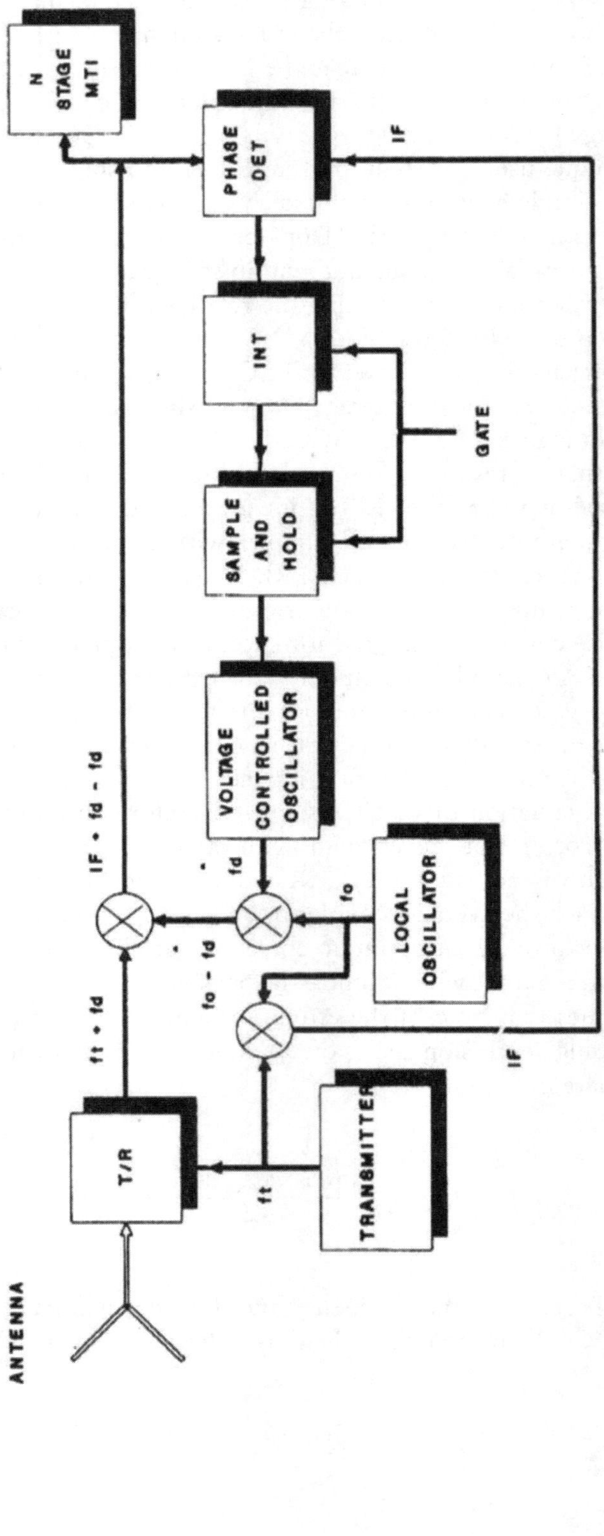

Figure 3.30 TACCAR block diagram. (*From:* [26, p. 3].)

where ϕ_R is the depression angle to each particular clutter cell. The peak error occurs at $\phi_R = \phi_{max}$ (see Figure 3.31), which results in a normalized peak error of

$$\hat{e}_n = \frac{f_e}{f_{ac} \cos \theta_a} = \frac{(\cos \phi_{max} - \cos \phi_{min})}{2} \tag{3.94}$$

The normalized TACCAR range error is depicted in Figure 3.32 as a function of the range normalized to the horizon range (R_{max}). In Figure 3.32, $R_{max} = \sqrt{h^2 + 2hr}$ with $R_{min} = R_{max}/10$, where h is the aircraft's height and r equals the 4/3 radius of the earth, approximately 4,600 nmi.

The TACCAR loss can be derived as

$$L_n = \left(\frac{f_e}{\sigma_c}\right)^{2N} + \sum_{k=1}^{N} \binom{2N}{2k} \left(\frac{f_e}{\sigma_c}\right)^{2(N-k)} \frac{1 \cdot 3 \ldots 2k - 1}{1 \cdot 3 \ldots N - 1} \tag{3.95}$$

where N is the number of delay sections, f_e is the frequency offset, and σ_c is the rms clutter width [26]. This function is plotted in Figure 3.33 for single, double, and triple cancellers. The improvement factor is then given by $I_N = I/L_N$, where I is the improvement factor as derived by assuming perfect motion compensation.

To illustrate the effect of TACCAR losses on the improvement factor, consider an airborne early warning radar with the following parameters: f = 425 MHz,

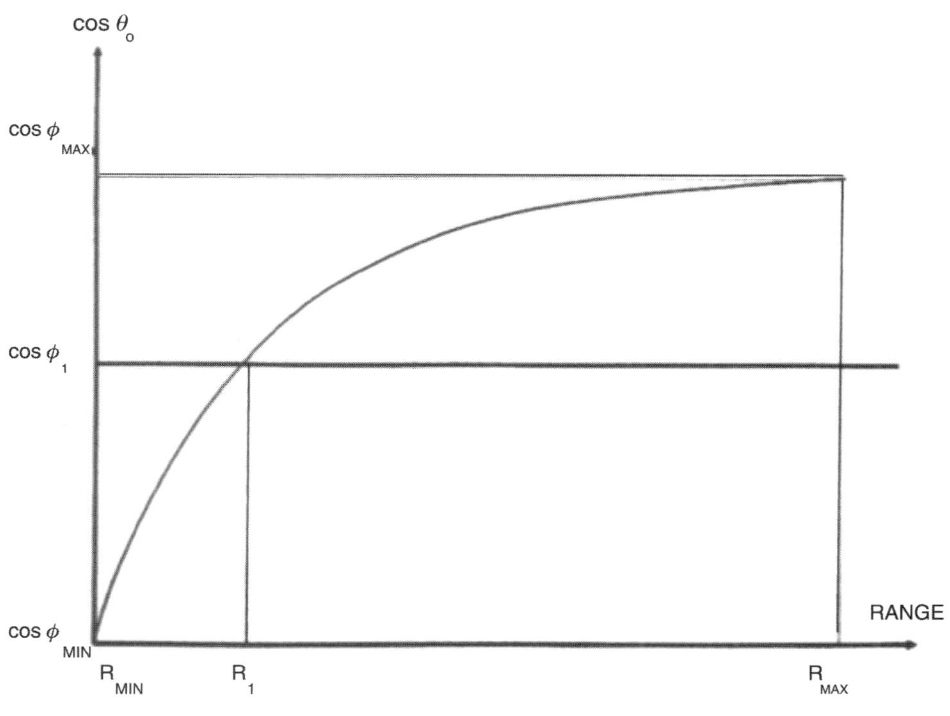

Figure 3.31 TACCAR range interval estimates. (*From:* [26, p. 15].)

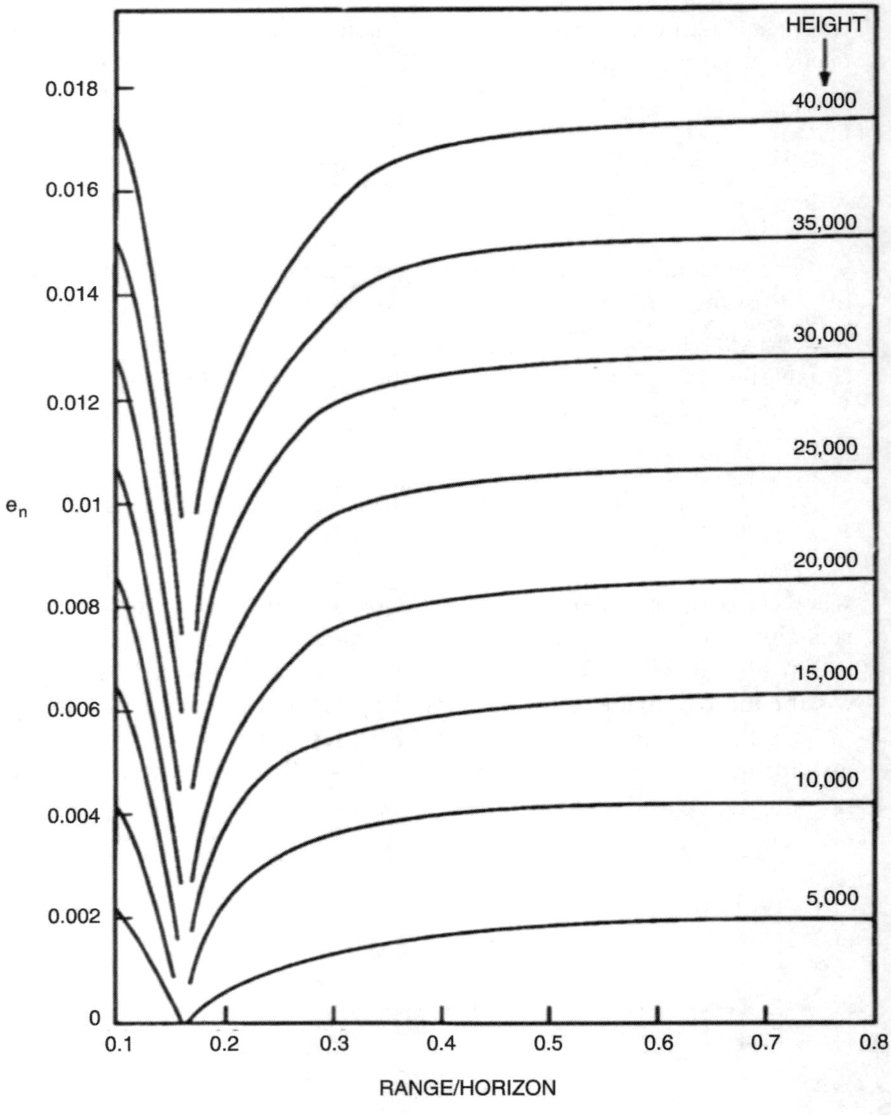

Figure 3.32 Normalized TACCAR errors. (*From:* [26, p. 7].)

$\alpha = 12$ rev/min, $D = 25$ ft, $f_r = 300$ pulses/s, and $v_{ac} = 300$ knots [9, 27]. The AEW radar is to operate over land with 25 knot surface winds at an altitude of 25,000 ft. The AEW radar uses a coherent double-canceller type of MTI with single-loop TACCAR motion compensation.

Consider the situation in which the radar antenna points along the aircraft's ground track ($\theta_a = 0°$). Internal clutter motion causes an rms velocity spread $\sigma_w = 0.0033 v_w^{1.261} = 0.2$ m/s, which results in an rms clutter frequency spread of $\sigma_i = 0.57$ Hz. The scanning clutter spread is given by $\sigma_s = 0.833\alpha/\pi\theta_B = 4.12$ Hz ($\alpha = 72°$/s, $\theta_B = 4.63°$). The combined rms clutter spread is then $\sigma_f = \sqrt{\sigma_c^2 + \sigma_s^2} = 4.16$ Hz, which results in an improvement factor $I_2 = 2/(\sigma_c T)^4$

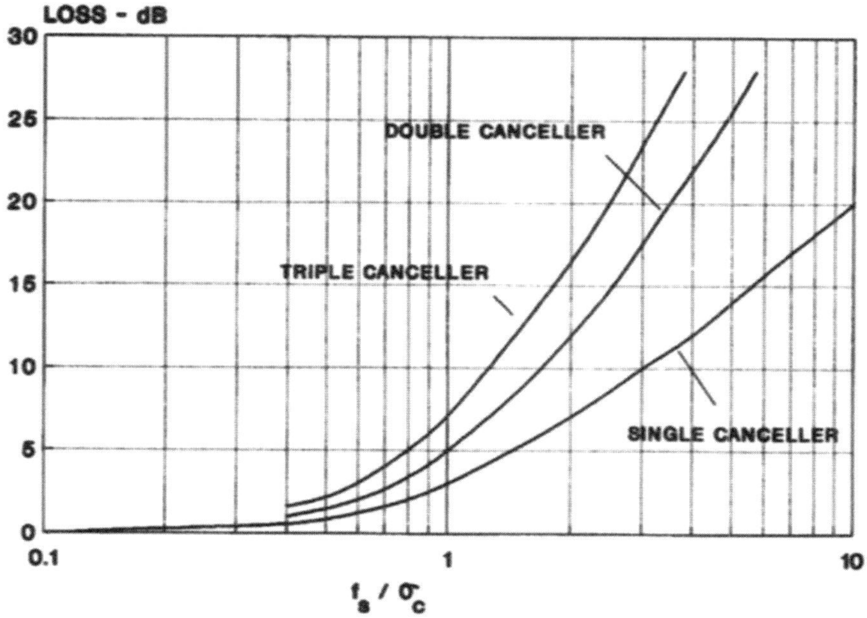

Figure 3.33 TACCAR losses for single, double, and triple MTI cancellers. (*From:* [26, p. 12].)

of 45.4 dB with perfect TACCAR motion compensation. Note that the frequency spread due to platform motion is negligible when the antenna points along the aircraft's ground track ($\sigma_{pm} \rightarrow 0$).

From Figure 3.32, the normalized TACCAR range error (e_n) is on the order of 0.01 over most of the radar's range coverage. This results in a frequency offset error of $f_e = e_n f_{ac} \cos \theta_a = 0.01 \cdot 438 = 4.38$ Hz when the radar's antenna points along the aircraft's ground track. The ratio $f_e/\sigma_c = 4.38/4.16 = 1.05$, which, from Figure 3.33, results in a TACCAR improvement factor loss for the double canceller ($N = 2$) of $L_2 = 3$ dB. The overall improvement factor is then given by $I_N = 45.4 - 3 = 41.4$ dB.

As the radar antenna points away from the aircraft's ground track, the frequency spread due to platform motion (see Section 3.2.3) tends to dominate, reaching its maximum value at $\theta_a = 90°$. For example, at $\theta_a = 90°$, the spectrum spread due to platform motion is $\sigma_{pm} = 0.528 v_{ac} \sin \theta_a / D = 10.7$ Hz, whereas, at $\theta_a = 30°$, the spread becomes $\sigma_{pm} = 5.35$ Hz. By combining with the other spreading factors, the total spread $\left(\sigma_f = \sqrt{\sigma_c^2 + \sigma_s^2 + \sigma_{pm}^2}\right)$ becomes $\sigma_f = 11.5$ Hz at $\theta_a = 90°$, resulting in an improvement factor of $I_2 = 37$ dB, the TACCAR range error is $f_e = 3.8$ Hz, the TACCAR loss is $L_2 = 1$ dB, and the resulting overall improvement factor thus is $I_N = 37 - 1 = 36$ dB.

The preceding example illustrates that only modest improvement factors are available from conventional AEW radars of this type. More advanced systems require improvement factors on the order of 60 dB. To accomplish this, several improvements are necessary:

- An integrated two-way main lobe-to-sidelobe antenna power ratio of 60 dB;
- System stability factors of 10^{-9};

- Use of triple instead of double cancellers;
- Use of MTI-coherent integrator combinations (16-pulse);
- TACCAR correction at three points in range;
- Improved platform motion cancellation using displaced phase center antennas (DPCAs).

The effect of the platform motion on the antenna aperture can be resolved into a normal component ($v_n = v_{ac} \cos \theta_0 \cos \phi_0$) and a tangential or rotational component ($v_t = v_{ac} \sin \theta_0 \cos \phi_0$). The normal component causes a shift in the mean frequency of the clutter spectrum that is corrected over a particular range interval by using a TACCAR loop as discussed earlier. The rotational component causes a widening of the clutter spectrum, which, as the example illustrated, limits the MTI improvement factor as the antenna points away from the aircraft's ground track. This effect can be corrected by using an antenna array that electronically displaces the phase center of the antenna by an amount equal and opposite to the physical displacement of the antenna caused by platform motion. The DPCA technique in conjunction with TACCAR thus provides complete motion compensation for an airborne MTI.

► Example 10 Motion Compensating Adaptive Digital MTI [28]

A common situation experienced in both land-based radar, operated in proximity to the sea, and naval radar, operated in proximity to land, is the occurrence of both land and sea clutter within its field of view. Ground clutter has zero velocity, whereas sea clutter has a variable mean velocity of up to 5 m/s (see Figure 1.20) [29]. In land-based radar the MTI canceller clutter notch is typically set to zero velocity to attenuate the heavy ground clutter, resulting in a decreased improvement factor when targets are detected over sea clutter. For naval radars a form of TACCAR is employed to tune the MTI canceller to the mean velocity of the sea resulting in improvement factor degradation over ground clutter. Figure EX-10.1 depicts a motion compensating adaptive digital MTI that allows detection of targets

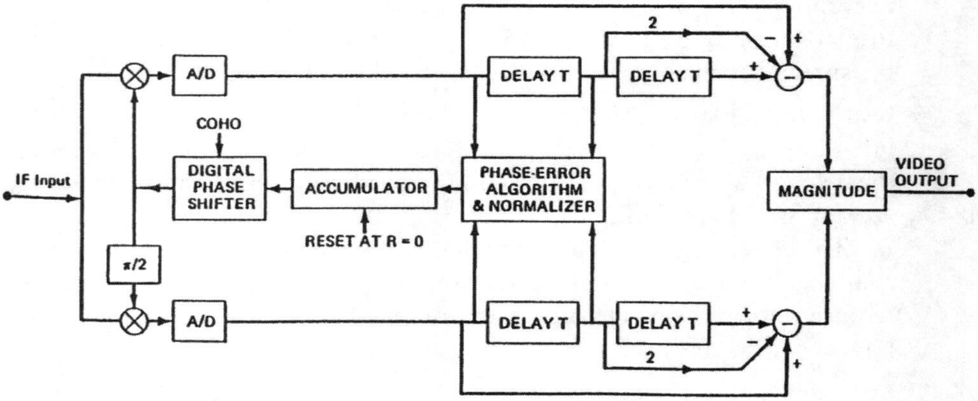

Figure EX-10.1 Motion compensating adaptive digital MTI. (*From:* [28, p. 15.61].)

over both sea and ground clutter. It adapts the MTI clutter notch within each range cell by estimating the clutter's Doppler frequency within that cell.

The phase error associated with the clutter's Doppler frequency is estimated from the relationship given by

$$d\phi = 2\pi f_d T$$

This phase error is then stored in an accumulator to determine the time (T_d) associated with a 2π radian phase shift. The Doppler frequency can then be estimated as

$$f_d = \frac{1}{T_d}$$

This Doppler frequency is then applied to the COHO through a digital phase shifter so that the Doppler shift from phase-to-phase is removed. This operation is depicted in Figure EX-10.2. The phase error averaging must be performed separately in both I and Q channels on the measured phase in each range cell due to the 2π ambiguity of the phase representation. Providing 5-bit resolution in the digital phase shifter ensures that the spurious sidebands will be at least 30 dB down from the primary response [30].

Examination of the diagram indicates that this form of motion compensation is appropriate for compensating mean clutter Dopplers that have velocities less than the velocity equivalent of the radar's PRF. Otherwise the phase error is ambiguous.

Further the compensation is ineffective when clutters with different velocities are present in the same range cell. This results for land–sea boundaries where

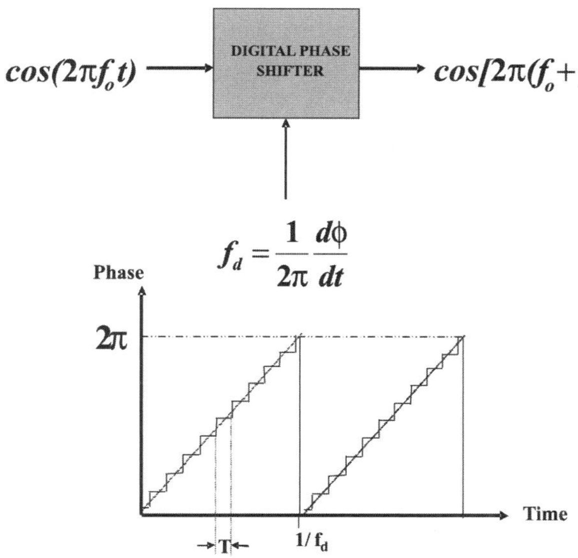

Figure EX-10.2 Serrodyning using a digital phase shifter.

different types of clutter are contained within the antenna's azimuth beamwidth. An approach that may be effective in this case is to employ a multiple pulse canceller whose notches are adaptively tuned to each mean clutter velocity [28].

Note that cells containing targets must be censored from the phase-measuring estimator depicted in Figure EX-10.1. This can be accomplished by censoring cells that have large phase changes on a pulse-to-pulse basis, indicating a target, and substituting phase measurements from surrounding cells not containing targets.

▶

The rotational Doppler shift caused by platform motion is given by $f_d = 2v_{ac}\,\theta \sin\,\theta_0 \cos\,\phi_0/\lambda$ (see Section 3.2.3 and Figure 3.17). The differential phase shift due to this component over an interpulse period (T) is given by

$$\eta = \int_0^T 2\pi f_d\,dt \approx 2\pi f_d T \tag{3.96}$$

or

$$\eta = \frac{4\pi v_{ac} T\theta \sin\,\theta_0 \cos\,\phi_0}{\lambda}$$

where v_{ac} is the aircraft's velocity, $T = 1/f_r$ is the radar's interpulse period, θ_0 is the antenna's azimuth angle, ϕ_0 is the antenna's depression angle, λ is the radar's wavelength, and θ is the azimuth angular displacement from the boresight. A vector diagram (see Figures 2.16 and 2.32) showing the phase rotation of the clutter vector due to the platform motion rotational Doppler component is depicted in Figure 3.34(a). Figure 3.34(b) shows how the phase advance can be compensated with the addition of two quadrature vectors (e_1 and e_2). The quadrature vectors are given by

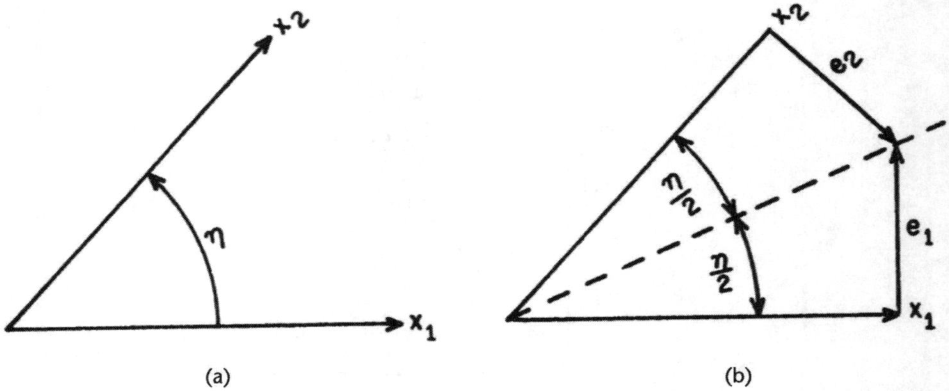

Figure 3.34 (a, b) Vectorial representation of DPCA platform motion compensation. (*From:* [31, p. 6].)

$$e_1 = jx_1 \tan\left(\frac{\eta}{2}\right) \tag{3.97}$$

$$e_2 = -jx_2 \tan\left(\frac{\eta}{2}\right)$$

where x_1 and x_2 are the clutter returns from successive transmitted pulses and η is the phase rotation. If the one-way power antenna pattern is $G(\theta)$, the amplitudes x_1 and x_2 are modulated by the two-way pattern $G^2(\theta)$, resulting in

$$e_1 = jG^2(\theta) \tan\left(\frac{\eta}{2}\right) \tag{3.98}$$

$$e_2 = jG^2(\theta) \tan\left(\frac{\eta}{2}\right) = \Delta(\theta)$$

These vectors can be realized by transmitting the pattern $G(\theta)$ and receiving pattern $G(\theta)$ plus an additional pattern $\Delta(\theta)$.

Figure 3.35 depicts the method used to apply DPCA compensation to the first section of a multistage MTI canceller [24, 31]. The returns received through the $\Delta(\theta)$ pattern are subtracted from the returns from the first pulse received through pattern $G(\theta)$ and added to the returns from the second pulse. The $\Delta(\theta)$ pattern is approximated by the difference pattern of a conventional amplitude monopulse antenna. Because the antenna pattern defined by $\Delta(\theta)$ generally must be approximated, the error in this pattern results in an imperfect DPCA correction and consequently in degraded MTI performance. Also, DPCA corrects only one canceller of a multiple-stage MTI, resulting in less than complete correction when double and higher order cancellers are employed.

The power transfer function for the canceller with DPCA correction depicted in Figure 3.35 is given by

$$|H_\eta(f)|^2 = \left(1 + \tan^2 \frac{\eta}{2}\right)[2 \sin \pi(f - f_d)T]^2 \, [2 \sin \pi f T]^{2N-2} \tag{3.99}$$

where η is the interpulse phase advance, f_d is the rotational component of the platform motion Doppler shift, T is the interpulse period, and $N = n - 1$ is the

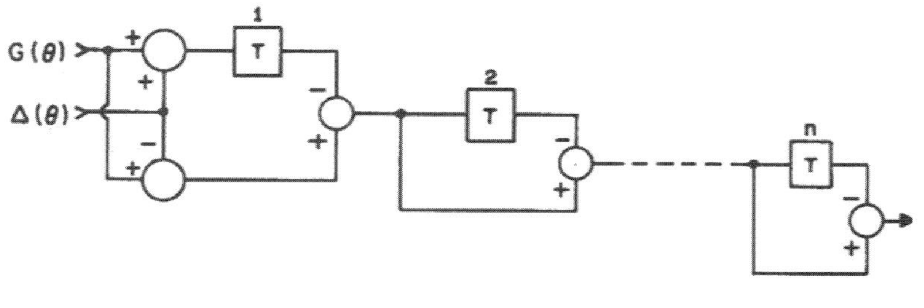

Figure 3.35 DPCA correction applied to the first stage of an N-stage MTI canceller. (*From:* [31, p. 14].)

number of MTI canceller stages [31]. This equation demonstrates that the null of the first canceller is shifted over the rotational component of the platform motion Doppler and the other canceller stages are not affected by the DPCA correction. The improvement factor with DPCA correction can be derived, assuming a Gaussian-shaped clutter spectrum, as follows:

$$I_N = 2^N\left(1 \cdot 3 \ldots 2N-1\right)\sqrt{2\pi\sigma_c}\Big) \int_{-\theta_0}^{\theta_0} \frac{G^4(\theta)\,d\theta}{n!} \int_{-\theta_0}^{\theta_0} G^4(\theta)(1+\tan^2 \pi f_c T) H_n(\theta)\,d\theta \tag{3.100}$$

where the integration is over the main beam (θ_0) of the antenna pattern [31]. The improvement factors derived from this equation for a (sin x)/x antenna pattern with double and triple cancellers is depicted in Figures 3.36 and 3.37, respectively, as a function of the fraction of the antenna aperture displaced per interpulse period ($x = v_p T \cos \phi_0 \sin \theta_a / D$). The dashed curves represent the MTI improvement factor with platform motion compensated by DPCA and the solid curves illustrate the platform motion limitation.

Comparison of the MTI improvement factor for a single canceller in conjunction with a (sin x)/x antenna pattern shows that if the fraction of antenna aperture displaced per interpulse period is less than $x = 1/10$, the MTI improvement factor almost perfectly compensates for the rotational platform motion [31]. Therefore, in this case, the improvement factor for surface clutter generally will be determined by the antenna scanning and internal motion limitations (see Sections 3.2.1 and 3.2.2). Antenna scanning modulation of the clutter return can also be compensated by using an implementation similar to that employed for DPCA [24]. Scanning motion compensation adds one-half the interpulse differential amplitude caused by scanning the two-way power antenna pattern, $\Delta G^2(\theta)$, to one pulse and then

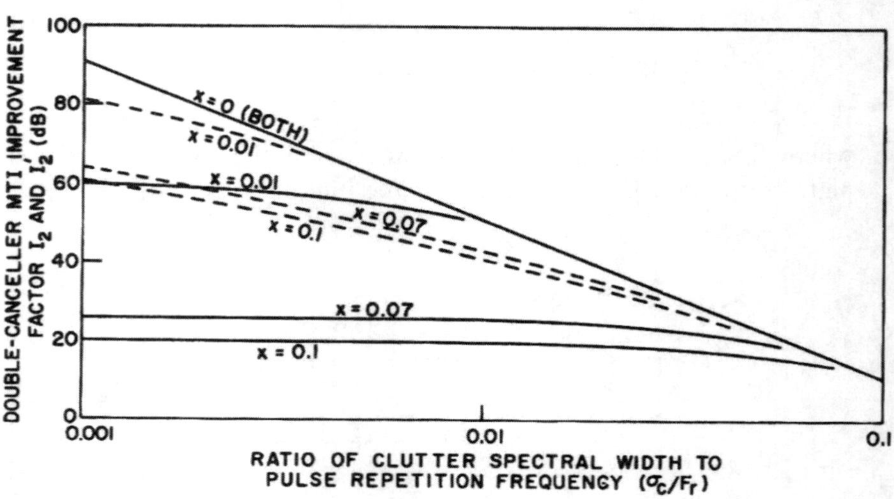

Figure 3.36 Comparison of the MTI improvement factor for a double canceller with and without DPCA correction. (*From:* [31, p. 8].)

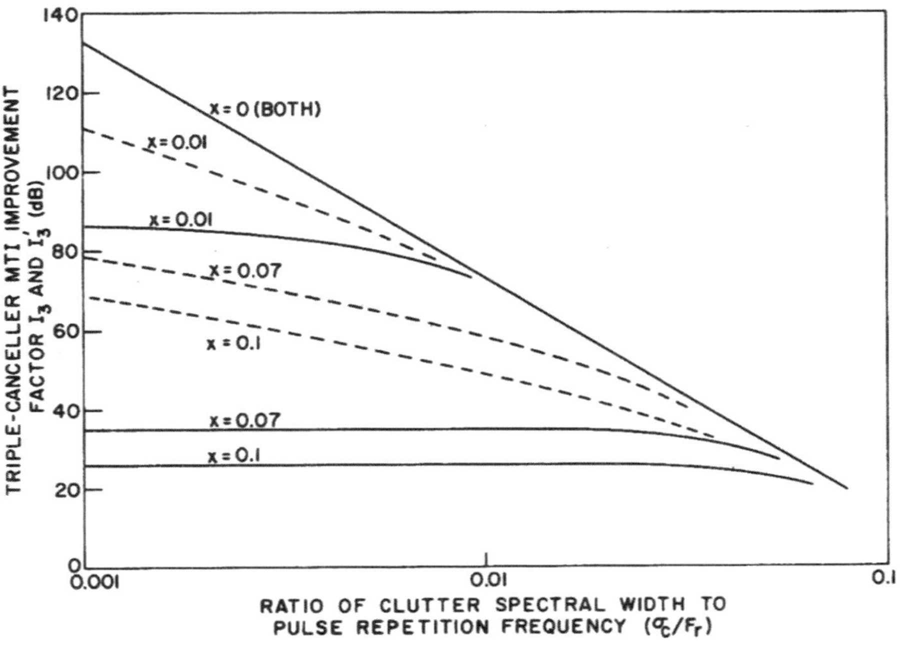

Figure 3.37 Comparison of the MTI improvement factor for a triple canceller with and without DPCA correction. (*From:* [31, p. 9].)

subtracts it from the next pulse, using the same implementation depicted for DPCA (see Figure 3.35).

Figures 3.36 and 3.37 illustrate that DPCA compensation of double and triple cancellers is not perfect, but there is appreciable correction, especially for ground clutter, where the intrinsic spectral widths are small. These figures also illustrate the additional improvement that can be achieved by a more complete compensation. Perfect compensation for motion is represented by the curves labeled $x = 0$, and comparing these curves with the dashed curves indicates the additional improvement that ideal motion compensation may provide.

3.2.8 Combining Improvement Factor Effects

The previous sections each derived an improvement factor limitation on the assumption that all other effects were absent. In practice, many of the effects will be simultaneously present and the composite improvement factor must consider the interaction of all such effects.

The basic assumption in combining the individual effects is that all are statistically independent and each can be represented as a separate modulation on a correlated carrier. When this condition exists, then the received clutter voltage is given by $v_t = a_t b_t c_t d_t \ldots \cos \omega t$, which results in a demodulated autocorrelation function, given by $\rho_c(\tau) = \rho_a(\tau)\rho_b(\tau)\rho_c(\tau) \ldots$. The power spectral density is found by Fourier transforming the ACF as

$$S_c(\omega) = S_a(\omega) * S_b(\omega) * S_c(\omega)\ldots \qquad (3.101)$$

where * indicates the convolution procedure

$$S_1(\omega) * S_2(\omega) = \int_{-\infty}^{\infty} \frac{S_1(\omega) S_2(\omega - y)\, dy}{2\pi}$$

The procedure is particularly convenient when the individual effects can be represented by a Gaussian-shaped PSD $S(\omega) = P_c \exp\left(-\omega^2/2\sigma_c^2\right)/\sqrt{2\pi}\sigma_c$, where σ_c is the standard deviation of the clutter spectrum (rad/s) related to the 3-dB bandwidth (ω_3 – rad/s) as $\sigma_3 = \omega_3/\sqrt{\ln 4} = 0.849\omega_3$. Note that when synchronous detection is involved, the 3-dB video bandwidth (B_3) is given by $B_3 = \sqrt{\ln 4}\sigma_c/\pi = (\sigma_c/2)/1.334$.

When individual Gaussian-shaped power spectral densities are convolved, the resulting clutter power spectral density is also Gaussian shaped, but has a variance σ_c^2, which is the sum of the variances of each of the individual effects:

$$\sigma_c^2 = \sigma_1^2 + \sigma_2^2 + \ldots + \sigma_n^2 \tag{3.102}$$

For example, if the situation involves clutter spreading due to internal motion (σ_i), scanning modulation (σ_s), platform motion (σ_{pm}), and transmitter instability (σ_x), the variance of the Gaussian-shaped clutter spectrum is

$$\sigma_c^2 = \sigma_i^2 + \sigma_s^2 + \sigma_{pm}^2 + \sigma_x^2 \tag{3.103}$$

If noncoherent MTI using square-law detection were involved, the clutter variance would be given by $\sigma_{nc}^2 = 2\sigma_c^2$, although the spectrum would still be Gaussian shaped. In a coherent MTI, the target spectrum also is spread by antenna scanning modulation as with the clutter spectrum and often is represented by a Gaussian-shaped spectral density.

With a noncoherent MTI, the predetection target spectrum is convolved with the clutter spectrum to form the video target spectrum (i.e., $\sigma_t^2 = 2\sigma_s^2 + \sigma_i^2 + \sigma_{pm}^2 + \sigma_x^2$). The video coherent and noncoherent MTI target and clutter spectra for both fixed and airborne radars are tabulated in Table 3.7.

The improvement factor can be found by considering the composite clutter spectrum with its associated composite variance. Alternatively, the improvement factor can be found by computing the individual improvement factor for each additive noiselike effect and combining these as

$$I^{-1} = I_1^{-1} + I_2^{-1} + \ldots + I_N^{-1} \tag{3.104}$$

This relationship can be developed by considering $I = \overline{G}\,\text{CA}$, where $\overline{G}$ is the MTI gain and CA is the clutter attenuation. The clutter attenuation is given by

Table 3.7 Video Target and Clutter Spectrum Variances for Fixed and Airborne Radars

Coherent MTI	*Noncoherent MTI*
Fixed:	
Target σ_s^2	$2\sigma_s^2 + \sigma_x^2 + \sigma_i^2$
Clutter $\sigma_s^2 + \sigma_x^2 + \sigma_i^2$	$2\sigma_s^2 + 2\sigma_x^2 + 2\sigma_i^2$
Airborne:	
Target σ_s^2	$2\sigma_s^2 + \sigma_x^2 + \sigma_i^2 + \sigma_{pm}^2$
Clutter $\sigma_s^2 + \sigma_x^2 + \sigma_i^2 + \sigma_{pm}^2$	$2\sigma_s^2 + 2\sigma_x^2 + 2\sigma_i^2 + 2\sigma_{pm}^2$
Variances:	
σ_i^2 = Internal motion	σ_s^2 = Scanning modulation
σ_x^2 = Transmitter instability	σ_{pm}^2 = Platform motion

$$\mathrm{CA} = \frac{\int_0^{\infty} \sum_{L=1}^{N} S_n(\omega)\, d\omega}{\int_0^{\infty} \sum_{L=1}^{N} S_N(\omega)\, |H(\omega)|^2\, d\omega} \tag{3.105}$$

where $S_N(\omega)$ is the individual effect clutter spectral density and $H(\omega)$ is the MTI filter response. If we interchange summation and integration, and note that the numerator always equals the clutter power, we have a relationship combining the clutter attenuation for the individual effects (CA_N) as $(\mathrm{CA})^{-1} = (\mathrm{CA}_1)^{-1} + (\mathrm{CA}_2)^{-1} + \ldots + (\mathrm{CA}_N)^{-1}$. Multiplication of the clutter attenuation by the MTI gain then provides the relationship for the improvement factor:

$$\left(\overline{G} \cdot \mathrm{CA}\right)^{-1} = \left(\overline{G} \cdot \mathrm{CA}_1\right)^{-1} + \left(\overline{G} \cdot \mathrm{CA}_2\right)^{-1} + \ldots + \left(\overline{G} \cdot \mathrm{CA}_N\right)^{-1} \tag{3.106}$$

When additive noiselike disturbances (i.e., system instabilities) are combined with multiplicative effects (i.e., clutter spreading), these effects must be considered separately before finding the overall improvement factor. For example, consider primary instability effects due to transmitter frequency change, STALO frequency change, COHO frequency change, pulse timing jitter, pulsewidth jitter, and pulse amplitude change, where each individual instability causes an improvement factor limitation of 50 dB. The composite improvement factor due to system instabilities would then be equal to $I_s = I_{IND}/6$ or 42.2 dB.

Now, assume that the improvement factor limitation due to clutter spectral spreading is caused by antenna scanning and internal clutter motion effects, where the limitation due to each effect is 43 dB when a double canceller MTI is used. The improvement factor for a double binomial canceller using Gaussian-shaped clutter is given by $I_c = 2/(\sigma_{c1} T)^4$ and results in $\sigma_{c1} T = 0.1$ radian for the example. The improvement factor considering both effects is then given by

$I_c = 2/\left(\sigma_{c1}^2 + \sigma_{c2}^2\right)^2 T^4$, which reduces to $I_c = 1/2\,(\sigma_{c1} T)^4$ or 37 dB. The resulting overall improvement factor is hence given by $I = \left(I_c^{-1} + I_s^{-1}\right)^{-1}$ or 35.9 dB for the example.

3.3 Oscillator Effects on Doppler Radar Performance

The performance of a coherent Doppler radar system often is limited by phase noise generated in the STALO or master oscillator. In the master oscillator power amplifier system (see Figures 1.15 and 2.20), the same STALO is used as the reference oscillator in the receiver and, with an IF offset, as the transmitted signal. When phase noise is present on the STALO, the resulting transmitted signal contains noise sidebands that extend into the frequency region occupied by Doppler-shifted targets [38]. When this noise-transmitted signal is reflected from clutter scatterers, it induces a noise background that is sometimes characterized as "*transmitted clutter*." This noise is both a function of the strength of the clutter scatterer returns and the level of sideband phase noise measured with respect to the carrier amplitude. The situation is further complicated when the return signal is mixed with the noisy STALO in the radar receiver to produce an IF signal that is subsequently synchronously detected by using the offset oscillator. The phase noise in the STALO is thus correlated between transmission and reception with the degree of correlation depending on the range of the scatterers with respect to the radar. This correlation produces a cancellation effect for short-range clutter scatterers (which pose the most severe problem), while multiplying the noise power by a factor of 4 for the most unfavorably positioned scatterers.

STALO phase noise, although important for low-PRF MTI radars, becomes the critical performance factor for high-PRF PD radars, because in high-PRF PD radars the highly ambiguous range response causes close-in clutter returns to compete with long-range target returns. The resulting clutter attenuation required for a high-PRF PD system is then on the order of 80 to 100 dB compared to the 50 to 60 dB required for low-PRF Doppler systems.

Figure 3.38 depicts the block diagram of a typical Doppler radar transmitter-receiver system. If the phase noise associated with the microwave reference oscillator (STALO) is given by ϕ_t, then the transmitted signal (T_0) can be represented as $v_t = \cos(\omega_t t + \phi_t)$. The local oscillator is given by $v = \cos(\omega t + \phi_t)$, where the phase noise from the IF offset oscillator is assumed to be negligible compared to that from the microwave STALO due to the large frequency ratio between the two sources. The return from a clutter scatterer at range R is then given by $v_c = \cos[\omega_t(t - t_d) + \phi_t - t_d]$, where $t_d = 2R/c$ is the two-way scatterer return time delay. The receiver output is obtained by mixing the clutter signal with the local oscillator signal as $v_r = \cos(\omega_{IF} t - \omega_t t_d + \phi_c)$, where $\phi_c = \phi_{t-t_d} - \phi_t$. If the phase noise is assumed to be given by $\phi_t = \Delta\phi \sin \omega_m t$, where ω_m is the noise radian frequency, the resulting output of the coherent Doppler radar synchronous phase detector can be represented as $v_p = \Delta\phi[\sin \omega_m(t - t_d) - \sin \omega_m t]$, which reduces to $v_p = \Delta\phi 2 \sin(\omega_m t_d/2) \cos(\omega_m t + \phi_0)$, where the voltage correlation factor $2 \sin(\omega_m t_d/2)$ directly multiplies the phase noise voltage spectral density, $S_\phi(\omega)$. The PSD seen by the Doppler processor due to phase noise is therefore given by

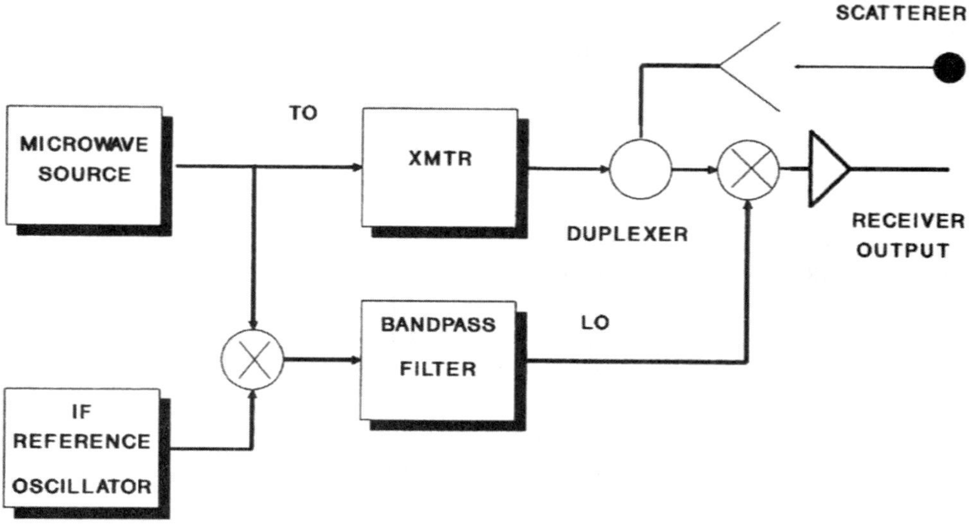

Figure 3.38 Block diagram of common oscillator Doppler radar transmitter-receiver. (*From:* [33, p. 161].)

$S_p(\omega) = 4 \sin^2 (\omega_m t_d/2) S_\phi(\omega)$, where $S_\phi(\omega)$ is the master oscillator (STALO) phase noise PSD, and t_d is a time delay, which depends on the range to the clutter scatterer. The correlation factor is plotted in Figure 3.39 for $t_d = 25\ \mu s$, where a peak (+6 dB) is reached at $f_m = 1/2t_d$ (20 kHz in Figure 3.39), followed by a null at $f_m = 1/t_d$ (40 kHz in Figure 3.39).

In a low-PRF coherent MTI using a binomial canceller, the clutter attenuation can be expressed as

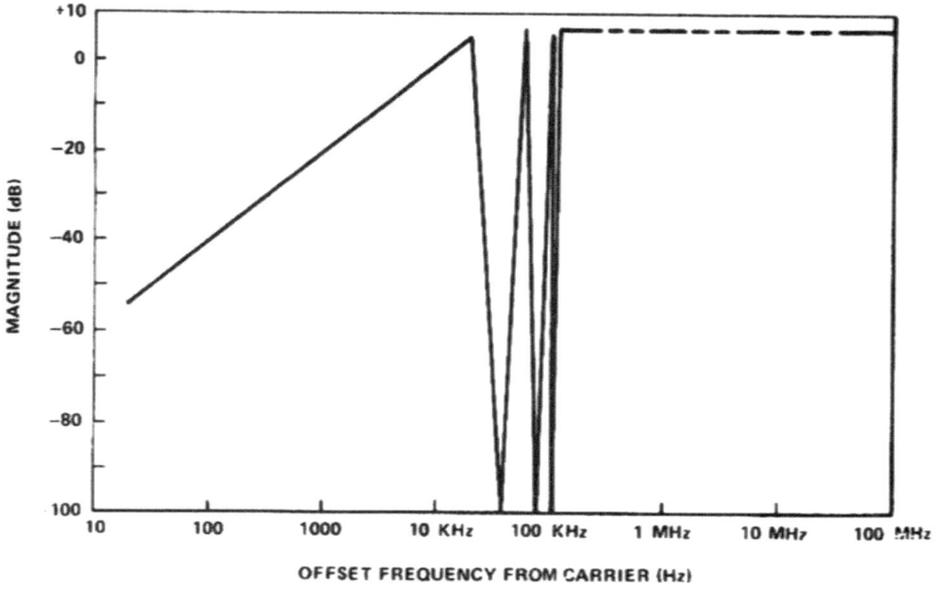

Figure 3.39 Correlation factor for master oscillator phase noise ($t_d = 25\ \mu s$). (*From:* [33, p. 92].)

$$\mathrm{CA} = 8 \int_{0}^{\infty} \sin^2(\pi f t_d) \sin^{2N}(\pi f T) S_\phi(f)\, df \tag{3.107}$$

where N is the number of canceller stages ($n = N + 1$ pulses processed), t_d is the two-way range delay, T is the radar's interpulse interval, and $S_\phi(f)$ is the master oscillator phase noise power spectrum normalized to the carrier power level. Figure 3.40 depicts several permissible phase noise master oscillator spectra ($S_\phi(f)$) for a single canceller MTI that provide the indicated level of clutter attenuation at an approximately 20 nmi range (t_d = 250 μs) with a radar PRF of 1,000 pulses/s (T = 1,000 μs). The oscillator phase noise spectrum is composed of three distinct components [34]: flicker noise, the phase noise spectrum of which is proportional to $1/f^3$ at small values of f_m; external additive oscillator noise, the phase spectrum of which is "white" or flat with frequency; and internal oscillator noise, the phase spectrum of which is proportional to $1/f^2$.

Several measured master oscillator phase noise spectral densities are shown superimposed on the permissible phase noise spectra in Figure 3.41. Generally, cavity-stabilized oscillators (e.g., dielectric resonant oscillators, electrostatically focused klystrons, cavity-stabilized klystrons) have less phase noise than multiplied (e.g., varactor multiplied, step recovery) crystal oscillators, which generally have a low fundamental frequency, under 100 MHz. Also, in general, lower fundamental frequency master oscillators have less phase noise than higher fundamental frequency oscillators, resulting in lower frequency MTI radars having greater performance potential than higher frequency MTI radars.

To understand the situation for high-PRF Doppler radars, we must first consider the clutter attenuation requirement for such radars. The requirements are different for surface-based and airborne PD radars due to both the geometrical and platform-motion aspects of these radars. The clutter return for surface-based high-PRF radars

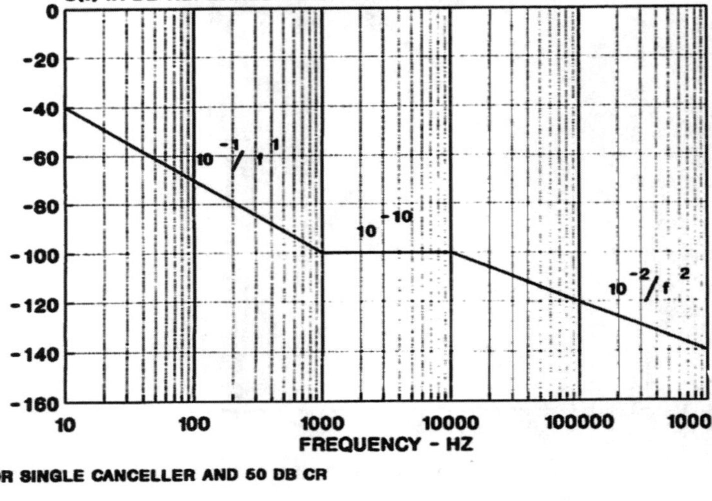

Figure 3.40 Permissible master oscillator phase noise spectra (t_d = 250 μs). (*From:* [32, p. 163].)

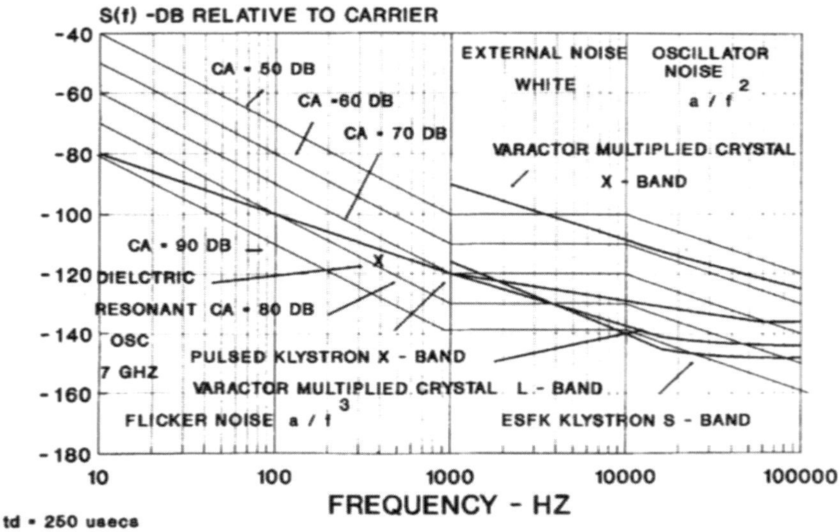

Figure 3.41 Permissible master oscillator phase noise spectral densities ($t_d = 250\ \mu s$). (*From:* [32, p. 163].)

is similar to that for a CW radar except that the CW clutter return is multiplied by the duty factor ($D_u = \tau / T$, where τ = pulsewidth, and T = interpulse period) of the radar. This produces a large main beam close-in clutter return that is a function of the radar's elevation antenna pattern with respect to the clutter scatterers in proximity to the radar.

The clutter return for airborne high-PRF radar is a function of the platform's speed and altitude as well as the antenna's radiation pattern. The main beam clutter return is modulated by the transmitter's phase noise into the frequency band covered by the Doppler filter bank, where it then competes with the Doppler frequencies returned from targets. The main beam clutter return for airborne radar is less than that of surface-based radar because the minimum range returning ground clutter always must be greater than the altitude of the aircraft. However, the main beam intercepts the clutter patch at an increased grazing angle, which has higher reflectivity than the lower grazing angles of surface-based radars.

The clutter power returned to surface-based CW radar can be expressed as

$$C = \left(\frac{P_t G^2 \lambda^2 \theta_a}{(4\pi)^3 L} \right) \cdot \int_0^\infty \left(\frac{\sigma^0 F^4}{r^3} \right) dr \tag{3.108}$$

where P_t is the transmitter power, θ_a is the azimuth beamwidth (rad), σ^0 is the clutter backscatter coefficient, F^4 is the two-way power antenna pattern propagation factor, L is the transmission and beam shape loss, and r is the range to the clutter scatterer [35]. If a Gaussian-shaped antenna beam is assumed, $F^4 = \exp\left(-5.54\phi^2 / \phi_e^2\right)$, where ϕ_e is the 3-dB elevation beamwidth. Further, using the small angle approximation $\phi = h / r$, where h is the height of the antenna's phase center above the surface, and assuming that σ^0 is independent of the grazing angle results in

$$\int_0^\infty \left(\frac{\sigma^0 \exp\left(\frac{-5.54h^2}{\phi_e r^2}\right)}{r^3} \right) dr = \frac{\sigma^0 \phi_e}{(3.33h)^2} \tag{3.109}$$

The CW clutter power can then be expressed as

$$C = \frac{P_T G^2 \lambda^2 \theta_a \phi_e^2 \sigma^0}{(4\pi)^3 (3.33h)^2 L} \tag{3.110}$$

Further simplifications can be made by substituting $G = 4\pi/(\theta_a \phi_e) L_n$ and $\theta_a = 1.5\lambda/D_a$, where D_a is the azimuth antenna aperture and $L_n = 1.2$ is the antenna pattern loss experienced when aperture weighting is used to provide low antenna sidelobes. The resulting CW clutter power is given by

$$C = \frac{P_t \sigma^0 \lambda D_a}{400h^2} \tag{3.111}$$

where P_t is the transmitter power, σ^0 is the clutter backscatter coefficient, λ is the radar wavelength, D_a is the antenna's azimuth width, h is the antenna's phase center height above the surface, and a beam shape loss of $L_B = 1.33$ is assumed. The high-PRF PD clutter average power return can then be determined as

$$C = \frac{P_t d_u \sigma^0 \lambda D_a}{400h^2} \tag{3.112}$$

where d_u is the PD radar's duty cycle. If the clutter power returned from the transmitter carrier (center line of spectrum) is desired, then $C = P_t d_u^2 \sigma^0 \lambda D_a / 400h^2$.

A further refinement of the PD radar clutter return equation can be obtained by using the constant reflectivity (γ) model in which the backscattering coefficient is given by $\sigma^0 = \gamma/\sin\phi$ (see Section 1.3.2.1). If we use the approximation $\sin\phi = h/r$, the integral in the clutter power equation is

$$\int_0^\infty \left(\frac{\sigma^0 F^4}{r^3} \right) dr = \int_0^\infty \frac{\gamma h \exp\left(\frac{-5.54h^2}{\phi_e^2 r^2}\right)}{r^4} dr \tag{3.113}$$

which equals $\gamma\phi_e^3/29.4h^2$. Substituting this into the CW clutter power equation results in

$$C = \frac{P_t G^2 \lambda^2 \theta_a \phi_e^3 \gamma}{29.4(4\pi)^3 L h^2} \tag{3.114}$$

which can be further simplified ($\phi_e = 1.5\lambda / D_e$, where D_e is the antenna's elevation aperture), resulting in

$$C = \frac{P_t d_u \lambda^2 \gamma D_a}{533 D_e h^2} \tag{3.115}$$

where the terms are as previously defined.

As an example, assume an X-band radar with a circular aperture ($D_a = D_e$) operating at a height $h = 3$m above a ground surface with $\gamma = 0.1$. For a transmitted power of 10 kW and a duty cycle of $d_u = 0.1$ ($P_{\text{ave}} = 1$ kW), the average clutter return power is $C = 1.9 \times 10^{-5}$ W, whereas the clutter power associated with the central line is 1.9×10^{-6} W [23]. If a Doppler filter bank using filters with a Doppler bandwidth of 1 kHz ($B_D = 1$ kHz, $t_0 = 10^{-3}$ sec) were employed and if the total clutter energy were concentrated in the filter, then it would equal 1.9×10^{-8} W-s. The noise energy level assuming a noise temperature $T_s = 3{,}000$ K would be $4 \cdot 10^{-20}$ W/Hz. The clutter-to-noise ratio would then be C/N = 4.75×10^{11} = 117 dB. Hence, the clutter attenuation requirement (to reduce the clutter's rms level to the noise level) for this example would be on the order of 120 dB, thus establishing a stringent requirement on the master oscillator phase noise stability and the Doppler filter sidelobe capability. Note that the correlation factor must be taken into account when establishing the master oscillator's allowable phase noise spectrum.

In the airborne case, the ground clutter return for high-PRF PD radar is spread over a velocity range equal to $\pm 2v_{ac}/\lambda$ as shown in Figure 3.42 (see Section 1.1.4). The prominent features of the clutter return are: (1) the main lobe clutter return, which is most severe when the radar is in a look-down mode and occurs at a mean frequency equal to $f_{ML} = 2v_{ac} \cos\theta_a \cos\phi_e / \lambda$; (2) the altitude line clutter, which represents the specular-like reflections received from the scatterers directly under the aircraft returned through the radar's sidelobes at a mean frequency of $f = 0$; and (3) the sidelobe and backlobe clutter returns, which range in frequency from $f_d = 2v_{ac}/\lambda$ for the forward-directed sidelobes to $f_d = -2v_a/\lambda$ for the backward-directed backlobes. Also shown in Figure 3.42 are the frequency placements of the Doppler filter bank: in the clutter-free region to detect closing aircraft or missile targets; in the forward clutter sidelobe region to detect chased targets or for Doppler beam sharpening of ground returns.

Figure 3.43 also depicts the effects of transmitter phase noise, which spreads the clutter return spectrum into the frequency region occupied by the Doppler filter bank. This effect can be visualized by considering the transmitted spectrum as a bank of frequencies about each individual line spectrum (due to phase noise), rather than as a single line spectrum. When this spectrum is convolved with each individual clutter scatterer (to form the total clutter return), the effect is to modulate each frequency component of the pure clutter return spectrum (no phase noise) by the phase noise spectrum. Because the phase noise spectrum has considerable frequency

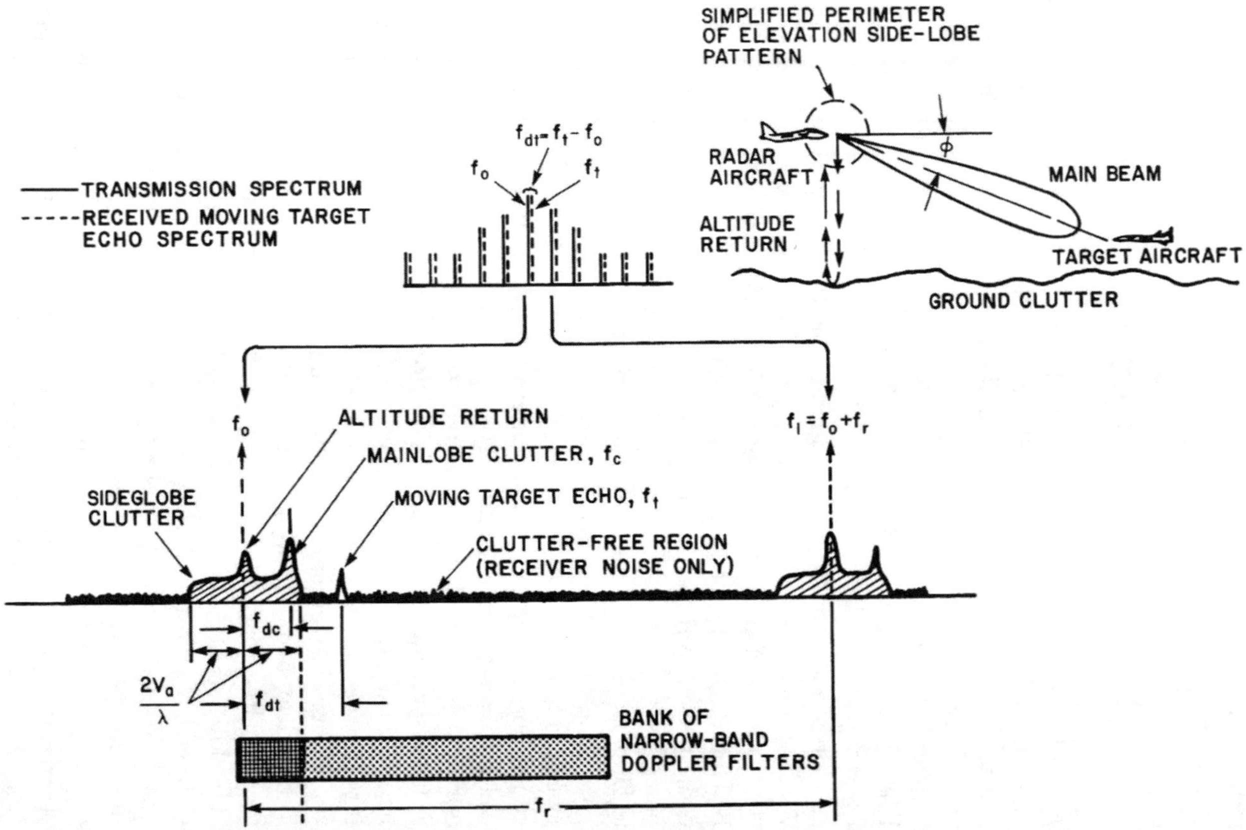

Figure 3.42 Clutter return and Doppler filter placement for high-PRF PD radar. (*From:* [34, p. 166].)

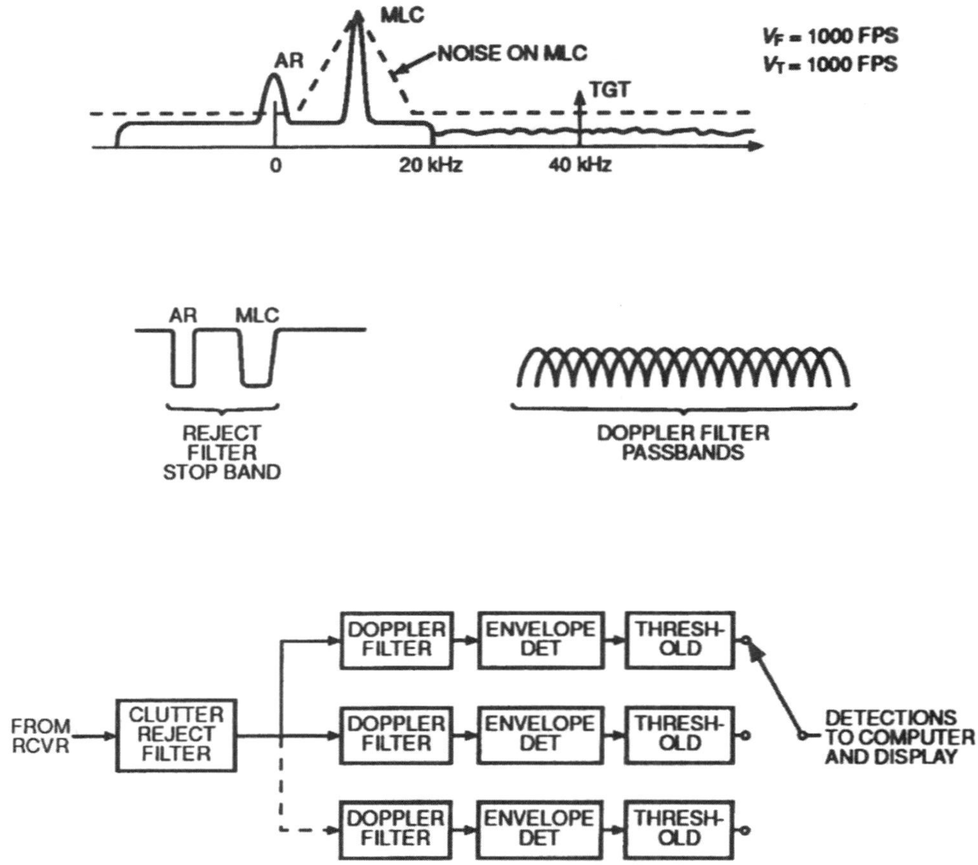

Figure 3.43 High-PRF PD search processing. (*From:* [32, p. 167].)

extent, it translates components of the main lobe clutter return from the notch rejection filter into the frequency region occupied by the Doppler filter bank. The objective is then to hold the phase noise spectrum to a level with respect to the carrier level such that the translated clutter spectrum has an rms value in each Doppler filter that is less than the receiver noise level. When this objective cannot be accomplished (generally in the Doppler filters closest to the main lobe clutter frequency), an adaptive CFAR is needed to desensitize the Doppler filter to this clutter leakage.

To consider the effect of main lobe clutter leakage, we need to determine the magnitude of the main lobe clutter return. Assuming that the airborne radar uses a pencil beam, the approximate equivalent clutter radar cross section is given by $\sigma_c = \sigma^0 \pi(\theta R)^2/4$, where θ is the two-way power antenna beamwidth $\left(\theta = \theta_{3\,\mathrm{dB}} / \sqrt{2}\right)$. Substituting this into the radar equation results in

$$C_{ML} = \frac{P_t d_u G^2 \theta_3^2 \sigma^0 \lambda^2}{\left(\sqrt{2}\,16\pi\right)^2 R^2 L} \tag{3.116}$$

where d_u is the radar's duty factor, θ_3 is the antenna's one-way power 3-dB beamwidth, R is the range to the MLC patch, and L is a system loss term. Substituting $G = \pi^2/\theta_3^2 = (\pi D/\lambda)^2$, where D is the circular antenna aperture diameter, results in $C_{ML} = P_t d_u \sigma^0 D^2/50R^2 L$ [23].

Assuming the same parameters as for the surface PD radar (i.e., $P_t = 10$ kW, $d_u = 0.1$, $\sigma^0 = 0.1$) with $D = 0.7$ m, $L = 1$, and $R = 10$ km, results in an average main lobe clutter power return equal to $C_{ML} = 10^{-8}$ W or a return associated with the central spectral line of 10^{-9} W. The clutter energy concentrated in a 1 kHz Doppler filter is 10^{-11} W-s and the receiver noise energy ($T_s = 3{,}000$ K) is $N = 4 \times 10^{-20}$ W/Hz, resulting in a C/N = 2.5×10^8 = +84 dB, requiring a clutter attenuation of the order of 80 dB. This results in a less stringent requirement on oscillator phase noise for the airborne case than the surface radar case. A typical master oscillator phase noise spectrum is shown in Figure 3.44.

As an additional example, consider X-band airborne intercept high-PRF PD radar. The parameters are $P_t = 2.5$ kW, $d_u = 0.5$, $\theta_3 = 2.2°$, $NF = 5$ dB, $B_d = 200$ Hz, $G = 37$ dB, $L = 3.9$ dB, and $\sigma^0 = 0.15$. The main lobe clutter return in the central line at $R = 100$ nmi then equals $C_{ML} = -78$ dBm (the central line average power is $d_u^2 P_t$). The noise power in the 200 Hz Doppler bandwidth equals $P_N = (-174 - 3 + 23) = -149$ dBm, where the noise power in the Doppler filter is reduced by the duty factor ($d_u = -3$ dB) because the receiver is gated on only half the time. The resulting C/N at 100 nmi is +71 dB, which implies that the master oscillator phase noise spectrum must be on the order of 80 dB below the carrier power level over the frequency region covering the Doppler filter bank. At 100 nmi, the central line power returned from a 5 m^2 target is about -149 dBm, so that $R_0 = 100$ nmi (R_0 is the range where S/N = 1). The main lobe clutter and target power levels for this example are plotted in Figure 3.45.

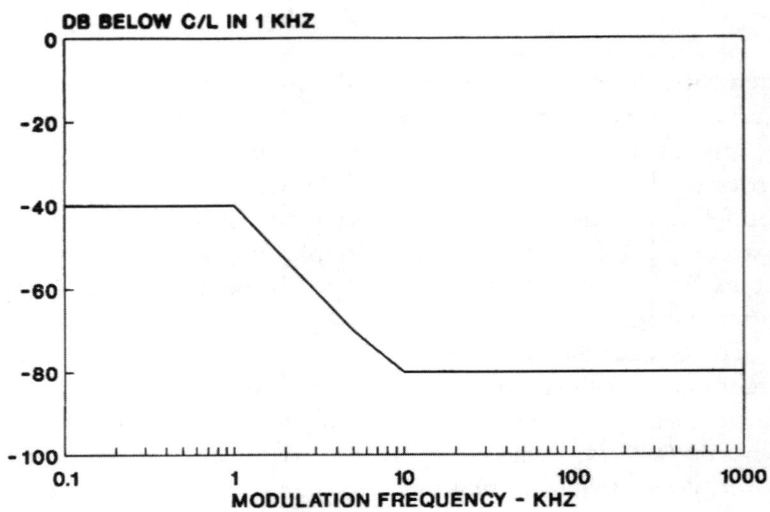

Figure 3.44 Typical high-PRF PD phase noise spectrum. (*From:* [32, p. 169].)

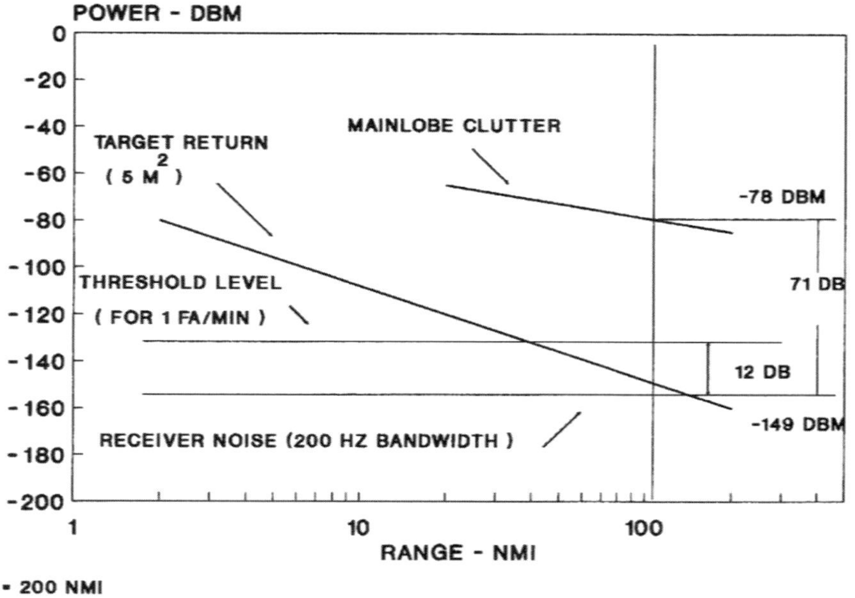

Figure 3.45 Airborne intercept PD radar main lobe clutter and noise levels. (*From:* [32, p. 169].)

References

[1] Marcum, J., "A Statistical Theory of Target Detection by Pulsed Radar," *IRE Trans.*, Vol. IT-6, No. 2, April 1960, pp. 209–211; also in S. Haykin, (ed.), *Detection and Estimation*, Halstead Press, 1976.

[2] Swerling, P., "Probability of Detection for Fluctuating Targets," *IRE Trans.*, Vol. IT-6, No. 2, April 1960, pp. 269–308; also in S. Haykin, (ed.), *Detection and Estimation*, Halstead Press, 1976.

[3] Schleher, D. C., *Introduction to Electronic Warfare*, Norwood, MA: Artech House, 1986.

[4] Barton, D. K., (ed.), *The Radar Equation*, Vol. 2 of *Radars*, Dedham, MA: Artech House, 1974.

[5] Schleher, D. C., "Performance Comparison of MTI and Coherent Doppler Processors," *Proc. IEEE Int. Radar Conference*, London, November 1982, pp. 154–158.

[6] Kanter, I., "A Generalization of the Detection Theory of Swerling," *Proc. IEEE EASCON Conf.*, Washington, D.C., September 1974, pp. 198–205; also in S. Haykin, (ed.), *Detection and Estimation*, Halstead Press, 1976.

[7] Dillard, G., and J. Richard, "Performance of an MTI Followed by an Incoherent Integrator for Nonfluctuating Signals," *Proc. IEEE Int. Radar Conference*, Washington, D.C., April 1980, pp. 194–199.

[8] Emerson, R., "First Probability Densities for Receivers with Square Law Detectors," *J. Applied Physics*, Vol. 24, No. 9, September 1953, pp. 1168–1176.

[9] Schleher, D. C., (ed.), *MTI Radar*, Dedham, MA: Artech House, 1978.

[10] Emerson, R., *Some Pulsed Doppler MTI and AMTI Techniques*, R-274, Rand Corporation, March 1954; also in D. C. Schleher, (ed.), *MTI Radar*, Dedham, MA: Artech House, 1978, pp. 77–142.

[11] Andrews, G., *Optimal Doppler Radar Processing*, Report 7727, Naval Research Laboratory, Washington, D.C., May 21, 1974.

[12] Skolnik, M. I., *Introduction to Radar Systems*, 2nd ed., New York: McGraw-Hill, 1980.

[13] Berkowitz, R., *Modern Radar*, New York: Wiley, 1965.

[14] Kretschmer, F., F. Lin, and B. Lewis, "A Comparison of Noncoherent and Coherent MTI Improvement Factors," *IEEE Trans. on Aerospace and Electronic Systems*, Vol. AES-19, No. 3, May 1983, pp. 398–404.

[15] Skolnik, M. I., (ed.), "MTI Radar," Chapter 17 in *Radar Handbook*, 1st ed., New York: McGraw-Hill, 1970.

[16] Stimson, G., *Introduction to Airborne Radar*, Hughes Aircraft Company, El Segundo, CA, 1983.

[17] Nathanson, F., *Radar Design Principles*, New York: McGraw-Hill, 1969.

[18] Fishbein, W., S. Graveline, and O. Rittenback, *Clutter Attenuation Analysis*, ECOM-2808, U.S. Army Electronics Command, Fort Monmouth, NJ, March 1967; also in D. C. Schleher, (ed.), *MTI Radar*, Dedham, MA: Artech House, 1978.

[19] Skolnik, M. I., *Introduction to Radar Systems*, 3rd ed., New York: McGraw-Hill, 2001.

[20] Billingsley, J., and J. Larrabee, *Measured Spectral Extent of L- and X-Band Radar Reflections from Wind-Blown Trees*, Report CMT-57, MIT Lincoln Laboratory Lexington, MA, February 1987.

[21] Ward, H., and W. Shrader, "MTI Performance Degradation Caused by Limiting," *IEEE EASCON Record*, Washington, D.C., September 1968, pp. 168–174.

[22] Ward, H., "The Effect of Bandpass Limiting on Noise with a Gaussian Spectrum," *Proc. IEEE*, Vol. 57, No. 11, November 1969, pp. 2089–2090c.

[23] Barton, D. K., *Modern Radar System Analysis*, Norwood, MA: Artech House, 1988.

[24] Skolnik, M. I., (ed.), "Airborne MTI," Chapter 18 in *Radar Handbook*, 1st ed., New York: McGraw-Hill, 1970.

[25] Muehe, C., et al., "New Techniques Applied to Air-Traffic Control Radars," *Proc. IEEE*, Vol. 62, No. 6, June 1974, pp. 716–723.

[26] Andrews, G., *Airborne Radar Motion Compensation Techniques, Evaluation of TACCAR*, Report NRL-7407, Naval Research Laboratory, Washington, D.C., April 1972.

[27] Ap Rhys, T., and G. Andrews, *AEW Radar Antennas*, AGARD Conference Preprint No. 139, *Antennas for Avionics*, 1974; also in D. C. Schleher, (ed.), *MTI Radar*, Dedham, MA: Artech House, 1978.

[28] Skolnik, M. I., (ed.), "MTI Radar," Chapter 15 in *Radar Handbook*, 2nd ed., New York: McGraw-Hill, 1990.

[29] Pidgeon, W., "The Doppler Dependence of Radar Sea Return," *J. of Geophysical Research*, February 1968, pp. 1333–1341.

[30] Schleher, D. C., *Electronic Warfare in the Information Age*, Norwood, MA: Artech House, 1999.

[31] Andrews, G., *Airborne Radar Motion Compensation Techniques, Evaluation of DPCA*, Report 7426, Naval Research Laboratory, Washington, D.C., July 1972.

[32] Schleher, D. C., *MTI and Pulsed Doppler*, Norwood, MA: Artech House, 1991.

[33] Goldman, S., "Oscillator Phase-Noise Proves Important to Pulse Doppler Radar Systems," *Microwave Systems News*, Vol. 14, No. 2, February 1984, p. 92.

[34] Raven, R., "Requirements on Master Oscillators for Coherent Radar," *Proc. IEEE*, Vol. 54, No. 2, February 1966, pp. 237–243.

[35] Barton, D. K., (ed.), *CW and Doppler Radar*, Vol. 7 of *Radars*, Dedham, MA: Artech House, 1974.

[36] Brennan, L., and I. Reed, "Quantization Noise in Digital Moving Target Indication Systems," *IEEE Trans. AES,* AES-2, November 1966, pp. 655–658.

[37] Nathanson, F., *Radar Design Principles,* 2nd ed., New York: McGraw-Hill, 1999.

[38] Everett, C. L., "Phase Noise Contamination in Doppler Spectra," *Microwave Journal,* September 1996, pp. 105–122.

CHAPTER 4

Clutter Properties and Data

Radar clutter is defined as unwanted echoes, typically from the ground, sea, rain or other precipitation, chaff, birds, insects, and aurora. The designer of a Doppler radar is generally interested in discriminating against these unwanted returns; hence, we must understand their properties to provide an effective radar design.

Distributed clutter is defined as radar signals backscattered from a spatially continuous distribution of scatterers with no subset of scatterers predominating. This type of clutter is generally observed from sea return, precipitation return, and returns from most undeveloped land terrain (excluding mountainous terrain). When the radar's wavelength is very small compared to the spatial resolving power of the radar, a reasonable approach is to model distributed clutter as the summation of phase-independent returns from a large number of small scatterers. This leads to representing the clutter as a random process having the statistical properties of Gaussian noise. Hence, the expected (mean) clutter power received from any incremental resolution element (surface or volume region) depends solely on the physical area of that region and a proportionality constant, σ^0, which represents the local clutter cross section per unit area, or η, which represents the local clutter cross section per unit volume (see Section 1.5.2.1). The significance of this result is that, if σ^0 (or η) is specified at each point of the clutter surface, we can completely determine the temporal, spectral, and statistical properties of the clutter input to any radar. Consequently, any description that permits σ^0 (or η) to be realistically specified over the expected clutter surface can be regarded as a suitable model for distributed clutter [1].

The clutter backscattering coefficient σ^0 for distributed surface clutter is determined by a number of factors, which include the radar's wavelength, the transmitted and received polarizations, the grazing angle of the incident wave to the surface, and the type of surface or terrain. The grazing angle to the terrain surface can never be known exactly and, hence, must be specified in terms of the mean surface. This will cause σ^0 to fluctuate for a fixed grazing angle due to local variations in the surface slope. Similarly, to isolate and identify every conceivable environment is not possible, resulting in the use of broad classifications for the surface type (e.g., desert, farmland, sea state number). Within each broad environmental type, σ^0 will vary further because of the large variety of possible subclasses contained within each classification. As is apparent from this discussion, σ^0 is not given by a precise number for each surface classification, but is a random variable characterized by a probability distribution. In Section 1.5.1, we described the Gaussian or Rayleigh clutter model, which generally represents the statistical properties of σ^0

for distributed clutter. With this model, the complete first-order statistical properties of σ^0 are determined by the specification of $\overline{\sigma^0}$, the mean value of σ^0.

A considerable amount of experimental data is available on the measurement of $\overline{\sigma^0}$ for various surfaces. Unfortunately, these data vary widely, which generally reflects the complex nature of the surface and the difficulty in extracting the precise condition under which the data were obtained. This has led to the use of highly averaged values for $\overline{\sigma^0}$ that represent the combination of data obtained through many experiments under supposedly similar conditions, but with wide variability in the data [2].

The concept of distributed clutter represents a consistent theory whereby a physical model results in an analytical model (i.e., the Gaussian random process), which can then be used to design and analyze the performance of a Doppler radar. With this model, the clutter's amplitude pdf, its correlation with time and spectral density, and its variation over time and location is completely specified (see Section 1.5). Further, the form of the clutter process is independent of the radar's parameters (e.g., pulsewidth and antenna pattern) and homogeneous over the complete clutter field. With these significant advantages, the distributed clutter model is not surprisingly used as the basis for most clutter descriptions and the objective of most experimental programs is to measure the parameters (e.g., $\overline{\sigma^0}$, amplitude pdf, correlation function) associated with this model.

Unfortunately, distributed clutter very rarely exists in nature. Generally, clutter is not stationary; it varies both temporally and spatially. Clutter is sensitive to the parameters of the radar used to induce it and cannot in general be represented by a stationary random process. There is much confusion between its short- and long-term temporal properties and its spatial distribution.

To illustrate some of these points, consider the situation for land clutter, which represents one of the most stringent environments for a Doppler radar (see Figures 1.20 and 1.21). Figure 4.1 depicts how land clutter appears on ground-based search radar with several azimuth beamwidths (1.3° and 6°) and pulsewidths (2 and 10 μs) [3]. The spatial variation of the clutter is apparent, as is its sensitivity to the radar's parameters. In addition, the dynamic range of land clutter generally varies over a much wider dynamic range than is predicted by the Rayleigh clutter distribution (e.g., 20 dB from the 5% to the 99% probability points for Rayleigh clutter) [4]. Because of this, many investigators have used the log-normal or the Weibull amplitude probability distributions to describe land and sea clutter with wide dynamic ranges (see Section 1.5.4).

One problem with representing land clutter as distributed clutter is the presence of the many discrete scatterers over developed land terrain. Discrete clutter, sometimes called specular or *point clutter*, is backscattered from physically isolated dominant reflecting objects on the terrain. This clutter includes radar signals backscattered from both developed areas (e.g., cities and urban areas, where the discrete scatterers occur in clusters) and physically isolated structures (e.g., water towers and silos). Radar returns from mountainous terrain may also be regarded as a form of discrete clutter in which the dominant reflectors are the scattering facets of the mountain slopes.

Discrete clutter signals [1, 5] are distinguished from distributed clutter in that they generally exhibit larger amplitudes (e.g., the RCS for an individual scatterer

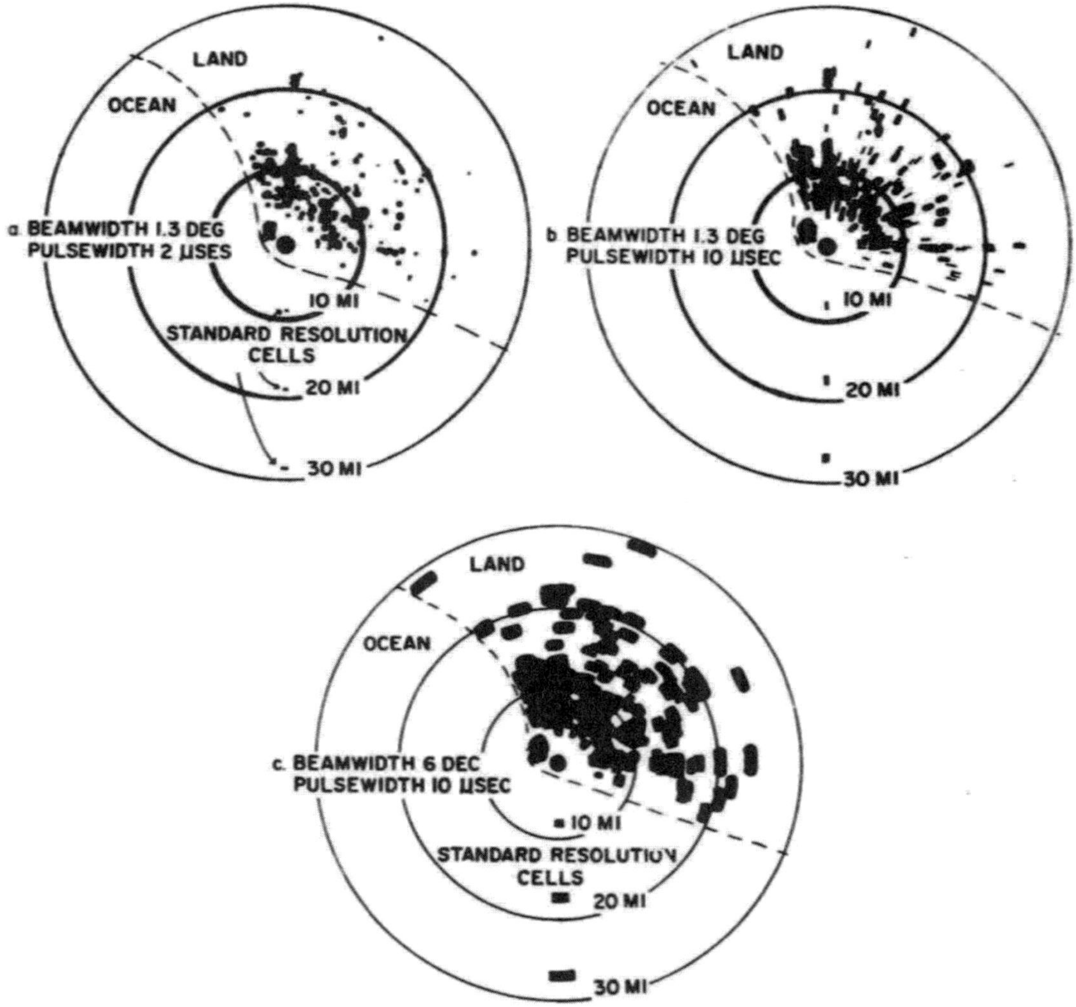

Figure 4.1 Land clutter for several azimuth beamwidths and pulsewidths. (*From:* [3, p. 296].)

might range from 10^2 to 10^6 m^2), fluctuate more slowly, and cannot be described by Gaussian statistics. Experimental evidence shows that discrete scatterers exist in sufficient numbers over land to have a significant effect on radar performance. In general, land clutter can be viewed as a composite between distributed and discrete clutter. The uncertainty of the discrete field of scatterers over land terrain is generally the mechanism that imparts a wide dynamic range to land clutter and its large spatial variability. Note that most experimental clutter measurements over developed land include composite discrete and distributed clutter, which cannot be separated into component parts, and hence measurement statistics of parameters such as $\overline{\sigma^0}$ and the probability distribution of σ^0 represent combined statistical data.

Other anomalies occur when land clutter is viewed at low grazing angles such as with ground-based radar. For relatively flat terrain at low grazing angles, the clutter reflectivity can be viewed as the product of the factor $\sigma^0 \mathbf{F}_c^4$ [6]. The first is the distributed clutter backscatter coefficient (σ^0) and the second represents the two-way propagation factor for the radar, clutter, and path geometry. For lower grazing angles, below the critical angle, the propagation factor (F_c^4) produces a significant effect and, hence, must be evaluated to determine the clutter reflectivity. For higher grazing angles, F_c^4 generally approaches a value of 1. The wide dynamic range observed for low grazing angles of both land and sea clutter can be predicted by the dependence of the propagation factor on the fourth power of the clutter source height. A difference of 1m or 2m of height can result in a change in tens of decibels in observed reflectivity [6].

Backscatter from clutter returns can be either coherent or noncoherent [4]. If the phase of a wave is constant or varies in a deterministic manner, it is called *coherent*. If the phase of a wave is random and uniformly distributed over a 2π interval of phase, it is called *noncoherent*. The clutter cross section can be expressed as

$$\sigma = \left| \sum_{i=1}^{n} \sqrt{\sigma_i} e^{j\phi_i} \right|^2 \tag{4.1}$$

where σ_i is the individual scatterer RCS and ϕ_i is its phase. The phase factor $e^{j\phi_i}$ includes phase change on reflection and phase delay due to the range of the radar to the scatterer. As an example, if all scatterers were of equal strength ($\sigma_i = \sigma$), for noncoherent scattering

$$\overline{\sigma}_c = \sum_{i=1}^{n} \sigma = n\sigma \tag{4.2}$$

whereas for coherent scattering with all scatterers in phase

$$\overline{\sigma} = \left| \sum_{i=1}^{n} \sqrt{\sigma} \right|^2 = |n\sqrt{\sigma}|^2 = n^2 \sigma \tag{4.3}$$

Hence, if there were ten $\sigma = 1\text{-m}^2$ scatterers in a clutter patch, the RCS could vary from 10 to 100 m^2 depending on the type of scattering involved.

In general, the scattering properties of clutter return are dependent on the incident polarization and the properties of the scattering surface. The rectangular scattering matrix can be expressed in terms of the clutter RCS as

$$\mathbf{S}_c = \begin{pmatrix} (\sigma_{HH})^{1/2} e^{j\phi_{HH}} & (\sigma_{HV})^{1/2} e^{j\phi_{HV}} \\ (\sigma_{VH})^{1/2} e^{j\phi_{VH}} & (\sigma_{VV})^{1/2} e^{j\phi_{VV}} \end{pmatrix} \tag{4.4}$$

where the subscripts denote horizontal (H) and vertical (V) polarization on transmitting and receiving, respectively. For monostatic radar $(\sigma_{HV})^{1/2} = (\sigma_{VH})^{1/2}$ and $\phi_{HV} = \phi_{VH}$, resulting in a symmetrical matrix. Scattering surfaces many times depolarize the incident wave. For example, measurements on deciduous or pine trees showed ratios of $\overline{\sigma}_{HH}/\overline{\sigma}_{HV}$ and $\overline{\sigma}_{VV}/\overline{\sigma}_{VH}$ in the range of 3 to 10 dB [4].

4.1 Surface Clutter

Clutter return backscattered from the land or sea is caused by the various scatterers located within the radar resolution cell of the radar. The scatterers are characterized by the normalized parameter σ^0, which represents the radar cross section per unit area of surface. Thus, the RCS can be expressed as

$$\sigma_{cx} = \sigma^0 A_c \tag{4.5}$$

where σ^0 (m^2/m^2) is the normalized clutter reflectivity per unit area and A_c (m^2) is the illuminated surface area within the radar resolution cell. The clutter power returned from the clutter patch with RCS of σ_{cx} can be expressed as

$$P_c = \frac{P_t G_t^2 \lambda^2 F_c^4 \sigma_{cx}}{(4\pi)^3 R_c^4 L_c} \tag{4.6}$$

where P_t is the transmitter power, G_t is the antenna gain, λ is the radar wavelength, L_c is the clutter transmission and beam shape losses, F_c^4 is the clutter pattern propagation factor, and R_c is the range to the clutter patch. For range-ambiguous radars, the clutter RCS must be weighted by the range and pattern propagation factor to the various clutter patches resulting in

$$P_c = \frac{P_t G_t^2 \lambda^2}{(4\pi)^3 L_c} \sum_{i=1}^{n} \frac{\sigma_{cxi} F_{ci}^4}{R_{ci}^4} \tag{4.7}$$

where σ_{ci}, F_{ci}^4, and R_{ci}^4 are the respective values associated with each of the ambiguous clutter elements. Note that, in a range-ambiguous radar (e.g., PD radar), the clutter elements closest to the radar generally contribute the most incremental clutter power. Also the pattern propagation factor is most significant for low grazing angles and often can be represented as the two-way power antenna radiation pattern:

$$F_c^4 = \exp\left(\frac{-5.54\phi^2}{\phi_{el}^2}\right) \tag{4.8}$$

for a Gaussian-shaped beam as described in Section 3.3.

The surface clutter area (A_c) illuminated by the radar can be represented by the two cases depicted in Figure 4.2 [7]. The first case (beamwidth-limited) applies for large grazing angles where the range associated with the pulsewidth ($R_t = c\tau/2$) is large compared to the projected length in the radial direction of the beam. The second, more usual case (pulse-length-limited) applies for low and medium grazing

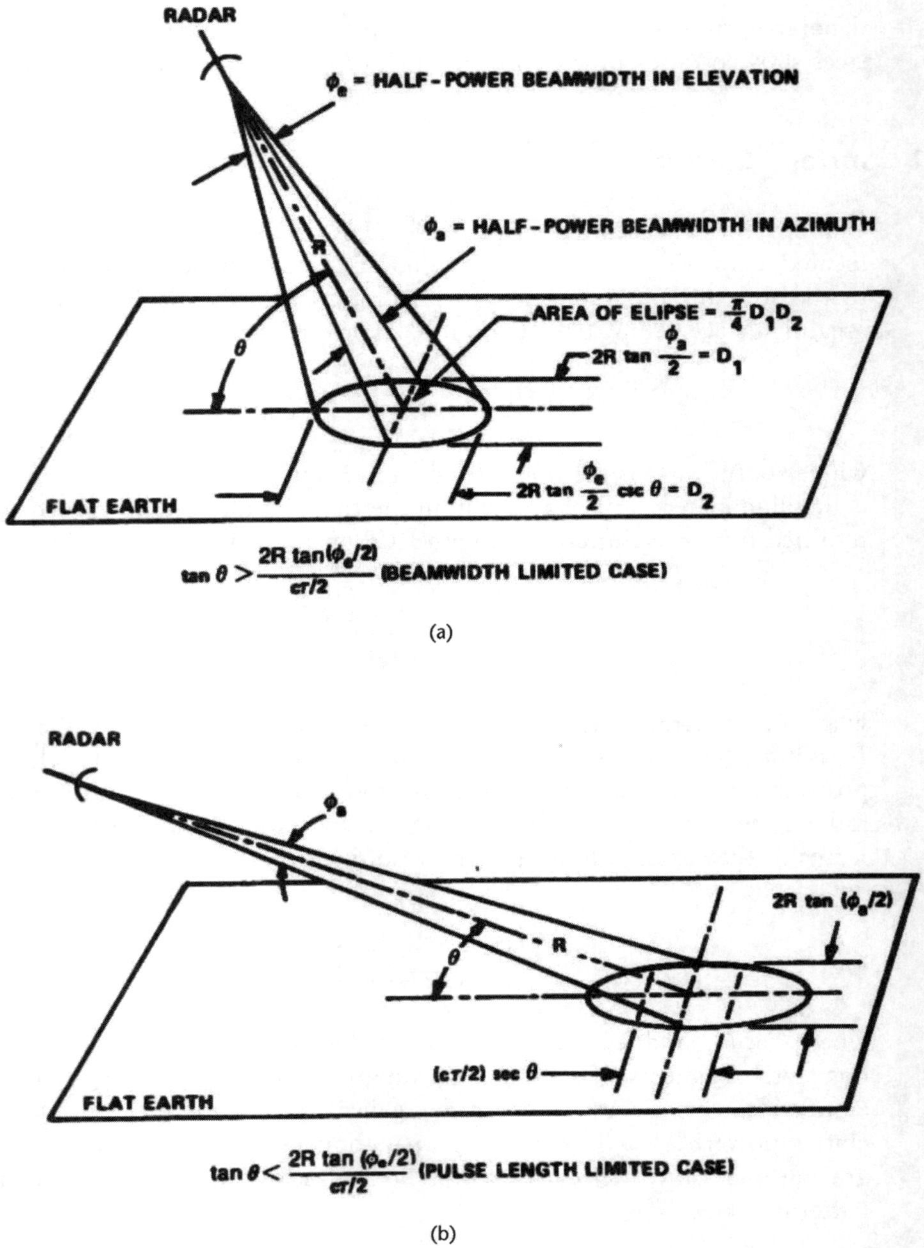

Figure 4.2 (a, b) Surface clutter area illuminated by pulsed radar. (*From:* [4, p. 32].)

angles where the pulse length derived range is less than the projected radial beam length. For small two-way power azimuth (θ_{az}) and elevation (ϕ_{el}) beamwidths less than 10° the clutter area can be expressed for the beam-limited case as

$$A_c = \frac{\pi R^2}{4} \theta_{az} \phi_{el} \csc \psi; \qquad \left(\tan \psi > \frac{\phi_{el} R}{R_\tau} \right) \tag{4.9}$$

and for the pulsewidth-limited case as

$$A_c = R\theta_{az}\frac{c\tau}{2}\sec\psi; \qquad \left(\tan\psi < \frac{\phi_{el}R}{R\tau}\right) \tag{4.10}$$

Equation (4.10) generally applies to surface-based radars, where the clutter patch is simply the projection of the area formed by the azimuth beam dimension at the clutter's range ($R\theta_{az}$) and the projected range associated with the radar pulsewidth ($c\tau\sec\psi/2$). Also note that the one-way power beamwidths for θ_{az} and ϕ_{el} are often used in place of the more accurate two-way power antenna beamwidths. The resulting difference in (4.5) and (4.6) is less than 1.5 dB, which generally is insignificant when compared to the uncertainty in the associated clutter data parameters used to compute the effective clutter RCS.

As an example, consider S-band surface-based radar (i.e., ASR-9 with τ = 1.05 μs and θ_{az} = 1.3°), which views land clutter at a range of 10 nmi with a corresponding grazing angle of ψ = 1°. The pulse-limited clutter area is then found from (4.10) as $A_c = 6.6 \times 10^4$ m^2. If the clutter consisted of wooded terrain, the backscatter coefficient would be approximately σ^0 = –30 dB, so that the clutter RCS ($\sigma_c = \sigma^0 A_c$) would be σ_c = 66 m^2. Over desert terrain, σ^0 = –45 dB, so that the clutter RCS would be σ_c = 2 m^2. For sea clutter (e.g., sea state 3), the backscatter coefficient is further reduced to σ^0 = –50 dB at 1° grazing angle. The RCS for sea clutter with this parameter would then be σ_c = 0.66 m^2.

As another example, consider X-band low-PRF airborne radar operating at an altitude of 20,000 ft and looking down on a low-altitude target at a 10 nmi slant range. The pertinent parameters of the radar are a pulsewidth of 0.5 μs and an azimuth beamwidth of 3°. First, the grazing angle to the ground clutter patch is given by

$$\psi = \sin^{-1}\left[\frac{h}{R}\left(1 + \frac{h}{2r_e}\right) - \frac{R}{2r_e}\right] \approx \sin^{-1}\left(\frac{h}{R}\right) \tag{4.11}$$

where h is the antenna's height, R is the slant range, and r_e = 4,587 nmi is the 4/3 earth radius. The grazing angle can be calculated as 19°. Substituting into (4.10) results in $A_c = 7.7 \times 10^4$ m^2. For wooded terrain at ψ = 19°, σ^0 = –20 dB. The clutter RCS is then equal to σ_c = 770 m^2.

In the preceding examples, if we assume that both the target and clutter return are correlated completely over the period that the antenna dwells on the target and the clutter power is much greater than the noise power, the signal-to-clutter power ratio can be expressed as

$$\frac{S}{C} = \frac{P_s}{P_c} = \frac{\sigma_t}{\sigma^0 R\theta_{az}\left(\frac{c\tau}{2}\right)\sec\psi} \tag{4.12}$$

The maximum clutter-limited range is then given by

$$R_{max} = \frac{\sigma_t}{\left(\frac{S}{C}\right)_{min} \sigma^0 \theta_{az} \left(\frac{c\tau}{2}\right) \sec \psi} \tag{4.13}$$

Note that the clutter-limited range is generally independent of the transmitted power. Also, equation (4.13) can be used for partially correlated targets and clutter if $(S/C)_{min}$ is defined as the minimum signal to clutter required for the actual correlation. Note that the substitution of the constant-γ model into (4.13), at angles above the critical angle, leads to clutter that is constant with range, depending instead on radar antenna height.

As a further example, consider the case where the radar observes surface clutter at near-vertical incidence. The value of σ^0 will depend on the surface reflectivity and any quasispecular reflections from generally randomly tilted surface facets. For a 10° antenna beam, a value for σ^0 on the order of +17 dB has been predicted [8]. Experimental data taken at 3.8 GHz indicate values of σ^0 over water on the order of +20 dB, whereas for land the values varied from about 0 to 10 dB depending on the terrain involved [4]. The clutter power can be calculated by substituting the value for the clutter area given by (4.5) into (4.3) and substituting $G = \pi^2 / \theta_{az} \phi_{el}$, resulting in

$$P_c = \frac{\pi P_t A_e \sigma^0}{64 R^2 \sin \psi} \tag{4.14}$$

where

$$A_e = \frac{G_t \lambda^2}{4\pi} \tag{4.15}$$

is the antenna's capture area, R is the range to the surface, and ψ is the near-vertical incidence grazing angle. The clutter power thus varies inversely as the square of the range to the surface.

In airborne radar, the strong reflection from near-vertical incidence results in "altitude line" clutters (see Figure 2.9 and Example 27), the Doppler frequency spectrum of which is centered about zero frequency. The clutter power can generally be computed from (4.14) as modified to account for reception of the clutter return through the antenna's sidelobe structure (i.e., $P_{al} = P_c / G_{SL}^2$, where G_{SL} is the ratio of the peak antenna gain to the sidelobe gain in the direction of the clutter). This equation applies in the beam-limited case, where the altitude or beamwidth is such that the trailing edge of the pulse does not reach the ground before the leading edge passes out of the antenna beam. Otherwise, the pulse-limited case applies, where the clutter return is now inversely proportional to the third power of the range, resulting in a weaker clutter return.

From a radar scattering viewpoint, the range of possible grazing angles can be divided into three distinct regions: the near grazing incidence region, the plateau region, and the near-vertical incidence region, as depicted in Figure 4.3 [4]. Within each of these regions, the dependence of σ^0 on the grazing angle and wavelength

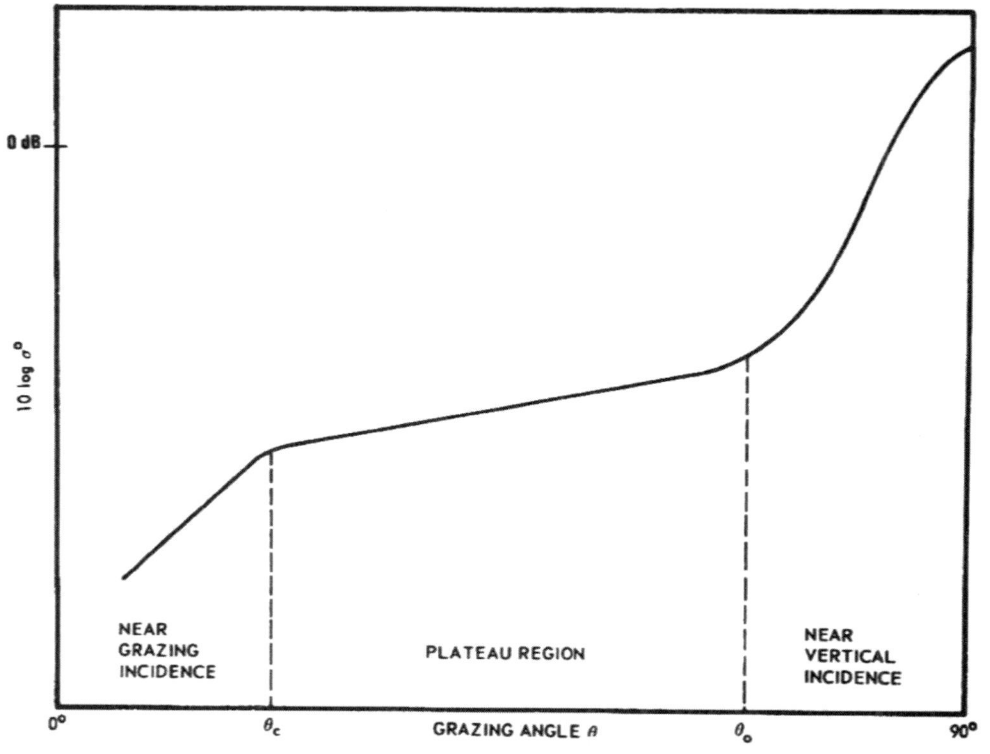

Figure 4.3 General dependence of the radar backscattering coefficient σ^0 on grazing angle for surface clutter. (*From:* [4, p. 213].)

can be characterized to some extent. However, the boundaries of the three regions change with wavelength, surface conditions, and polarization.

In the near grazing incidence region, σ^0 varies rapidly with the grazing angle. It generally decreases rapidly with a decreasing grazing angle and longer wavelengths. When the surface in a scattering sense is rough, σ^0 is consistently larger for all grazing angles and frequencies compared to a smooth surface. Generally, roughness can be defined (i.e., Rayleigh criterion) as

$$\sigma_h \sin \psi \geq \frac{\lambda}{8} \tag{4.16}$$

where σ_h is the rms height of the surface irregularities, ψ is the grazing angle, and λ is the radar's wavelength. Thus, roughness in the scattering sense depends on both surface irregularities and the frequency and grazing angle of the incident radiation. For example, at a grazing angle of 1° and a frequency of 2 GHz (λ = 0.15 m) a surface with irregularities on the order of $\Delta h \geq 1.07$m will appear rough. Generally, rough surfaces tend to backscatter more of the incident radiation than smooth surfaces. In the limit, a perfectly smooth surface at less than near-vertical incidence will forward scatter all incident radiation and σ^0 will then be zero.

The interference concept is frequently used to explain certain properties of surface clutter observed in the low-grazing-angle region [9]. This concept postulates that the return from each scatterer can be modeled as the combined return from a direct reflection from the scatterer and an indirect reflection scattered from the underlying clutter surface. The clutter backscatter is then modified by the pattern propagation factor F_c^4 that applies for multipath phenomena and generates a lobe pattern caused by destructive and constructive interference [4]. The critical angle, ψ_{cl} (rad), which defines this region, is approximately equal to the angle for the first maximum of the lobe pattern given by

$$\psi_{cl} = \sin^{-1}\left(\frac{\lambda}{4\pi\sigma_h}\right) \tag{4.17}$$

where σ_h is the rms deviation of the surface from the mean surface. Equation (4.17) generally applies only for sea clutter illuminated with horizontally polarized radiation (i.e., when the reflection coefficient $\Gamma = -1$). However, the general form of the equation

$$\psi_{cl} = \sin^{-1}\left(\frac{\lambda}{k\sigma_h}\right) \tag{4.18}$$

(with k a constant depending on polarization and surface type) could be expected to apply to other situations [10]. The backscatter coefficient for grazing angles less than the critical angle then equals

$$\sigma^0 = F_c^4 \sigma_p^0 \tag{4.19}$$

where F_c^4 is the pattern propagation factor in the region below the first interference lobe maximum:

$$F_c^4 = 16 \sin^4\left(\frac{2\pi\sigma_h \sin\psi}{\lambda}\right) \tag{4.20}$$

and σ_p^0 is the value of σ^0 at the critical angle. Equation (4.13) predicts that the backscattering coefficient σ^0 for surface clutter will vary approximately proportional to ψ^4 and λ^{-4} below the critical angle (ψ_c). Some experimental evidence confirms this behavior in accordance with interference theory [4].

Interference theory can also be used to explain the difference in σ^0 for sea clutter between horizontal and vertical polarization, the difference lying in their respective reflection coefficients. For horizontal polarization, the reflection coefficient for perfectly conducting surfaces (i.e., sea clutter) has a magnitude of 1 and a phase shift of 180° (i.e., $\Gamma = -1$). This results in destructive interference at grazing angles below the critical angle (ψ_{cl}) given in (4.17). However, for vertical polarization the situation is quite different. The magnitude and phase of the reflection coefficient each vary with both grazing angle and frequency. The grazing angle, for which the phase is $\pi/2$ and the magnitude is a minimum, is know as *Brewster's angle*. Above Brewster's angle, the reflection phase shift becomes small (i.e., $\phi \to 0$); hence, destructive interference can occur only well below Brewster's angle.

Therefore, the critical angle for vertical polarization depends on the grazing angle at which the frequency-dependent Brewster's angle occurs and may become independent of the surface roughness.

The shape of the backscattering coefficient curve (σ^0) for horizontal and vertical polarization is depicted in Figure 4.4 for UHF. In this frequency range, Brewster's angle is expected to be on the order of 3° to 4°, whereas the critical angle for horizontal polarization and a moderate sea (σ_h = 4 ft) is on the order of $\psi_c = 9°$. Hence, the backscattering coefficient (σ^0) would be expected to be generally higher for vertical polarization than for horizontal polarization at low frequencies. As the radar's frequency is increased, Brewster's angle increases and the critical angle decreases to become approximately equal (i.e., $\psi_c = 5°$) at 1 GHz. For further increases in frequency, Brewster's angle becomes much greater than the critical angle so that the backscattering coefficients (σ^0) at near-grazing angle incidence would be expected to coalesce for horizontal and vertical polarization. For example, at 10 GHz, Brewster's angle is 8° and the critical angle is usually less than 1° [4].

In general, the surface clutter return for vertical polarization tends to be greater than for horizontal polarization. The return for vertical polarization tends to be stronger because the phase lag of the reflection coefficient is such that more energy exists at points close to the surface. As λ decreases or the scatterers are raised

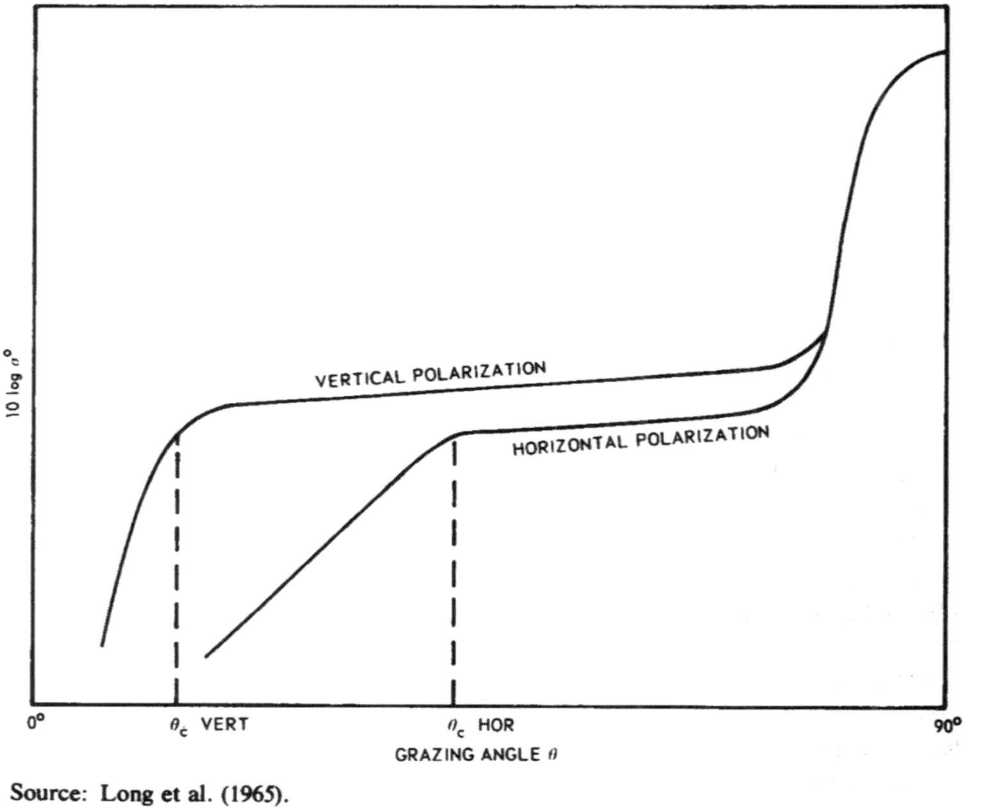

Source: Long et al. (1965).

Figure 4.4 General appearance of radar backscattering coefficient (σ^0) at UHF. (*From:* [4, p. 278].)

higher above the surface (corresponding to a rougher surface), more of the interference pattern tends to be averaged and its value at points close to the surface becomes less significant. For very small λ or a very rough surface, the phase of the reflection coefficient becomes unimportant and σ^0 depends only on the average field strength (i.e., the magnitude of the reflection coefficient). Finally, at very small grazing angles, only those scatterers with exceptional height will be illuminated, causing quasispecular reflections [9]. For sea clutter, this results in spiky returns that are characterized by non-Gaussian probability distributions (see Section 1.3.4).

In the plateau region, the change of σ^0 with grazing angle is slow and generally can be approximated by the constant gamma model [4], given by

$$\sigma^0 = \gamma \sin \psi \tag{4.21}$$

where γ is a constant and ψ is the grazing angle. In this region, most surfaces generally appear rough in a scattering sense [see (4.11)] and the scattering is incoherent. The backscattering coefficient (σ^0) changes slowly with wavelength (i.e., σ^0 is proportional to λ^{-1} in the microwave region), whereas its magnitude generally increases with the surface roughness. The transition angle between the plateau and near-vertical incidence region is on the order of 60°, but this angle varies with the surface roughness, whereas the transition is gradual rather than sharp. The amplitude probability distribution of the surface clutter in the plateau region is generally characterized by Rayleigh statistics.

At near-vertical incidence, the backscatter is larger when the surface is smooth than when it is rough. This is in contrast to the plateau region, where the opposite condition applies. Clutter returns in this region become significant for airborne radars that experience strong returns from terrain directly underneath the aircraft. This type of return generally enters the radar through its antenna sidelobes and has zero Doppler frequency. The clutter return from near-vertical incidence is strongest over smooth water or land (e.g., asphalt or concrete roads) and weakest over rough terrain (e.g., sea clutter with high sea states and wooded or cultivated land) [4].

4.1.1 Land Clutter

Land clutter is one of the most difficult types of clutter to characterize because it generally lacks a homogeneous structure. The large variation of natural and human-made objects over most terrain makes this clutter difficult to classify into type or obtain consistent experimental data. Under apparently similar conditions, experimental measurements obtained by using airborne radar will differ from those of ground radar. One explanation is that airborne measurements tend to smooth the large spatial variations prevalent in land clutter data (thereby producing a spatial average), whereas fixed ground measurements tend to accentuate shadowing and geometric anomalies, which are then averaged over time to produce smoothed statistical clutter data [7]. The result of this uncertainty in land clutter data is that most characterizations are composed of highly smoothed data that must be used with caution in the design of Doppler radars.

The general approach in describing land clutter is to state its properties in terms of an average backscatter coefficient (σ^0), an amplitude probability distribution, and a power spectral density. However, because of the spatial variability of land clutter, we also need to specify its spatial distribution. One measure of the spatial distribution is to consider σ^0 as a random variable, the probability distribution of which represents values of σ^0 that can be found in independent spatial cells [11]. The specification as to what might constitute an independent spatial cell is an open question. One study of this effect, based on high-resolution radar data, considered triangular correlations consisting of a short correlation distance (i.e., no correlation after 50 ft) and a long correlation distance (i.e., no correlation after 1,000 to 2,000 ft), which was a function of the type of terrain involved [1].

Some general characteristics of land clutter can be deduced from the general properties of surface clutter. The properties of smooth terrain [see (4.16)], such as pavement, desert, or very flat land, follow the general characteristics of surface clutter illustrated in Figure 4.3, which depicts the three regional variations of σ^0 with grazing angle. Within this class is an increase of σ^0 with frequency usually no faster than linear with the radiation frequency. In the interference region, the reflection coefficient of terrain is tempered by the dielectric properties of the surface, which reduces its magnitude (i.e., $\rho < 1$). Because this factor determines the degree of destructive interference caused by the reflected wave, land clutter could be expected to exhibit a slower variation of σ^0 with grazing angle (ψ) in this region and also be less sensitive to polarization changes (e.g., horizontal versus vertical) than for a perfectly reflecting surface.

As the surface roughness of the terrain increases, the various regions tend to merge into a single region. In this region, the diffusion law applies and the backscattering coefficient (σ^0) exhibits a dependence on grazing angle given by the constant-γ relationship ($\sigma^0 = \gamma \sin \psi$). However, σ^0 does not tend to zero as the grazing angle approaches zero, but rather approaches some finite value that apparently depends on the frequency and terrain [12]. The backscattering coefficient exhibits some variation with polarization, but no systematic effects are definable [7]. Frequency variations are, at most, a linear increase of σ^0 with frequency [13], but the experimental data defining this variation have such a large variance that it is difficult to define a systematic change with frequency or state that any difference really exists [1]. In urban areas, certain building surfaces may be more effective flat-plate reflectors when the surface irregularities are small compared to the radiation wavelength, which might cause an increase in σ^0 for a decrease in frequency [10].

Examples of rough terrain are mountains, cities, and urban areas with a significant number of human-made structures. However, other terrains might be considered rough (in the scattering sense) if their irregularities would become significant when compared to the wavelength of the incident radiation (i.e., $\sigma_h \sin \psi \geq \lambda/8$). For example, at X-band and a grazing angle of 30°, a surface would be considered rough if $\sigma_h \geq \lambda/4 = 0.75$ cm. At this frequency, most land surfaces would appear rough, and the rough terrain model characteristics would apply.

For surface-based radar, land clutter is generally returned from scatterers with a grazing angle of less than 1°. Limited experimental data are available in this region to determine the magnitude and extent of land clutter. An analytical model developed for this region (called the *horizon and transhorizon region*) has been used to determine the magnitude of clutter returns at X-band for both level and hilly terrain [6]. The essence of this model is to express the backscatter coefficient as the product of the two-way pattern propagation factor (F_c^4) and the constant-γ backscatter coefficient ($\sigma^0 = \gamma \sin \psi$). The X-band clutter reflectivity predicted over level terrain, as depicted in Figure 4.5, indicates that due to diffraction low clutter levels may be returned from beyond the geometrical horizon. These returns may be significant in overpowered or sensitive radars designed to detect small RCS targets. A similar curve for clutter reflectivity over rolling or hilly terrain is depicted in Figure 4.6. Note that the clutter is extended to greater ranges and the decline beyond the geometrical horizon is less steep. Figure 4.7 depicts the effect of frequency on the clutter reflectivity at low grazing angles for both level and hilly terrain. The effect at UHF is most pronounced where the increased scatterer height (from σ_h = 1m to 10m) can increase the clutter strength by the order of 60 dB. At these low grazing angles, the high sensitivity to terrain height irregularities will cause large variations in clutter reflectivity, and the clutter's amplitude probability distribution will generally be described by non-Gaussian statistics.

4.1.1.1 Average σ^0 for Various Terrains

The backscatter coefficient (σ^0) for land clutter is difficult to specify because of the wide variety types and spatial variability of this parameter. Nevertheless, σ^0 is a useful parameter for describing extended target areas because the radar cross section is normalized with respect to the surface area illuminated by the radar. The wide variability of the data makes the use of an average value of σ^0 an appropriate

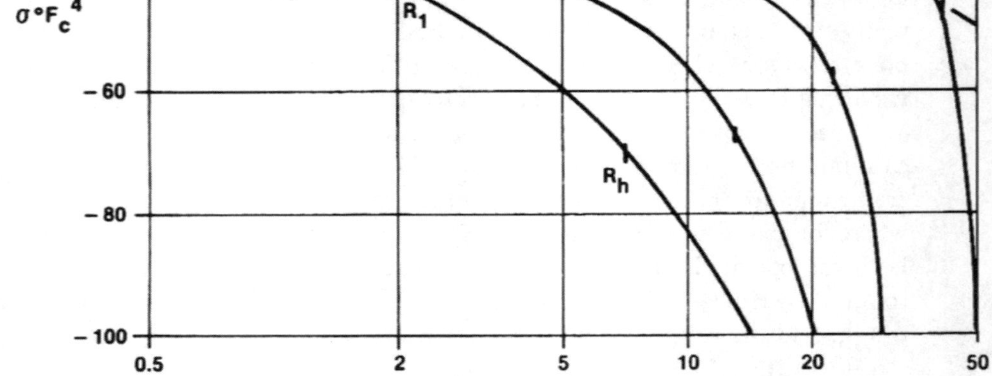

Figure 4.5 Land clutter backscatter coefficient for low grazing angles at X-band over level terrain. (*From:* [8, p. 130].)

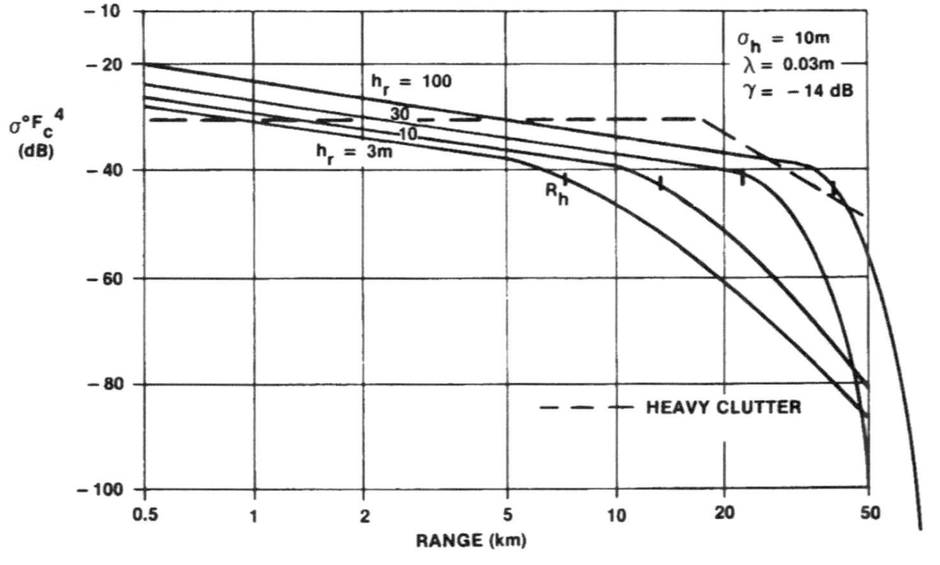

Figure 4.6 Land clutter backscatter coefficient for low grazing angles at X-band over rolling or hilly terrain. (*From:* [8, p. 130].)

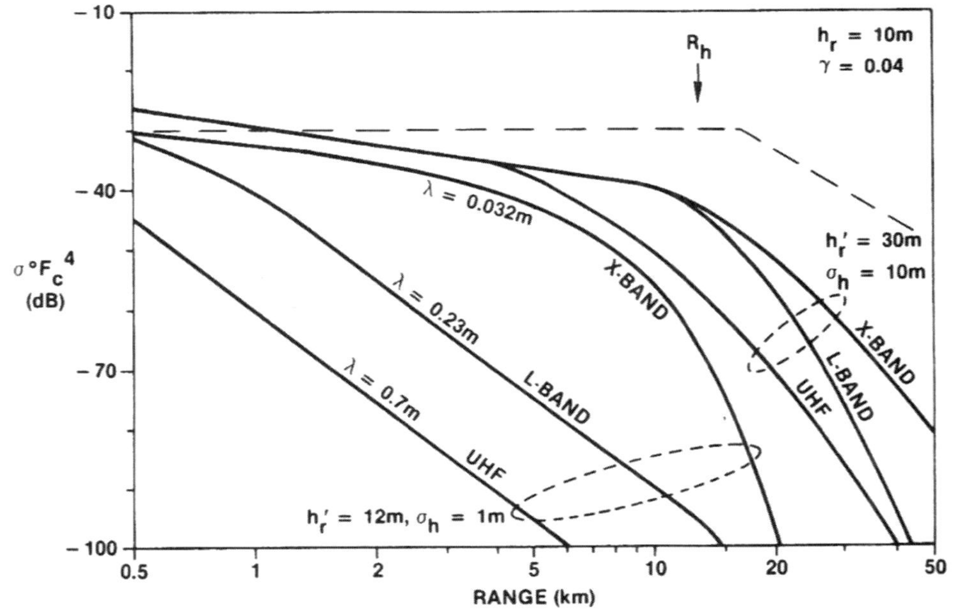

Figure 4.7 Land clutter backscatter coefficients for low grazing angles as a function of frequency. (*From:* [8, p. 130].)

parameter, which will again be designated $\overline{\sigma^0}$. For airborne measurement data, $\overline{\sigma^0}$ naturally becomes a spatial average. For ground-derived data, $\overline{\sigma^0}$ becomes a temporal average for high-resolution data and a composite (both temporal and spatial) average for low-resolution data. Further, many clutter bases represent heavily averaged data that include measurements from many experimental programs that are normalized and then averaged.

In general, $\overline{\sigma^0}$ depends on grazing angle, frequency, polarization, surface roughness, and the type of terrain involved; polarization effects are not pronounced. The wind does not generally affect $\overline{\sigma^0}$, but does affect the spectra of some types of land clutter. Reflectivity values for land clutter are generally higher than for any other form of radar clutter.

Different terrain types are categorized into generic classes to systematize the data into a usable form. Note that the reflection characteristics vary greatly within each terrain type. A general classification of types is as follows:

- Relatively flat land, deserts;
- Rural rolling countryside, farmland;
- Forests, heavy woods, jungle;
- Mountainous regions;
- Urban areas with a high density of artificial structures [7].

Over the years, a number of major efforts have been conducted to collect and provide a definitive model for the backscatter coefficient over land. The most recent was conducted by Lincoln Laboratory and measured low grazing angle clutter from surface instrumentation radar located on an adjustable 30- to 100-ft tower to ranges of 1 to 50 km [14, 15]. Frequencies covered include VHF, UHF, L-band, S-band, and X-band. A wide variety of terrain was examined from 42 different sites with good ground truth using a wide 80 dB range receiver with greater than 2 dB accuracy. Range resolution was 36 m or 150 m for VHF and UHF, respectively, with 50 m or 150 m for the other frequencies.

The study provided mean and median backscattering coefficients using detailed terrain descriptions of the clutter types measured. Median values are provided in Table 4.1 [2]. Note that at VHF forest clutter is 10 to 15 dB stronger than at microwaves, whereas farmland clutter is 20 to 30 dB less at VHF than at microwaves. In general, the researchers found that the spatial distribution of the clutter examined can be represented by a Rayleigh distribution at depression angles above 6 to 8° and a Weibull distribution for lower angles. The difference between horizontal and vertical polarizations was typically 1 or 2 dB.

However, the major contribution of this measurement program was in the temporal characterization of low grazing angle ground clutter. Basically land clutter can be grouped into those scatterers that move with the wind and those that stay fixed such as tree trunks, rocks, bare ground, and other stationary objects [73]. It was found that the moving component spectra can be characterized by a two-sided exponential distribution [72]. This has implications both for MTI (see Example 3 in Chapter 1) and STAP [16] signal processing.

The two-sided exponential ground clutter model given in (1.42) describes the PSD for windblown trees and other vegetation over a frequency range from UHF

Table 4.1 Grouped Median Values of Land Clutter Backscatter by Terrain Class

Terrain	*VHF*	*UHF*	*L-Band*	*S-Band*	*X-Band*
Urban	−20.9	−16	−12.6	−10.1	−10.8
Mountains	−7.6	−10.6	−17.5	−21.4	−21.6
Forest/high relief					
Depression angle > 1°	−10.5	−16.1	−18.2	−23.6	−19.9
Depression angle < 0.2°	−19.3	−16.8	−22.6	−24.6	−25
Forest/low relief					
Depression angle > 1°	−14.2	−15.7	−20.8	−29.3	−26.5
Depression angle 0.4°–1°	−26.2	−29.2	−28.6	−32.1	−29.7
Depression angle < 0.3°	−43.6	−44.1	−41.4	−38.9	−35.4
Agriculture/high relief					
Terrain slope > 2°	−32.4	−27.3	−26.9	−34.8	−28.8
Agriculture/low relief					
1° < terrain slope > 2°	−27.5	−30.9	−28.1	−32.5	−28.4
Terrain slope > 1°	−56	−41.1	−31.6	−30.9	−31.5
Desert					
Depression angle > 1°	−38.2	−39.5	−39.6	−37.9	−25.6
Depression angle < 0.3°	−66.8	−74	−68.6	−54.4	−42

to X-band. The ratio between the dc or zero-Doppler term and the fluctuating or ac term can be approximated as

$$r = 394 w^{-1.55} f_{\text{GHz}}^{-1.21} \tag{4.22}$$

where w is the wind speed in knots and f_{GHz} is the frequency in gigahertz.

The dc term of the covariance function is given by

$$R(0) = \frac{r}{r+1} \tag{4.23}$$

whereas the ac term is given by

$$R(\tau) = \frac{1}{r+1} \frac{(\beta\lambda)^2}{(\beta\lambda)^2 + (4\pi\tau)^2} \tag{4.24}$$

with the shape parameter β approximated as follows:

$$\beta = \frac{1}{0.105(\log(w) + .476)} \tag{4.25}$$

Alternatively, measured values of β can be found in Table 1.5.

The correlation time is important to data collection and processing. It was found that the correlation time for windblown trees is a function of frequency, whereby the longer the wavelength, the longer the correlation time. Typical correlation times for clutter from windblown trees are given in Table 4.2 [16–18, 72].

The significance of these measurements is that the shape of the land clutter spectrum from windblown trees depends primarily on wind speed but is invariant

Table 4.2 Typical Correlation Time for Windblown Clutter

Frequency	*VHF*	*UHF*	*L-Band*	*S-Band*	*X-Band*
Typical correlation time (sec)	5.04	0.94	0.95	0.081	0.049

when expressed in velocity units from VHF to X-band. The description of the PSD by the two-sided exponential distribution extends the lower level spectral components beyond those expected from the Gaussian-shaped spectrum traditionally used for this situation. This results in a lower MTI improvement factor than was previously predicted (see Example 3). Also the model predicts that the dc component of the spectra will be much larger at high frequencies than at X-band [2].

Another model generated for the Overland Radar Program provides data appropriative for airborne radar. Note that most summaries use a similar database but differ in their interpretation of the data or extrapolations to fill voids in them. Also measurements taken from airborne instrumentation radar generally have difficulty obtaining accurate data at low grazing angles. The difference between airborne and surface instrumentation lies in the range to the radar horizon. Surface instrumentation might have a range of 10-to-20 km while airborne instrumentation ranges are 100-to-200 km. The resulting difference in dynamic range for a scatterer near the horizon is then the order of 40 dB requiring receiver dynamic ranges of 80 dB or more for accurate airborne measurements.

Figure 4.8 depicts the average backscattering coefficient ($\overline{\sigma^0}$) for five terrains associated with distributed land clutter. The curves represent an average of about 24 different experimental programs that measured σ^0 associated with the types of terrain listed. The data utilized span a frequency range from UHF to X-band and involve both horizontal and vertical polarizations, with grazing angles from 1/4°

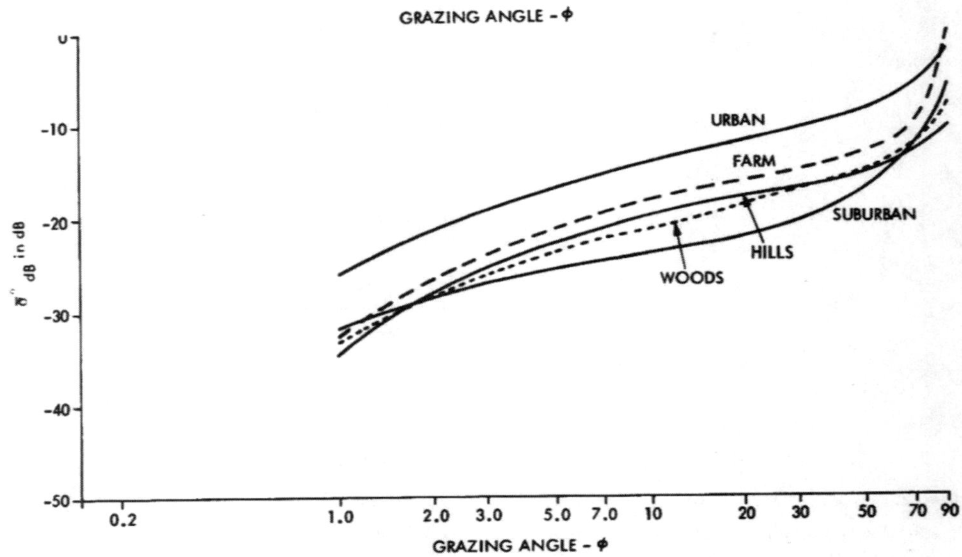

Figure 4.8 Mean backscattering coefficient $\overline{\sigma^0}$ dB for five terrains. (*From:* [20, p. 60].)

to 70° [1, 19, 20]. These data, with the exception of those concerned with desert terrain, represent an average of all frequencies and polarizations associated with a particular type of terrain. The rationale for this approach is that the data revealed no systematic trend with frequency or polarization and, hence, represented fluctuations about a mean value.

The backscatter coefficient data illustrated in Figure 4.8 extend down to a grazing angle of 1° and up to 90°. However, the data utilized provided high statistical confidence only over a range of 10° to 90° by using a constant-γ reflectivity model (i.e., $\sigma^0 = \gamma \sin \psi$). The data points outside this region (i.e., 1° to 10°), although sparse, tended to confirm the theoretical model used to extend the higher angle data into this region [1].

The curves for the backscattering coefficient presented in Figure 4.8 represent the average for many data points. These curves are intended to represent the statistical spatial average for $\overline{\sigma^0}$ within the spatially independent resolution patch that will be given later in Table 4.4 for the various terrains listed. Further, the clutter is assumed to be log-normally distributed within each independent elemental clutter patch. The random variable σ^0_{dB} (σ^0 plotted on a decibel scale) for each elemental clutter patch will then be normally distributed. Further, σ^0_{dB} for the radar's surface clutter cell (formed by a number of elemental clutter patches) will also be normally distributed. The standard deviation for each elemental clutter patch of the backscattering coefficient σ^0_{dB} is depicted in Figure 4.9 for farmland

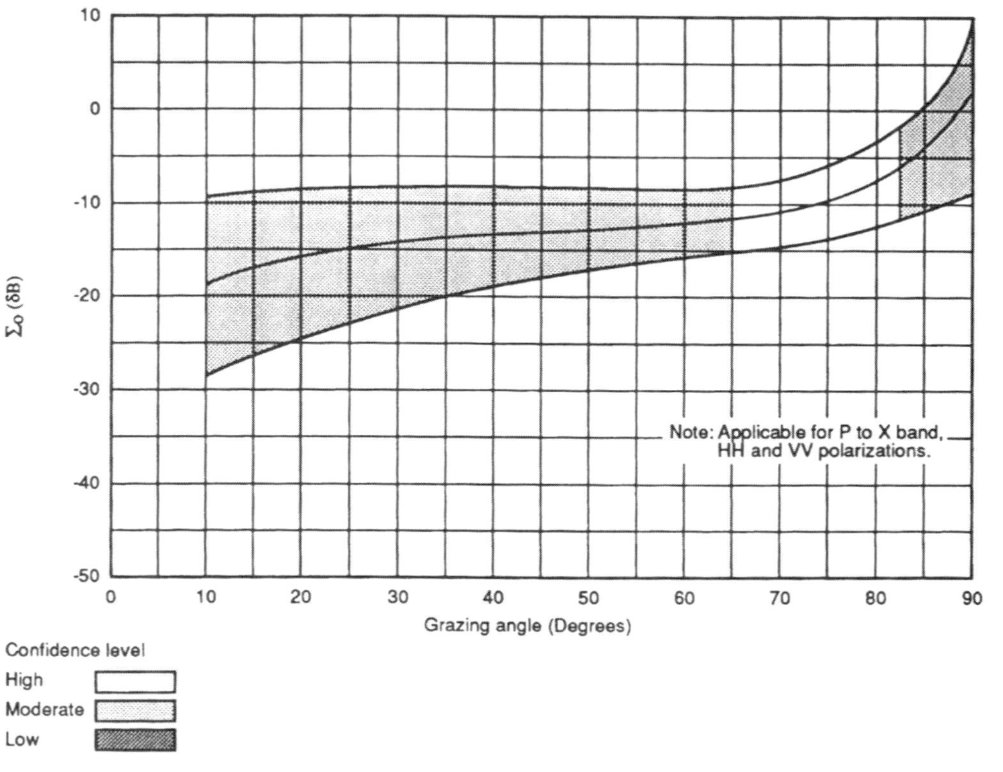

Figure 4.9 Range of backscatter coefficients for farmland clutter. (*From:* [1, p. 9].)

clutter and in Figure 4.10 for woodland clutter (see also Figure 4.52 for other terrains).

The surface clutter resolution cell will, in general, be larger than the resolution patch given in Table 4.3, resulting in a number (n) of independent clutter patches located within the overall surface clutter resolution cell. The spatial amplitude distribution of σ^0_{dB} will then be normally distributed with standard deviation

$$s_2 = \Delta\sigma^0_{\text{dB}} = \frac{\sum_{i=1}^{n} \left(\Delta\sigma^0_{\text{dBi}}\right)}{n} \quad (4.26)$$

For example, if nine independent cells ($A_c = 10^6$ ft^2) were located within the surface clutter resolution cell for woodland clutter ($\sigma^0_{\text{dB}} = -20$ dB and $s_2 = 7.5$ dB at 10°), then the standard deviation of the clutter patch would be 2.5 dB. Thus (see Figure 4.10), at the 84th percentile (the one σ point), the value of σ^0_{dB} would be less than −17.5 dB, and at the 99th percentile (the two σ point), the value of σ^0_{dB} would be less than −15 dB. Note that the temporal variation (e.g., Rayleigh distribution with mean σ^0_{dB}), which is important for dynamic range considerations in Doppler radars, would be superimposed on the spatial amplitude distribution.

Figure 4.11 depicts the average backscatter coefficient ($\overline{\sigma^0}$) associated with desert terrain. These curves show that σ^0 increases with increasing frequency. The

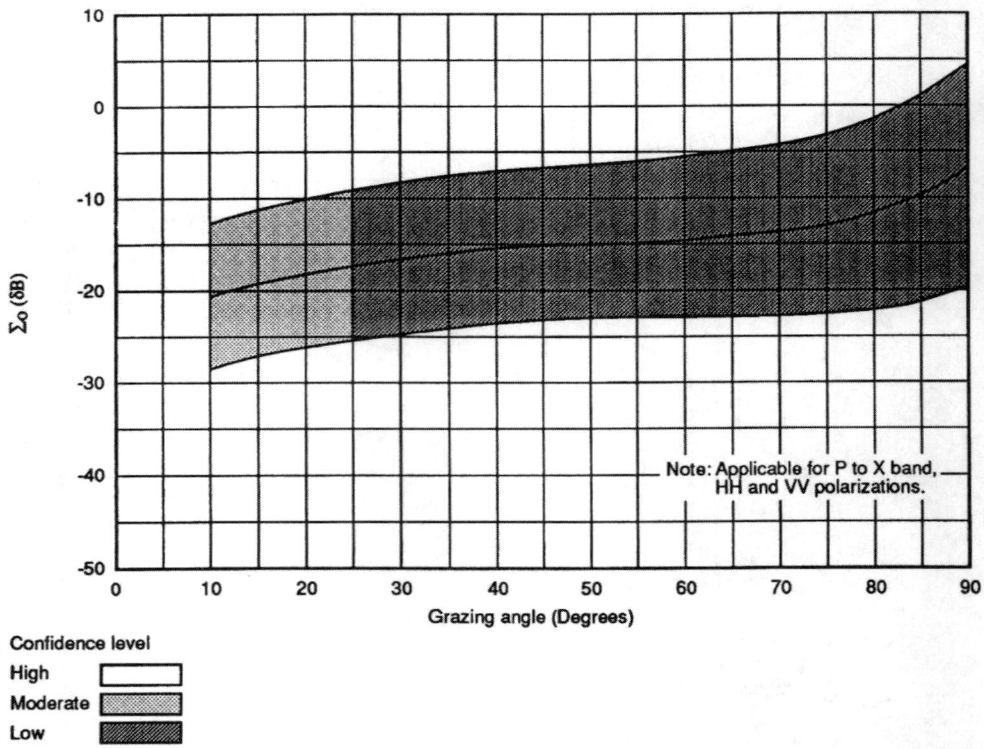

Figure 4.10 Range of backscatter coefficients for woodland clutter. (*From:* [1, p. 10].)

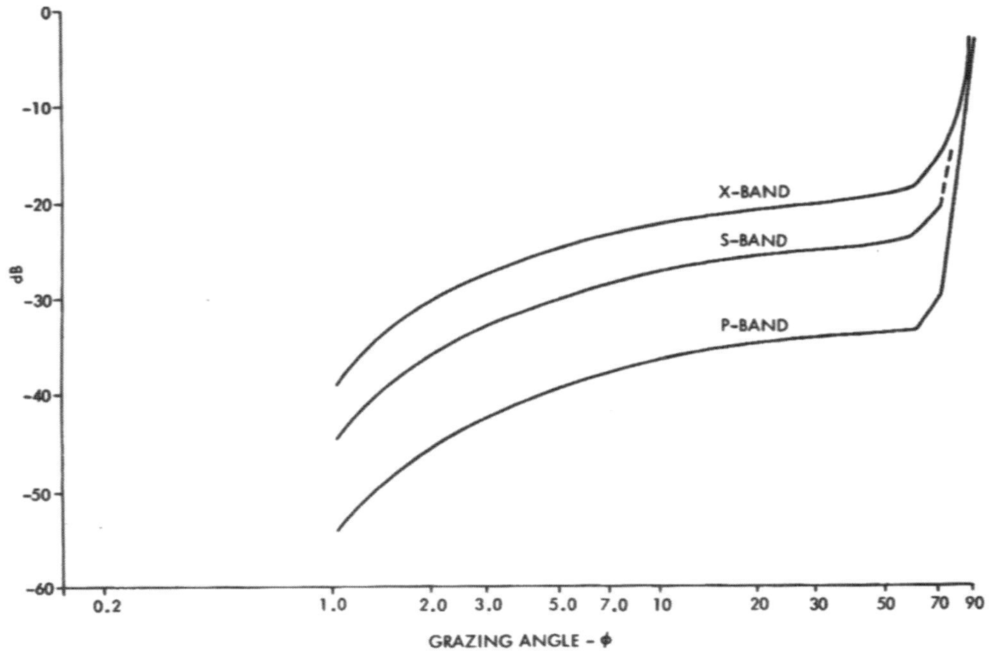

Figure 4.11 Mean backscatter coefficients $\overline{\sigma^0}$ dB for desert terrain. (*From:* [19, p. 54].)

curve at X-band was derived from high-fidelity data available from a 10° to 70° grazing angle. Available data at other frequencies indicated frequency dependence, but were not of sufficient fidelity to allow a complete determination of $\overline{\sigma^0}$ versus grazing angle. Therefore, the X-band curve was modified by the factor 40 – 10 log [f (MHz)] dB to obtain the curves at S-band and UHF. These curves were extended into the grazing angle region below 10° by using a constant-γ model ($\sigma^0 = \gamma \sin \psi$). The available data at near-vertical incidence showed no frequency dependence and, hence, the curves at all frequencies were linearly interpolated from their value at 70° into a single value at 90°, which was determined from the X-band data [1].

The confidence interval for desert terrain that contains 90% of the data points is depicted in Figure 4.12 for X-band data. These X-band data should be scaled by the factor 40 – 10 log f_o (MHz) to provide confidence intervals for other frequency bands. The data analyzed showed no polarization dependence for desert terrain and, hence, the curves apply for both horizontal and vertical polarization.

Another set of data that characterizes the mean backscattering coefficient ($\overline{\sigma^0}$) at S-band ($\lambda = 0.1$m) is depicted in Figure 4.13 [13]. These data include curves for mountainous terrain and smooth pavement not provided in the Overland Radar Program data. In the diffuse region where the σ^0 (dB) curve is virtually a straight line, the data may be scaled to other microwave frequencies by using the following relation:

$$\sigma^0 = \sigma^0[3\ \text{GHz}] \cdot \frac{f\,(\text{GHz})}{3} \tag{4.27}$$

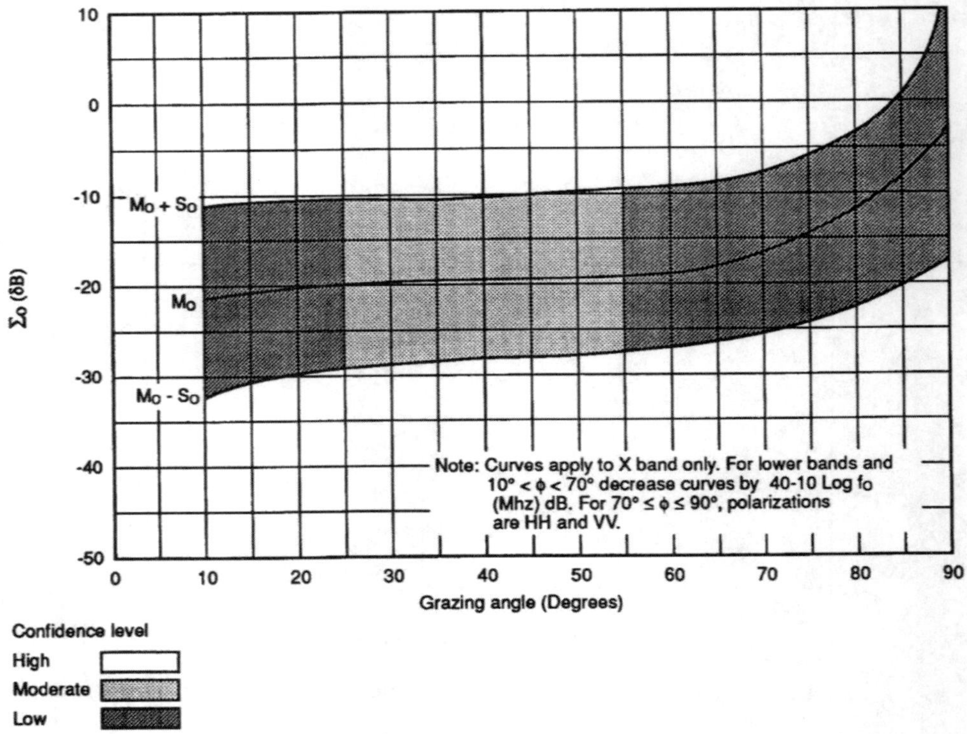

Figure 4.12 Range of backscatter coefficients for desert terrain at X-band. (*From:* [1, p. 11].)

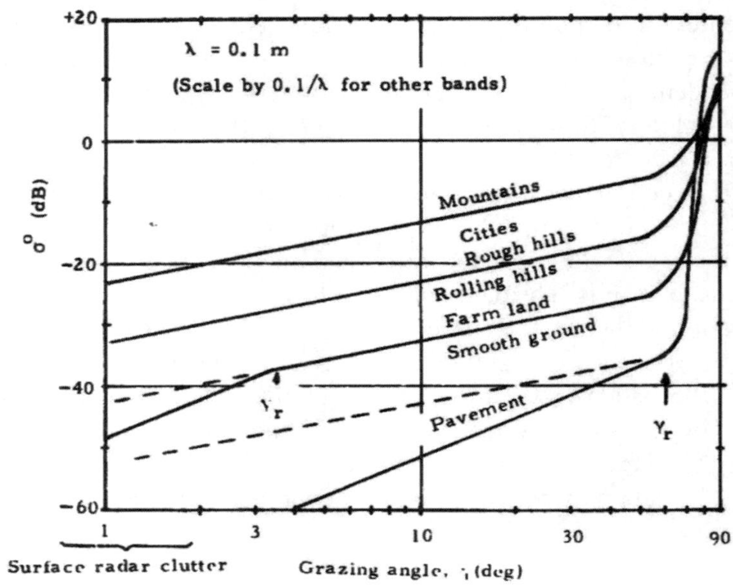

Figure 4.13 Land clutter reflectivity data at S-band. (*From:* [13, p. 160].)

Further, the critical angle (ψ_c) can be scaled to other microwave frequencies by way of the relation:

$$\psi_c = \psi_c\,[3\text{ GHz}] \cdot \frac{3}{f\,(\text{GHz})} \tag{4.28}$$

The data given in Figure 4.13 indicate a generally smaller value of $\overline{\sigma^0}$ for city and farmland terrain than that given by the Overland Radar Program data. The curves exhibit the three grazing angle regions characteristic of surface clutter. At high grazing angles, there is a quasispecular region for which the reflectivity is very high and is higher for smooth than rough terrain. In the intermediate-angle diffuse scattering region, the grazing angle dependence is given by a constant-γ model ($\sigma^0 = \gamma \sin \psi$). At very low grazing angles, there is an interference region for which the grazing angle dependence is much steeper than in the diffuse region. In the quasispecular region at near-vertical incidence grazing angles where the curves for $\overline{\sigma^0}$ rise steeply, there is essentially no frequency dependence.

The linear frequency dependence of land clutter postulated for the model given in Figure 4.13 is generally considered to apply throughout the microwave region if the surface is rough relative to a wavelength [see (4.11)]. If the same variation were applied throughout the millimeter-wave (MMW) region, very large values of σ^0 would be expected. Limited data [21], depicted in Figures 4.14, 4.15, and 4.16 for backscatter from wooded terrain, indicate that the linear variation does not apply for these higher frequencies. However, the data show that the average backscatter coefficient ($\overline{\sigma^0}$) is generally larger in the MMW region, but with an apparent

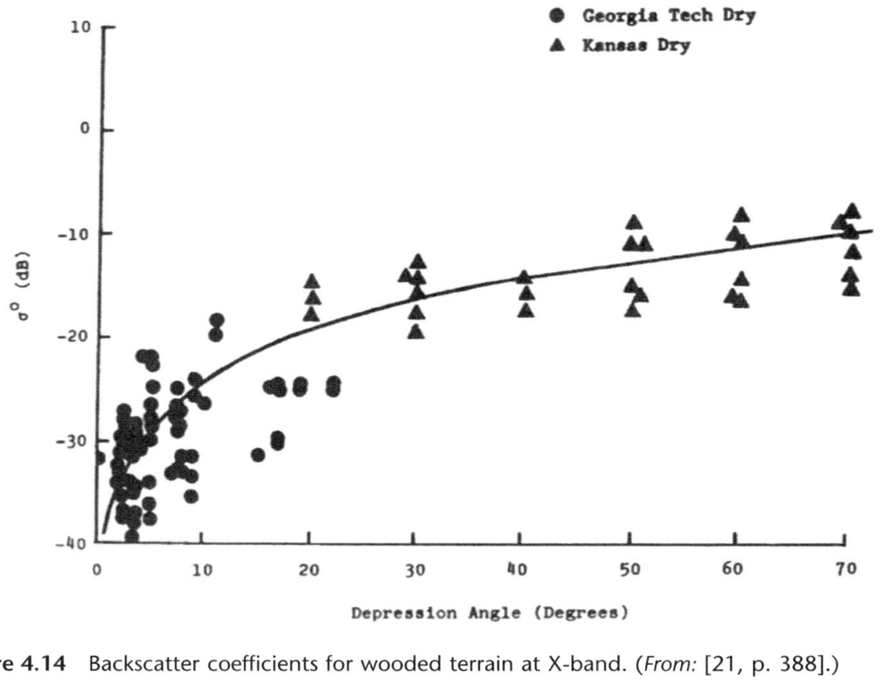

Figure 4.14 Backscatter coefficients for wooded terrain at X-band. (*From:* [21, p. 388].)

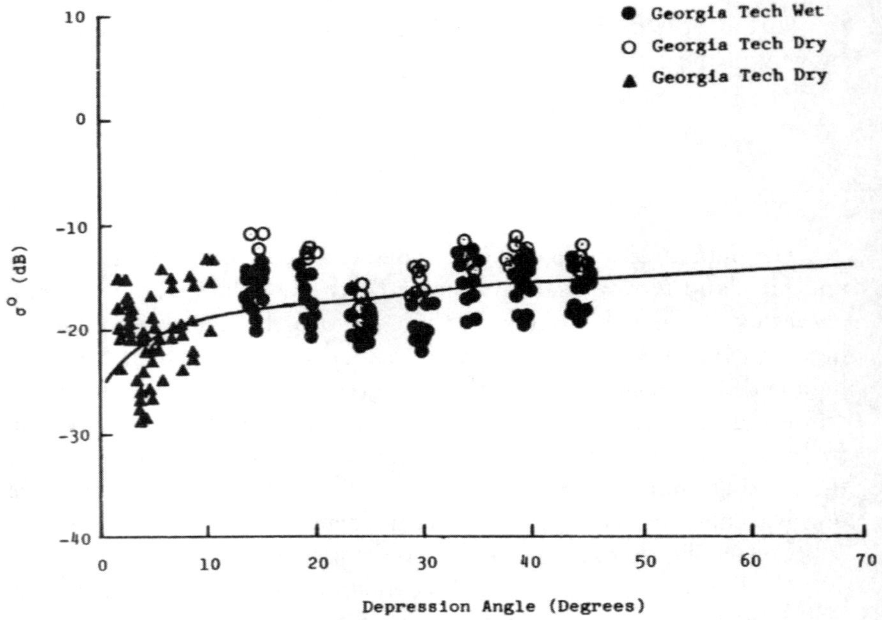

Figure 4.15 Backscatter coefficients for wooded terrain at 35 GHz. (*From:* [21, p. 388].)

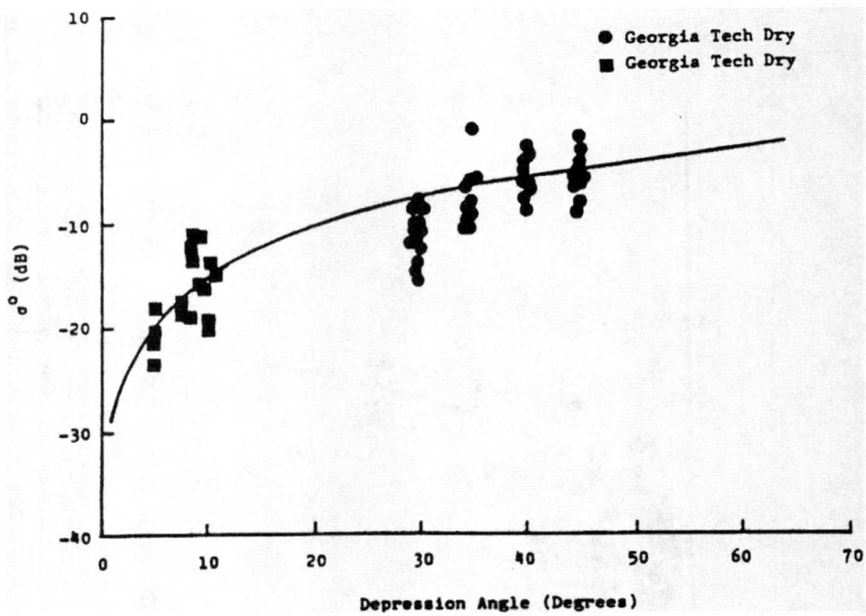

Figure 4.16 Backscatter coefficients for wooded terrain at 95 GHz. (*From:* [21, p. 388].)

anomaly at low grazing angles, where the frequency variation seems to reverse itself.

▶ Example 11 Land Clutter Backscattering Coefficient Models

Land clutter is a result of the reflection due to backscatter from ground surface scatterers within the radar's resolution cell. If the surface is smooth, then the backscatter is small, whereas a rough surface may produce a large amount of return. Generally roughness can be defined by the following inequality using the Rayleigh criteria:

$$\sigma_h \sin \psi \geq \frac{\lambda}{8}$$

where σ_h is the standard deviation of the rms height of the surface irregularities, ψ is the grazing angle and λ is the radar's wavelength. Thus the backscatter is generally a function of the surface structure, grazing angle, and radar frequency.

At low grazing angles backscatter from the land is generally greater than that from the sea due to the greater reflectivity of cultural objects. This complicates the detection of small low-flying missiles and aircraft from both land- and ship-based radars [72]. However a compromising feature is that the spectrum of land clutter is generally narrow, allowing MTI and PD radars to provide high improvement factors that enhance targets with significant motion relative to most land objects.

The characterization of land clutter is a difficult problem because of the many varieties of natural and man-made objects that may be intercepted by the radar beam. The general procedure is to group various types of terrain into generic classes as indicated in Table EX-11.1, which shows both its surface roughness and gamma value [6]. For example, at an X-band frequency, desert terrain is rough down to a grazing angle of 0.2° and at UHF it is rough down to 4.8°.

The large variation of natural and man-made objects over most terrain makes land clutter difficult to classify by type. This variation also makes it hard to obtain consistent experimental data. An approach followed by Nathanson [7, 23] is to average the data from about 50 experimenters that compiles the mean backscattering coefficient (σ^0) taken from similar terrains. These data includes grazing angles from 1° to 60° over frequencies from 1 to 35 GHz. Terrains include desert, farmland,

Table EX-11.1 Surface Roughness and Gamma for Generic Terrains [22]

Terrain Type	γ *(dB)*	σ_h *(m)*
Mountains	–5	100
Urban	–5	10
Wooded hills	–10	10
Rolling hills	–12	10
Farmland	–15	3
Desert	–20	1
Smooth surface	–25	0.3

rolling-wooden hills, and urban. Many of the data points are annotated with a note that a 5 dB error is not unlikely. Some values were obtained by extrapolation and interpolation and there are significant gaps and inconsistencies in the data.

The Nathanson land clutter data are collected in matrices embedded in the MATLAB program nathlanda.m where the columns are aligned by frequency for grazing angle data contained in the rows. The program provides a polynomial least-mean-square fit that smooths the data and an interpolation program that allows data to be calculated for frequencies between the various data points. The smoothed data indicate that an adjustment within the 5 dB tolerance would allow more consistent curves. This is contained in MATLAB program nathlanda.m whose output for desert terrain is depicted in Figure EX-11.1. These data were obtained using a fourth-order polynomial least mean square fit.

The Georgia Institute of Technology (GIT) has developed an empirical model to predict the mean backscatter of land clutter at grazing angles up to about 70°. The equation for the model is given by [21, 24]

$$\sigma^0 = A(\theta + C)^B \exp\left(\frac{-D}{1 + \dfrac{0.1\sigma_h}{\lambda}}\right)$$

where θ is the depression angle in radians; parameters A, B, C, and D are empirically defined constants; λ is the radar wavelength in meters; and σ_h is the standard deviation of the surface roughness in meters. The parameters A, B, C, and D are tabulated in the MATLAB program git_landclt.m for soil/sand, grass, tall grass crops, trees, urban, wet snow, and dry snow for frequencies of 3 to 95 GHz. The constant D is generally zero for all but flat terrains such as desert or roads. Also the parameters are not available for all frequencies. A sample graph obtained from the model for urban terrain at X-band is depicted in Figure EX-11.2. For comparison

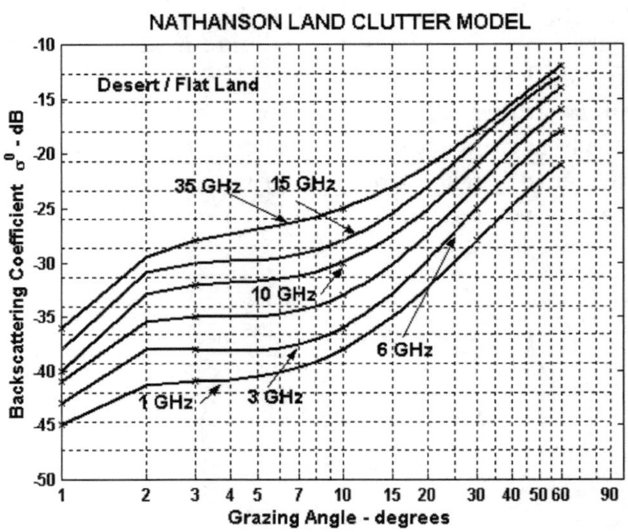

Figure EX-11.1 Nathanson desert clutter data.

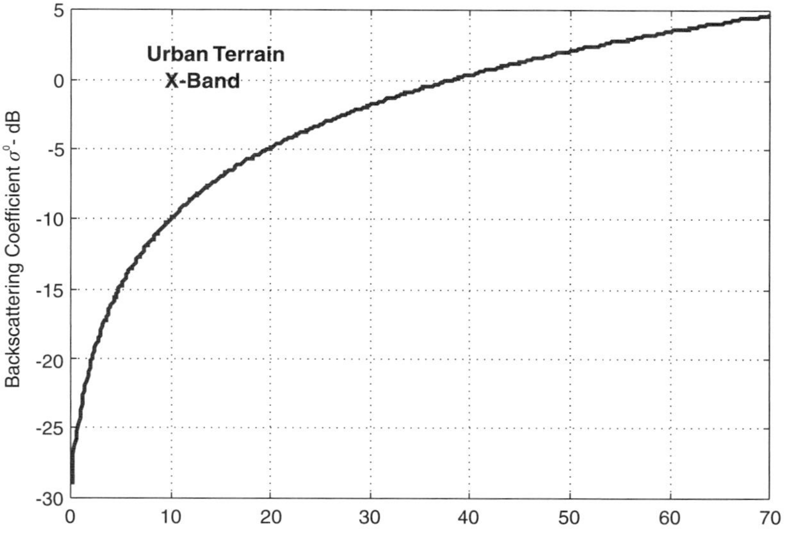

Figure EX-11.2 GIT backscatter model for urban terrain.

the MATLAB program also provides a plot of the backscatter estimated from the constant gamma model using the parameters given in Table EX-11.1 as modified by an adjustment in frequency suggested by Nathanson [7].

An airborne land clutter model proposed by Morchin [25, p. 150] extends Barton's data [6] and is appropriate primarily for surface-based radars. The model postulates three regions. The first is a low grazing angle or interference region that applies to desert terrain or other flat surfaces. The upper limit of this region is defined by a critical angle in radians as

$$\theta_e = \sin^{-1}\left(\frac{\lambda}{4\pi h_e}\right)$$

where λ is the wavelength and h_e is the surface roughness given in meters by

$$h_e = 9.3\beta_o^{2.2}$$

The parameters for the model including β_o are defined in Table EX-11.2.

The critical angle factor in decibels that applies only to desert terrain is given by

Table EX-11.2 Parameters for Airborne Land Clutter Model

Terrain	*A*	*B*	β_o
Desert	–29	$\pi/2$	0.14
Farmland	–24	$\pi/2$	0.2
Wooded hills	–19	$\pi/2$	0.4
Mountains	–14	1.24	0.5

$$\sigma_c^0 = -10 \log\left(\frac{\theta_c}{\theta_g}\right)$$

where θ_g is the grazing angle when $\theta_g < \theta_c$. The absorption factor u in decibels is given by

$$u = -10 \log\left(\frac{4.7}{\sqrt{f_{\text{GHz}}}}\right)$$

The mean backscatter coefficient for the low grazing angle and plateau regions is then given by

$$\sigma_{\text{dB}}^0 = A - 10 \log \lambda + 10 \log(\sin \theta_g) + u + \sigma_c^0$$

The mean backscatter coefficient for the specular region in decibels due to Beckman and Spizzichino [26] is given by

$$\sigma_s^0 = 10 \log\left\{\cot^2 \beta_o \exp\left[\frac{-\tan^2 (B - \theta_t)}{\tan^2 \beta_o}\right]\right\}$$

where $B = \pi/2$ for all terrains, except 1.24 for mountains, and is added to the low grazing angle and plateau regions component when $\sigma_s^0 > 0$ to form the composite model.

The data for the model are contained in the MATLAB program landclt_morchina.m and apply to frequencies from UHF to X-band. A plot for S-band is given in Figure EX-11.3, which also contains comparable Nathanson data for desert terrain that show reasonable agreement.

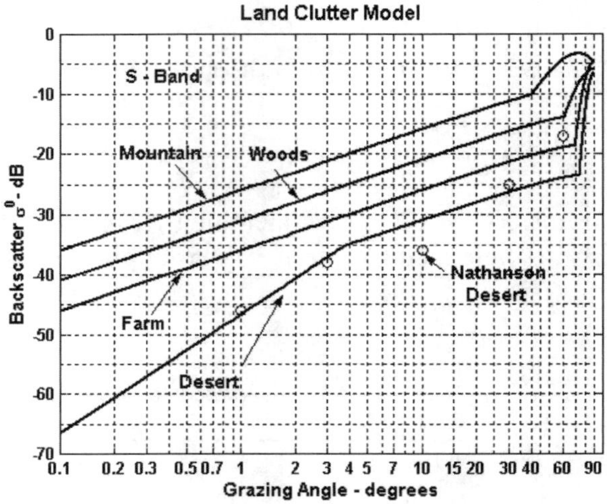

Figure EX-11.3 Airborne land clutter model for various terrains.

4.1.1.2 Amplitude Probability Distribution Function

Two types of amplitude statistics are useful in describing land clutter: temporal and spatial amplitude distributions. Temporal amplitude statistics refer to fluctuations over time in the amplitude of the clutter return due to the relative motion of the individual scatterers. The relative scatterer motion can be increased by either the physical movement of the radar platform or antenna scanning, which induces apparent scatterer motion due to the changing aspect angle between the scatterers and the radar. This temporal fluctuation can be described in terms of an amplitude probability distribution function, where the fluctuation rate determines the frequency spectrum of the clutter. The temporal amplitude statistics give information on the percentage of time during which the returns have a given range of values, and the frequency spectrum provides information on how rapidly the values change.

If ground-based radar illuminates a particular clutter patch, any clutter amplitude fluctuation observed will be due to intrinsic temporal fluctuation of the scatterers due to wind effects. Now, if the radar is pointed at another land clutter patch containing scatterers unrelated to the first patch, a different temporal amplitude distribution will generally be observed. This effect, which is prevalent for land clutter, is due to significant physical variations in the clutter field from point to point. This variability results from the inhomogeneities in the clutter and can be described in terms of a spatial probability distribution that defines the underlying statistics for each point in the clutter field.

When land clutter consists primarily of a large number of randomly distributed scatterers, approximately equal in size, the varying amplitude v_c of the clutter can be represented by a Rayleigh pdf:

$$p(v_c) = \frac{v_c \exp\left[-v_c^2/2\sigma^0\right]}{\sigma^0}; \qquad v_c \geq 0 \tag{4.29}$$

where the average clutter power $P_c = 2\sigma_{cx}^2$. Because the instantaneous clutter power is given by $P = v^2 = k\sigma^0$ the clutter's backscatter coefficient (σ^0) can be represented by the exponential pdf given by (see Section 1.5.1)

$$p(\sigma^0) = \frac{\exp\left(-\sigma^0/\overline{\sigma^0}\right)}{\overline{\sigma^0}} \tag{4.30}$$

where $\overline{\sigma^0}$ is the mean clutter backscattering coefficient.

Rayleigh statistics are generally considered to apply to distributed land clutter viewed at high grazing angles ($\psi > 5°$) with a medium- or low-resolution radar (pulsewidth > 0.5 μs) [19]. When discrete clutter (e.g., buildings or other artificial structures) is present in addition to distributed Rayleigh clutter, the clutter's Rician pdf can be written as [19]

$$p(P) = \frac{(1+m^2)\exp\left(\dfrac{-m^2 - P(1+m^2)}{\overline{P_c}}\right) I_o\left(2m\sqrt{\dfrac{(1+m^2)P}{\overline{P_c}}}\right)}{\overline{P_c}} \tag{4.31}$$

where $m^2 = S^2/P_o$ is the ratio of the discrete (S^2) to the distributed (P_o) clutter power and $I_0(\cdot)$ is the modified Bessel function. In this case, the total clutter power $\overline{P_c} = S^2 + P_o$ is proportional to the clutter's mean backscattering coefficient $\overline{\sigma^0}$.

Data related to the spatial variation of land clutter are relatively scarce. One report that summarized data from mountainous terrain at S-band, wooded terrain at L-band, and both wooded and cultivated terrain at X-band recommended the use of the Weibull probability distribution [74] to represent the spatial statistics of land clutter [11]. Figure 4.17 depicts X-band data for cultivated land, which show an increasing dynamic range with decreasing grazing angle. These data are fitted with a Weibull probability distribution function (see Section 1.5.4), given by

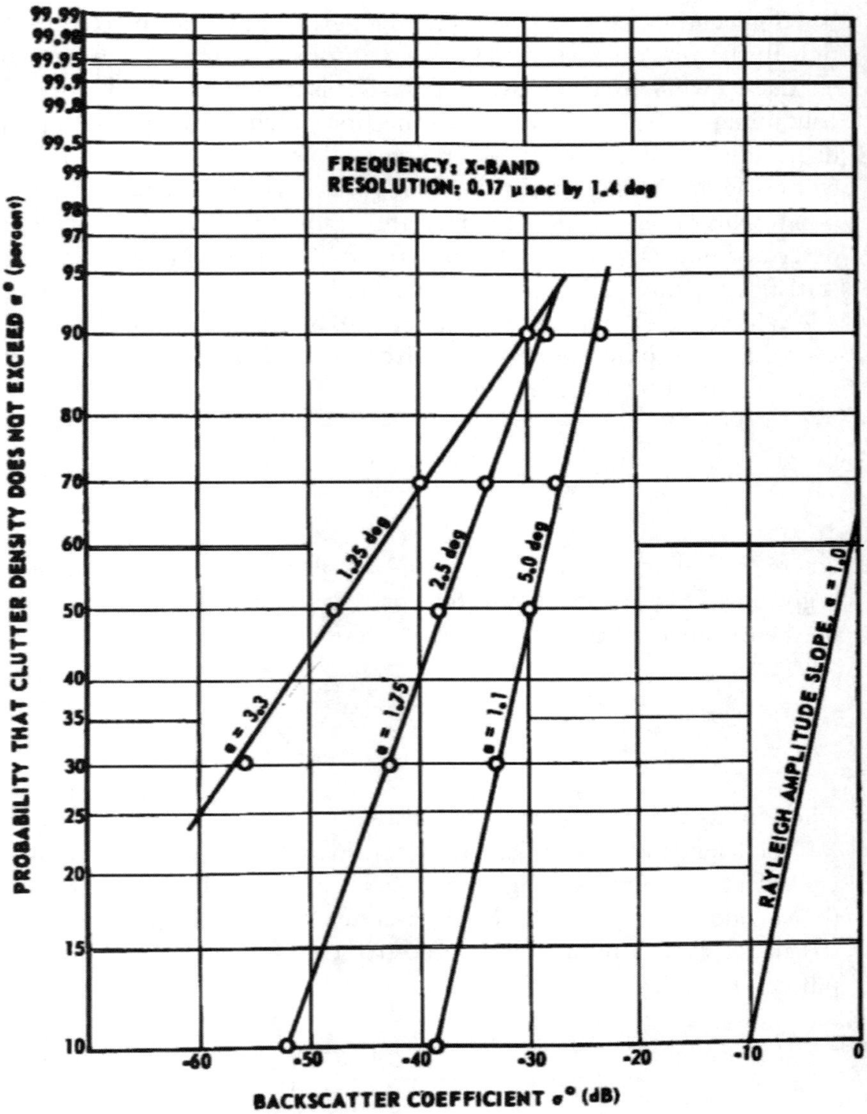

Figure 4.17 Land clutter spatial distribution for cultivated land at X-band. (*From:* [11, p. 11].)

$$P(\sigma^0) = 1 - \exp\left[-\ln 2\left(\frac{\sigma^0}{\sigma_m^0}\right)^\beta\right] \tag{4.32}$$

where $\beta = \alpha/2$ is the Weibull parameter and σ_m^0 is the median clutter backscattering coefficient that can be related to the mean backscattering coefficient $(\overline{\sigma^0})$ as

$$\overline{\sigma^0} = \frac{\sigma_m^0 \Gamma\left(1 + \frac{1}{\beta}\right)}{(\ln 2)^{1/\beta}} \tag{4.33}$$

Table 4.3 lists some values of β and σ_m^0 that apply to several different types of terrain [11]. Note that the Weibull distribution reduces to the exponential probability distribution (Rayleigh amplitude) for $\beta = 1$.

Another distribution used to represent the spatial distribution of land clutter is the log-normal distribution (see Section 1.3.4) [1]. The log-normal probability distribution function for the clutter's backscatter coefficient (σ^0) is given by

$$P(\sigma^0) = 1 + \frac{1}{2}\operatorname{erf}\left[\frac{\ln\left(\frac{\sigma^0}{\sigma_m^0}\right)}{\sqrt{2}\sigma_p}\right] \tag{4.34}$$

where σ_p is the standard deviation of the underlying normal distribution and σ_m^0 is the median backscatter coefficient. The mean backscatter coefficient is given by

$$\overline{\sigma^0} = \rho\sigma_m^0 \tag{4.35}$$

where ρ is the mean-to-median ratio given by

$$\rho = \exp\left(\frac{\sigma_p^2}{2}\right) \tag{4.36}$$

Table 4.4 lists some parametric values for the land clutter log-normal distribution [19].

Table 4.3 Weibull Parameters for Land Clutter Distributions [11, p. 7]

	Frequency Band	*Grazing Angle (°)*	*Resolution (μs by°)*	σ_m^0 *(dB)*	β
Rocky mountains	S	—	2.0 × 1.5	–46.3	0.256
Forest (spring)	X	0.7	0.17 × 1.4	–42.5	0.253
Forest (winter)	X	0.7	0.17 × 1.4	–36.4	0.266
Cultivated land (spring)	X	1.25	0.17 × 1.4	–47.8	0.303
Cultivated land (spring)	X	2.5	0.17 × 1.4	–38	0.571
Cultivated land (spring)	X	5.0	0.17 × 1.4	–29.8	0.909
Cultivated land (summer)	X	1.25	0.17 × 1.4	–46.5	0.352
Cultivated land (summer)	X	2.5	0.17 × 1.4	–37.8	0.429
Cultivated land (summer)	X	5.0	0.17 × 1.4	–25.3	1.0

Table 4.4 Log-Normal Parameters for Land Clutter Distributions

Terrain	*Frequency Band*	*Grazing Angle* (°)	σ_p
Discrete	S	Low	3.916
Distributed	S	Low	1.380
Various	UHF-K_a	10–70	0.728–1.584

A study [27] to measure the worst case amplitude statistics experienced by minimally attended L-band air search radar, deployed in Alaska, determined that the distribution functions range from contaminated normal to log-normal [28]. The measurements were taken with depression angles from the radar to the clutter that varied from –0.8° to 1.2° with pulsewidths of 0.5, 1, 3, and 6 μs. The worst case clutter statistics obtained assuming a log-normal distribution for mountains, hills, and valleys are given in Table 4.5. These values were used in the design specification for the ARSR-4 air route surveillance radar [29].

The spatial distributions are of interest for scanning radars where the returns from several points on the ground are generally processed before a detection decision is made. Of interest is the depth of correlation associated with each clutter cell before adjacent cells can be considered decorrelated. An autocorrelation analysis of sidelooking airborne radar (SLAR) data provided the results depicted in Figure 4.18, which indicate the distance required to decorrelate various types of terrain [20].

The decorrelation function in Figure 4.18 can be approximated by two correlation functions with associated decorrelation distances. The first decorrelation

Table 4.5 Worst-Case Land Clutter Model at Low Grazing Angles: Alaskan Environment

Terrain	*Vegetation*	*Height* (*ft*)	σ^0 *Values Median*	σ^0 *Values 84th Percentile*
Mountains	Bare	4,000	–20 dB	–14 dB
Wooded hills	Small trees, grass	1,000–4,000	–19 dB	–12 dB
Valleys, muskeg	Heavily wooded	1,000	–20 dB	–13 dB
Composite (15–100 nmi)		—	–23.5 dB	–17.5 dB

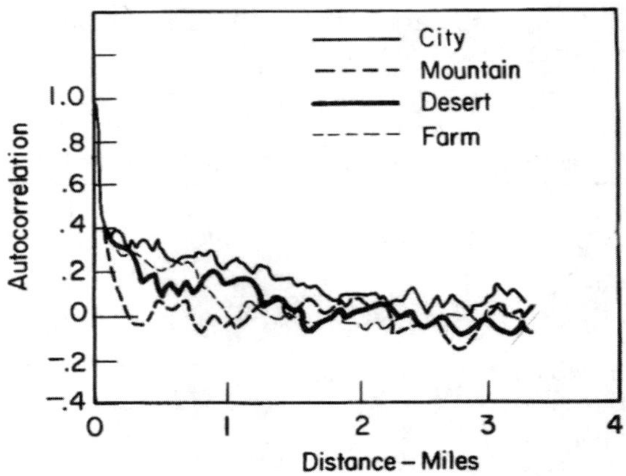

Figure 4.18 Spatial decorrelation distances for various types of land clutter. (*From:* [30, p. 199].)

distance (d_1) provides a rapid decorrelation to about the 0.4 level, and the second decorrelation distance (d_2) is much longer and represents a long-term correlation. The fast decorrelation distance was found to be on the order of 50 ft for all terrain types; the values for the long decorrelation distances are given in Table 4.6. Note that the value of the long decorrelation distance is significantly larger for more or less homogeneous terrains such as farms and woodland than for urban or mountainous terrains, which have sharp facets in the terrain structure.

In general, land clutter displays non-Gaussian amplitude statistics (i.e., wide clutter dynamic ranges) when viewed at low grazing angles or with a high-resolution radar. The low-grazing-angle effect, which is important for surface-based radars, is generally attributed to shadowing effects in which significant portions of the clutter may be obscured from the radar, depending on its aspect with respect to the radar. As the aspect changes, different parts of the clutter send strong reflections back to the radar, causing large amplitude fluctuations. The same sort of effect generally occurs for mountainous or city clutter, except that this effect can also occur for high grazing angles and low-resolution radar. Non-Gaussian amplitude statistics for high-resolution radar are generally caused by the radar resolving individual clutter scatterers with returns that vary widely in amplitude. Figure 4.19 depicts several non-Gaussian land clutter distributions obtained for surface-based radars at low grazing angles, showing the wide dynamic range associated with this type of clutter.

4.1.1.3 Land Clutter Spectra

The intrinsic spectrum of land clutter results from the relative motion of wind-driven scatterers (e.g., tree branches, leaves, vegetation). As the wind increases, the motion increases and the spectrum broadens. In general, the intrinsic spectral width is almost directly proportional to the transmitted frequency throughout the microwave region. Note that further spectral spreading of the clutter occurs due to antenna scanning, platform motion, transmitter instability, and clutter-limiting effects. These effects are covered in Sections 3.2.2 through 3.2.5.

The intrinsic spectral width for wooded land clutter is depicted in Figure 4.20 as a function of wind speed (see also Table 3.8). The standard deviation of the spectrum expressed in velocity units (σ_v – m/s) given in the figure is intended to be used in a Gaussian-shaped spectrum. The Gaussian-shaped spectrum is given by

$$S(f) = \frac{P_c \exp\left(f^2/2\sigma_f^2\right)}{\sqrt{2\pi}\sigma_f} \tag{4.37}$$

Table 4.6 Decorrelation Distances for Land Clutter

Terrain Type	*Long Decorrelation Distance (ft)*
Woodland	500–1,000
Desert	1,000–2,000
Farmland	1,000
Mountains, hills	500
Urban, suburban	500

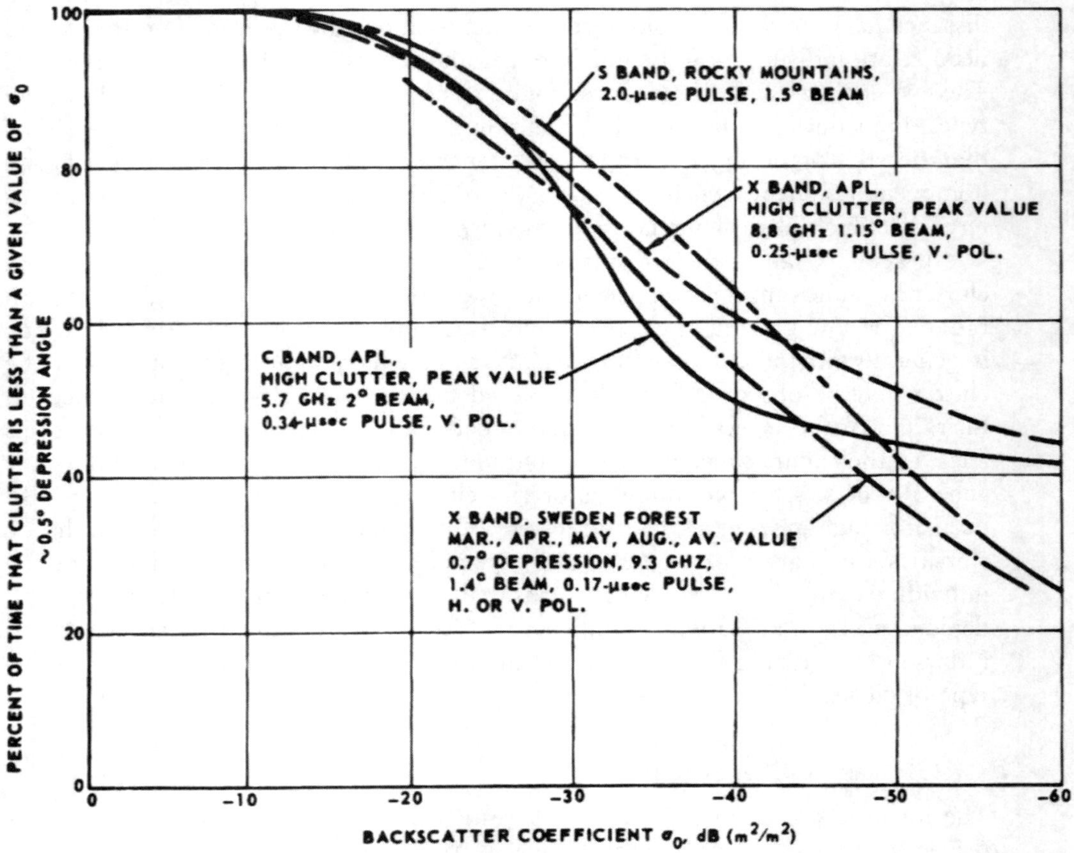

Figure 4.19 Land clutter backscatter distributions from surface radars at low grazing angle. (*From:* [23, p. 216].)

where $\sigma_f = 2\sigma_v/\lambda$ is the standard deviation of the spectrum and P_c is the clutter power. Also shown is the ratio of the fixed discrete clutter component to the fluctuating distributed clutter component [e.g., see (4.19); $m^2 = S^2/P_o$]. Equation (4.37) applies for the spectrum of the fluctuating component, whereas the discrete clutter component is represented by an impulse function $\delta(f)$ at zero frequency.

Experimental data indicate that the Gaussian-shaped spectrum given by (4.37) may underestimate the higher frequency components of the land cutter spectrum. A two-sided exponential has been suggested as a better fit to experimental X-band data than the Gaussian-shaped spectrum [14, 72]. Data at other frequencies suggested that a more general relationship for the clutter spectrum, given by

$$S(f) = \frac{k}{\left[1 + \left(\frac{f}{f_c}\right)^n\right]} \tag{4.38}$$

is appropriate. Here, f_c is the corner frequency proportional to the wind speed and n is an exponent that varies with the radar's transmitted frequency [27, 69].

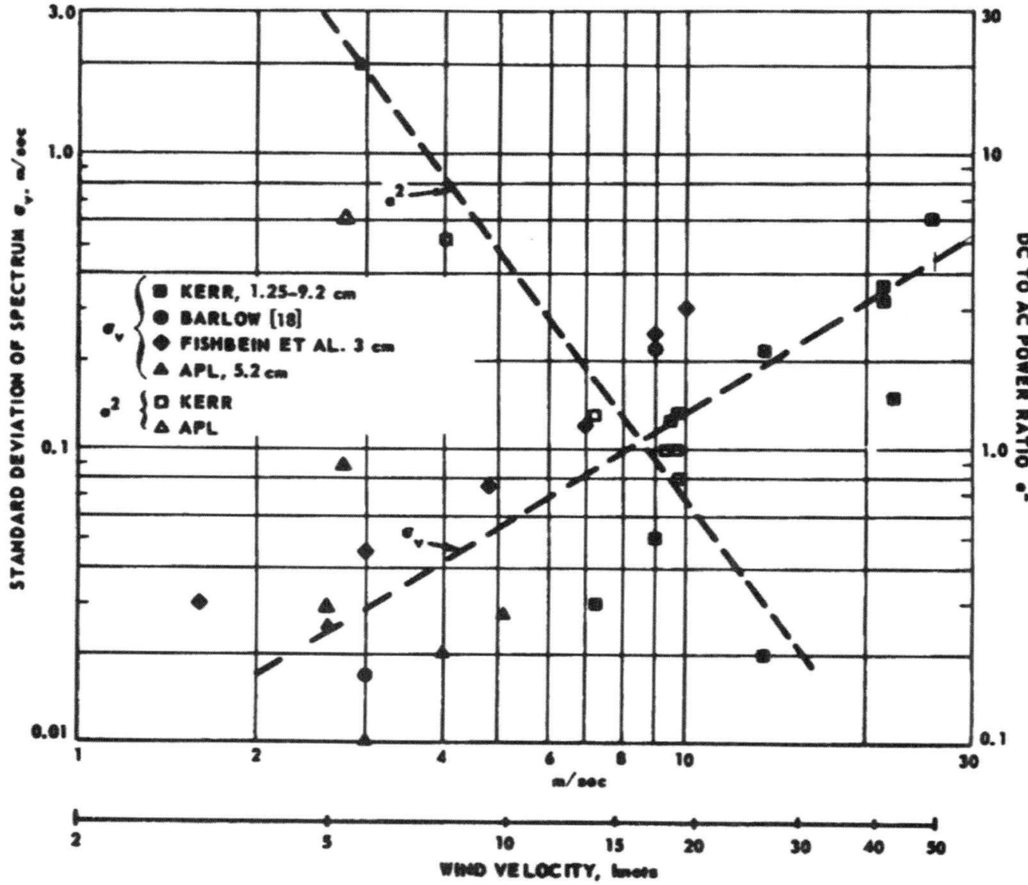

Figure 4.20 Spectral widths for wooded terrain as a function of wind speed. (*From:* [23, p. 274].)

Parameters for this relationship with 6 to 15 mi/h wind speeds are given in Table 4.7.

4.1.2 Sea Clutter

Sea clutter depends on radar parameters such as frequency, polarization, grazing angle, pulsewidth, and azimuth beamwidth. With the exception of near-vertical incidence, the reflectivity of sea clutter would be negligible were it not for wind-generated waves and capillaries (small waves caused by wind ripples superimposed

Table 4.7 Power Spectrum Function Parameters for Land Clutter at Various Frequencies

	Frequency			
Parameter	*9.5 GHz*	*16 GHz*	*35 GHz*	*95 GHz*
n	3	3	2.5	2
f_c (Hz)	9	16	21	35

on main waves). The properties of sea clutter are thus related to the sea-surface configuration, the most prominent feature of which are sea waves. The sea condition is generally described in terms of sea state, which relates to the height of the waves.

The sea-surface condition is related to the interaction of two somewhat independent disturbances. Distant winds cause swell, a fairly regular surface of nearly sinusoidal waves. Local winds cause steeper, less regular waves that occur in a random manner over the sea surface. Local winds blowing for an appreciable period of time produce a fully arisen sea, which achieves a quasi steady-state condition with recognizable upwind, crosswind, and downwind characteristics. The conventional description of sea state relates it to wind speed and assumes a fully arisen condition. In general, backscatter from the sea depends on sea state (i.e., wave height and wind speed) and the direction of the waves relative to that of the radar beam.

The reflectivity characteristics of the sea are generally related to sea state number in the same manner as land clutter reflectivity is related to terrain type. The sea states listed in Table 4.8 use the Douglas scale, which relates the sea condition to the wave height [4]. Because the wave height is instrumental in establishing the backscattering properties of sea clutter, the Douglas scale is an appropriate measure.

The statistical wave height parameter used in the Douglas scale $h_{1/3}$ denotes the average crest-to-trough sea height of the highest one-third of all of the waves. If the wave heights are assumed to have a Gaussian amplitude distribution with standard deviation σ_h, we can deduce that $h_{1/3} = 4\sigma_h$ and $h_{av} = 2.6\sigma_h$ [10].

The theory of sea clutter attributes part of the reflection to the existence of capillary waves (ripples) generated by local wind gusts of short duration. The capillary waves, which have wavelengths on the order of an inch, are superimposed on the larger wave structure and, because of their small size, assume greater importance as the frequency of the radar is increased. In this theory, the capillary waves are assumed to act approximately like flat-plate reflectors of varying size and orientation. This capillary wave structure apparently causes the interference effects at low grazing angles that result in σ^0 being sensitive to the radar's frequency and polarization. The capillary waves, in effect, become scatterers above the average sea surface that reflect both a direct and an indirect wave (off the sea surface), resulting in destructive interference at low grazing angles [4].

Sea clutter is generally described by its backscatter coefficient (σ^0), amplitude probability distribution, and power spectral density. This type of clutter differs

Table 4.8 Douglas Sea State Levels

Sea State	*Description*	*Wave Height ($h_{1/3}$ ft)*	*Approximate Wind Speeds (knots)*
1	Smooth	0–1	0–6
2	Slight	1–3	6–12
3	Moderate	3–5	12–15
4	Rough	5–8	15–20 (whitecaps)
5	Very rough	8–12	20–25
6	High	12–20	25–30 (gale)
7	Very high	20–40	30–50

from land clutter in its generally homogeneous structure, its average velocity proportional to wind velocity, which affects the central frequency of its spectrum, and the effects caused by the different dielectric properties of the sea as compared to land terrain, which affect the magnitude and phase of the reflection coefficient. The nonzero Doppler shift associated with sea clutter requires using motion compensation techniques in MTI radars.

The variation of average backscattering coefficient ($\overline{\sigma^0}$) with grazing angle (ψ) for sea clutter follows the three scattering regions of general surface clutter, depicted earlier in Figure 4.3. For microwaves, the only strong frequency dependence is associated with the interference effect, which determines the location of the critical angle (ψ_c) and makes σ^0 a strong function of λ for angles less than ψ_c. For a given sea state, ψ_c is proportional to the wavelength (i.e., $\psi_c = \sin^{-1}\lambda/4\pi\sigma_h$). For horizontal polarization, the curve of σ^0 versus ψ for low frequencies is expected to be similar to that for microwaves, except that the critical angle will be larger. This creates an expected variation, as depicted in Figure 4.21, in which the low-grazing-angle region [i.e., ψ^4 dependence; see (4.20)] is extended and the plateau region is foreshortened as the frequency is lowered into the UHF, VHF, and HF regions [4].

For vertical polarization, σ^0 would not be expected to vary as ψ^4 in the low-grazing-angle region because the magnitude of the reflection coefficient is generally less than 1 and the phase angle is not 180°. The curve for vertical polarization is expected to resemble that for horizontal polarization while differing in several

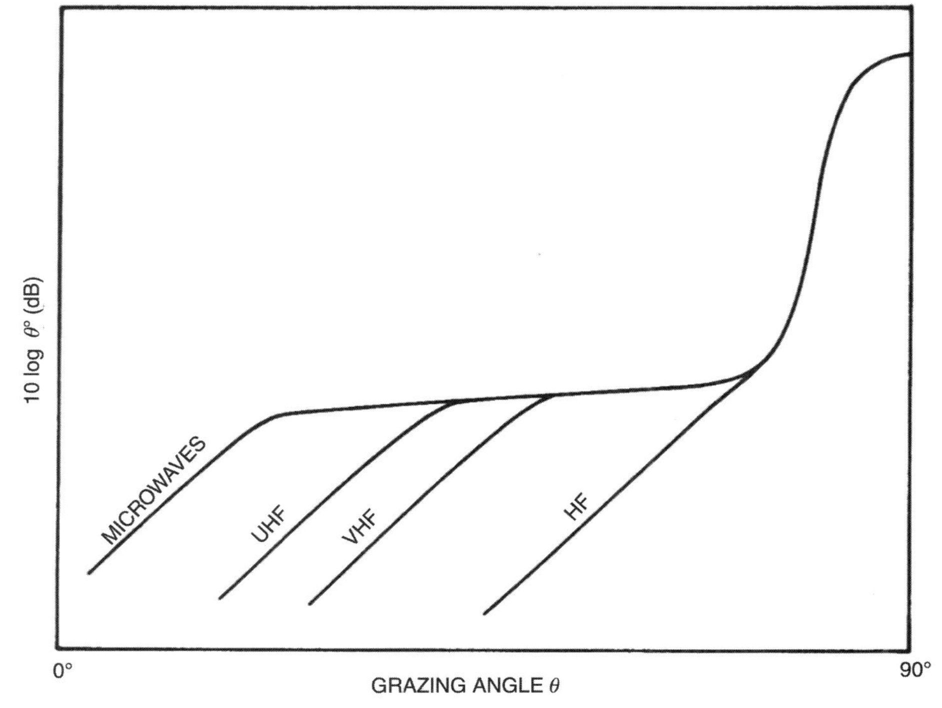

Figure 4.21 Backscatter coefficients σ^0 as a function of grazing angle for horizontal polarization at various frequencies. (*From:* [4, p. 276].)

important respects. First, σ_{VV}^0 will generally be larger than σ_{HH}^0 for most grazing angles. Second, the critical angle ψ_c will not be proportional to wavelength. Third, the functional dependence of σ^0 on ψ_c in the low-grazing-angle region will not be proportional to ψ^4.

For vertical polarization, destructive interference can occur only at grazing angles below Brewster's angle, where the reflection coefficient approaches $\Gamma = -1$. At higher frequencies (e.g., X-band), the critical angle generally occurs at grazing angles less than Brewster's angle. For example, at X-band frequencies the critical angle might be 1°, whereas Brewster's angle would be about 8° [4]. Thus, as the grazing angle is decreased at high microwave frequencies, destructive interference would generally be expected to occur at about the same angle for both horizontal and vertical polarizations. This condition persists until the frequency approaches about 1 GHz, where Brewster's angle and the critical grazing angle are approximately equal for low and moderate sea states. As the frequency is further decreased, Brewster's angle is less than the critical grazing angle and the shape of the σ^0 curve for vertical polarization will become significantly different from that for horizontal polarization.

The situation for vertical polarization and low frequencies (UHF) was depicted earlier in Figure 4.4. For low frequencies, as the grazing angle decreases, destructive interference will not be expected until the grazing angles become very small (on the order of 1° at 30 MHz). Therefore, the plateau region of the σ_{VV}^0 curve is expected to extend to much smaller angles than for σ_{HH}^0. Further, the critical angle for low frequencies and vertical polarization will be generally independent of surface irregularities. The critical angle will be determined primarily by the Brewster angle effect, which is dependent on frequency. As the grazing angle approaches zero, σ_{VV}^0 should approach σ_{HH}^0 because the vertical polarized reflection coefficient approaches unity with a phase shift of 180° [4].

The backscatter from the sea depends on the direction of the wind relative to the direction of the radar antenna beam. Backscattering is generally higher when the radar beam looks into the wind than when looking downwind or crosswind. There might be a variation as great as 5 to 10 dB in σ^0 as the antenna scans 360° in azimuth. The increased backscatter into the wind is apparently due to the asymmetrical nature of sea waves, which tend to be steeper on the upwind side.

A large number of radars that operate over water are horizontally polarized to take advantage of the apparently lower values of σ^0 with this polarization. However, when horizontally polarized short-pulse radar ($\tau \leq 0.5$ μs) is used at low grazing angles (typical for surface-based radars), the clutter backscatter return becomes spatially nonuniform. The clutter return then contains occasional spiky quasispecular echoes that can often persist for several seconds. The random spiky returns have a radar cross section at X-band that is on the order of 1 m^2 for an individual spike. To resolve individual spikes, a radar pulsewidth on the order of tens of nanoseconds is required. Sea spikes are apparently associated with breaking waves that form whitecaps. Sea spikes are also observed with vertically polarized short-pulse radars, but the effect is much less pronounced than for horizontal polarization.

Thus, as the resolution cell size of the radar is reduced, the nature of the sea clutter changes. The effect of sea spikes on radar performance can be profound, causing a large increase in false alarms, or a reduction in sensitivity to desired targets as the threshold is increased to reduce the false alarms. The resolution of radars operating in sea clutter conditions is therefore an important design parameter to be optimized for maximum detection performance.

4.1.2.1 Average σ^0 for Various Sea States

Because sea clutter is a fluctuating quantity, experimental σ^0 data must be evaluated statistically. The usual statistical measure used to quantify σ^0 is the time average, which we will specify as $\overline{\sigma^0}$. In some cases [4], the median (value of σ^0 for which the pdf has a value of 50%) is used to specify the clutter's backscattering coefficients ($\overline{\sigma^0}$) average value. If the probability distribution of the fluctuations is known, we are able to relate the median to the mean value. For example, for a Rayleigh distribution, the mean-to-median ratio is given by 10 log $(\sigma^0/\sigma^0_m) = 1.6$ dB (see Section 1.5.4). We should note that the correct value to use in the radar range equation for σ^0 is the mean value, and standard detection curves (computed for thermal noise interference) can be used to specify signal-to-clutter ratios, provided the sea clutter amplitude follows a Rayleigh pdf [10].

Most compilations of $\overline{\sigma^0}$ show wide variations due to the problems of defining sea conditions, the experimental difficulties of sea clutter measurement, and the many factors on which sea clutter backscatter depends. One approach [7] is to average the available data over many of the variables, which provides an indication of trends and can be used for initial system design. A compilation obtained by averaging and extending data on $\overline{\sigma^0}$ acquired from a large number of experiments for grazing angles between 0.1° and 30° is given in Tables 4.9 to 4.14 [7, 32]. These data represent (1) a time average over several tens of milliseconds; (2) an average of upwind, crosswind, and downwind values; and (3) radar pulsewidths in the 0.5- to 10-μs region, where Rayleigh statistics apply and the resolution does not allow resolving the individual wave structure. Note that for the small grazing angles tabulated (i.e., $\psi = 0.1°$ and 0.3°), the depression angle of the radar antenna beam may be different due to the earth's curvature [see (4.11)].

Data are given in Tables 4.9 to 4.14 for the mean backscattering coefficient $\overline{\sigma^0}$ as a function of grazing angle for sea states 0 through 5. Horizontal and vertical polarization data are plotted in Figures 4.22 and 4.23. These curves employ some smoothing of the data points to provide continuous curves. The sea state 3 data were chosen as representative of typical or average sea conditions [10].

The Naval Research Laboratory (NRL) has developed a four-frequency airborne radar system capable of making surface clutter measurements at frequencies ranging from UHF (428 MHz) to X-band (8,910 MHz) [33]. The coherent pulsed radar is capable of transmitting a sequence of four frequencies (8,910, 4,455, 1,228, and 428 MHz) alternately on horizontal and vertical polarization. Pulsewidths range from 0.1 to 2 μs, allowing almost simultaneous measurement of surface clutter over a wide frequency range with both horizontal and vertical polarization. The advantage of this measurement system is that it removes the surface condition variability, yet allows a parametric study of frequency and polarization sensitivity.

Tables 4.15 and 4.16 show sea clutter backscatter data taken under rough sea conditions (i.e., wave height = 16.4 ft, approximately sea state 6), which compares

Table 4.9 Normalized Sea Backscatter Coefficient σ^0 for a Grazing Angle of 0.1° [7, 32]

		Reflection Coefficient (dB) Below 1 m^2/m^2 at Frequency (GHz)						
		UHF	*L*	*S*	*C*	*X*	K_u	K_a
Sea State	*Polarization*	*0.5*	*1.25*	*3.0*	*5.6*	*9.3*	*17*	*35*
0	V	94	87	80	75	67	61	55
	H	103	96	89	84	76	70	64
1	V	89	82	75	70	62	56	50
	H	96	89	82	77	69	63	57
2	V	92	83	74	65	56	47	37
	H	96	87	78	69	59	50	51
3	V	86	76	67	57	47	38	28
	H	90	80	71	61	51	42	32
4	V	71	65	59	53	47	41	35
	H	72	66	60	54	48	42	36
5	V	71	63	56	48	41	34	26
	H	71	63	56	48	41	34	26

Table 4.10 Normalized Sea Backscatter Coefficient σ^0 for a Grazing Angle of 0.3° [7, 32]

		Reflection Coefficient (dB) Below 1 m^2/m^2 at Frequency (GHz)						
		UHF	*L*	*S*	*C*	*X*	K_u	K_a
Sea State	*Polarization*	*0.5*	*1.25*	*3.0*	*5.6*	*9.3*	*17*	*35*
0	V	84	78	73	67	61	55	49
	H	94	89	83	78	72	67	61
1	V	76	71	66	61	56	51	46
	H	89	83	76	70	63	57	50
2	V	77	70	64	57	50	44	37
	H	84	76	69	61	53	46	38
3	V	70	64	57	50	44	37	30
	H	76	69	61	53	45	38	30
4	V	62	57	53	48	43	38	33
	H	62	57	52	47	42	37	31
5	V	58	53	48	43	38	34	29
	H	62	56	49	42	36	29	23

Table 4.11 Normalized Sea Backscatter Coefficient σ^0 for a Grazing Angle of 1° [7, 32]

		Reflection Coefficient (dB) Below 1 m^2/m^2 at Frequency (GHz)						
		UHF	*L*	*S*	*C*	*X*	K_u	K_a
Sea State	*Polarization*	*0.5*	*1.25*	*3.0*	*5.6*	*9.3*	*17*	*35*
0	V	75	69	64	59	53	48	43
	H	85	80	75	70	65	60	55
1	V	68	64	59	55	50	45	41
	H	81	74	66	59	52	45	38
2	V	62	58	53	49	45	41	37
	H	74	67	60	54	47	40	33
3	V	57	53	49	45	41	37	32
	H	67	61	55	49	42	36	30
4	V	51	47	44	40	37	34	30
	H	53	49	45	41	36	32	28
5	V	43	41	39	36	34	32	30
	H	49	52	46	39	33	26	20

Table 4.12 Normalized Sea Backscatter Coefficient σ^0 for a Grazing Angle of 3° [7, 32]

		Reflection Coefficient (dB) Below 1 m^2/m^2 at Frequency (GHz)						
		UHF	*L*	*S*	*C*	*X*	K_u	K_a
Sea State	*Polarization*	*0.5*	*1.25*	*3.0*	*5.6*	*9.3*	*17*	*35*
0	V	64	61	59	57	54	52	50
	H	76	72	67	63	59	54	50
1	V	58	55	52	49	46	43	40
	H	68	63	59	54	50	45	41
2	V	55	52	49	46	42	39	36
	H	63	59	54	50	45	41	36
3	V	45	43	41	40	38	36	35
	H	58	54	50	45	41	37	33
4	V	39	38	37	36	34	33	32
	H	51	47	44	41	37	34	31
5	V	38	37	35	34	32	31	29
	H	50	45	41	36	32	27	23

Table 4.13 Normalized Sea Backscatter Coefficient σ^0 for a Grazing Angle of 10° [7, 32]

		Reflection Coefficient (dB) Below 1 m^2/m^2 at Frequency (GHz)						
		UHF	*L*	*S*	*C*	*X*	K_u	K_a
Sea State	*Polarization*	*0.5*	*1.25*	*3.0*	*5.6*	*9.3*	*17*	*35*
0	V	48	47	46	46	45	45	44
	H	62	59	57	54	52	49	47
1	V	46	45	44	43	42	41	39
	H	58	56	53	50	48	45	42
2	V	38	38	37	36	36	35	34
	H	55	52	49	46	44	41	38
3	V	34	34	33	33	32	32	32
	H	51	48	44	41	37	34	30
4	V	31	31	30	30	30	30	30
	H	47	44	41	37	34	31	28
5	V	27	27	27	27	27	26	26
	H	44	41	38	35	31	28	25

Table 4.14 Normalized Sea Backscatter Coefficient σ^0 for a Grazing Angle of 30° [7, 32]

		Reflection Coefficient (dB) Below 1 m^2/m^2 at Frequency (GHz)						
		UHF	*L*	*S*	*C*	*X*	K_u	K_a
Sea State	*Polarization*	*0.5*	*1.25*	*3.0*	*5.6*	*9.3*	*17*	*35*
0	V	43	43	42	42	41	41	40
	H	47	48	48	49	50	50	51
1	V	39	38	38	38	38	37	37
	H	45	45	45	45	45	45	45
2	V	34	33	33	32	32	31	31
	H	42	41	41	40	40	39	39
3	V	29	29	28	28	28	27	27
	H	40	39	38	36	35	33	32
4	V	29	28	27	26	24	23	22
	H	39	38	36	35	33	32	31
5	V	21	21	20	20	20	20	19
	H	39	35	32	29	26	23	20

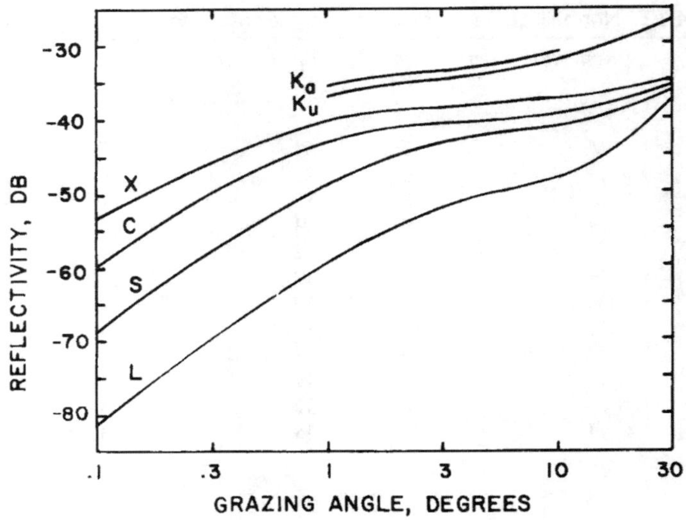

Figure 4.22 Mean backscattering coefficient σ^0 as a function of grazing angle for horizontal polarization at various frequencies (sea state 3). (*From:* [10, p. 316].)

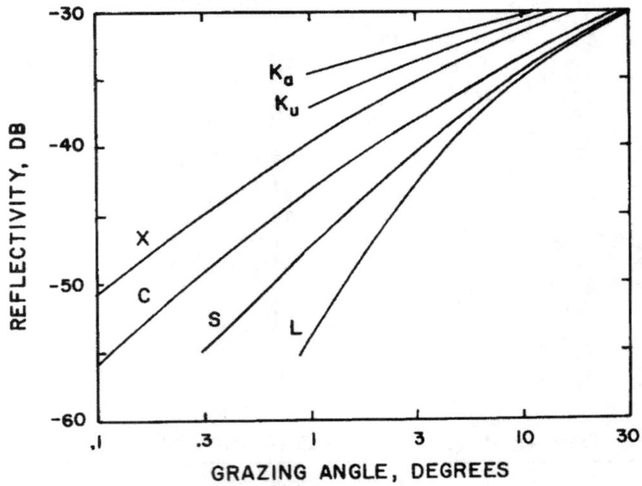

Figure 4.23 Mean backscattering coefficient σ^0 as a function of grazing angle for vertical polarizations at various frequencies (sea state 3). (*From:* [10, p. 317].)

sea clutter return for a wide range of frequencies, polarizations, and relative wind directions. Note that the calibration accuracy for these data is ±2 dB, and the backscattering data are given in median values rather than mean values (i.e., 1.6 dB should be added to the median values to provide mean values).

Table 4.15 provides rough sea backscatter data versus grazing angle for vertically and horizontally polarized transmissions and receptions (i.e., σ_{VV} and σ_{HH}), different wind directions (upwind, downwind, and crosswind), and various frequencies between UHF and X-band. The data indicate no apparent backscatter frequency sensitivity for vertical polarization, whereas the frequency sensitivity for horizontal

Table 4.15 Median Values of the Normalized Radar Cross Section σ^0 for Directly Polarized Signals Measured at X-, C-, L-, and P-Bands [34]

Depression Angle (degrees)	*Wind Direction*[b]	X_{VV}	X_{HH}	C_{VV}	C_{HH}	σ^0 (dB)[a] L_{VV}	L_{HH}	P_{VV}	P_{HH}
5	U	−35.5	−37.5	−34.5	−39.5	−40.5	−45.0	−41.0	−47.0
	D	−38.5	−43.0	−40.0	−44.0	−43.0	—	−43.0	—
	C	−37.0	−40.5	−39.5	−41.5	−44.0	−48.5	−42.5	—
10	U	−31.5	−35.0	−28.5	−38.5	−36.0	−48.0	−36.0	−46.5
	D	−33.5	−42.5	−31.5	−43.5	−34.5	−51.0	−36.0	−49.0
	C	−34.5	−39.5	−33.5	−41.5	−35.5	−48.5	−36.0	−49.0
20	U	−27.0	−33.5	−25.0	−35.0	−31.0	−41.5	−30.0	−47.5
	D	−29.5	−38.5	−28.0	−40.0	−29.0	−42.5	−28.5	−46.5
	C	−30.0	−38.5	−30.0	−40.0	−30.5	−43.5	−30.5	−45.5
30	U	−25.5	−30.5	−22.0	−32.5	−28.0	−37.5	−28.5	−37.0
	D	−27.0	−35.5	−23.5	−36.5	−26.0	−36.5	−28.0	−36.5
	C	−29.0	−35.5	−25.5	−35.5	−26.5	−33.0	−28.0	−37.5
45	U	−20.5	−24.0	−19.0	−27.0	−27.0	−32.0	−25.5	−31.0
	D	−23.5	−28.5	−21.5	−31.0	−25.0	−32.0	−24.5	−30.5
	C	−25.5	−29.0	−23.5	−30.0	−27.0	−34.5	−25.5	−32.5
60	U	−17.5	−18.5	−14.5	−19.0	−20.5	−22.0	−23.0	−25.5
	D	−18.5	−20.0	−16.0	−21.0	−19.0	−21.5	−22.0	−24.5
	C	−19.5	−20.5	−17.5	−21.0	−22.0	−24.0	−21.0	−24.0
75	—	−9.5	−10.0	−7.0	−9.0	—	—	−13.0	−15.0
90	—	−1.0	−1.0	+2.0	+1.0	—	—	+0.5	−2.0

[a]February 20, 1969: wind velocity, 29 knots; wave height, 16.4 ft.
[b]U = upwind, D = downwind, and C = crosswind.

Table 4.16 Median Values of the Normalized Radar Cross Section σ^0 for Cross-Polarized Signals Measured at X-, C-, L-, and P-Bands [34]

Depression Angle (degrees)	*Wind Direction*[b]	X_{VH}	X_{HV}	C_{VH}	C_{HV}	σ^0 (dB)[a] L_{VH}	L_{HV}	P_{VH}	P_{HV}
5	U	−44.0	−47.5	−42.0	—	−47.0	—	−47.0	—
	D	−46.0	−51.0	−43.0	—	−48.5	—	−48.0	—
	C	−44.5	−47.5	−43.5	—	—	—	−47.5	—
10	U	−42.5	−43.5	−41.0	−43.0	−48.5	−50.0	−45.5	−51.0
	D	−46.0	−46.5	−44.5	−46.0	−48.0	−48.5	−46.0	−51.0
	C	−45.5	−47.0	−44.0	−47.0	−47.5	−48.5	−46.5	−51.5
20	U	−39.0	−40.0	−38.0	−39.0	−42.5	−45.0	−42.0	−46.0
	D	−42.0	−43.5	−40.5	−40.5	−42.0	−41.5	−42.0	−45.0
	C	−43.0	−43.5	−41.0	−43.0	−43.5	−45.0	−42.0	−45.0
30	U	−38.0	−39.0	−36.5	−36.0	−41.0	−41.5	−40.0	−43.5
	D	−40.0	−41.0	−38.0	−38.0	−39.5	−39.0	−40.5	−44.0
	C	−40.0	−40.5	−38.0	−38.5	−39.0	−39.5	−40.5	−43.5
45	U	−36.0	−35.0	−34.5	−33.5	−41.0	−40.5	−40.0	−42.5
	D	−38.0	−37.5	−36.0	−36.0	−38.5	−38.0	−38.5	−41.5
	C	−37.5	−37.0	−36.5	−37.0	−41.5	−42.0	−40.0	−43.0
60	U	−35.0	−33.5	—	—	—	—	−38.5	−40.5
	D	−35.0	−34.0	—	—	—	—	−37.5	−39.5
	C	−34.5	−33.5	—	—	—	—	−37.0	−39.5

[a]February 20, 1969: wind velocity, 29 knots; wave height, 16.4 ft.
[b]U = upwind, D = downwind, and C = crosswind

polarization exhibits a λ^{-1} to $\lambda^{-1/4}$ dependence. As expected, vertical polarization backscatter generally exceeds that of crosswind or downwind. Another effect noticed with these sets of experimental measurements was a saturation effect, whereby the backscatter tends to have an upper limit as a function of grazing angle [33]. This upper limit is attributed to the tilting effect of the composite wave surface (i.e., capillary waves superimposed on sea surface).

Table 4.16 lists cross-polarization backscatter values (i.e., σ_{VH} and σ_{HV}), where transmission is on one polarization (e.g., vertical) and reception is on the orthogonal polarization (e.g., horizontal). The data show agreement between values for σ_{HV} and σ_{VH}, indicating a symmetrical scattering matrix. The values for the cross-polarized components generally range on the order of 10 dB below the vertically polarized component. However, for horizontally polarized backscatter, the values range from on the order of 10 dB above the cross-polarized component at high frequencies (X-band) to about that of the cross-polarized component at low frequencies (L-band and UHF) and low grazing angles ($\psi < 20°$).

► Example 12 Sea Clutter Backscattering Coefficient Models

The detection performance of radar systems depends on an accurate prediction of the background clutter. Four sea clutter models are in common use. The foundation for several of the models is Nathanson's data, which were compiled from about 60 independent experiments [7, 23]. These data compile the mean backscattering coefficient (σ^0) as a function of sea state (SS), grazing angle, frequency and polarization. Nathanson averages values over all wind directions relative to the look angle of the radar. Some values were obtained by extrapolation and interpolation and there are significant gaps and inconsistencies in the data.

Both the Technology Service Corporation (TSC) and Hybrid (HYB) models are based on the Nathanson data [36, 40]. The Georgia Institute of Technology (GIT) uses a theoretical multipath model to predict the mean backscatter at low grazing angles that may under-estimate the actual values due to ducting effects. Above a 1° grazing angle, the GIT model is consistent with the other models. The Sittrop model is based on actual data for upwind and crosswind conditions but has limited frequency coverage (X–K_u-band).

The characteristics of these models are shown in Table EX-12.1. Algorithms that compute the values of these models can be found in several references [35–43]. The models generally use either Douglas sea state or wind speed as an input to describe the sea surface. Other inputs are grazing angle, look angle of the wind

Table EX-12.1 Characteristics of Sea Clutter Models

Model	*GIT*	*HYB*	*TSC*	*Sittrop*	*Nathanson*
Frequency (GHz)	1–100	0.5–35	0.5–35	X–K_u-band	0.5–35
Sea description	Wind	Douglas	Douglas	Wind	Douglas
Conditions	Speed (m/s)	SS 0–5	SS 0–5	Speed (kt)	SS 0–5
Grazing angle (°)	0.1–10	0.1–30	0.1–90	0.2–10	0.1–30
Look angle (°)	0–180	0–180	0–180	Cross/upwind	Average
Polarization	H, V	H, V	H, V	H, V	H, V

with respect to radar antenna pointing, and polarization. MATLAB computer programs that calculate the mean sea clutter backscatter are given in sea_clt_gitl.m, sea_clt_hybridl.m, sea_clt_tsc2.m, sea_clut_sittrop.m, nathseal.m, nath_sea2.m, and seaclt_barton.m.

A comparison of these models for a frequency of 9.3 GHz, SS-3, and horizontal polarization is depicted in Figure EX-12.1. The plots for the GIT, TSC, and Hybrid models are given for upwind conditions. As expected the GIT underestimates the mean backscatter for low grazing angles. The Nathanson data points average the values over all look angles and, hence, are somewhat lower than the models using upwind conditions. Note that an average of the Sittrop upwind and crosswind models proves the best match to the Nathanson data. The rationale for using upwind values is that it is prudent to design radars for worst case conditions.

Other examples (see Figures EX-12.2 through EX-12.4) cover the frequency range from L-band to K_u-band. These MATLAB programs are contained in sea_clt_cmp.m, sea_clt_comp_kuvs.m, sea_clt_comp_lv5.m, and sea_clt_comp_sh3a.m. In general, the Hybrid model overestimates the mean backscatter for grazing angles between 3 and 30 degrees. The TSC model appears to provide the best correlation with the Nathanson data over the range of grazing angles between 0.1 and 30°.

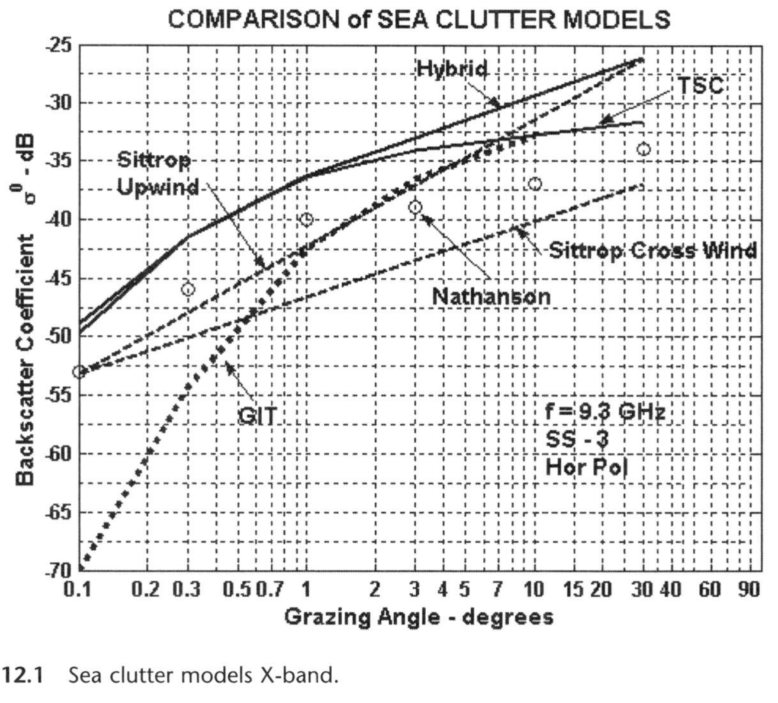

Figure EX-12.1 Sea clutter models X-band.

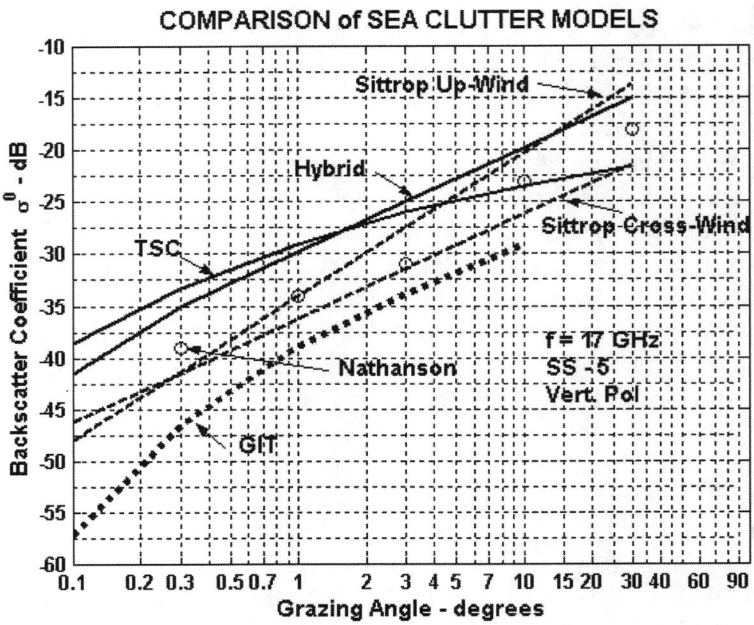

Figure EX-12.2 Sea clutter models for Ku-band.

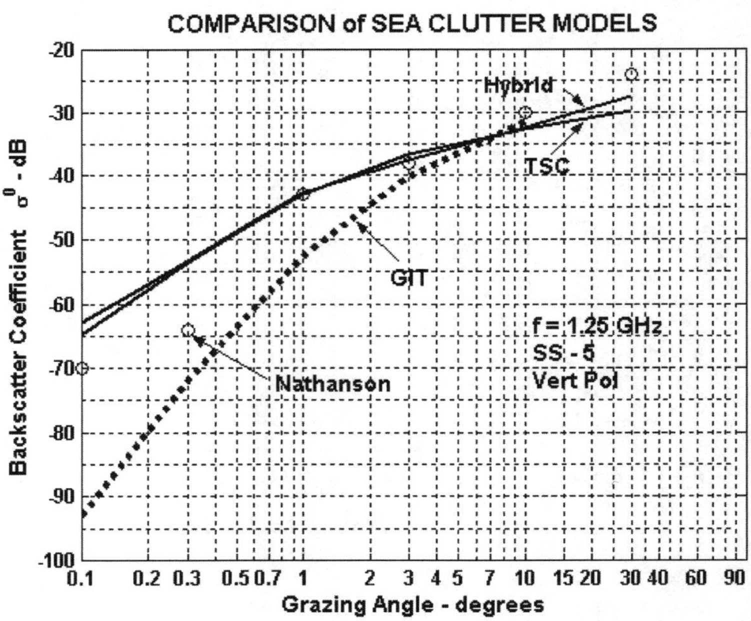

Figure EX-12.3 Sea clutter models for L-band.

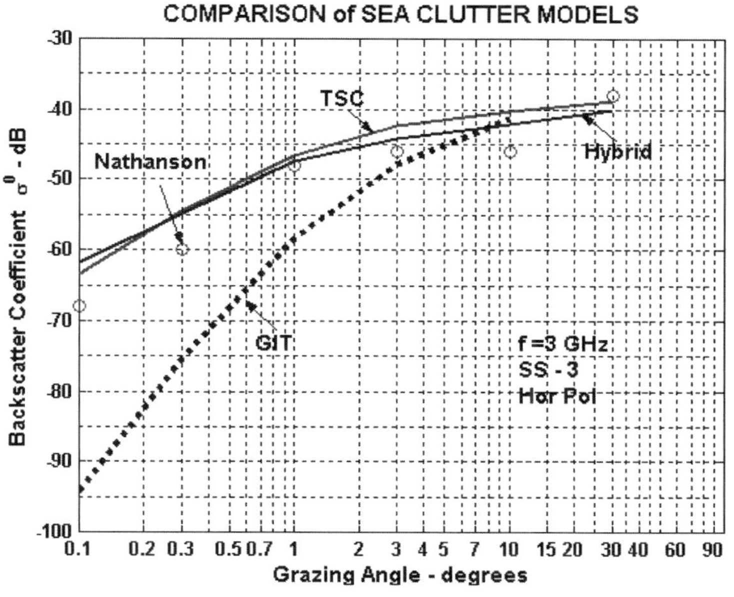

Figure EX-12.4 Sea clutter models for S-band.

4.1.2.2 Amplitude Probability Distribution Functions for Sea Clutter

The amplitude of sea clutter fluctuates in a statistical manner that can be characterized in terms of a pdf. When the return from sea clutter can be modeled as resulting from a number of independent random scatterers with no one scatterer predominating, the envelope-detected amplitude fluctuation voltage can be described by the Rayleigh amplitude pdf given by

$$p(v) = \frac{v}{\sigma^2} \exp\left(-\frac{v^2}{2\sigma^2}\right) \tag{4.39}$$

which is a function of only the clutter power given by $P_c = 2\sigma^2$ [see also (4.30) for the exponential power pdf]. The Rayleigh pdf applies to sea clutter when the resolution cell (proportional to the pulsewidth and azimuth beamwidth) or area illuminated by the radar is relatively large. Experimental data show that this distribution does not apply for high-resolution radar that illuminates only a small area of the sea.

In general, the Rayleigh distribution applies for sea clutter when the size of the area of ocean illuminated instantaneously by the radar is large compared to the wave structure. In a typical sea, as has been estimated, if the area has linear dimensions of about 250 ft, a sufficient number of scatterers (i.e., 5 to 10) contribute to the backscatter to justify the use of Rayleigh statistics. This corresponds to a pulsewidth of 0.5 μs or larger and an azimuth beamwidth of 2.5° at 1 nmi or 0.25° at 10 nmi.

One measure used to quantify the departure from Rayleigh statistics is the ratio of the standard deviation to the mean value of σ^0, which equals 1 for Rayleigh-distributed clutter [this follows from (4.30), where $E(\sigma^0) = \overline{\sigma^0}$ and $E[(\sigma^0)^2] = 2(\overline{\sigma^0})^2$]. Figure 4.24 depicts this measure when applied to sea return from both C-band and X-band surface radars covering pulse lengths between 0.008 and 3 μs with an azimuth beamwidth of 0.9° to 3° [7]. These data indicate an increasing deviation from Rayleigh statistics as the pulsewidth is decreased, especially on horizontal polarization. The increased probability of high amplitudes is generally attributed to the presence of spikes or specular reflections in the sea clutter return.

Sea clutter may also be non-Rayleigh at very low grazing angles, even when the resolution cell is large enough to result in Rayleigh clutter at higher grazing angles. This is due to a shadowing effect that results in fewer significant scattering elements in a given area of ocean. This effect also was found to be more pronounced for horizontal than vertical polarization [10].

Figure 4.25 depicts the cumulative probability distribution measured for sea clutter by using short-pulse (0.4 μs) X-band, surface-based radar with a narrow (0.8°) beamwidth [4]. These data are plotted on a normal probability scale against a decibel scale, where a straight line represents a log-normally distributed probability function [see (4.34)]. That the measured sea clutter data can be approximated in the tail region by a log-normal distribution (i.e., a straight line approximation) is apparent from inspection of the data shown on the curve. Also apparent is the longer tail of the distribution associated with horizontal as compared to vertical

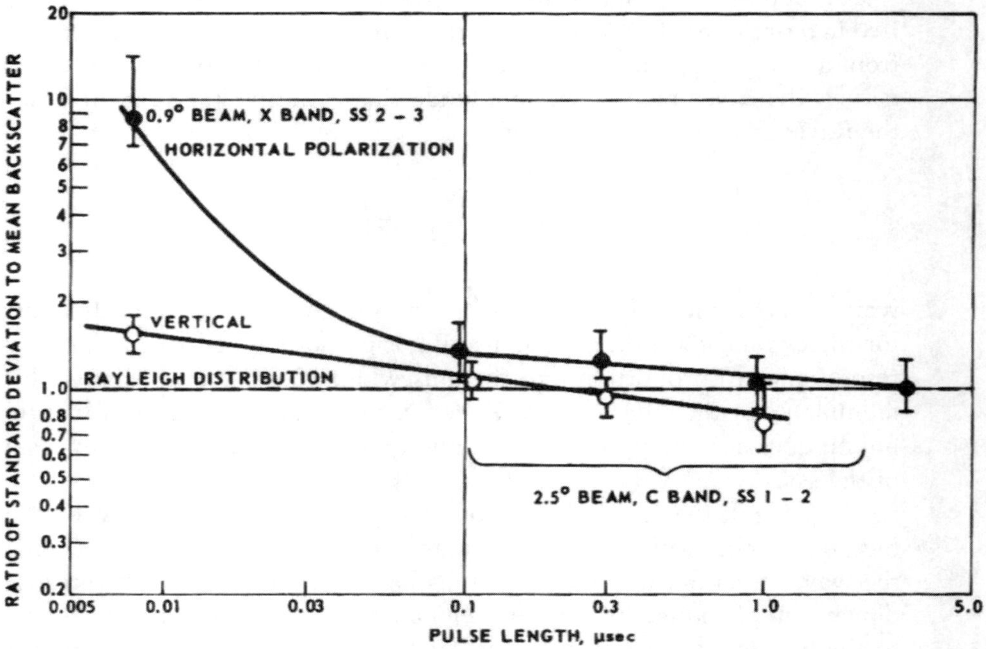

Figure 4.24 Distribution of the standard deviation to mean for short-pulse sea backscatter σ^0. (*From:* [23, p. 255].)

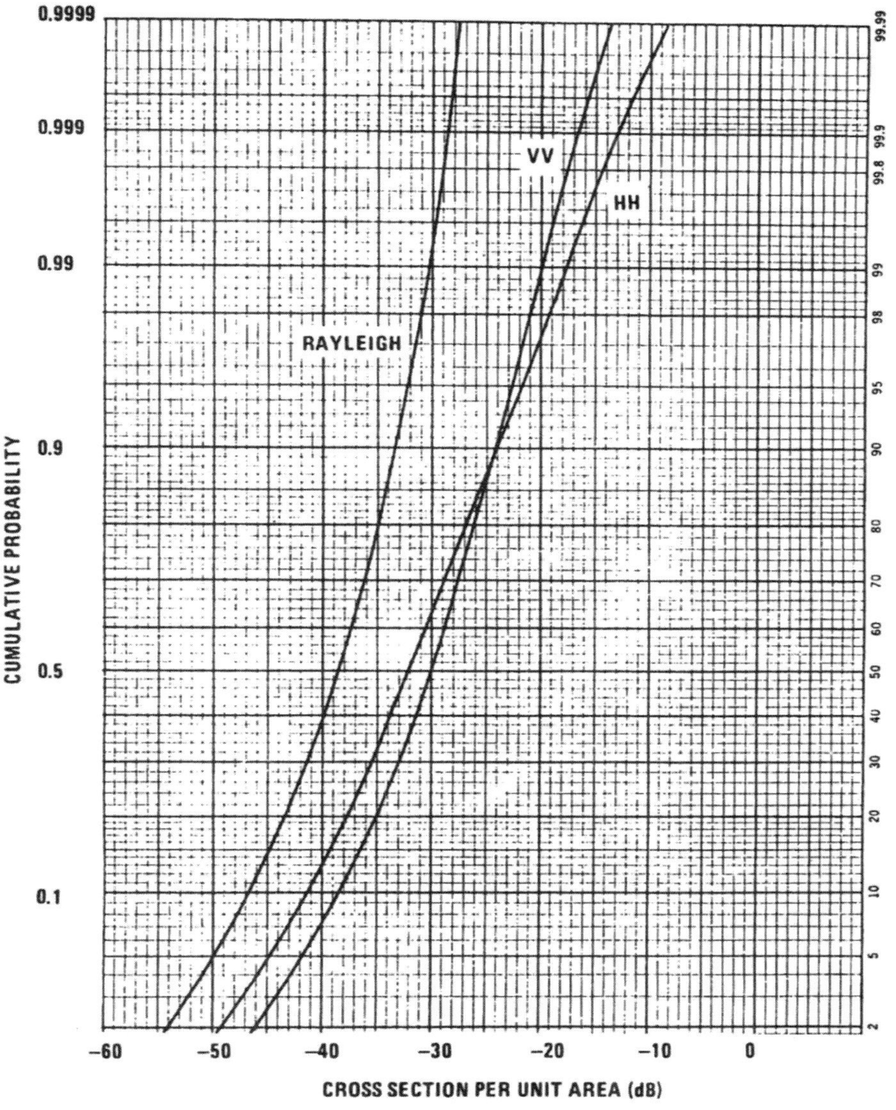

Figure 4.25 Cumulative amplitude pdf for an X-band sea backscatter coefficient. (*From:* [4, p. 185].)

polarization. Measured standard deviations were 7 ± 0.6 dB for horizontal polarization and 4 ± 0.6 dB for vertical polarization [4]. A reference Rayleigh distribution is also shown in Figure 4.25, which allows determination of the increased dynamic range associated with log-normal clutter at the higher cumulative probability values.

Figure 4.26 depicts experimental high-resolution (pulsewidth = 20 ns, beamwidth = 0.5°) sea clutter data taken at a grazing angle of 4.7° for vertical polarization [28]. The amplitude probability data generally fall between the Rayleigh and log-normal cumulative probability distribution functions shown for comparison on the curves. This situation is typical for most high-resolution radars, where the Rayleigh distribution tends to underestimate the amplitude distribution. As shown in the

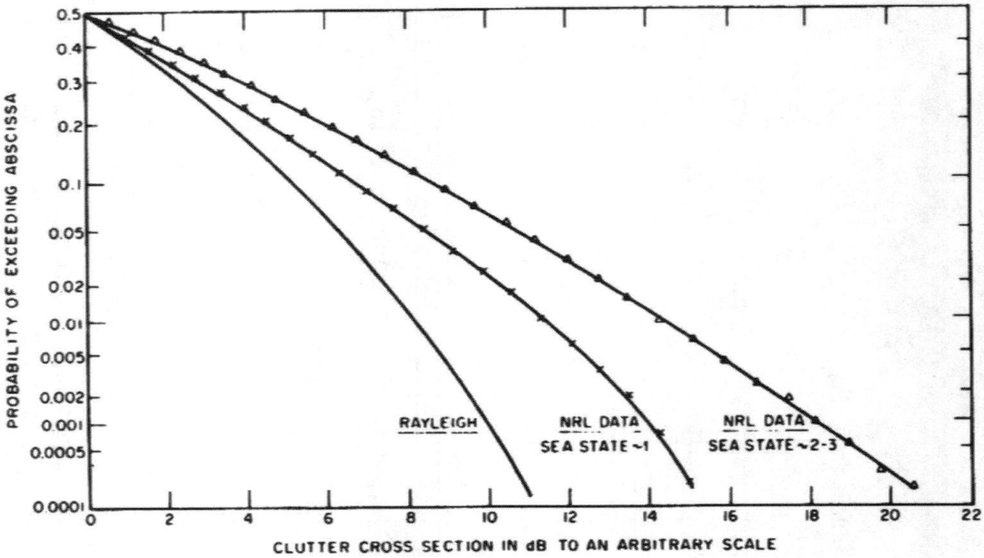

Figure 4.26 Cumulative probability distribution function for vertically polarized, high-resolution sea clutter data. (*From:* [28, p. 8].)

curve, two effects are associated with increasing sea roughness (i.e., increasing sea state). First, the tails of the distribution generally increase with increasing sea state. In addition, the tails tend to approach a log-normal distribution as the sea becomes rougher. Both effects combine to increase the dynamic range of the sea clutter, which requires the threshold detector to be set as a higher value than if the clutter were Rayleigh distributed to avoid excessive false alarms.

The log-normal amplitude voltage pdf representing envelope-detected sea clutter is given by

$$p(v) = \frac{1}{\sqrt{2\pi}\sigma_v v} \exp\left[\frac{-\ln^2\left(\frac{v}{v_m}\right)}{2\sigma_v^2}\right]; \quad v > 0 \tag{4.40}$$

where σ_v is the standard deviation of the underlying normal distribution and v_m is the median value of the log-normal distribution. Transforming this distribution to represent the distribution for clutter power ($P = v^2$), which is proportional to the backscatter coefficient (see Section 1.3.1), results in

$$p(\sigma^0) = \frac{1}{\sqrt{2\pi}\sigma_p \sigma^0} \exp\left[\frac{-\ln^2\left(\frac{\sigma^0}{\sigma_m^0}\right)}{2\sigma_p^2}\right] \tag{4.41}$$

where σ_m^0 is the median value of σ^0 and $\sigma_p = 2\sigma_v$ is the standard deviation of the underlying normal generating distribution (i.e., standard deviation of $\ln \sigma^0$). The mean-to-median ratio $\rho = \overline{\sigma^0}/\sigma_m^0$ is frequently used to characterize the tails of a probability distribution (see Section 1.5.4). The higher this ratio, the longer the tails of the distribution. For the log-normal distribution, the mean-to-median ratio is given by (4.36) as $\rho = \exp(\sigma_p^2/2)$. Figure 4.27 depicts values of ρ expressed on a decibel scale (i.e., $\rho_{dB} = 10 \log \rho$) as a function of grazing angle and sea state [19]. These data were obtained experimentally by using a K_u-band radar transmitting a short pulse (pulsewidth = 0.1 μs). The values given for ρ allow various log-normal fits to be synthesized for characterizing sea clutter at various grazing angles and sea states.

Because the log-normal distribution generally overestimates the real clutter's amplitude probability distribution (see Figure 4.26), several other distributions have been proposed for a closer match. The Weibull distribution, which was originally developed for land clutter, has also been used to model sea clutter data [44]. This distribution generally provides a fit closer to sea clutter data than the log-normal distribution, especially in the tail region, and has the added advantage that the Rayleigh distribution is a member of the Weibull family of distributions. Figure 4.28 depicts amplitude probability distributions plotted for sea clutter data (SS-3, Ku-band, horizontal polarization, 0.1-μs pulsewidth, into the wind) using a log-Rayleigh versus decibel scale in which a Weibull distribution plots as a straight line (see Section 1.5.4). The data show a fit close to a Weibull distribution with longer tails at the lower grazing angles. Note that the distribution approaches a Rayleigh clutter distribution at a grazing angle $\psi = 30°$.

The Weibull pdf for the clutter backscatter (σ^0) is given by

$$p(\sigma^0) = \frac{\beta}{\sigma_m^0} \cdot \ln 2 \left(\frac{\sigma^0}{\sigma_m^0}\right)^{\beta-1} \exp\left[-(\ln 2)\left(\frac{\sigma^0}{\sigma_m^0}\right)^{\beta}\right] \tag{4.42}$$

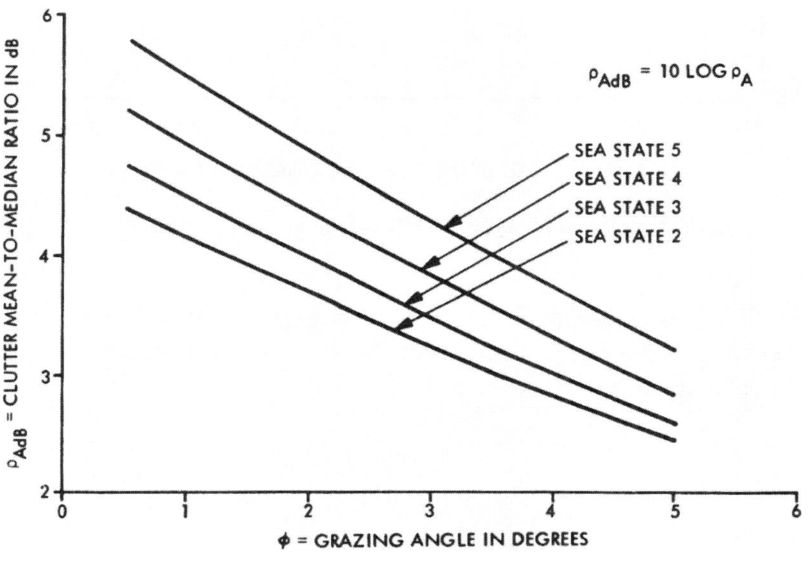

Figure 4.27 Sea clutter mean-to-median ratios for log-normally distributed sea clutter for various sea states and grazing angles. (*From:* [19, p. 43].)

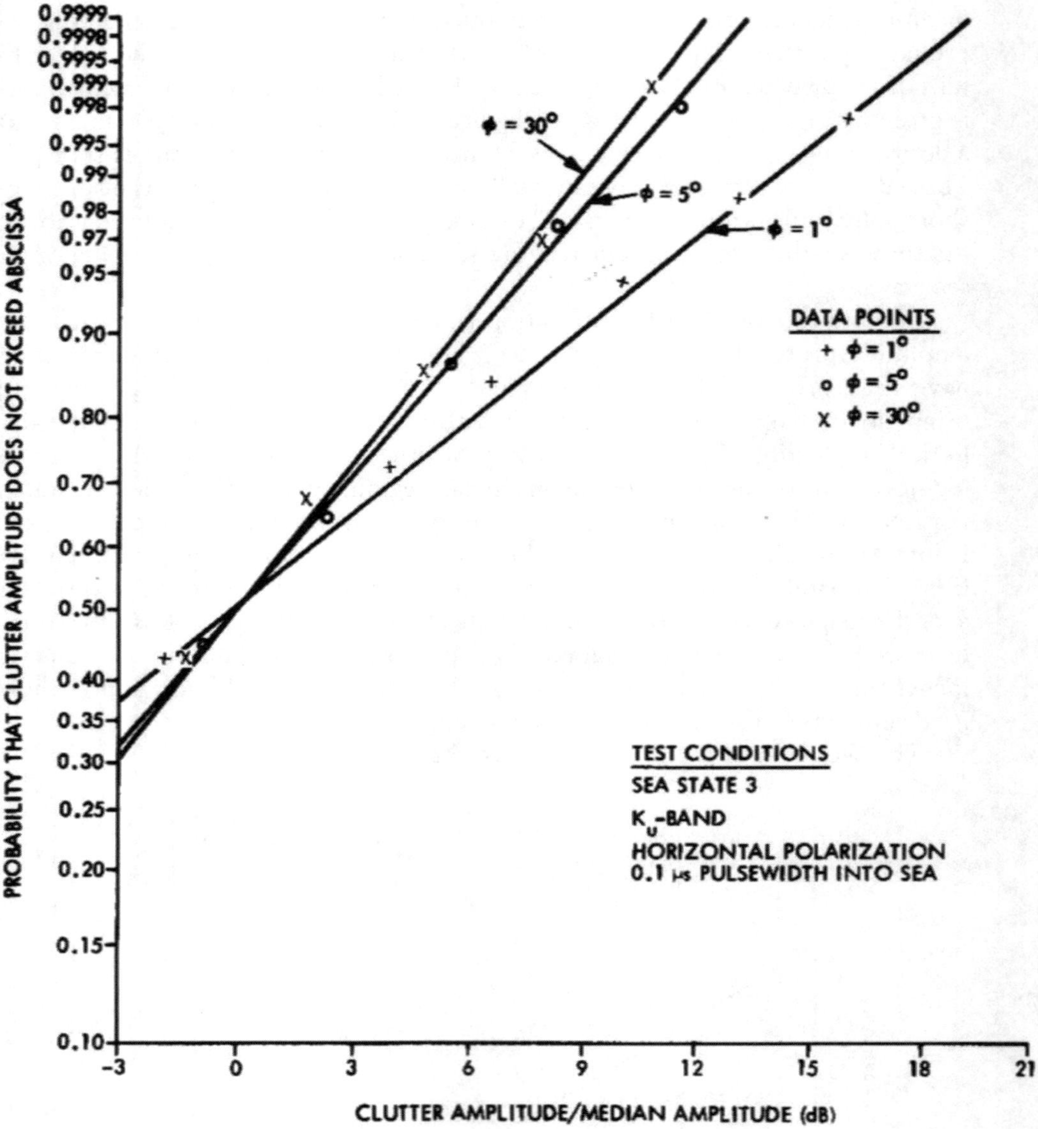

Figure 4.28 Weibull distribution for sea clutter at several grazing angles. (*From:* [19, p. 49].)

where σ_m^0 is the median value of the clutter and β is a parameter relating to the skewness of the distribution [19]. The mean-to-median ratio $(\overline{\sigma^0}/\sigma_m^0)$ for Weibull clutter is given by

$$\rho = \frac{\Gamma\left(1 + \frac{1}{\beta}\right)}{(\ln 2)^{1/\beta}} \tag{4.43}$$

Note that for $\beta = 1$ the Weibull distribution reverts to the exponential density function $\rho(\sigma^0) = \exp(-\sigma^0/\overline{\sigma^0})/\overline{\sigma^0}$, which characterizes Rayleigh clutter. Table

4.17 lists some values obtained experimentally for the Weibull skewness parameter β [19].

The K-distribution sea clutter model (see Section 1.5.4) was developed to provide a mathematical description for the composite model sea surface; the model theorizes that sea clutter is due to the compound effect of two fluctuating components. The speckle component due to capillary waves is modeled as a classical Rayleigh-distributed clutter (see Section 1.5.1). This component has a relatively short decorrelation time (on the order of 10 ms) that can be further decorrelated by using frequency agility (but this is not appropriate for MTI radar). The other component is associated with the sea swell structure of a fully arisen sea, which has a long correlation time (on the order of seconds) and is not affected by frequency agility. This component is modeled with a chi-distribution obtained by using an empirical fit to experimental sea clutter data [45].

The K-distribution is derived by assuming that the mean value of the Rayleigh-distributed component is a random variable with chi-distributed probability distribution. The resulting voltage amplitude pdf is found from the relation

$$p(v) = \int_0^\infty p(v/y)\, p(y)\, dy \tag{4.44}$$

where $p(v/y)$ is the Rayleigh probability density conditioned on its mean value y and $p(y)$ is a chi pdf. The mean value (y) of the Rayleigh pdf [see (4.39)] is equal to $y = \sqrt{\pi/2}\sigma$ so that the Rayleigh pdf conditioned on its mean level y is given by

$$p(v/y) = \frac{\pi v}{2y^2} \exp\left(-\frac{\pi v^2}{4y^2}\right); \qquad v \geq 0 \tag{4.45}$$

The chi pdf that represents the fluctuation of the mean level is given by

$$p(y) = \frac{2b^{2v}}{\Gamma(v)} y^{2v-1} \exp(-b^2 y^2); \qquad y \geq 0 \tag{4.46}$$

The resulting envelope-detected voltage amplitude density function of the sea clutter can then be evaluated from (4.34) as

$$p(v) = \frac{4c(cv)^v K_{v-1}(2cv)}{\Gamma(v)} \tag{4.47}$$

Table 4.17 Weibull Skewness Parameter for Several Grazing Angles and Sea States

Sea State	*Frequency Band*	*Beamwidth (degrees)*	*ψ (degrees)*	*Pulsewidth (μs)*	*β*
SS-1	X	0.5	4.7	0.02	0.726
SS-3	K_u	5	1–30	0.1	0.58–0.8915

where $c = \sqrt{\pi/4b}$ is a scale parameter, v is a shape parameter, and $K_{v-1}(\cdot)$ is the modified Bessel function of order $v = 1$. The nth moments of $p(v)$ are given by

$$E(v^n) = \frac{\Gamma\left(v + \frac{n}{2}\right)\Gamma\left(1 + \frac{n}{2}\right)}{c^n \Gamma(v)} \tag{4.48}$$

so that the clutter power is given by $P_c = E(v^2) = v/c^2$. If we use the transformation $\sigma^0 = kv^2$ and substitute for the clutter mean value $\overline{\sigma^0} = v/c^2$, we obtain the sea clutter's backscattering coefficient (σ^0) pdf:

$$p(\sigma^0) = \frac{2v\left(\frac{v\sigma^0}{\overline{\sigma^0}}\right)^{v-1/2} K_{v-1}\left(2\sqrt{\frac{v\sigma^0}{\overline{\sigma^0}}}\right)}{\Gamma(v)\,\overline{\sigma^0}} \tag{4.49}$$

The probability distribution function for K-distribution clutter can then be found as

$$P(\sigma^0) = \frac{1 - 2\left(\frac{v\sigma^0}{\overline{\sigma^0}}\right)^{v} K_v\left(2\sqrt{\frac{v\sigma^0}{\overline{\sigma^0}}}\right)}{\Gamma(v)} \tag{4.50}$$

where $\overline{\sigma^0}$ is the mean backscattering coefficient, v is a shape parameter that determines the dynamic range of the clutter and K_v is the modified Bessel function of the second kind.

The K-distribution backscatter coefficient (σ^0) mth-order moment can be deduced from (4.48) as

$$E[(\sigma^0)^m] = \frac{\Gamma(v + m)\Gamma(m + 1)}{c^{2m}\Gamma(v)} \tag{4.51}$$

which in normalized form $\left(M_m = \overline{(\sigma^0)^m} / (\overline{\sigma^0})^m\right)$ becomes equal to [46]

$$M_m = \frac{\Gamma(v + m)\Gamma(m + 1)}{v^m \Gamma(v)} \tag{4.52}$$

The second-order moment $M_2 = 2(v + 1)/v$ then becomes a measure of the dispersion. Figure 4.29 depicts sea clutter data fitted with a K-distribution. Run 2 on the illustration denotes SS-1 to SS-2, 10 to 15 knots upwind, 1.5° grazing angle, circular polarization, a 270-ns pulsewidth, and $M_2 = 2.97$. Run 18 denotes SS-3, 18 knots downwind, with 1° grazing angle, vertical polarization, and $M_2 = 2.88$. Figure 4.30 depicts wide dynamic-range sea clutter data fitted with a K-distribution. Run 9 on the figure denotes SS-2, 15 knots upwind, 1° grazing angle, horizontal polarization, a 70-ns pulsewidth, and $M_2 = 5.98$. Both Figures 4.29 and 4.30 have

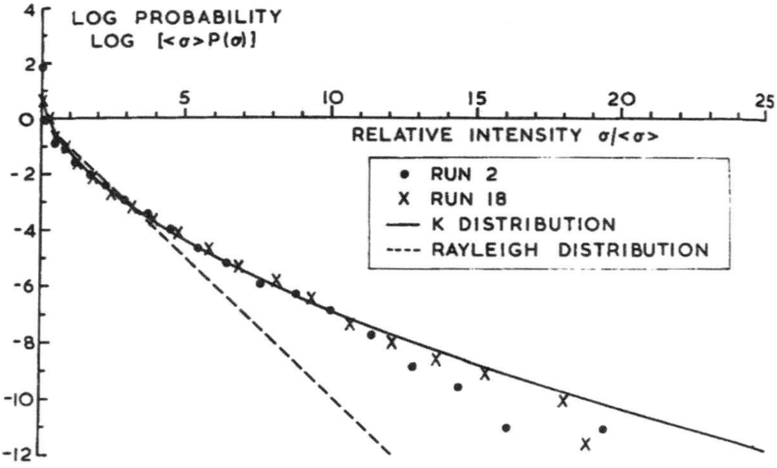

Figure 4.29 K-distributed sea clutter for circular and vertical polarization. (*From:* [46, p. 108].)

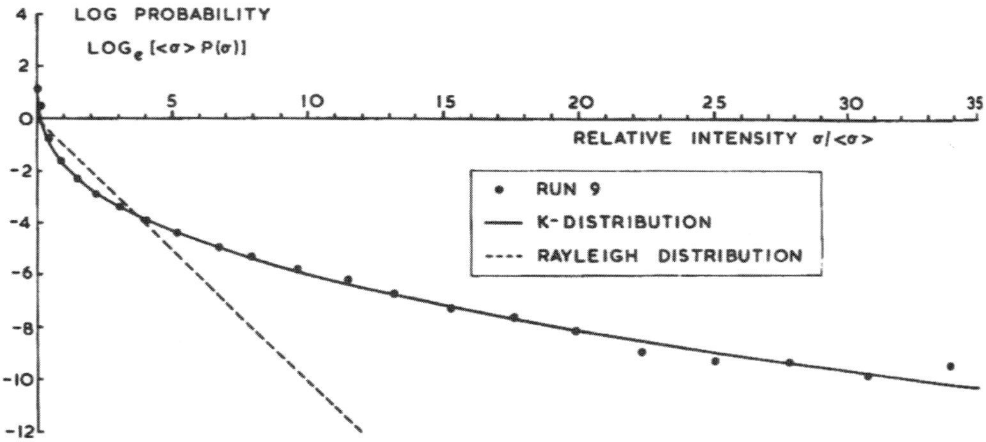

Figure 4.30 K-distributed sea clutter for horizontal polarization. (*From:* [46, p. 108].)

Rayleigh distributions superimposed to illustrate the extent of the tail distribution. In general, sea clutter distributions at grazing angles between 1° to 1.5° measured for pulsewidths of 70 and 270 ns at SS-1 to SS-5 with horizontal, vertical, and circular polarizations had values of v ranging from 0.1 to 3 [46].

The K-distribution provides a clutter tail distribution very similar to that provided by the Weibull distribution. However, its properties are more difficult to quantify analytically because of the unavailability of formulations for the modified Bessel function $K_v(\cdot)$. The value of the K-distribution lies primarily in the area of devising signal processing configurations based on its compound structure (i.e., processing the Rayleigh speckle component). The signal processor can be configured to process the Rayleigh-distributed speckle component with the understanding that its mean value fluctuates with a chi distribution. For example, if the probability of detection $P_d(\overline{\sigma^0})$ is evaluated for Rayleigh-distributed clutter, then the overall probability of detection is found from

$$P_d = \int_0^{\infty} P_d(\overline{\sigma^0})\, p(\sigma^0)\, d\sigma^0 \tag{4.53}$$

4.1.2.3 Sea Clutter Spectra

The relative motion of the sea surface with respect to the radar causes an intrinsic Doppler frequency shift of the return signal from the individual scatterers. Two effects are of interest to designers of MTI and PD radars. The first is the spectral width and shape, which result from the spread of velocity of the elemental scatterers of the sea surface. The second, a mean Doppler frequency shift of the entire spectrum, can be either up or down from the central frequency of the sea clutter return. The amount of the shift reflects the average radial velocity of the waves relative to the radar, which is typically a few knots or less. This mean or average Doppler shift is given by

$$f_m = \frac{2v_w}{\lambda} \tag{4.54}$$

where v_w is the average radial velocity of the waves with respect to the radar. Note that v_w can be either positive or negative, depending on wind conditions that generate the sea surface conditions. This shift is sometimes important in MTI radars, requiring motion compensation techniques to adjust the MTI filter to coincide with the center frequency of the Doppler spectrum (see Section 3.2.7). In other cases, the shift is so small relative to the MTI filter's passband (the MTI filter's passband is approximately equal to a band of Doppler frequencies corresponding to the radar's PRF) that it falls within the MTI clutter filter rejection notch, and no motion compensation is required.

The spectrum of sea clutter often is assumed to have a Gaussian shape. There is some question whether this is correct [10], but it is an analytically convenient assumption. With a Gaussian shape, the spectral width is characterized by its standard deviation σ_f, which can be converted to velocity units as

$$\sigma_v = \frac{\lambda}{2}\sigma_f \tag{4.55}$$

The width of the velocity spectrum varies with the agitation of the sea surface, becoming broader as the sea state increases. Expressing the width in velocity units makes the spread independent of frequency, thus allowing data from different experiments to be compared on an equivalent basis.

Figure 4.31 depicts the variation of bandwidth (double-sided half-power spectral width) with sea state (see Table 4.6), which is compiled from many experiments involving horizontal, vertical, and circular polarizations. Note that these data

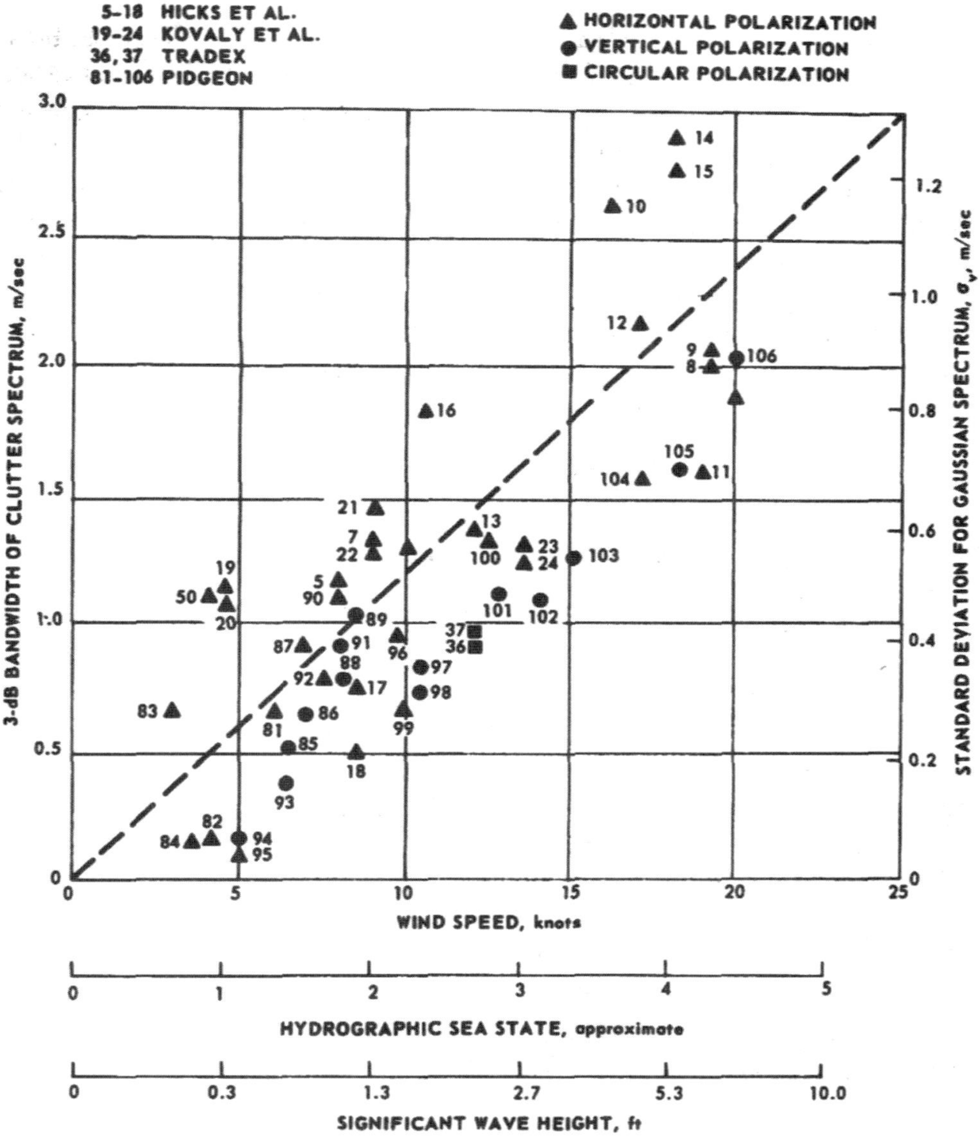

Figure 4.31 Bandwidth of sea clutter spectrum for coherently detected sea clutter. (*From:* [23, p. 243].)

represent the spectrum spread prior to envelope detection (i.e., coherent detection) and involve measurements taken over a frequency range of 220 MHz to 9.5 GHz [7]. The double-sided half-power spectral width can be related to the standard deviation of a Gaussian spectrum by using the following relation:

$$\sigma_v = 0.425\Delta v \tag{4.56}$$

where Δv is the half-power width. The linear relationship for the coherent half-power bandwidth indicated by the dotted line can be expressed as

$$\Delta f_{\mathrm{Hz}} = 3.6 f_{\mathrm{GHz}} S \tag{4.57}$$

where f_{GHz} is the radar's frequency in gigahertz and S is the sea state (see Table 4.5).

Figure 4.32 depicts the variation of bandwidth for envelope-detected sea clutter as a function of sea state, where experimental data are taken with both horizontal

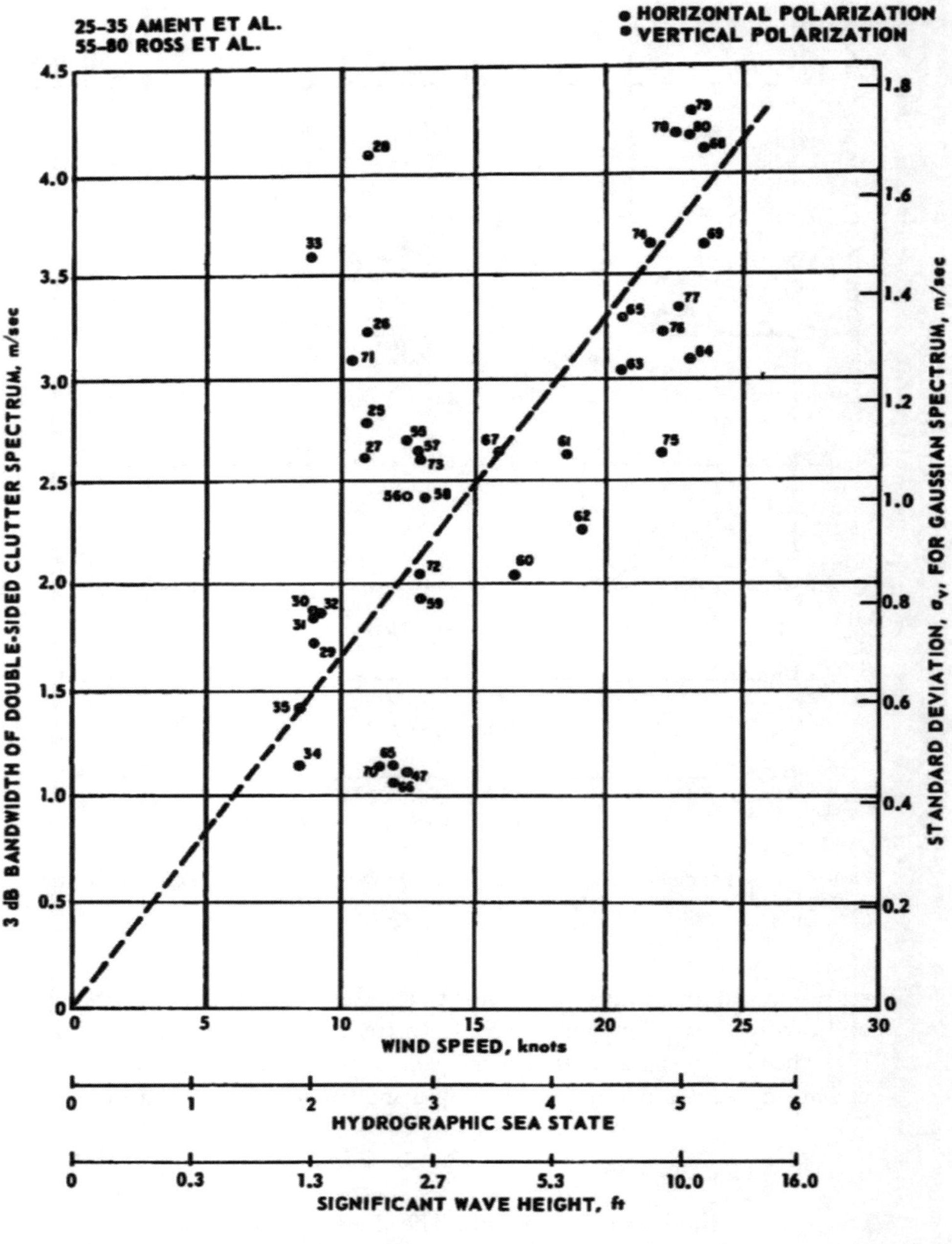

Figure 4.32 Bandwidth of envelope-detected sea clutter spectrum. (*From:* [23, p. 244].)

and vertical polarization [7]. The spectral width of the envelope-detected video can be derived as the convolution of the coherent spectrum, which itself results in

$$\sigma_{\text{video}}^2 = 2\sigma_{\text{Doppler}}^2 \tag{4.58}$$

where $\sigma_{\text{Doppler}}^2$ is the coherent spectral variance. This relationship indicates that the half-power width of the video spectrum is greater by the factor $\sqrt{2}$ than the coherent half-power spectral width.

Figure 4.33 depicts the upwind-downwind mean Doppler shift for sea clutter as a function of wind speed. These data show an upper limit on the spectral shift on the order of 7 knots and a difference between horizontally and vertically polarized data. The difference is apparently caused by the sensitivity of the vertically polarized return to just wave height, whereas sensitivity of the horizontally polarized return is to both wave height and wind speed [7].

The basic cause for the sea clutter spectra is the relative motion of the individual scattering areas on the surface. This motion is caused by both the large swell waves and the short capillary waves that ride on the swell surface. One theory of the backscattering mechanism attributes it to a Bragg resonance from the capillary wave structure tilted by the large swell waves. This model accounts for both a frequency dependence (the resonant water wavelength $\lambda_r = 2\lambda_w \cos \psi$, where λ_w is the water wavelength) and a polarization dependence [2].

A typical envelope-detected autocorrelation function for sea clutter is depicted in Figure 4.34 [4]. Note that this function consists of a long decorrelation time related to the period of the individual waves passing through the radar clutter patch and a short decorrelation time due to the effect of the wind on the wave structure. The fast fluctuations have a spectral width on the order of 100 Hz, whereas the slow fluctuation spectrum is on the order of 1 Hz [4].

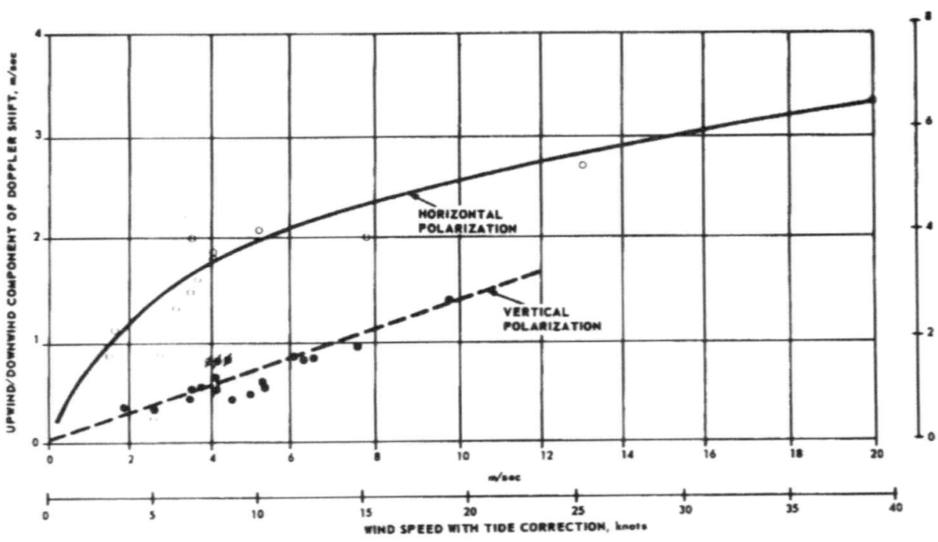

Figure 4.33 Mean Doppler shift for sea clutter [23, p. 248].

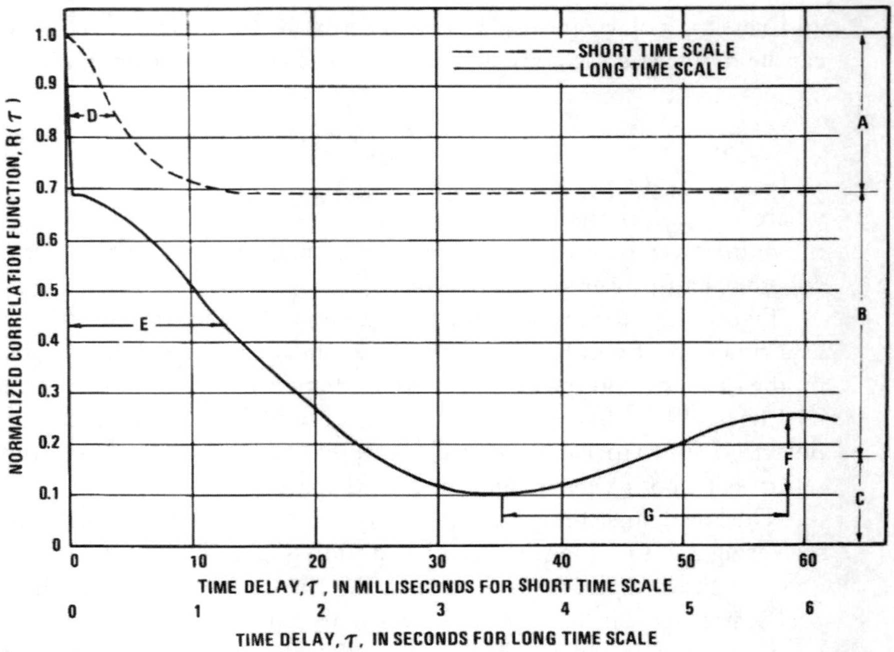

Source: Long (1974).

Figure 4.34 Typical sea cutter autocorrelation function. (*From:* [4, p. 145].)

Of significance is the ratio of the power contained in the fast fluctuating component (A) compared to that in the slow fluctuating component (B). Typical ratios for B/A are 0.7, 0.07, and 0.2, respectively, for *HH*, *VV*, and *HV* polarizations [4]. The significant difference of these values for different polarizations indicates a basic difference in the scattering mechanism associated with various polarizations. This also can be observed from the character of the echo return, which appears spiky for *HH* polarization and relatively noiselike for vertical polarization.

4.2 Atmospheric Clutter

A key attribute of radar is its ability to function in all weather at long ranges. However, as the frequency of the radar is increased, the reflection from precipitation generally increases as the fourth power of frequency throughout the microwave region, becoming significant for conventional radars that operate above UHF frequencies. High- or medium-PRF Doppler radars generally can separate the precipitation echoes from those of other targets contained within the same resolution cell. This property can be used to discriminate against precipitation clutter or, in the case of Doppler weather radar, to analyze the weather return.

The properties of chaff clutter are similar to those of precipitation clutter. However, the reflectivity of chaff, which consists of many lightweight metallic dipoles, generally is uniform over a wide frequency band. This can be achieved by using dipoles of various lengths that are resonant throughout the frequency coverage band of the chaff.

Clutter return backscattered from precipitation or chaff is caused by the various scatterers generally assumed to be uniformly distributed within the radar's volumetric resolution cell. The scatterers are characterized by the normalized parameter η, which represents the radar cross section per unit volume. Thus, the RCS of volume-filling clutter can be expressed as

$$\sigma_{cx} = \eta V_c \tag{4.59}$$

where η is the volume clutter reflectivity (m^2/m^3) and Vc is the radar's volume resolution cell. The volume resolution cell depicted in Figure 4.35 is given by

$$V_c = \frac{\pi}{4}\theta_{az}\phi_{el}R^2\left(\frac{c\tau_e}{2}\right) \tag{4.60}$$

where θ_{az} is the one-way azimuth beamwidth in radians, ϕ_{el} is the one-way elevation beamwidth in radians, R is the range in meters to the resolution cell, τ_e is the effective radar pulsewidth in seconds, and c is the velocity of light in meters per second. The power returned to the radar from the clutter is then equal to [2, p. 443; 7, p. 69]

$$P_c = \frac{P_t G_t G_r \lambda^2 \eta \theta_{az} \phi_{el} c\tau_e}{1024(\ln 2)\pi^2 R^2 L} \tag{4.61}$$

where P_t is the transmitted power; G_t and G_r are the transmitting and receiving Gaussian beam-shaped antenna gains, respectively; L is a loss term; and the other terms are as previously defined. Note that this equation contains a two-coordinate beam shape loss equal to $8 \ln 2/\pi$ that is only appropriate if this factor is not included in η.

Other atmospheric clutter consists of reflections from birds and insects. Reflections from individual birds are generally small but the cumulative effect from migrating birds or flocks of birds is significantly greater. Table 4.18 gives the RCS

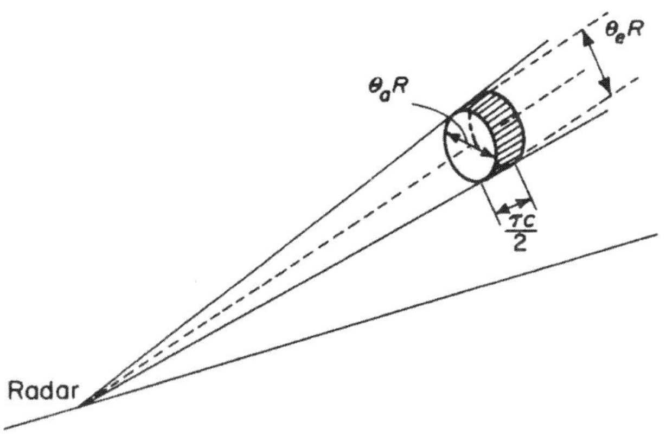

Figure 4.35 Resolution cell for volume clutter. (*From:* [47, p. 97].)

Table 4.18 Mean Radar Cross Section (dBsm) of Birds

Bird	*UHF*	*S-Band*	*X-Band*
Grackle	–42	–26	–28
Sparrow	–57	–29	–38
Pigeon	–30	–21	–28

taken with vertical polarization of some individual birds at UHF, S-band, and X-band frequencies. The increased RCS at S-band is due to a resonant effect from the bird's size. Birds typically fly at speeds from 15 to 40 knots, which can be troublesome for MTIs with low blind speeds.

Returns from single birds at S-band can range from 10^{-4} to 10^{-2} m^2. The return is principally from the body with very little from the wings. For large birds, the body is resonant near L-band and is in the Rayleigh (steady-state) region at UHF. Typically, anywhere from one to several hundred birds may occupy a resolution cell. Although the mean return from a typical flock of birds may be low (i.e., on the order of 10^{-2} m^2), the tail of the distribution has been observed to return up to 10 m^2. Birds usually fly at altitudes less than 7,000 ft with air speeds between 15 and 45 knots [19]. Returns from birds viewed by L-band search radar with a display range of 50 km are depicted in Figure 4.36.

The RCSs for several different kinds of insects taken with vertical polarization are listed in Table 4.19 [7]. As with birds, their RCS increases rapidly as the resonance region is reached. The RCS of some insects (especially the hawk moth) are large at X-band frequencies.

Swarms of insects have been observed by meteorologists using powerful radars. The swarms may cover a large area and, in general, drift with the winds. The large number of insects within a radar resolution can substantially increase the mean

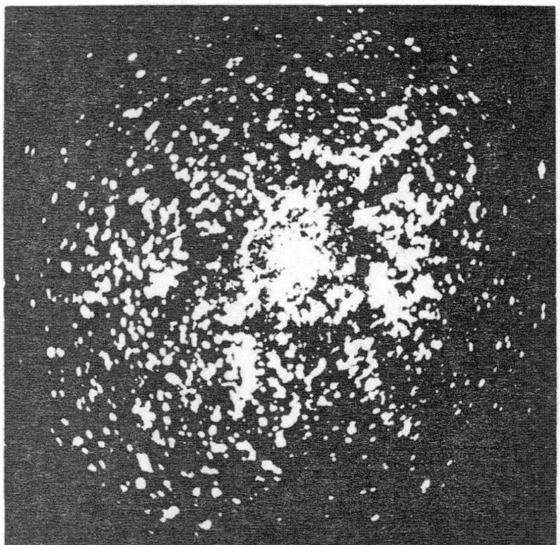

Figure 4.36 Birds viewed from an *L*-band search radar. (*From:* [5, p. 167].)

Table 4.19 Typical Mean RCS (dBsm) of Individual Insects

Insect	Frequency Band UHF	S	X
Hawk moth (5 cm)	−54	−30	−18
Worker bee (1.5 cm)	−52	−37	−28
Dragonfly	−52	−44	−30

value over that of an individual insect. In addition, the RCS of both birds and insects have been reported to follow a log-normal probability distribution, resulting in large tail returns of up to 10 m^2.

4.2.1 Precipitation Clutter

The backscattering properties of rain depend on the transmitted frequency, polarization, and number and size of the raindrops. Because the raindrops are approximately spherical, scattering in the Rayleigh region occurs when the ratio of the raindrops' diameter to the wavelength is small. In this region, the RCSs of the raindrops exhibit a fourth power variation with frequency (i.e., $\sigma_{cx} = kf^4$). The Rayleigh scattering theory approximation generally applies for rain below C-band and, except for the heaviest rain, represents a good approximation for X-band. Figure 4.37 depicts the variation of the backscattering coefficient η (m^2/m^3) as a function of the radar frequency and precipitation rate for rain and snow [8]. These curves apply to frequencies up to the order of 100 GHz, where the backscatter becomes relatively independent of frequency, but rather depends on the number of rain scatterers contained in a resolution cell. Figure 4.38 depicts the variation of the backscatter coefficient through the MMW frequency region for different rain rate

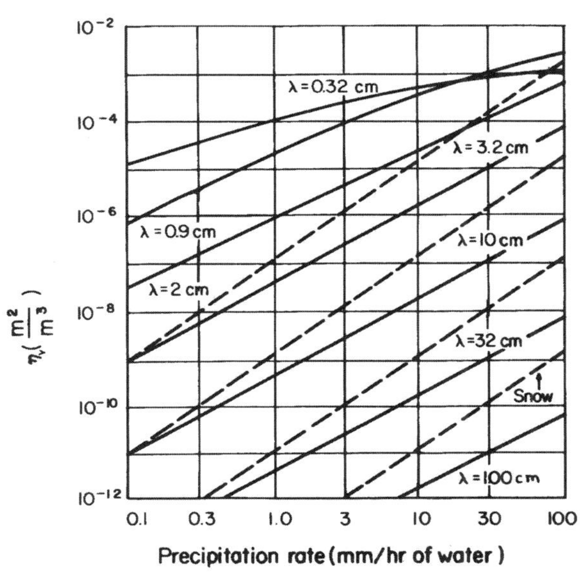

Figure 4.37 Backscattering coefficients η for rain and snow. (*From:* [8, p. 134].)

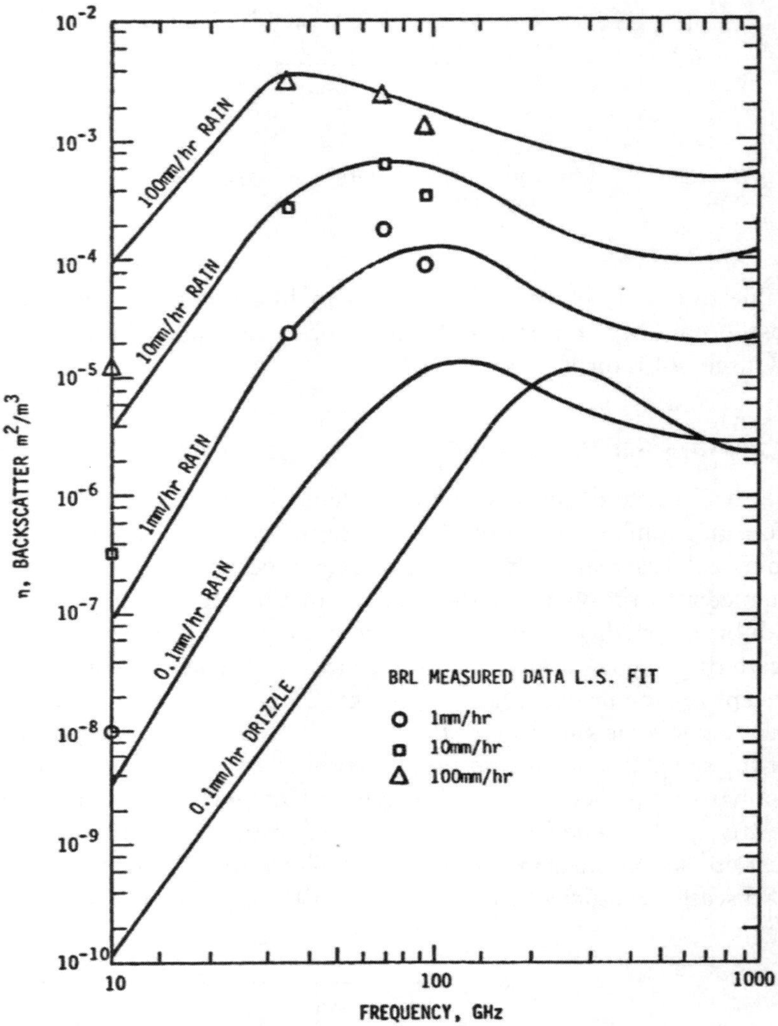

Figure 4.38 Calculated rain backscatter coefficients versus frequency. (*From:* [48, p. 287].)

intensities. The curves given in Figures 4.37 and 4.38 apply to linearly polarized radar or circularly polarized radar using an opposite sense for transmission and reception. When same-sense circular polarization is used, an attenuation of approximately 20 dB occurs in the backscattering coefficient.

The problem of calculating radar performance in precipitation clutter is complicated by the variability of raindrop intensity within a rainstorm, both horizontally and vertically. The vertical extent of rain may be as high as 25,000 ft for thunderstorms, but usually is not above 12,000 ft for steady rain. This can result in a considerable variation in rain reflectivity. In one rainstorm, for example, a 10 dB variation was observed within a range of 1 nmi [7]. More typical variations for moderate rain are a variation of ±1 dB in the mean value over a 10 nmi distance. The approximate extent of various rain rates is tabulated in Table 4.20 [23].

Table 4.20 Extent of Various Rain Rates

Rate	*Diameter (nmi)*	*Maximum Altitude (ft)*
Light (0–1 mm/h)	Unlimited	10,000
Moderate (4 mm/h)	30–50	10,000–15,000
Heavy (16 mm/h)	20	20,000
Intense (50 mm/h)	2	40,000

When precipitation starts as snow or hail at a high altitude and then melts as it drops, a phenomenon is created called the *bright band effect*. This effect results from the increased reflectivity (about 7 dB) of the melting snow or hail at the melting altitude, typically around 11,000 to 12,000 ft in a band of about 1,000-ft thickness [7].

4.2.1.1 Amplitude Probability Distribution Function for Rain Clutter

The backscatter from rain clutter results from the cumulative effect of a large number of independent scatterers (i.e., raindrops). Thus, in accordance with the central limit theorem, the rain clutter should be described by a Gaussian distribution. Thus, the received RF clutter voltage has a two-dimensional normal pdf, and the envelope-detected voltage has a Rayleigh distribution. This has been confirmed by experimental measurement performed at X-band frequencies [49]. The Rayleigh envelope distribution should apply even for high-resolution radars because these contain a large number of individual scatterers within a radar resolution cell.

Other measurements indicate that the envelope-detected amplitude pdf has longer tails than indicated by a Rayleigh distribution. One set of data taken at 95 GHz with vertically polarized radar and a rain rate of 5 mm/h was closely fitted with a log-normal distribution [50]. This distribution applied to rain rates between 2 and 60 mm/h and overall frequencies measured (9.375, 35, 70, and 95 GHz). Typical standard deviations for these measurements with vertical and circular polarization and rain rates from 5 to 20 mm/h were 7.5 to 8.5 dB at 9.375 GHz, 2.5 to 3.0 dB at 35 GHz, and 2.9 to 3.2 dB at 95GHz. The standard deviations were most sensitive to frequency change (i.e., 9.375 to 35 GHz) and insensitive to polarization or rain rate change.

Another set of data fitted the return from rain and clouds with a Weibull distribution [51]. These data were obtained with an L-band (1.3 GHz) ARSR radar. Weibull-shaped parameters between $\beta = 0.68$ and 1 were used to fit the observed data. Unfortunately, the small number of data samples collected did not permit the fit to be used with high confidence in the tail region (see Section 1.5.4).

4.2.1.2 Precipitation Clutter Spectra

The relative motion of the scatterers (raindrops) within the weather clutter's resolution cell produces a spread of velocities and Doppler frequencies, resulting in a frequency spectrum of width related to the spread of velocity. The rain clutter spectrum is spread about some mean value determined by the wind velocity. This is particularly significant in an MTI radar operating in the microwave frequency region (above L-band), where the weather clutter reflectivity is sufficient to result in an appreciable clutter return and the first MTI blind speed is such that higher speed targets are aliased (translated from a multiple-PRF zone) into the same frequency range as the weather clutter. This situation becomes progressively worse as the radar's frequency is increased. A radar processor specifically designed to cope with this situation (i.e., the MTD) is described in Section 1.6. An estimate of the mean velocity extent of rain clutter (e.g., up to about 30 m/s) and its spread (e.g., from 1 m/s at short range to 5 m/s at 200 nmi) was given in the environmental diagram of Figure 1.20 for an air traffic control radar.

In general, four atmospheric effects contribute to the total spectral width of rain clutter [7]. Wind shear involves a change in wind speed with altitude, which results in a distribution of radial velocities over the vertical extent of the beam, as depicted in Figure 4.39. This component is generally the most significant of all the contributing factors in determining the spectrum of weather clutter and most pronounced for radars with broad elevation beams (e.g., cosecant-squared antenna beams). The other components of the weather clutter's spectrum are beam broadening, turbulence, and falling velocity distribution. Beam broadening is due to the finite width of the radar beam, which causes a spread of radial velocity components when the radar is looking crosswind. Turbulence is caused by the fluctuating currents of the wind, which causes a radial velocity distribution centered about the mean wind velocity. The falling velocity distribution involves a spread in the velocity components among the reflecting particles that constitutes the weather clutter located within the radar's resolution cell. For a Gaussian-shaped power spectral density given by

$$S(f) = \frac{P_c}{\sqrt{2\pi\sigma_f^2}} \exp\left(-f^2/2\sigma_f^2\right) \tag{4.62}$$

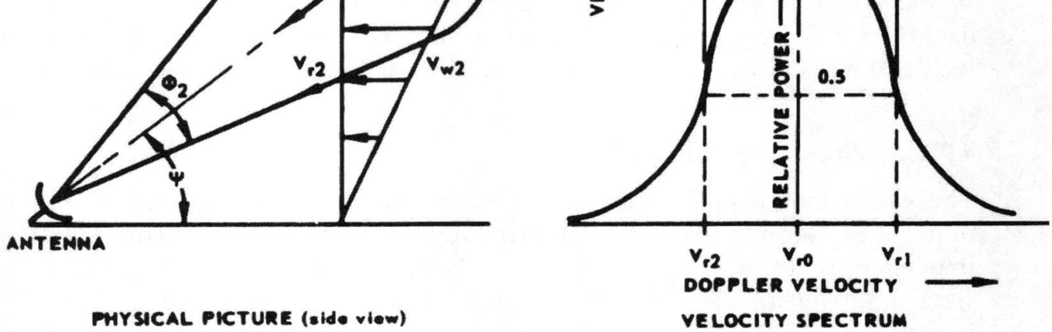

Figure 4.39 Effects of wind shear on the Doppler spectrum. (*From:* [23, p. 207].)

the frequency spectral variance (σ_f^2) is related to the Doppler velocity spectral variance (σ_v) as $\sigma_f = 2\sigma_v/\lambda$. The variance of the Doppler velocity spectrum can be represented as the sum of the variances of each contributing factor [7]:

$$\sigma_v^2 = \sigma_{\text{shear}}^2 + \sigma_{\text{beam}}^2 + \sigma_{\text{turb}}^2 + \sigma_{\text{fall}}^2 \tag{4.63}$$

The standard deviation of the wind shear component is given by

$$\sigma_{\text{shear}} = 0.42kR\phi_{el2} \tag{4.64}$$

where k is the wind shear velocity gradient and ϕ_{el2} is the two-way, half-power radar's elevation beamwidth. The two-way elevation beamwidth is used in this case since it was used in developing the theoretical model and confirmed experimentally [52, pp. 511–512]. A value $k = 5.7$ (4 m/s-km) is the maximum gradient and for an arbitrary azimuth $k = 4$ (4 m/s-km) is appropriate. Care must be exercised in applying (4.64) to ensure that the elevation beam is filled with precipitation. In practice, σ_{shear} is limited to about 6 m/s for beamwidths of 2.5° or less because the precipitation is limited to some altitude ceiling (e.g., 15,000 to 20,000 ft) [7].

The beam-broadening velocity spectral component has a standard deviation:

$$\sigma_{\text{beam}} = 0.42v_o\,\theta_{az}\sin\beta \tag{4.65}$$

where v_o is the wind velocity at wind center, θ_{az} is the two-way half-power azimuth beamwidth, and β is the azimuth angle relative to the wind velocity at beam center. The beam-broadening component is generally small (e.g., less than 0.5 m/s for a wind speed of 60 knots and a beamwidth of 2°).

The standard deviation of falling velocity for rain is approximately $\sigma_{\text{fall}} =$ 1 m/s for the vertical velocity component and is nearly independent of the rain intensity. At elevation angle ψ, the approximate standard deviation is given by [7, p. 246]

$$\sigma_{\text{fall}} = \sin\psi \quad [\text{m/s}] \tag{4.66}$$

For small elevation angles, this component is insignificant compared to that due to shear effects.

Fluctuations about the mean value of the spectrum are called *turbulence*. The standard deviation due to turbulence is nearly independent of height and generally varies over a range of 0.5 to 2 m/s with an average value on the order of $\sigma_{\text{turb}} =$ 1 m/s.

The Doppler velocity spectral width is generally a function of the range, elevation, and azimuth beamwidths, wind speed at beam center, and relative direction of the wind with respect to the radar beam. Experimental data for various ranges and an elevation beamwidth of 1.4° is depicted in Figure 4.40.

The normalized frequency spectrum for rain clutter at 9.375 GHz with vertical polarization is depicted in Figure 4.41 [53]. These data are fitted with a spectrum of the form $S(f) = P_o/1 + (f/f_c)^n$, where f_c is the corner frequency and n is an

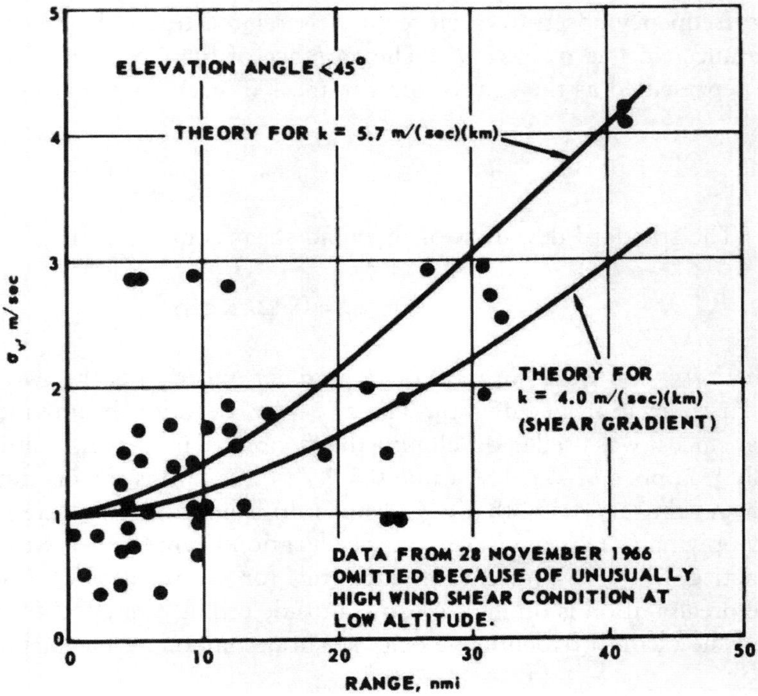

Figure 4.40 Experimental rain clutter standard deviation for 1.4° elevation beamwidth. (*From:* [23, p. 211].)

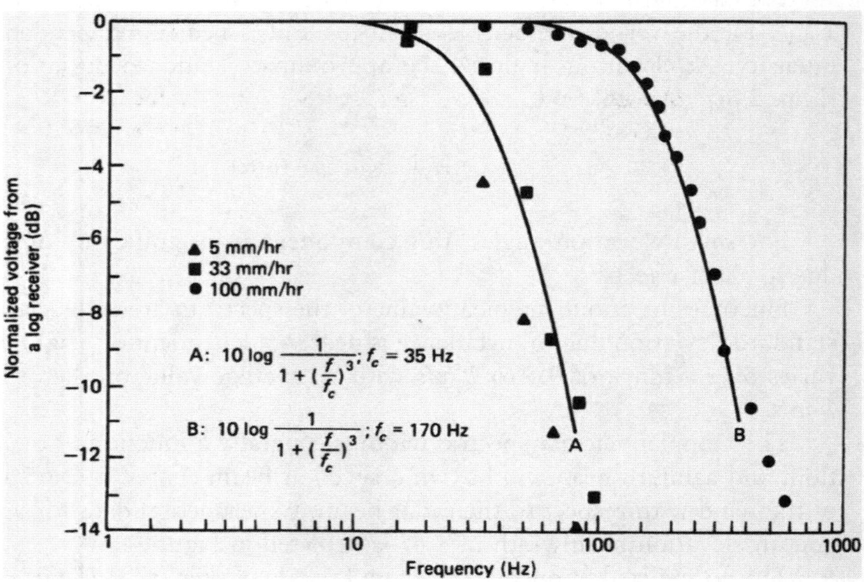

Figure 4.41 Normalized frequency spectrum of rain return at 9.375 GHz and vertical polarization. (*From:* [24, p. 300].)

Table 4.21 Measured Rain Spectrum Parameters

Frequency (GHz)	*Rate (mm/h)*	f_c *(Hz)*	n
10	5	35	3
	100	70	3
35	5	80	2
	100	120	2
70	5	175	2
	100	500	2
95	5	140	2
	100	500	2

exponent equal to either 2 or 3. Measured values for f_c and n for frequencies of 10, 35, 70, and 95 GHz with rain rates of 5 and 100 mm/h are given in Table 4.21.

► Example 13 Range in Precipitation Clutter and Chaff

Surface-based air surveillance radar must generally operate in a multiclutter environment consisting of land, sea, chaff, and precipitation clutter. The multiclutter environment is illustrated in Figure 1.20 for air traffic control radar. Note that land and sea clutter generally extend to the radar horizon, whereas precipitation and chaff clutter extend over the full operating range of the radar. Also the magnitude of precipitation clutter is dependent on the fourth power of frequency, resulting in a more difficult problem for radars operating above L-band frequencies. Chaff is generally cut to respond at the frequency of the radar it is trying to jam.

One solution to this problem is the MTD processor that is covered in Section 1.6. This Doppler processor in its basic form attenuates ground or sea clutter using an MTI canceller followed by an FFT that censors precipitation and chaff. A potential problem for this type of design is illustrated in Figure EX-6.5 of Example 6. If uniform weighing is employed in the FFT, then strong clutter returns may cross-couple through the 13.2-dB sidelobes into the active Doppler filters. This requires application of a weighting window that reduces the sidelobes by 30 dB while increasing the Doppler frequency bandwidth by 26%. Examination of Figure EX-6.5 shows that when the clutter is at the peak of one filter it is about at the −13-dB points in adjacent filters. Thus a strong AGC action must be applied in these adjacent filters to reduce the clutter, thereby attenuating target returns in these filters.

For civil radars operating at L-band frequencies and below the magnitude of precipitation, clutter is small and can generally be handled using a circularly polarized antenna. Cancellation of rain clutter in light rain using CP provides between 25 and 32 dB improvement depending on the range to the clutter cell. In heavy rain the cancellation is reduced to 2 to 19 dB, which results from a change in the raindrop shape from spherical to elliptical. Depolarization of aircraft targets with CP can further reduce the S/C ratio by 2 to 6 dB. Multipath reflections can further distort the apparent raindrop shape, causing a further reduction in improvement factor [7]. Military radars that must contend with chaff must use another technique since CP has limited effect on chaff dipoles.

The RCS of rain can be expressed as [2, p. 442]

$$\sigma_c = \frac{\eta \pi \theta_{az} \phi_{el} \left(\frac{c\tau_e}{2} \right) R^2}{8 \ln 2}$$

where η is the volumetric backscattering coefficient (m^2/m^3); θ_{az} and ϕ_{el} are the one-way azimuth and elevation beam widths in radians, respectively; τ_e is the effective pulsewidth; and R is range to the target. Note that this equation contains a beam shape factor loss equal to 8 ln $2/\pi$ that is appropriate only if this factor is not included in the experimentally determined value for η [54, p. 15.10].

The backscattering coefficient can be approximated as

$$\eta = 7 f_{\mathrm{GHz}} r^{1.6} 10^{-12}$$

where r is the rain rate in mm/h. The equivalent RCS of rain clutter for the S-band ASR-9 radar (see Table 1.1) is shown in Figure EX-13.1 for light (1 mm/h), moderate (4 mm/h), and heavy (16 mm/h) rain.

The S/C ratio for precipitation clutter is given by

$$\frac{S}{C} = \frac{\sigma_t I_c}{\eta \theta_{az} \phi_{el} \left(\frac{c\tau_e}{2} \right) R^2}$$

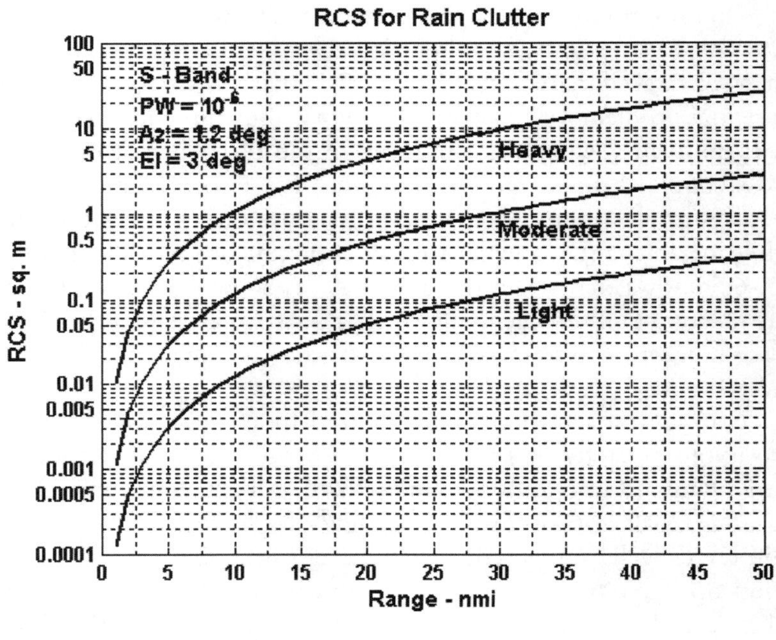

Figure EX-13.1 Equivalent RCS of rain clutter for ASR-9 radar.

where σ_t is the target RCS and I_c is the improvement factor that enhances the target signal with respect to the clutter. Note that we have assumed that beam shape losses are equal for both the target and clutter [7, p. 234]. In general, radars operating at L-band and greater frequencies require some form of Doppler processing to allow detection of targets in precipitation clutter.

The two appropriate Doppler processors are the MTD and the cascaded MTI canceller (shown earlier in Figure 1.13). In the cascaded MTI processor, the zeros of the cancellers are placed over the mean velocities of the respective clutters present. With respect to the ASR-9 a fixed double canceller would be used to eliminate ground clutter followed by an adaptive canceller as described in Example 10 to eliminate the precipitation clutter or chaff. In the following discussion, we calculate the number of MTI cancellers needed to suppress the precipitation clutter in the adaptive canceller. Note that the fixed MTI canceller is only needed for the first 10 to 20 nmi of range where the precipitation clutter RCS is less than at maximum range. For example, a fixed double canceller followed by an adaptive single canceller can be used for the first 10 to 20 nmi in range. The cascaded MTI can then be adapted to using a triple adaptive canceller to cover the remaining ranges out to the maximum range.

The maximum range in precipitation clutter can be found as [54, p. 15.10]

$$R = \left(\frac{\sigma_t I_c}{\eta \theta_{az} \phi_{el} (c\tau_e/2)(S/C)L} \right)^{1/2}$$

where the S/C ratio is evaluated from the MS curves for the particular target type and L is the system losses associated with the target with respect to the clutter. The standard deviation of rain clutter spectra varies from 1 to 5 m/s depending on wind conditions and range. At S-band this corresponds to frequencies of 20 to 100 Hz and correlations of 10 to 50 ms. Hence, generally little or no integration gain is available to enhance the S/C ratio.

The improvement factor for binomial cancellers is given by

$$I_c = \frac{2^{n-1}}{(n-1)!(\sigma_f T)^{2(n-1)}}$$

where n is the number of pulses processed, σ_f is the radian standard deviation of the spectra, and T is the period of the radar. The spectral spread for rain clutter in meters per second is given by

$$\sigma_v = \sqrt{\sigma_s^2 + \sigma_{\text{shear}}^2 + \sigma_{\text{beam}}^2 + \sigma_{\text{fall}}^2 + \sigma_{\text{turb}}^2}$$

where σ_s is the standard deviation due to antenna scanning in meters per second and σ_{shear} is given by

$$\sigma_{\text{shear}} = 0.42 k R_{\text{km}} \phi_{el2}$$

where $k = 4$ m s^{-1} km^{-1}, R is the range in kilometers, and ϕ_{el2} is the two-way elevation beam width. Approximating $\sigma_{\text{fall}} = \sigma_{\text{turb}} = 1$ and converting to radian frequency results in

$$\sigma_f^2 = \sigma_{\text{srad}}^2 + \left(\frac{4\pi}{\lambda}\right)^2 \left[2 + \sigma_{\text{beam}}^2 + (1.68 \cdot 10^{-3} \phi_{EL} R)^2\right]$$

where $\sigma_c = 2\pi\sigma_f$ and R is in meters. Substituting this expression into the formula for the improvement factor indicates that it varies with range. The MATLAB program precip_rng_mtic.m calculates the maximum detection range in rain clutter for cancellers that process $n = 2, 3, 4$, and 5 pulses, using Newton's method, while the same program can be used to calculate the maximum range in chaff by substituting the appropriate value for η for chaff. The standard deviation of turbulence and fall is estimated at 0.7 m/s for chaff as discussed in [7]. The reduction from the parameters for rain is due to the reduced fall velocity of chaff. The MATLAB program precip_rng_cpa.m calculates the detection range in rain clutter where both a circular polarized (CP) antenna and MTI processing are employed.

The analysis involving a CP antenna is complicated in that the shape of the raindrops changes from a spherical to an elliptical shape depending upon the precipitation rate. The cancellation ratio provided by the CP can be determined as

$$CR_{dB} = 20 \log\left(\frac{1 - E^2}{1 + E^2}\right)$$

where E is the ellipticity of the raindrop. Note that [7, p. 155] states that practical system values of E range from 0.96 to 0.98 so that infinite cancellation is not possible. The ellipticity of a particular size raindrop is given in [68, p. 507]. The Atwood-Burrows probability distribution of raindrop sizes for various rain rates is tabulated in [67, p.23.9]. This can be used to determine the ellipticity probability distribution for various rain rates. The mean value of the ellipticity can then be found and used to find the cancellation ratio for various values of the rain rate. MATLAB program cr_cirpola.m provides a plot of cancellation ratio versus rain rate for ideal CP ($E = 1$) and practical values of $E = 0.98$ and 0.96.

Another consideration that limits the cancellation ratio of CP is due to the elongation of the raindrop in the horizontal plane. This causes greater attenuation for the horizontal polarized component as compared to the vertical polarized component. This differential attenuation increases the ellipticity of the received CP wave that decreases the effectiveness of the CP cancellation as the transmitted frequency increases [7, p. 155]. MATLAB program cp_attenb.m plots cancellation ratio against rain rate and includes frequency, raindrop distortion, and practical CP system ($E = 0.98$) effects. For heavy rain (16 mm/hr), there is little effect below 3 GHz, but serious detoriation occurs above X-band [7, p. 156].

The ranges provided against a 5 m^2 Swerling Type 1 target in rain and chaff for an S-band radar with ASR-9 parameters are given in Table EX-13.1. For rain MTI cancellers of order 1–4, with and without CP are considered. For chaff

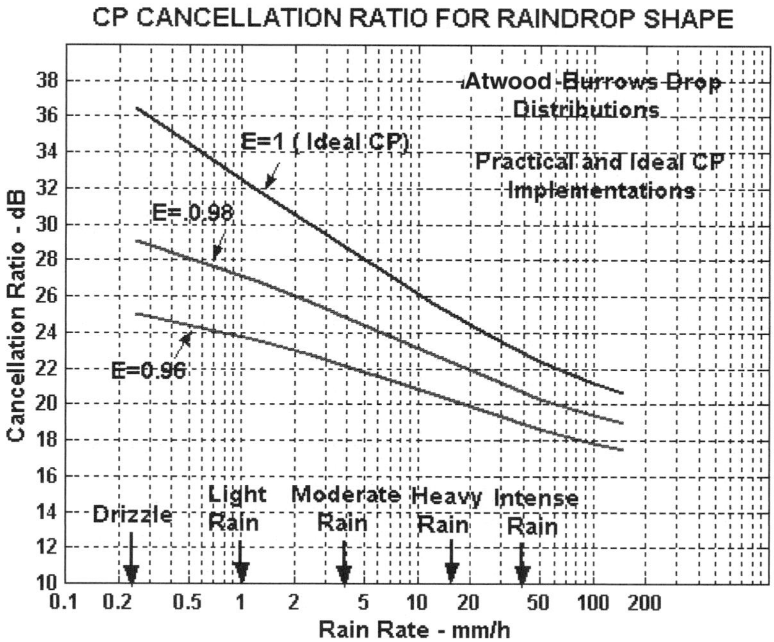

Figure EX-13.2 Cancellation ratio as function of rain rate (E = 1, 0.98, and 0.96).

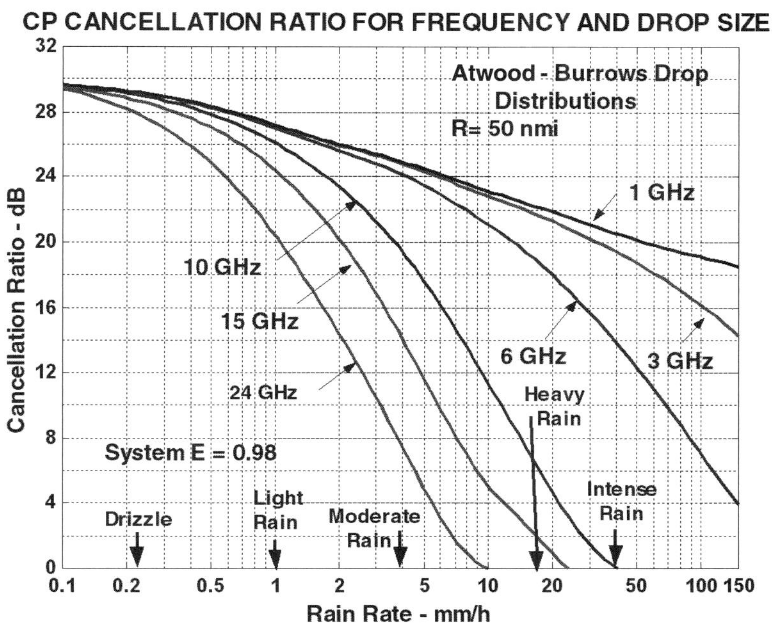

Figure EX-13.3 CP cancellation ratio as function of frequency and raindrop size.

Table EX-13.1 Range in Precipitation and Chaff with MTI Cancellers and CP

	Range (nmi)							
Cancellers	*1*	*2*	*3*	*4*	*1CP*	*2CP*	*3CP*	*4CP*
Drizzle (0.25 mm/hr)	21.74	29.85	33.69	34.69	37.46	37.46	37.46	37.46
Light rain (1 mm/hr)	10.57	19.1	24.35	26.8	37.33	37.33	37.33	37.33
Medium rain (4 mm/hr)	4.19	11.08	16.88	20.22	22.53	30.56	34.28	35.18
Heavy rain (16 mm/hr)	1.43	5.22	10.71	14.59	8.67	17	22.45	25.15
Intense rain (40 mm/hr)	0.69	2.75	7.21	11.25	3.37	9.68	15.5	18.98
Light chaff ($\eta = 10^{-8}$)	3.82	10.8	16.53	19.95				
Moderate chaff ($\eta = 10^{-7}$)	1.27	5.42	10.96	14.78				
Heavy chaff ($\eta = 10^{-6}$)	0.4	2.06	6.36	10.37				

MTI canceller processing is considered since CP is not appropriate. MTI motion compensation is assumed beyond the surface clutter horizon extent. Note that CP is generally effective in improving detection ranges and provides adequate detection for the most probable rain rates below 4 mm/hr [7, p. 219].

▶

4.2.2 Chaff

Chaff is a form of volumetric radar clutter composed of distributed metallized reflectors, dispensed into the atmosphere to interfere with and confuse radar operation. The chaff usually consists of a large number of dipoles designed to resonate at the frequencies of the radars that they are attempting to confuse. Chaff can be used in either a corridor-laying or self-protection mode. Chaff dispensed from an aircraft at a steady rate over a fairly long period is used to form a corridor that conceals following aircraft. Self-protection involves launching relatively small quantities of chaff in controlled bursts to cause weapon-associated tracking radar to point at the chaff rather than at the protected vehicle. The design of a chaff dispensing and deployment system is quite different for each of the two modes of operation [48]. Types of chaff distribution are depicted in Figure 4.42.

An elemental piece of chaff can be considered a dipole antenna whose output terminals are short circuited. When the chaff dipole is excited with RF energy, the shortest length just long enough to permit an electrical charge to propagate from one end to the other and back during an RF cycle is on the order of one-half the wavelength of the incident energy. When the chaff is cut for this resonance condition (typically, 0.46 to 0.48λ), the RCS of the elemental chaff dipole average over all possible orientations is given by

$$\sigma_c = 0.15\lambda^2 \tag{4.67}$$

where λ is the radar wavelength. Resonance can occur at harmonic frequencies (shorter wavelengths) as shown in Figure 4.43, but have lower absolute RCSs due to the dependence on wavelength of the relationship given by (4.67) [48].

If a number of elemental chaff dipoles are combined to form a chaff cloud with spacing between elements on the order of two wavelengths or greater, the average RCS of the cloud is given by

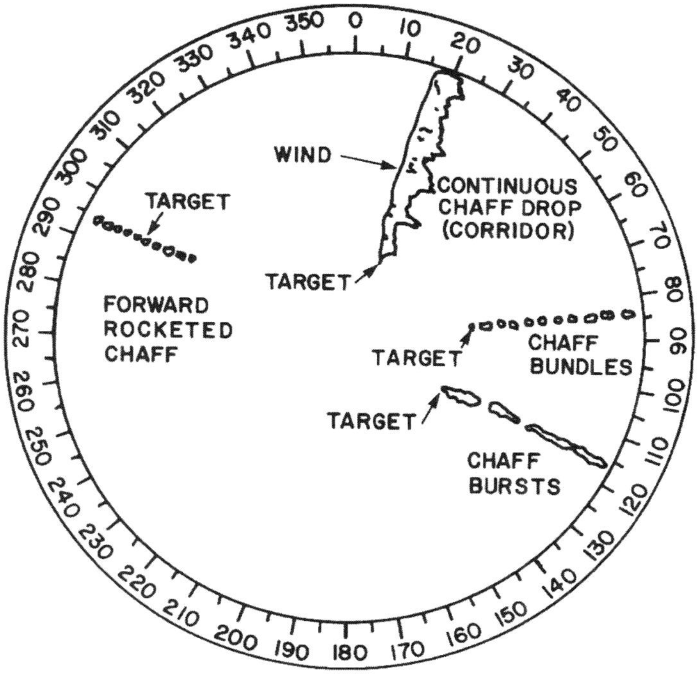

Figure 4.42 Types of chaff distributions. (*From:* [55, p. 414].)

$$\sigma_c = 0.15N\lambda^2 \tag{4.68}$$

where N is the number of effective chaff dipoles. The bandwidth of an elemental dipole is a function of the length-to-diameter (L/d) ratio for circular dipoles and the length-to-width ($4L/w$) ratio for rectangular dipoles. The chaff bandwidth ranges from 10% to 25% of the resonant frequency for L/d of 100 to 10,000, where the higher bandwidths are associated with the lower ratios. A typical design provides a bandwidth approximately equal to 15% of the center frequency for an L/d of 1,000. Figure 4.44 shows the RCS versus frequency for a representative design of a broadband chaff package covering the range from 2 to 12 GHz, where various elements are stagger-tuned to cover the whole band of interest. Note that the number of dipoles associated with the higher frequency chaff elements is substantially greater than that associated with the lower frequency chaff elements. Also, the RCS response at any particular frequency is the sum of the responses of all chaff dipoles resonant at any value lower than the design frequency. Thus, the RCS associated with a particular group of elements cut to the same length (e.g., typically $L = 0.48\lambda$) is generally lower than the actual RCS at this frequency, except for the group with the longest cut lengths. Note that the weight of 25×50-μm aluminum foil chaff to achieve the multiband RCS of 40 to 50 m^2 depicted in Figure 4.44 is approximately 0.1 lb [48].

The reflectivity of chaff depends on the deployment doctrine (i.e., corridor or self-protection missions) employed and whether the chaff has bloomed fully or is in a transient state [48]. In the corridor saturation mission, a large amount of chaff

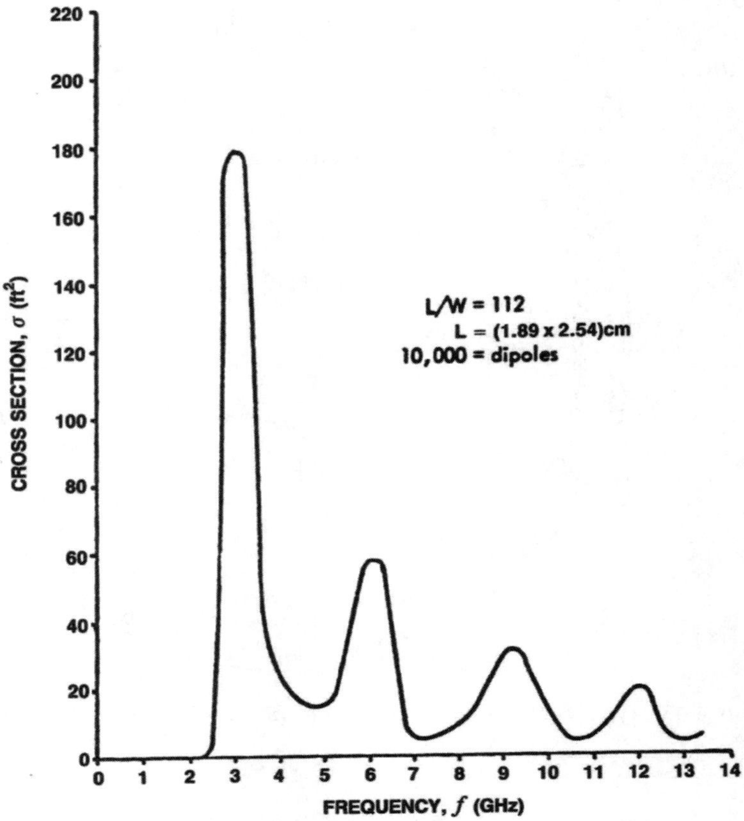

Figure 4.43 Frequency response of chaff dipole RCSs. (*From:* [48, p. 186].)

is dispensed throughout the corridor such that any aircraft entering the corridor are shielded from radar detection and tracking. Although the primary function of the chaff corridor is to prevent radar acquisition or tracking of aircraft within the corridor, a secondary function is to provide protection for the chaff-deploying aircraft against radar-directed weapons [56].

For effectively denying information regarding strike composition and timing, a chaff corridor should originate outside the radar detection range and be uninterrupted throughout its length. The chaff RCS in the radar resolution cells should be sufficient to prevent detection of the shielded aircraft, taking into consideration any chaff discrimination capabilities of the radar. The chaff RCS initially deployed by the chaff-dispensing aircraft must be greater than the chaff RCS subsequently required for corridor use by the penetrating aircraft. This is necessary to allow for chaff cloud growth and chaff diffusion across the resolution cell boundaries. How much greater the initial chaff RCS should be compared to the chaff RCS per resolution cell at time of use would depend on several factors, such as the chaff-dispensing system, the RCS of the aircraft to be shielded within the corridor, the corridor growth and chaff fall characteristics, the length and use time required of the corridor, and the resolution cell's volume.

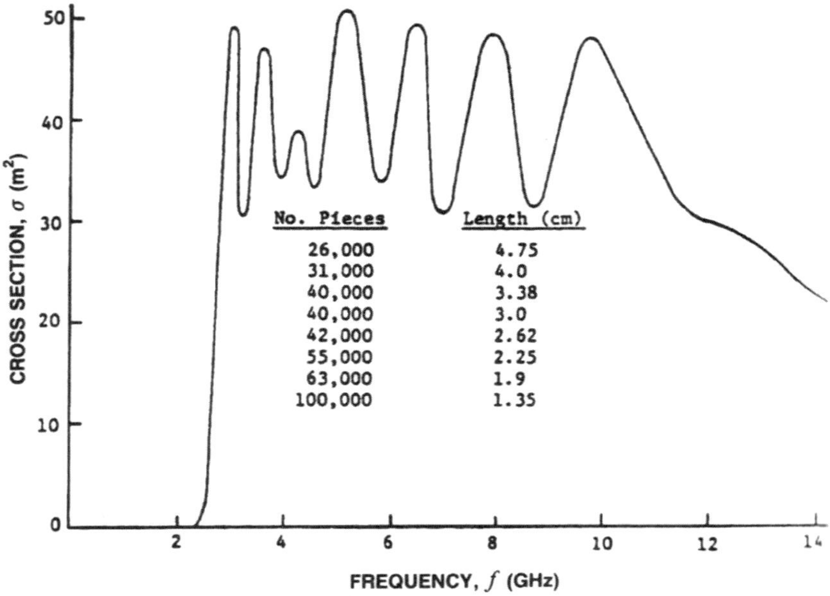

Figure 4.44 Frequency response of chaff dipole RCSs. (*From:* [48, p. 241].)

As an example, consider the ASR-9 radar ($\lambda = 0.1$ m, $\theta_{az} = 1.3°$, $\phi_{el} = 4.8°$, $\tau = 1.05$ μs) that at a range of 50 nmi has a volume resolution cell of $V = 10^9$ m^3 [see (4.59)]. Assume that we want to generate a chaff corridor with an RCS of 50 m^2 within this volume cell. The chaff reflectivity will then equal $\eta = 5 \times 10^{-8}$ m^2/m^3. The number of chaff dipoles contained within the radar's resolution cell can be found from (4.68) as $N = 3.33 \times 10^4$ dipoles. The density of dipoles within the resolution cell is then 3.33×10^{-5} dipoles/m or 1 dipole per 31 m^3. The most efficient chaff is made by aluminizing cylindrical glass dipoles. The diameter of each chaff dipole is 25 μm and its density is 2550 kg/m^3 [48]. The volume of each dipole is given by $V = 0.48\pi d^2\lambda/4 = 0.377\pi d^2$. The weight of each dipole is thus $w_1 = 961 d^2\lambda$ kg. The total weight of the aluminized glass dipoles required to provide an RCS = 50 m^2 within the radar's volume resolution cell at a wavelength $\lambda = 0.1$ m is then $w = Nw_1/e$ where e is an efficiency factor that allows for breakage, mistuning, mismatching, and other chaff inefficiencies. For a chaff efficiency of 50%, the total weight of chaff to fill a resolution cell would hence equal 4×10^{-3} kg.

The preceding analysis can be expressed as a relationship of the chaff's RCS with respect to the weight of chaff required to produce a given RCS at a specified frequency. This relationship is given by

$$\sigma_c = \frac{k w_{\text{lb}}}{f_{\text{GHz}}} \tag{4.69}$$

where w_{lb} is the weight of the chaff in pounds, f_{GHz} is the radar's frequency in gigahertz, and k is a constant that depends on the properties of the type of chaff utilized. The value of k ranges from 17,000 for aluminized glass chaff to on the order of 3,000 for aluminum foil chaff [48].

▶ Example 14 Chaff Fundamentals [55]

Chaff consists of a large number of shortened tuned dipole antennas. Tuned dipoles are cut to the first resonance point (length = $\lambda/2$) since this provides the maximum RCS for the least amount of material. The capture area of the tuned dipole is given by $G\lambda^2/4\pi$, where G is the dipole gain. The captured signal is reflected from the short so that it essentially doubles the voltage or generates four times the power at the antenna terminals. This is further reradiated with gain G to produce an equivalent RCS equal to

$$\sigma = \frac{\lambda^2 G^2}{\pi}$$

The dipole gain varies with angle and for vertical polarization can be expressed as

$$G = \frac{2\left[\dfrac{\cos(kl\cos\theta) - \cos kl}{\sin\theta}\right]^2}{\displaystyle\int_0^{\pi}\left[\frac{\cos(kl\cos\theta) - \cos kl}{\sin\theta}\right]^2 \sin\theta\, d\theta}$$

where $k = 2\pi/\lambda$ and l is half the dipole length. The gain is maximum normal to the antenna ($\theta = \pi/2$) where

$$G = \frac{2[1 - \cos kl]^2}{\displaystyle\int_0^{\pi}\left[\frac{\cos(kl\cos\theta) - \cos kl}{\sin\theta}\right]^2 d\theta}$$

The integral in the denominator equals 1.21888 for half-wave dipoles, 3.3181 for full-wave dipoles, and 1.7582 for $3\lambda/2$ length dipoles. The corresponding maximum gains are 1.64 (2.15 dB) for half-wave dipoles, 2.41 (3.82 dB) for full-wave dipoles, and 1.41 (0.56 dB) for $3\lambda/2$ length dipoles. Substituting this into the above equation results in the maximum RCS from a single vertically polarized dipole as

$\sigma = 0.86\lambda^2$ half-wave dipole

$\sigma = 1.85\lambda^2$ full-wave dipole

$\sigma = 0.41\lambda^2$ $3\lambda/2$ dipole

In a chaff cloud, the dipoles are randomly oriented with respect to the illuminating radar. An approximate relationship that expressed the RCS of a randomly oriented dipole is

$$\overline{\sigma} = (1 + \cos kl)^2 \frac{G^2(\theta, kl)\lambda^2}{\pi}$$

Evaluation of this relationship is accomplished in the MATLAB program dipole_chaff.m. The response shown in Figure EX-14.1 illustrates the harmonic structure of the dipole response. The average RCS ($\overline{\sigma}$) for randomly oriented half-wave (typically 0.46 to 0.48λ) dipoles at resonance is generally given as $\sigma = 0.15\lambda^2$, which includes efficiency, breakage, and polarization losses not considered in the preceding equation.

A number of elemental chaff dipoles with spacing greater than two wavelengths can be combined to form a chaff cloud whose total RCS is given by

$$\sigma_t = 0.15kN\lambda^2$$

where N is the total number of dipoles and k is an efficiency factor generally estimated as 0.8.

A typical chaff pod is the ALE-43 whose characteristics are listed in Table EX-14.1. The chaff dispenser consists of eight chaff rovings that contain a total of 320 lb of chaff. Each roving consists of 3,000 strands of continuous glass fibers half coated with aluminum. The twisted strands from each roving are fed through separate channels to the cutting mechanism where they are pulled between a platen and a pressure plate. The selected cutter head rotates and cuts the strands against the platen, to the proper dipole length, and then ejects them at about 25 fps into

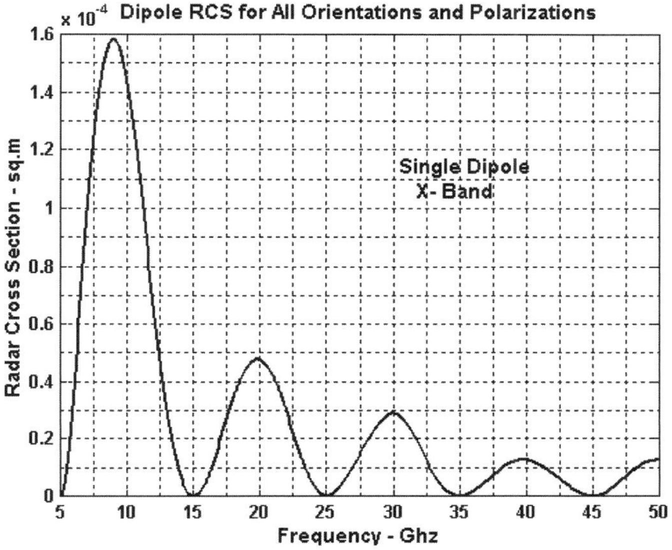

Figure EX-14.1 Harmonic response of single X-band tuned dipole.

Table EX-14.1 ALE-43 Chaff Pod Characteristics

Characteristic	*Parameter*
Pod weight empty	340 lb
Total chaff payload	320 lb
Roving bundles	8
Dispensing rate	0.48 lb/s per roving bundle
Continuous dispensing mode	11 minutes (max)
Pulse dispensing mode	On 1–9 sec; off 1–9 sec
Cutter rollers	Select 1 of 3 in flight
Pod dimension	Length = 11 ft; diameter = 19 in.
Performance	550 knots (max) to 50 K ft

the turbulent air stream. Strands from each of the eight rovings are cut simultaneously and dipoles are continuously ejected from the pod into the turbulent air. Initially they are randomly tumbling and moving at the speed of the aircraft dropping them. They slow down to about 87% of their initial speed almost immediately as they clear the air turbulence caused by the pod.

The initial width of the chaff cloud is no wider than the outlet of the ALE-43 when the dipoles are deployed but migrate as soon as they are dispensed into the turbulent air stream. In general, the width of the corridor expands to about 100 feet within 1 sec. Once the aircraft moves away from the chaff, the chaff continues to spread horizontally at a rate of about 1 ft for each 1.5 ft it drops. As an example, suppose a corridor is dispensed at 5,000 ft of altitude by an aircraft with a velocity of 450 knots (750 feet/s). The fall rate of chaff at this altitude is about 60 ft/min. In 17 minutes the chaff falls 1,000 ft and the spread is 667 + 100 = 767 ft. During this period the dispensing aircraft has moved about 127.5 nmi away from the point at which the aircraft's dispenser was turned on.

To illustrate the use of the continuous dispensing mode, assume that the ALE-43 is to lay a radial corridor against an X-band radar that operates at 9.9 GHz with a pulsewidth of 0.8 μs. At this frequency the ALE-43 has the capacity in the continuous mode to dispense an RCS of 704 m^2/s. If the dispensing aircraft is flying at 420 knots (700 ft/s), it transverses the 400-ft radial dimension of the radar resolution cell in 0.57 sec. This places about 400 m^2 of RCS in the radar's resolution cell. If the mission is to protect four F/A-18 strike aircraft within a radar resolution cell (RCS = 20 m^2 per aircraft), then this would be more than adequate, but cannot be reduced due to the operating limitation of the dispenser.

A further example assumes that the ALE-43 is to lay a tangential corridor against the radar described above that has a 5° azimuth beamwidth and a maximum range of 50 nmi. The path through the tangential dimension of the radar resolution cell is 26,500 ft ($5\pi/180 \cdot 50 \cdot 6{,}080$) long. The dispensing aircraft at 700 ft/s would be within the radar resolution cell for 37.86 sec. Using the *Pulse Dispensing Mode* with 1 sec on and 9 sec off would dispense 4 sec of chaff in the radar resolution cell, resulting in an RCS of 2,816 m^2 that far exceeds the protection requirement.

▶

4.2.2.1 Amplitude Probability Distribution Function for Chaff

The large numbers of randomly moving dipoles that make up the chaff cloud ideally satisfy the conditions for Gaussian-distributed clutter (see Section 1.5.1).

In fact, experimental analysis of chaff echo returns shows full agreement with the Gaussian clutter theory [49]. This means that the chaff's envelope-detected amplitude pdf follows a Rayleigh distribution, $p(v) = v \exp(-v^2/2\sigma^2)/\sigma^2$; $P_c = 2\sigma^2$, whereas the RCS for chaff has an exponential pdf, $p(\eta) = \exp(-\eta/\overline{\eta})/\overline{\eta}$.

Note that, although this distribution is valid when the chaff has reached its steady-state conditions (i.e., the rate of change of $\overline{\eta}$ with time is small), the probability distribution changes significantly when the chaff cloud is growing in a transient state. The preceding reflects the difficulty of modeling a nonstationary process (the growing chaff cloud) by a stationary Gaussian random process.

4.2.2.2 Chaff Clutter Spectra

The main *Electronic Protection* (EP) technique used to combat chaff is Doppler processing. In MTI radar, the techniques used in the MTD processor (see Section 1.6) are applicable. These techniques generally allow the rejection of both ground clutter and chaff (or precipitation) clutter, although these two types of clutter occupy different parts of the frequency spectrum (see Section 8.6). High-PRF PD radar is considered to be an even better method for eliminating chaff because its unambiguous Doppler response allows a clutter rejection filter to be set above the velocity response of the chaff. However, care must be exercised in radars operated in extended clutter with range ambiguities (e.g., high- and medium-PRF PD) because the spectral response at each range-ambiguous clutter cell will be different (both center frequency and spread), and the total clutter spectrum is the sum of the individual clutter spectra. This is illustrated in Figure 4.45 for four separate range ambiguities, where the chaff clutter spectrum is spread over a significantly wider range than that associated with an individual range cell [8].

The voltage autocorrelation function for chaff obtained from experimental data is depicted in Figure 4.46 [49]. The curve shows an autocorrelation different from one at $\tau = 0$ due to the effect of receiver noise. Also, the curve approaches $\rho = 0$ very slowly, which is indicative of a slow secular variation of the average clutter power. This variation has been attributed to the change of orientation of the chaff dipoles as they fall, which causes an accompanying change in the chaff's RCS. When these anomalies are removed from the experimental autocorrelation function, we find that it can be well approximated by a Gaussian-shaped function. Thus, the power spectral density will also be approximated by a Gaussian-shaped function.

Chaff dipoles have high aerodynamic drag that causes their velocity to drop to that of the local wind conditions a few seconds after being dispensed. This means that the horizontal velocity of the chaff will be determined primarily by the wind speed in the radar resolution cell containing the chaff. Its vertical velocity (falling rate) depends on the dipole dimensions and material, with a falling velocity of 0.3 m/s attributed to aluminized glass at sea level.

The width of the Doppler spectra for chaff can be divided into four components in a manner similar to that for precipitation clutter [see (4.63)]. The components due to wind shear and beam broadening due to the finite widths of the beam should be identical to those for precipitation [see (4.64) and (4.65)]. The value for turbulence is on the order of $\sigma_{turb} = 1$ m/s below 12,000 ft of altitude and 0.7 m/s

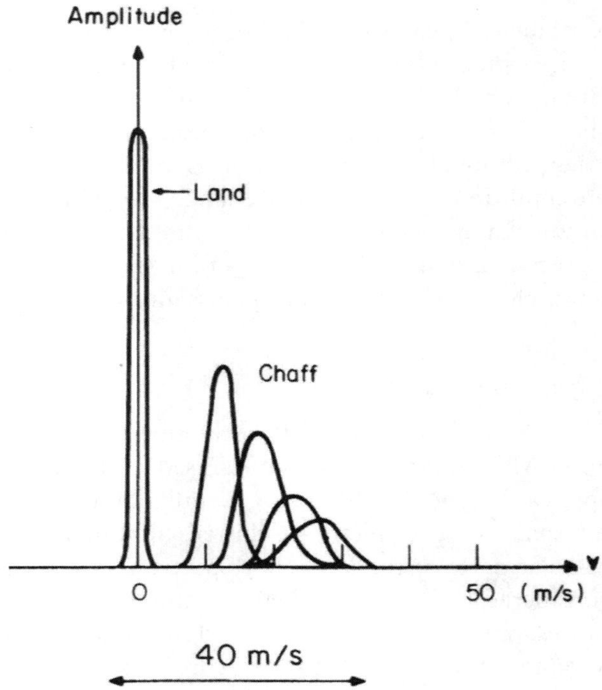

Figure 4.45 Spectrum of range ambiguous chaff clutter. (*From:* [8, p. 137].)

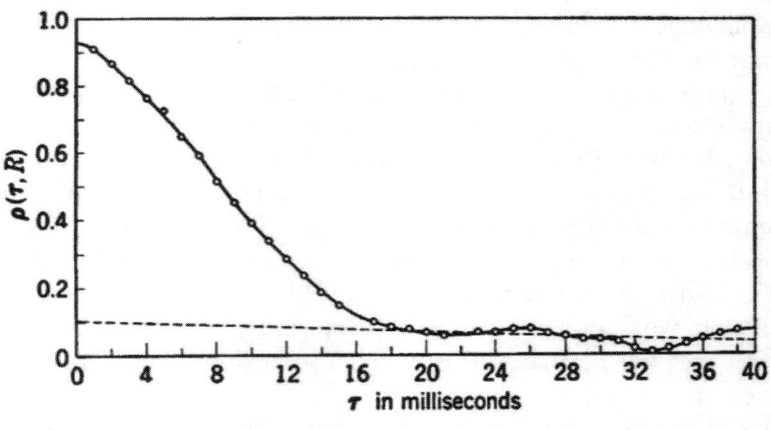

Figure 4.46 Voltage autocorrelation of chaff at X-band. (*From:* [49, p. 567].)

above that altitude. The final component of the chaff spectra is the Doppler spread due to the variation in falling velocity of the dipoles. Because chaff dipoles fall more slowly than raindrops, this component is estimated at $\sigma_{\text{fall}} = 0.45 \sin \psi$ m/s [7].

As with precipitation, the spectra of chaff clutter is generally determined by the wind shear component, which is given by $\sigma_{\text{shear}} = 0.42kR\phi_{el}$, where k is the velocity gradient (4 m/s-km is suggested for arbitrary azimuth angles), R is the

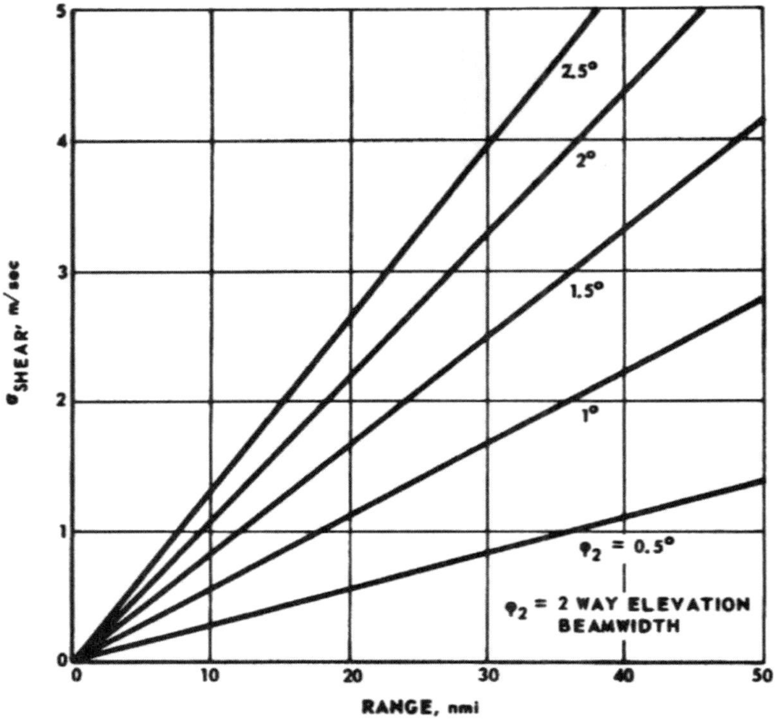

Figure 4.47 Spectral width of chaff due to wind shear (average over 360°). (*From:* [23, p. 208].)

slant range to the clutter resolution cell in kilometers, and ϕ_{el} is the two-way elevation antenna beamwidth in radians. However, in contrast with precipitation, chaff altitude is not generally limited to 20,000 ft, and spectra spreads greater than 6 m/s may be observed when chaff is dispensed above this altitude. Spectral widths for this component are depicted in Figure 4.47 [7]. Note that observed spectral widths for chaff are apparently not affected by the age of the chaff (assuming the chaff has bloomed) and, when normalized to wavelength, are independent of frequency (i.e., σ_v does not depend on frequency) [49].

► Example 15 Range in Ground and Chaff Clutter Using FFT Processing

FFT processing can be used to track moving targets in a multiclutter environment consisting of ground, precipitation, and chaff clutter. An example of such an FFT processor is that used in the SA-15 target engagement radar described in Chapter 1 [75–78]. Basically the TER phased-array radar functions to track both the target and the Gauntlet missile. The TER is cued by the Target Acquisition Radar (TAR), which measures the designated targets range, angles, and Doppler. To acquire the target, the TER is first mechanically rotated to the target's azimuth angle. Then the limited scan phased array is electronically scanned to cause the target tracker to lock onto the designated target. The target's Doppler is then tuned so that the target's spectra are centered between PRF lines. Because the TER is of the low-PRF type, the spectra of the ground and chaff clutter are also aliased into this same region between the PRF lines. Selection of an appropriate PRF then results in the clutter spectrums occurring in different Doppler filters than the target. Detailed parameters of the TER are not generally available in the open literature so educated estimates are tabulated in Table EX-15.1. The frequency, known to be in K_u-band, was selected to provide a Doppler shift of 50 Hz/knot. The first blind speed is given by

$$v_{bs} = \frac{c\,\mathrm{PRF}}{2f}$$

resulting in 41.152 m/s or 80 knots.

When the TER is tracking a 600-knot radial target, its Doppler is 30 kHz. The target spectrum is then aliased so that it is centered between the first PRF lines. The mean of the ground clutter spectrum is then at zero Doppler, while the mean chaff Doppler is located below the target spectrum in accordance with its radial velocity. The 32 Doppler filters formed by the FFT are each separated by 125 Hz and equally spaced between the PRF lines.

For low-altitude targets the magnitude of the ground clutter is much stronger than that of the chaff clutter and the target. Its spectrum is relatively narrow and is located in the main band of, at most, two Doppler filters. The ground clutter can then be suppressed through appropriate AGC action in the affected filters. However, the ground clutter may be cross-coupled into the other Doppler filters through their sidelobes. The solution to this problem is to window the FFT. One of the more versatile windows is the Chebyshev window that allows a particular sidelobe level to be specified while providing the minimum 3 dB bandwidth consistent with this sidelobe level (see Example 6). Unfortunately windowing also has the undesirable effect of spreading all of the filters' 3 dB bandwidths so that additional Doppler filters might be affected more by the clutter than if windowing was not used.

The point improvement factor for the windowed FFT is given by [19]

Table EX-15.1 Estimated TER Parameters

Parameters	*Values*
Peak transmitter power	15 kW
Average transmitter power	600W
Frequency	14.58 GHz
Pulsewidth	10 μs
Antenna gain	43 dB
PRF	4,000 Hz
PC bandwidth	2.5 MHz
Azimuth beamwidth	1°
Elevation beamwidth	1°
Losses	12 dB
Doppler processor	32-point FFT
Doppler sidelobes	–35 dB
Antenna sidelobes	–18 dB

$$I_c = \frac{\left(\sum_{i=0}^{N-1} \mathbf{a}_i\right)^2}{\sum_{i=0}^{N-1} \sum_{j=0}^{N-1} \mathbf{a}_i \mathbf{a}_j \cos\left(2\pi(i-j)\frac{k}{N}\right) \rho_c[(i-j)]}$$

where $\mathbf{a}_{i,j}$ are the window weights, N is the number of FFT points, k is the filter number and

$$\rho_C(i) = \exp\left(-\frac{i^2\Omega^2}{2}\right); \qquad \Omega = 2\pi\sigma_f T$$

is the normalized correlation coefficient while σ_f is the standard deviation of the Gaussian-shaped spectrum.

The range in chaff can be found using the expression (see Example 13)

$$R_{\text{chaff}} = \left[\frac{\sigma_t I_C}{\eta\,\theta_{az}\,\phi_{el}\left(\frac{c\tau_e}{2}\right)\left(\frac{S}{C}\right)}\right]^{1/2}$$

where $\eta = 10^{-6}$, 10^{-7}, and 10^{-8} for heavy, medium, and light chaff, respectively. The other terms are defined in Example 13 as is the spectrum spread in radians per second:

$$\sigma_C^2 = \left(\frac{4\pi}{\lambda}\right)^2 \left[2 + \sigma_{\text{beam}}^2 + (1.68 \cdot 10^{-3}\,\phi_{el2}\,R)^2\right]$$

This indicates that I_c is a function of range and that the range in chaff must be solved using Newton's method.

The range in ground clutter may be solved directly as

$$R_{\text{ground}} = \frac{\sigma_t I_C}{\sigma^\circ\,\theta_{az}\left(\frac{c\tau_e}{2}\right)\left(\frac{S}{C}\right)}$$

where θ_{az} is defined as the azimuth one-way power beamwidth. The spectrum spread in radians per second is due mostly to wind-driven vegetation and is given by

$$\sigma_w = \frac{0.0414 \cdot v_w^{1.261}}{\lambda}$$

where v_w is in knots [19].

The MATLAB program fft_range.m performs the necessary calculations and finds the range provided by the FFT processor at the peak of each FFT filter in chaff and ground clutter. The program also calculates the equivalent RCS of the ground clutter, the gain of the FFT with windowing, and the S/C in ground clutter. Figure EX-15.1 shows the ranges achieved using the FFT processor in ground and chaff clutter. The target is assumed to be a 600-knot cruise missile flying at an altitude of 10m whose Doppler is aliased into the 16th Doppler filter. The radial component of the chaff is assumed to be 25 knots centered in the 10th filter while the ground clutter is at 0 knots in the 1st filter. The chaff spectrum is spread into filters 7 through 13 due to the shear component of the chaff spectrum. The bandwidth of each Doppler filter is 159 Hz while its voltage gain is 19.68.

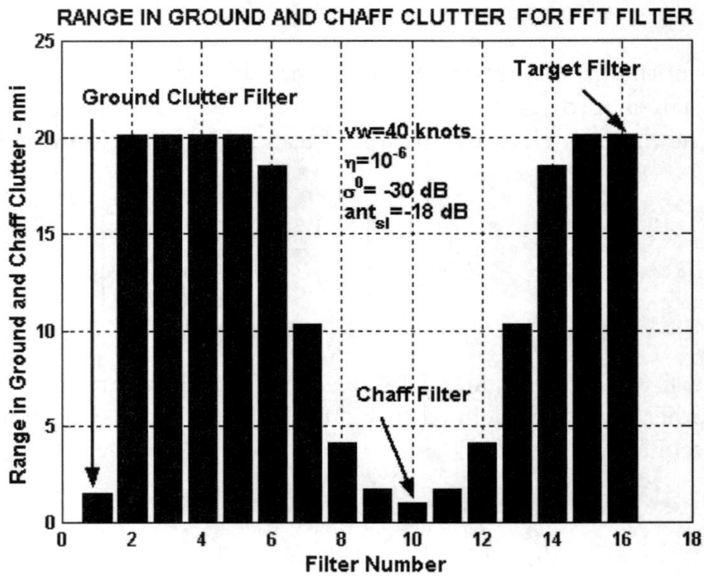

Figure EX-15.1 TER range in heavy chaff and ground clutter.

▶

4.3 Approximations to Clutter Amplitude Statistics

Determination of the performance of MTI or PD radars in clutter generally requires the representation of the clutter as a random process. The Gaussian clutter model (see Section 1.5.1) is an example of a physical model developed into an analytic formulation that allows meaningful representation of the clutter's properties. In this case, the clutter process is completely specified by its mean value and covariance function (i.e., spectral characteristics). However, the mean value, which is proportional to the clutter's average backscattering coefficient ($\overline{\sigma^0}$ or $\overline{\eta}$), is a function of many variables (e.g., terrain type, grazing angle, frequency, polarization). Therefore, we want analytic parametric expressions that allow determining the clutter's mean as a function of the variables associated with a particular representation. The parametric representation of the clutter's backscattering coefficient is usually obtained by using an approximation procedure (e.g., a least mean square fit) to a set of experimental data.

When the clutter cannot be described by using a Gaussian clutter model, then a complete representation requires the specification of the clutter's multidimensional pdf (see Section 1.5.4). Experimental data generally are not available to accomplish this, and a compromise is to specify the first pdf of the clutter's backscattering coefficient $p(\sigma^0)$ and the clutter's covariance function. This involves an empirical fit, where one criterion applied is that the resulting amplitude pdf has analytic properties that allow the detection performance in clutter to be determined. The most popular non-Gaussian amplitude pdfs employed are the log-normal, Weibull, and K-distributions.

4.3.1 Approximate Average Backscatter Coefficient ($\overline{\sigma^0}$)

The following models are provided for the average backscattering coefficient of rain, land, and sea clutter. These models generally are based on averaged experimental data obtained from a number of sources that is then combined and fitted with a parametric model.

4.3.1.1 Parametric Rain Clutter Model

Based on an analytic model, the backscatter coefficient per unit volume of rain clutter as a function of frequency and rainfall rate can be determined as

$$\overline{\eta} = 7 f_{\mathrm{GHz}}^4 \, r^{1.6} \times 10^{-12} \qquad \left[\frac{\mathrm{m}^2}{\mathrm{m}^3}\right] \tag{4.70}$$

where f_{GHz} is the radar frequency in gigahertz and r is the rainfall rate in millimeters per hour [2].

An empirical fit to rain clutter data in the high microwave and MMW region provides the following parametric relationship for the backscattering coefficient of precipitation clutter:

$$\overline{\eta} = A r^B \qquad \left[\frac{\mathrm{m}^2}{\mathrm{m}^3}\right] \tag{4.71}$$

where r is the rain rate in millimeters per hour and A and B are as given in Table 4.22 [50].

4.3.1.2 Parametric Land Clutter Model

A parametric model that provides an empirical fit to a large sample of experimental land clutter data for the average backscatter coefficient (σ^0) is given by

Table 4.22 Rain Backscatter Model Coefficient

Frequency (GHz)	*A*	*B*
9.4	1.3×10^{-8}	1.6
35	1.2×10^{-6}	1.6
70	4.2×10^{-5}	1.1
95	1.5×10^{-5}	1.0

$$\overline{\sigma^\circ} = A(\psi + C)^B \exp\left[\frac{-D}{\left(1 + \frac{0.1\sigma_h}{\lambda}\right)}\right] \tag{4.72}$$

where ψ is the grazing angle (rad), σ_h is the standard deviation of the surface (cm), λ is the radar wavelength (cm), and A, B, C, and D are empirically derived constants [24]. The model provides $\overline{\sigma^0}$ for various types of land clutter including soil, grass, trees, sand, rocks, urban areas, and snow-covered ground as a function of grazing angle, surface roughness, and frequency. The values for the parameters A, B, C, and D are given in Table 4.23. Note that $\overline{\sigma^0}$ variations with surface roughness occur in the model for only soil or sand ($\sigma_h \approx 0.2$ to 2.5 cm) at frequencies below 15 GHz. The parameters for trees have been adjusted to match available data at

Table 4.23 Land Clutter Model Constants

Constant	*Frequency*	*Soil, Sand*	*Grass*	*Tall Grass, Crops*	*Trees*	*Urban*	*Wet Snow*	*Dry Snow*
A	3	0.0045	0.0071	0.0071	0.0028	0.362	—	—
	5	0.0096	0.015	0.015	0.0047	0.779	—	—
	10	0.025	0.039	0.039	0.0095	2.0	0.0246	0.195
	15	0.05	0.079	0.079	0.019	2.0	—	—
	35	—	0.125	0.301	0.036	—	0.195	2.45
	95	—	—	—	0.046	—	1.138	3.6
B	3	0.83	1.5	1.5	0.64	1.8	—	—
	5	0.83	1.5	1.5	0.64	1.8	—	—
	10	0.83	1.5	1.5	0.64	1.8	1.7	1.7
	15	0.83	1.5	1.5	0.64	1.8	—	—
	35	—	1.5	1.5	0.64	—	1.7	1.7
	95	—	1.5	1.5	1.1	—	0.83	0.83
C	3	0.0013	0.012	0.012	0.002	0.015	—	—
	5	0.0013	0.012	0.012	0.002	0.015	—	—
	10	0.0013	0.012	0.012	0.002	0.015	0.0016	0.0016
	15	0.0013	0.012	0.012	0.002	0.015	—	—
	35	—	0.012	0.012	0.012	—	0.008	0.0016
	95	—	0.012	0.012	0.012	—	0.008	0.0016
D	3	2.3	0.0	0.0	0.0	0.0	—	—
	5	2.3	0.0	0.0	0.0	0.0	—	—
	10	2.3	0.0	0.0	0.0	0.0	0.0	0.0
	15	2.3	0.0	0.0	0.0	0.0	—	—
	35	—	0.0	0.0	0.0	—	0.0	0.0
	95	—	0.0	0.0	0.0	—	0.0	0.0

X-band [57] and 95 GHz [21]. In addition, parameters for soil or sand, grass, and crops have been adjusted to be consistent with the model at other frequencies.

Figure 4.48 illustrates a model estimating $\overline{\sigma^0}$ developed for the Overland Radar Program (AWACS) by averaging a wide compilation of X-band farmland clutter data obtained from various airborne experimental programs [1]. The variability in the data is evident as is the lack of data below 10°. The recommended procedure for extrapolating the model into the low-angle region is to use a constant gamma model ($\sigma^0 = \gamma \sin \psi$) with gamma chosen to provide continuity with the 10° to 90° model [1, p. 49]. Data variability is attributed to natural effects rather than instrumentation error. Data at L-band were found to fit the X-band model leading to the conclusion that the model is valid from X-band to at least L-band.

4.3.1.3 Parametric Sea Clutter Model

A number of the unique characteristics of sea clutter are illustrated in a sea clutter model developed for use with airborne radars. The backscattering coefficient (σ^0)

Ohio State Data (X Band)

GRAPH	TERRAIN	POL.
A	4' Oats	VV
B	3' Soybeans	VV
C	12" Wheat stubble	VV
D	Disked field	VV
E	Plowed field	HH
F	1/2" Green wheat	VV

Other Data (X Band)

GRAPH	AGENCY	TERRAIN	POL.
G	Goodyear	Arizona Irrigated farmland	HH
H	Goodyear	Arizona Dry grassland	HH
J	NRL	New Mexico Pecan orchards	HH
K	NRL, Grant & Yoolee	Green grass Sandy Loam	VV

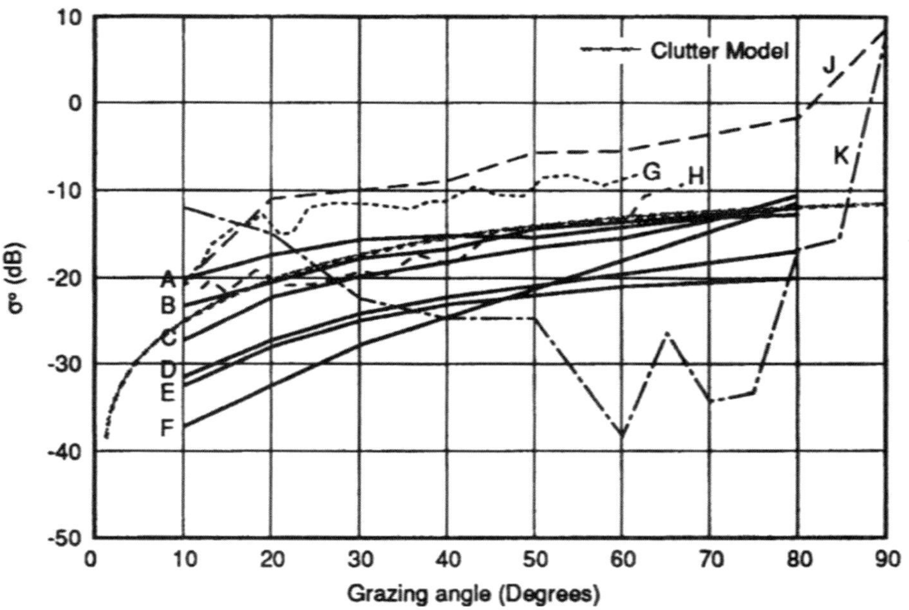

Figure 4.48 Land clutter model at X-band for crops versus experimentally determined farmland clutter. (*From:* [1, p. 44].)

for sea clutter is dependent on frequency, grazing angle, sea and wind aspect, sea state and wind speed, and polarization [19].

The variation with polarization is the most difficult to quantify. For moderate to low sea states, horizontal polarization has yielded consistently lower sea returns than vertical polarization in the upwind and downwind directions. However, in crosswind, vertical polarization may give a lower return. In rough seas, the differences between the two linear polarizations may diminish or reverse at large grazing angles. A disadvantage of horizontal polarization is the appearance of highly correlated "sea spikes," especially upwind. Because of this uncertainty, a model for horizontal polarization is developed first and then modified for vertical polarization.

The general shape of the curve for the backscattering coefficient (σ^0) as a function of grazing angle was shown earlier in Figure 4.3. The three major regions of dependence are the small-angle region, the plateau region, and the large-angle region.

The critical angle (ϕ_c), where the curves for the plateau and small-angle regions intersect, is given by

$$\phi_c = \frac{\lambda}{2.5 h_w} \tag{4.73}$$

where ϕ_c = critical angle in radians, λ = wavelength in feet, and h_w = peak-to-trough wave height in feet of the highest 10% of the waves. An approximate relationship for h_w (in feet) is given by

$$h_w = 1.05 \exp(0.495 s) \quad \text{[ft]} \tag{4.74}$$

where s is the Douglas sea state.

The variation of σ^0 in the plateau region is taken as proportional to $\phi^{0.4}$ and the variation in the small-angle region is $\phi^{5.0}$. The aspect dependence is modeled as

$$\sigma^0_{\text{dB}}(\beta) = \sigma^0_{\text{u dB}} - 4 + 4 \cos \beta \tag{4.75}$$

where β = angle between antenna boresight and upwind direction and $\sigma^0_{\text{u dB}}$ is the backscatter coefficient for the upwind direction. A frequency dependence of $\lambda^{-1/2}$ was chosen as best representing the data for the plateau region.

A calibration point is needed to quantify the model. This was taken as λ = 3 cm, ϕ = 10°, s = 4, and σ^0 = –42 dB. The value of σ^0_p in the plateau region is then

$$\sigma^0_p = \frac{k \phi^{0.4} \exp(0.792 s)}{\lambda^{1/2}} \tag{4.76}$$

where s = Douglas sea state, ϕ = grazing angle (degrees), λ = wavelength (cm), and $k = 1.83 \times 10^{-6}$. The value of σ^0_s in the small-angle region is given by

$$\sigma_s^0 = \frac{\phi^5 \exp(3.069 s)}{c \lambda^{5.1}} \tag{4.77}$$

where $c = 1.17 \times 10^5$.

The value of σ_L^0 in the large-angle region was not modeled, but the transition grazing angle appears to lie in the 30° to 60° region.

Analytical expressions were not obtained for vertical polarization. Grazing angle and frequency dependence were observed to be the same as for horizontal polarization. The dependence on sea state is very complicated.

4.4 Analytical Clutter Representations

Analytical clutter models are useful in the design and analysis of MTI and PD radars. These models attempt to represent the random clutter in such a manner that system performance can be determined analytically. Many of the characteristics of these models are discussed elsewhere in this text (e.g., see Section 1.5). This section collects the many relationships involved with the various models and presents them in a consistent manner.

The simplest clutter model is the Gaussian model that represents the random clutter process as a multidimensional Gaussian process. The model has been employed almost exclusively in MTI analyses. An attractive feature of this model is that the clutter process is specified completely by its mean value and covariance function. The assumption of the Gaussian model leads to an optimal MTI processor that is linear.

The Gaussian clutter assumption generally leads to an optimistic evaluation of MTI performance. At present, none of the analytical techniques available allow a determination of the performance of MTI systems in the presence of non-Gaussian clutter. The principal method of attacking this problem is through a Monte-Carlo computer simulation, in which the actual MTI system is exercised by applying statistical samples of the postulated clutter and target processes.

4.4.1 The Gaussian Process Representation

Distributed clutter that is returned from a spatially continuous distribution of scatterers with no subset of scatterers predominating can generally be represented by a Gaussian process. Such clutter is usually associated with weather clutter, chaff, sea clutter observed with a low-resolution radar (pulsewidth $\tau > 0.5$ μs) or with a high-resolution radar at high grazing angles ($\phi > 5°$), and land clutter observed from high grazing angles ($\phi > 5°$) over undeveloped terrain.

The random clutter process is represented at the clutter carrier frequency (f_c) as

$$c_t = x_t \cos \omega_c t - y_t \sin \omega_c t \tag{4.78}$$

where x_t and y_t are zero mean, identically distributed, low-pass independent normal processes with variance σ^2. The clutter process c_t is strict sense stationary and, hence, its characteristics are independent of the time origin.

The voltage envelope of c_t is given by

$$v_t = \sqrt{x_t^2 + y_t^2} \tag{4.79}$$

and has a Rayleigh first pdf:

$$p_v(v) = \frac{v}{\sigma^2} \exp\left(\frac{-v^2}{2\sigma^2}\right); \qquad v \geq 0 \tag{4.80}$$

independent of time due to the stationarity, and is a function of only the clutter power:

$$P_c = 2\sigma^2 \tag{4.81}$$

The power amplitude distribution function can be found by applying the transformation $P = v^2$:

$$P_p(P) = \frac{1}{P_c} \exp\left(\frac{-P}{P_c}\right) \tag{4.82}$$

The clutter cross-sectional area (backscatter coefficient · intercepted area) is proportional to the power envelope, resulting in an exponential pdf for the clutter cross section (σ):

$$p_p(\sigma) = \frac{1}{\overline{\sigma}} \exp\left(\frac{-\sigma}{\overline{\sigma}}\right) \tag{4.83}$$

Here, $\overline{\sigma}$ is the mean clutter cross section, given by

$$\overline{\sigma} = \overline{\sigma^0} A_c \tag{4.84}$$

where $\overline{\sigma^0}$ is the mean backscattering coefficient and A_c is the area intercepted by the radar cell.

The power spectral density associated with the randomly fluctuating portion of the clutter cross section is usually assumed to be Gaussian shaped in that it is the result of several independent effects. The PSD of the power envelope is given by

$$S_p(f) = \frac{P_c}{\sqrt{2\pi\sigma_f^2}} \exp\left(\frac{-f^2}{2\sigma_f^2}\right) \tag{4.85}$$

where P_c is the clutter power and σ_f is the standard deviation of the spectrum. The standard deviation of the spectrum (σ_f) is related to the rms velocity clutter spread (σ_v) by

$$\sigma_c = \frac{2\sigma_v}{\lambda} \tag{4.86}$$

The autocorrelation function associated with the power envelope is given by

$$R_p(\tau) = P_c \exp\left(\frac{-\tau^2}{2\sigma_\tau^2}\right) \tag{4.87}$$

where

$$\sigma_\tau = \frac{1}{2\pi\sigma_f} \tag{4.88}$$

In many applications, to work with the normalized autocorrelation function is convenient:

$$\rho_p(\tau) = \frac{R_p(\tau)}{P_c} = \exp\left(\frac{-\tau^2}{2\sigma_\tau^2}\right) \tag{4.89}$$

Of interest in MTI work and other coherent processing systems are the power spectral density and autocorrelation function of x_t and y_t. The relationship between the normalized autocorrelation functions of P_t and x_t is known [49] to be given by

$$\rho_p(\tau) = \rho_x^2(\tau) = \rho_y^2(\tau) \tag{4.90}$$

Thus, the spectral density of the power envelope can be found from the power spectral density of the coherent quadrature components by

$$S_p(\omega) = S_x(\omega) * S_x(\omega) \tag{4.91}$$

where * represents convolution. Hence, for a Gaussian-shaped spectrum, all power spectral densities and autocorrelations are Gaussian shaped, and the standard deviation of the power envelope can be found from those of the coherent quadrature components by using

$$\sigma_p^2 = 2\sigma_x^2 = 2\sigma_y^2 \tag{4.92}$$

To summarize, for MTI work, the power spectral density or autocorrelation functions are given by (4.85), (4.87), and (4.89), using the appropriate standard deviation given by (4.92). When working with experimental or published data, care must be exercised to ascertain whether the standard deviations referenced are the coherent (σ_x) or noncoherent (σ_p) standard deviations.

Some MTI applications require modeling clutter that has relative motion with respect to the radar. This is accomplished through modifying (4.85) by the mean clutter Doppler shift ($\bar{f}_d$), resulting in

$$S_p(f) = \frac{P_c}{\sqrt{2\pi\sigma_f^2}} \exp\left[\frac{-(f - f_d)^2}{2\sigma_f^2}\right] \tag{4.93}$$

When the clutter process (4.78) is sampled by pulsed radar, the probability density of the samples is given by a multidimensional Gaussian distribution:

$$p_{x_1 \ldots x_n}(x_1 \ldots x_n) = \frac{\exp\left(\frac{-\mathbf{x}^T \mathbf{M}_x^{-1} \mathbf{x}}{2}\right)}{(2\pi)^{n/2} |\mathbf{M}_x|^{1/2}} \tag{4.94}$$

where $\mathbf{x} = (x_1, x_2, \ldots, x_n)^T$, and the covariance matrix is given by

$$\mathbf{M}_x = \sigma_x^2 \begin{pmatrix} 1 & \rho_{12} \ldots & \rho_{1n} \\ \rho_{21} & \ddots & \rho_{2n} \\ \rho_{n1} & \ldots & 1 \end{pmatrix} \tag{4.95}$$

where

$$\rho_{ij} = \exp\left[\frac{-(i-j)^2 \Omega^2}{2}\right]; \qquad \Omega = 2\pi\sigma_f T \tag{4.96}$$

for a Gaussian-shaped distribution.

If the clutter is passed through a linear MTI, the Gaussian form of the distribution is preserved. The pdf at the output of the MTI can be found by using (4.94) with the covariance matrix modified by the effect of the MTI filter on the clutter process.

For analytical convenience, the clutter process is sometimes assumed to be a Gaussian Markov process. This process has an exponential normalized function given by

$$\rho_{\alpha\beta} = \exp(-\gamma |\alpha - \beta| T) \tag{4.97}$$

and a power spectral density given by

$$S(\omega) = \frac{2\gamma P_c}{(\gamma^2 + \omega^2)} \tag{4.98}$$

The inverse covariance matrix ($\mathbf{M}_x^{-1}$) has a particularly simple form:

$$\mathbf{M}_x^{-1} = \frac{1}{1-\rho_{12}^2} \begin{bmatrix} 1 & -\rho_{12} & 0 & \ldots & 0 & 0 \\ -\rho_{21} & 1+\rho_{12}^2 & -\rho_{12} & \ldots & 0 & 0 \\ \vdots & \vdots & \vdots & \ddots & \vdots & \vdots \\ 0 & -p_{12} & 1-\rho_{12}^2 & \ldots & 0 & 0 \\ 0 & 0 & 0 & \ldots & 1+\rho_{12}^2 & -\rho_{12} \\ 0 & 0 & 0 & \ldots & -\rho_{21} & 1 \end{bmatrix} \tag{4.99}$$

4.4.2 The Rician Clutter Representation

This model is similar to the Gaussian process representation except that a predominantly steady scatterer (S) is added to the distributed clutter. The first pdf corresponding to (4.82) becomes

$$p_p(P) = \frac{1 + m^2}{\overline{P}} e^{-m^2} e^{-[P/\overline{P}(1+m^2)]} I_0\left(2m\sqrt{\frac{(1+m^2)P}{\overline{P}}}\right) \tag{4.100}$$

where the ratio of the steady (S^2) to distributed (P_0) power ratio is given by

$$m^2 = \frac{S^2}{P_0} \tag{4.101}$$

and the total power ($\overline{P}$) proportional to the cross-sectional area is given by

$$\overline{P} = S^2 + P_0 \tag{4.102}$$

For MTI work, the clutter process at the carrier frequency is obtained by modifying (4.78) to

$$c_t = (S + x_t) \cos \omega_c t - y_t \sin \omega_c t \tag{4.103}$$

The autocorrelation function for the in-phase (I) process and Gaussian-shaped distributed clutter is

$$R(\tau) = P_0\left[m^2 + \exp\left(\frac{-\tau^2}{2\sigma_\tau^2}\right)\right] \tag{4.104}$$

The resulting power spectral density is

$$S(f) = m^2 P_0\, \delta(f) + \frac{P_0}{\sqrt{2\pi\sigma_f^2}} \exp\left(\frac{-f^2}{2\sigma_f^2}\right) \tag{4.105}$$

which illustrates that the MTI radars operating from fixed platforms in this clutter must be designed to provide a null at zero frequency.

4.4.3 The Log-Normal Clutter Representation

The log-normal clutter representation has been used to model high-resolution (pulsewidth $\tau < 0.5\ \mu$s) sea-clutter data, where the sea clutter is observed at grazing angles of less than 5°. In addition, ground clutter observed at low grazing angles has also been characterized by using the log-normal model.

The log-normal distribution is highly skewed, resulting in a relatively high probability of large clutter amplitudes. False alarm probabilities are controlled by raising the detection thresholds, thereby causing a reduced probability of detection (P_d) as compared to that achieved in Rayleigh clutter.

The log-normal model generally tends to overestimate the dynamic range of the real clutter distribution, whereas the Rayleigh model tends to underestimate the dynamic range. Thus, analyses with the log-normal model produce conservative results, whereas optimistic results are produced with the Rayleigh model. A good design procedure is to design for the range of distributions between Rayleigh and log-normal.

The log-normal sea clutter model is associated with sea spikes, a quasispecular reflection from wave crests during the formation of white caps. The log-normal model for ground clutter seems to be associated with large directive scatterers that, depending on viewing aspect, produce bright ridges and shadowing, resulting in a large clutter dynamic range.

The log-normal character of sea clutter is shown in Figure 4.49 for data taken with harbor surveillance radar [58]. As is apparent, the data above the median can be closely approximated by straight lines, indicating the close fit provided by the log-normal model.

The log-normal character of ground clutter data is apparent for clutter returned from cities as shown in Figure 4.50. These data were synthesized from high-resolution synthetic aperture radar data [20].

The experimental log-normal data depicted in Figures 4.49 and 4.50 represent long-time averages of the clutter cross section in a resolution cell. From an MTI standpoint, the short-time characteristics of the clutter are important, and two alternative interpretations of the data may be conjectured.

The first interpretation represents the random clutter process as a wide-sense stationary log-normal process. This is obtained analytically by passing the multi-dimensional Gaussian process represented by (4.94) through an exponential non-

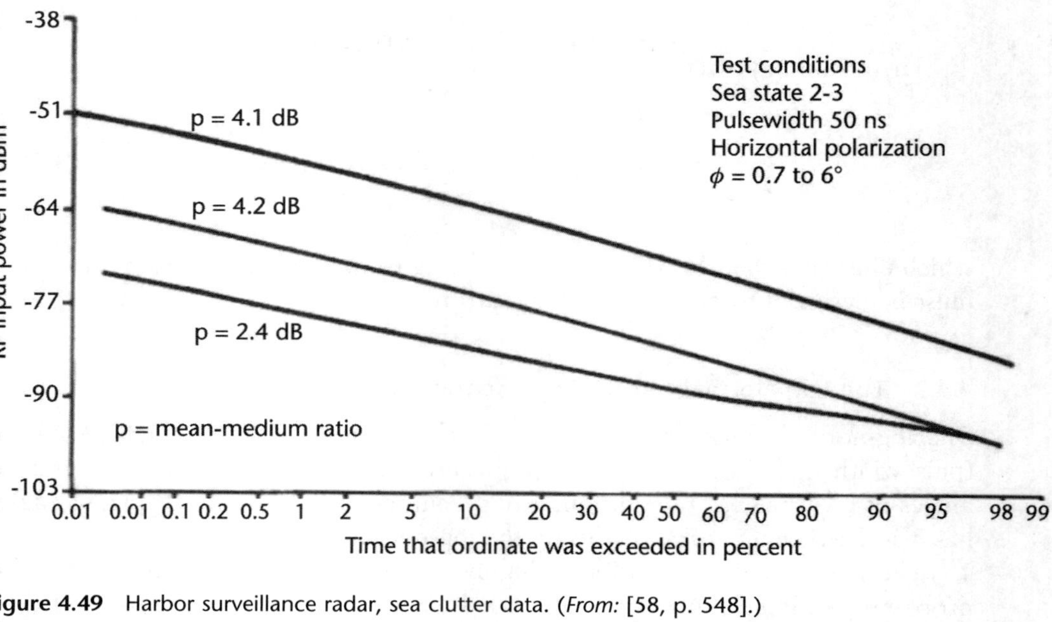

Figure 4.49 Harbor surveillance radar, sea clutter data. (*From:* [58, p. 548].)

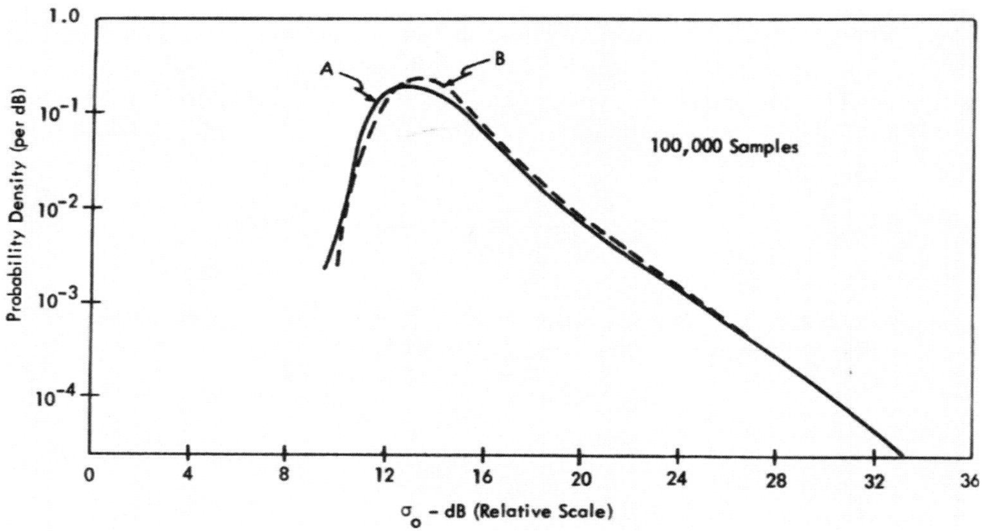

Figure 4.50 Probability density function of two cities. (*From:* [20, p. 26].)

linearity. To achieve the desired stationarity, the x_t and y_t processes of (4.78) must be uncorrelated. This interpretation has the advantage that it is a straightforward extension of the Gaussian clutter model, and many of the properties of this log-normal model can be deduced from corresponding properties of the Gaussian model.

The second interpretation models the clutter as a time-varying process having parameters that are stationary within a radar observation period over a specified spatial correlation distance. The conditional pdf $p_v(v\,|\,\sigma^0)$ is Rayleigh or Rician over the spatial correlation distance associated with each spatial correlation block that varies statistically in accordance with a log-normal distribution. Parameters of this model are discussed later in Section 4.5.1 on the IIT clutter model.

Both models lead to the same approximate result if uncorrelated (either temporally or spatially) samples are taken from the distribution. This situation occurs in fast-scanning radars that sample the clutter on a scan-to-scan basis. For MTI systems, the two interpretations can lead to different results, but no guidelines are currently available to indicate a clear choice.

By using the first interpretation, the log-normal clutter model is formed by expressing (4.78) as

$$c_t = v_t \cos(\omega_c t + \phi_c) \tag{4.106}$$

where v_t is the envelope of (4.79) and ϕ_c is a uniformly distributed random variable. If the quadrature components $x_t = v_t \cos \phi_c$ and $y_t = v_t \sin \phi_c$ are zero mean, uncorrelated, and have identical band-limited spectral densities, c_t is a wide-sense stationary process. The first pdf of the envelope is given by

$$p_v(v) = \frac{1}{\sqrt{2\pi}\sigma_v v} \exp\left[-\frac{\ln^2 (v/v_m)}{2\sigma_v^2}\right]; \qquad v \geq 0 \tag{4.107}$$

where σ_v is the standard deviation of the underlying normal distribution and v_m is the median value of the log-normal distribution.

The power envelope density function (proportional to the clutter cross-sectional area) is found by applying the transformation $A = v^2$ to (4.107) resulting in

$$p_p(\sigma) = \frac{1}{\sqrt{2\pi}\sigma_p \sigma} \exp\left[-\frac{\ln^2 (\sigma/\sigma_m)}{2\sigma_p^2}\right]; \qquad \sigma \geq 0 \tag{4.108}$$

where $\sigma_p = 2\sigma_v$, and σ_m is the median value of the clutter cross section. The mean value of the clutter cross section is given by

$$\overline{\sigma} = \rho_A \sigma_m \tag{4.109}$$

where ρ_A is the mean-to-median ratio:

$$\rho_A = \exp\left(\frac{\sigma_p^2}{2}\right) \tag{4.110}$$

The distribution function for the cross-sectional area is given by

$$P_p(\sigma) = \frac{1}{2}\left[1 + \operatorname{erf}\left(\frac{1}{\sqrt{2}\sigma_p} \ln \frac{\sigma}{\sigma_m}\right)\right] \tag{4.111}$$

On log-probability paper, this function plots as a straight line, as depicted in Figure 4.51 ($A_{\text{dB}} = 10 \log \sigma$). The straight line is easily determined for either the voltage envelope or cross-sectional area using the relationships provided in Figure 4.51 for the 0.5 and 0.9999 cumulative probability points. The mean or median cross-sectional area is determined from knowledge of the clutter's backscattering coefficient.

The characteristics of log-normal clutter are often given in terms of its underlying normal distribution. We can see from Figure 4.51 that the log-normal clutter distribution can be expressed as

$$p_N(A_{\text{dB}}) = \frac{1}{\sqrt{2\pi}\sigma_{\text{dB}}} \exp\left[-\frac{A_{\text{dB}}^2}{2\sigma_{\text{dB}}^2}\right] \tag{4.112}$$

where

$$A_{\text{dB}} = 10 \log_{10} \frac{\sigma}{\sigma_m} \tag{4.113}$$

Transforming (4.112) by using (4.113) results in the log-normal distribution of (4.108) with

$$\sigma_p = 0.2303\sigma_{\text{dB}} \tag{4.114}$$

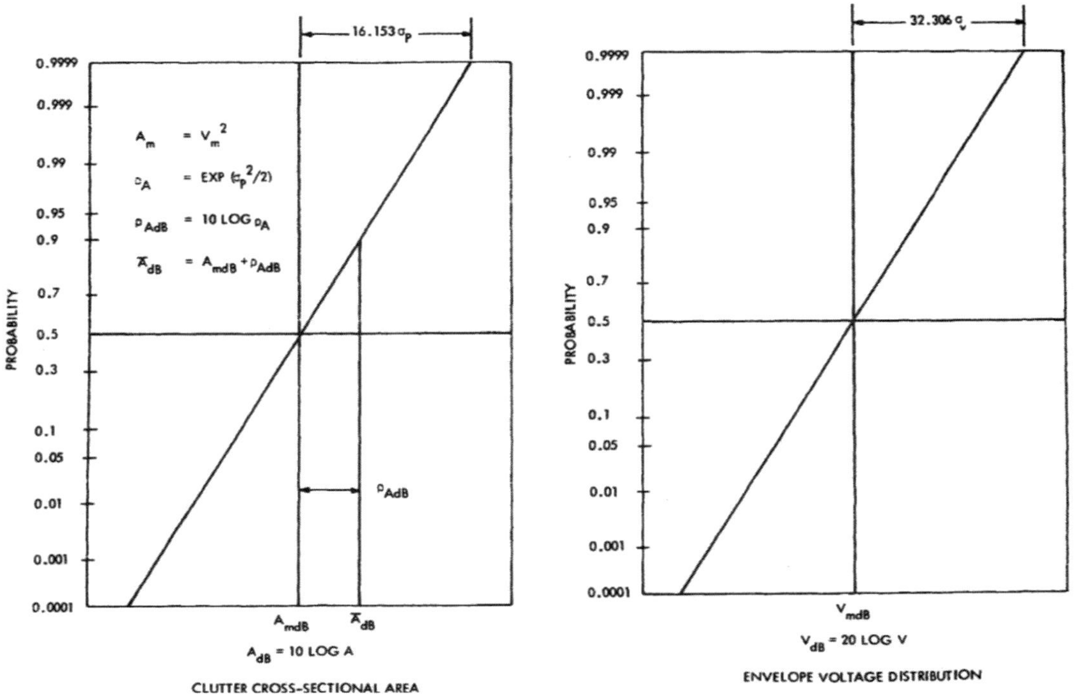

Figure 4.51 Log-normal clutter model. (*From:* [19, p. 46].)

Equation (4.114) can be used to relate the log-normal clutter parameters in natural units (σ_{dB}).

Table 4.24 gives some reported values of σ_p for various types of clutter. For MTI work, the correlation properties of the sample log-normal clutter model are important. A complete characterization of the log-normal model can be obtained by passing the multidimensional Gaussian distribution of (4.86) through an exponential nonlinearity. This is discussed in [59] and involves specification of the mean and covariance matrices associated with the multidimensional Gaussian distribution. The elements of the Gaussian covariance matrix ($\mathbf{m}_{xij}$) are related to the elements of the log-normal covariance matrix, $\mathbf{m}_{\ln(ij)}$, by

Table 4.24 Log-Normal Clutter Parameters

Terrain or Sea State	*Band Frequency*	ϕ	σ_p
Sea State 2–3	X	4.7°	1.382
Sea State 3	K_u	1–5°	1.440–1.960
Sea State 4	X	0.24°	1.548
Sea State 5[a]	K_u	0.50°	1.634
Land clutter (discrete)	S	Low	3.916
Land clutter (distributed)	S	Low	1.380
Land clutter[a]	P to K_a	10°–70°	0.728–2.584
Rain clutter	X and 95 GHz	—	0.680
Radar angels	—	—	1.352–1.620

[a]Estimated value.

$$\mathbf{m}_{xij} = \ln\left(1 + \frac{\mathbf{m}_{\ln(ij)}}{\overline{x_i}\,\overline{x_j}}\right) \tag{4.115}$$

where $\overline{x}_{i,j}$ denotes the elements of the mean vectors of the log-normal clutter. Several relationships useful in converting from Gaussian (x_t) to log-normal (z_t) variables through an exponential nonlinearity, $A\,\exp(\cdot)$ are

$$E(z_t) = A\,\exp\left[\frac{R_x(0)}{2}\right] \tag{4.116}$$

$$R_z(\tau) = A^2\,\exp[R_x(0) + R_x(\tau)] \tag{4.117}$$

$$C_z(\tau) = A^2\,\exp[R_x(0)]\{\exp[R_x(\tau)] - 1\} \tag{4.118}$$

where $R_x(\tau)$ is the ACF of the Gaussian variable, and $R_z(\tau)$ is the ACF of the log-normal variable, and $C_z(\tau)$ is the covariance function of the log-normal variable.

4.4.4 The Weibull Clutter Representation

The Weibull clutter representation has properties that lie between the Rayleigh (a member of the Weibull family) and log-normal clutter models. The Weibull model has been used to model both land and sea clutter and offers the potential to accurately represent the real clutter distribution over a much wider range of conditions than either the log-normal or Rayleigh model.

Figure 4.28 showed experimental sea clutter distributions plotted on log-Rayleigh probability paper, and Figure 4.17 presented experimental land clutter data. The close fits provided by the straight lines indicate that the data can be closely modeled by Weibull statistics. In addition, the decreasing skewness with increasing grazing angle indicates that the distribution is approaching a Rayleigh distribution.

The two clutter model interpretations that apply to the log-normal model also apply to the Weibull clutter model. For MTI systems, the two interpretations can lead to different results, but no guidelines are currently available to indicate a clear choice.

In a Weibull clutter model, the first probability distribution of the voltage output of an envelope detector normalized to the median v_m is given by

$$p_v(R) = \alpha \cdot \ln 2 \cdot R^{\alpha-1}\,\exp(-\ln 2 \cdot R^{\alpha}); \qquad R > 0 \tag{4.119}$$

where $R = v/v_m$ and α is a parameter that relates to the skewness of the distribution. This is related to the amplitude pdf of the clutter cross section σ by the transformation $\sigma = v^2$, resulting in

$$p_A(\sigma_c) = \beta \cdot \ln 2 \cdot \sigma_c^{\beta-1}\,\exp\left(-\ln 2 \cdot \sigma_c^{\beta}\right); \qquad \sigma_c > 0 \tag{4.120}$$

where $\sigma_c = \sigma/\sigma_m$, $\beta = \alpha/2$, and σ_m is the median of the clutter cross-sectional area. The mean of the clutter cross section is given by

$$\overline{\sigma} = \frac{\sigma_m \Gamma\left(1 + \frac{1}{\beta}\right)}{(\ln 2)^{1/\beta}} \tag{4.121}$$

The distribution function representing the clutter cross-sectional area on a decibel scale is given by

$$\sigma_{\text{dB}} = \sigma_{m_{\text{dB}}} + \frac{1.592}{\beta} + \left(\frac{10}{\beta}\right) \log\left\{\ln\left[\frac{1}{(1 - P_A(\sigma))}\right]\right\} \tag{4.122}$$

This function plots as a straight line on log-Rayleigh probability paper. The straight line is easily determined by plotting the 0.5 probability point as

$$\sigma_{\text{dB}} = \sigma_{m_{\text{dB}}} \tag{4.123}$$

and the 0.9999 probability point as

$$\sigma_{\text{dB}} = \sigma_{m_{\text{dB}}} + \frac{11.235}{\beta} \tag{4.124}$$

Some reported values of the Weibull model parameters are given in Table 4.25. The Weibull clutter distribution is sometimes defined in terms of a slope parameter (α), given by

$$\alpha = \frac{1}{\beta} \tag{4.125}$$

The correlation properties of the sample Weibull clutter model are important for MTI work. The correlation matrix of the Weibull distribution can be related to the correlation matrix of a Gaussian distribution by the following set of transformations [60]. A Rayleigh vector ($\mathbf{v}$) is formed from two zero-mean independent Gaussian vectors $\mathbf{x}$ and $\mathbf{y}$ by

$$\mathbf{v}_i = \sqrt{\mathbf{x}_i^2 + \mathbf{y}_i^2} \tag{4.126}$$

Table 4.25 Weibull Clutter Parameters

Terrain or Sea State	*Band Frequency*	*Beamwidth (degrees)*	*Grazing Angle (degrees)*	*Pulsewidth (μs)*	β
Rocky mountains	S	1.5	—	2	0.256
Wooded hills	L	1.7	≈0.5	3	0.313
Forest	X	1.4	0.7	0.17	0.253–0.266
Cultivated land	X	1.4	0.7–5	0.17	0.303–1
Sea state 1	X	0.5	4.7	0.02	0.726
Sea state 3	K_u	5	1–30	0.1	0.58–0.8915

The Rayleigh vector is transformed into a Weibull vector ($\mathbf{R}$) by

$$r_i = v_i^{2/\alpha} \tag{4.127}$$

Elements of the correlation matrix ($\mathbf{m}_{xij}$) of the Gaussian vectors are related to elements of the correlation matrix of the Weibull vector ($\mathbf{s}_{ij}$) by

$$\mathbf{s}_{ij} = \left\{ \frac{\Gamma^2\left(1+\frac{1}{\alpha}\right)}{\Gamma\left(1+\frac{2}{\alpha}\right) - \Gamma^2\left(1+\frac{1}{\alpha}\right)} \right\} \cdot \left[{}_2F_1\left(-\frac{1}{\alpha}, -\frac{1}{\alpha}, \mathbf{m}_{xij}^2\right) - 1 \right] \tag{4.128}$$

where ${}_2F_1(\cdot)$ is the hypergeometric function.

4.4.5 The *K*-Distribution Clutter Representation

A representation of K-distribution clutter has been used to model the envelope-detected return from both land and sea clutter [61]. The properties of the K-distribution are similar to those of the Weibull distribution (see Section 4.1.2.2) with values in the tail region ranging from that of the Rayleigh distribution (a member of the K-distribution family, as is the Weibull) to approach that of the log-normal distribution. The K-distribution representation is derived from a composite structure that assumes that the clutter can be represented as a Gaussian process with a varying mean. The mean is further assumed to fluctuate slowly so that its value is fixed over the processing time in the radar signal processor. This allows analyzing the system performance by using Gaussian (Rayleigh envelope) statistics, which are then averaged over the ensemble of mean values to determine the detection performance for Swerling fluctuating targets. Thus, although the ensemble probability distribution of the clutter's backscatter coefficient is represented by a K-distribution, the actual analysis is generally performed on the compound distributions that consist of the Rayleigh envelope for the fast fluctuating component and the chi-square distribution for the slowly fluctuating mean.

The Rayleigh pdf of the envelope-detected K-distribution voltage conditioned on its mean value $\left(y = \sqrt{\pi/2}\sigma\right)$ is given by

$$p(v \mid y) = \frac{\pi v}{2y^2} \exp\left(-\frac{\pi v^2}{4y^2}\right); \qquad v \geq 0 \tag{4.129}$$

where $\sqrt{2}\sigma$ is the standard deviation of the clutter power associated with this component. The chi pdf that represents the fluctuation of the mean level is given by

$$p(y) = \frac{2b^{2v}}{\Gamma(v)} y^{2v-1} \exp\left(-b^2 y^2\right); \qquad y \geq 0 \tag{4.130}$$

where v is a shape parameter and b is a scale parameter. The resulting K-distributed pdf of the envelope-detected clutter is then given by

$$p(v) = \frac{4c(cv)^{v} K_{v-1}(2cv)}{\Gamma(v)} \tag{4.131}$$

where $c = \sqrt{\pi/4b}$ is a scale parameter, v is a shape parameter, and $K_{v-1}(\cdot)$ is the modified Bessel function of order $v - 1$. The nth moments of $p(v)$ are given by

$$E(v^n) = \frac{\Gamma\left(v + \frac{n}{2}\right)\Gamma\left(1 + \frac{n}{2}\right)}{c^n \Gamma(v)} \tag{4.132}$$

so that the clutter power is given by $P_c = E(v^2) = v/c^2$. The pdf for the clutter power can be derived by using the transformation $p = v^2$ as

$$p(p) = \frac{2c^{v+1} p^{v-1/2} \mathbf{K}_{v-1}\left(2c\sqrt{p}\right)}{\Gamma(v)} \tag{4.133}$$

The cumulative probability distribution function for the clutter's power that is proportional to the clutter's backscattering coefficient (σ^0) is then given by

$$P(p) = \frac{1 - 2c^{v} p^{v/2} \mathbf{K}_{v}\left(2c\sqrt{p}\right)}{\Gamma(v)} \tag{4.134}$$

For most clutter situations, the value of the shape parameter (v) varies over the range $0.1 < v < \infty$, where the smaller values (i.e., $v \to 0.1$) correspond to clutter with high tails, and as v approaches ∞, the clutter distribution approaches a Rayleigh distribution. Values of v between 0.1 and 3 have been experimentally determined to approximate high-resolution (τ = 70–270 ns) sea clutter data at low grazing angles (ϕ = 1°–5°) [46]. We can also show that the moments of K-distributed clutter range from that of the Rayleigh distribution to approach those of the log-normal distribution [61].

▶ Example 16 Simulation of Radar Clutter Distributions [62]

A clutter model assumption is implicit in the design or analysis of MTI radar. The function of the clutter model is to represent the random clutter process in such a manner that the system performance can be determined. Ideally, the model accurately reflects the clutter process in a realistic manner. Practically, the clutter database is rarely adequate to achieve this objective, and the art of modeling the clutter involves a compromise between accurate representation and analytic convenience.

The traditional clutter representation uses the Rayleigh model. However, experimental data have shown that real clutter exhibits a much wider dynamic range than this model. Generally, Weibull, log–normal, and K-distribution models have been used to model radar clutter. The Weibull and K-distribution models have the advantage that the Rayleigh model is a member of both these families. The log-normal model generally overestimates the dynamic range of the clutter.

The cumulative probability distributions for the clutter RCS that is described by these models are given in Table 1.6. MATLAB program rad_clt_plot.m provides a plot of these distributions, which is depicted in Figure EX-16.1. The parameters used in these models are appropriate for high-resolution SS-3 sea clutter. The Weibull distribution ($\beta = 0.5$) and K-distribution ($\nu = 0.3$) provide very similar results. As expected the log-normal distribution provides the widest dynamic range.

Most random variables used in radar simulations can be generated by the transformation method [63]. To apply this method, the cumulative probability distribution $P(x)$ of the random variable x must generally be available and invertible. To understand this method, first note that $P(x)$ is a nondecreasing function with values that range from 0 to 1. If we envision a graph of $P(x)$ as a function of x where the ordinate is probed with a uniform distributed random variable u with support 0 to 1, then each randomly selected value of u can be used to find an equivalent value of the random variable x. The random value x is then distributed in accordance with the cumulative probability distribution $P(x)$.

As an example of this method, consider an exponential distribution given by

$$P(x) = 1 - \exp(-\alpha x)$$

Then letting $P(x) = u$ we can solve for x as

$$x = \frac{-\ln(u)}{\alpha}$$

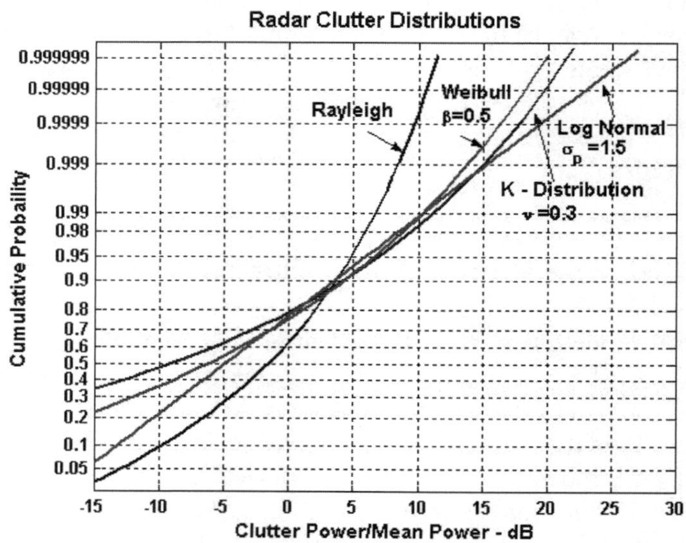

Figure EX-16.1 Probability distributions for clutter models.

where we recognize that both $(1 - u)$ and u are uniformly distributed random variables with support 0 to 1. Table EX-16.1 describes a number of radar clutter distributions that can be simulated using uniformly distributed random variables u and v with support 0 to 1 [62]. In MATLAB a uniformly distributed random variable can be generated by the function rand, while a Gaussian distribution can be generated by randn.

K-distribution clutter is derived by assuming that a Rayleigh variate is conditioned on its mean value represented by a chi or root gamma distribution. Unfortunately, the chi distribution is not invertible so that it cannot be generated using the transformation method. However, because the probability density is available, it can be generated using the rejection method [64, p. 551].

In the rejection method, an invertible comparison function $f(t)$ is selected which everywhere lies above the pdf $p(t)$ of the variable to be generated. Then a test random variable (t) is generated from the comparison function and the ratio $q = f(t)/p(t)$ is formed. Next, a uniform deviate $0 - 1$ is generated and if less than q, the test random variable is accepted and otherwise rejected.

An algorithm for generation of a gamma random deviate of order v is given in MATLAB program knoise.m. The algorithm first generates a gamma random variable of integer order k using the logarithmic method and a second gamma random variable of order $a = v - k < 1$ using the rejection method. The overall gamma deviate is then the sum of the two gamma random variables. The efficiency of the simulation is enhanced by partitioning the comparison function into subregions such that

$$f_1(t) = \frac{t^a - 1}{\Gamma(a)}; \qquad 0 \le t \le 1$$

$$f_2(t) = \frac{e^{-t}}{\Gamma(a)}$$

Table EX-16.1 Simulation of Radar Clutter Using Transformation Method

Distribution	*Parameters*	*Simulated Values*
Gaussian	$N(0, \sigma)$	$x = \sigma[-2\ln(u)]^{1/2}\cos(2\pi v)$
	$E(x, y) = 0$	$y = \sigma[-2\ln(u)]^{1/2}\sin(2\pi v)$
	$N(m, \sigma)$	$x_m = m + \sigma[-2\ln(u)]^{1/2}\cos(2\pi v)$
Rayleigh	σ	$x_R = \sigma[-2\ln(u)]^{1/2}$
Exponential	α	$x_E = \dfrac{[-\ln(u)]}{\alpha}$
Weibull	R_m, β	$x_W = R_m\left[\dfrac{-\ln(u)}{\ln(2)}\right]^{1/2\beta}$
Log-normal	σ	$x_{LN} = \exp(x)$

The chi or root gamma random variable u is then obtained using the transform

$$u = \frac{\sqrt{t}}{b}$$

where the clutter power is given by $P_c = 2v/b^2$.

To illustrate the ease with which complicated radar detection problems can be simulated, we describe the simulation of radar, which functions to detect small targets in sea clutter [62]. Most radars of this class work on the same principle. They transmit a wide bandwidth providing high range resolution matched as closely as possible to the physical dimensions of the target. A rapid scanning antenna (150 to 300 r/min) is used to sample target and clutter returns on a scan-to-scan basis. Because the clutter returns decorrelate on a scan-to-scan basis, while the target returns remain correlated, it is possible to extract targets using scan-to-scan integration.

The most difficult component to model is the radar clutter. As the resolution of the radar increases, reflections from individual wave facets can return high amplitudes called sea spikes, which significantly exceed target returns. The composite surface model postulates that sea clutter is due to the compound effect of reflections from the many small wave facets modulated by the underlying sea swell structure. Each of these components responds differently to the use of frequency agility by the radar.

Although this form of sea clutter representation accurately represents experimentally determined sea clutter statistics, its properties are difficult to quantify analytically. Fortunately, it is relatively straightforward to simulate this form of sea clutter by emulating its physical model. Using this approach provides the added advantage that the effects of both frequency agility and high range resolution on the maximum detection range are automatically embedded in the model.

To simulate this detection problem, the K-distributed clutter voltage in receiver noise is simulated using MATLAB code by:

$$r_n = \sqrt{\frac{ux^2 + P_n}{2}} * (-2 * \ln(\text{rand}))^{1/2}$$

where rand is a uniformly distributed random variable 0 to 1, ux is a root gamma (chi) variable, and P_n is the noise power. The in-phase and quadrature components of the interference using MATLAB are given by

$$x_i = r_n * \cos(2\pi * \text{rand})$$

$$x_q = r_n * \sin(2\pi * \text{rand})$$

resulting in a signal-plus-interference random variable given by

$$r_s = \sqrt{(v_s - x_i)^2 + x_q^2}$$

where v_s is the steady or fluctuating target random variable.

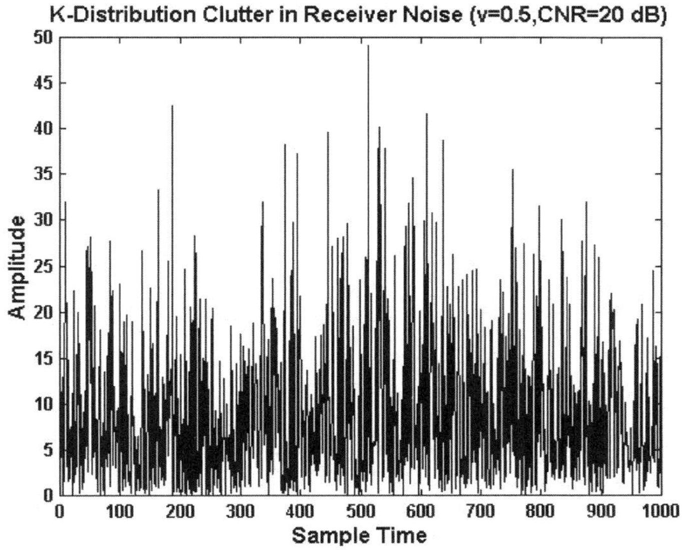

Figure EX-16.2 Simulation of *K*-distribution clutter in receiver noise.

Figure EX-16.2 depicts a MATLAB simulation of the time series generated for *K*-distributed clutter and receiver noise. The spiky nature of the simulated sea clutter is apparent. Also apparent is the difficulty of detecting small fluctuating targets whose magnitudes are significantly less than that of the sea spikes.

▶

4.5 Clutter Model Environments

This section describes two clutter model environments developed to represent those seen by airborne search radar (the IIT model) and ground-based search radar.

4.5.1 The IIT Radar Clutter Model [20]

Representing one of the more comprehensive clutter models, the IIT model was developed for use with *Overland Airborne Radars* on the basis of extensive data analysis. The model provides for the spatial distribution of the clutter and does not explicitly address intrinsic or induced variations in the clutter return.

The IIT model treats radar ground clutter as consisting of two additive components: *distributed* clutter and *discrete* clutter. The former is indicative of returns from natural terrains; the latter corresponds to returns from physically isolated and dominant artificial scatterers.

The complete model is intended for use in computer simulations and is far too complex for manual calculations. In the following section, a general description of the model is given to provide insight into clutter modeling and a basis for approximations that may be used for simple calculations. The reader is referred to [20] for a complete description of the model.

The distributed clutter backscattering cross section per unit area (σ^0) is a function of frequency, polarization, grazing angle, terrain, and environmental factors. The first four factors are deterministic in nature, whereas environmental factors are random and account for variations in grazing angle.

Figures 4.8 and 4.11 depicted average backscattering coefficients for various terrains as a function of grazing angle. The data given in Figure 4.8 were independent of frequency and polarization because uncertainties in the database would be larger than variations due to these parameters. For desert terrain, the variation with frequency was significant and, hence, should be included in the model.

The instantaneous clutter power returned to the radar is proportional to the effective clutter cross-sectional area, the distribution of which is given by

$$p_p(\sigma) = \frac{1}{\overline{\sigma}} \exp\left(\frac{-\sigma}{\overline{\sigma}}\right) \tag{4.135}$$

The factor $\overline{\sigma}$ is a random variable having a value that is a function of the spatial properties of the clutter distribution, given by

$$\overline{\sigma} = \overline{\sigma^0}(x, y) A_c \tag{4.136}$$

where $\overline{\sigma^0}(x, y)$ is the spatially dependent clutter backscatter coefficient and A_c is the area intercepted by the radar resolution cell.

For any fixed set of spatial parameters, (x_0, y_0) is selected from a log-normal distribution by

$$p_N\left(\sigma^0_{\text{dB}}\right) = \frac{1}{\sqrt{2\pi} s_0} \exp\left[-\frac{1}{2}\left(\frac{\sigma^0_{\text{dB}} - \overline{\sigma^0_{\text{dB}}}}{s_0}\right)^2\right] \tag{4.137}$$

where

$$\sigma^0_{\text{dB}} = 10 \log_{10} \sigma^0 \tag{4.138}$$

and s_0 is the standard deviation of σ^0_{dB}. The term $\overline{\sigma^0}$ was given earlier in Figures 4.8 and 4.11, whereas s_0 is given by

$$s_0 = \sqrt{s_1^2 + s_2^2} \tag{4.139}$$

where s_1 and s_2 are defined in Figures 4.52 and 4.53. Both $\overline{\sigma^0_{\text{dB}}}$ and s_0 are functions of the grazing angle and type of terrain.

As an intermediate step in the development of the model, consider that the terrain is divided into a checkerboard kind of pattern. Each cell of the checkerboard is defined by a correlation distance (d) over which the value of $\sigma^0(x_0, y_0)$ applies. The average value of the backscatter coefficient, $\overline{\sigma^0}(x, y)$, to be used in (4.136) is found by integrating over those cells of the checkerboard contained within the

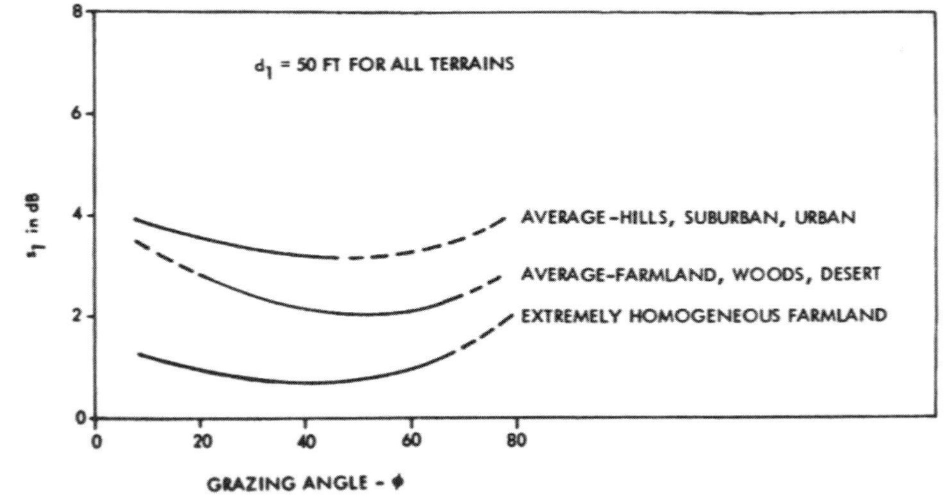

Figure 4.52 Standard deviation of short-range spatial correlation for various terrains. (*From:* [20, p. 64].)

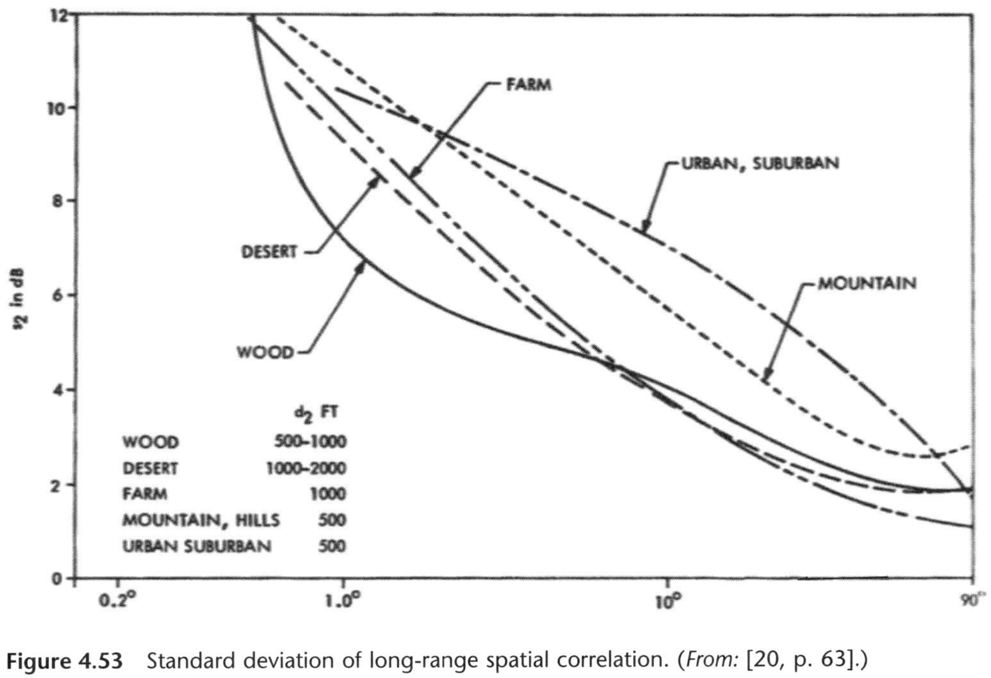

Figure 4.53 Standard deviation of long-range spatial correlation. (*From:* [20, p. 63].)

window formed by the radar's resolution cell. For high accuracy, the radar's two-way antenna pattern is used as an amplitude weighting for one dimension of the window.

As the radar antenna is scanned, or when the radar is on a moving platform, new cells of the checkerboard enter the radar resolution window and old cells are discarded. Thus, the radar induces a temporal variation in the clutter statistics that accounts for the correlation properties of the clutter.

The next refinement of the model adds a triangular correlation function across each cell of the checkerboard. This spatial ACF is given by

$$r(\Delta x, \Delta y) = \frac{\exp\left[(0.2303 s_0)^2 \rho_{12}^2(\Delta x, \Delta y)\right] - 1}{\exp\left[(0.2303 s_0)^2\right] - 1} \tag{4.140}$$

where

$$\rho_{12}^2(\Delta x, \Delta y) = \rho(\Delta x)\rho(\Delta y) \tag{4.141}$$

$$\begin{aligned} \rho(\Delta z) &= 1 - \left|\frac{\Delta z}{d}\right|; \qquad \left|\Delta z = \sqrt{(\Delta x)^2 + (\Delta y)^2}\right| \le d \\ &= 0; \qquad |\Delta z| > d \end{aligned} \tag{4.142}$$

and d is the correlation distance.

For computer simulation, the model at any point in the terrain field can be generated by

$$\sigma_{\mathrm{dB}}^0(x, y) = \overline{\sigma_{\mathrm{dB}}^0} + s_0 U_0(x, y) \tag{4.143}$$

where $U_0(x, y)$ is a correlated, zero-mean, unit variance, sample function of a two-dimensional Gaussian random process. Standard computer programs are available for generating $U_0(x, y)$. For this model, values of $d = d_2$ from Figure 4.54 should be used.

Analysis of synthetic aperture radar (SAR) data [20] indicates that a further refinement of the ACF of the model is required for accurately depicting the clutter return. The revised ACF depicted in Figure 4.54 is a two-segment approximation to an exponential function. The first correlation distance (d_1) provides a rapid decorrelation, and the second correlation distance (d_2) is much longer and represents a long-term correlation. The value of d_2 is much longer for homogeneous terrain such as farms and woodland than for urban or mountainous terrain where sharp changes occur.

For computer simulation, (4.143) is modified so that

$$\sigma_{\mathrm{dB}}^0(x, y) = \overline{\sigma_{\mathrm{dB}}^0} + s_1 U_1(x, y) + s_2 U_2(x, y) \tag{4.144}$$

where s_1 and s_2 are given in Figures 4.52 and 4.53, and $U_1(x, y)$ and $U_2(x, y)$ are sample functions from correlated, zero-mean, unit variance, two-dimensional Gaussian processes.

The amplitude probability distribution of discrete clutter is given by a log-normal distribution defined as

$$p_N(A_{\mathrm{dB}}) = \frac{1}{\sqrt{2\pi} s_d} \exp\left[-\frac{1}{2}\left(\frac{A_{\mathrm{dB}} - m_d}{s_d}\right)^2\right] \tag{4.145}$$

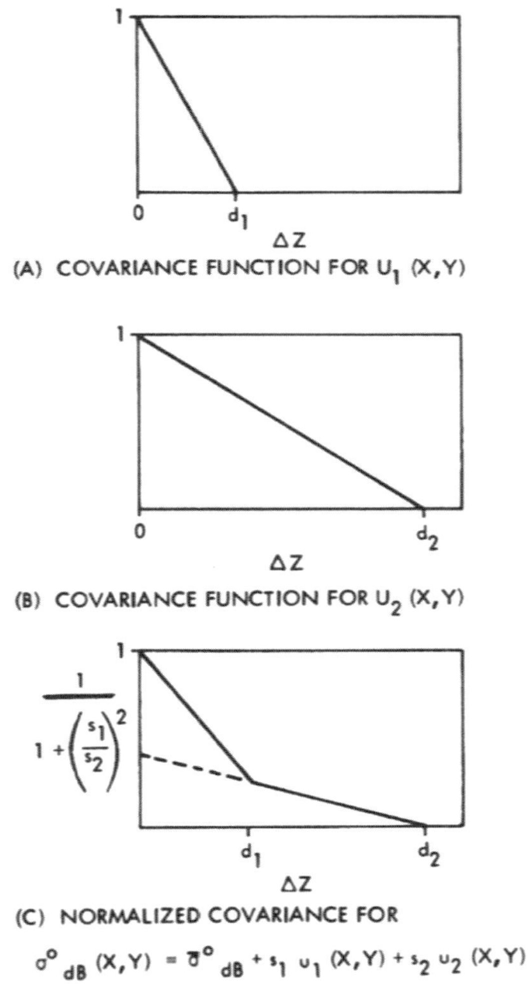

Figure 4.54 (a–c) Covariance function for long- and short-range spatial covariance. (*From:* [20, p. 65].)

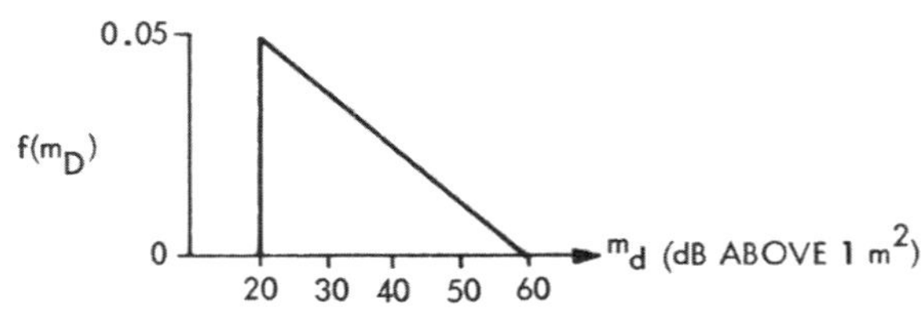

Figure 4.55 Probability distribution law for discrete clutter. (*From:* [1, p. 25].)

where

$$A_{\mathrm{dB}} = 10 \log A_d \qquad [\mathrm{dBsm}] \tag{4.146}$$

s_d is the standard deviation of A_{dB}, and m_d is the median of A_d with respect to a cross section of 1 m^2. The value of s_d is approximately equal to 5 dB. The value of m_d, given in Figure 4.55, ranges from 10^2 to 10^6 m^2 with a triangular probability distribution.

4.5.2 A Search Radar Model Environment

Several environmental models have been postulated that apply to ground-based air search radar in a benign environment [5, 65]. These models generally consist of an environment that generates a collection of radar returns corresponding to those most commonly observed on the display of ground-based air search radar. The environment comprises land and sea clutter, precipitation clutter, and spurious targets such as birds and ground vehicles that contaminate the radar display.

The model environment allows various competing radar designs to be compared against a standard reference. The environment is such that there are points in the coverage where desired targets cannot be detected and others where false alarms are indicated. False alarms are generally considered to be a far more serious problem because they tend to overload the data processing function of the radar (i.e., radar display or automatic detection equipment), thus limiting its effectiveness. For this reason, false alarm elimination is generally a high priority.

A model environment for air search radar is listed in Table 4.26 [5]. This model is intended to stress the radar's signal processor rather than to be representative of a typical environment. In the model, land clutter is represented by point surface reflectors, which vary in amplitude from 10 to 10^4 m^2. The points are spread randomly over a sector from 0° to 180° in azimuth and 40 km in range. If the radar is completely surrounded by a land mass throughout its operating range,

Table 4.26 Air Search Radar Model Environment

Return	*Cross Section*		*Azimuth (°)*	*Range (km)*	*Radial Velocity (m/s)*	*Height (m)*
Land:	Number	σ of each (m^2)	0–180	0–40	0	(surface)
Fixed points	450	10^4				
uniformly	900	10^3				
distributed	4,500	10^2				
in area	6,800	10				
Sea (sea state 4)	$\sigma^0 = 3 \times 10^{-5}$	(m^2/m^2)	180–360	0–20	2–4	(surface)
Rain (mm/h):	Frequency	η (m^2/m^3)	−25–25	20–40	0–30	500–2,000
	1,300	2×10^{-10}				
	3,000	5×10^{-9}				
	5,000	5×10^{-8}				
	10,000	5×10^{-7}				
Birds	Number of birds	σ of each (m^2)				
	2,000	10^{-2}	0–360	0–50	15	200
Cars	Number of cars	σ of each (m^2)				
	50	1	0–180	0–40	15	(surface)

Interference: Four identical radars at a distance $R_{max}/2$ from the radar under analysis; there is a clear line of sight to each; PRFs differ slightly, whereas center frequencies are displaced by 2%, 4%, 6%, and 8% of f_0.

then the coverage would be extended to 360° in azimuth with the same reflector density.

An alternate model [65] represents land clutter in a distributed manner, where the RCS per resolution cell depends on the size of the cell given by the half-power azimuth beamwidth (θ_{az}) in degrees multiplied by the pulse length (τ) in seconds. With θ_{az} in the range 0.4° to 4° and τ in the range 0.1 to 10 μs, the probability distribution is log-normal and has the parameters for linear polarization given in Table 4.27. In each case, the distribution is truncated at 50 dBsm. Circular polarization reduces the RCS by 3 dB, and the clutter return is independent of frequency. The velocity spectrum is Gaussian in shape with zero-mean velocity and a standard deviation of 0.3 m/s.

In the model, a second 180° azimuth sector is devoted to sea clutter and extends to the radar horizon (antenna height is 20 m) at approximately 20 km. A sea state 4 condition is postulated with a spectral spread corresponding to radial velocities of 2 to 4 m/s.

Rain falling at a rate of 4 mm/h in a 30° sector from 20 to 140 km is used to represent a storm in the radar's coverage. Spectrum spreads corresponding to radial velocities from 0 to 30 m/s (depending on the wind direction) are used with the highest rain densities occurring at altitudes from 500 to 2,000m. Note that the backscattering coefficient (η) is a strong function of the radar transmitted frequency so that different values are given for the various radar bands.

Alternately [65], the dimensions of the radar coverage volume affected by the rain as a function of precipitation rate are given in Table 4.28. The duration is the maximum expected in any one place in a year. The backscattering coefficient for rain is given by $\overline{\eta} = 7 \times 10^{-48} \cdot f^4 \cdot r^{1.6}$ [m^2/m^3], where f is the frequency in hertz and r is the rain rate in millimeters per hour. The velocity spectrum of the rain is given by $[1 + (7.3 \cdot 10^{-3} \cdot k \cdot R \cdot \phi_{el2})]^{1/2}$ [m/s], where k is the wind shear

Table 4.27 Model Parameters for Log-Normal Ground Clutter

$\theta_{az}\tau$ (° -s)	*Median (dBsm)*	*Standard Deviation (dB)*	*Number of Resolution Cells Affected by Clutter*
10^{-5}	0	20	5×10^3
10^{-6}	−15	20.1	5×10^4
10^{-7}	−30	20.8	5×10^5

Table 4.28 Precipitation Volume as a Function of Rain Rate

Precipitation Rate (mm/h)	*Diameter (km)*	*Ceiling (km)*	*Maximum Duration (hr)*
2	300	4	13
4	45	4	5
8	35	8	1.8
16	20	8	0.6
32	8	8	0.2
64	1	8	0.06

across the vertical beamwidth in m/s -km, R is the range in meters, and ϕ_{el2} is the vertical two-way half-power beamwidth in degrees. The mean velocity of the precipitation is determined by the wind speed, which varies with height, as tabulated in Table 4.29.

Birds are a source of moving clutter for low-PRF air search radar. The model uses an arbitrary number of 2,000 birds uniformly distributed in all azimuths over a range from 0 to 50 km. The small RCS of each bird is 0.01 m^2, so only sensitive radars that search for small targets will be bothered by this source of clutter. A random radial velocity distributed between ±15 m/s with a maximum height of 200m is assumed for each bird.

The distribution of the number of birds expected within a particular distance from the radar has been modeled by using ornithological data [66]. The density of birds is a random variable that depends on many factors, including the migratory pattern of the bird flocks. The model for the bird distributions is given in Table 4.30. Note that the number of birds used in the alternative search model (i.e., 2,000) approximately corresponds to the 50th percentile of the bird distribution model.

Moving ground vehicles such as cars are another source of radar returns that may bother low-PRF air search radar. Fifty cars are specified, which are uniformly distributed throughout the land clutter region and have a random uniformly distributed ±15 m/s radial velocity. Return from cars as viewed by an L-band search radar using MTI and a display range of 50 km is depicted in Figure 4.56.

Interference is specified as that which would exist if radiation from identical radars with a small variation of parameters were received by the radar under evaluation. For this purpose, four interfering radars are specified, each located within the LOS at a distance equal to half the maximum range. PRFs differ slightly, whereas center frequencies are displaced 2%, 4%, 6%, and 8% from the carrier frequency. This type of interference produces so-called *rabbits*, which move at a rate determined by the difference between the PRF of the radar under evaluation and that of the interfering radar. Return from an interfering radar and land clutter, as viewed on L-band search radar with a 50 km range, is depicted in Figure 4.57.

Table 4.29 Wind Speed Variation with Height

	Wind Speed Exceeded (% Time)		
Height (km)	*50% (m/s)*	*10% (m/s)*	*1% (m/s)*
Surface	5	9	13.5
2	11.5	21	31
4	13.5	24.5	36
6	18.5	34	50
8	23.5	43	63

Table 4.30 Bird Distribution Model

Observation Percentile (%)	*Bird Density (per mi^2)*
36.2	None
50	3
90	48
99	400

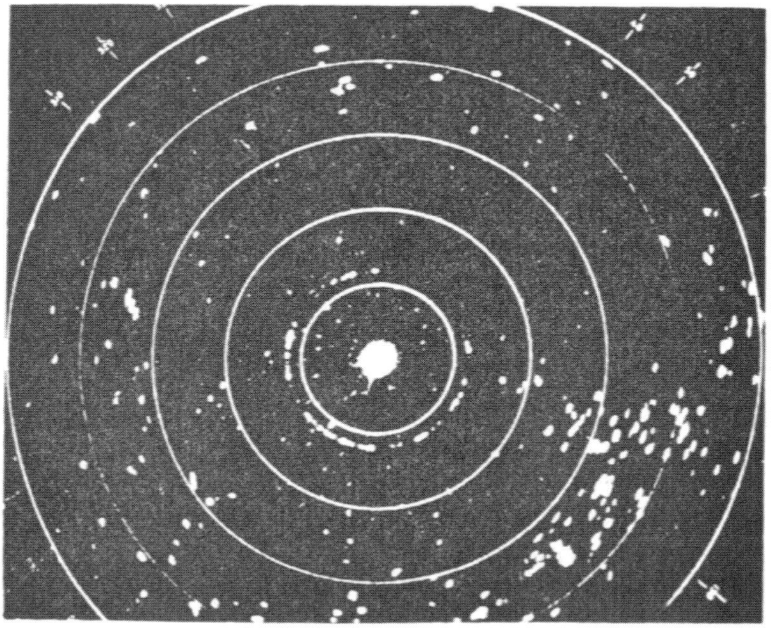

Figure 4.56 Cars viewed by L-band search radar using MTI and a display range of 50 km. (*From:* [5, p. 167].)

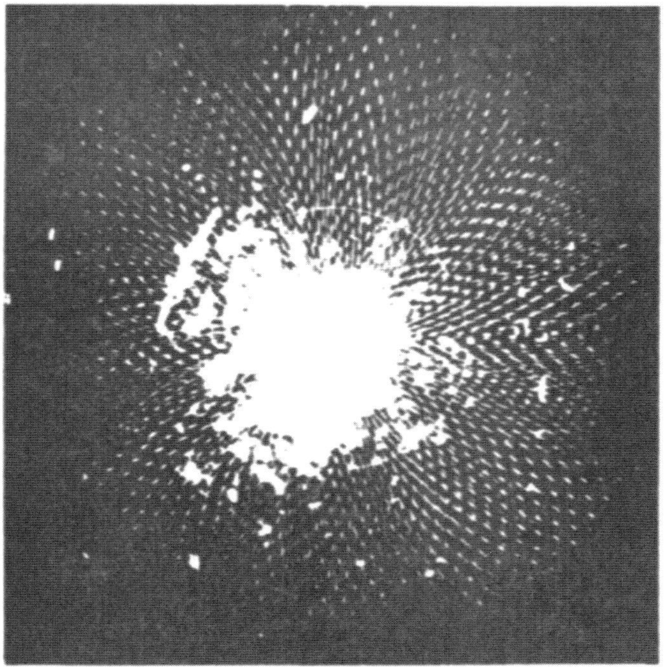

Figure 4.57 Interference and land clutter viewed on L-band search radar. (*From:* [5, p. 167].)

References

[1] Greenstein, L., A. Brindley, and R. Carlson, *A Comprehensive Ground Clutter Model for Airborne Radar*, Overland Radar Technology Program Report, IIT Research Institute, Chicago, September 1969.

[2] Skolnik, M. I., *Introduction to Radar Systems*, 3rd ed., New York: McGraw-Hill, 2001.

[3] Barton, D. K., and W. Shrader, "Interclutter Visibility in MTI Systems," *IEEE EASCON Record*, Washington, D.C., 1969, pp. 294–297; also in D. K. Barton, (ed.), *Radar Clutter*, Vol. 5 of *Radars*, Dedham, MA: Artech House, 1977.

[4] Long, M. W., *Radar Reflectivity of Land and Sea*, Dedham, MA: Artech House, 1983.

[5] Ward, H., "A Model Environment for Search Radar Evaluation," *IEEE EASCON Record*, Washington, D.C., 1971, pp. 164–171; also in D. K. Barton, (ed.), *Radar Clutter*, Vol. 5 of *Radars*, Dedham, MA: Artech House, 1977.

[6] Barton, D. K., "Land Clutter Models for Radar Design and Analysis," *Proc. IEEE*, Vol. 73, No. 2, February 1985, pp. 198–204.

[7] Nathanson, F., *Radar Design Principles*, 2nd ed., Mendham, NJ: SciTech, 1999.

[8] Barton, D. K., *Modern Radar System Analysis*, Norwood, MA: Artech House, 1988.

[9] Durlach, N., *Influence of the Earth's Surface on Radar*, Technical Report 373, AD 627635, MIT Lincoln Laboratory, January 1965.

[10] Blake, L. V., *Radar Range-Performance Analysis*, Artech House, Norwood, MA, 1986.

[11] Boothe, R., *The Weibull Distribution Applied to Ground Clutter Backscatter Coefficient*, Report RE-TR-69-15, U.S. Army Missile Command, Huntsville, AL, June 1969; also in D. C. Schleher, (ed.), *Automatic Detection and Radar Data Processing*, Dedham, MA: Artech House, 1980.

[12] Hayes, R., and F. Dyer, *Land Clutter Characteristics for Computer Modeling of Fire Control Radar Systems*, EES/GIT Project A-1485, AD 912490, Georgia Institute of Technology, Atlanta, GA, May 1973.

[13] Barton, D. K., (ed.), *Radar Clutter*, Vol. 5 of *Radars*, Dedham, MA: Artech House, 1975.

[14] Posner, F., "Spiky Sea Clutter at High Range Resolutions and Very Low Grazing Angles," *IEEE Trans. on Aerospace and Electronic Systems*, Vol. AES-38, No. 1, January 2003, pp. 58–73.

[15] Baker, C., "K-Distributed Coherent Sea Clutter," *IEE Proc.*, Vol. 138, Pt. F, No. 2, April 1991, pp. 89–92.

[16] Billingsley, J., et al., "Impact of Experimentally Measured Doppler Spectrum of Ground Clutter on MTI and STAP," *Radar-97, IEE Int. Radar Conference*, Edinburgh, Scotland, 1997, pp. 290–294.

[17] Lombardo, P., and J. Billingsley, "A New Model for the Doppler Spectrum of Windblown Radar Ground Clutter," *Proc. IEEE National Radar Conference*, Boston, MA, April 1999, pp. 142–147.

[18] Billingsley, J., *Low Angle Land Clutter Models*, Norwich, NY: William Andrew Publishing, 2002.

[19] Schleher, D. C., (ed.), *MTI Radar*, Dedham, MA: Artech House, 1978.

[20] Kazel, S., et al., *Extensions to the ORT Clutter Model*, IIT Research Institute, Chicago, June 1971.

[21] Currie, N. C., and S. Zehner, "Millimeter Wave Land Clutter Model," *Proc. IEE Int. Radar Conference*, London, October 1982, pp. 385–389.

[22] Barton, D., and W. Barton, *Modern Radar System Analysis Software, Version 2.0*, Norwood, MA: Artech House, 1993.

[23] Nathanson, F., *Radar Design Principles*, 1st ed., New York: McGraw-Hill, 1969.

[24] Eaves, J., and E. Reedy, *Principles of Modern Radar*, New York: Van Nostrand Reinhold, 1987.

[25] Morchin, W., *Airborne Early Warning Radar*, Norwood, MA: Artech House, 1990.

[26] Hou, X., and N. Morinaga, "Detection Performance in K-Distributed and Correlated Rayleigh Clutter," *IEEE Trans. on Aerospace and Electronic Systems*, Vol. AES-25, No. 5, September 1989, pp. 634–642.

[27] Simkins, W., V. Vannicola, and J. Ryan, *Seek Igloo Radar Clutter Study*, RADC-TR-338, ADA047897, Griffiss AFB, New York, October 1977.

[28] Trunk, G., and S. George, "Detection of Targets in Non-Gaussian Sea Clutter," *IEEE Trans. on Aerospace and Electronic Systems*, Vol. AES-6, No. 5, September 1970, pp. 620–628; also in D. K. Barton, (ed.), *Radar Clutter*, Vol. 5 of *Radars*, Dedham, MA: Artech House, 1977.

[29] Lay, R., J. Taylor, and G. Brunins, "ARSR-4: Unique Solutions to Long-Recognized Radar Problems," *Proc. IEEE Int. Radar Conference*, Washington, D.C., May 1990, pp. 6–11.

[30] Schleher, D. C., *MTI and Pulse Doppler Radar*, Norwood, MA: Artech House, 1991.

[31] Currie, N. C., F. Dyer, and R. Hayes, *Radar Land Clutter Measurements at 9.5, 16, 35 and 95 GHz*, Technical Report 3 DAA 25-73-0256, Georgia Institute of Technology, Atlanta, GA, March 1975.

[32] Petts, G., "Radar Systems," in *Handbook of Electronic System Design*, C. Harper, (ed.), New York: McGraw-Hill, 1980.

[33] Guinard, N., and J. Daley, "An Experimental Study of a Sea Clutter Model," *Proc. IEEE*, Vol. 58, No. 4, April 1970, pp. 543–550; also in D. K. Barton, (ed.), *Radar Clutter*, Vol. 5 of *Radars*, Dedham, MA: Artech House, 1977.

[34] Daley, J., W. Davis, and N. Mills, *Radar Sea Returns in High Sea States*, Report 7142, Naval Research Laboratory, September 25, 1970.

[35] Antipov, I., *Simulation of Sea Clutter Returns*, DSTO-TR-0679, Defence Science and Technology Organisation, Salisbury, South Australia, 1998.

[36] Choong, P. L., *Modeling Airborne L-Band Sea and Coastal Land Clutter*, DSTO-TR-0945, Defence Science and Technology Organisation, Salisbury, South Australia, March 2000.

[37] Chan, H. C., "Radar Sea Clutter at Low Grazing Angles," *IEE Proc.*, Vol. 137, Pt. F, No. 2, April 1990, pp.102–111.

[38] Sittrop, H., "On the Sea Clutter Dependency on Windspeed," *Radar-77, IEE Conf. Proc.*, No. 155, London, England, 1977, pp. 110–114.

[39] Watts, S., "A Practical Approach to the Prediction and Assessment of Radar Performance in Sea Clutter," *Proc. IEEE Radar Conference*, Washington, D.C., May 1995, pp. 181–186.

[40] Reilly, J. P., and G. D. Dockery, "Influence of Evaporation Ducts on Radar Sea Return," *IEE Proc.*, Vol. 137, Pt. F, No. 2, April 1990, pp. 80–88.

[41] Dockery, G. D., "Method for Modeling Sea Surface Clutter in Complicated Propagation Environments," *IEE Proc.*, Vol. 137, Pt. F, No. 2, April 1990, pp. 73–79.

[42] Horst, M., F. Dyer, and M. Tuley, "Radar Sea Clutter Model," *Digest of Int. IEEE AP/S URSI Symposium*, College Park, MD, 1978.

[43] Olin, I., "Amplitude and Temporal Statistics of Sea Spike Clutter," *Radar-82, IEE Conf. Proc.*, No. 216, London, England, 1982, pp. 198–202.

[44] Schleher, D. C., "Radar Detection in Weibull Clutter," *IEEE Trans. on Aerospace and Electronic Systems*, Vol. AES-12, No. 6, November 1976, pp. 736–743; also in D. C. Schleher, (ed.), *Automatic Detection and Radar Data Processing*, Dedham, MA: Artech House, 1980.

[45] Watts, S., "Radar Detection Prediction in Sea Clutter Using the Compound K-Distributed Model," *IEE Proc.*, Vol. 132, Pt. F, No. 7, December 1985, pp. 613–620.

[46] Jakeman, E., and P. Pusey, "Statistics of Non-Rayleigh Sea Echo," *Proc. IEE Int. Radar Conference*, London, November 1977, pp. 105–109.

[47] Barton, D. K., *Modern Radar System Analysis*, 1st ed., Dedham, MA: Artech House, 1964.

[48] Schleher, D. C., *Introduction to Electronic Warfare*, Dedham, MA: Artech House, 1986.

[49] Kerr, D., *Propagation of Short Radar Waves*, New York: McGraw-Hill, 1951.

[50] Currie, N. C., F. Dyer, and R. Hayes, "Some Properties of Radar Returns from Rain at 9.375, 35, 70 and 95 GHz," *Proc. IEEE Int. Radar Conference*, Washington, D.C., 1975, pp. 215–220.

[51] Sekine, M., et al., "On Weibull-Distributed Weather Clutter," *IEEE Trans. on Aerospace and Electronic Systems*, Vol. AES-15, No. 6, November 1979, pp. 824–830.

[52] Nathanson, F., and J. Reilly, "Radar Precipitation Echoes," *IEEE Trans. on Aerospace and Electronic Systems*, Vol. AES-4, No. 4, July 1968, pp. 505–514.

[53] Currie, N. C., F. Dyer, and R. Hayes, *Analysis of Radar Rain Return at Frequencies of 9.5, 35, 70 and 95 GHz*, Technical Report 2, Army Contract DAA 25-73-0256, Georgia Institute of Technology, Atlanta, February 1975.

[54] Skolnik, M. (ed.), "MTI Radar," Chapter 15 in *Radar Handbook*, 2nd ed., New York: McGraw-Hill, 1990.

[55] Schleher, D. C., *Electronic Warfare in the Information Age*, Norwood, MA: Artech House, 1999.

[56] Mitchell, P., and R. Short, "How to Plan a Chaff Corridor," in *International Countermeasures Handbook*, Los Altos, CA: EW Communications, 1979, pp. 382–392.

[57] Katzin, M., D. Ringwalt, and T. Weaver, *Evaluation of Airborne Overland Radar Techniques and Testing*, Vol. II, *Clutter Data Appendixes*, RTD-TR-65, Air Force Systems Command, Wright Patterson AFB, Ohio, September 1965.

[58] Schleher, D. C., "Harbor Surveillance Radar Detection Performance," *IEEE J. Oceanic Engineering*, Vol. OE-2, No. 4, October 1977, pp. 318–325; also in D. C. Schleher, (ed.), *Automatic Detection and Radar Data Processing*, Dedham, MA: Artech House, 1980.

[59] Peebles, P., "The Generation of Correlated Log-Normal Clutter for Radar Simulations," *IEEE Trans. on Aerospace and Electronic Systems*, Vol. AES-7, No. 6, November 1971, pp. 1215–1217.

[60] Szajnowski, W., "The Generation of Correlated Weibull Clutter for Signal Detection Problems," *IEEE Trans. on Aerospace and Electronic Systems*, Vol. AES-13, No. 5, September 1977, pp. 536–540.

[61] Fante, R., "Detection of Multiscatter Targets in K-Distributed Clutter," *IEEE Trans. on Antennas and Propagation*, Vol. AP-32, No. 12, December 1984, pp. 1358–1362.

[62] Schleher, D. C., "Solving Radar Detection Problems Using Simulation," *IEEE AES Systems Magazine*, April 1995.

[63] Press, W., et al., *Numerical Recipes*, New York: Cambridge University Press, 1989.

[64] Knuth, D., *The Art of Computer Programming*, Reading, MA: Addison-Wesley, 1981.

[65] Edgar, A., E. Dodsworth, and M. Warden, "The Design of a Modern Surveillance Radar," *Proc. IEE Int. Radar Conference*, London, October 1973, pp. 8–13.

[66] Pollon, G., "Distributions of Radar Angels," *IEEE Trans. on Aerospace and Electronic Systems*, Vol. AES-8, No. 6, November 1972, pp. 717–721.

[67] Skolnick, M., (ed.), "Meteorological Radar," Ch. 23 in *Radar Handbook,* 2nd ed., New York, McGraw-Hill, 1990.

[68] Jones, D., "The Shape of Rain Drops," Circular 77, *Journal of Meteorology,* Vol. 16, No. 5, October 1959, pp. 504–510.

[69] Fishbein, W., S. Graveline, and O. Rittenback, *Clutter Attenuation Analysis,* Tech. Rep. ECOM-2808, U.S. Army Electronics Command, Fort Monmouth, NJ, March 1967 (in D. C. Schleher, (ed.), *Automatic Detection and Radar Data Processing,* Dedham, MA: Artech House, 1980.).

[70] Jakeman, E., and P. Pusey, "A Model for Non-Rayleigh Sea Echo," *IEEE Trans. on Antennas and Propagation,* Vol. AP-24, No. 6, November 1976, pp. 806–814.

[71] *Land Clutter Effects on Shipboard Radar,* PEO on Theater Air Defense, Department of Navy, Washington, D.C., April 11, 1996.

[72] Billingsley, J., and J. Larrabee, *Measured Spectral Extent of L- and X-Band Radar Reflections from Wind-Blown Trees,* MIT Lincoln Laboratory Rep. CMT-57, Lexington, MA, February 1987.

[73] Dong, Y., *Models of Land Clutter vs Grazing Angle, Spatial Distribution—L-Band VV Polarization Perspective,* DSTO-RR-0273, Edinburgh, South Australia, March 2004.

[74] Schleher, D. C., *Automatic Detection and Radar Data Processing,* Dedham, MA: Artech House, 1980.

[75] Fiszer, M., and J. Gruszczynski, "Russia's Roving SAMs," *Journal of Electronic Defense,* July 2002, pp. 47–56.

[76] http://enemyforces.com/missiles/tor.htm.

[77] http://globalsecurity.org/military/worlds/russia/sa-15.htm.

[78] http://greekmilitary.net/airdefence.htm.

CHAPTER 5

Theory of Optimum Radar Doppler Processing

Most surface and airborne radars operate in an environment where the clutter return obscures targets of interest. If the target is moving relative to the clutter, we can filter out the undesirable clutter return by exploiting the differential Doppler shift produced by the target relative to clutter radial motion. Doppler signal processors are used to extract the Doppler-shifted target from the background interference, which typically consists of radar clutter plus receiver noise.

In the design of Doppler signal processors, we may be initially tempted to ignore the receiver noise component of the background interference and concentrate on the interfering clutter. This leads to a design that attenuates the portion of the Doppler spectral region of interest, which contains the clutter, and passes those signals that lie outside the attenuation region. This type of Doppler processor, called a moving target indicator, provides a comb filter type of response that ideally has a flat passband in the region of expected targets and a stopband in the region of heavy clutter returns.

The MTI is effective in improving the system signal-to-clutter ratio, but has no capability in improving the system signal-to-noise ratio. This outcome limits detection performance of the basic MTI to the single-pulse performance determined by using the standard Marcum-Swerling radar detection analysis [1]. A natural extension of the MTI Doppler processor is the addition of an integrator to improve its detection performance through enhancing the S/N ratio [2].

The integrator can be implemented either coherently or incoherently (after detection). A bank of coherent integration filters is conveniently formed by using the FFT algorithm. The cascaded MTI-coherent integrator bank Doppler processor is usually called a *pulsed Doppler processor*, which means that the processor's filter response is matched to the target spectrum. The term *MTI processor* means that the processor's Doppler response is open to all possible target Doppler frequencies.

A processor that uses considerably less hardware than the PD processor is the cascaded MTI-incoherent integrator. This processor is implemented by envelope detection of the MTI output, followed by a video integrator. However, the MTI correlates receiver noise that, when incoherently integrated, causes a detection loss [3–6]. The loss increases as the number (m) of pulses processed by the MTI increases [5]. This loss is compounded by any residual clutter passed by the MTI, which, in the extreme, can completely negate any improvement provided by the video integration [6].

As is apparent from this discussion, detection performance of the MTI-coherent integrator Doppler processor is superior to that of the cascaded MTI-incoherent integrator processor. Also, a careful examination of any MTI-integrator cascade using batch processing shows that only one of the m pulses employed in the MTI is effective in building up the processor's S/N ratio, which implies an inherent detection loss [6]. Hence, none of the previously described Doppler processors provide the optimal performance, which is the motivation for development of a general theory of optimal Doppler processors.

The fundamental problem to be considered is the detection of n coherently related, Doppler-shifted target pulses embedded in a background of Gaussian-distributed interference, consisting of clutter and receiver noise. The initial RF phase of the coherent pulse train returned from the target is uncertain and hence will be randomly distributed with respect to the phase of the coherently transmitted waveform.

If all of the parameters associated with the signal and interference were known, we could apply optimal radar signal processing techniques [7] resulting in an optimal processor consisting of a complex-weight (amplitude and phase processing) transversal filter, followed by an envelope detector. The processor synthesis problem is then to find the optimal weights of the transversal filter. Because the optimum weights involve the assumed signal distribution, separate solutions would be natural in accordance with the a priori assumption of the Doppler region occupied by the target. Direct application of the theory results in a processor that is optimum at a single Doppler frequency. This will be called the *point Doppler processor*. Because the target Doppler shift is not known in most applications, a bank of transversal filters can be used to cover the Doppler frequency band of interest. Each filter is then optimum at its center frequency, but mismatched otherwise. In this chapter, we explore the question of straddling losses that occur between the optimal point responses of this type of Doppler processor. The cascaded MTI-coherent integrator compares most directly with the point Doppler processor, using less hardware, but with only one of the m pulses that forms the MTI being effective in building up the S/N ratio.

The optimal theory can also be applied to develop a Doppler processor that is optimum for a signal with a Doppler shift that has uniform probability for any value within the detection region. This criterion results in an optimal quadrature channel MTI Doppler processor [8–11]. The characteristics of the optimal MTI will be derived in this chapter and its performance compared with conventional MTI cancellers derived by using binomial weights.

The Doppler point and MTI processors each represent an extreme in their assumptions relative to the Doppler signal distribution. A third type of optimization criterion would be to divide the Doppler region of interest into equal intervals and provide a separate optimal processor for each region. We call this the *equal-interval processor* and compare its performance with that of the point Doppler processor.

5.1 Optimal Radar Doppler Processors

The structure of the optimal radar Doppler processor consists of a complex-weight transversal filter, followed by an envelope detector, as depicted in Figure 5.1. This

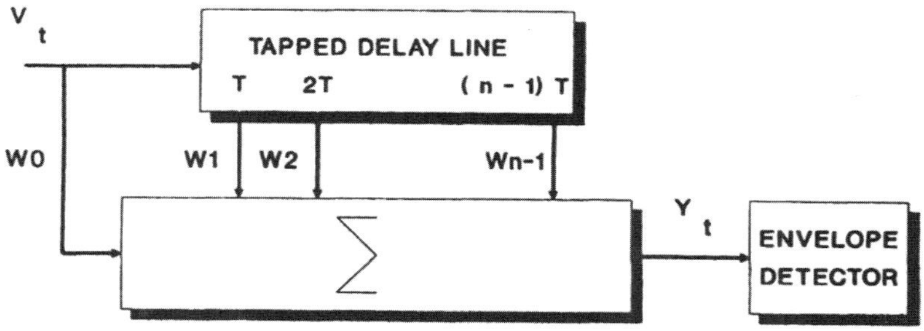

Figure 5.1 Optimal transversal filter processor. (*From:* [7, p. 281].)

structure was derived by using statistical detection theory [7] and, as shown, can be derived equivalently by maximizing the signal-to-interference ratio. The linear nature of the solution (i.e., a transversal filter) is a result of the Gaussian distribution of the interference, and the envelope detector is a consequence of the initially unknown phase of the coherent signal pulse train. The objective of the synthesis procedure for optimum radar Doppler processors is to choose the complex weights of the transversal filter so that the output signal-to-interference ratio is maximized for a specified Doppler input signal distribution.

In our study of optimal radar Doppler signal processors, we first develop some general properties of transversal filters including performance definitions. Then we develop formulas for the performance and complex weights of point, equal-interval, and MTI Doppler processors.

5.1.1 Transversal Filters

The complex output of the transversal filter depicted in Figure 5.2 is given by

$$r = \sum_{i=0}^{n-1} w_i v_i = \mathbf{w}^T \mathbf{v} = \mathbf{v}^T \mathbf{w} \tag{5.1}$$

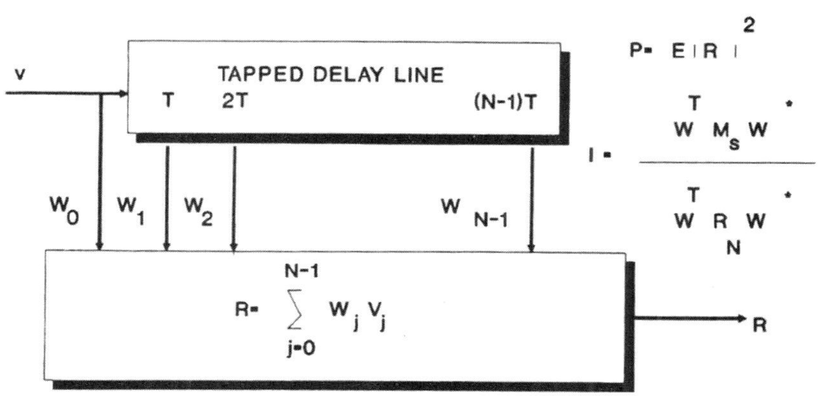

Figure 5.2 Transversal filter. (*From:* [7, p. 282].)

where $\mathbf{w}^T = (w_0, w_1, \ldots, w_{n-1})$ is the complex weight vector, $\mathbf{v}^T = (v_0, v_1, \ldots, v_{n-1})$ is the complex input signal vector, and the superscript T represents a matrix transpose operation. The output power of the transversal filter is given by

$$P_0 = E|\mathbf{r}|^2 = E(\mathbf{r}\mathbf{r}^*) = \mathbf{w}^T E(\mathbf{v}\mathbf{v}^{*T})\mathbf{w}^* \tag{5.2}$$

where the asterisk indicates a complex conjugate operation. When the input is interference the output power is given by

$$P_n = \mathbf{w}^T \mathbf{R}_n \mathbf{w}^* \tag{5.3}$$

where $\mathbf{R}_n$ is the interference's covariance matrix given by

$$\mathbf{R}_n = E(\mathbf{N}\mathbf{N}^{*T}) \tag{5.4}$$

and $\mathbf{N}^T = (N_0, N_1, \ldots, N_{n-1})$ is the complex input noise vector. When the input is a complex signal vector given by $\mathbf{s}e^{j\phi}$, where ϕ is randomly distributed 0 to 2π, and $\mathbf{s}^T = (s_0, s_1, \ldots, s_{n-1})$, the output power is given by

$$P_s = \mathbf{w}^T E(\mathbf{s}e^{j\phi}\mathbf{s}^{*T}\mathbf{e}^{-j\phi})\mathbf{w}^* = \mathbf{w}^T \mathbf{M}_s \mathbf{w}^* \tag{5.5}$$

where $\mathbf{M}_s$ is the signal covariance matrix:

$$\mathbf{M}_s = E(\mathbf{s}\mathbf{s}^{*T}) \tag{5.6}$$

The improvement factor of a Doppler processor is defined as the output S/N ratio divided by the input S/N:

$$I_f = \frac{\dfrac{P_{os}}{P_{on}}}{\dfrac{P_{is}}{P_{in}}} = \frac{\left(\dfrac{\mathrm{S}}{\mathrm{N}}\right)_o}{\left(\dfrac{\mathrm{S}}{\mathrm{N}}\right)_i} \tag{5.7}$$

The power gain of the processor is given by

$$G_s = \frac{P_{os}}{P_{is}} \tag{5.8}$$

and the interference cancellation is given by

$$G_n = \frac{P_{in}}{P_{on}} \tag{5.9}$$

so that the improvement factor can be written as

$$I_f = G_s G_n \tag{5.10}$$

The normalized improvement factor of a transversal filter ($P_{is} = P_{in}$) is given by

$$I_f = \frac{\mathbf{w}^T \mathbf{M}_s \mathbf{w}^*}{\mathbf{w}^T \mathbf{R}_n \mathbf{w}^*} \tag{5.11}$$

The performance of the transversal filter is optimized by finding solutions for the weights ($\mathbf{w}$) of (5.11) that maximize the improvement factor I_f. When I_f is maximized, the signal-to-interference ratio at the output of the transversal filter is maximized. This results in a maximum P_d when the interference is Gaussian distributed. The relationship between P_d and I_f will be developed later in this section.

The maximum improvement factor occurs at a stationary point in (5.11) for which

$$\mathbf{M}_x \mathbf{w}^* = \gamma \mathbf{R}_n \mathbf{w}^* \tag{5.12}$$

where γ is a scalar. Equation (5.12) can be written as a matrix equation:

$$\left(\mathbf{R}_n^{-1} \mathbf{M}_s - \gamma \mathbf{I}\right) \mathbf{w}^* = 0 \tag{5.13}$$

where $\mathbf{R}_n^{-1}$ is the matrix inverse of the covariance matrix and $\mathbf{I}$ is the identity matrix. Nontrivial solutions for (5.13) exist only when the determinant equals zero, as follows:

$$\left|\mathbf{R}_n^{-1} \mathbf{M}_s - \gamma \mathbf{I}\right| = 0 \tag{5.14}$$

In general, for a Doppler processor that processes n pulses there are n solutions for γ, which are called the *eigenvalues* of the matrix $\mathbf{R}_n^{-1} \mathbf{M}_s$. The value of the improvement factor is given by

$$I_f = \gamma_{\max} \tag{5.15}$$

where $\gamma_{\max}$ is the largest eigenvalue. Each eigenvalue has an associated eigenvector that can be found from (5.13). The eigenvector associated with $\gamma_{\max}$ allows determination of the optimum weights ($\mathbf{w}_0$). Equations (5.13) and (5.14) will be used later to find the optimum weights and improvement factors for the point, equal-interval, and MTI Doppler processors.

In general, the weights associated with optimal Doppler processors are complex. This means that both the phase and amplitude of each signal sample must be appropriately weighted in the transversal filter. Frequently, we want to implement the complex transversal filter by using only real operations. This can be accomplished by using a quadrature-channel processor, where both the real and imaginary components of the weights operate on the signal in each channel. Note that MTI Doppler processing requires only real weights, as will be derived in a later section.

Consider the complex transversal filter depicted in Figure 5.3, which is driven by the complex envelope signal $\tilde{v}_t$, and has a complex envelope output $\tilde{r}_i$ and complex weights $\tilde{w}$. Recall that the real input signal v_t is related to the complex envelope by

$$v_t = \mathrm{Re}[\tilde{v}_t e^{j\omega t}] = v_{ct} \cos \omega t + v_{st} \sin \omega t \tag{5.16}$$

where v_{ct} and v_{st} are quadrature baseband signal components. The complex envelope in terms of the quadrature components is given by

$$\tilde{v}_t = v_{ct} - jv_{st} \tag{5.17}$$

and the complex weights are given by

$$\tilde{\mathbf{w}} = \mathbf{w}_R - j\mathbf{w}_I \tag{5.18}$$

where $\mathbf{w}_R$ is the real component and $\mathbf{w}_I$ is the imaginary component. The complex envelope of the output is given by [12]

$$\tilde{r} = \tilde{\mathbf{w}}^T \tilde{\mathbf{v}} \tag{5.19}$$

where $\tilde{\mathbf{w}}^T = (\tilde{w}_0, \tilde{w}_1, \ldots, \tilde{w}_{n-1})$ and $\tilde{\mathbf{v}}^T = (\tilde{v}_0, \tilde{v}_1, \ldots, \tilde{v}_{n-1})$ are vector samples of the complex envelope. Substitution of (5.17) and (5.18) into (5.19) results in the complex-envelope output of the transversal filter in terms of the quadrature input signals and the real and imaginary weight vectors:

$$\tilde{r} = \left(\mathbf{w}_R^T \mathbf{v}_c - \mathbf{w}_I^T \mathbf{v}_s\right) - j\left(\mathbf{w}_R^T \mathbf{v}_s + \mathbf{w}_I^T \mathbf{v}_c\right) \tag{5.20}$$

The square-law envelope-detected output of the transversal filter is given by

$$|\tilde{r}|^2 = \left(\mathbf{w}_R^T \mathbf{v}_c - \mathbf{w}_I^T \mathbf{v}_s\right)^2 + \left(\mathbf{w}_R^T \mathbf{v}_s + \mathbf{w}_I^T \mathbf{v}_c\right)^2 \tag{5.21}$$

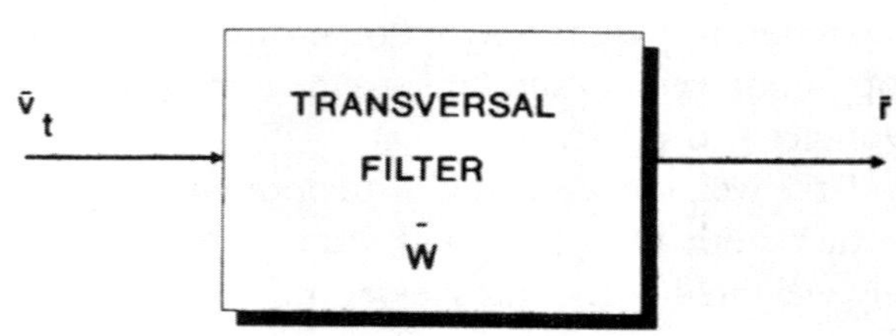

Figure 5.3 Complex transversal filter. (*From:* [7, p. 284].)

A block diagram of the complex transversal filter implemented with real components is depicted in Figure 5.4.

Of considerable interest is the detection performance of transversal filter Doppler processors. For maximum utility, the performance should be based on a *probability of detection* measure, because this will allow comparison with other types of Doppler and non-Doppler processors in a framework of the Marcum-Swerling radar detection analysis. In addition, the relationship between the Doppler processor's improvement factor and its probability of detection should be developed due to the extensive use of the improvement factor to measure the quality of the processor. As previously discussed, the probability of detection is increased as the improvement factor is increased [6].

Consider the real implementation of a complex transversal filter, as depicted in Figure 5.4. The input signal can be represented as

$$v_t = \sqrt{2P_s}\,\cos[(\omega_c + \omega_d)t + \phi] + x_t \cos \omega_c t - y_t \sin \omega_c t \tag{5.22}$$

where x_t and y_t include clutter plus noise components, P_s is the signal power, σ_c^2 is the clutter power, σ_n^2 is the receiver noise power, and ω_d is the target's radian Doppler frequency. The in-phase and quadrature components of the input are given by

$$i_t = \sqrt{2P_s}\,\cos(\omega_d t + \phi) + x_t \tag{5.23}$$

$$q_t = \sqrt{2P_s}\,\sin(\omega_d t + \phi) + y_t \tag{5.24}$$

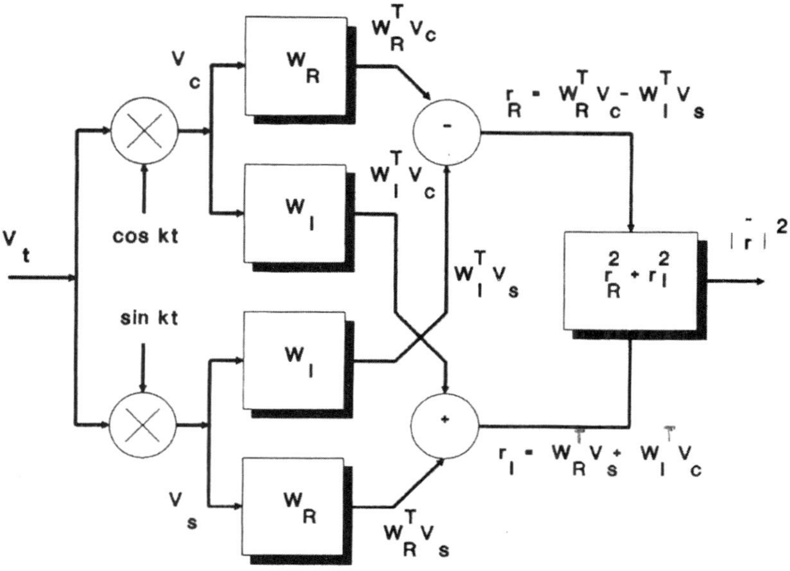

Figure 5.4 Block diagram of complex transversal filter implemented with real operations. (*From:* [7, p. 285].)

The quadrature channel outputs of the transversal filter are given by

$$\alpha = \sum_{j=0}^{n-1} w_{Rj}\, i(t - jT) - \sum_{j=0}^{n-1} w_{Ij}\, q(t - jT) \tag{5.25}$$

$$\beta = \sum_{j=0}^{n-1} w_{Rj}\, q(t - jT) + \sum_{j=0}^{n-1} w_{Ij}\, i(t - jT) \tag{5.26}$$

Now, both α and β are Gaussian-distributed random variables because they result from linear operations on Gaussian-distributed random processes. Therefore, the mean and variance of α and β will specify the joint probability density function $p(\alpha, \beta \mid s)$. The expected values of α and β can be found as

$$\begin{aligned} s_x = E(\alpha) = &\sum_{j=0}^{n-1} w_{Rj}\sqrt{2P_s}\cos[\omega_d(t - jT) + \phi] \\ &- \sum_{j=0}^{n-1} w_{Ij}\sqrt{2P_s}\sin[\omega_d(t - jT) + \phi] \end{aligned} \tag{5.27}$$

$$\begin{aligned} s_y = E(\beta) = &\sum_{j=0}^{n-1} w_{Rj}\sqrt{2P_s}\sin[\omega_d(t - jT) + \phi] \\ &+ \sum_{j=0}^{n-1} w_{Ij}\sqrt{2P_s}\cos[\omega_d(t - jT) + \phi] \end{aligned} \tag{5.28}$$

The variance of α or β is given by

$$\sigma_{\alpha\beta}^2 = \sigma_n^2 \sum_{j=0}^{n-1} \left(w_{Rj}^2 + w_{Ij}^2\right) + \sigma_c^2 \sum_{j=0}^{n-1}\sum_{i=0}^{n-1} (w_{Rj}w_{Ri} + w_{Ij}w_{Ii})\rho(i - j) \tag{5.29}$$

where ρ is the clutter's correlation coefficient, which for a Gaussian-shaped spectrum is

$$\rho(i - j) = \exp\left[\frac{(i - j)^2\Omega^2}{2}\right] = \rho(1)^{(i-j)^2} \tag{5.30}$$

and

$$\Omega = 2\pi\phi_f T \tag{5.31}$$

The power out of the transversal filter is given by

$$P_o = E(\tilde{\mathbf{r}}\tilde{\mathbf{r}}^*) = 2P_s\left[\sum_{j=0}^{n-1} \left(w_{Rj}^2 + w_{Ij}^2\right)\right] \tag{5.32}$$

The average power gain from (5.8) can be found as

$$\overline{G}_s = \sum_{j=0}^{n-1} \left(w_{Rj}^2 + w_{Ij}^2\right) \tag{5.33}$$

The interference cancellation [see (5.9)] for only the clutter component of the interference, called *clutter attenuation*, CA, is given by

$$\text{CA} = \sum_{j=0}^{n-1} \sum_{i=0}^{n-1} (w_{Rj} w_{Ri} + w_{Ij} w_{Ii}) \rho(i-j) \tag{5.34}$$

The normalized improvement factor for clutter alone ($2P_s = \sigma_c^2$) using (5.10) can be found as

$$I_{fc} = \frac{\sum_{j=0}^{n-1} \left(w_{Rj}^2 + w_{Ij}^2\right)}{\sum_{j=0}^{n-1} \sum_{i=0}^{n-1} (w_{Rj} w_{Ri} + w_{Ij} w_{Ii}) \rho(1-j)} \tag{5.35}$$

We further define the clutter-to-noise power ratio as

$$F = \frac{\sigma_c^2}{\sigma_n^2} \tag{5.36}$$

and we have

$$\sigma_{\alpha\beta}^2 = \sigma_n^2 \overline{G}_s \left(1 + \frac{F}{I_{fc}}\right) \tag{5.37}$$

Equation (5.37) states that the variance of the Gaussian-distributed interference at the output of the transversal filter is simply the sum of the receiver noise power plus that of the residual clutter power multiplied by the power gain of the processor. The residual clutter power can be found by dividing the input clutter power by the clutter improvement factor, I_{fc}. Because the processor power gain (G_s) multiplies both the signal and interference powers, we would expect this factor to drop out when the signal-to-interference ratio is formed.

An interesting relation is the normalized improvement factor (I_{fc}) for clutter alone as a function of that for the total interference. This relationship can be developed as

$$I_f = \frac{1+F}{F} \frac{I_{fc}}{1 + \frac{I_{fc}}{F}} \tag{5.38}$$

The overall improvement factor I_f tends to approach that for clutter alone (I_{fc}) as F becomes large. For I_{fc} that is very large compared to F, the overall improvement factor I_f approaches $F + 1$, indicating a maximum improvement relative to the clutter-to-noise power ratio.

The next step in the development of the probability of detection for the complex transversal filter is to determine the correlation between the quadrature channel outputs α and β. This can be determined as

$$\rho_{\alpha\beta} = E[(\alpha - s_x)(\beta - s_y)] = 0 \tag{5.39}$$

which follows from the basic property of the interference representation $E(x_t y_t) = 0$. Equation (5.39) specifies that the random variables α and β are uncorrelated, and as they are Gaussian, they are independent. Therefore, the joint density function can be written as

$$p(\alpha, \beta \mid s) = \frac{1}{2\pi\sigma_{\alpha\beta}^2} \exp\left[-\frac{(\alpha - s_x)^2 + (\beta - s_y)^2}{2\sigma_{\alpha\beta}^2}\right] \tag{5.40}$$

The envelope (r) is found by using $\alpha = r \cos\theta$ and $\beta = r \sin\theta$ resulting in

$$p(r \mid s) = \frac{r}{\sigma_{\alpha\beta}^2} \exp\left(-\frac{r^2 + s^2}{2\sigma_{\alpha\beta}^2}\right) I_0\left(\frac{rs}{\sigma_{\alpha\beta}^2}\right) \tag{5.41}$$

The square-law detected output is found by applying the transformation $z = r^2$, resulting in

$$p(z \mid s) = \frac{1}{2\sigma_{\alpha\beta}^2} \exp\left(-\frac{z + s^2}{2\sigma_{\alpha\beta}^2}\right) I_0\left(\frac{\sqrt{z}s}{\sigma_{\alpha\beta}^2}\right) \tag{5.42}$$

where $s^2 = s_x^2 + s_y^2 = 2\overline{G}_s P_s$.

The detection threshold T_d is found from (5.42) with $s = 0$ to be

$$T_d = 2\sigma_{\alpha\beta}^2 \ln\left(\frac{1}{P_{fa}}\right) \tag{5.43}$$

where P_{fa} is the false alarm probability. The probability of detection can be found from (5.40), (5.41), and (5.37) to be

$$P_d = \mathrm{Q}\left\{\left[\frac{2P_s}{\sigma_n^2\left(1 + \frac{F}{I_{fc}}\right)}\right]^{1/2}, \left[2 \ln\left(\frac{1}{P_{fa}}\right)\right]^{1/2}\right\} \tag{5.44}$$

where $\mathrm{Q}\{\cdot,\cdot\}$ is Marcum's Q function.

Equation (5.44) relates the complex transversal filter's probability of detection to the improvement factor for clutter alone, given by (5.35). Examination of (5.44) shows that P_d increases as the improvement factor increases. Also, this relation reveals the intuitively satisfying result that the detection probability depends on the ratio of signal to residual interference, where the residual interference is the receiver noise power plus the clutter power divided by the clutter improvement factor.

An alternative solution for the complex transversal filter's probability of detection can be found by solving (5.38) for I_{fc} in terms of I_f and substituting into (5.44):

$$P_d = Q\left\{\left[\frac{2P_s}{\sigma_n^2(1+F)/I_f}\right]^{1/2}, \left[2\ \ln\left(\frac{1}{P_{fa}}\right)\right]^{1/2}\right\} \tag{5.45}$$

This equation illustrates that the complex transversal filter is effective against both clutter and noise. Equation (5.45) is appropriate when the Doppler band of interest is limited, thereby improving the S/N ratio, whereas (5.44) is used for MTI transversal filters, for which there is no S/N ratio improvement.

5.1.2 The Point Doppler Processor

The point Doppler transversal filter is optimized at a point contained within the desired Doppler processing band and is nonoptimum or mismatched at all other points in the Doppler band. A Doppler processor is formed by using a number of point transversal filters to fill the Doppler region of interest. Any number of transversal filters can be used, although in practice the band is filled with n equally spaced filters, where n is the number of coherent pulses processed. This situation is depicted in Figure 5.5, and an example is presented in this section.

The complex input signal to the point Doppler transversal filter is given by

$$s_i = Ae^{j\omega_d t} \sum_{i=0}^{n-1} \delta(t - iT) \tag{5.46}$$

where A is the amplitude, ω_d is the radian Doppler frequency, n is the number of pulses processed, and T is the radar's interpulse interval. The signal vector is then $\mathbf{s}^T = (s_1, s_2, \ldots, s_n)$, where

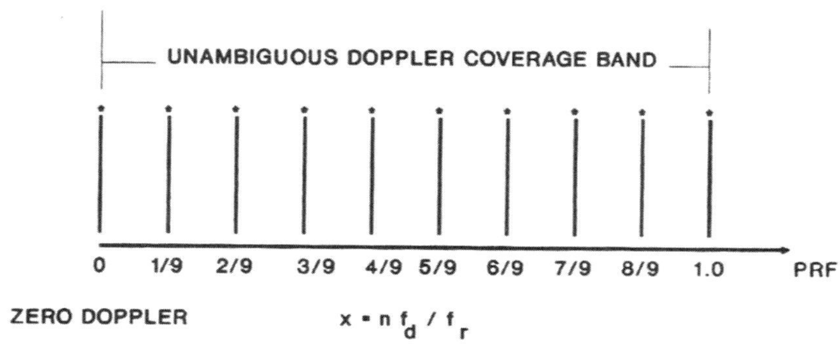

Figure 5.5 Design criterion for a point-optimized Doppler processor. (*From:* [7, p. 291].)

$$s_k = Ae^{j\omega_d(k-1)T}; \qquad k = 1, 2, \ldots, n \tag{5.47}$$

The normalized (divided by A^2) signal covariance matrix can be found by using (5.6) as

$$\mathbf{M}_s = \begin{pmatrix} 1 & e^{-j\omega_d T} & \ldots & e^{-j\omega_d(n-1)T} \\ & \ddots & & \vdots \\ e^{j\omega_d T} & & \ddots & \vdots \\ e^{j\omega_d(n-1)T} & \ldots & \ldots & 1 \end{pmatrix} \tag{5.48}$$

An element (m_s) of the normalized covariance matrix can be found as

$$m_{s_{kj}} = e^{j\omega_d T(k-1)} e^{-j\omega_d T(j-1)} = e^{j\omega_d(k-j)T} \tag{5.49}$$

where k is the row number ($k = 1, 2, \ldots, n$) and j is the column number ($j = 1, 2, \ldots, n$).

The optimal weights can be found as the eigenvector that corresponds to the maximum eigenvalue of $\mathbf{R}_n^{-1}\mathbf{M}_s$ [see (5.13)]. However, solution of (5.13) is not necessary because it can be determined, using statistical detection theory, that the optimum weights are

$$w_o = \mathbf{R}_n^{-1}\mathbf{s}^* \tag{5.50}$$

We can test this solution for the optimal weights by confirming that those given by (5.50) provide the maximum improvement factor. Substitution of (5.50) into (5.11) for the improvement factor and use of the solution $\left(\mathbf{R}_n^{-1}\mathbf{M}_s = \gamma_{\max}\mathbf{I}\right)$ will result in

$$I_f = \frac{\mathbf{w}_o^T\mathbf{M}_s\mathbf{w}_o^*}{\mathbf{w}_o^T\mathbf{R}_n\mathbf{w}_o^*} = \frac{\mathbf{w}_o^T\overbrace{\mathbf{M}_s\left(\mathbf{R}_n^{-1}\mathbf{s}^*\right)^*}^{I\gamma_{\max}}}{\mathbf{w}_o^T\underbrace{\mathbf{R}_n\left(\mathbf{R}_n^{-1}\mathbf{s}^*\right)^*}_{\mathbf{I}}} = \gamma_{\max} \tag{5.51}$$

Equation (5.51) confirms that the maximum likelihood weights given by (5.50) do, in fact, provide the maximum improvement factor. This is just another demonstration that, for Gaussian interference, the maximum likelihood and the maximum signal-to-interference ratio criteria are equivalent.

In addition to the optimal weights, we need to determine an expression for the improvement factor. An expression can be found by using a number of matrix theorems [6]. First, observe that the rank of the signal covariance matrix $\mathbf{M}_s$ is 1 because this is the order of the largest nonzero determinant in the matrix. This means that the rank of $\mathbf{R}_n^{-1}\mathbf{M}_s$ is one because the rank of a product of two matrices is equal to or less than the rank of either matrix. Because the number of nonzero eigenvalues is equal to the rank of a matrix, $\mathbf{R}_n^{-1}\mathbf{M}_s$ has only one nonzero eigenvalue that is real (complex eigenvalues are in conjugate pairs). The trace of a matrix is equal to the sum of all its eigenvalues, and because in this case there is only one eigenvalue, it must be equal to the improvement factor; therefore,

$$I_f = \text{Trace}\left(\mathbf{R}_n^{-1}\mathbf{M}_s\right) \tag{5.52}$$

Note that the improvement factor given by (5.52) applies only for the point at which the transversal filter is matched.

A general expression for the optimum weights and improvement factor of the point-optimized transversal filter can be derived in terms of the inverse interference covariance matrix $\mathbf{R}_n^{-1}$. Generally, we must invert the interference covariance matrix $\mathbf{R}_n$, which is given by

$$\mathbf{R}_n = \sigma_n^2 \begin{pmatrix} 1+F & F\rho_1 & \cdots & \cdots & F\rho_{n-1} \\ F\rho_1 & \ddots & & & \vdots \\ \vdots & & \ddots & & \vdots \\ \vdots & & & \ddots & \vdots \\ F\rho_{n-1} & \cdots & \cdots & \cdots & 1+F \end{pmatrix} \tag{5.53}$$

using an appropriate computer program. However, for a small number of samples ($n \leq 3$), we can obtain an analytic solution.

If the resulting elements of the inverse covariance matrix are given by α_{ij}:

$$\mathbf{R}_n^{-1} = \begin{pmatrix} \alpha_{11} & \alpha_{12} & \cdots & \cdots & \alpha_{ln} \\ \alpha_{12} & \ddots & & & \vdots \\ \vdots & & \ddots & & \vdots \\ \vdots & & & \ddots & \vdots \\ \alpha_{1n} & \cdots & \cdots & \cdots & \alpha_{11} \end{pmatrix} \tag{5.54}$$

then the optimum complex weights given by (5.50) are

$$w_{i-1} = \sum_{j=1}^{n} \alpha_{ij} \cos(j-1)\,\omega_d T - j \sum_{i=1}^{n} \alpha_{ij} \sin(j-1)\,\omega_d T \tag{5.55}$$

whereas the improvement factor given by (5.11) is

$$I_f = \frac{\sum_{i=0}^{n-1} |w_i|^2 + 2 \sum_{i=0}^{n-2} \sum_{j=i+1}^{n-1} \left[\text{Re}\left(w_i^* w_j\right) \cos(j-1)\,\omega_d T - \text{Im}\left(w_i^* w_j\right) \sin(j-i)\,\omega_d T\right]}{\sum_{i=0}^{n-1} |w_i|^2 + 2 \sum_{i=0}^{n-2} \sum_{j=i+1}^{n-1} \rho(i-j)\, \text{Re}\left(w_i^* w_j\right)} \tag{5.56}$$

where

$$\text{Re}\left(w_i^* w_j\right) = w_{Ri} w_{Rj} + w_{Ii} w_{Ij} \tag{5.57}$$

$$\text{Im}\left(w_i^* w_j\right) = w_{Ri} w_{Ij} - w_{Ii} w_{Ri} \tag{5.58}$$

Equation (5.56) is a general expression that allows computing the improvement factor at any Doppler frequency ω_d. However, remember that the weights in (5.56) must be computed at the point Doppler frequency where the transversal filter is optimized. Equation (5.52) can be utilized to find the improvement factor at the optimum point:

$$I_{fo} = \sum_{i=1}^{n} \alpha_{ij} + 2 \sum_{i=1}^{n-1} \sum_{j=i+1}^{n} \alpha_{ij} \cos(j-i)\,\omega_d T \tag{5.59}$$

Figure 5.6 depicts the results of an improvement factor calculation for a nine-pulse point-optimized Doppler processor. The curve labeled *optimum envelope* is derived using (5.59) and represents the maximum improvement factor that can be obtained by an optimized transversal filter centered at that Doppler frequency. The improvement factors are provided for two transversal filters obtained using (5.56). The filters are optimized at Doppler frequencies equal to 2/9 and 3/9 of the radar's PRF. The two transversal filters cross over at a point where the improvement factor is approximately –3 dB below the peak. The probability of detection for the transversal filter at each Doppler frequency can be found from (5.45) by using the appropriate improvement factor. Note that when a significant response occurs in

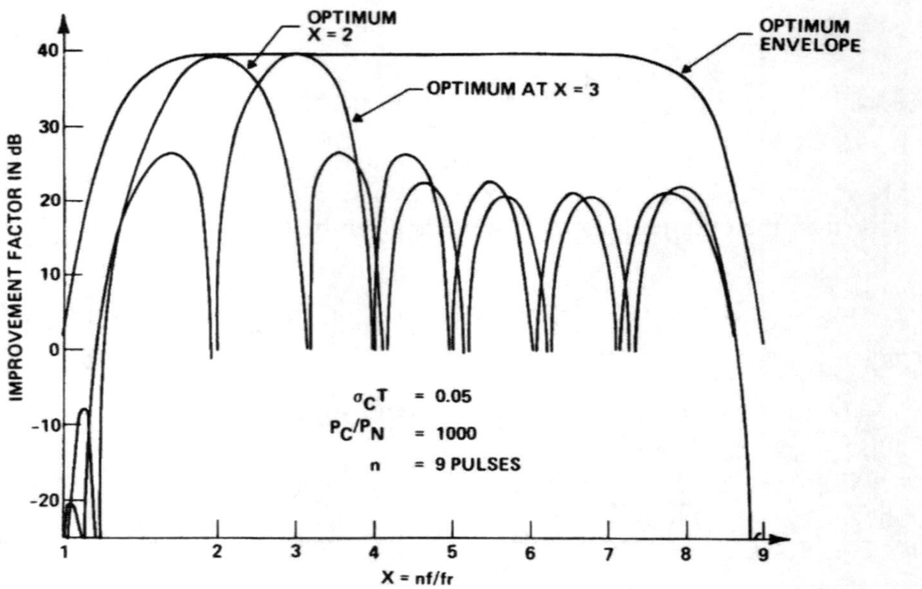

Figure 5.6 Improvement factor for point-optimized transversal filters ($x = 2, 3$). (*From:* [7, p. 293].)

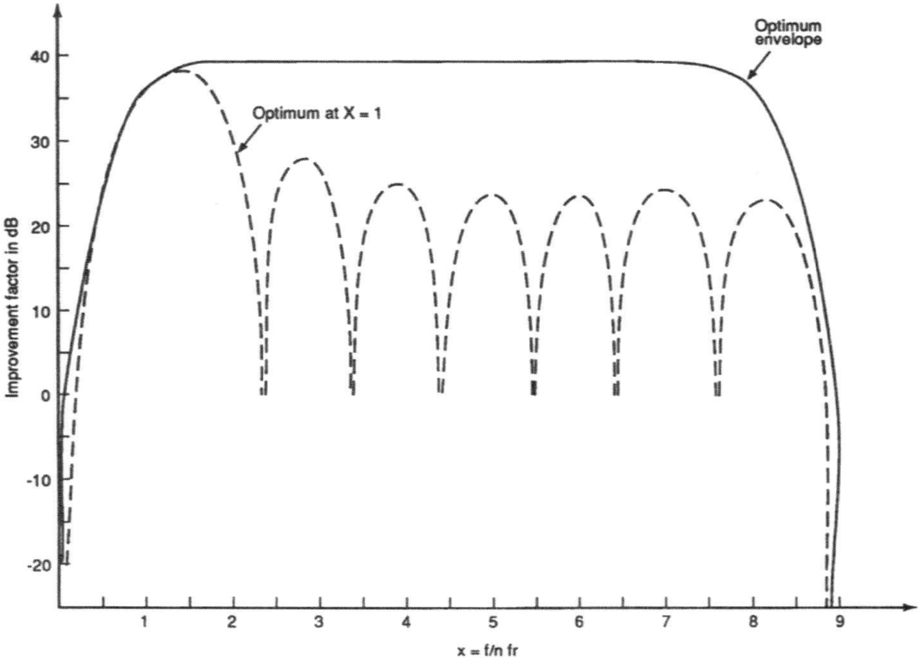

Figure 5.7 Improvement factor for point-optimized transversal filter ($x = 1$). (*From:* [7, p. 294].)

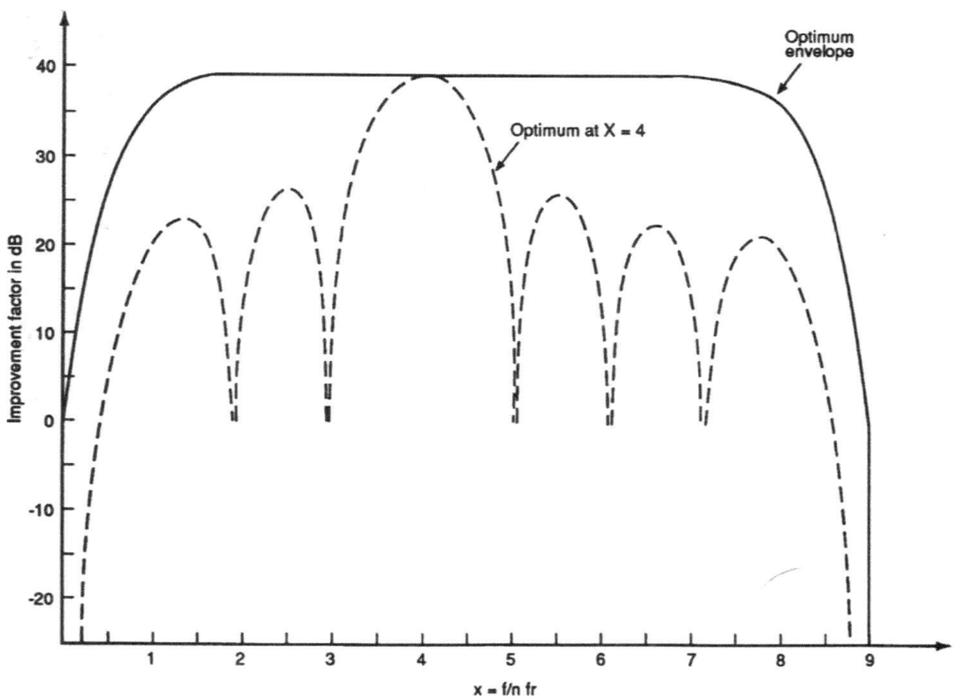

Figure 5.8 Improvement factor for point-optimized transversal filter ($x = 4$). (*From:* [7, p. 296].)

two contiguous filters, the joint probabilities of detection must be considered in determining the probability of detection at that point.

Figure 5.6 illustrates the characteristics of transversal filters that provide both S/C ratio and S/N ratio improvement. Note the deep nulls provided in the region about the PRF lines, where the heaviest clutter components are concentrated. Also, the width of the main lobe is approximately matched to the signal spectrum, which is assumed (in the design of the filter) to be centered about the point where the transversal filter is optimized.

Figure 5.7 illustrates a design optimized about a Doppler frequency that is 1/9 the radar's PRF. Note that the filter's response lies within the optimal improvement factor envelope, as expected. However, the conflicting requirements of maximizing the signal response in a region close to the heavy clutter region while minimizing the clutter response have caused a compromise in the target response: The main lobe width is reduced and the peak improvement factor is reduced as well.

Figure 5.8 illustrates a design optimized at a Doppler frequency that is 4/9 the radar's PRF. Note that the filter's sidelobe response in the vicinity of the heavy clutter region is reduced further from that of the previously illustrated designs (Figures 5.6 and 5.7), which were optimized at points close to the heavy clutter region.

5.1.3 The Optimized MTI Processor

The basic assumption of the MTI processor is that there is no *a priori* knowledge of the Doppler shift of the target return. The resulting target Doppler probability distribution is given by

$$p(f_d) = \begin{cases} T, & 0 < f_d < \dfrac{1}{T} \\ 0, & \text{otherwise} \end{cases} \tag{5.60}$$

The signal covariance matrix can be found by averaging the signal covariance matrix for each Doppler frequency [see (5.49)] with respect to the probability distribution of expected Doppler frequencies [see (5.60)]. The resulting normalized signal covariance matrix element is given by

$$m_{s_{kj}} = e^{j\pi(k-j)} \frac{\sin \pi(k-j)}{\pi(k-j)} \tag{5.61}$$

where k is the row number and j is the column number. Equation (5.61) for $m_{s_{kj}}$ is equal to 1 when $k = j$ and 0 when $k \neq j$. Therefore, the signal covariance matrix for the MTI case is

$$\mathbf{M}_s = \mathbf{I} \tag{5.62}$$

where $\mathbf{I}$ is the identity matrix.

The solution for the optimum weights is then given by (5.12), which can be written in the form:

$$(\mathbf{R}_n - \lambda \mathbf{I})\mathbf{w}^* = 0 \tag{5.63}$$

where λ are the scalar eigenvalues of $\mathbf{R}_n$. The improvement factor in this case is obtained from (5.11) and is given by

$$I_f = \frac{1}{\lambda_{\min}} \tag{5.64}$$

where the minimum eigenvalue is chosen to maximize I_f. The optimum weights are then found by solving (5.63), using the minimum eigenvalue $\lambda_{\min}$. To summarize, the optimum weights ($\mathbf{w}_o$) constitute the eigenvector that corresponds to the minimum eigenvalue:

$$(\mathbf{R}_n - \lambda_{\min}\mathbf{I})\mathbf{w}_o^* = 0 \tag{5.65}$$

The eigenvalues of $\mathbf{R}_n$ must be real because $\mathbf{R}_n$ is a real symmetric matrix [13]. Furthermore, this means that the optimum weights ($\mathbf{w}_o$) can always be taken as real [14], which simplifies the structure of the transversal filter depicted in Figure 5.4 to the form with only real weights depicted in Figure 5.9. This form is commonly referred to as a *quadrature channel MTI*.

The interference covariance matrix $\mathbf{R}_n$ given by (5.53) can be written as the sum of a clutter covariance matrix, $\mathbf{R}_c$, and a receiver noise matrix:

$$\mathbf{R}_n = \mathbf{R}_c + \sigma_n^2\mathbf{I} \tag{5.66}$$

where

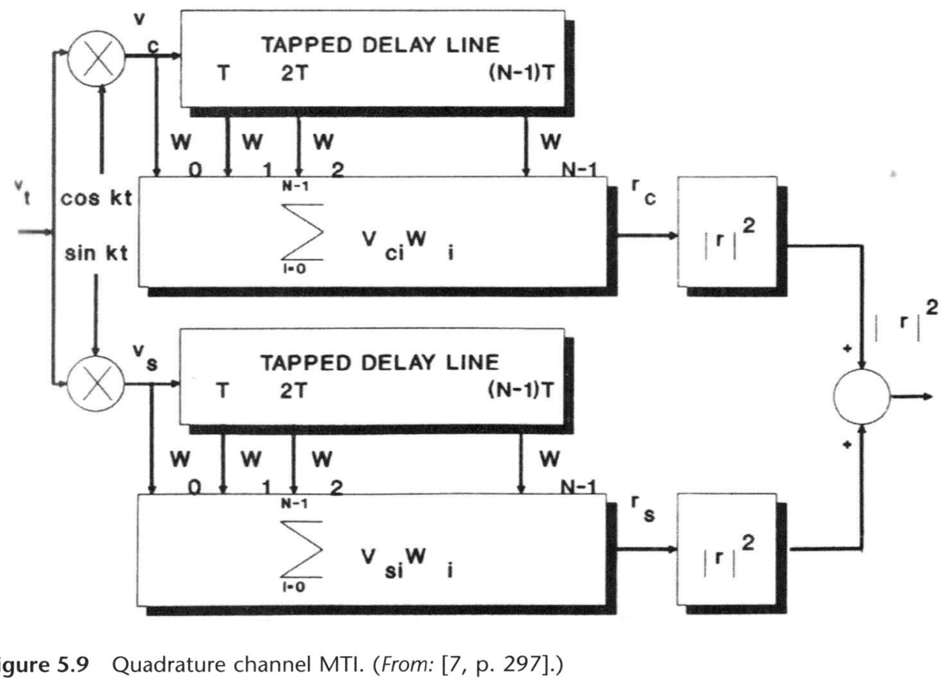

Figure 5.9 Quadrature channel MTI. (*From:* [7, p. 297].)

$$\mathbf{R}_c = \sigma_c^2 \begin{pmatrix} 1 & \rho_1 & \cdots & \cdots & \rho_{n-1} \\ \rho_1 & \ddots & & & \vdots \\ \vdots & & \ddots & & \vdots \\ \vdots & & & \ddots & \vdots \\ \rho_{n-1} & \cdots & \cdots & \cdots & 1 \end{pmatrix} \tag{5.67}$$

Substitution of (5.66) into the solution (5.65) results in

$$(\mathbf{R}_c - \lambda_c \mathbf{I})\mathbf{w}_o^* = 0 \tag{5.68}$$

where the eigenvalues of $\mathbf{R}_n$ are

$$\lambda_n = \frac{1}{F+1}(F\lambda_c + 1) \tag{5.69}$$

and λ_c are the eigenvalues of $\mathbf{R}_c$. The improvement factor given by (5.64) can be rewritten in terms of (5.69) as

$$I_f = \frac{F+1}{\dfrac{F}{I_{fc}} + 1} \tag{5.70}$$

From (5.70), we can see that the weights that maximize I_{fc} also maximize I_f. Hence, solving (5.68) for the optimum weights by using the clutter covariance matrix is equivalent to solving (5.38) for the optimum weights by using the interference covariance matrix. Also, as F approaches 0, I_f approaches 1, indicating that the MTI provides no improvement against receiver noise. In addition, as the clutter improvement factor becomes large, the overall improvement factor approaches $F + 1$.

The preceding discussion indicates that the MTI design can be optimized on the basis of only clutter considerations (5.68), without any loss of generality. Therefore, we will use this approach in determining the MTI weights and improvement factors that follow. The approach is consistent with much of the literature on MTI and focuses attention on the main function of the MTI, which is to eliminate clutter.

The improvement factor for the quadrature channel MTI is given by (5.35) specialized to real weights:

$$I_{fc} = \frac{\sum_{j=0}^{n-1} w_j^*}{\sum_{i=0}^{n-1}\sum_{j=0}^{n-1} w_i w_j \rho(i-j)} \tag{5.71}$$

Specialized formulas for the optimal weights and associated improvement factors can be derived by using (5.63) for $n = 2, 3, 4$ [8, 10, 11]; these are given in

Table 5.1. Beyond $n = 4$, solving (5.68) analytically becomes difficult, and computer solutions thus become appropriate.

The solutions given by (5.68) for the optimum MTI have a number of undesirable qualities that tend to restrict their practical utility [15]. First, the MTI filter's transfer function becomes increasingly narrow as the number of processed pulses increases, asymptotically approaching an impulse function at the minimum value of the clutter spectrum. This tendency is depicted in Figure 5.10, which illustrates transfer functions for up to 10 pulses. The improvement factor also increases with the number of processed pulses. This feature is illustrated in Figure 5.11, which depicts improvement factors for up to 16 processed pulses and various clutter spectral spreads.

Second, to have the sum of the filter weights equal to 0 is desirable, which eliminates fixed, discrete targets. Table 5.1 indicates that this condition is satisfied only when the optimized MTI processes an even number of pulses.

An alternative optimization procedure using quadratic programming [16] allows excess improvement factor to be traded for the flexibility of specifying a desired MTI velocity response [15]. An example is depicted in Figure 5.12, which provides an improvement factor in excess of 30 dB with a significantly widened filter response that is essentially flat in the MTI's passband. The quadratic programming optimization procedure also allows for specifying that the sum of MTI filter weights equal 0 regardless of the number of processed pulses.

The improvement factor for binomial weights is given in Figure 5.11 in addition to that for the optimized MTI design. For a narrow clutter spectrum, the performance achieved with binomial weights approaches that for the optimized design. The discrepancy becomes greater as the number of processed pulses and the clutter spectrum increase. For a small number of processed pulses ($n \leq 5$), the simplicity of using binomial weights is usually traded for the relatively modest improvement factor gain that can be achieved by using an optimized MTI design.

The binomial weights for an MTI design are given by

$$w_k = (-1)^k \binom{n}{k}; \qquad k = 0, 1, \ldots, n-1 \tag{5.72}$$

Table 5.1 MTI Optimum Weights and Improvement Factors

Pulse	*Weights*	*Improvement Factor*
2	$1, -1$	$\dfrac{1}{1-\rho_1}$
3	$1; -\dfrac{\rho_2 + \sqrt{\rho_2^2 + 8\rho_1^2}}{2\rho_1}; 1$	$\dfrac{1}{1 - \dfrac{\rho_2}{2}\left[\sqrt{1 + 8\left(\dfrac{\rho_1}{\rho_2}\right)^2} - 1\right]}$
4	$w_1 = \dfrac{\sqrt{\dfrac{(\rho_1-\rho_3)^2}{4} + (\rho_1-\rho_2)^2} + \dfrac{(\rho_1-\rho_3)}{2}}{\rho_1-\rho_2}$ $1; w_1; -w_1; -1$	$\dfrac{1}{1 - \dfrac{\rho_1+\rho_2}{2} - \sqrt{\dfrac{(\rho_1-\rho_3)^2}{4} + (\rho_1-\rho_2)^2}}$

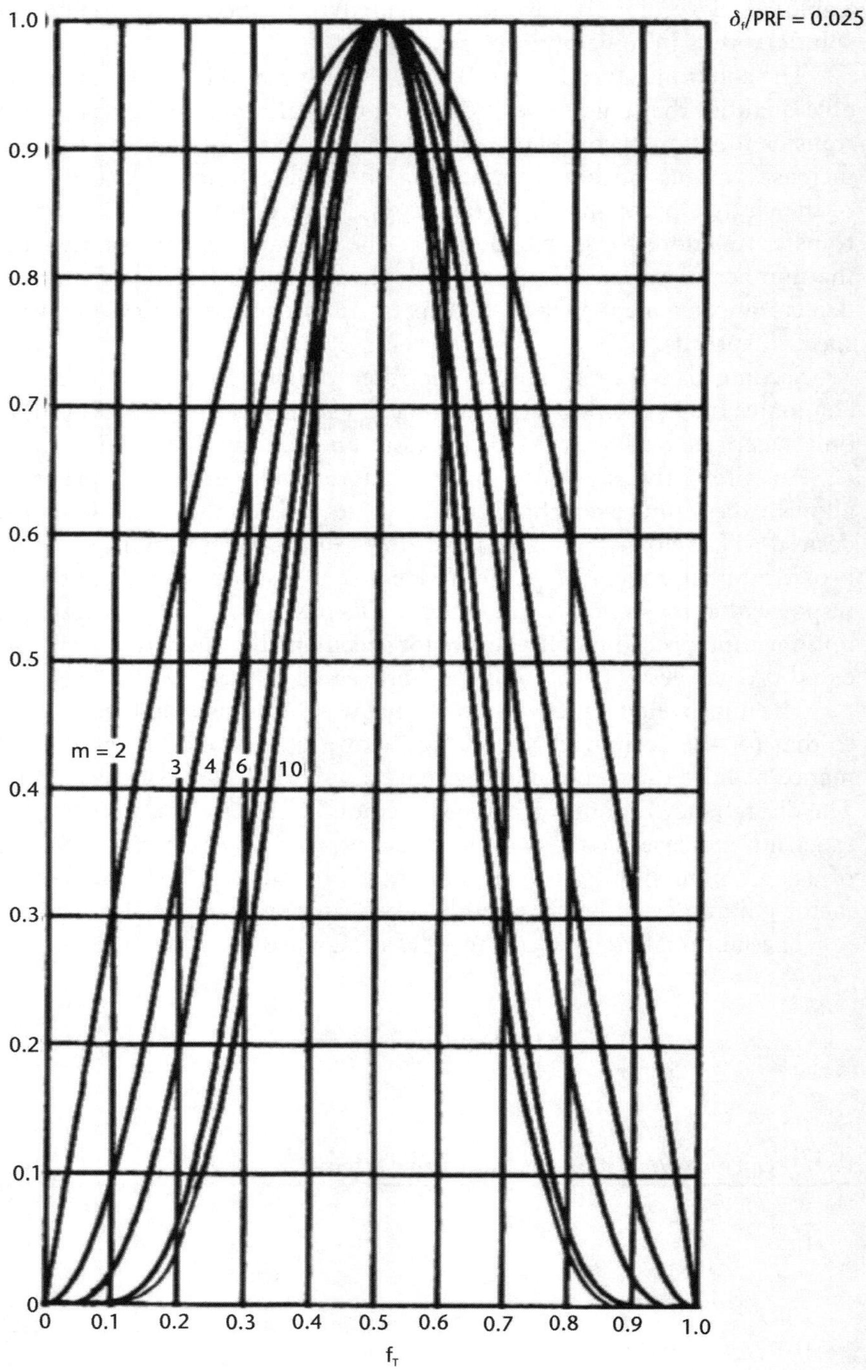

Figure 5.10 Velocity responses for optimum MTI. (*From:* [7, p. 300].)

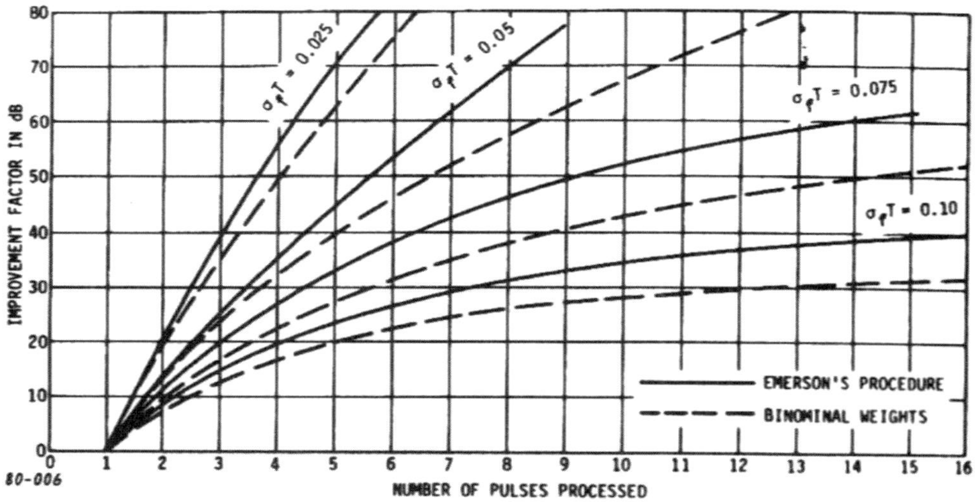

Figure 5.11 Improvement factor for optimum and binomial MTI. (*From:* [6, p. 302].)

The improvement factor is given by

$$I_{fc} = \frac{\sum_{j=0}^{n-1} \binom{n}{j}^2}{\sum_{j=0}^{n-1} \sum_{k=0}^{n-1} (-1)^{k+j} \binom{n}{j} \binom{n}{k} \rho(j-k)} \tag{5.73}$$

which also can be written as [8]

$$I_{fc} = \frac{1}{1 - 2\frac{n}{n+1}\rho_1 + 2\frac{n}{n+1}\frac{n-1}{n-2}\rho_2 + \ldots} \approx \frac{2^n}{n!(\Omega)^{2n}} \tag{5.74}$$

where the approximation applies for $\Omega = 2\pi\sigma_f T \ll 1$. The average power gain is given by

$$\overline{G} = 1 + n^2 + \left[\frac{n(n-1)}{2!}\right]^2 + \left[\frac{n(n-1)(n-2)}{3!}\right]^2 + \ldots \tag{5.75}$$

The binomial weights represent an MTI design that consists of a cascade of $n - 1$ single MTI cancellers (single delay and subtract MTIs). A popular Doppler processor design uses a single or double MTI canceller to attenuate the clutter, followed by a coherent or incoherent integrator to build up the S/N ratio.

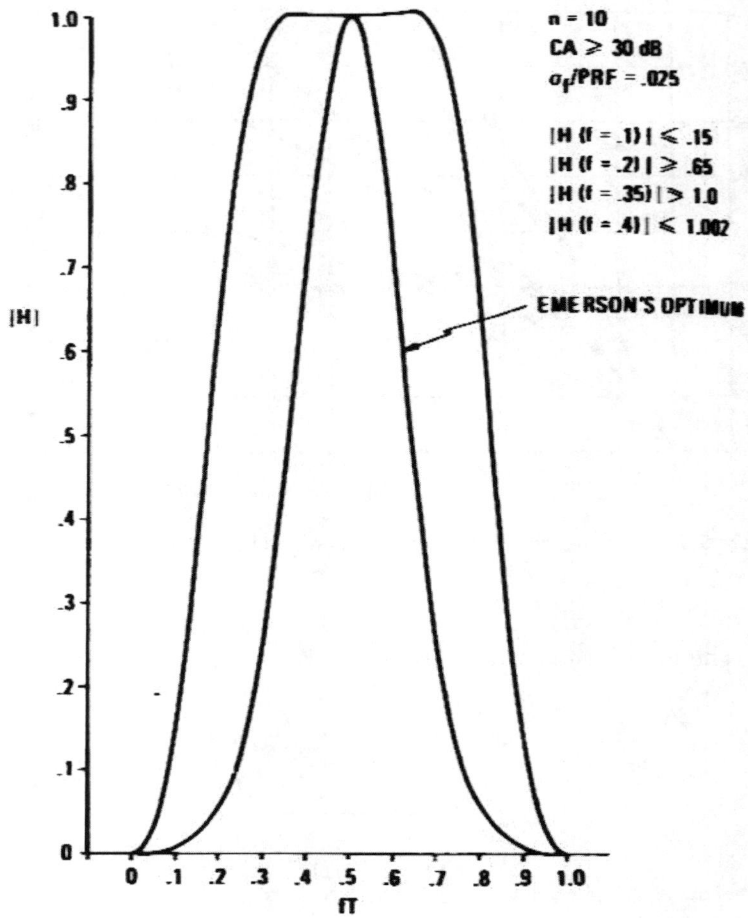

Figure 5.12 MTI velocity response using quadrature programming. (*From:* [15, p. 176].)

► Example 17 MTI Improvement Factor for Optimum and Binomial Weights

MTI binomial weights are given by

$$w_k = (-1)^k \frac{n!}{k!(n-k)!}$$

where n is the number of MTI delay sections and k is the weight number 0, 1, 2, . . . , n. The MTI improvement factor for binomial weights is given by

$$I = \frac{\mathbf{w}^T \mathbf{M}_S \mathbf{w}^*}{\mathbf{w}^T \mathbf{R}_N \mathbf{w}^*} \qquad \mathbf{w}^T = [w_0, w_1, \ldots, w_n]$$

$$\mathbf{M}_S = \begin{bmatrix} 1 & 0 & 0 \\ 0 & 1 & 0 \\ 0 & 0 & 1 \end{bmatrix} \qquad \mathbf{R}_N = \begin{bmatrix} 1 & \rho_1 & \rho_2 \\ \rho_1 & 1 & \rho_1 \\ \rho_2 & \rho_1 & 1 \end{bmatrix}$$

$$\rho(iT) = \exp\left(\frac{-i^2\Omega^2}{2}\right) \qquad \Omega = \frac{2\pi\sigma_f}{\text{PRF}}$$

where $\mathbf{w}^T$ is the transpose of the weight vector, σ_f is the standard deviation of the Gaussian-shaped clutter spectrum, PRF is the pulse repetition frequency of the radar, $\mathbf{M}_S$ is the signal covariance matrix and $\mathbf{R}_N$ is the clutter's covariance matrix.

The optimal weights can be determined by finding the minimum eigenvalue of $\mathbf{R}_n$. This can be accomplished by using the MATLAB function $\lambda = \min(\text{eig}(\text{Rn}))$. Then the optimal improvement factor is given by

$$I = \frac{1}{\lambda}$$

The MATLAB program impfac_comp.m plots improvement factors for both binomial and optimum weights as a function of $x = \sigma_f/\text{PRF}$ for $n = 1$ to 5. The plot is depicted in Figure EX-17.1.

The performance limitation for surface-based radars is generally due to antenna scanning that spreads the clutter spectrum (see Section 3.2.2). The improvement factor for this case is given by:

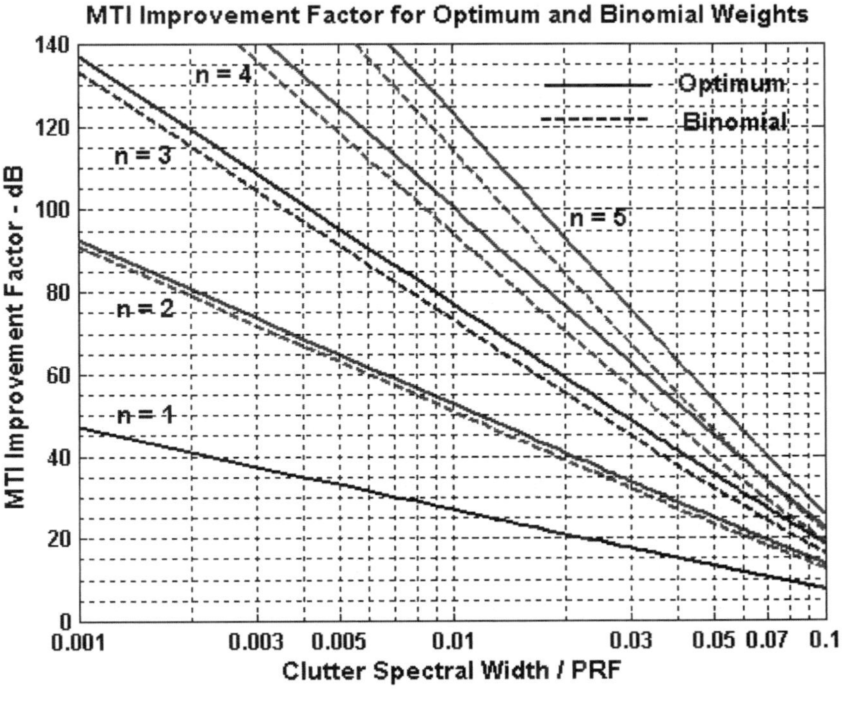

Figure EX-17.1 Improvement factor for binomial and optimum MTI cancellers.

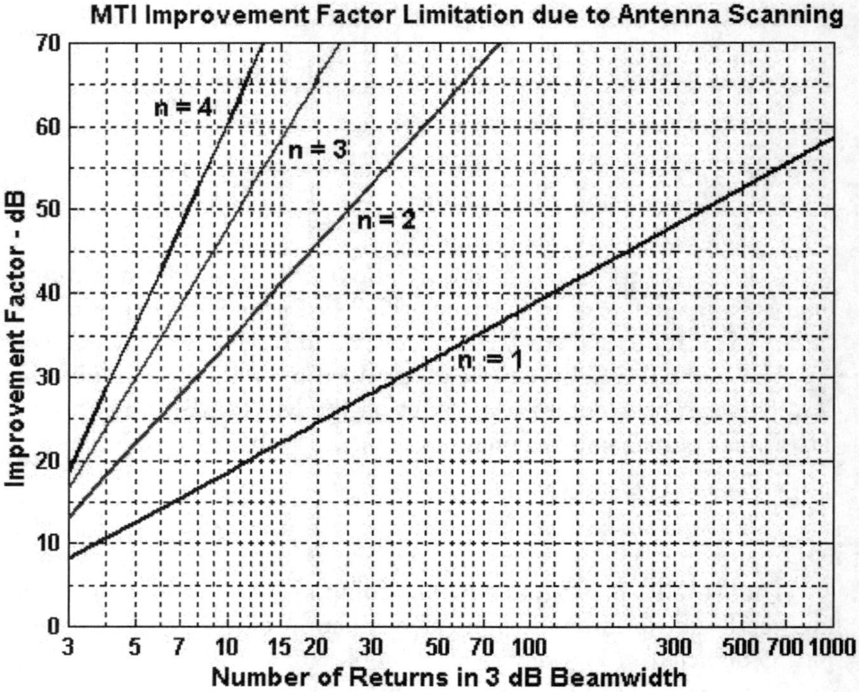

Figure EX-17.2 MTI performance due to scanning clutter.

$$I_n = \frac{n_B^{2n}}{(1.388)^n n!}$$

where n_B is the number of hits in the 3 dB antenna beamwidth. MATLAB program impfac_antscan.m plots the improvement factor as a function of $n_B = 1$ to 4. This plot is shown in Figure EX-17.2.

▶

5.1.4 The Equal-Interval Doppler Processor

The equal-interval Doppler transversal filter optimized the detection performance within an interval that is a subset of the Doppler processing band. A Doppler processor is formed by dividing the Doppler coverage band into equal intervals and employing an optimized transversal filter in each. This criterion is depicted in Figure 5.13: The Doppler band is divided into n intervals, where n is the number of processed pulses.

The equal-interval Doppler processor reflects a balanced assumption relative to the distribution of the input signals Doppler shift. This processor is optimized over a wider region than the point Doppler processor and provides S/N ratio improvement not available with the MTI processor. However, the computation of optimum weights and improvement factor is more difficult than for the other types of processors.

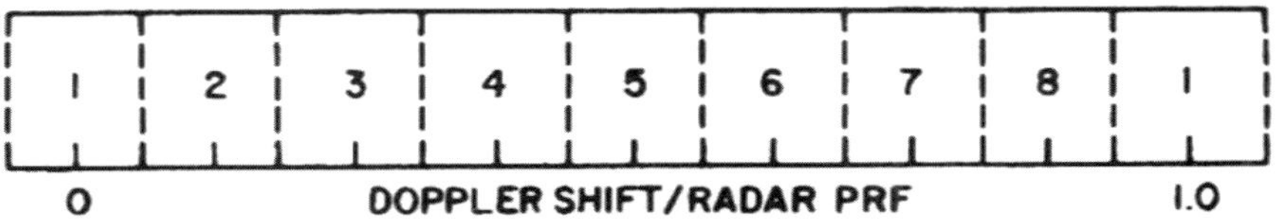

Figure 5.13 Design criterion for equal-interval optimized transversal filter. (*From:* [17, p. 6].)

For the equal-interval Doppler processor, the target Doppler probability distribution is given by

$$p(f_d) = \begin{cases} nT, & \dfrac{(2k-1)}{2nT} \le f_d \le \dfrac{(2k+1)}{2nT} \\ 0, & \text{otherwise} \end{cases} \tag{5.76}$$

where $k = 0, 1, 2, \ldots, n-1$ is the filter number. The signal covariance matrix is found by averaging (5.49) for each Doppler frequency with respect to the probability distribution of expected Doppler frequencies [see (5.76)]. The resulting normalized signal covariance matrix element is given by

$$m_{s_{il}} = e^{j2\pi k(i-l)/n} \frac{\sin[\pi(i-l)/n]}{\pi(i-l)/n} \tag{5.77}$$

where i is the row number, l is the column number, k is the filter number, and n is the number of equal intervals (number of processed pulses).

The optimal weights can be found as the eigenvector corresponding to the maximum eigenvalue of $\mathbf{R}_n^{-1}\mathbf{M}_s$ [see (5.13) and (5.14)]. The improvement factor can be found as the value of the maximum eigenvalue ($\gamma_{\max}$) of $\mathbf{R}_n^{-1}\mathbf{M}_s$ [see (5.15)]. This solution is more difficult than for the point-optimized processor because $\mathbf{M}_s$ is no longer a singular matrix. Thus, the improvement factor is not equal to the trace of the matrix $\mathbf{R}_n^{-1}\mathbf{M}_s$ [see (5.52)] and, in addition, a simple expression for the weights is not available [see (5.50)]. In fact, the solution using (5.14) is somewhat involved, because the matrix $\mathbf{R}_n^{-1}\mathbf{M}_s$ is the product of a real symmetric matrix and a Hermitian matrix, which results in a matrix that has complex elements and apparently no simple properties. Therefore, a solution using the general relations (5.13), (5.14), and (5.15) is most suitable for a computer program that finds eigenvalues and eigenvectors for complex matrices or when a small number of pulses are processed, for which an analytic solution can be found.

We now proceed in a somewhat intuitive way that provides a simple solution that ought to approach the general solution. The solution for the optimal weights at any point in the Doppler domain is given by $\mathbf{w}_o = \mathbf{R}_n^{-1}\mathbf{s}^*$. A quasi-optimum solution over an interval in the Doppler domain is obtained by using the average of the optimum weights over that interval. This, in turn, can be obtained by averaging the point-optimized weights of (5.50), using the probability distribution of (5.76). The resulting equal-interval-optimized complex weights are given by

$$\overline{w}_{i-1} = \sum_{j=1}^{n} \left[\alpha_{ij} \cos(j-1)\varpi_d T \frac{\sin(j-1)\pi/n}{(j-1)\pi/n} - j\alpha_{ij} \sin(j-1)\varpi_d T \frac{\sin(j-1)\pi/n}{(j-1)\pi/n} \right] \quad (5.78)$$

where α_{ij} are the elements of the inverse covariance matrix [see (5.54)], n is the number of equal intervals, ϖ_d is the Doppler radian frequency at the center of the interval ($\varpi_d T = k/n$), and T is the radar's interpulse interval. Note that (5.78) can also be interpreted as using (5.50) for the optimum-point weights with an averaged signal vector $\overline{\mathbf{s}}^*$ substituted for the monochromatic signal $\mathbf{s}^*$. The averaged signal vector is found by averaging s^* using the probability distribution of (5.76). The corresponding improvement factor is given by (5.56), using the averaged weights of (5.78):

$$I_f = \frac{\sum_{i-0}^{n-1} |\overline{w}_i|^2 + 2\sum_{i=0}^{n-2}\sum_{j=i+1}^{n-1} \left[\mathrm{Re}\left(\overline{w}_i^* \overline{w}_j\right)\cos(j-1)\overline{\omega}_d T - \mathrm{Im}\left(\overline{w}_i^* \overline{w}_j\right)\sin(j-i)\overline{\omega}_d T\right]}{\sum_{i=0}^{n-1} |\overline{w}_i|^2 + 2\sum_{i=0}^{n-2}\sum_{j=i+1}^{n-1} \rho(i-j)\,\mathrm{Re}\left(\overline{w}_i^* \overline{w}_j\right)} \quad (5.79)$$

Figure 5.14 depicts the results of an improvement factor calculation for a nine-pulse equal-interval-optimized Doppler processor. The improvement factors are

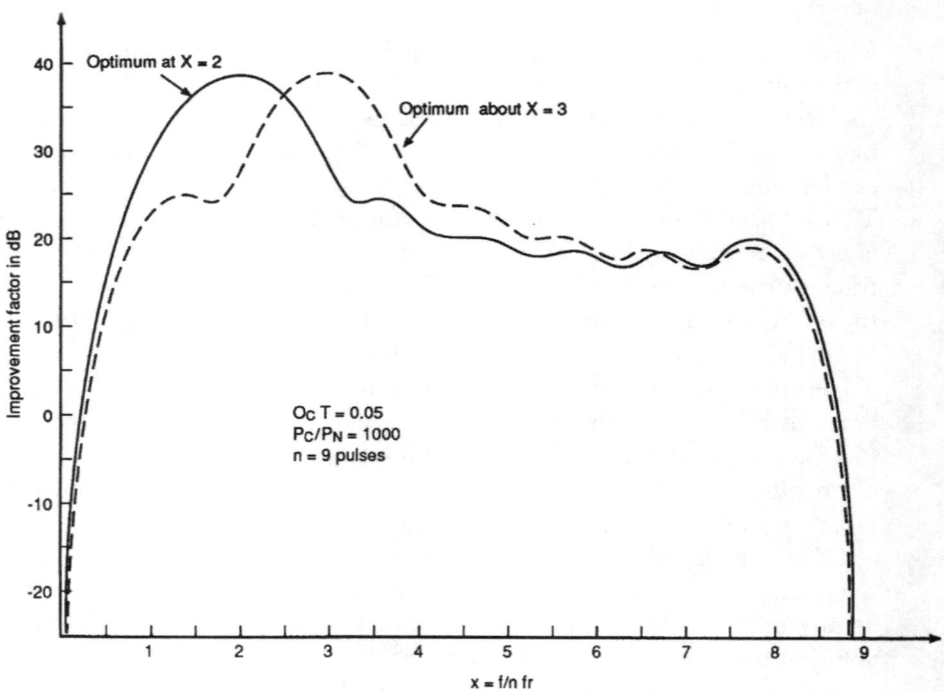

Figure 5.14 Improvement factor for equal-interval optimized transversal filters (x = 2, 3). (*From:* [7, p. 305].)

obtained by using (5.79) for two transversal filters optimized about Doppler frequencies centered at 2/9 and 3/9 of the radar's PRF. The response of these filters should be compared with that of equivalent filters shown in Figure 5.6, which were optimized by using the point Doppler criterion. The principal differences are that the peak response of the interval-optimized transversal filter is –1 dB below the point-optimized filter and the interval-optimized filters cross over at 2 dB rather than the 3 dB of the point-optimized filters. The result is that the equal-interval-optimized transversal filter will provide a more uniform response than the point-optimized design for an equal number of distributed filters. Nonetheless, the point-optimized design will have improvement factor peaks throughout the Doppler coverage region that exceed the peak of the interval-optimized filter.

Another advantage claimed for the interval-optimized transversal design is a better response when the filter is optimized about zero Doppler shift [17]. The filter response for this condition is depicted in Figure 5.15. Note that this filter also has a response about the PRF line at $x = 9$ because the optimized interval is ambiguous about ±1/2 unit around the PRF lines that include both the zero Doppler frequency and the first PRF line ($x = 9$).

Figure 5.16 depicts the improvement factor response for transversal filters that are point optimized at $x = 0$ and $x = 1/2$. The design optimized at $x = 0$ has a poor response due to difficulty in coping with the conflicting requirements of a maximum response for a signal at zero Doppler frequency and maximum attenuation for clutter centered about the same frequency. Therefore, the interval-optimized design has a better response than the point-optimized design when the design point is a

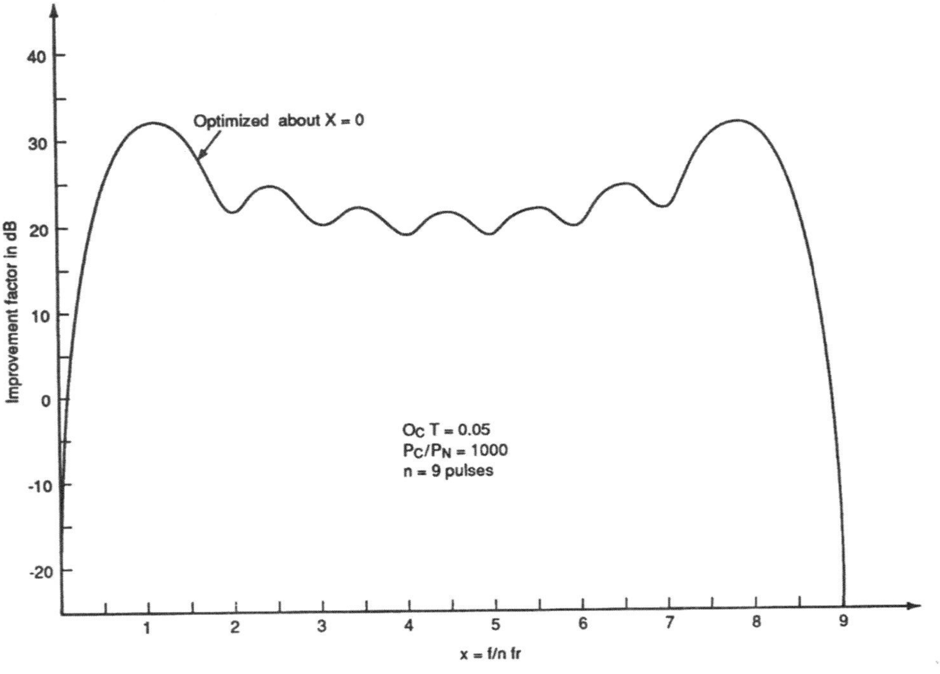

Figure 5.15 Improvement factor for equal-interval optimized transversal filter ($x = 0$). (*From:* [7, p. 306].)

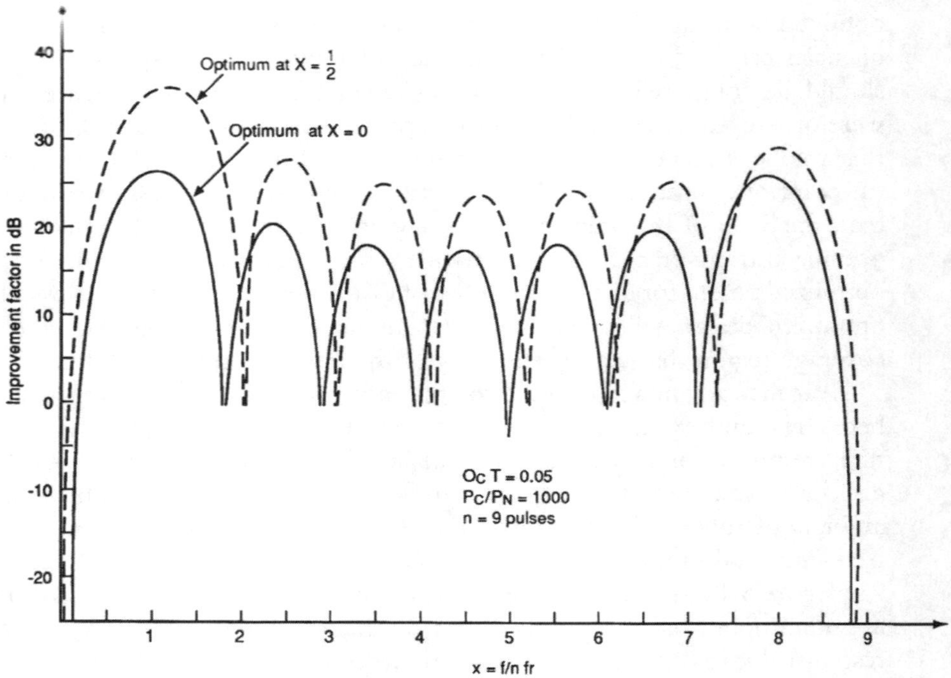

Figure 5.16 Improvement factor for point-optimized transversal filters (x = 0, 1/2). (*From:* [7, p. 307].)

zero Doppler shift. However, note that this situation is easily corrected by point optimizing the transversal filter about $x = 1/2$, which, in effect, removes the conflicting requirements.

5.2 Cascaded MTI-Coherent Integrator [18, 19]

The theory for the optimal radar Doppler processor design presented in the previous section prescribed the use of transversal filters, the weights of which are determined by the assumed target Doppler distribution and interference covariance matrix. An almost optimum alternative Doppler processor uses an MTI followed by a bank of coherent integrators [19].

The two-stage cascaded MTI-coherent integrator uses the MTI to enhance the S/C ratio and the coherent integrator primarily to enhance the S/N ratio. The design criterion for each element is relatively straightforward. The MTI is designed to attenuate the worst expected clutter spectrum, and the coherent integrator is matched to the expected dwell time on target.

However, the MTI-coherent integrator provides an inherently less efficient structure than the optimal transversal filter when the interference covariance matrix is known. The inefficiency occurs in the MTI portion of the filter, where only one of the m pulses used in the MTI effectively builds up the S/N ratio. Therefore, if a total of n radar pulses are available for Doppler processing, only $n - m + 1$ pulses are available for coherent integration. As a general design principle, the higher the order of the MTI filters, the fewer pulses available for coherent integration and the less enhancement of the S/N ratio available in the processor. Practical designs of this processor usually employ single or double MTI cancellers for this reason.

A popular version of a cascaded MTI-coherent integrator uses the quadrature channel MTI, depicted earlier in Figure 5.9, followed by a coherent integrator implemented with the FFT algorithm. This processor is shown in Figure 5.17. The complex form of the MTI output (quadrature channel) is ideally suited to the FFT processor, which accepts complex inputs. The outputs of the FFT coherent integrator as illustrated in Figure 5.18 are n envelope detected channels that span the Doppler region between PRF lines. Each channel can have a separate gain or CFAR control allowing suppression of a portion of the Doppler band that contains excess interference. This capability is important in controlling weather clutter, chaff returns, and electronic interference.

In this section, the essential characteristics of the cascaded MTI-coherent integrator are derived. Initially, a time-domain approach will be followed because this allows derivation of the weights of an equivalent transversal filter. These weights could then be substituted into (5.11) to determine the improvement factor for the cascaded processor. However, this procedure results in somewhat complex multiple

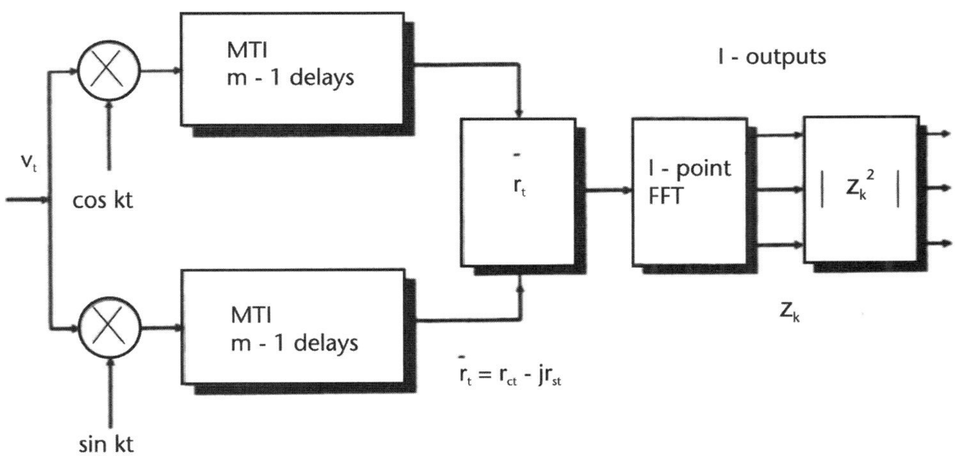

Figure 5.17 Cascaded MTI-FFT coherent integrator. (*From:* [7, p. 308].)

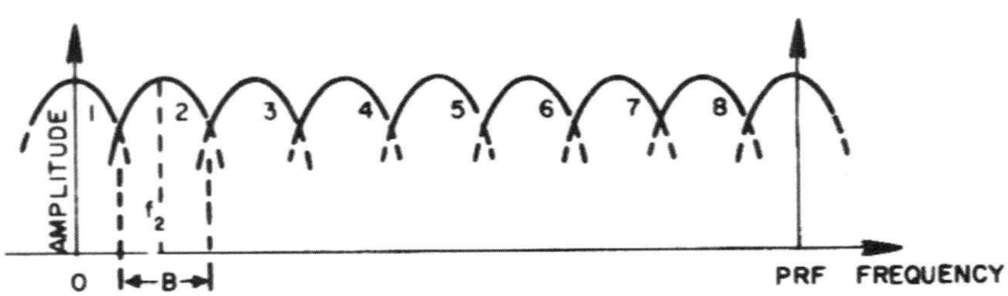

Figure 5.18 FFT coherent integrator filter bank. (*From:* [19, p. 5].)

summations [19, 20], and as an alternative we provide a relatively straightforward frequency-domain analysis that allows calculation of the improvement factor.

The purpose of pursuing the time-domain analysis is that it illustrates a distinctive feature of the cascade processor, which is important in the actual design of this type of Doppler processor. As is apparent by referring to the quadrature channel MTI shown in Figure 5.9, the first MTI output is obtained after m pulses have entered the MTI. Subsequent MTI outputs, which are applied to the l-point FFT coherent integrator, are obtained as long as input pulses to the processor are available. The output of the processor is then a random variable z, obtained after $m + l - 1$ pulses are applied to the cascaded processor, but where only the last l of the MTI output pulses are applied to the integrator. Analytically, all of this is accommodated by a frequency-domain analysis, thereby significantly reducing the possibility of error, for which several examples can be found in the published literature.

The discrete Fourier transform (DFT) of a sequence of complex input samples $\tilde{r}_i = (\tilde{r}_0, \tilde{r}_1, \ldots, \tilde{r}_{l-1})$ obtained from the output of the MTI is given by [21, 22]

$$\tilde{s}(k) = \sum_{i=0}^{l-1} \tilde{r}(i) e^{-j(2\pi ik/l)} \tag{5.80}$$

where $k = 0, 1, 2, \ldots, l - 1$ is the output filter number and l is the number of coherently integrated pulses. The similarity to (5.1), which describes the output of a transversal filter, is apparent, where the complex weight vector is identified as

$$\alpha_k^i = e^{-j(2\pi k/l)} \tag{5.81}$$

The equivalent transversal filter that represents the DFT function is depicted in Figure 5.19.

The impulse response of the l-point DFT as represented by the transversal filter of Figure 5.19 is given by

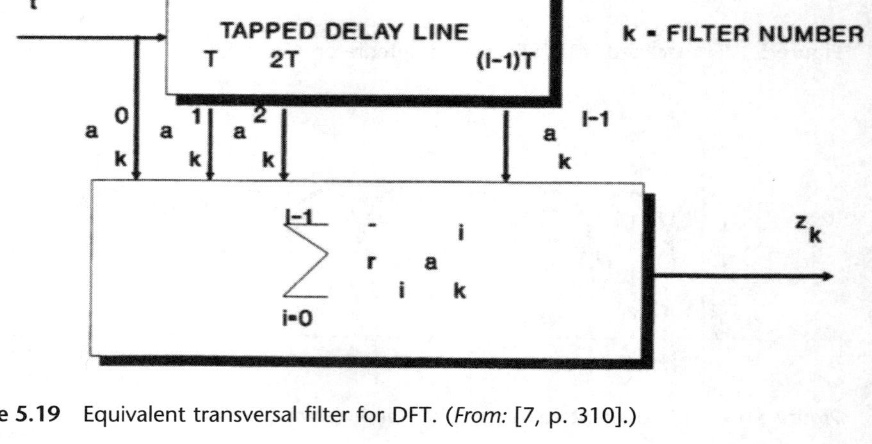

Figure 5.19 Equivalent transversal filter for DFT. (*From:* [7, p. 310].)

$$h(t) = \sum_{i=0}^{l-1} \alpha_k^i \, \delta(t + iT) \tag{5.82}$$

The transfer function is obtained as the Fourier transform of the impulse response:

$$H(j\omega) = \sum_{i=0}^{l-1} \alpha_k^i e^{+j\omega_d T} = \sum_{i=0}^{l-1} \left[e^{-j(2\pi k/l - \omega_d T)} \right]^i \tag{5.83}$$

which can be written as

$$H(j\omega) = \frac{e^{-j(2\pi k/l - \omega_d T)^l} - 1}{e^{-j(2\pi k/l - \omega_d T)} - 1} \tag{5.84}$$

where the series $\Sigma_{i=0}^{l-1} b^i = (b^l - 1)/(b - 1)$ is used. The power transfer function is obtained from (5.84) as

$$|H(j\omega)|^2 = \left| \frac{\sin\left(\frac{l\omega_d T}{2} - k\pi\right)}{\sin\left(\frac{\omega_d T}{2} - \frac{\pi k}{l}\right)} \right|^2 \tag{5.85}$$

The power transfer function for the first filter ($k = 0$) is given by

$$|H(\omega)|^2 = \left| \frac{\sin\left(\frac{l\omega_d T}{2}\right)}{\sin\left(\frac{\omega_d T}{2}\right)} \right|^2; \qquad k = 0 \tag{5.86}$$

which has zeros located at

$$\omega_d T = \frac{2\pi p}{l}, \qquad p = 1, 2, \ldots, l \tag{5.87}$$

and maxima at

$$\omega_d T = \frac{2\pi}{l}\left(p + \frac{1}{2}\right), \qquad p = 1, 2, \ldots, l \tag{5.88}$$

The value at $\omega_d = 0$ is given by

$$|H(0)|^2 = \lim_{\omega_d \to 0} \left| \frac{\sin\left(\frac{l\omega_d T}{2}\right)}{\sin\left(\frac{\omega_d T}{2}\right)} \right|^2 = l^2 \tag{5.89}$$

Therefore, the normalized power transfer function as found by expanding the numerator of (5.85) and dividing by the maximum value is given by

$$|H(j\omega)|_n^2 = \left| \frac{\sin(l\omega_d T/2)}{l \sin(\omega_d T/2 - \pi k/l)} \right|^2 \quad (5.90)$$

which is depicted in Figure 5.20 for a 16-point DFT. The generalized maximum of the power transfer function of the l-point DFT filter occurs at

$$\omega_d T = \frac{2\pi k}{l}, \qquad k = 0, 1, \ldots, l - 1 \quad (5.91)$$

with normalized value equal to 1.

Thus, a bank of filters can be produced as depicted in Figure 5.18 by using the l available output ports of the DFT [see (5.80)]. Each output is formed by an equivalent transversal filter of the form shown in Figure 5.19, which has a maximum at a normalized Doppler frequency $f_d/f_r = k/l$ and a power transfer function given by (5.90).

The DFT is efficiently implemented by using the FFT algorithm [21, 22]. The FFT digital filter allows formation of the full coherent-integration bank to minimize the number of multiplication and addition operations. This will be discussed more fully in the section on general MTI Doppler filters, but note that this efficient

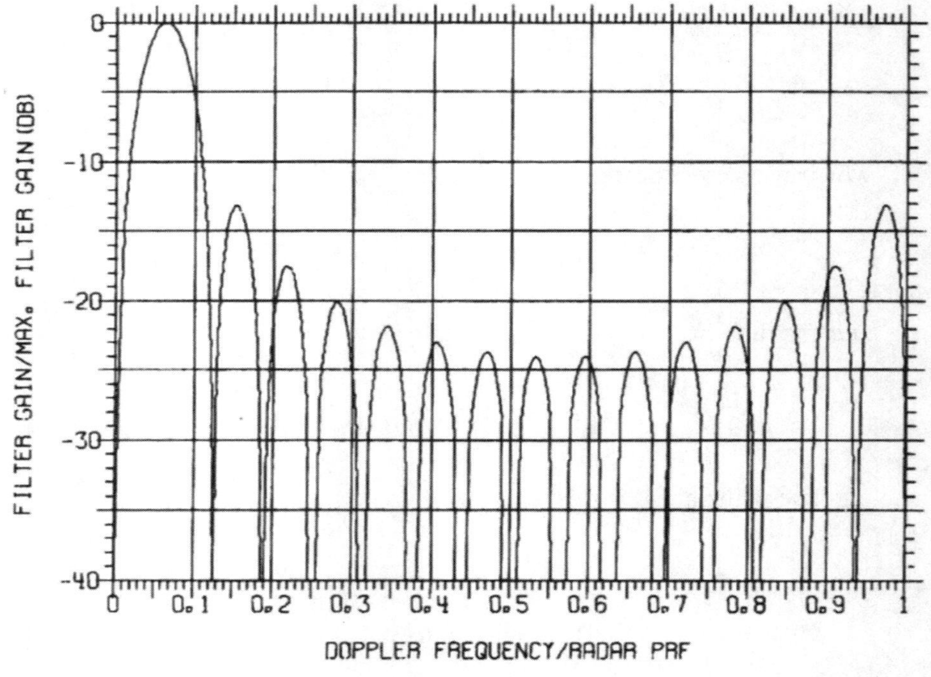

Figure 5.20 Power transfer function for DFT coherent integrator. (*From:* [19, p. 9].)

implementation is the principal reason for the popularity of the cascaded MTI-coherent integrator processor discussed in this section.

The power transfer function for the FFT coherent integrator given by (5.85) can be approximated by replacing the sin(·) function in the denominator with its argument, resulting in

$$|H(j\omega)|^2 \approx \frac{\sin^2(l\omega_d T/2 - k\pi)}{(\omega_d T/2 - k\pi/l)^2} \tag{5.92}$$

The $(\sin^2 x)/x^2$ form of (5.92) illustrates two characteristics of the FFT coherent filter bank. First, the sidelobe structure associated with each coherent integration filter is relatively high (see Figure 5.20). The analogy is to the pattern of a uniformly illuminated antenna aperture. The relative amplitude of the power transfer function sidelobes from (5.85) and (5.92) are listed for reference purposes in Table 5.2. Note that the approximation of (5.92) is not as good in the higher order sidelobes as the argument x increases [23].

Second, the uniformity of the overall response of the FFT coherent-integration Doppler filter bank is poor. From (5.92), we can see that adjacent filters cross over when the argument $x = \pi/2$. The magnitude of each individual filter is approximately −3.9 dB below its peak response at this point. The effect of this lack of uniformity is a relatively large straddling loss, the maximum value of which occurs when the target's Doppler frequency is located in the region midway between the peak responses ($\omega_d = 2\pi k/lT$) of the coherent integration filter bank.

The sidelobe and crossover characteristics of the FFT coherent integrator are due to the use of a rectangular window where all input samples within the data window are accepted with equal weight. By analogy to antenna radiation pattern theory, we know that sidelobes can be reduced and the main pattern lobe broadened by using an appropriate amplitude taper applied to the illumination function. This can be accomplished by using one of the standard window functions employed in the design of digital filters [24]. We shall concentrate on two particular types of windows, the Hamming and Hann (or hanning) windows, which are depicted in Figure 5.21. Both of these windows can be represented by the following equation:

$$g(t) = (1 - a) + a \cos\left(\frac{2\pi t}{lT}\right) \tag{5.93}$$

Table 5.2 Peak Sidelobe Response of Power Transfer Function for a Coherent Integration Filter

Sidelobe Number	*Peak Amplitude*	*(sin x)/x Approximation*
1	−13.3 dB	−13.5 dB
2	−17.6 dB	−17.9 dB
3	−20.8 dB	−20.8 dB
4	−21.8 dB	−23.0 dB
5	−23.0 dB	−24.8 dB
6	−23.7 dB	−26.2 dB

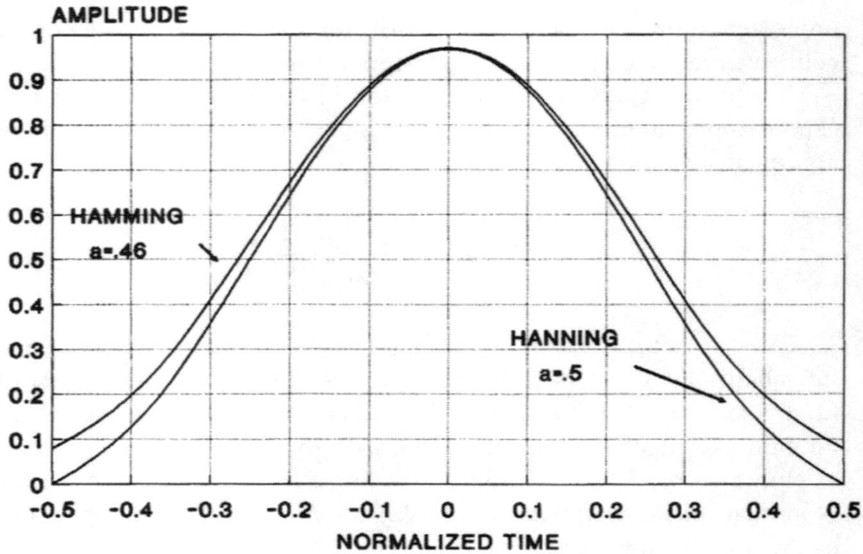

Figure 5.21 Hamming and Hanning window functions. (*From:* [2, p. 314].)

where $-lT/2 \le t \le lT/2$ and $a = 0.5$ for the hanning window and $a = 0.46$ for the Hamming window [25].

The application of a window or amplitude weighting function modifies the expression for the DFT of (5.80) to

$$\tilde{s}_w(k) = \sum_{i=0}^{l-1} \tilde{r}(i) g(i) e^{-j[2\pi ik/l]} \tag{5.94}$$

where $g(i)$ is the sampled data equivalent to the continuous window function of (5.93). The impulse response of the DFT given by (5.82) can be written as

$$h_w(t) = g(t) \sum_{i=0}^{l-1} a_k^i \, \delta(t + iT) = g(t) h(t) \tag{5.95}$$

where $g(t)$ is a continuous periodic version of (5.93) that assumes a value only at the sampling instants where $\delta(t + iT)$ is nonzero. The Fourier transform of $g(t)$ is given by

$$G(\omega) = 2\pi(1 - a)\,\delta(\omega) + a\pi\left[\delta\left(\omega - \frac{2\pi}{lT}\right) + \delta\left(\omega + \frac{2\pi}{lT}\right)\right] \tag{5.96}$$

The voltage transfer function of the DFT coherent integrator with a Hamming or hanning window can be found by Fourier transforming (5.95), which prescribes the convolving of the frequency functions [26] given by (5.84) and (5.96) $[H(\omega) * G(\omega)]$, resulting in

$$H_w(j\omega) = 2\pi(1 - a) H_k(j\omega) + \pi a [H_{k+1}(j\omega) + H_{k-1}(j\omega)] \tag{5.97}$$

where the subscripts on the transfer functions indicate that the k, $k+1$, and $k-1$ transfer functions are to be combined as indicated. Equation (5.97) can be simplified to

$$H_w(j\omega) = H_k(j\omega) + \alpha[H_{k+1}(j\omega) + H_{k-1}(j\omega)] \tag{5.98}$$

where the weights are $\alpha = 0.5$ for hanning weighting and $\alpha = 0.426$ for Hamming weighting.

The interpretation of (5.98) is that an effective weighting function window can be applied to the input data of the FFT coherent integrator by using symmetrical postcombining weighting of the adjacent filters combined with the filter under consideration. The sidelobes [24] with hanning weighting (–31 dB) and Hamming weighting (–41 dB) are considerably better than for uniform weighting (–13 dB). For hanning weighting, adjacent filters cross over at –0.4 dB and for Hamming weighting at –0.84 dB, which is a considerable improvement with respect to uniform weighting (–3.9 dB). Note that the broadening of the coherent integrator's main lobe response with weighting increases the amount of receiver noise power in the filter's output, which has the deleterious effect of decreasing the output S/N ratio. This will be considered in more detail at the end of this section when we compare the detection performance of various Doppler processors.

Consider now the MTI portion of the cascaded MTI-coherent integrator. From Section 5.1 on optimal radar Doppler processors, we know that the MTI is a specialized form of transversal filter that has only real weights. The MTI is completely specified by defining these real weights. The optimal weights for a two-pulse MTI are $w = (1, -1)$ resulting in an impulse response given by

$$h_{\mathrm{MTI}}(t) = \delta(t) - \delta(t - T) \tag{5.99}$$

and a voltage transfer function of

$$|H_{\mathrm{MTI}}(\omega)| = \left|2 \sin\left(\frac{\omega_d T}{2}\right)\right| \tag{5.100}$$

For the cascaded MTI-coherent integrator, we are interested in the lowest order m-pulse MTI that provides the desired improvement factor. Therefore, to use binomial weights is appropriate because the loss of improvement factor for a low-order MTI is minimal compared to optimal weights (see Figure 5.11) and the design procedure is simplified. Binomial weights [see (5.72)] can be obtained by successive convolution of the optimal weights for a two-pulse MTI (single canceller). Thus, the power transfer function for an MTI utilizing binomial weights can be found by simply multiplying (5.100) with itself $m-1$ times and squaring, resulting in

$$|H_{\mathrm{MTI}}(j\omega)|^2 = \left|2 \sin\left(\frac{\omega_d T}{2}\right)\right|^{2(m-1)} \tag{5.101}$$

The peak power gain of the binomial MTI is found by evaluating (5.101) at $\omega_d T = \pi$, resulting in $\overline{G}_s = 2^{2(m-1)}$. Normalization of (5.101) to its peak value results in the normalized power transfer function for a binomial MTI:

$$|H_{\text{MTI}}(j\omega)|^2 = \left|\sin\left(\frac{\omega_d T}{2}\right)\right|^{2(m-1)} \tag{5.102}$$

where m is the number of pulses processed by the MTI and T is the radar's interpulse interval.

The next step in the development is to derive an expression for the weights of a transversal filter that represents the kth output port of the combined MTI-FFT coherent integrator. Consider the cascade of an m-pulse MTI processor with an l-point FFT coherent integrator, depicted in Figure 5.22. The l-point FFT coherent integrator is represented by an equivalent transversal filter that provides the kth output port. In accordance with (5.80), the coherent integrator output z_k is given by

$$z_k = \sum_{i=m-1}^{l+m-2} \alpha(i-m+1)\tilde{r}(i-m+1) = \mathbf{a}_k^T \tilde{\mathbf{r}} \tag{5.103}$$

where $\mathbf{a}_k^T = (a_0, a_1, a_2, \ldots, a_{l-1})$ are the complex weights of the kth output of the FFT; $\tilde{\mathbf{r}} = (\tilde{r}_0, \tilde{r}_1, \tilde{r}_2, \ldots, \tilde{r}_{l-1})$ are the complex outputs of the MTI that occur in sequence after m input pulses applied to the MTI-coherent integrator cascade. Each output of the MTI in this index system is given by

$$\tilde{r}(j) = \sum_{i=j}^{m+j-1} w(i-j)\tilde{v}(i) = \mathbf{w}^T\tilde{\mathbf{v}} \tag{5.104}$$

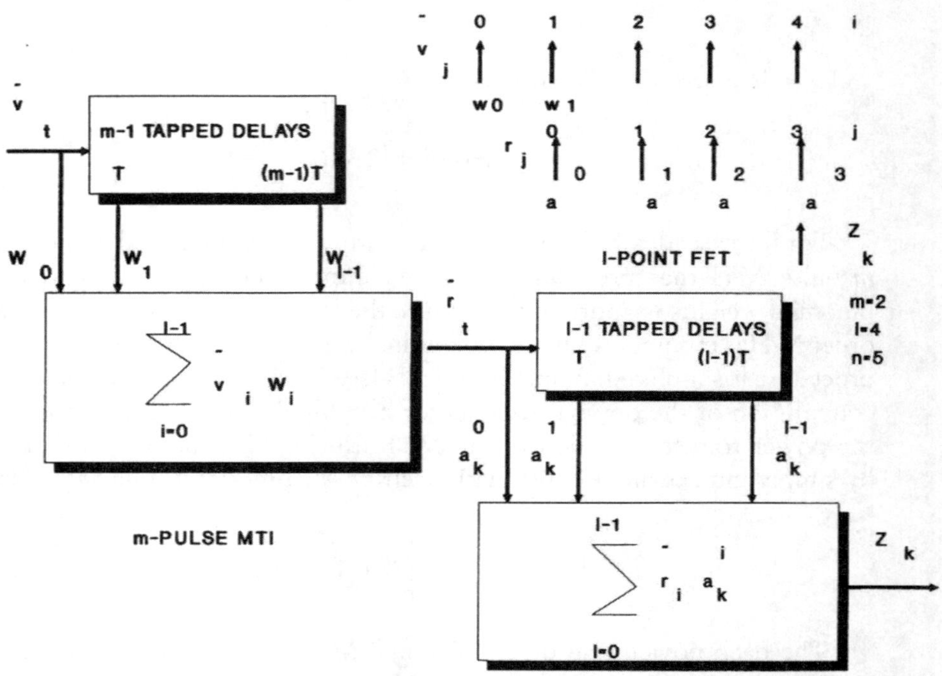

Figure 5.22 Cascaded MTI-equivalent FFT transversal filter. (*From:* [7, p. 318].)

so that the total output of the cascaded MTI-FFT coherent integrator is given by

$$z_k = \sum_{i=m-1}^{n-1} \sum_{j=i-m+1}^{i} \alpha(i-m+1)w(j-i+m-1)v(j) = \mathbf{b}^T\tilde{\mathbf{v}} \quad (5.105)$$

where $n = l + m - 1$ is the total number of input pulses applied to the processor, $\tilde{\mathbf{v}} = (\tilde{v}_0, \tilde{v}_1, \ldots, \tilde{v}_{n-1})$ is the complex input voltage and $\mathbf{b}^T = (b_0, b_1, \ldots, b_{l+m-1})$ is the weight vector of an equivalent transversal filter for the total processor. From (5.105), the equivalent weight vector can be expressed as [19]

$$\mathbf{b} = \hat{\mathbf{A}}\mathbf{a}_k \quad (5.106)$$

where $\alpha_k^T = (\alpha_0, \alpha_1, \ldots, \alpha_{l-1})$ is the FFT weight vector and $\hat{\mathbf{A}}^T$ is an $m + l - 1$ by l matrix formed from the MTI weight vector as follows:

$$\hat{\mathbf{A}}_1 = \begin{bmatrix} w_o \\ \\ w_{m-1} \\ 0 \\ 0 \\ 0 \end{bmatrix} \left.\vphantom{\begin{matrix}0\\0\\0\end{matrix}}\right\} \begin{matrix} l-1 \\ \text{zeros} \end{matrix}, \quad \hat{\mathbf{A}}_2 = \begin{bmatrix} 0 \\ w_o \\ \\ w_{m-1} \\ 0 \\ 0 \end{bmatrix}, \quad \hat{\mathbf{A}}_l = \begin{bmatrix} 0 \\ 0 \\ 0 \\ \\ w_o \\ w_{m-1} \end{bmatrix} \left.\vphantom{\begin{matrix}0\\0\\0\end{matrix}}\right\} \begin{matrix} l-1 \\ \text{zeros} \end{matrix} \quad (5.107)$$

As an example, consider a single-canceller MTI ($m = 2$), followed by a four-point FFT coherent integrator ($l = 4$). The four weight vectors of the FFT are given by $\mathbf{a}_k = (1, 1, 1, 1)$, $(j, 1, -j, -1)$, $(-1, 1, -1, 1)$, and $(-j, 1, j, -1)$, whereas the weight vector of the MTI is given by $\mathbf{w} = (1, -1)$. If we select the second FFT weight vector ($k = 1$), we have

$$\hat{\mathbf{A}} = \begin{bmatrix} 1 & 0 & 0 & 0 \\ -1 & 1 & 0 & 0 \\ 0 & -1 & 1 & 0 \\ 0 & 0 & -1 & 1 \\ 0 & 0 & 0 & -1 \end{bmatrix} \quad (5.108)$$

$$\mathbf{b} = \begin{bmatrix} 1 \\ -1-j \\ j-1 \\ 1+j \\ -j \end{bmatrix} \quad (5.109)$$

which is the weight vector of the equivalent transversal filter, the output of which is given by (5.105). This equivalent transversal filter is depicted in Figure 5.22, which also shows the timing involved with the various indices used in (5.103), (5.104), and (5.105).

Alternatively, the normalized power transfer function of the combined MTI-coherent integrator is given by the product of (5.90) and (5.102):

$$|H(j\omega)|^2 = \frac{[\sin(\omega_d T/2)]^{2(m-1)} \sin^2(l\omega_d T/2 - k\pi)}{l \sin(\omega_d T/2 - k\pi/l)} \tag{5.110}$$

The improvement factor can be found from

$$I_f = \frac{|H(j\omega)|_N^2 \int_0^{\omega_r} S_I(\omega)\, d\omega}{\int_0^{\omega_r} S_I(\omega)\, |H(j\omega)|_N^2\, d\omega} \tag{5.111}$$

where the interference power spectral density for a Gaussian-shaped clutter spectrum and receiver noise is given by

$$S_I(\omega) = \frac{P_n}{\omega_r} + \frac{P_c}{\sqrt{2\pi}\sigma_c}\left[e^{-\omega^2/2\sigma_c^2} + e^{-(\omega-\omega_r)^2/2\sigma_c^2}\right]; \qquad 0 \le \omega \le \omega_r \tag{5.112}$$

where P_n is the receiver noise power, P_c is the clutter power, σ_c^2 is the variance of the Gaussian-shaped clutter spectrum's radian frequency response, and ω_r is the pulse repetition radian frequency. A convenient way to normalize (5.112) is to divide by the receiver noise power P_n and convert to f_d as the independent parameter, resulting in

$$I_f = \frac{|H(jf)|^2 \int_0^{f_r} S_I(f)\, df}{\int_0^{f_r} S_I(f)\, |H(jf)|_N^2\, df} \tag{5.113}$$

$$S_I(f) = \frac{1}{f_r} + \frac{F}{\sqrt{2\pi}\sigma_f}\left[e^{-f_d^2/2\sigma_f^2} + e^{-(f_d-f_r)^2/2\sigma_f^2}\right] \tag{5.114}$$

where F is the clutter-to-noise power ratio, f_r is the radar's pulse repetition frequency, and σ_f^2 is the variance of the clutter's Gaussian-shaped spectrum.

Figure 5.23 depicts the improvement factor for a cascaded single-canceller, MTI 8-point FFT coherent integrator with no integrator weighting. Responses for adjacent filters centered at $k = 2$ and $k = 3$ are shown. This figure illustrates the relatively high crossover loss and high sidelobes associated with a rectangular window coherent integrator. The improvement factor at the integrator center frequency is 38.5 dB, which compares to 39.1 dB for a transversal filter design (see Figure 5.6) with the same number of nine processed pulses. Note, however, that the peak improvement factor, which is skewed to the right due to the effect of the MTI response, is on the order of 39 dB [27].

Figure 5.24 depicts the improvement factor for a cascaded single-canceller, MTI 8-point FFT coherent integrator with Hamming weighting ($\alpha = 0.426$). Again, responses for adjacent filters centered at $k = 2$ and $k = 3$ are shown. Note the broadened response that reduces the crossover loss and the reduced sidelobe structure. However, the improvement factor is reduced to 37.9 dB at the integrator center frequency, whereas the peak improvement factor is reduced to 38.3 dB. Therefore, the effect of the coherent-integrator weighting function is to provide a more uniform overall Doppler frequency response with minimal sidelobes at the expense of a reduced peak improvement factor.

A comparison between the point-optimized transversal filter Doppler processor; a cascaded single-canceller MTI 8-point FFT coherent integrator; and a cascaded single-canceller, MTI 8-point FFT coherent integrator with Hamming weighting is given in Table 5.3. The basis of the comparison is on the averaged improvement

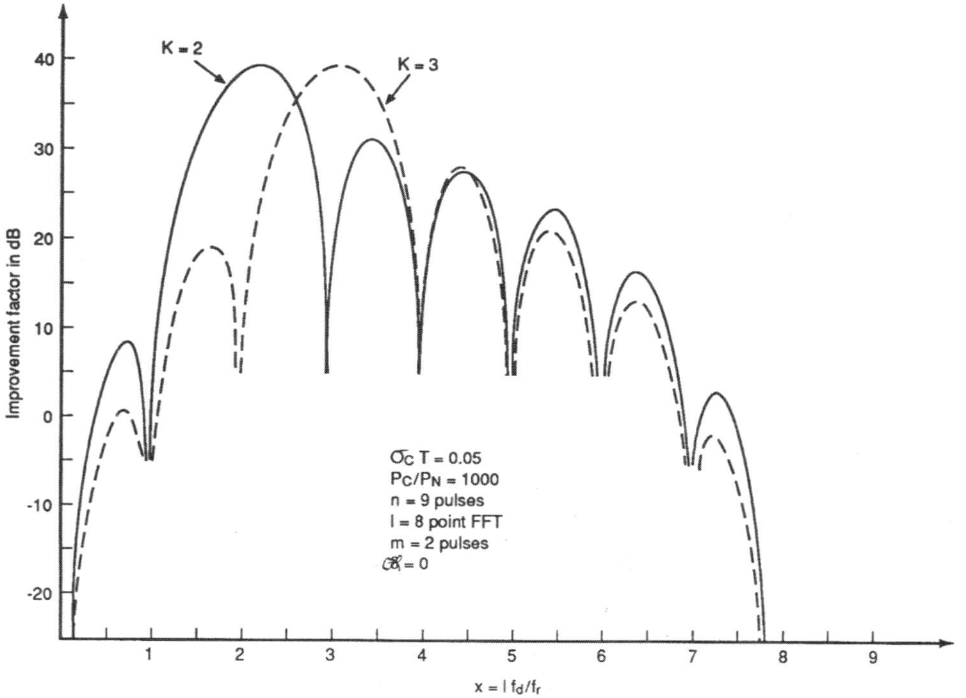

Figure 5.23 Improvement factor for a single-canceller, MTI-coherent integrator. (*From:* [7, p. 320].)

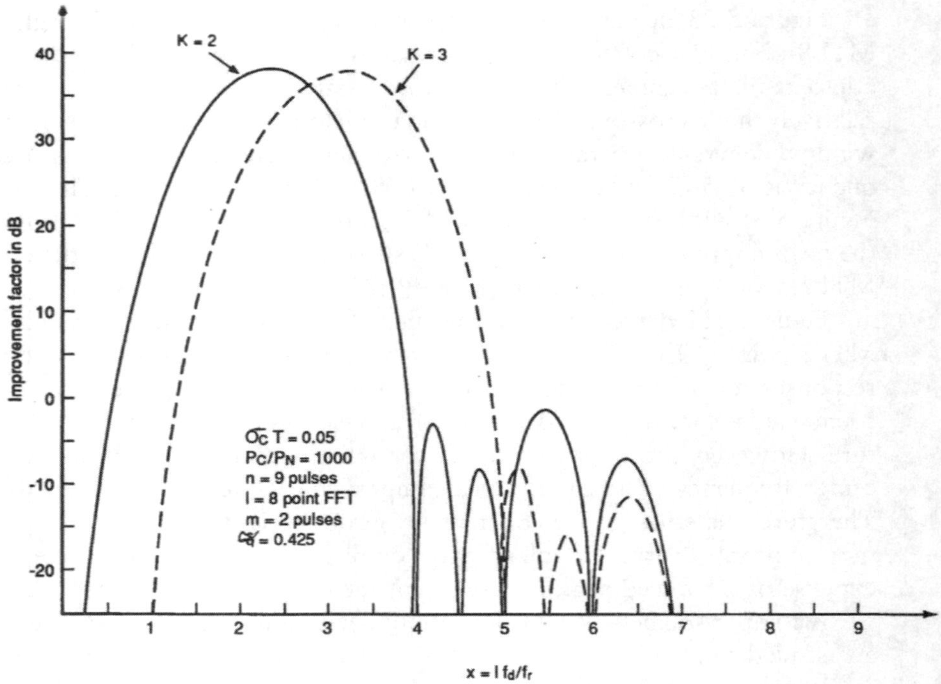

Figure 5.24 Improvement factor for single-canceller, MTI-coherent integrator with Hamming weighting. (*From:* [7, p. 321].)

Table 5.3 Performance Comparison of Doppler Processors

Doppler Processor	*Average Improvement Factor Between Crossovers*
Point-optimized transversal filter	38.03 dB
MTI-coherent integrator FFT	37.63 dB
MTI-weighted coherent integrator	37.61 dB

Note: $n = 9$, $m = 2$, $l = 8$, $\alpha = 0.426$, $\sigma_c T = 9.05$, $F = 1{,}000$.

factor of a single output port ($k = 2$) where the average is taken between crossover points of adjacent filters. We can see that, as expected, the point-optimized transversal filter provides an advantage of approximately 0.4 dB for the situation considered, whereas the Hamming weighting results in minimal loss of performance.

The significance of the 0.4 dB loss in improvement factor can be determined by using (5.45), which converts the improvement factor into detection performance on the basis of probability of detection. Equation (5.45) is applicable for any transversal filter in Gaussian-distribution interference. Because we have seen that the cascaded MTI-FFT has a transversal filter representation [see (5.106)] the applicability of (5.45) is established. As an example, if the transversal-filter Doppler processor provides a $P_d = 0.9$ with a $P_{fa} = 10^{-6}$, the MTI-coherent integrator provides a $P_d = 0.83$ for the same parameters.

5.3 Cascaded MTI-Noncoherent Integrator

A Doppler processor used extensively in practice is an MTI followed by an incoherent integrator. The significant feature of this processor, as compared to the cascaded MTI-coherent integrator described in the previous section, is a considerable reduction in hardware. However, the processor can have severe detection losses compared to the Doppler processors previously described, the performance of which approached optimal limits.

The detection loss associated with the cascaded MTI-incoherent integrator is due to the number of pulses processed in the MTI and the residual clutter that passes through the MTI. Initially, let us consider the effect of receiver noise on the processor, because this establishes the maximum performance that can be obtained. Also available for this case is an approximate analysis that allows a simple calculation of detection performance [3, 28].

The cascaded MTI-incoherent integrator is depicted in Figure 5.25. This implementation uses a quadrature MTI (see Figure 5.9), although a single-channel MTI can be used with additional detection loss [29]. The MTI correlates the receiver noise, which has the effect of reducing the number of independent noise samples available for integration. This effect reduces the integration gain of the integrator, which is proportional to the number of independent samples processed.

An approximate solution can be obtained by assuming that the detected output (z) of the MTI depicted in Figure 5.25 has a Gaussian distribution [3]. The output of the integrator then has a Gaussian distribution due to its linear nature. The detection performance of the processor is then determined by the S/N ratio at the output of the integrator.

To emphasize the linear nature of the integrator, it is depicted in Figure 5.26 as a transversal filter with weights $\mathbf{a}^T = (1, 1, 1, \ldots, 1)$, which is excited by the Gaussian-distributed vector $\mathbf{z}^T = (z_0, z_1, \ldots, z_{l-1})$. The S/N ratio at the integrator's

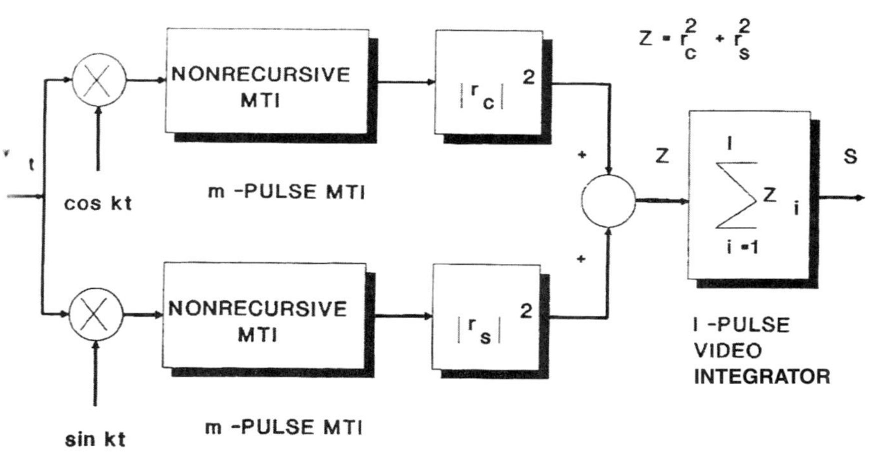

Figure 5.25 Cascaded MTI-incoherent integrator. (*From:* [7, p. 323].)

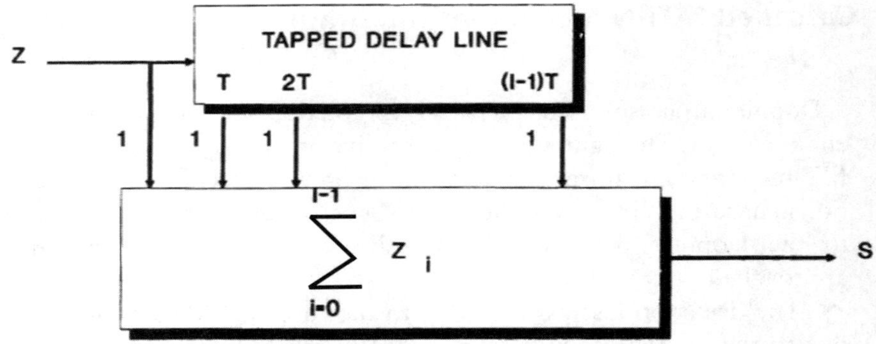

Figure 5.26 Transversal filter integrator. (*From:* [7, p. 323].)

output can be determined by applying the relationships for the transversal filter developed in Section 5.1.

The signal power output of the transversal filter integrator can be found by applying (5.5) with normalized covariance matrix [see (5.48) with $\omega_d = 0$]:

$$\mathbf{M}_s = \begin{pmatrix} 1 & \cdots & 1 \\ \vdots & & \vdots \\ 1 & \cdots & 1 \end{pmatrix} \tag{5.115}$$

whereas the noise power output is found by using (5.3) with normalized covariance matrix [see (5.67)]:

$$\mathbf{R}_n = \begin{pmatrix} 1 & \rho_1^2 & \cdots & \rho_{l-1}^2 \\ & \ddots & & \vdots \\ \rho_1^2 & & \ddots & \vdots \\ \rho_{l-1} & \cdots & \cdots & 1 \end{pmatrix} \tag{5.116}$$

where $\rho^2(j)$ and $j = 1, 2, \ldots l - 1$, are the correlation coefficients of the squared output of the MTI. The factor by which the integrator improves the output S/N ratio is given by

$$n_e = \frac{l^2}{l + 2 \sum_{j=1}^{l-1} (l - j)\rho^2(j)} \tag{5.117}$$

where

$$\rho(j) = \frac{\sum_{i=0}^{m-1} \sum_{k=0}^{m-1} w_i w_k \, \delta(k - i + j)}{\sum_{i=0}^{m-1} w_i^2}; \qquad \delta(0) = 1 \tag{5.118}$$

and $\mathbf{w}^T = (w_0, w_1, \ldots, w_{m-1})$ is the MTI weight vector. The factor n_e is called the *equivalent number of independent samples* because it determines the integration gain in an equivalent manner, just as if n_e independent samples were applied to the integrator. The integration loss for the approximate Gaussian statistics is given by [5]

$$L = 10 \log\left(\frac{l}{n_e}\right)^{1/2} \tag{5.119}$$

If we examine (5.117) for the extreme condition of uncorrelated samples ($\rho = 0$), we find that $n_e = l$, which is appropriate for independent Gaussian samples. At the other extreme of completely correlated samples ($\rho = 1$), we find, as expected, that $n_e = 1$ and the integrator has no effect. For a single canceller $\mathbf{w} = (1, -1)$, $n_e/l \approx 2/3$ ($L = 0.88$ dB), whereas for a double canceller $\mathbf{w} = (1, -2, 1)$, $n_e/l \approx 18/35$ ($L = 1.444$ dB) [3].

The approximate analysis using Gaussian statistics can be expected to provide accurate results only when l is large, which causes the input distribution to the integrator to approach a Gaussian distribution. Table 5.4 lists the integration loss for m = 2-, 3-, 4-, and 5-pulse MTI binomial cancellers obtained by using importance-sampling simulation techniques for a linear envelope detector averaged for $l = 4$, 8, 16, and 32 pulses [5]. Also shown is the approximate loss computed by an n_e calculation using standard detection curves with identical parameters.

The approximate analysis just described indicates that the correlating effect of the MTI on receiver noise causes a detection loss when using video integration: The noise increases with the number of pulses processed by the MTI. However, the function of the MTI is to allow detection of targets embedded in clutter, and a complete analysis must consider the effect of an interfering background that consists of both receiver noise and clutter. An exact analysis of this problem is available [6, 30, 31] and will follow.

The square-law-detected, video-integrated output of the cascaded MTI-incoherent integrator depicted in Figure 5.25 is given by

$$s = \sum_{i=0}^{l-1} z_i = \sum_{i=0}^{l-1} r_{ci}^2 + r_{si}^2 \tag{5.120}$$

This problem is identical in form to the analysis developed for the detection of a steady target in receiver noise, except that in this case the noise is correlated

Table 5.4 MTI Noise Integration Loss ($P_d = 0.9$, $P_{fa} = 10^{-6}$)

MTI Pulses	*Average Integration Loss (dB)*	*Loss Calculated Using n_e (dB)*
2	1.0	1.1
3	1.8	1.8
4	2.2	2.4
5	2.5	2.8

due to both the MTI action and the clutter component of the interfering background. The quadrature output vectors of the MTI, $\mathbf{r}_c = (r_{c0}, r_{c1}, \ldots, r_{cl-1})$ and $\mathbf{r}_s = (r_{s0}, r_{s1}, \ldots, r_{sl-1})$, are Gaussian distributed because they are the result of a linear process on a Gaussian vector. As is well known [32], the correlated Gaussian vectors $\mathbf{r}_c$ and $\mathbf{r}_s$ can always be represented by an equivalent set of uncorrelated components obtained through a transformation that diagonalizes the covariance matrix. Application of the uncorrelated representation of vectors $\mathbf{r}_c$ and $\mathbf{r}_s$ to (5.120) determines that the random variable s has a noncentral chi-squared distribution when a signal is present and a chi-squared distribution for the case of no signal [33]. Application of the statistics of the integrator output (s) in a standard probability of detection calculation allows determination of the detection performance of the cascaded MTI-incoherent integrator.

The two l-dimensional Gaussian pdfs of the quadrature channel MTI are given by

$$p(\mathbf{r}_c, \mathbf{r}_s) = \frac{\exp\left\{-\frac{1}{2}[(\mathbf{r}_c - \mathbf{s}_x)^T\mathbf{C}^{-1}(\mathbf{r}_c - \mathbf{s}_x) + (\mathbf{r}_s - \mathbf{s}_y)^T\mathbf{C}^{-1}(\mathbf{r}_s - \mathbf{s}_y)]\right\}}{(2\pi)^l|\mathbf{C}|} \tag{5.121}$$

where the quadrature mean vectors $\mathbf{s}_x^T = (s_{x0}, s_{x1}, \ldots, s_{xl-1})$ and $\mathbf{s}_y^T = (s_{y0}, s_{y1}, \ldots, s_{yl-1})$ are given by (5.27) and (5.28) (with $P_s = A^2/2$ and $\phi = 0$):

$$s_{xk} = A \sum_{j=0}^{m-1} w_j \cos[\omega_d(t_k - jT)] \tag{5.122}$$

$$s_{yk} = A \sum_{j=0}^{m-1} w_j \sin[\omega_d(t_k - jT)] \tag{5.123}$$

The l by l covariance matrix $\mathbf{C}$ is given by [7]:

$$\mathbf{C} = \begin{pmatrix} C_{11} & \cdots & C_{1l} \\ \vdots & \ddots & \vdots \\ C_{l1} & \cdots & C_{ll} \end{pmatrix} \tag{5.124}$$

with these elements:

$$c_{kq} = \sum_{i=0}^{m-1} \sum_{j=0}^{m-1} w_i w_j E[v_{ci+k-1} v_{cj+q-1}] \tag{5.125}$$

where $\mathbf{w}^T = (w_0, w_1, \ldots, w_{m-1})$ is the MTI weight vector and $\mathbf{v}_c = (v_{c0}, v_{c1}, \ldots, v_{cl+m-2})$, or, equivalently, $\mathbf{v}_s = (v_{s0}, v_{s1}, \ldots, v_{sl+m-2})$ is the quadrature MTI input interference vectors.

The key step in the development is to rotate the coordinate system using a linear transformation ($\mathbf{L}$) such that $\mathbf{C}$ has elements only on the main diagonal. The rotated MTI quadrature outputs then become $\mathbf{r}_c' = \mathbf{L}\mathbf{r}_c$ and $\mathbf{r}_s' = \mathbf{L}\mathbf{r}_s$ where $\mathbf{L}$ is formed by the eigenvectors $\mathbf{e}_k^T = (e_{k1}, \ldots, e_{kl})$ of $\mathbf{C}$:

$$\mathbf{L} = \begin{pmatrix} e_{11} & e_{21} & \ldots & e_{l1} \\ \vdots & \vdots & & \vdots \\ e_{1l} & e_{2l} & \ldots & e_{ll} \end{pmatrix} \tag{5.126}$$

and the diagonal matrix ($\boldsymbol{\lambda}$) has elements corresponding to the eigenvalues (λ_k) of $\mathbf{C}$ [32]:

$$\boldsymbol{\lambda} = \begin{pmatrix} \lambda_1 & 0 & 0 \\ 0 & \ddots & 0 \\ 0 & 0 & \lambda_l \end{pmatrix} \tag{5.127}$$

The elements of (5.126) and (5.127) are found by solving the following matrix equation:

$$(\mathbf{C} - \boldsymbol{\lambda}_k \mathbf{I})\mathbf{e}_k = 0; \qquad k = 1, 2, \ldots, l \tag{5.128}$$

and can be checked by using

$$\boldsymbol{\lambda} = \mathbf{L}^T \mathbf{C} \mathbf{L} \tag{5.129}$$

The quadrature mean vectors at the MTI output are also rotated by the transformation, resulting in $\mathbf{s}_x' = \mathbf{L}\mathbf{s}_x$ and $\mathbf{s}_y' = \mathbf{L}\mathbf{s}_y$.

Equation (5.121) can now be written in terms of the rotated coordinate system as

$$p\left(\mathbf{r}_c', \mathbf{r}_s'\right) = \frac{\exp\left[-\dfrac{1}{2}\displaystyle\sum_{k=1}^{l} \dfrac{\left(r_{ck-1}' - s_{xk-1}'\right)^2 + \left(r_{sk-1}' - s_{yk-1}'\right)^2}{\lambda_k}\right]}{(2\pi)^l \displaystyle\prod_{k=1}^{l} \lambda_k} \tag{5.130}$$

where the rotation has decorrelated the quadrature MTI output vectors, which in turn has made the components independent; hence, the overall pdf is simply the product of the individual component density functions. The characteristic function of the square-law-detected, video-integrator output (s) of the cascaded processor can then be found from (5.130) by using standard techniques (see [33] for derivation) as

$$M(p) = \prod_{k=1}^{l} \exp\left(-\frac{pg_k^2}{1 + 2p\lambda_k}\right) \Big/ (1 + 2p\lambda_k) \tag{5.131}$$

where $p = -j\omega$ and g_k^2 are the projections of the signal vector onto the kth eigenvector:

$$g_k^2 = \sum_{i=1}^{l} \sum_{j=1}^{l} (s_{xi-1}s_{xj-1} + s_{yi-1}s_{yj-1})e_{ki}e_{kj} \tag{5.132}$$

Detection thresholds can be found by transforming (5.132) with $g_k = 0$ to obtain the no-signal pdf ($s \geq 0$):

$$p(s) = \sum_{k=1}^{l} C_k \exp(-s/2\lambda_k)/2\lambda_k \tag{5.133}$$

where

$$C_k = \prod_{\substack{j=1 \\ j \neq k}}^{l} \left(\frac{\lambda_k}{\lambda_k - \lambda_j} \right) \tag{5.134}$$

Integration of (5.133) to find the false alarm probability in terms of the detection thresholds (s_T) results in [6, 30, 31]

$$P_{fa} = \sum_{k=1}^{l} \left\{ \prod_{\substack{j=1 \\ j \neq k}}^{l} \left(\frac{\lambda_k}{\lambda_k - \lambda_j} \right) \right\} \exp\left(\frac{-s_T}{2\lambda_k} \right) \tag{5.135}$$

Normalized detection thresholds for a single canceller MTI ($m = 2$) and an 8-pulse video integrator ($l = 8$) are plotted in Figure 5.27 for various Gaussian-shaped clutter spectra ($\sigma_c T = 0.01, 0.05, 0.1$) and clutter-to-receiver noise powers ($F = 0$, 5,000, 10,000, and 20,000).

The procedure for finding the probability of detection is more complex than that for finding the detection thresholds. The difficulty arises because direct inversion of the characteristic function (5.131) to obtain the pdf is apparently not available in a closed form. Therefore, a numerical solution uses the Gram-Charlier series approach.

The cumulant generating function can be defined as [34]

$$h(p) = \ln M(p) \tag{5.136}$$

The cumulants are then obtained by evaluating

$$\chi_k = (-1)^k \frac{\partial^k}{\partial p^k} [h(p)]|_{p=0} \tag{5.137}$$

A convenient way to evaluate the cumulants is to express (5.131) as

$$M(p) = \prod_{k=1}^{l} e^{-s_k} e^{\alpha_k s_k / p + \alpha_k} \cdot \frac{\alpha_k}{p + \alpha_k} \tag{5.138}$$

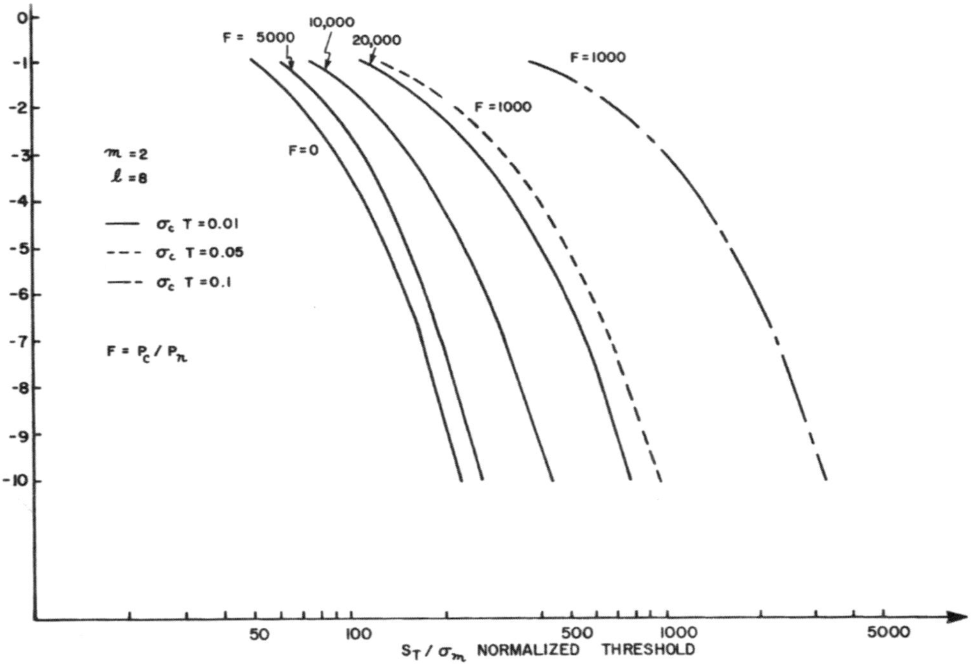

Figure 5.27 Normalized thresholds for cascaded MTI-incoherent integrator. (*From:* [7, p. 329].)

where $s_k g_k^2/2\lambda_k$ and $\alpha_k = 1/2\lambda_k$. The cumulant generating function thus becomes

$$h(p) = \sum_{k=1}^{l} C_k + \frac{\alpha_k s_k}{p + \alpha_k} - \ln(p + \alpha_k) \tag{5.139}$$

and the cumulants therefore become

$$\chi_i = \sum_{k=1}^{l} (2\lambda_k)^{i-1}(i-1)!\left[ig_k^2 + 2\lambda_k\right] \tag{5.140}$$

The cumulants are now used to obtain a Gram-Charlier series [32] for the detection probabilities of the cascaded MTI-incoherent integrator. A three-term series with Edgeworth grouping was found to provide good accuracy for the case of a single-canceller in cascade with an 8-pulse incoherent integrator [6].

The general procedure for obtaining the detection characteristics of the cascaded MTI-incoherent integrator is approached as follows:

- Form the covariance matrix using (5.125), which is a function of the MTI order (m), the number of pulses incoherently integrated (l), and the spectral characteristics of the clutter $\left(\sigma_c^T\right)$.
- Evaluate the eigenvalues (λ_k) and eigenvectors $\left(\mathbf{e}_k^T\right)$ associated with covariance matrix $\mathbf{C}$.

- Find the detection thresholds from (5.135), which involve the eigenvalue (λ_k), the desired false alarm probability (P_{fa}), and the clutter-to-receiver noise power ratio (F).
- Evaluate the transformed signal magnitude g_k^2 from (5.132), which involves the eigenvectors $\left(\mathbf{e}_k^T\right)$, the signal amplitude ($A$), the Doppler frequency (ω_d), and the MTI weights ($\mathbf{w}^T$). Use (5.122) and (5.123) to evaluate the quadrature MTI signal vectors $\mathbf{s}_x^T$ and $\mathbf{s}_y^T$.
- Use (5.140) to evaluate the processor cumulants from g_k^2 and λ_k and then substitute the cumulants into a Gram-Charlier series to evaluate detection probabilities.

► Example 18 Integration Gain for MTI-Noncoherent Integrator

The block diagram of an MTI followed by a noncoherent integrator is depicted in Figure EX-18.1. The performance of this processor is discussed in Section 5.3. When noise- alone is present, a loss occurs due to the correlating action of the MTI that varies from 1 to 2.5 dB depending on the order of canceller employed (single to fourth order). However, when clutter is present, this loss increases. This loss depends on the magnitude of the clutter power with respect to the noise power and the improvement factor provided by the MTI.

The first step in analyzing the cascaded MTI integration system is to determine the detection thresholds. A plot of the thresholds from MATLAB program mti_nci_thresh.m is depicted in Figure EX-18.2 for a single canceller followed by an 8-pulse noncoherent integrator with $\sigma_c T = 0.01$. The thresholds are obtained by plotting

$$P_{fa} = \sum_{k=1}^{l} \left\{ \prod_{\substack{j=1 \\ j \neq k}}^{l} \left(\frac{\lambda_k}{\lambda_k - \lambda_j} \right) \right\} \exp\left(\frac{-s_T}{2\lambda_k} \right)$$

for clutter-to-noise power ratios of $F = 0$, 5,000, 10,000, and 20,000.

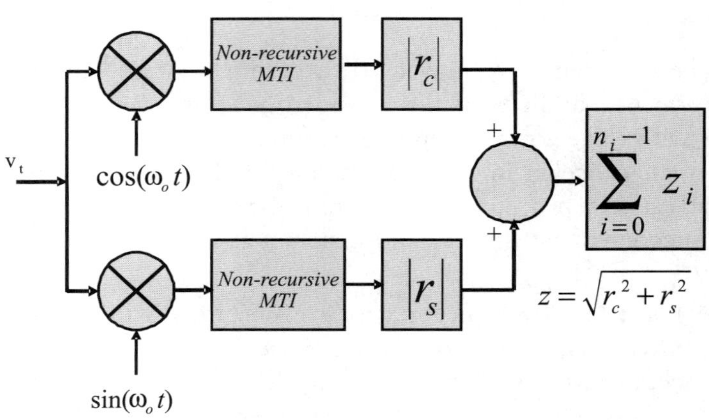

Figure EX-18.1 Nonrecursive MTI with noncoherent integrator.

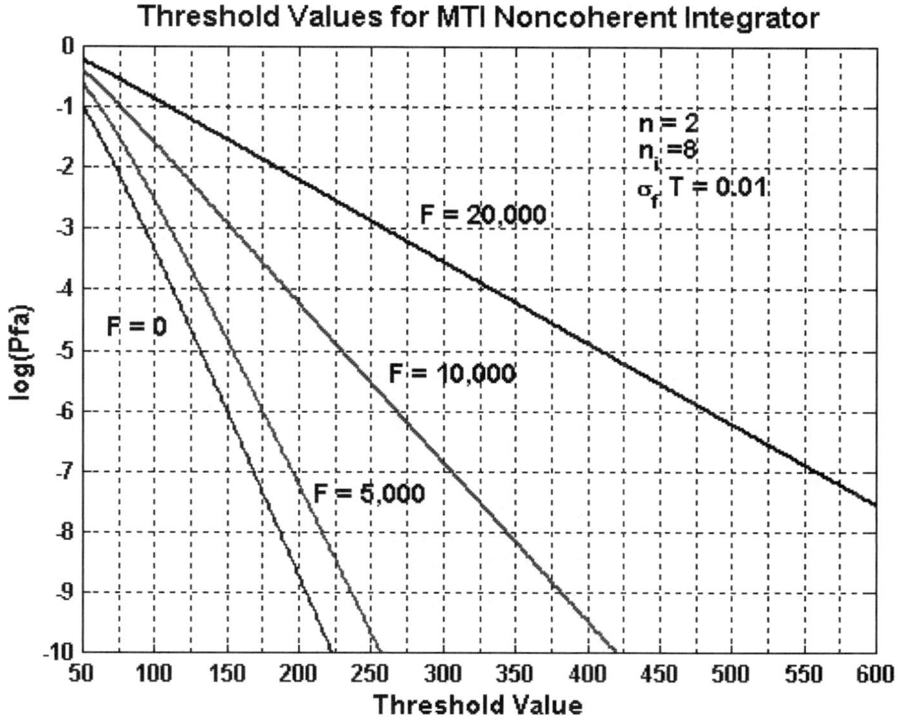

Figure EX-18.2 Threshold values for cascaded MTI noncoherent integrator.

More precise values are found in MATLAB program thresh_mti_nci.m and thresh_mti_ncia.m using Newton's iteration method to solve for the detection threshold α_T for particular values of P_{fa}, n_i, and F. Table EX-18.1 provides a short list of threshold values for a single canceller MTI with $n_i = 8$ and $2\pi\sigma_f/\text{PRF} = 0.01$.

The procedure for determining detection probabilities starts with finding the noise power normalized covariance matrix whose elements are given by

$$C_{kl} = \sum_{i=0}^{m-1} \sum_{j=0}^{m-1} w_i w_j \rho^*(i + k - j - l)$$

with weights $\mathbf{w}^T = [w_o, \ldots, w_{m-1}]$ and $k, l = 1, 2, \ldots, n_i$ while

Table EX-18.1 Threshold Values for MTI and Noncoherent Integrator ($m = 2$, $n_i = 8$, $2\pi\sigma_f/\text{PRF} = 0.01$)

P_{fa}	$F = 0$	$F = 1{,}000$	$F = 5{,}000$	$F = 10{,}000$	$F = 20{,}000$
10^{-3}	93.0124	94.9316	110.4373	153.2723	259.0955
10^{-4}	112.4733	114.4202	132.4501	191.3548	333.8704
10^{-5}	131.4242	133.3903	153.8058	229.4294	408.6453
10^{-6}	150.0613	152.0420	174.7568	267.5034	483.4202
10^{-7}	168.4914	170.4834	195.4424	305.5773	558.1952
10^{-8}	186.7786	188.7797	215.9466	343.6512	632.9701
10^{-9}	204.9637	206.9722	236.3234	381.7251	707.6421

$$\rho^*(0) = F + 1 \qquad \rho^*(ii) = F\rho(ii) \qquad F = \frac{P_c}{P_n}$$

where P_n is the noise power, P_c is the clutter power, n_i is the number of pulses integrated, and F is the clutter power-to-noise power ratio. The clutter correlation function is given by

$$\rho(ii) = \exp\left(\frac{-ii^2}{2}\frac{(2\pi\sigma_f)^2}{\text{PRF}}\right) = \rho(1)^{ii^2}$$

where σ_f is the clutter spectrum standard deviation.

The next step in the development is to find the eigenvalues (λ_k) and eigenvector ($\mathbf{e}_k$) of the matrix **C** that allows the output vector of the MTI to be decorrelated. This can be conveniently accomplished numerically using the eig(**C**) command in MATLAB that returns both the eigenvalues and eigenvectors of **C**. This allows the characteristic function of the probability density function to be determined as

$$\Phi(p) = \exp\left[\sum_{k=1}^{n_i} \frac{pg_k^2}{(1 - 2p\lambda_k)}\right] \prod_{k=1}^{n_i} \frac{1}{(1 - 2p\lambda_k)}$$

where $p = j\omega$ and

$$g_k^2 = \sum_{i=1}^{n_i} \sum_{j=1}^{n_i} 4A^2 \sin^2\left(\frac{\omega_d T}{2}\right) \cos[\omega_d T(i-j)]\, \mathbf{e}_{ki}\mathbf{e}_{kj}$$

Detection curves for an MTI followed by a noncoherent integrator can be developed from the characteristic function of the system. The characteristic function of the integrated MTI output can be found as

$$\Phi_o(p) = \Phi^{n_i}(p)$$

Taking the inverse FFT of the complex conjugate of $\Phi_o(p)$ provides the pdf of the integrated output

$$p(z) = ifft[\Phi_o(p^*)]$$

The probability of detection can then be found as

$$P_d = \int_{\alpha_T}^{\infty} p(z)\, dz$$

where α_T is the detection threshold.

The characteristic function of an MTI-noncoherent integrator output is then found as

$$\Phi_o(p) = \left\{ \exp\left[\sum_{k=1}^{n_i} \frac{p g_k^2}{(1 - 2p\lambda_k)} \right] \prod_{k=1}^{n_i} \frac{1}{(1 - 2p\lambda_k)} \right\}^{n_i}$$

Using MATLAB program pd_mti_ncia.m, the inverse FFT of the complex conjugate of this expression is used to determine the integrator's pdf. Then the pdf from α_T to ∞ is integrated using the thresholds determined above, thereby finding P_d. The P_d plots for a single canceller ($m = 2$) followed by a noncoherent integrator ($n_i = 8, 32$) with $2\pi\sigma_f/\text{PRF} = 0.01$ and $F = 0$, 5,000, 10,000, and 20,000 are depicted in Figures EX-18.3 and EX-18.4. The MATLAB program pd_mti_ncia_8.m and pd_mti_ncia_32.m provide these figures.

The resulting detection curve for integrating 8 pulses using a single canceller MTI followed by a noncoherent integrator is depicted above. Note that the improvement factor for this example can be found as 43 dB, resulting in the noise power and residual clutter power being equal for F = 20,000. For this condition the integration gain is about 1 dB since the target and residual clutter are both integrated by the same amount.

The integration gain for a single-canceller MTI followed by a noncoherent integrator is depicted in Figure EX-18.5 for n_i = 3, 8, 16, 32, and 64 pulses. This curve illustrates that a high improvement factor and hence a low residual clutter-to-noise ratio is necessary to achieve a significant integration gain in clutter.

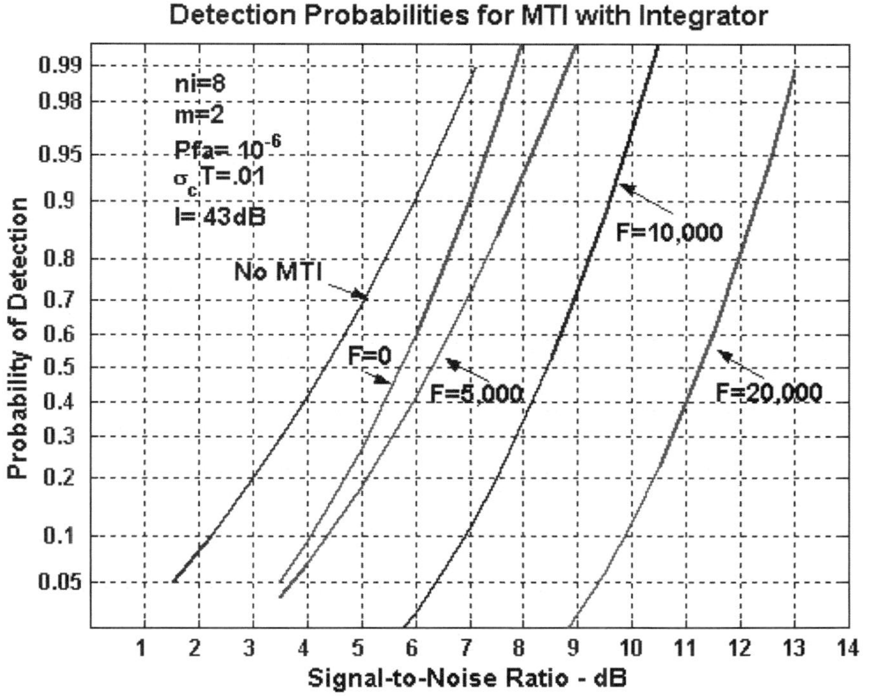

Figure EX-18.3 Detection probability for MTI with noncoherent integrator for $n_i = 8$.

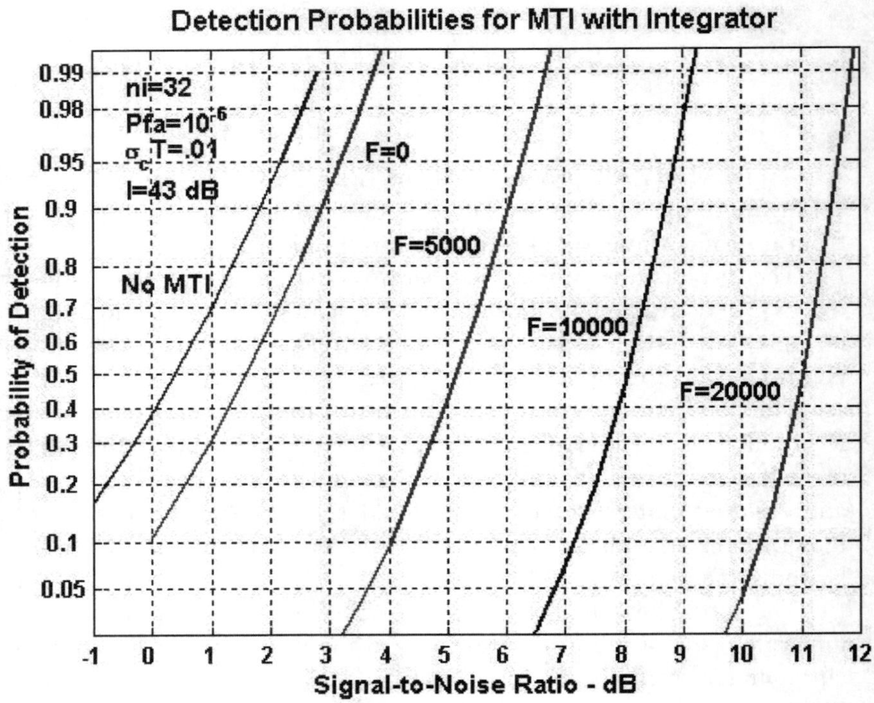

Figure EX-18.4 Detection probability for MTI with noncoherent integrator for $n_I = 32$.

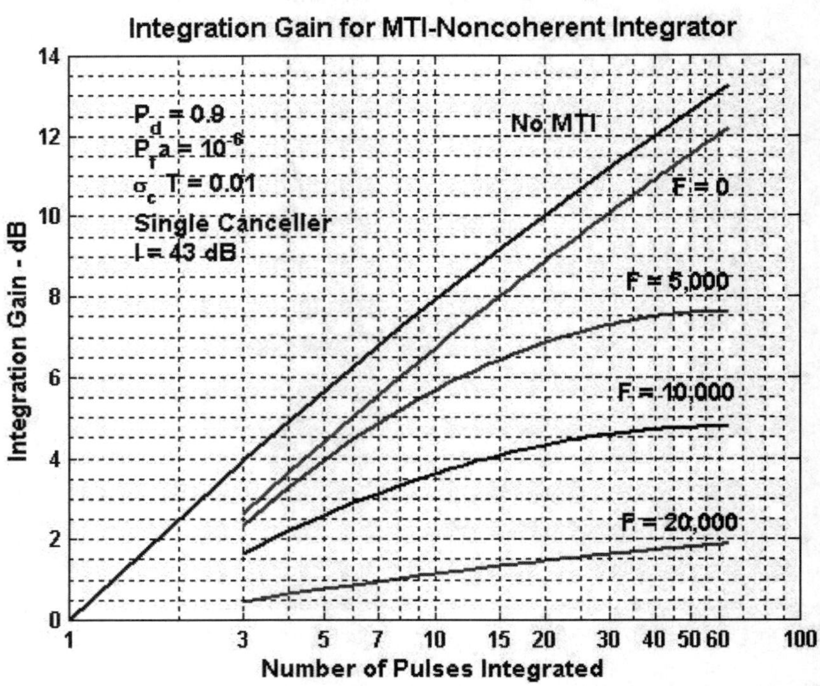

Figure EX-18.5 MTI integration gain with noncoherent integrator.

▶

The evaluation of the detection probabilities of the cascaded MTI-incoherent integrator is best accomplished by using a computer program generated from this procedure. A computer library program that generates eigenvalues and eigenvectors from a covariance matrix input is a very useful tool in implementing a computer program for the procedure. Before giving results of detection calculations for the MTI-incoherent integrator processor, we present several illustrations of calculations involving these equations in addition to a special case for which a closed-form solution of eigenvalues and eigenvectors is available.

Consider the case where a single-canceller quadrature channel MTI ($m = 2$) is followed by a four-pulse video integrator ($l = 4$). The covariance matrix is given by

$$\mathbf{C} = \sigma_n^2 \begin{bmatrix} 2 + 2F(1 - \rho_1) & -1 + F(2\rho_1 - \rho_2 - 1) & F(2\rho_2 - \rho_1 - \rho_3) & F(2\rho_3 - \rho_2 - \rho_4) \\ -1 + F(2\rho_1 - \rho_2 - 1) & 2 + 2F(1 - \rho_1) & -1 + F(2\rho_1 - \rho_2 - 1) & F(2\rho_2 - \rho_1 - \rho_3) \\ F(2\rho_2 - \rho_1 - \rho_3) & -1 + F(2\rho_1 - \rho_2 - 1) & 2 + 2F(1 - \rho_1) & -1 + F(2\rho_1 - \rho_2 - 1) \\ F(2\rho_3 - \rho_2 - \rho_4) & F(2\rho_2 - \rho_1 - \rho_3) & -1 + F(2\rho_1 - \rho_2 - 1) & 2 + 2F(1 - \rho_1) \end{bmatrix} \tag{5.141}$$

where σ_n^2 is the receiver noise power, the $\rho(i)$ denote the clutter correlation coefficients, and F is the clutter-to-receiver noise power ratio. The covariance matrix is simplified when no clutter is present ($F = 0$):

$$\mathbf{C} = \sigma_n^2 \begin{pmatrix} 2 & -1 & 0 & 0 \\ -1 & 2 & -1 & 0 \\ 0 & -1 & 2 & -1 \\ 0 & 0 & -1 & 2 \end{pmatrix} \tag{5.142}$$

For this case, an analytic solution is available for the eigenvalues and eigenvectors associated with the covariance matrix [30, 35]:

$$\lambda_k = 4 \sin^2 \frac{k\pi}{2(l + 1)} \tag{5.143}$$

$$e_{kj} = \sqrt{\frac{2}{l + 1}} \sin \frac{kj\pi}{l + 1}; \qquad j = 1, 2, \ldots, l \tag{5.144}$$

The quadrature mean vectors from (5.122) and (5.123) are

$$s_{xk} = -2A \sin\left(\frac{\omega_d T}{2}\right) \sin\left[\omega_d T\left(k - \frac{1}{2}\right)\right] \tag{5.145}$$

$$s_{yk} = 2A \sin\left(\frac{\omega_d T}{2}\right) \cos\left[\omega_d T\left(k - \frac{1}{2}\right)\right] \tag{5.146}$$

The signal amplitude from (5.132) then is

$$g_k^2 = \sum_{i=1}^{l} \sum_{j=1}^{l} 4A^2 \sin^2\left(\frac{\omega_d T}{2}\right) \cos[\omega_d T(i-j)] e_{ki} e_{kj} \tag{5.147}$$

Detection curves for $\omega_d T = 1/4$ are evaluated in [31].

Figures 5.28, 5.29, and 5.30 depict detection curves for a cascaded single-canceller MTI and an 8-pulse video integrator ($\omega_d T = 1/4$) with narrow ($\sigma_c T = 0.01$), medium ($\sigma_c T = 0.05$), and wide ($\sigma_c T = 0.1$) Gaussian-shaped clutter spectra, respectively [6]. Also shown for comparison purposes are detection curves for an optimal nine-pulse transversal filter and a quasi-optimum cascaded single-canceller MTI 8-pulse coherent integrator.

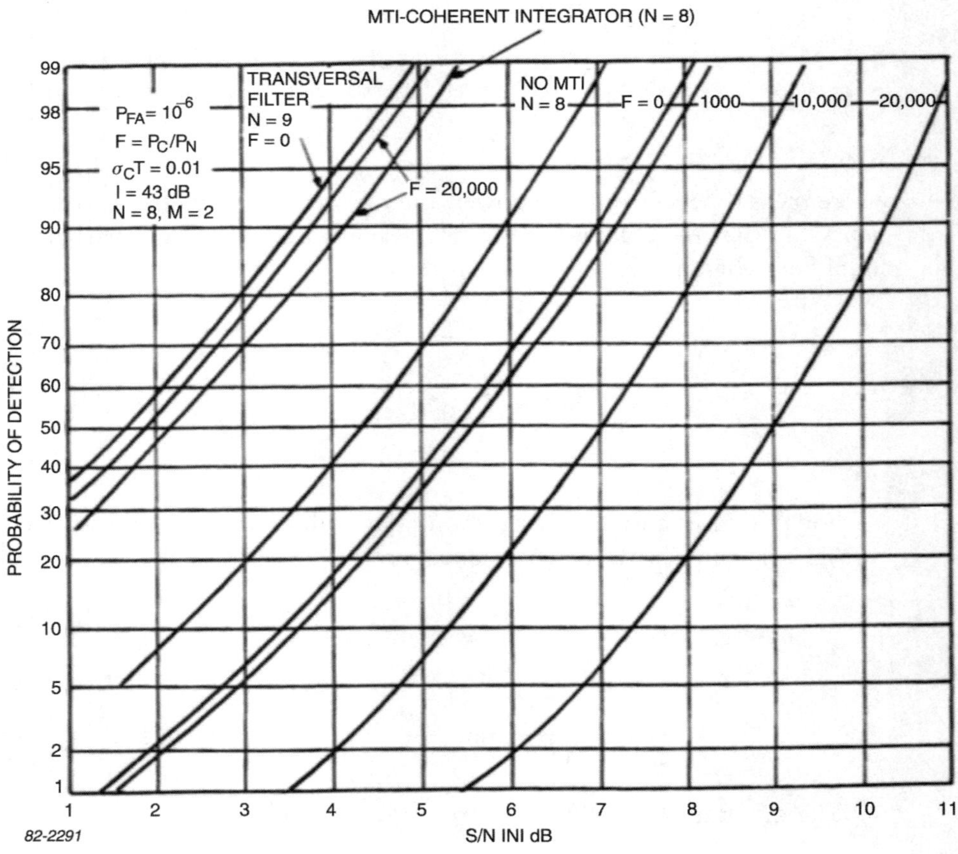

Figure 5.28 Probability of detection for a single-canceller quadrature MTI with video integration ($N = 8$). (*From:* [6, p. 157].)

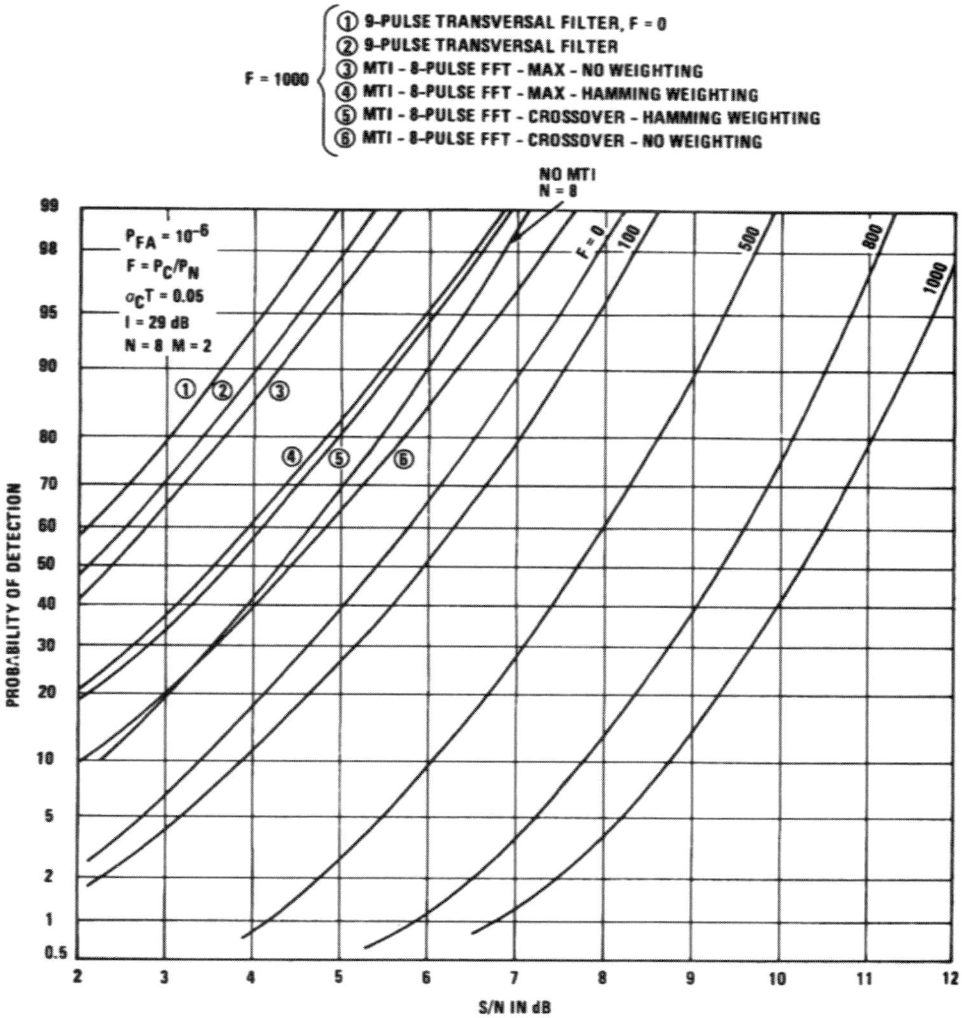

Figure 5.29 Probability of detection for a single-canceller quadrature MTI with video integration ($N = 8$). (*From:* [6, p. 158].)

In Figure 5.28, for a narrow clutter spectrum ($\sigma_c T = 0.01$), the MTI improvement factor (see Table 5.1) is 20,000 (43 dB). The detection loss for the optimal transversal filter is on the order of 0.2 dB between no clutter ($F = 0$) and a clutter-to-noise power ratio of $F = 20{,}000$. The detection performance in heavy clutter ($F = 20{,}000$) of the MTI-coherent integrator at maximum response is essentially that of an 8-pulse coherent integrator, indicating that the processor was effective in removing the clutter. The effect of the MTI and clutter on the incoherent integrator results in a loss of 1.1 dB with noise alone and 4.3 dB when the clutter residual power equals the noise power.

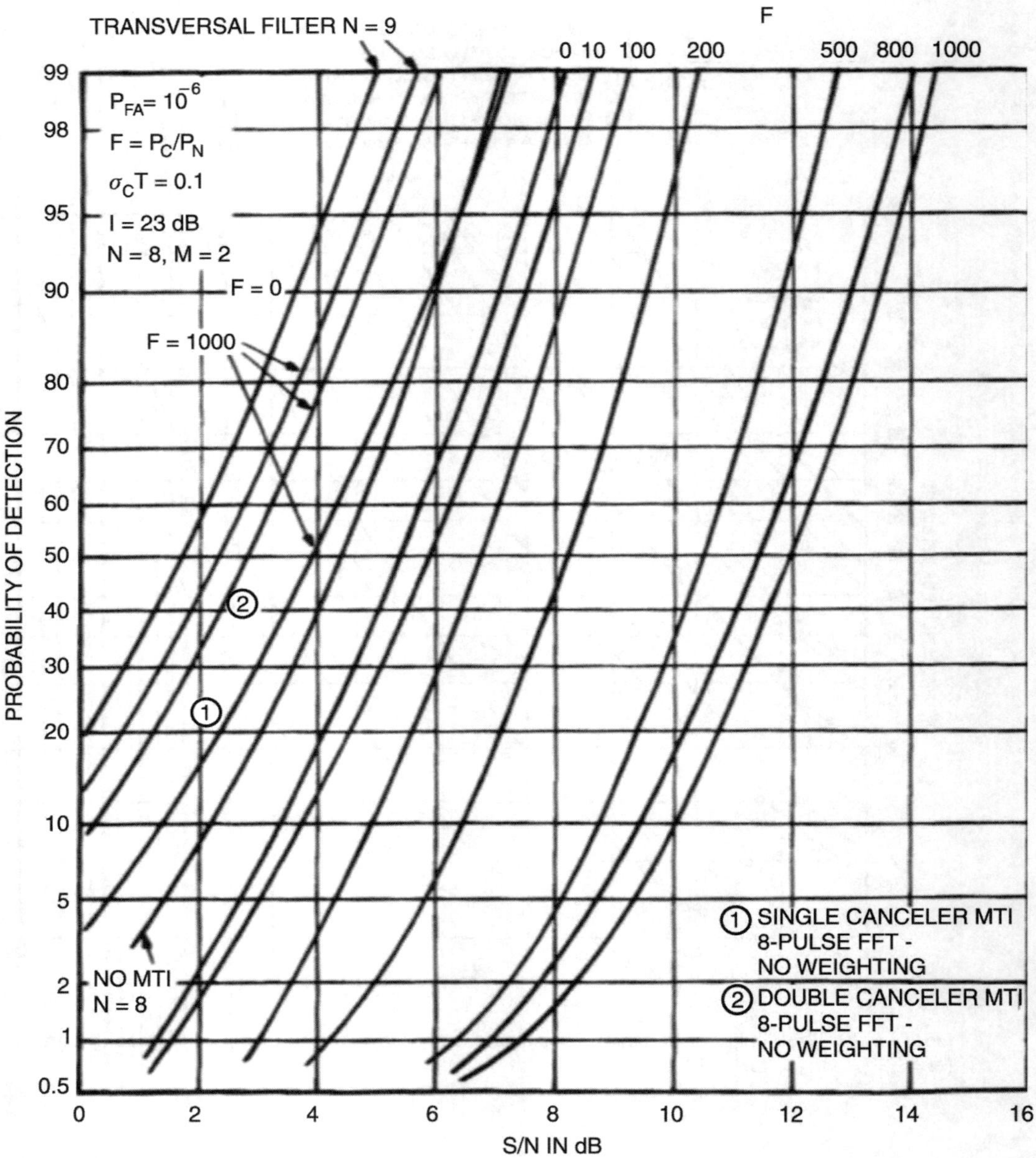

Figure 5.30 Probability of detection for a single-canceller quadrature MTI with video integration ($N = 8$). (*From:* [6, p. 158].)

► Example 19 Comparison Transversal Filter and Cascaded MTI-FFT

In Example 3 we analyzed the improvement factor and transfer functions provided by a cascaded MTI-FFT processor. The analysis was enabled by converting the cascade to an equivalent transversal filter and indicated the need to prevent cross coupling by using Chebyshev windowing to reduce FFT sidelobes. In this example we extend this analysis to examine the probability of detection of the cascaded

MTI-FFT and find that this provides a large improvement over the MTI-noncoherent integrator combination examined in Example 18. We then explore the performance provided by a pure point-optimized bank of transversal filters and find that this configuration provides an advantage over the cascaded MTI-FFT both in an enhanced probability of detection and clutter processing capability. As a note of interest, the original MTD processor employed a cascaded MTI-FFT (see Section 1.6) while later improved versions converted to transversal filter processors that provided enhanced performance.

In Example 3 the equivalent transversal filter weights of the cascaded MTI-FFT using a double canceller MTI and N-point FFT were shown to be given by

$$\mathbf{w}_{\text{MTI–FFT}} = [\mathbf{w}_{\text{FFT}}, 0] - [0, \mathbf{w}_{\text{FFT}}]$$

where

$$\mathbf{w}_{\text{FFT}} = \exp\left[\frac{-2\pi j(p-1)(K-1)}{N}\right]$$

and K is the filter number and $p = 1, 2, \ldots, N$. Further the clutter covariance matrix ($\mathbf{R}_n$) and signal covariance matrix ($\mathbf{M}_s$) were calculated. The improvement factor for clutter is then given by

$$I_{fc} = \frac{\mathbf{w}^T \mathbf{M}_s \mathbf{w}^*}{\mathbf{w}^T \mathbf{R}_N \mathbf{w}^*}$$

The improvement factor of the MTI against noise provides no gain as can be shown by evaluating the relationship

$$I_N = \frac{\sum_{j=0}^{n-1} w_j^2}{\sum_{j=0}^{n} \sum_{k=0} w_j w_k \boldsymbol{\rho}_n(j-k)}$$

where the w_j are the MTI weights and $\boldsymbol{\rho}_n$ is the identity matrix. The improvement factor of the FFT against noise is equal to the number of FFT points (N) when evaluated at the peak of its response and otherwise from (5.85). Hence, the integration gain (IG) of the MTI-FFT cascade against noise when evaluated at the peak response of the FFT is equal to N.

Both the MTI and FFT are linear systems as is their cascade. The application of a Gaussian distributed process to a linear system results in another Gaussian distributed process at its output. Hence, we can apply Marcum's analysis for a square-law envelope detector to determine the probability of detection for the cascaded MTI-FFT. This is given by

$$P_d = Q\left\{\left[\frac{2\text{snr}_{in} IG}{\left(1+\frac{F}{I_{fc}}\right)}\right]^{1/2}, \left[2\ln\left(\frac{1}{P_{fa}}\right)\right]^{1/2}\right\}$$

where Q(a, b) is Marcum's Q function, F is the clutter-to-noise power ratio, P_{fa} is the desired probability of false alarm, and snr_{in} is the input signal-to-noise ratio. The MATLAB program pd_mti_ffta.m provides an output plot of P_d on a normal probability scale. The program applies for any order binomial MTI and FFT points with uniform, Chebyshev, or Hamming weighting. It uses the MATLAB function marcumq (a, b) to compute the Q function. The program output for a single canceller followed by an 8-point FFT with uniform weighting is depicted in Figure EX-19.1. Note the deterioration of performance as the clutter power increases. This can be improved by using a higher order MTI canceller at the expense of less performance for $F = 0$.

The point-optimized transversal filter, discussed in Section 5.1.2, provides an attractive alternative to the cascaded MTI-FFT processor. In this processor a number of point-optimized transversal filters are generally equally spaced through the Doppler region of interest. The spacing is generally selected so that the transversal filter passbands cross over at their 3 dB points.

The interference (clutter and noise) covariance matrix is given by

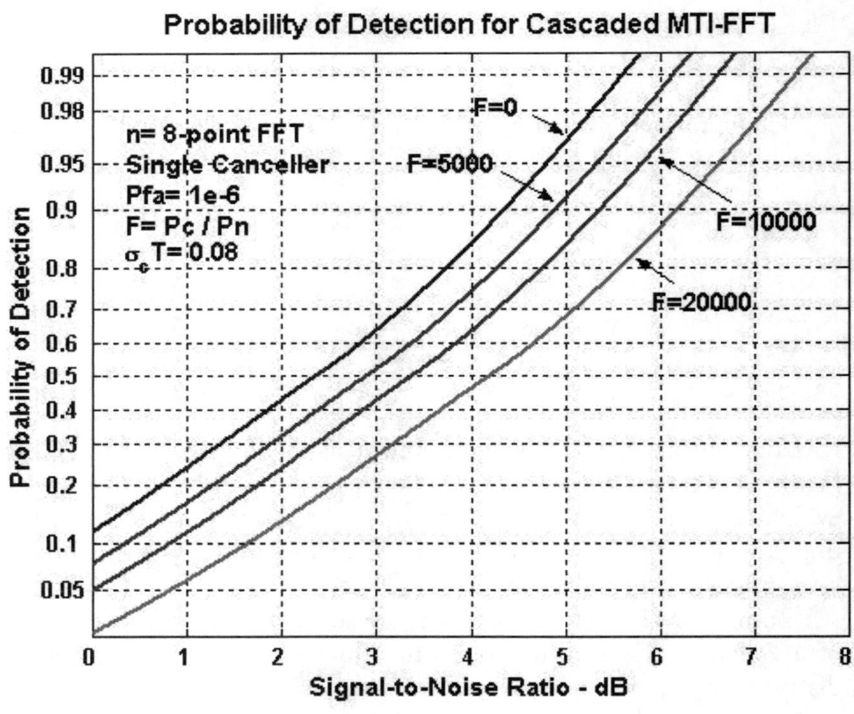

Figure EX-19.1 Probability of detection for cascaded MTI-FFT.

$$\mathbf{R}_n = \begin{pmatrix} 1+F & F\rho_1 & \cdots & F\rho_{n-1} \\ F\rho_1 & \ddots & a_{23} & \vdots \\ \vdots & & \ddots & \vdots \\ F\rho_{n-1} & \cdots & \cdots & 1+F \end{pmatrix}$$

where F is the clutter-to-noise power ratio and ρ_i is the clutter correlation coefficient. The filter is tuned to the desired point by selecting the signal vector as

$$s_k^T = e^{j\omega_d(k-1)/\text{PRF}} \qquad k = 1, 2, \ldots, n$$

where n is the number of points and f_d is the Doppler frequency of interest. The optimum transversal filter weights can then be found as

$$\mathbf{w}_o = \mathbf{R}_n^{-1}\mathbf{s}^*$$

where $\mathbf{R}_n^{-1}$ is the inverse covariance matrix. The envelope of the optimum improvement factors can be found from the relationship

$$I_{fo} = \text{trace}\left(\mathbf{R}_n^{-1}\mathbf{M}_s\right)$$

where $\mathbf{M}_s$ is the signal covariance matrix evaluated at all possible Doppler frequencies from 0 to PRF. The optimum point improvement factor frequency response is evaluated from

$$I_f = \frac{\mathbf{w}_o^T\mathbf{M}_s\mathbf{w}_o^*}{\mathbf{w}_o^T\mathbf{R}_N\mathbf{w}_o^*}(1+F)$$

where $\mathbf{w}_o$ are the optimum weights and $\mathbf{M}_s$ is evaluated over the Doppler frequency range. These values are evaluated in MATLAB program trans_filt_impfacbx.m and shown in Figure EX-19.2 for a 9-point filter tuned to PRF/3 and $F = 5{,}000$.

The next step in the development is to determine the probability of detection for the optimized point transversal filter followed by a square-law envelope detector evaluated at the tuned Doppler frequency. This can be found from Marcum's analysis as

$$P_d = Q\left\{[2\text{snr}_{in} I_f]^{1/2}, \left[2\ln\left(\frac{1}{P_{fa}}\right)\right]^{1/2}\right\}$$

The resulting probability of detection for the transversal filter is depicted in Figure EX-19.3.

The MATLAB program compare.m plots a comparison between the transversal filter and cascaded MTI-FFT processors. This is shown in Figure EX-19.4 for a 9-point FFT and an MTI followed by an 8-point FFT. When no clutter is present, the snr to provide a 0.9 probability of detection is within about 0.5 dB. However,

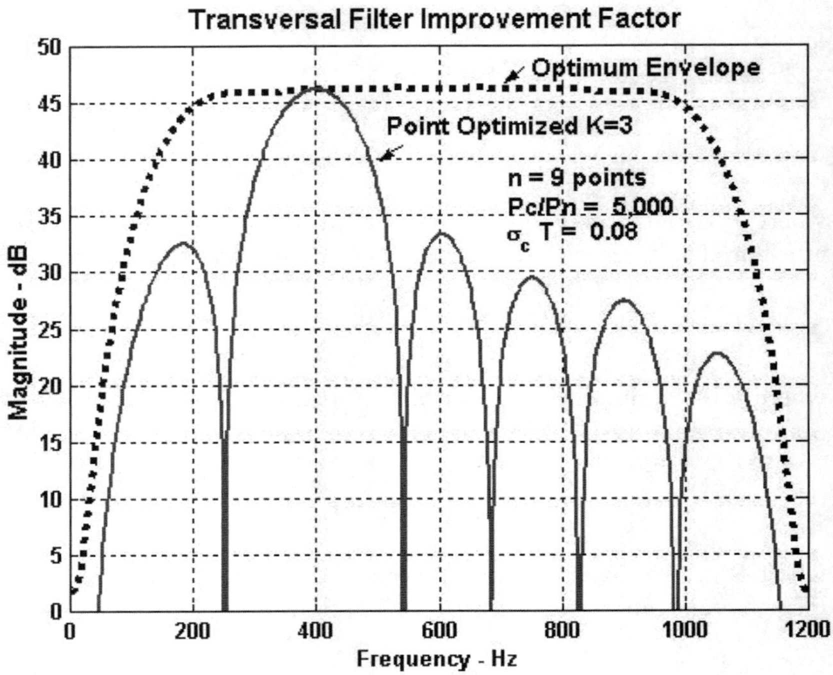

Figure EX-19.2 Transversal filter improvement factor and envelope.

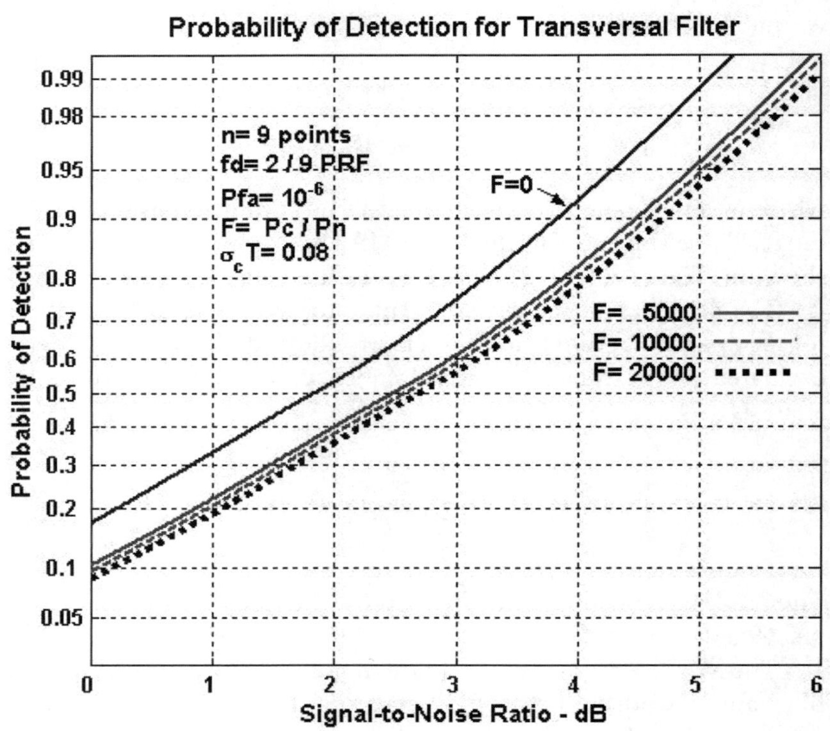

Figure EX-19.3 Probability of detection for transversal filter.

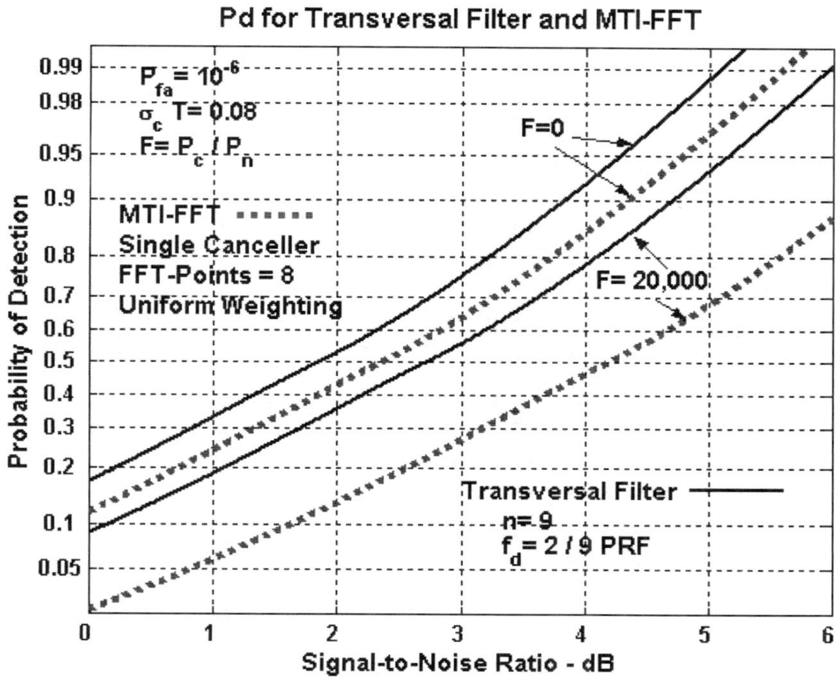

Figure EX-19.4 Probability of detection for transversal filter and MTI-FFT.

when heavy clutter is present ($F = 20{,}000$), the margin is 1.25 dB in favor of the transversal filter.

In Figure 5.29, the detection performance is shown for a medium clutter spectrum ($\sigma_c T = 0.05$). The single-canceller MTI improvement factor for this case is 29 dB. The detection loss of the optimum transversal filter is on the order of 0.4 dB between no clutter ($F = 0$) and a clutter-to-noise power ratio of $F = 1{,}000$. The peak detection performance of the cascaded MTI-coherent integrator is within the order of 0.15 dB of an 8-pulse coherent integrator in noise, which indicates good performance in clutter. However, at the crossover between adjacent Doppler filters, the composite performance (considering detection probabilities in both filters) is almost –2 dB below the peak. Use of Hamming weighting of the coherent integrator response considerably smooths the overall Doppler response, but the overall performance for $F = 1{,}000$ is uniformly on the order of –1.4 dB below that of an ideal 8-pulse coherent integrator in noise. The detection loss of the MTI-incoherent integrator is again 1.1 dB for noise alone and 3.5 dB when residual clutter power is equal to noise power.

Figure 5.30 depicts the detection performance for a wide clutter spectrum ($\sigma_c T = 0.1$). The single-canceller MTI improvement factor for this case is 23 dB. Performance of the optimum transversal filter again is good over strong clutter ($F = 1{,}000$) with the reduction on the order of 0.8 dB with respect to noise-only performance ($F = 0$). The performance of the single-canceller coherent integrator for F = 1,000 is poor, being on the order of −2.7 dB below that of an 8-pulse coherent integrator in noise, which indicates that a better MTI filter is required. Use of a double canceller, $\mathbf{w}^T$ = (1, −2, 1), instead of a single canceller, reduces the detection loss for F = 1,000 to within 0.5 dB of an 8-pulse coherent integrator in noise. Detection loss for the MTI-incoherent integrator ranges from 1.1 dB for noise alone to 7.5 dB for F = 1,000. This loss is plotted in Figure 5.31 as a function of residual clutter power (F/1). The curve indicates that residual clutter power must be made significantly less than receiver noise power if good detection performance is to be realized by using the MTI-incoherent integrator Doppler processor.

The performance comparison in Figures 5.28 through 5.30 can be summarized as follows:

- The optimum complex weight transversal filters provide the best detection performance, especially in heavy clutter environments.
- The performance of cascaded MTI-coherent integrators is inherently somewhat lower than that of transversal filters for an equal number of processed pulses.
- Coherent integrator weighting is required to provide a uniform overall Doppler processing response.
- Higher order MTI filters in combination with coherent integrators are required for wide clutter spectra to provide good detection performance.
- MTI-incoherent integrator Doppler processors are inherently lossy. For a single-canceller MTI with 8-pulse incoherent integration, losses range from 1.1 dB for noise-only, to 3.5 to 4 dB when residual clutter equals noise, and to more than 7.5 dB when residual clutter is five times the noise power.

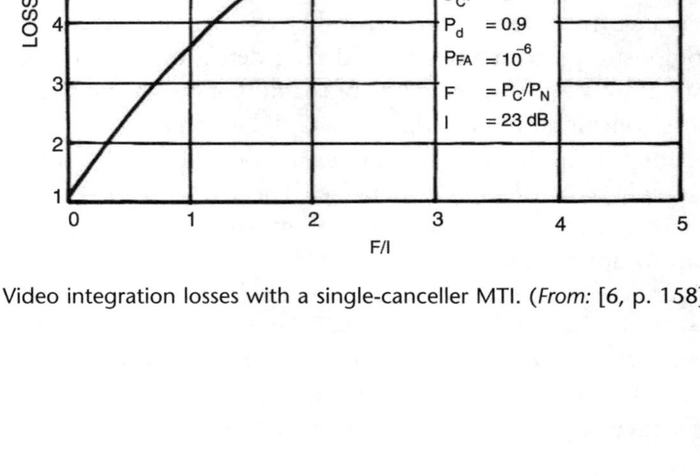

Figure 5.31 Video integration losses with a single-canceller MTI. (*From:* [6, p. 158].)

References

[1] DiFranco, J., and W. Rubin, *Radar Detection*, Dedham, MA: Artech House, 1980.

[2] Skolnik, M., (ed.), *Radar Handbook*, 2nd ed., New York: McGraw-Hill, 1990.

[3] Hall, W., and H. Ward, "Signal-to-Noise Loss in Moving Target Indicator," *Proc. IEEE*, Vol. 56, No. 2, February 1968, pp. 233–234.

[4] Trunk, G., "MTI Integration Loss," *Proc. IEEE*, Vol. 65, No. 11, November 1977, pp. 1620–1621.

[5] Trunk, G., *MTI Noise Integration Loss*, Report 8132, Naval Research Laboratory, Washington, D.C., July 1977.

[6] Schleher, D. C., "Performance Comparison of MTI and Coherent Doppler Processors," *Proc. IEE Int. Radar Conference*, London, November 1982, pp. 154–158.

[7] Schleher, D. C., *MTI and Pulsed Doppler Radar*, 1st ed., Norwood, MA: Artech, 1989.

[8] Schleher, D. C., ed., *MTI Radar*, Artech House, Dedham, MA, 1978.

[9] Emerson, R., *Some Pulsed Doppler MTI and AMTI Techniques*, Report R-274, Rand Corporation, March 1954; also in D. C. Schleher, (ed.), *MTI Radar*, Dedham, MA: Artech House, 1978.

[10] Capon, J., "Optimum Weighting Functions for the Detection of Sampled Signals in Noise," *IEEE Trans. on Information Theory*, Vol. IT-10, No. 2, April 1964, pp. 152–159; also in D. C. Schleher, (ed.), *MTI Radar*, Dedham, MA: Artech House, 1978.

[11] Andrews, G., *Optimum Radar Doppler Filtering Techniques*, Report 7727, Naval Research Laboratory, Washington, D.C., May 1974.

[12] Van Trees, H., *Detection, Estimation, and Modulation Theory*, Part III, New York: Wiley, 1971.

[13] Ayres, F., *Theory and Problems of Matrices*, New York: Schaum Publishing, 1962.

[14] R. Bellman, *Introduction to Matrix Analysis*, New York: McGraw-Hill, 1970.

[15] Schleher, D. C., and D. Schulkind, "Optimization of Nonrecursive MTI," *Proc. IEE Int. Radar Conference*, London, October 1977, pp. 182–185; also in D. C. Schleher, (ed.), *MTI Radar*, Dedham, MA: Artech House, 1978.

[16] Dantzig, G., *Linear Programming and Extensions*, Princeton, NJ: Princeton University Press, 1963.

[17] Andrews, G., *Comparison of Radar Doppler Filtering Techniques*, Report 7811, Naval Research Laboratory, Washington, D.C., October 1974.

[18] McAulay, R., *A Theory of Optimum Moving Target Indicator* (MTI) Digital Signal Processing, Supplement 1, MIT Lincoln Laboratory, Lexington, MA, October 31, 1972.

[19] Andrews, G., *Performance of Cascaded MTI and Coherent Integration Filters in a Clutter Environment*, Report 7533, Naval Research Laboratory, Washington, D.C., March 1973.

[20] Dillard, G., "Signal-to-Noise Ratio Loss in an MTI Cascaded with Coherent Integration Filters," *Proc. IEEE Radar Conference*, Washington, D.C., April 1975, pp. 117–122.

[21] Brigham, E., and R. Murrow, "The Fast Fourier Transform," *IEEE Spectrum*, Vol. 4, No. 12, December 1967, pp. 63–70.

[22] Bergland, G., "A Guided Tour of the Fast Fourier Transform," *IEEE Spectrum*, Vol. 6, No.7, July 1969, pp. 41–52.

[23] M. Skolnik, *Introduction to Radar Systems*, 3rd ed., New York: McGraw-Hill, 2001.

[24] Oppenheim, A., and R. Schafer, *Digital Signal Processing*, Englewood Cliffs, NJ: Prentice-Hall, 1975.

[25] Rabiner, L., and B. Gold, *Theory and Application of Digital Signal Processing*, Englewood Cliffs, NJ: Prentice-Hall, 1975.

[26] Schwarz, M., and L. Shaw, *Signal Processing: Discrete Spectral Analysis, Detection and Estimation*, New York: McGraw-Hill, 1975.

[27] Muller, B., "MTI Loss with Coherent Integration of Weighted Pulses," *IEEE Trans. on Aerospace and Electronic Systems*, Vol. AES-17, July 1981, pp. 549–552.

[28] Nathanson, F., *Radar Design Principles*, 2nd ed., Mendham, NJ: SciTech Publishing, 1991.

[29] Weiss, M., and I. Gertner, "Loss in Single-Channel MTI with Post-Detection Integration," *IEEE Trans. on Aerospace and Electronic Systems*, Vol. AES-18, No. 2, March 1982, pp. 205–208.

[30] Kanter, I., "A Generalization of the Detection Theory of Swerling," *Proc. IEEE EASCON Conference*, Washington, D.C., September 1974, pp. 198–205; also in S. Haykin, (ed.), *Detection and Estimation*, Halstead Press, 1976.

[31] Dillard, G., and J. Richard, "Performance of an MTI Followed by Incoherent Integration for Nonfluctuating Signals," *Proc. IEEE Int. Radar Conference*, Washington, D.C., April 1980, pp. 194–199.

[32] Cramer, H., *Mathematical Methods of Statistics*, Princeton, NJ: Princeton University Press, 1946.

[33] Whalen, A., *Detection of Signals in Noise*, New York: Academic Press, 1971.

[34] Kendall, M., and A. Stuart, *The Advanced Theory of Statistics*, Vol. 11, New York: Hafner Publishing, 1963.

[35] Jain, A., "Fast Karhunen-Loeve Transform for a Class of Random Processes," *IEEE Trans. on Communications*, Vol. COM-24, September 1976, pp. 1023–1029.

CHAPTER 6

MTI Systems

Functionally, MTI radar allows the detection of targets imbedded in clutter. The basic principle involves filtering the undesired clutter return by exploiting the differential Doppler frequency shift that exists when there is relative radial motion between the target scatterers and the clutter scatterers. The precise meaning of the term MTI *radar* is somewhat ubiquitous, but it is generally defined as follows.

First, an MTI radar generally operates at a sufficiently low PRF such that unambiguous range measurements are possible. At microwave frequencies, Doppler measurements are highly ambiguous, creating so-called blind speeds, which occur in small regions about the radial target velocities corresponding to multiples of the radar's PRF. The objective of the MTI is to provide a comb filter response in the Doppler frequency domain that ideally has a stopband in regions of heavy clutter concentration and passes unattenuated all other target velocities. As shown in Section 5.1.3, this comb filter is the optimal linear filter when all target Doppler frequencies are assumed equally likely.

In general, the objective of the MTI processor is to remove the clutter, thereby improving the signal-to-clutter ratio. However, it does not improve the single-hit S/N ratio. As described in Sections 5.2 and 5.3, it is possible to follow the MTI with either a coherent or noncoherent (after detection) integrator. The MTI-noncoherent integrator can either improve or deteriorate the single-hit S/N ratio, depending on the situation and must be applied with caution.

6.1 MTI Configurations

A classic form of a ground-based surveillance radar type of coherent MTI is depicted in Figure 6.1. The coherent transmission is formed by mixing a stable local oscillator reference source with a coherent oscillator at an intermediate frequency (IF). Transmission then occurs at the upconverted sum of the STALO and COHO frequencies. The received target signals are downconverted to IF by mixing with the STALO where they are further converted into baseband signals by comparison with the COHO in a phase detector. The output of the phase detector is a bipolar video signal at the target's Doppler frequency (see Section 2.3.1).

The ground and weather clutter spectrum for this coherent ground-based MTI is depicted in Figure 6.2. Note that the ground clutter spectral lobes (which are spread primarily by scanning modulation) appear at multiples of the radar's PRF. This is due to the sampled data aspect of the pulse radar waveform. Weather

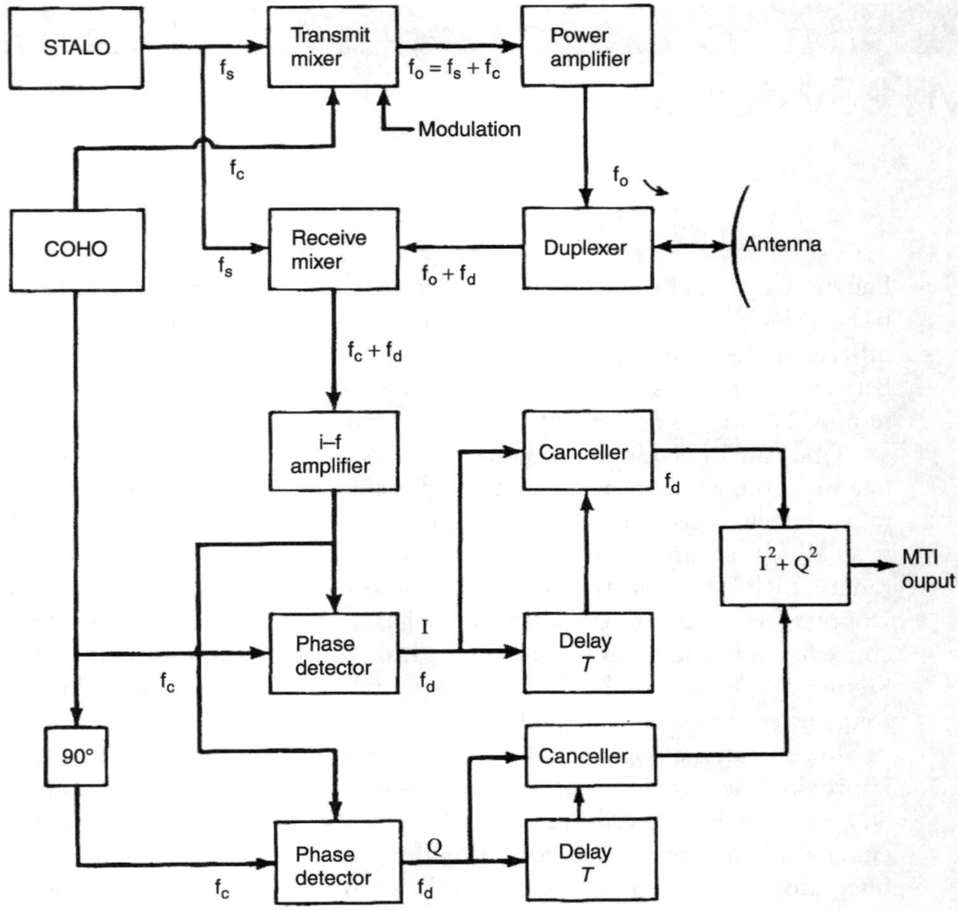

Figure 6.1 Ground-based coherent MTI system. (*From:* [1, p. 238].)

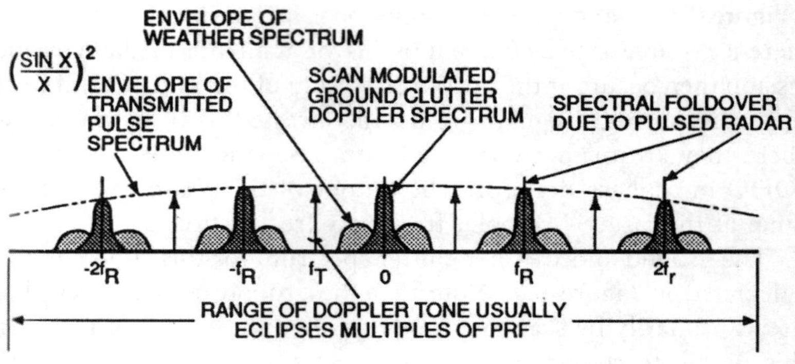

Figure 6.2 Ground and weather clutter spectrum for a coherent ground-based MTI. (*From:* [2, p. 341].)

clutter, which in general has a nonzero radial velocity, is shown between the PRF lines with greater spectral width (due to wind shear) than for ground clutter (see Figure 1.20). The range of target Doppler frequencies usually encompasses multiple PRF regions, thereby causing the target Doppler to be ambiguous. When the target Doppler falls into those regions occupied by the ground clutter lobes, the MTI cannot detect the target. This is the blind speed effect in MTI systems, where the MTI is unable to detect targets with radial velocities that create Doppler frequencies corresponding to multiples of the radar's PRF. Also note that the weather clutter spectrum is also spread into multiple PRF regions, thereby potentially causing target responses with higher ambiguous radial speeds to be aliased by the weather clutter.

A quadrature type of MTI canceller that employs in-phase and quadrature phase detector channels is shown in Figure 6.1. This type of MTI processor preserves the coherent nature of the signals and conforms to the structure of the optimal MTI described in Section 5.1.3. An alternative configuration that uses a single-phase detector is sometimes employed, particularly in older designs. With this type of design, both positive and negative frequencies are folded into a band between zero frequency and half the PRF ($f_r/2$). This results in the blind phase phenomenon in which signals in quadrature with the COHO reference are rejected.

The use of a single-channel coherent MTI results in a sensitivity loss against receiver noise as compared to fully coherent MTI systems that process both in-phase and quadrature channels. The most severe loss occurs for nonfluctuating targets with high detection probabilities when no postdetection integration is used (see Section 2.3.1). The asymptotic loss using postdetection integration with a large number of pulses approaches 1.5 dB. The number of degrees of freedom for a fluctuating target is reduced by a factor of 2, which increases the fluctuation loss when compared to fully coherent processing.

The function of the MTI canceller is to attenuate the clutter returns by applying a comblike filter response, as depicted in Figure 6.3 for single- and double-cancellation units. The type of response shown has deep nulls about the PRF lines and, hence, is appropriate for use against ground clutter. In general, the greater the number of pulses processed in the MTI canceller, the wider the rejection notch and the greater the clutter attenuation (see Figures 2.25 and 3.5). Also note that in ground-based surveillance radar, the width of the ground clutter lobe is determined

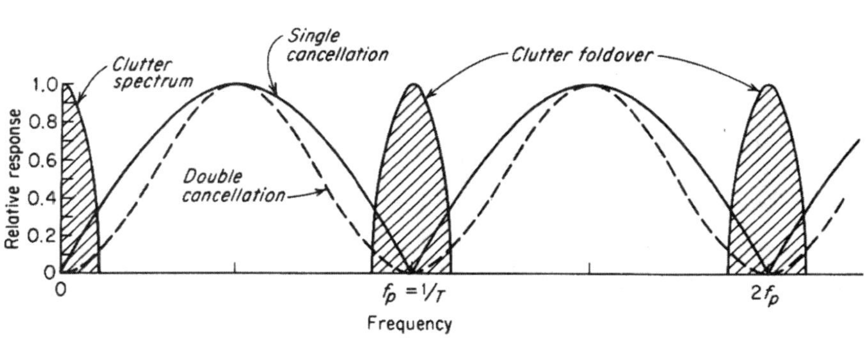

Figure 6.3 MTI filter response for single and double cancellers. (*From:* [3, p. 116].)

primarily by antenna scanning modulation (see Section 3.2.2). Thus, the slower the antenna scans, the narrower the clutter lobe and the greater the clutter attenuation for a given canceller complexity (i.e., number of pulses processed).

A second type of coherent MTI applicable for radars using power oscillator-type transmitters (e.g., magnetrons) functions by phase-locking the COHO to each transmitter pulse (see Figure 2.21). The COHO can be locked by synchronizing its phase to the transmitter phase for each transmitted pulse. Alternatively, a continuously oscillating COHO can be utilized, and the phase of the transmitted pulse may be measured and used to compensate the phase detector output, as depicted in Figure 6.4 [1]. The latter approach is generally preferred in digital MTI implementations (using a digital phase shifter) because a continuously running COHO is generally more stable than one that must be continuously synchronized.

The coherent-on-receive MTI is incapable of cancelling multiple-time-around clutter returns that occur at ranges greater than those corresponding to a basic pulse repetition interval. Thus, this type of MTI is not appropriate to situations where multiple-time-around clutter returns are anticipated (e.g., from cities, mountains, et cetera). When this occurs, the fully coherent (see Figure 6.1) MTI must be used. Also, if PRF staggering is employed to compensate for blind speeds, block-

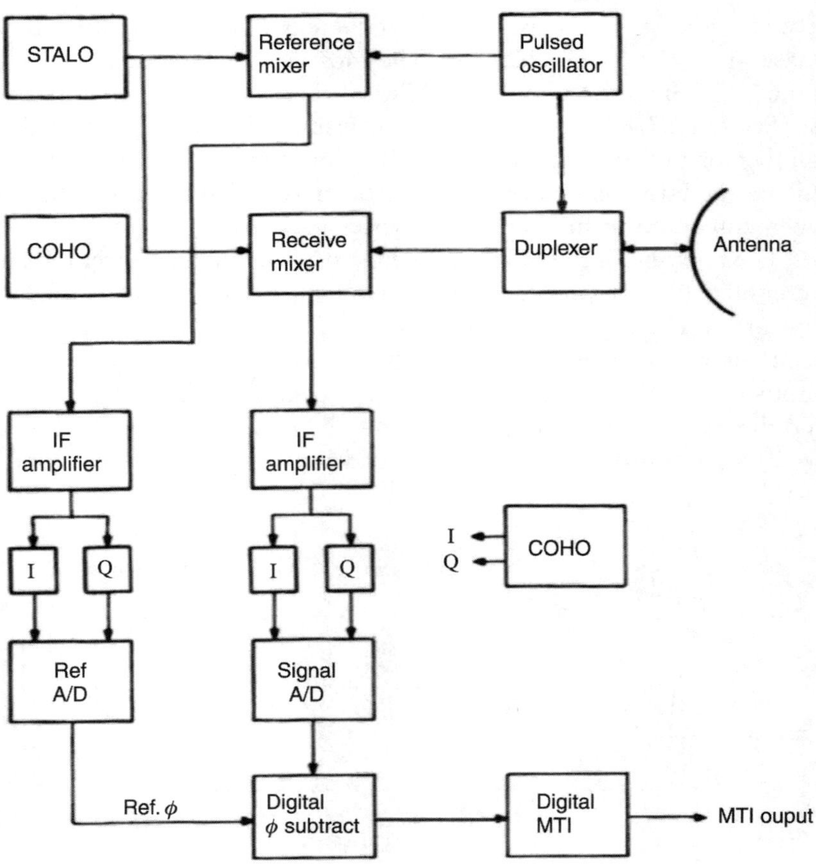

Figure 6.4 Coherent-on-receive MTI with digital phase storage. (*From:* [1, p. 242].)

to-block staggering must be used rather than pulse-to-pulse staggering to cancel ambiguous clutter [4].

Frequently, an MTI must cope with clutter having spectra with a nonzero central Doppler frequency. This occurs for fixed-based radars that experience wind-driven airborne clutter such as that associated with precipitation or chaff and also when sea clutter is present. Doppler frequency shift of the clutter also occurs when the MTI is located on a moving platform such as a naval ship or an airborne or space vehicle. In these cases, some form of reference frequency adjustment is required to move the null of the coherent MTI canceller over the clutter spectra. This problem is compounded when clutter spectra with different central frequencies occupy the same range resolution cell (e.g., weather clutter over ground clutter). Solutions to this problem include (1) cascading MTI cancellers with different null frequencies (e.g., see Section 1.3.3) and (2) following the MTI with a bank of Doppler filters in which the filters containing moving clutter returns are desensitized, thereby removing the undesired clutter (see Section 1.6; the MTD).

Compensation for platform motion can be accomplished in the coherent MTI by offsetting the COHO's frequency as depicted in Figure 6.5. With this approach, the COHO frequency is mixed with the output of a compensating VCO, the

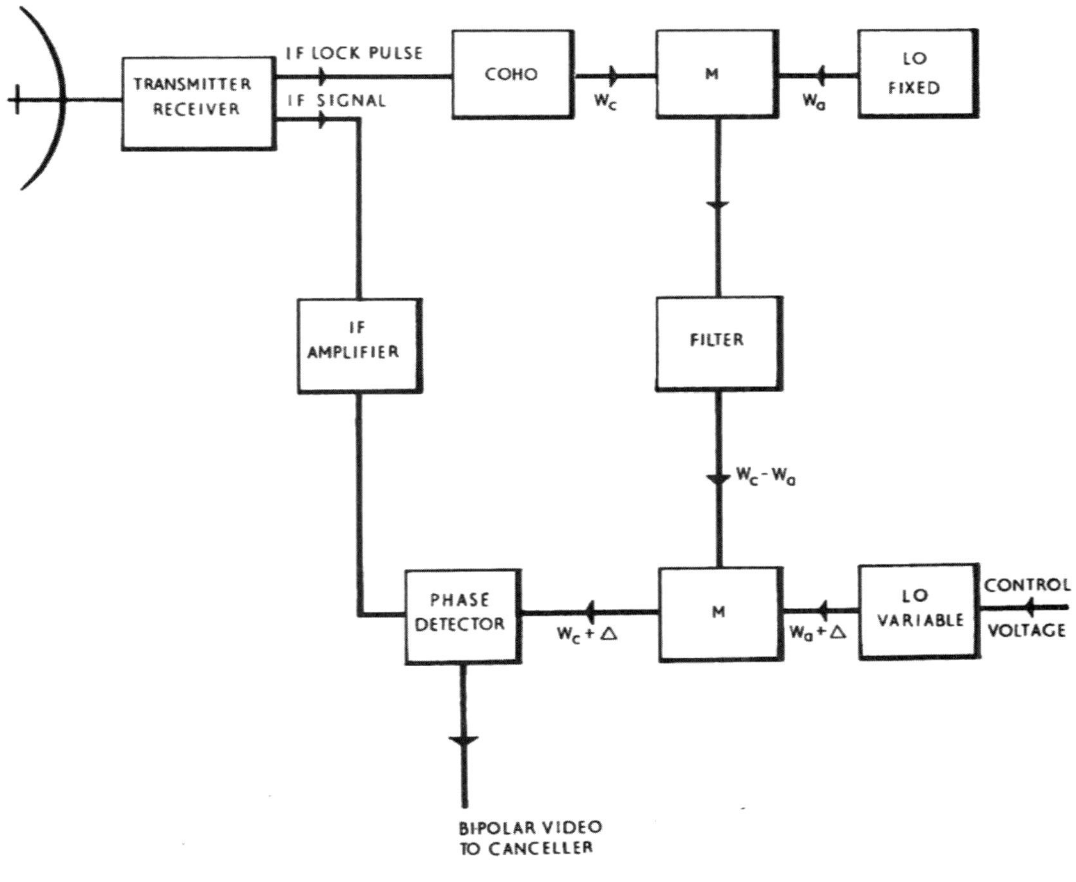

Figure 6.5 Velocity compensated MTI. (*From:* [5, p. 155].)

frequency of which is adjusted to bring the center frequency of the clutter spectra to zero. The difficulty of this method lies in deriving an accurate control signal for the VCO. In some applications (e.g., naval surface radar), we can derive an open-loop correction from the antenna's azimuth angle, the ship's true motion from its navigation system, and the antenna's pitch and roll motion from gyro sensors. However, in most applications we need an adaptive correction derived by locking the offset oscillator to the clutter's center frequency. In principle, the clutter locking must be accomplished in each range cell (see Example 10) because the velocity vector from the radar to the clutter patch generally differs for each cell. However, in practice, we face the problem of finding an uncontaminated sample of the clutter signal to lock the frequency. This generally restricts the number of clutter-locking points in the range dimension to three or four, which provides adequate performance (see Section 3.2.7 and Figure 3.33) in most applications.

The TACCAR technique provides an architecture often used in clutter-locking situations. Figure 6.6 depicts the general form of this type of system (also see Figure 3.30) as applied to a coherent-on-receive MTI using an IF canceller. The chief feature of TACCAR is the use of a VCO in a phase-locked loop where the correction for the clutter Doppler is obtained from the averaged measurement of the clutter Doppler frequency within a sampled range interval.

TACCAR automatically centers the clutter rejection notch at the averaged Doppler frequency of the clutter within the sampled range interval. TACCAR can

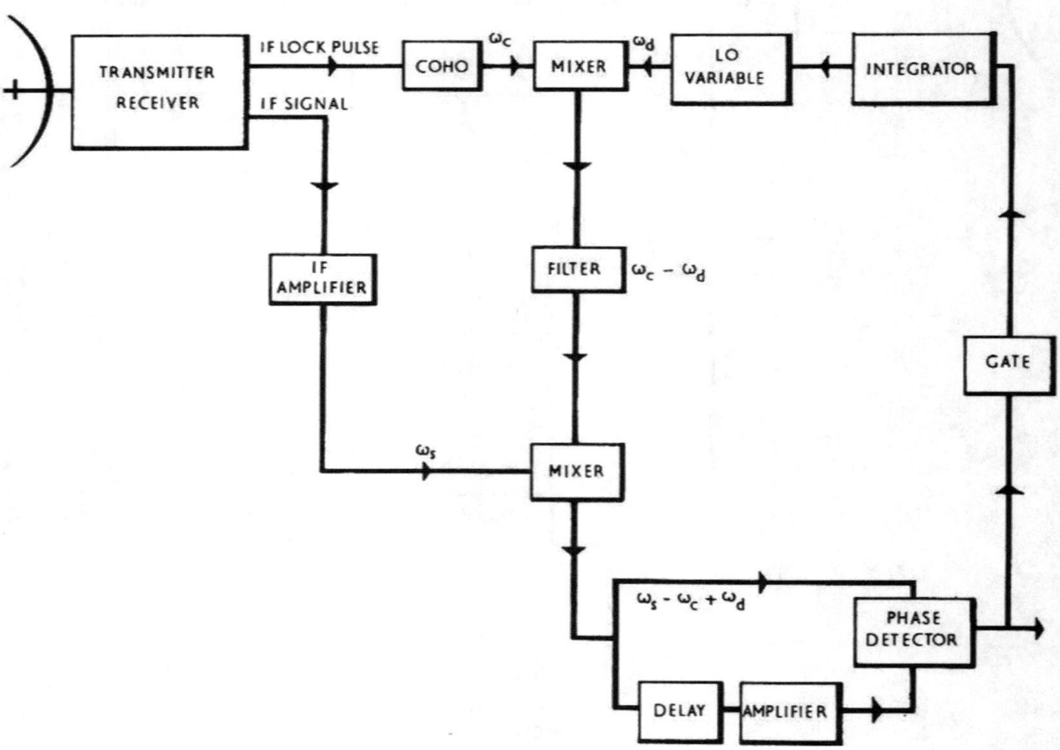

Figure 6.6 Clutter-locking MTI system. (*From:* [5, p. 156].)

be applied to cancel any fixed or moving clutter source and, hence, is capable of cancelling ground, sea, weather, and chaff clutter within any gated range interval for radars that are either fixed (e.g., land based) or on a moving platform (e.g., naval or airborne). The system cannot, however, simultaneously cancel two different clutter sources within a particular range interval. For example, if weather clutter is superimposed on ground clutter return, the TACCAR cannot simultaneously cancel both signals.

The primary application of TACCAR is in airborne MTI (AMTI) or when MTI is used on naval radar where in both cases the primary clutter component (i.e., land or sea clutter) is relatively narrow in spectral width and the Doppler frequencies are to be cancelled one at a time. Another application [6] is used on a land-based MTI system where ground clutter is cancelled close in, while weather and chaff clutter that occurs at different Dopplers is cancelled at ranges beyond the ground clutter.

An alternative method for controlling the null frequency of a coherent MTI canceller is to insert a phase shifter in the direct channel as depicted in Figure 6.7. The phase shifter for constant velocity clutter (v_c) will be adjusted to a value given by

$$\Delta\phi = \frac{4\pi v_c}{\lambda f_r} \tag{6.1}$$

where λ is the radar's wavelength and f_r is the radar's PRF. The phase shifter, by changing the phase of each radar return pulse by an amount $\Delta\phi$, effectively tunes the canceller null frequency to the average Doppler frequency of the clutter.

The coherent MTI uses the phase fluctuations in the return signal to recognize the Doppler component produced by a moving target (see Section 2.3.1). The operation of this type of MTI depends on a reference signal at the radar receiver that is coherent with the transmitter signal. Alternatively, a noncoherent MTI (see Section 2.3.1) that processes the envelope-detected target return signal can be employed. This type of MTI is also called a *clutter-referenced MTI* because the relative rotation of the target phaser with respect to the clutter phaser provides the essential action that allows this type of MTI to function (see Figure 2.16).

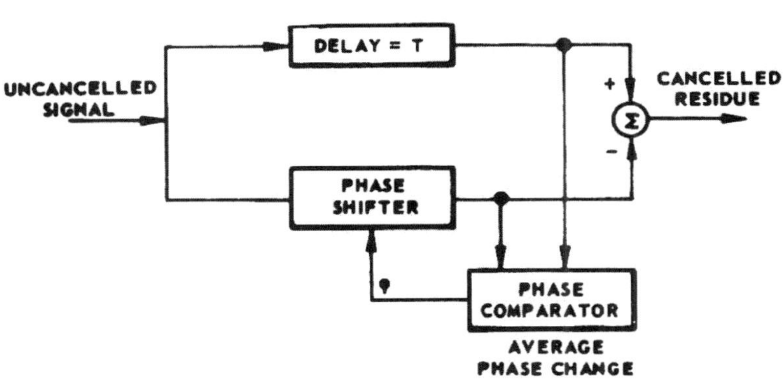

Figure 6.7 Phase-shifter clutter-locking MTI canceller. (*From:* [7, p. 395].)

Hence, for this MTI, the clutter return itself, in the vicinity of the target return, becomes the reference that replaces the function of the COHO in a coherent-type MTI. A limitation of this technique is that moving target echoes in a region not containing clutter do not induce the characteristic modulation onto the received envelope at the target's Doppler frequency, which allows the target to be detected. Because of this, the noncoherent MTI is sometimes described as being "blind in the clear." One solution is to use a clutter gate that switches the output from normal operation (no MTI) to MTI operation in the presence of clutter. The switching process, however, is difficult to control properly when the clutter appears intermittently.

A block diagram of a typical noncoherent MTI system is depicted in Figure 6.8. The transmitter may be either a power amplifier or a pulsed oscillator, and a STALO is used in the receiver to translate the signal to IF. The stability of the transmitter, for a given MTI performance level, must be comparable to that of a coherent MTI, although the frequency stability of the STALO can be somewhat less. The IF signals are envelope detected in a *linear-logarithmic* (lin-log) receiver that provides a dynamic range compatible with that of the MTI canceller. Alternatively, *sensitivity time control* (STC) can be employed separately or in conjunction with the lin-log characteristic to compress the dynamic range of the system. The transition between the linear and logarithmic characteristics of the receiver is set equal to the expected value of the improvement factor of the MTI. In theory, for Rayleigh-distributed clutter, the use of the logarithmic characteristics results in a constant residue, because the fluctuation component of the clutter (i.e., rms value) is proportional to its amplitude (i.e., mean value). However, when the clutter is significantly non-Gaussian or when significant stationary fixed targets (e.g., water towers, large fixed buildings) exist, the residue has significant fluctuations.

The envelope-detected signal of the noncoherent MTI is then processed through a conventional single-channel MTI canceller (the quadrature channel of a coherent MTI is not present) that filters the Doppler modulation produced by the beating of the target return and the clutter. Because this system depends on the beating of the target signal with the clutter signal, the difference in strength between the two component signals must be less than the subclutter visibility of the MTI for the

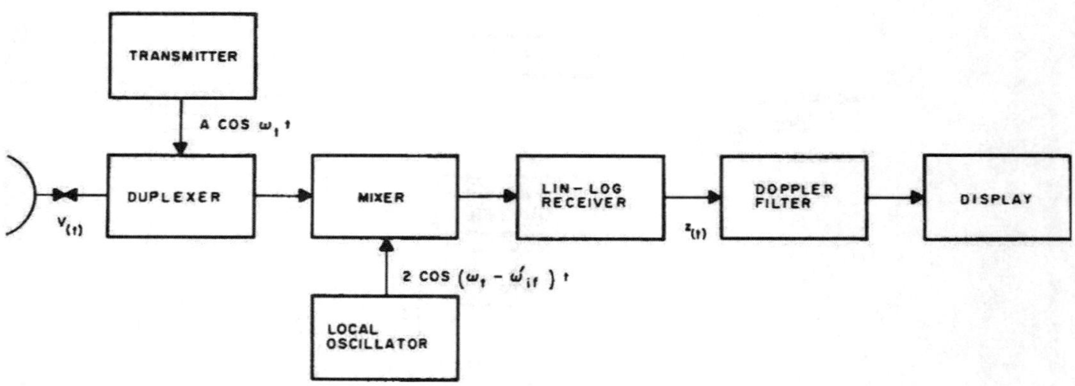

Figure 6.8 Noncoherent MTI. (*From:* [4, p. 202].)

system to function. For weak clutter signals, clutter gating partially overcomes this limitation.

The advantage of the noncoherent MTI is its simplicity (particularly that no motion compensation is required), which has led to its use in airborne applications where weight and size are at a premium. Its principal disadvantage with respect to a coherent MTI is its reduced performance (see Section 3.1.1), which is due in part to the lack of spectral purity of the clutter when used as a reference signal and also to the spatial patchiness of clutter, which causes gaps in the MTI's coverage. Other limitations on the performance of a noncoherent MTI involve the following:

- Intermodulation products of the clutter and target spectra in the envelope detection, causing spectral spreading;
- Blind phase limitations due to the use of a single MTI channel;
- The inability of cancelling multiple-time-around clutter if a power oscillator transmitter is used;
- Increased sensitivity to MTI stability limitations that cause amplitude jitter.

The block diagram of a phase-sensitive noncoherent MTI that removes the requirement for clutter gating is depicted in Figure 6.9 [6]. In this configuration, the reference signal for the MTI is derived by hard limiting the returns (consisting of clutter or receiver noise) from the range cell adjacent to that containing the target. If clutter is present, it provides a zero Doppler reference and target phase variation as the Doppler frequency is detected and passed to the MTI canceller. If only noise is present along with a target, the target output appears with a random phase and passes through the MTI canceller.

The phase-sensitive noncoherent MTI provides outputs for moving targets in clutter and point targets in the clear. Otherwise, the same limitations that apply for the conventional noncoherent MTI also apply to the phase-sensitive noncoherent MTI. A further anomaly associated with this type of system is that the three signal paths contained within the MTI can stretch the target when detected over clutter, which can lead to uncertainty in determining the range to the target.

6.2 MTI Performance Considerations

MTI performance is generally determined by two distinct factors. The first is related to the ability of the MTI to filter or attenuate the clutter return. This is determined

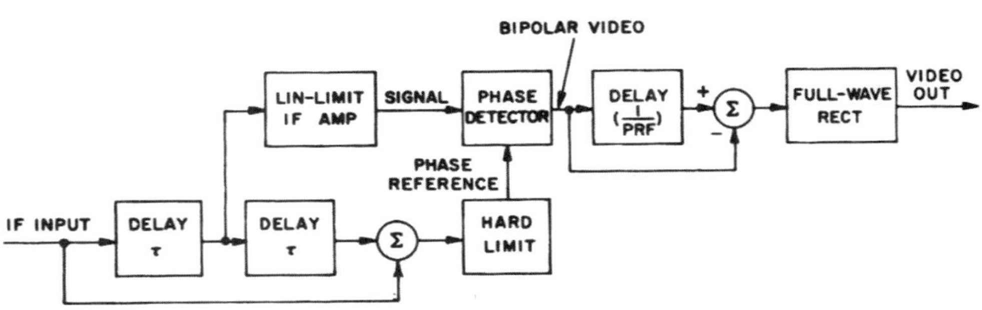

Figure 6.9 Phase-sensitive noncoherent MTI. (*From:* [6, p. 17-54].)

by the number of return pulses (n) processed by the MTI and the width and shape of the clutter spectrum. The second factor involves the overall system stability, which sets an upper limit on MTI system performance. This is a function of the many factors that result in pulse-to-pulse noise jitter in the system with the principal one being transmitter frequency instability (see Section 3.1.2).

The clutter spectral width generally is a function of antenna scanning modulation, wind-driven clutter motion, radar platform motion, and radar transmitter instability (see Section 3.2). For conventional ground-based air surveillance radar, the ground clutter lobe's spectral width is determined primarily by scanning modulation. Hence, the filtering performance of a nonrecursive ground-based MTI filter can be related to the time the antenna dwells on the target and the number of pulses processed by the MTI (see Section 3.2.2). For airborne radar, the ground clutter lobe's spectral width is determined primarily (particularly as the antenna scans off the aircraft's ground track) by the platform motion effect. This results in the airborne radar MTI filter's performance being determined by the aircraft's velocity, the azimuth scan angle, the antenna aperture in the azimuth dimension, and the number of pulses processed by the MTI (see Section 3.2.3). A further limitation imposed on the performance of an airborne MTI involves the Doppler sidelobe returns within the passband of the MTI, which are proportional in magnitude to the two-way antenna sidelobe levels (see Section 3.2.3).

6.2.1 MTI Performance for Ground-Based Surveillance Radar

The detection performance of an MTI is generally related to the MTI improvement factor (I_c), which is defined as the ratio of the output signal-to-clutter ratio of the MTI to the input signal-to-clutter ratio (see Sections 3.1 and 5.1.3). For a conventional ground-based surveillance radar, using a coherent MTI with an n-pulse binomial canceller, when the scanning modulation limitation dominates, the improvement factor (see Figure 3.7) can be approximated in terms of the number of pulses per antenna 3 dB beamwidth (n_B) as

$$I_c = \frac{n_B^{2(n-1)}}{(1.388)^{n-1}(n-1)!} \tag{6.2}$$

For a single canceller ($n = 2$), (6.2) results in an improvement factor $I_c = n_B^2/1.388$, whereas for a double canceller, the result is $I_c = n_B^4/3.844$.

As an example, the ASR-9 radar ($\dot{\theta} = 12.5$ rev/min, θ_3 dB $= 1.3°$, PRF $=$ 1,200 Hz, $n = 3$) results in $n_B = 20.8$ pulses per 3-dB beamwidth and an improvement factor of 46.9 dB.

The improvement factor is further reduced by the instabilities associated with the MTI system. To consider this case, the system instabilities are sometimes measured separately from the limitations caused by clutter spectral spreading resulting in an MTI cancellation ratio (CR) limitation. This can be combined (see Section 3.1.2) with the improvement factor limitation (called I_{ideal}) to result in an overall improvement factor (I_t) given by

$$I_t = \frac{1}{[(I_{\text{ideal}})^{-1} + (\text{CR})^{-1}]} \tag{6.3}$$

As an example, the ASR-9 has a system instability specified as CR = 60 dB, resulting in an overall improvement factor of I_t = 46.7 dB, which compares to a measured value of 45 dB [8].

6.2.2 MTI Performance for Airborne Surveillance Radar

The detection performance of low-PRF airborne MTI radar is generally dominated by platform motion considerations. The two primary effects are the spreading of the main lobe ground clutter platform motion component as the radar antenna scans away from the clutter's ground track and the translation of antenna sidelobe ground clutter into the MTI's passband (see Figures 2.4 and 2.5).

Generally, the result of increasing the spectral width of the ground clutter's main lobe component is to diminish the MTI's detection probability. Described in Section 2.2.1, this is a result from reducing the ratio of the clear region within which the MTI must operate, as compared to the blind region that contains main lobe clutter components spread by platform motion (see Figure 2.6). To examine the magnitude of this effect, consider the spectral spreading (in hertz) for the main lobe ground clutter platform motion component (see Section 3.2.3) given by

$$\sigma_{pm} = \frac{0.528 v_{ac} \sin \theta_0 \cos \phi_0}{D} \tag{6.4}$$

where v_{ac} is the aircraft's velocity (m/s), θ_0 is the azimuth angle with respect to the aircraft's ground track, ϕ_0 is the antenna's depression angle to the clutter patch, and D is the antenna's aperture width (m). A high-performance aircraft generally has a small antenna aperture and a large aircraft velocity, which tends to maximize the spectral spreading. The example given in Section 2.2.1 shows that for v_{ac} = 1,500 knots and D = 2.5 ft, the PRF must be increased to 4,000 Hz and the scan angle restricted to ±30° for equal blind and clear regions, which restricts the nonambiguous range to on the order of 20 nmi. The design of long-range AMTI radar with 360° scan capability involves using a larger antenna aperture in a lower speed aircraft (see Figure 2.6).

The improvement factor for an n-pulse coherent binomial MTI canceller with a platform motion limitation is given by

$$I_c = \frac{2^{n-1}}{(3.318 v_{ac} T \sin \theta_0 \cos \phi_0 / D)^{2(n-1)} (n-1)!} \tag{6.5}$$

where $T = 1/f_r$ is the radar's interpulse period. For a single canceller this equals

$$I_c = \frac{1}{5.5 (v_{ac} T \sin \theta_0 \cos \phi_0 / D)^2} \tag{6.6}$$

and for a double canceller it equals

$$I_c = \frac{1}{60.6(v_{ac}T \sin \theta_0 \cos \phi_0/D)^4} \tag{6.7}$$

For the high-performance aircraft AMTI example given in Section 2.2.1 ($v_{ac} = 1{,}500$ knots, $f_r = 4{,}000$ Hz, $\theta_0 = 30°$, $\psi_0 = 0°$, $D = 2.5$ ft), the improvement factor becomes 10.5 dB for a single canceller and 18.1 dB for a double canceller. This calculation illustrates the difficulty of achieving high detection performance with low-PRF AMTIs in a high-performance aircraft.

The severe limitations imposed by platform motion effects have led to the development of motion compensation techniques to provide increased performance in low-PRF AMTIs. The philosophy of these techniques is to remove the effects of platform motion before MTI processing. When this is accomplished, the limitations on MTI performance associated with fixed-based MTIs (i.e., antenna scanning modulation, transmitter instability) would also apply to AMTIs. However, we must recognize that, in the AMTI case, the returns from antenna sidelobes are aliased throughout the entire MTI frequency band, resulting in an improvement factor limitation equal to the ratio of the integrated energy entering the system through the main lobe to that entering through the sidelobes for the two-way antenna pattern.

Motion compensation techniques are generally applied in the coherent RF or IF portion of the MTI receiver to correct the spectral translation and spreading before MTI processing. It is natural to follow the motion compensation with a coherent MTI for maximum detection performance. TACCAR is used to compensate the coherent MTI for the mean Doppler shift induced by platform motion, and a *Displaced Phase-Center Antenna* (DPCA) is used to correct for platform motion spectral spreading (see Section 3.2.7).

For example, consider the case of the low-PRF E-2C AMTI radar designed to detect airborne targets over both water and land [4, 9]. The pertinent parameters for the AEW radar are a PRF of 300 Hz, an aperture width of 25 ft, a platform velocity of 300 knots, and a transmission frequency at UHF (e.g., 425 MHz). Figure 6.10 illustrates the MTI improvement factor for an N-delay canceller ($N = n - 1$) as a function of the clutter spectral spread (σ_c – Hz) to the PRF (f_r – Hz). The system characteristics that limit the MTI improvement factor can be related to an effective clutter spectral spreading and to a shift in the mean of the spectrum. The total clutter spectral spread can be determined (assuming statistical independence) by taking the square root of the sum of the squares of all the individual spectral widths (see Section 3.2.8).

The achievable MTI improvement factor for an AMTI system is ultimately limited by the internal motion of the clutter itself, the system stability, and the integrated main lobe-to-sidelobe ratio of the two-way antenna pattern. However, in practice the significant limitations arise from: (1) platform motion, (2) antenna sidelobe levels, (3) antenna scanning modulation, and (4) system stability. The expected values for each of these components for an AEW radar are superimposed on Figure 6.10 to allow the relative importance of each component to be determined [4, 9].

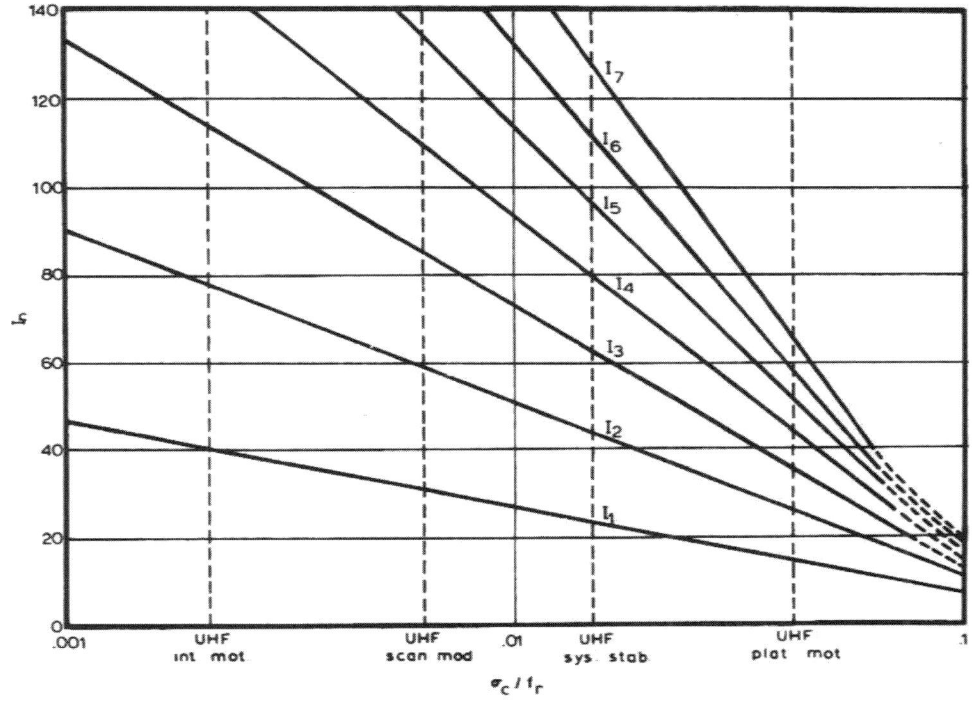

Figure 6.10 MTI improvement factor for an *N*-delay canceller with system limitations. (*From:* [9, p. 6; 2, p. 460].)

In Figure 6.10, a typical value for the spectral width due to internal clutter motion of $\sigma_v = 0.2$ m/s is used (see Section 3.2.1). At UHF, this results in an rms spectral spread of 0.57 Hz ($\sigma_f = 2\sigma_v/\lambda$) and a normalized value $\sigma_f/f_r = 0.002$. Figure 6.10 shows that the MTI improvement factor for this condition is 40 dB for a single canceller, 78 dB for a double canceller, and 114 dB for a triple canceller. If the system were limited by internal clutter motion, all practical requirements could be achieved easily by the use of a double canceller.

In general, platform motion can be resolved into two components, one parallel and one perpendicular to the antenna aperture (see Section 3.2.7). These components have the effect of both spreading the spectrum of the clutter returns [see (6.4)] and shifting the mean of the spectrum. Considering only the platform motion spreading effect, the rms value from (6.4) ($v_{ac} = 300$ knots, $\theta_0 = 90°$, $\phi_0 = 0°$, $D = 25$ ft) equals $\sigma_{pm} = 10.7$ Hz, which results in a normalized value $\sigma_{pm}/f_r = 0.0357$, using our E-2C example with $f_r = 300$ Hz. Figure 6.10 shows that the MTI improvement factor is about 15 dB for a single canceller, 25 dB for a double canceller, and 35 dB for a triple canceller with the antenna aperture parallel to the velocity vector. For this case, many practical requirements cannot be met with single, double, or triple cancellers, thereby determining the requirement for motion compensation processing.

With the antenna aperture perpendicular to the velocity vector, the dominant effect is the shifting of the mean of the clutter spectrum. In general, the mean of the Doppler spectrum is determined by the platform velocity, the transmitted frequency, and the angle to the clutter patch under consideration given by

$$f_c = \frac{2v_{ac} \cos \phi_0 \cos \psi_0}{\lambda} \tag{6.8}$$

Motion compensation of this effect is generally required for successful MTI performance.

Because the antenna sidelobes essentially cover 360°, the Doppler shifts received through the sidelobes are associated with velocities ranging from $-v_{ac}$ (v_{ac} is the aircraft's velocity) to $+v_{ac}$. This Doppler spread can cover the entire MTI ambiguous frequency band. As a result, these sidelobe returns are not cancelled and the MTI improvement factor is limited by the ratio of the integrated energy entering the system through the main lobe to that entering through the sidelobes for the two-way antenna pattern. At L-band an integrated one-way main lobe-to-sidelobe ratio of 30 dB is achievable, whereas at UHF the coupling of the antenna radiation pattern to the aircraft structure considerably degrades this ratio [4, pp. 456–457].

The effect of antenna scanning modulation is related to the antenna's scan rate and beamwidth with standard deviation given by

$$\sigma_f = \frac{0.265\dot{\theta}}{\theta_B} \tag{6.9}$$

This results in an antenna scanning modulation spectral spread for the example ($\dot{\theta}$ = 6 rev/min, θ_B = 6.4°) of σ_f = 1.5 Hz, which has a normalized value σ_f / f_r = 0.005. The scanning modulation spectral width is affected by the antenna scan rate, the antenna aperture, and the transmitted frequency, but generally most requirements can be met without compensation for this effect.

System stability is greatly simplified through the use of digital processing, which allows a single clock to control the complete system, including the MTI delays. However, the pulse-to-pulse phase stability of the transmitter is not eliminated by digital processing, so that the performance of the system is generally limited by the transmitter stability. For this type of system, the interpulse transmitter stability relative to the transmitter frequency is estimated at 10^{-8}, resulting in a normalized stability of $\Delta f / f_r = 0.0142$. The corresponding MTI improvement factor (see Figure 6.10) is 22 dB for a single canceller, 42 dB for a double canceller, and 62 dB for a triple canceller. Most requirements can be met with a stability of this magnitude.

The conclusion of the preceding discussion is that the factors limiting the improvement factor for this type of AMTI radar are the antenna sidelobes and platform motion. Motion correction techniques, therefore, are necessary for adequate performance in conjunction with a low-sidelobe antenna.

A TACCAR type of correction is generally required to correct for the mean shift in the Doppler spectrum due to platform motion. This correction is made by phase locking the MTI's COHO to the clutter's mean frequency. The time constant of this phase-locked loop must be large compared with the interpulse period of the radar to maintain the pulse-to-pulse coherency needed for successful MTI operation. This leads to a single correction for all ranges. However, for an aircraft flying above the earth's surface, the mean Doppler shift to each clutter cell is different because it is proportional to the angle of arrival, which is a function of range. Hence, the TACCAR correction is generally incomplete, although this could be improved through the use of multiple correction circuits (see Section 3.2.7).

The effect of single-point TACCAR correction results in a smaller MTI improvement factor that is a function of the correction error and the clutter spectral spread (see Figures 3.31 and 3.32). For a typical set of parameters, the limitations due to incomplete motion compensation when the antenna is pointed in the direction of the aircraft's velocity vector are depicted in Figure 6.11 [4, 9].

Use of a DPCA is a technique that compensates for the component of velocity parallel to the antenna aperture. This is accomplished by either physically or electrically displacing the phase center of the antenna in a direction opposite to the velocity component of the aircraft (see Section 3.2.7). A single canceller can be almost perfectly compensated by DPCA with essentially no reduction in improvement factor. For higher order cancellers (double, triple, and so on) only the first canceller stage is effectively compensated. This results in an imperfect compensation so that the effects of this platform motion component are not completely eliminated. Figure 6.12 illustrates typical parameters for the MTI improvement factor, including the effect of DPCA limitations for single, double, and triple cancellers. The MTI

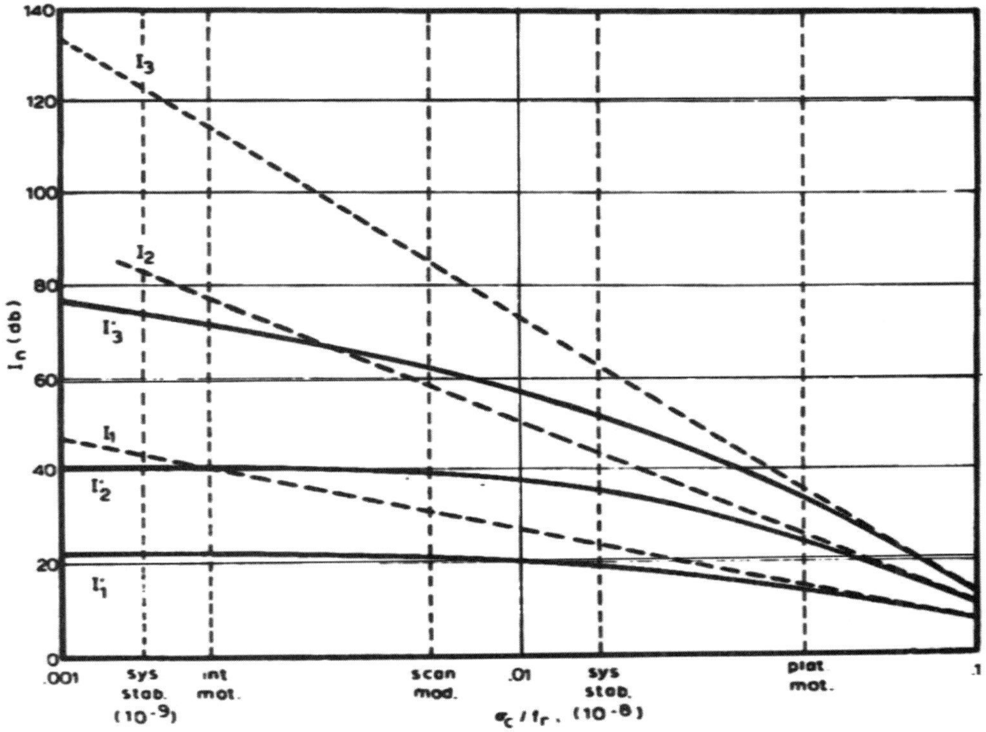

Figure 6.11 MTI improvement factor with TACCAR limitation for an *N*-delay canceller at UHF. (*From:* [4, p. 461; 9, p. 12-7].)

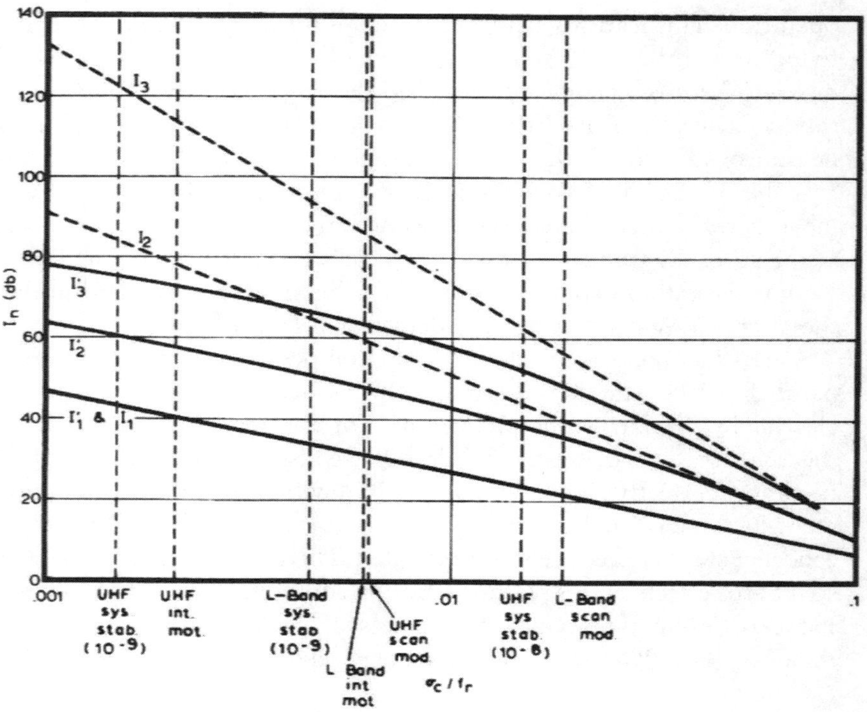

Figure 6.12 MTI improvement factor with DPCA limitations for an *N*-delay canceller. (*From:* [4, p. 462; 9, p. 12-8].)

improvement factor for a single canceller with DPCA coincides with the perfectly compensated curve. The improvement factors for double and triple cancellers are much less than the perfectly compensated curves.

Figure 6.12 indicates that for the example system with DPCA and TACCAR motion compensation, the limiting factor involves the effect of system instabilities. If system stability can be increased to 10^{-9}, then scanning modulation becomes the system limitation. However, the UHF transmitter frequency selected for the radar has the effect of increasing the antenna sidelobes due to excitation of the aircraft's structure in this frequency range [4, 9]. An increase to L-band (i.e., $f = 1{,}200$ MHz) would substantially reduce the antenna sidelobe level. However, a change to L-band would result in several changes to the system performance limitations depicted in Figures 6.10, 6.11, and 6.12.

First, at an L-band frequency, the scanning modulation component spread would increase in direct proportion to the change in frequency from UHF (see Section 3.2.2). As depicted in Figure 6.12, this would then become the dominant limitation, assuming that the system stability could be maintained at 10^{-9}. Notice that the internal clutter motion limitation would also be increased at an L-band frequency. However, a more serious limitation would be the increased TACCAR limitations that occur at the L-band frequency, as depicted in Figure 6.13. These occur due to the higher Doppler shifts at the increased transmitter frequency. For a double canceller, these limitations are about 25 dB for scanning modulation and

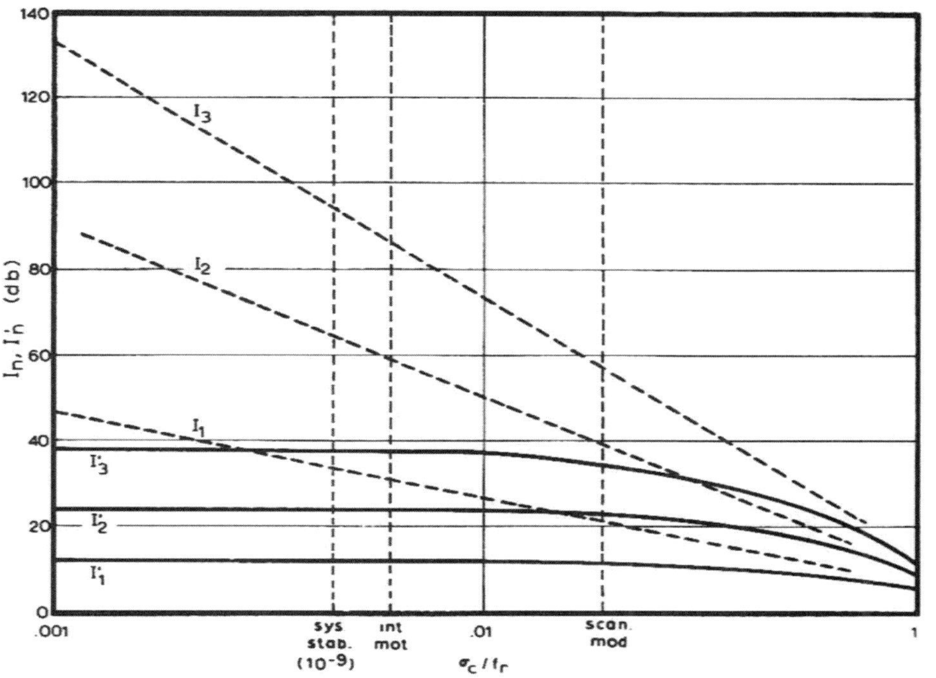

Figure 6.13 MTI improvement factor with TACCAR limitations for an *N*-delay canceller at *L*-band. (*From:* [9, p. 12-8; 2, p. 462].)

TACCAR losses (antenna at 90°). If the TACCAR losses and the DPCA limitations could be eliminated, then the limitation to the MTI improvement factor caused by scanning modulation at L-band would be about 55 dB for a triple canceller (see Figure 6.13). Reduction of these limitations consists of making TACCAR corrections at additional points in the range and applying a DPCA correction to all cancellers [10, 11].

To summarize, in order to ensure adequate performance of the example long–range, low-PRF AMTI radar, the following system features are desirable [4, 9]:

- Integrated main lobe-to-sidelobe ratios for the two-way antenna pattern on the order of –60 dB;
- Pulse-to-pulse transmitter stability factor equal to 10^{-9} of the transmitted frequency;
- Triple canceller MTI processor;
- TACCAR correction at multiple range points (three minimum);
- DPCA correction of all three cancellers.

► Example 20 DPCA Improvement Factor for MTI Cancellers [3, 11, 12]

The main limitation to the use of MTI cancellers in airborne radar is the spreading of the clutter spectrum due to platform motion. The standard deviation of the platform motion component of the clutter spectral spread is given by

$$\sigma_{pm} = \frac{0.528 v_p \cos \phi_{eo} \sin \theta_{azo}}{D}$$

where v_p is the platform velocity, ϕ_{eo} is the depression angle, θ_{azo} is the azimuth angle, and D is the aperture length of the antenna. The improvement factor is then given by

$$I_n = \frac{2^N}{N!} \left(\frac{1}{2\pi\sigma_{pm} T} \right)^{2N}$$

where N is the number of canceller stages and T is the radar interpulse period. DPCA electronically moves the phase center of the antenna in an equal and opposite direction to the physical movement, thereby causing the phase center of the antenna to be stationary in space. This compensation when applied in conjunction with a single canceller provides perfect compensation for platform motion. When multiple-stage cancellers are employed, the compensation is not perfect. In this example we develop a MATLAB program that calculates the improvement factor for airborne radars using single, double, and triple MTI cancellers with DPCA motion compensation.

The power transfer function for an N-stage MTI employing DPCA is

$$|H_\eta|^2 = \left(1 + \tan^2 \frac{\eta}{2}\right) \left[2 \sin\left(\pi\left(f - f_d'\right)T\right)\right]^2 [2 \sin(\pi f T)]^{2N}$$

where

$$\eta = 2\pi f_d' T$$

and

$$f_d' \approx \frac{2 v_p \cos \phi_{eo} \sin \theta_{azo} \theta}{\lambda}$$

where θ is the incremental azimuth angle about the antenna's boresight. Note that both η and f_d' are functions of θ. The received clutter power is modulated by the two-way antenna power gain $G^4(\theta)$. The spectrum of the clutter is given by

$$S(f, \theta) = G^4(\theta) \exp\left[-\frac{1}{2} \left(\frac{f - f_d'}{\sigma_{pm}} \right)^2 \right]$$

The total input clutter power is then

$$P_{ic} = \int_{-\pi}^{\pi} \int_{-\infty}^{\infty} S(f, \theta)\, df\, d\theta$$

The output clutter power is given by

$$P_{oc} = \int_{-\pi}^{\pi} \int_{-\infty}^{\infty} S(f, \theta)\, |H_\eta|^2 \, df \, d\theta$$

The MTI power gain is given by

$$\overline{G} = \frac{2^N}{N!} [1 \cdot 3 \cdot 5 \ldots (2N - 1)]$$

Then the improvement factor for an N-stage canceller with DPCA is given by

$$I_n' = \frac{\overline{G} P_{ic}}{P_{oc}}$$

Following the analysis in [11] a uniformly illuminated antenna is assumed with antenna pattern

$$G^4(x) = \left[\frac{\sin(x)}{x}\right]^4$$

where $x = \pi a \sin \theta / \lambda$.

The improvement factors for a single and double canceller with DPCA are given by

$$I_1' = I_1 \frac{G_0}{G_0 + G_1} \qquad I_2' = I_2 \frac{G_0}{G_0 + \frac{2}{3} I_1 G_1}$$

where

$$G_0 = \int_{-\pi}^{\pi} G^4(x) \, dx \qquad G_1 = \int_{-\pi}^{\pi} G^4(x) \tan^2(cx) \, dx$$

and $c = 2v_p \cos \phi_{eo} \sin \theta_{azo} T/D$.

The improvement factor for a triple canceller is given by

$$I_3' = I_3 \frac{G_0}{G_2 + \frac{36}{15} I_1 G_3 + \frac{8}{15} I_2 G_4}$$

where

$$G_2 = \int_{-\pi}^{\pi} G^4(x)\cos^2(cx)\,dx \qquad G_3 = \int_{-\pi}^{\pi} G^4(x)\sin^2(cx)\,dx$$

and

$$G_4 = \int_{-\pi}^{\pi} G^4(x)\sin^2(cx)\tan^2(cx)\,dx$$

The MATLAB program dpca_impfacb.m evaluates these equations using the trapz.m function to perform numerical integration and the sinc(t) function to evaluate $\sin(x)/x$ with singularity at $x = 0$. Curves are plotted in terms of motion parameter

$$x = \frac{v_p T}{D}\cos\phi_e \sin\theta_{az}$$

which represents the fraction of the antenna aperture displaced per interpulse period.

Figure EX-20.1 shows the performance of a single canceller with and without DPCA for $x = 0$, 0.01, 0.07, and 0.1 over a range of σ_f/PRF of 0.001 to 0.1. As

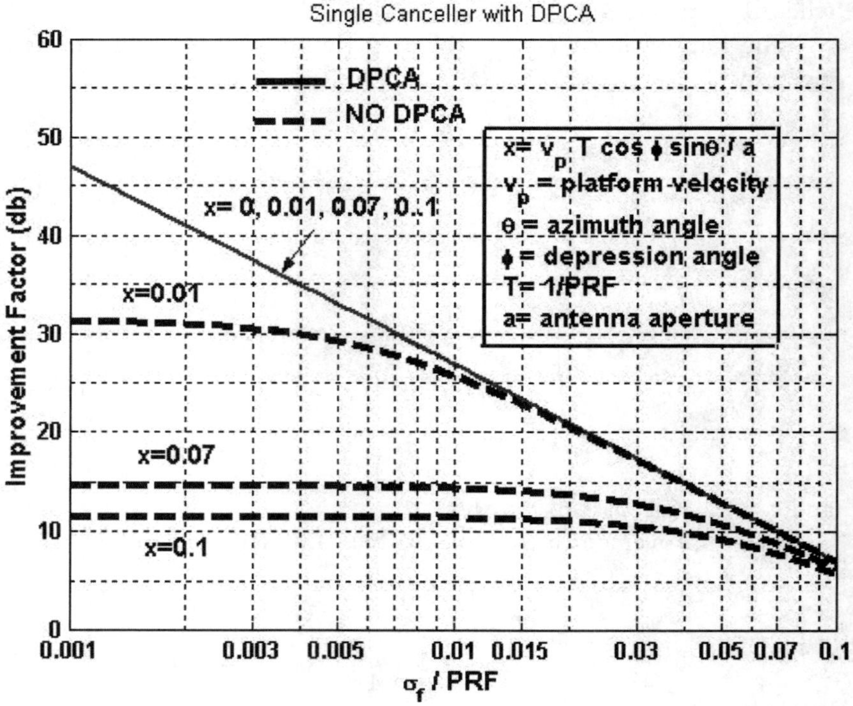

Figure EX-20.1 Improvement factor for a single canceller with and without DPCA.

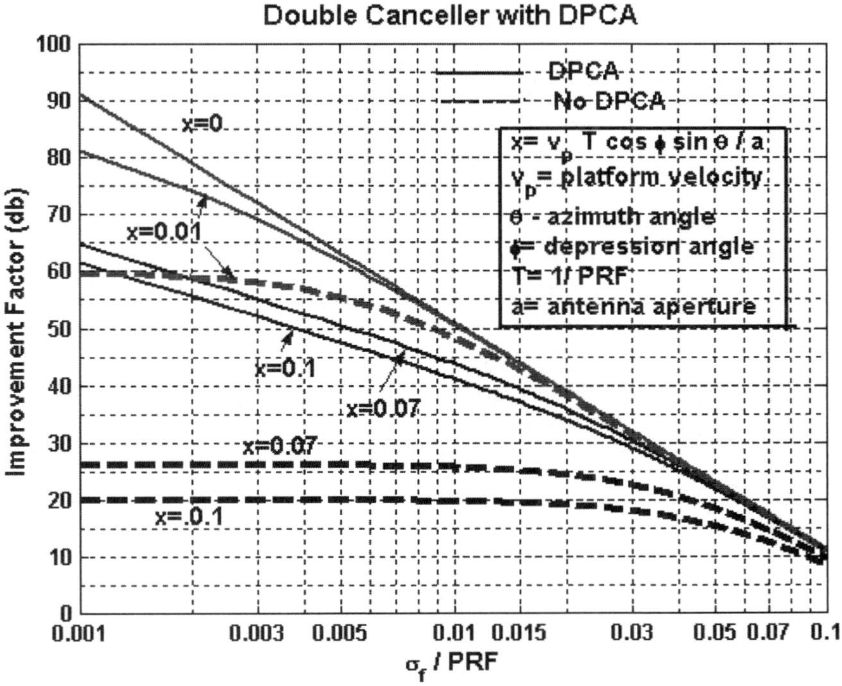

Figure EX-20.2 Improvement factor for double canceller with and without DPCA.

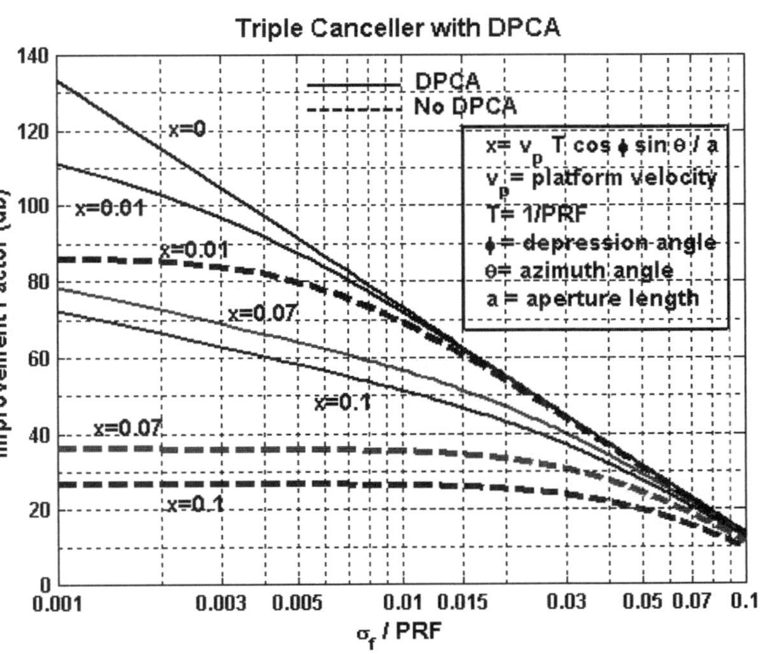

Figure EX-20.3 Improvement factor for triple canceller with and without DPCA.

expected DPCA provides complete motion compensation for this case. The curves without DPCA show that a single canceller without DPCA has limited performance capability.

Figure EX-20.2 shows the performance of a double canceller with and without DPCA for $x = 0, 0.01, 0.07$, and 0.1 over a range of σ_f/PRF for 0.001 to 0.1. The curve illustrates that DPCA is most effective for relatively small inherent spectral widths that might be associated with overland situations. Also the performance deteriorates as the platform velocity increases, causing a relatively larger fraction of the antenna aperture to be displaced per interpulse period.

Figure EX-20.3 shows the performance of a triple canceller with and without DPCA for $x = 0, 0.01, 0.07$, and 0.1 over a range of σ_f/PRF = 0.001 to 0.1. For small platform velocities and σ_f/PRF < 0.01, improvement factors of about 60 dB can be realized without DPCA. For higher platform velocities ($x = 0.1$), DPCA is required to achieve improvement factors in this range.

▶

6.2.3 MTI Performance for a Nonlinear Receiver

An important requirement for a PD radar receiver is to maintain linear operation over a wide dynamic range in order to minimize clutter spectral spreading (see Section 3.2.5). In general, the radar receiver is required to operate on very weak signals and convert them to a voltage sufficient to drive the Doppler processor. Typically, about 100 dB of gain is required to convert weak signals at the receiver input to an acceptable voltage at the output. Because the range of input signals (including clutter and target returns) may also exceed 100 dB, if truly linear, the receiver must amplify a 100 dB range of input signals by 100 dB, which, in general, is an impossible task. Therefore, receivers are nonlinear in the sense that they use some technique (e.g., limiter, log amplifier, AGC, or STC) to compress the total range of input signals to a usable range at the output. Because the process of amplification and detection in the receiver always results in some degree of distortion of the input signal, this is described in terms of spurious signal levels, compression, or other appropriate terminology.

In Section 3.2.5, the receiver was modeled as a linear device followed by a limiter, which was described by an error function transfer function. The transfer function for this type of receiver is plotted in Figure 6.14 on a decibel scale and represents a severe limiting action that may be employed, for example, if the receiver is followed by a digital processor employing an A/D converter (i.e., the dynamic range is approximately 6B dB, where B is the number of bits in the A/D converter). However, in general, receivers do not exhibit this severe limiting action, and in this section we develop MTI improvement factor relationships using traditional receiver distortion theory and intermodulation products characterized by third-order intercept points. An advantage of this approach is that standard methods are available to measure receiver nonlinearity, which then can be used to specify receiver characteristics in terms of achievable MTI improvement factors.

A general expression relating output to input for many electronic devices is a power series of the form:

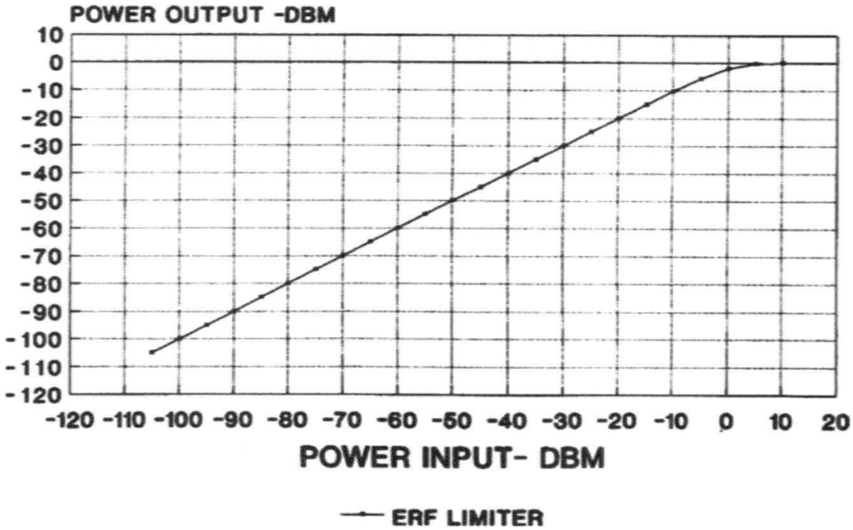

Figure 6.14 Nonlinear receiver with error function transfer characteristic. (*From:* [2, p. 357].)

$$e_0 = k_1 e_i + k_2 e_i^2 + k_3 e_i^3 + \ldots \tag{6.10}$$

This expression describes the transfer characteristics of virtually all electronic devices including mixers and amplifiers, which are the key elements of most receivers. When the input signal contains multiple frequency components (i.e., clutter), the input can be represented as

$$e_i = e_1 \cos \omega_1 t + e_2 \cos \omega_2 t + e_3 \cos \omega_3 t + \ldots \tag{6.11}$$

When this signal is applied to the receiver represented by (6.10), the resulting output consists of a large number of intermodulation terms (see [13] for all terms through the fifth order). Examination of the intermodulation terms reveals that the amplitudes of fundamental terms $\left(\text{e.g., } e_0 = k_1 e_1 \cos(\omega_1 t)\right)$ are directly proportional to the level of the input signal; the amplitudes of second-order responses $\left(\text{e.g., } e_0 = k_2 e_1 e_2 \cos(\omega_1 t - \omega_2 t)\right)$ are proportional to the square of the input signal's amplitudes; and the amplitudes of third-order responses $\left(\text{e.g., } e_0 = k_3 e_1 e_2^2 \cos(2\omega_1 t)\right)$ are proportional to the cube of the input signal's amplitudes [14].

A plot of these responses against the input signal level on a log-log graph will be straight lines with slopes corresponding to the order of the response, as depicted in Figure 6.15. The fundamental output will increase 1 dB for every decibel that the input increases. The second-order response when stimulated with two equal amplitude input tones differing in frequency will increase 2 dB for every decibel that the input tones are increased. The third-order response when stimulated with two equal-amplitude input tones differing in frequency will increase by 3 dB for every decibel that the input tones are increased.

The intersection of a projection of the straight-line portion of the second- or third-order response curves with a projection of the linear portion of the fundamen-

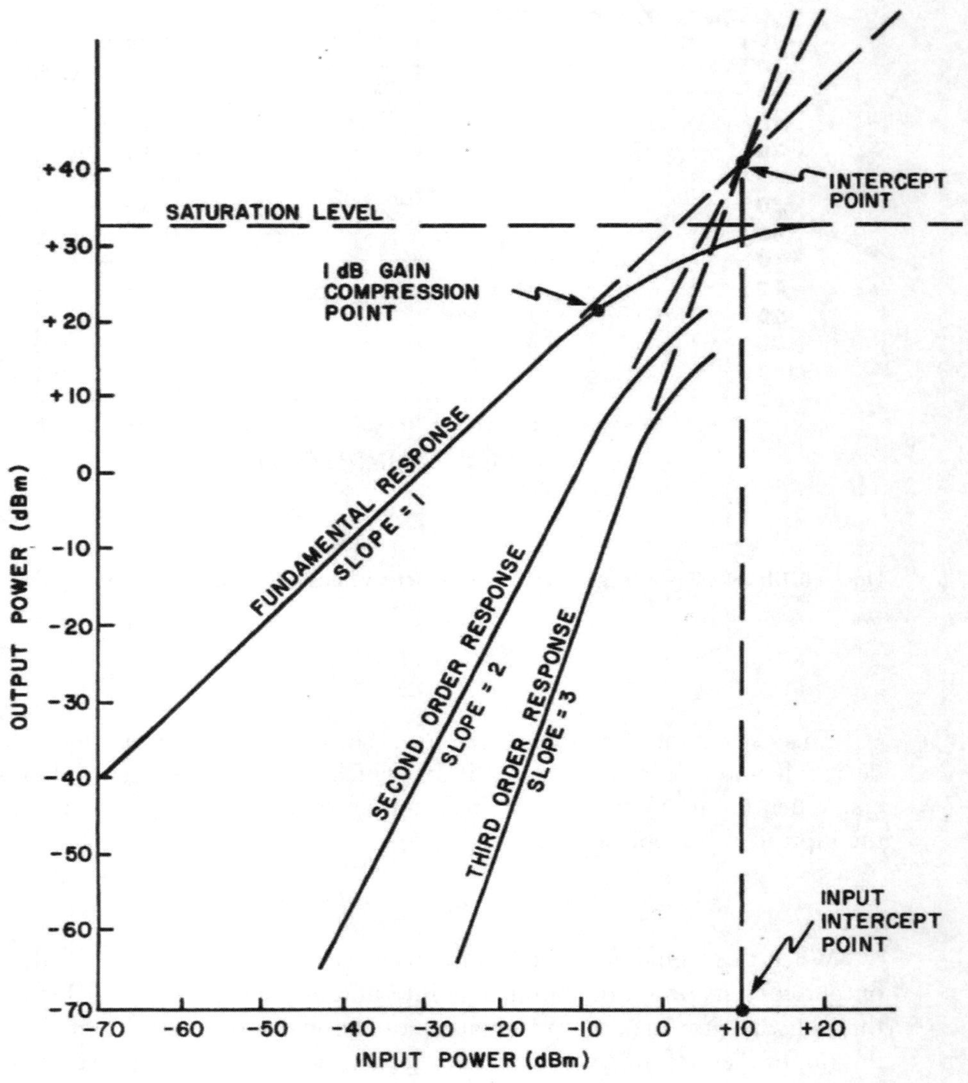

Figure 6.15 Nonlinear receiver second- and third-order intercept points. (*From:* [14, p. 71].)

tal response curve is called an *intercept point*. The second- and third-order plots may or may not intercept the fundamental line at the same point, depending on system design. These characteristics (the known slopes of the response curves and the second- and third-order intercept points) can be used to determine the intermodulation response of the receiver.

In many practical receivers, even-order responses are suppressed using balanced designs (e.g., balanced mixers) resulting in domination by the third-order response. For this situation, the receiver characteristic is given by

$$e_0 = k_1 e_i + k_3 e_i^3 \tag{6.12}$$

where the power gain is given by $G = k_1^2$ and k_3 is the third-order distortion coefficient. The autocorrelation function at the receiver output is given by

$$R_0(\tau) = E[e_0(t)e_0(t+\tau)] \tag{6.13}$$

resulting in

$$R_0(\tau) = GR_i(\tau) + 6k_3^2 R_i^3(\tau) \tag{6.14}$$

where $R_i(\tau)$ is the input ACF to the receiver. The normalized receiver output ACF can then be found from

$$\rho_0(\tau) = \frac{R_0(\tau)}{R_0(0)} \tag{6.15}$$

and assuming a Gaussian-shaped ACF at the receiver input given by

$$R_i(\tau) = P_c e^{-(\sigma_c \tau)^2/2} \tag{6.16}$$

where P_c is the clutter's power, σ_c (rad/s) is the clutter's spectral spread, and τ is the separation in time between samples. The resulting normalized ACF is then given by

$$\rho_0(\tau) = F_1 e^{-(\sigma_c \tau)^2/2} + F_2 e^{-3(\sigma_c \tau)^2/2} \tag{6.17}$$

where

$$F_1 = \frac{1}{1+\gamma} \tag{6.18}$$

$$F_2 = \frac{\gamma}{1+\gamma} \tag{6.19}$$

and where

$$\gamma = \frac{6k_3^2 P_c^2}{G} \tag{6.20}$$

The third-order intercept point at the output can be found by equating the fundamental and third-order responses, resulting in

$$P_{\text{IPO}} = e_0^2 = \frac{4k_1^3}{3k_3} \tag{6.21}$$

If we solve for k_3 and substitute into (6.20), we have

$$\gamma = 10.7\left(\frac{P_c G}{P_{\text{IPO}}}\right)^2 \tag{6.22}$$

which expresses the nonlinear receiver's normalized ACF in terms of the ratio of the clutter power level to the third-order intercept power level, both at the receiver output.

The improvement factor for a single canceller is given by

$$I_1 = \frac{1}{1 - \rho_0(T)} \tag{6.23}$$

which results in

$$I_{NL1} = \frac{1}{1 - F_1 e^{-(\sigma_c T)^2/2} + F_2 e^{-3(\sigma_c T)^2/2}} \tag{6.24}$$

For a $\sigma_c T \ll 1$, the resulting MTI improvement factor can be reduced to

$$I_{NL1} = \frac{I_1}{F_1 + 3F_2} \tag{6.25}$$

where $I_1 = 2(\sigma T)^2$ is the MTI improvement factor for a single-canceller MTI using a linear receiver. Similarly, the MTI improvement factor for a nonlinear receiver and a double canceller can be found to be

$$I_{NL2} = \frac{I_2}{F_1 + 9F_2} \tag{6.26}$$

where $I_2 = 2/(\sigma_c T)^4$. Extension to a triple canceller results in

$$I_{NL3} = \frac{I_3}{F_1 + 27F_2} \tag{6.27}$$

where $I_3 = 4/3(\sigma_c T)^6$.

Examination of (6.22) indicates that, as the third-order intercept point is made large compared to the clutter power, $\gamma \to 0$ causing the nonlinear receiver MTI improvement factors to approach those achieved with a linear receiver. Conversely, when the clutter power is large with respect to the third-order intercept point (i.e., the receiver is saturated), $F_1 \to 0$ and $F_2 \to 1$, which results in a reduction in MTI improvement factor of 4.8 dB for the single canceller, 9.5 dB for the double canceller, and 14.3 dB for the triple canceller.

In summary, the receiver nonlinear characteristic can be determined through measurement (and specification) of the intercept points of various orders. These can then be related to the MTI improvement factor in terms of the relative levels of clutter power to the intercept points for different orders of MTI cancellers and arbitrary clutter spectral shapes. In the preceding section, we have developed the appropriate relationships for a typical case where third-order distortion dominates the receiver characteristic and the clutter spectrum exhibits a Gaussian shape.

6.3 MTI Radar Filter Design

MTI filters can be either recursive or nonrecursive. Nonrecursive filters often are preferred because of their superior transient response.

6.3.1 Nonrecursive MTI Filters

The nonrecursive or transversal filter, as shown in Figure 6.16, consists of a tapped delay line, where the output is formed by the addition of weighted signals from each tap (see Sections 2.3.2 and 5.1.1). The impulse response of such a filter is given by

$$h(t) = \sum_{k=0}^{n-1} w_k \, \delta(t - kT) \tag{6.28}$$

where n is the number of pulses processed, w_k are the filter weights, and $\delta(\cdot)$ is an impulse function. The Laplace transform of the impulse response that provides the filter's transfer function is given by

$$H(p) = \sum_{k=0}^{n-1} w_k e^{-kpT} \tag{6.29}$$

where p is a complex variable and T is the radar's interpulse period. The z-transform is given by $z = \exp(pT)$, which results in

$$H(z) = \sum_{k=0}^{n-1} w_k z^{-k} = \frac{\sum_{k=0}^{n-1} w_k z^{n-1-k}}{z^{n-1}} \tag{6.30}$$

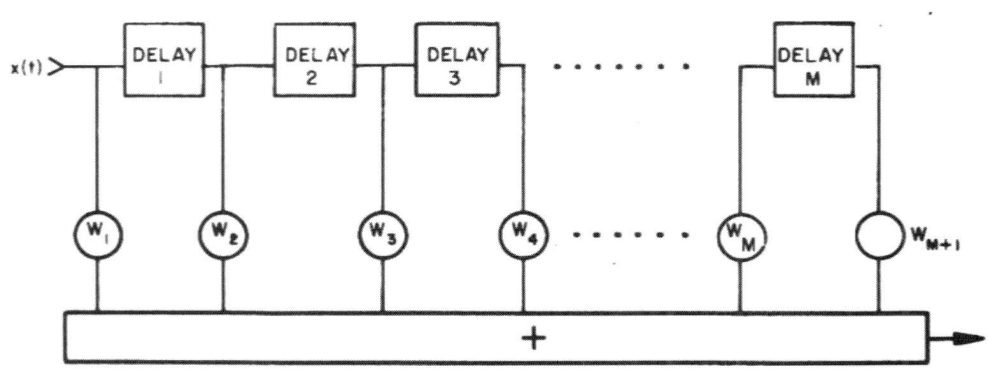

Figure 6.16 MTI transversal or nonrecursive filter. (*From:* [15, p. 2].)

indicating that the nonrecursive MTI has $n-1$ poles (equal to the number $N = n-1$ of delay sections) at the origin in the z plane.

The simplest form of transversal filter is the single canceller (see Figure 2.18) consisting of a single delay-and-subtract function with weights $\mathbf{w} = (1, -1)$ and $n = 2$. The transfer function of the single-canceller MTI (6.30) is

$$H(z) = \frac{z-1}{z} \tag{6.31}$$

indicating a pole at the origin and a zero located on the unit circle at $z = 1$, as depicted in Figure 6.17. A cascade of $N = n-1$ single cancellers, as depicted in Figure 6.18, has a transfer function given by

$$H(z) = \left(\frac{z-1}{z}\right)^{n-1} \tag{6.32}$$

indicating $n-1$ poles at the origin and $n-1$ zeros on the unit circle at $z = 1$. This type of MTI canceller is known as a *binomial canceller* because the weights have alternate binomial coefficients given by

$$w_k = (-1)^k \binom{n}{k} = \frac{(-1)^k n!}{k!(n-k)!} \tag{6.33}$$

The frequency response of the single canceller can be found from the diagram shown in Figure 6.19 by using the relationship for $|H(j\omega)|$, which is found by first

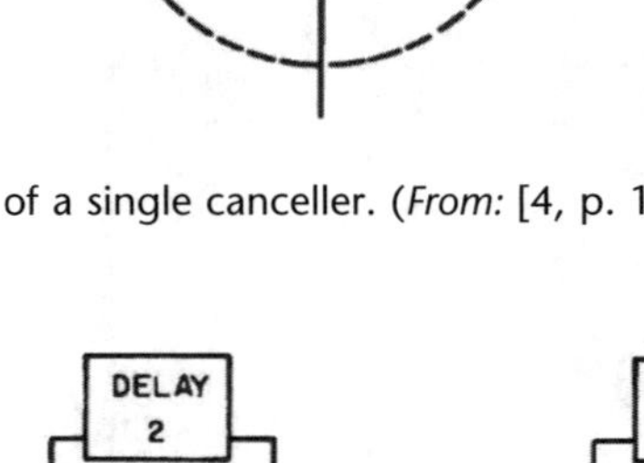

Figure 6.17 Pole zero pattern of a single canceller. (*From:* [4, p. 166].)

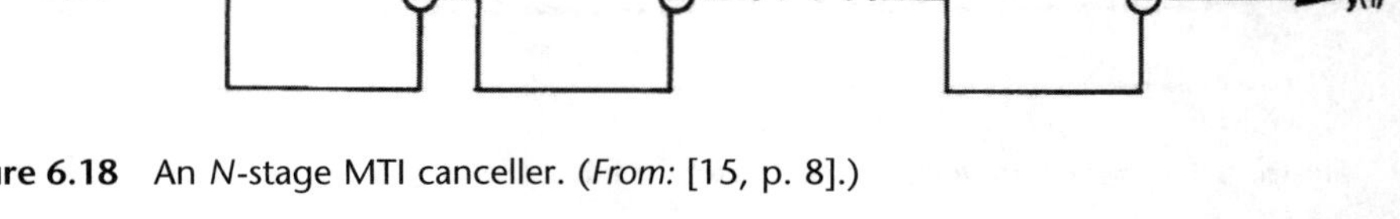

Figure 6.18 An *N*-stage MTI canceller. (*From:* [15, p. 8].)

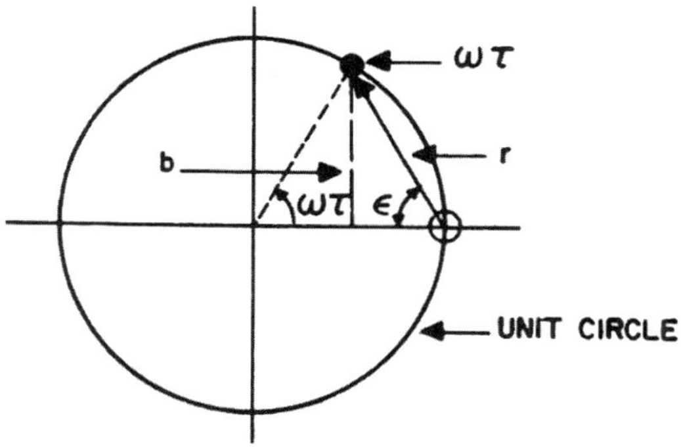

Figure 6.19 MTI frequency response of single canceller from the pole-zero diagram. (*From:* [4, p. 166].)

taking the product of vector lengths from zeros to points on the unit circle and then dividing by the product of vector lengths from poles to points on the unit circle. This results in a transfer function for a single canceller given by

$$|H(j\omega)| = \sqrt{\sin^2(\omega T) + [1 - \cos(\omega T)]^2} = 2\sin\left(\frac{\omega T}{2}\right) \tag{6.34}$$

whereas that of the general n-pulse binomial canceller is given by

$$|H(j\omega)| = \left|2\sin\left(\frac{\omega T}{2}\right)\right|^{n-1} \tag{6.35}$$

where n is the number of pulses processed by the MTI.

Another nonrecursive MTI canceller uses optimal weights rather than binomial weights [4, 16, 17]. The optimal weights are those which maximize the MTI improvement factor under the constraint that all target velocities have equal probability of occurrence (see Section 5.1.3). The MTI improvement factor (see Section 3.1) is given by

$$I_f = \frac{\sum_{j=0}^{n-1} w_j^2}{\sum_{j=0}^{n-1}\sum_{k=0}^{n-1} w_j w_k \rho_c(j-k)} \tag{6.36}$$

where w_j denotes the MTI weights and $\rho_c(\cdot)$ is the clutter's correlation coefficient. The power gain of the MTI is given by

$$\overline{G} = \sum_{j=0}^{n-1} w_j^2 \tag{6.37}$$

which represents an amplification of the target signal component by the MTI and must be normalized when assessing the MTI's clutter attenuation effect.

First, consider the optimal weights for a two-pulse MTI canceller. Equation (6.33) for the improvement factor can be specialized to

$$I_{f1} = \frac{w_0^2 + w_1^2}{w_0^2 + w_1^2 + 2\rho_1 w_0 w_1} \tag{6.38}$$

where ρ_1 is the single lag correlation coefficient. Next, the improvement factor can be maximized subject to the constraint that the MTI gain $\overline{G}$ is a constant by forming the following object function:

$$J = w_0^2 + w_1^2 + \rho_1 w_0 w_1 - \lambda\left(w_0^2 + w_1^2\right) \tag{6.39}$$

where λ is a Lagrangian multiplier [18]. Then, using the calculus of variations, the stationary points are found, where $\partial J/\partial w_0 = 0$ and $\partial J/\partial w_1 = 0$, resulting in the following linear equations:

$$\begin{aligned} w_0(1 - \lambda) + w_1\rho_1 &= 0 \\ w_1(1 - \lambda) + w_0\rho_1 &= 0 \end{aligned} \tag{6.40}$$

Elimination of w_0 and w_1 from these equations results in an expression for the Lagrangian multiplier as

$$\lambda^2 - 2\lambda + 1 - \rho_1^2 = 0 \tag{6.41}$$

The solution of the equation results in

$$\lambda = 1 - \rho_1 \tag{6.42}$$

for the value of λ that maximizes the improvement factor. Substitution of this value into (6.40) results in $w_0 = -w_1$ for the optimal weights, which, when normalized, produce a weight vector $\mathbf{w} = (1, -1)$ and an improvement factor $I_1 = 1/(1 - \rho_1)$. This demonstrates that the binomial MTI is optimum for the case when only a single-delay MTI is involved.

For higher order MTIs, the procedure used for the single-delay MTI can be followed, but the optimal weights are different than those of the binomial MTI and the solution becomes more complex. Rather than using Lagrangian multipliers, the procedure described in Section 5.1.3 is generally used to determine the optimal weights for higher order MTIs. This procedure solves the matrix equation:

$$(\mathbf{R_n} - \lambda_{min}\mathbf{I})\mathbf{w} = 0 \tag{6.43}$$

by determining the eigenvalues (λ_{min}) of

$$|\mathbf{R_n} - \lambda_{min}\mathbf{I}| = 0 \tag{6.44}$$

where $\mathbf{R}_n$ is the normalized covariance matrix and $\mathbf{I}$ is the identity matrix. Hence, the optimal weights are found as the eigenvector that corresponds to the minimum eigenvalue. The improvement factor is then given by

$$I_f = \frac{1}{\lambda_{\min}} \tag{6.45}$$

where the eigenvalues $\lambda_{\min}$ are identical to the Lagrangian multiplier in the calculus of variations approach.

To demonstrate the eigenvalue approach, the determinant of (6.44) for the single delay is given by

$$\begin{vmatrix} 1-\lambda & \rho_1 \\ \rho_1 & 1-\lambda \end{vmatrix} = 0 \tag{6.46}$$

This results in (6.41), which can be solved for $\lambda_{\min} = 1 - \rho_1$. The optimal weights can then be found from (6.40) as the eigenvector $\mathbf{w} = (1, -1)$. Solutions for optimum weights and improvement factors for double- and triple-delay MTI cancellers are given in Table 5.1.

The optimal weight MTI canceller provides a greater improvement factor than the binomial MTI canceller when greater than two pulses (one delay section) are processed. Figure 6.20 depicts the improvement factor provided with both optimal and binomial weights. If the standard deviation of the clutter spectrum is 0.01 times the PRF, an additional improvement factor can be achieved with optimal weights on the order of 2 dB for the double canceller (n = 3), 3.5 dB for the triple

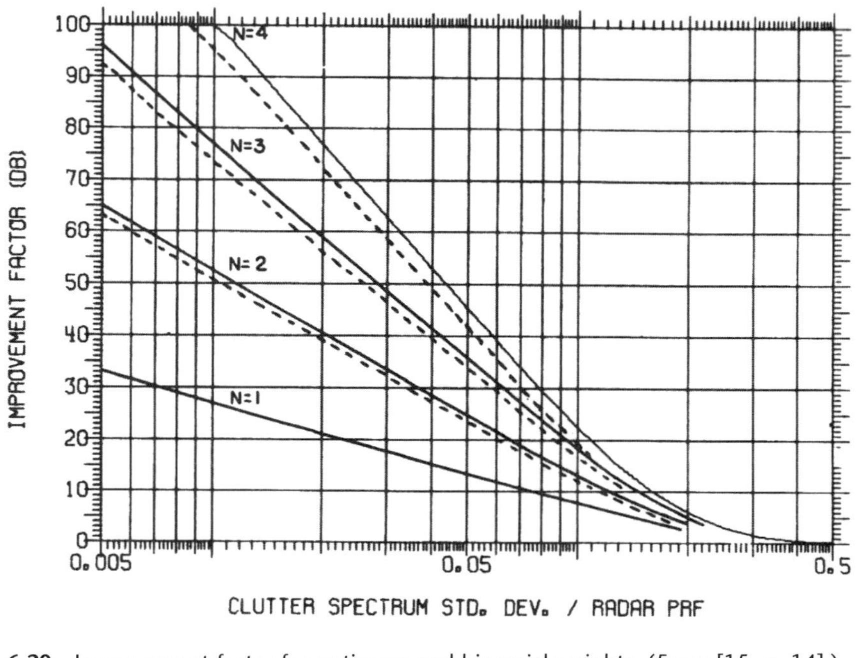

Figure 6.20 Improvement factor for optimum and binomial weights. (*From:* [15, p. 14].)

canceller ($n = 4$), and 5 dB for the quadruple canceller ($n = 5$). To achieve this performance, accurate weights are required and the clutter spectral width should be narrow with respect to the radar's PRF [15].

The z-plane transfer function of the nonrecursive MTI given by (6.30) can also be written in pole-zero form as

$$H(z) = k\frac{(z - z_1)(z - z_1^*)\ldots}{z^{n-1}} \tag{6.47}$$

when the number of delays (cancellers) is even, and

$$H(z) = k\frac{(z - z_0)(z - z_1)(z - z_1^*)\ldots}{z^{n-1}} \tag{6.48}$$

when the number of delays is odd, where z_0, z_1, and z_1^* are the zero locations in the z-plane, and the poles are located at the origin ($z = 0$). Note that for an even number of delays (e.g., a double canceller) the zeros generally occur in conjugate pairs except for the binomial canceller, where they coalesce at $z = 1$. For an odd number of delays, one or more zeros are placed at $z = 1$ (causing a null at zero frequency and multiples of the PRF), and the remaining zeros generally occur in conjugate pairs, except for the binomial canceller. Also, for a nonrecursive MTI canceller, the number of zeros and poles is equal to the number of delay sections.

To illustrate the design of nonrecursive MTI filters in the z-plane, consider the case of the double-delay ($n = 3$) MTI canceller. The normalized ($k = 1$) z-plane transfer function is given by

$$H_2(z) = \frac{z^2 - (z_1 + z_1^*)z + z_1 z_1^*}{z^2} \tag{6.49}$$

where z_1 and z_1^* are the complex conjugate zero locations and $|z_1|^2 = z_1 z_1^*$ is the vector length of the zero from the origin. Comparison of this representation with (6.30) allows the normalized weights to be determined in relation to the zero locations as $w_0 = 1$, $w_1 = -(z_1 + z_1^*) = -2 \text{ Re } z_1$, $w_2 = z_1 z_1^* = |z_1|^2$. Solution of the matrix equation $(\mathbf{R}_n - \lambda_{\min}\mathbf{I})\ \mathbf{w} = 0$ for the optimal weights (i.e., finding the eigenvector) leads to the conclusion that, for the single-delay canceller, $w_0 = -w_1$; whereas for the double canceller, $w_0 = w_2$; and for the triple canceller $w_0 = -w_3$ and $w_1 = -w_2$ (see Table 5.1). This can be generalized to the conclusion that the optimum filter weights are symmetrical for an even number of delay sections (e.g., double canceller) and antisymmetrical for an odd number of delay sections (e.g., single and triple cancellers). Thus, from (6.49), we can further deduce that, when the weights are normalized such that $w_0 = 1$, the largest indexed weight has absolute value $|w_{n-1}| = 1$, resulting in the zeros of the optimal weight MTI canceller being located on the unit circle in the z plane (i.e., $|w_{n-1}| = |z_{n-1}| = 1$), as depicted in Figure 6.21 for the double-delay canceller.

Another consequence of the weight pattern just described in that the transfer function of the nonrecursive MTI canceller can be determined from the filter weights as

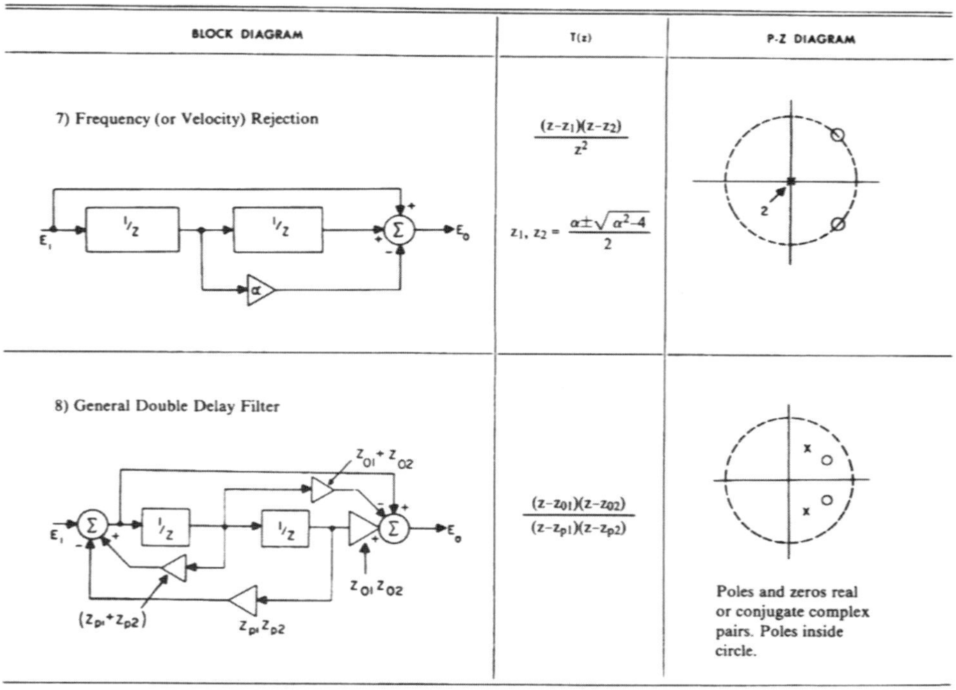

Figure 6.21 Pole-zero pattern for the double-delay canceller with optimal weights. (*From:* [4, p. 175].)

$$|H(j\omega)| = \left| \sum_{i=0}^{(N-1)/2} (-1)^i 2|w_i| \sin\left(\frac{N}{2} - i\right)\omega T \right| \tag{6.50}$$

for N odd (e.g., single or triple canceller) and

$$|H(j\omega)| = \left| (-1)^{N/2}|w_{N/2}| + \sum_{i=0}^{(N-2)/2} (-1)^i 2|w_i| \cos\left(\frac{N}{2} - i\right)\omega T \right| \tag{6.51}$$

for N even (e.g., double canceller), where N is the number of MTI delay sections. For example, the transfer function for a single canceller can be determined for (6.46) as $|H(j\omega)| = |-|w_1| + 2|w_0| \cos \omega T|$. For a binomial double canceller with weights $\mathbf{w} = (1, -2, 1)$ this becomes $|H(j\omega)| = 2(1 - \cos \omega T) = 4 \sin^2 (\omega T)/2$.

From Figure 6.21, the transfer function for the double-delay optimal weight MTI can be written as

$$H_2(z) = \frac{z^2 - 2\cos\phi z + 1}{z^2} \tag{6.52}$$

where $\phi = \omega T$. The weights of the MTI can be deduced from (6.30) as $\mathbf{w} = (1, -2\cos\phi, 1)$. The improvement factor is given by

$$I_f = \frac{1 + 2\cos^2\phi}{1 + 2\cos^2\phi - 4\rho_1\cos\phi + \rho_2} \tag{6.53}$$

where ρ_1 and ρ_2 are the correlation coefficients evaluated at $t = T$ and $2T$. If we let $\cos\phi = x$ and find the maximum I_f by setting $dI_f/dx = 0$, we have

$$x = \cos\phi = \frac{\rho_2 + \sqrt{\rho_2^2 + 8\rho_1^2}}{4\rho_1} \tag{6.54}$$

The optimal weight is then given by $\mathbf{w} = (1, w_1, 1)$ where

$$w_1 = -\left(\frac{r}{2} + \sqrt{\frac{2 + r^2}{4}}\right) \tag{6.55}$$

and $r = \rho_2/\rho_1$, which for a Gaussian-shaped correlation coefficient is $r = \exp[-3(\sigma_c T)^2/2]$. The optimum weights for a triple-delay MTI canceller (see Table 5.1) can be found in a similar manner, where the transfer function is given by

$$H_3(z) = \frac{(z-1)(z^2 - 2\cos\phi z + 1)}{z^3} \tag{6.56}$$

resulting in an optimal weight vector equal to $\mathbf{w} = [1, -(1 + 2\cos\phi), (1 + 2\cos\phi), -1]$.

There are several limitations to the optimum weight MTI canceller. First, note [see (6.48)] that in the design of MTI cancellers with an even number of delay section (e.g., double-delay canceller), the zeros occur at other than the zero frequency (i.e., $z = 1$). This means that fixed discrete targets are not being fully cancelled by the MTI filter. Also, as the number of pulses processed by the MTI increases, the design procedure results in significant variations in the MTI passband (see Figure 2.25) and a larger improvement factor than can be supported by MTI stability considerations. A quadratic programming algorithm has been developed that allows trading the excess improvement factor for greater velocity coverage consistent with the cancellation of fixed discrete targets [4, 19].

▶ Example 21 Transfer Function and Improvement Factor for MTI

The power transfer function for an MTI with binomial coefficients is given by

$$|H(\omega)| = \left|2\sin\left(\frac{\omega T}{2}\right)\right|^{2n}$$

where ω is the radian frequency, T is the period, and n is the number of MTI delay sections. MATLAB program mti_canceler.m plots the normalized MTI transfer function against the parameter f_d/PRF for n equal to 1 – 9. Note that the passband becomes increasingly narrow as the number of MTI delay sections is increased. This is an inherent problem with binomial cancellers generally limiting the number of delay sections to three or four. The transfer function is depicted in Figure EX-21.1.

MTI binomial weights are given by

$$w_k = (-1)^k \frac{N!}{k!(N-k)!}$$

where N is the number of MTI delay sections and k is the weight number 0, 1, 2, . . . , N. The MTI improvement factor is given by:

$$I = \frac{\mathbf{w}^T \mathbf{M}_S \mathbf{w}^*}{\mathbf{w}^T \mathbf{R}_N \mathbf{w}^*} \qquad \mathbf{w}^T = [w_0, w_1, \ldots, w_n]$$

where

$$\mathbf{M}_S = \begin{bmatrix} 1 & 0 & 0 \\ 0 & 1 & 0 \\ 0 & 0 & 1 \end{bmatrix} \qquad \mathbf{R}_N = \begin{bmatrix} 1 & \rho_1 & \rho_2 \\ \rho_1 & 1 & \rho_1 \\ \rho_2 & \rho_1 & 1 \end{bmatrix} \qquad \rho(iT) = \exp\left(\frac{-i^2\Omega^2}{2}\right) \qquad \Omega = \frac{2\pi\sigma_f}{\text{PRF}}$$

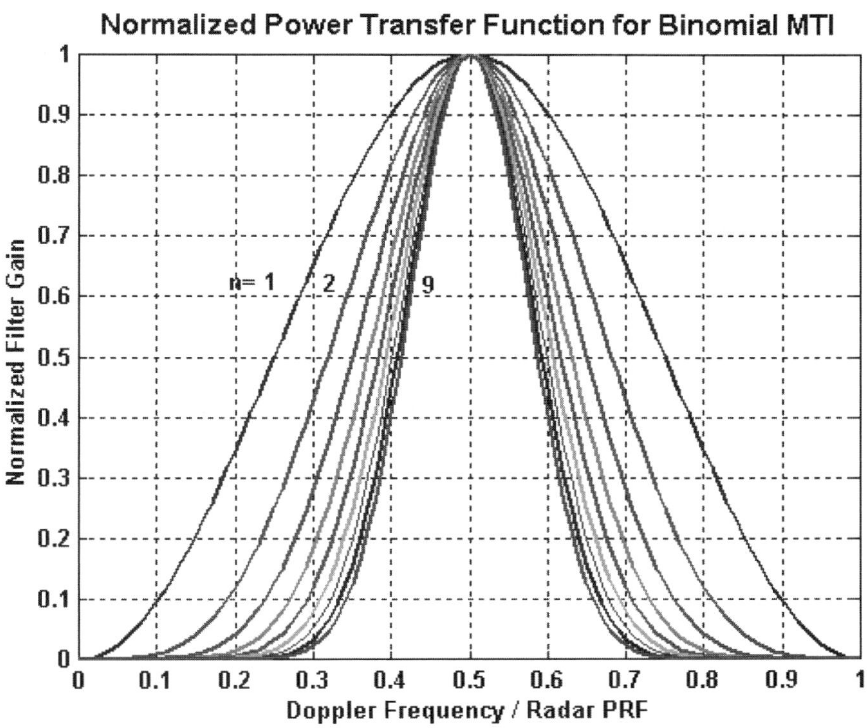

Figure EX-21.1 Power transfer function for binomial cancellers.

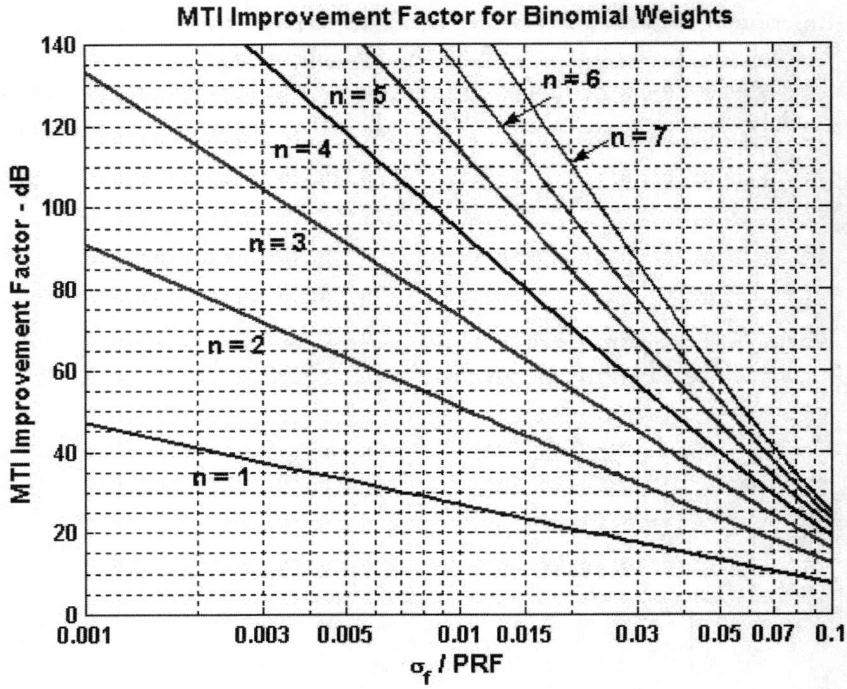

Figure EX-21.2 Improvement factor for binomial cancellers.

and $\mathbf{w}^T$ is the transpose of the weight vector, σ_f is the standard deviation of the Gaussian-shaped clutter spectrum, PRF is the pulse repetition frequency of the radar, $\mathbf{M}_S$ is the signal covariance matrix, and $\mathbf{R}_N$ is the clutter's covariance matrix.

MATLAB program impfac_binary.m plots the improvement factor as a function of $x = \sigma_f/\mathrm{PRF}$ for $N = 1$ through 7. Figure EX-21.2 depicts the improvement factor for binomial cancellers.

▶

6.3.2 Recursive MTI Filters

The most general form of MTI canceller is the recursive type (see Example 5), the block diagram of which is depicted in Figure 2.26. Recursive MTI filters employ feedback loops as well as the feed-forward configurations associated with nonrecursive filters. The feedback loops allow shaping of the MTI velocity response, which, in principle, makes possible the synthesis of almost any frequency response function. However, the recursive filters are susceptible to transients that may circulate through the MTI filter for many radar interpulse intervals, depending on the degree of shaping employed in the MTI. Transients in the MTI may be caused by large clutter returns that act as a step input, the use of step-scan antenna beam programming (e.g., in phased arrays), or by intentional or unintentional nonsynchronous interference. When recursive MTI filters are employed, some form of transient protection generally must be used as well [4, 20].

The z-plane transfer function for a recursive MTI is given by

$$H(z) = k\frac{z^{n-1} + a_0 z^{n-2} + \ldots + a_{n-1}}{z^{n-1} + b_0 z^{n-2} + \ldots + b_{n-1}} \tag{6.57}$$

where n is the number of pulses processed by the MTI filter and a_i, $i = 0, 1, 2, \ldots, n-1$ are the feed-forward coefficients and b_i, $i = 0, 1, 2, \ldots, n-1$ are the feedback coefficients. The transfer function can be written in pole-zero form as

$$H(z) = k\frac{(z - z_{01})(z - z_{02}) \ldots (z - z_{0n-1})}{(z - z_{p1})(z - z_{p2}) \ldots (z - z_{pn-1})} \tag{6.58}$$

where z_{0i}, $i = 1, 2, \ldots, n-1$ are the filter zeros and z_{pi}, $i = 1, 2, \ldots, n-1$ are the filter poles.

In practice, most recursive MTI filters are synthesized as cascades with each cascade having no more than two delay sections. Thus, no feedback or feed-forward path needs to span more than two delay sections.

A basic single-delay building block is depicted in Figure 6.22 [4, 21]. The transfer function for this filter is given by

$$H(z) = k_f\frac{\left(z + \frac{1}{k_f}\right)}{z - k_b} \tag{6.59}$$

where the zero is located at

$$z_0 = \frac{-1}{k_f} \tag{6.60}$$

and the pole at

$$z_p = k_b \tag{6.61}$$

For MTI filters used in fixed-based radars operating against ground clutter, we generally want a zero located at zero frequency, which is achieved in the single-delay configuration by setting $k_f = -1$.

The basic double-delay building block is depicted in Figure 6.23 [4, 21]. The transfer function for this filter is given by

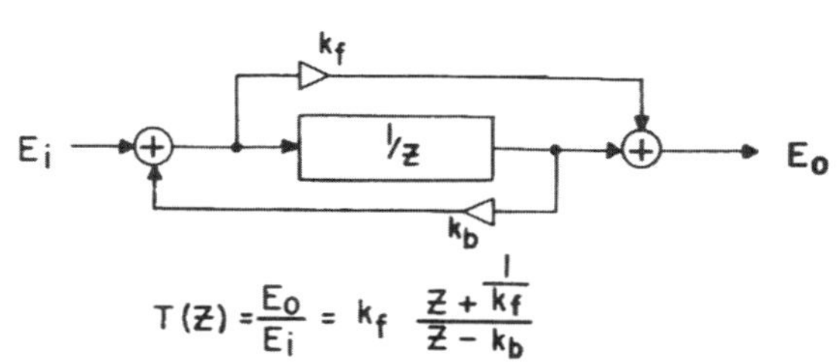

Figure 6.22 Single-delay recursive MTI filter with separate pole-zero control. (*From:* [21, p. 160].)

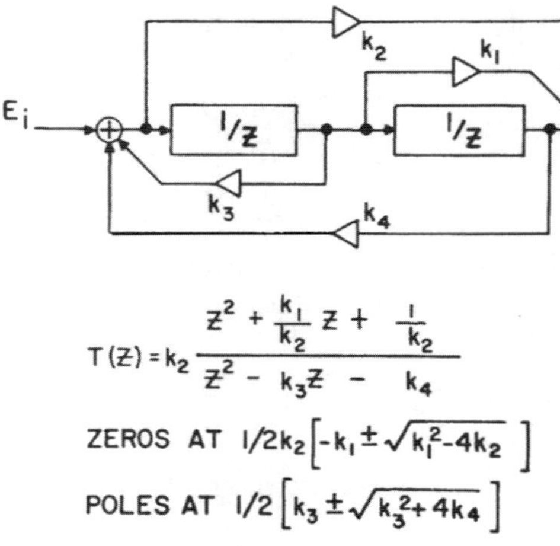

Figure 6.23 Double-delay recursive MTI Filter with separate pole-zero control. (*From:* [21, p. 160].)

$$H(z) = k \frac{z^2 + \left(\frac{k_1}{k_2}\right)z + \frac{1}{k_2}}{z^2 - k_3 z - k_4} \tag{6.62}$$

where the zeros are located at

$$z_0 = \frac{1}{2k_2}\left(-k_1 \pm \sqrt{k_1^2 - 4k_2}\right) \tag{6.63}$$

and the poles at

$$z_p = \frac{1}{2}\left(k_3 \pm \sqrt{k_3^2 + 4k_4}\right) \tag{6.64}$$

This double-delay configuration is the only one where the feedback factors involved in the location of the poles do not affect the zeros and vice versa. The pole locations are controlled solely by the feedback loops, whereas the two feed-forward loops govern only the zero locations. If, in the synthesis procedure of the overall MTI filter, the number of poles exceeds the number of zeros, the last poles can be obtained with the basic circuit by dropping either one or both of the zeros. One zero is lost by setting k_1 or $k_2 = 0$ and both zeros are dropped by letting $k_1 = k_2 = 0$.

An example of a triple-canceller MTI designed by using the basic single-delay and double-delay building blocks is depicted in Figure 6.24 [22]. The single-delay section is designed with $k_f = -1$ and $k_b = \alpha_4$, whereas the double-delay section has $k_1 = -2\alpha_1$, $k_2 = 1$, $k_3 = \alpha_2$, and $k_4 = \alpha_3$. The single-delay zero is then at $z_0 = 1$ and the pole is at $z_p = \alpha_4$ (see (6.60) and (6.61)). The double-delay zeros are located at $z_0 = \alpha_1 \pm \sqrt{\alpha_1^2 - 1}$, whereas the poles are located at

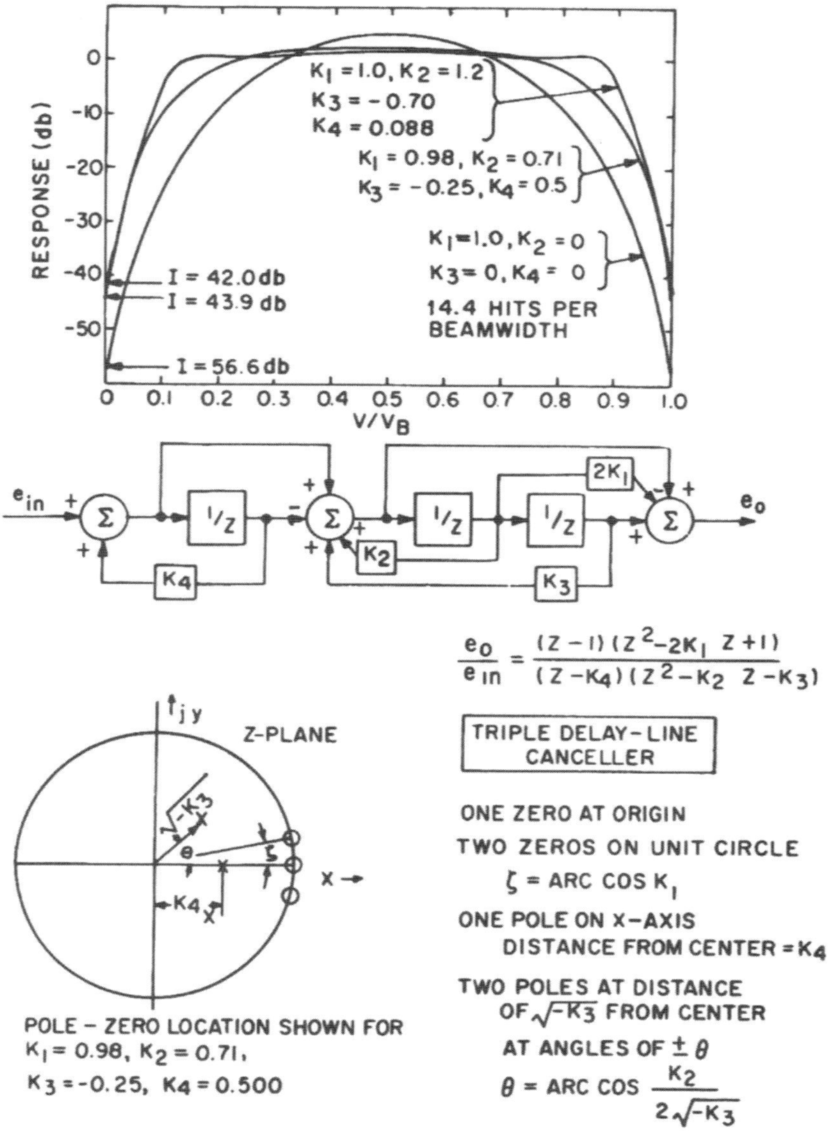

Figure 6.24 Triple MTI canceller design example. (*From:* [22, p. 15-26].)

$z_p = (1/2)\left(\alpha_2 \pm \sqrt{\alpha_2^2 + 4\alpha_3}\right)$ [see (6.63) and (6.64)]. The resulting transfer function is then given by

$$H(z) = \frac{(z-1)(z^2 - 2\alpha_1 z + 1)}{(z-\alpha_4)(z^2 - \alpha_2 z - \alpha_3)} \tag{6.65}$$

The pole-zero pattern is illustrated in Figure 6.24 and has three zeros located on the unit circle at frequencies $\omega_o T = 0, \pm\cos^{-1}\alpha_1$ with a single pole located on the x axis of the z plane at $x = \alpha_4$, and a conjugate pair of poles located at a radius equal to $\sqrt{\alpha_3}$ with angles $\theta = \pm\cos\, \alpha_2/2\sqrt{\alpha_3}$. The diagram illustrates how the velocity response of the filter can be widened at the expense of a reduced improvement factor by the appropriate adjustment of the filter's parameters. Note that the filter's transfer function can be found from the pole-zero diagram by finding the ratio of the product of the zero vectors to the product of the pole vectors for different points (i.e., $\theta = \cos\, \omega T$) around the unit circle. The improvement factor is then found from the relationship $I_f = \overline{G} \cdot \text{CA}$ as

$$I_f = \frac{|\overline{H(f)}|^2 \displaystyle\int_0^{\text{PRF}} S(f)\, df}{\displaystyle\int_0^{\text{PRF}} |H(f)|^2\, S(f)\, df} \tag{6.66}$$

where $H(f)$ is the filter's transfer function and $S(f)$ is the clutter spectrum.

Another recursive filtering configuration commonly found in analog-shaped canceller implementations is depicted in Figure 6.25. This configuration is similar to a binomial double canceller [see (6.29)], except that the velocity response can be widened by using the feedback paths, which move the poles from the origin in the z plane. The advantage of this configuration in analog implementations is that,

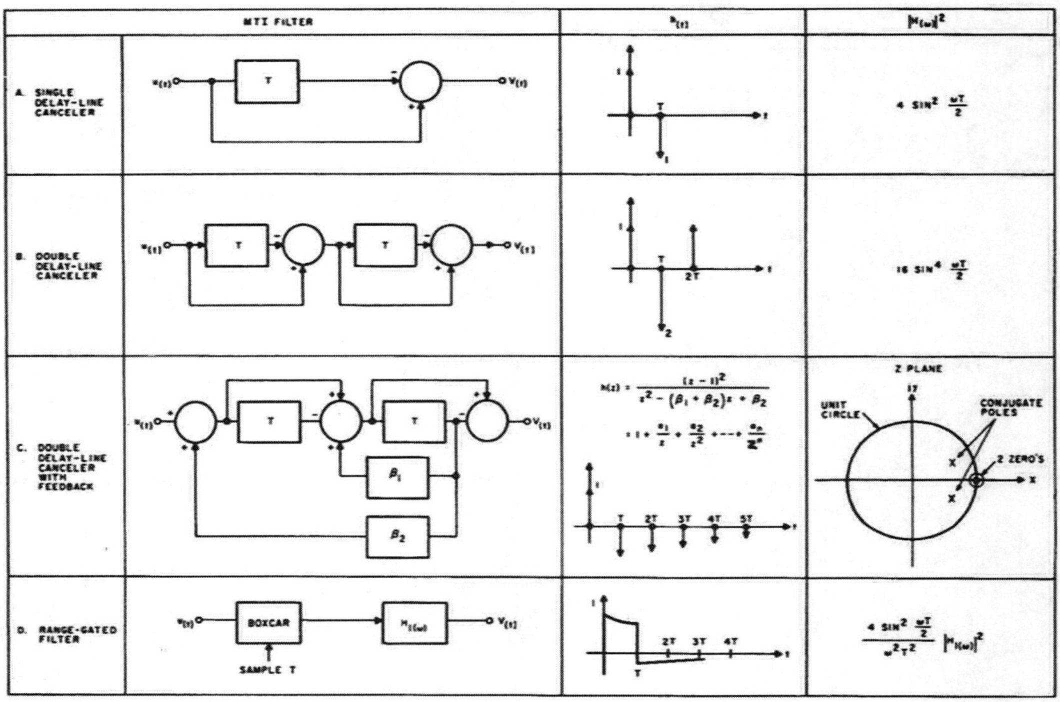

Figure 6.25 Shaped double-canceller MTI filter. (*From:* [4, p. 193].)

if one filter section drifts so that its zero in the z plane moves away from $z = 1$, the other filter section is still available to cancel fixed, discrete targets at zero frequency.

The normalized transfer function of the shaped double-delay canceller is given by

$$H(z) = \frac{(z-1)^2}{z^2 + (\beta_1 + \beta_2)z + \beta_2} \tag{6.67}$$

where a double zero is located at $z = 1$ and the poles are located at

$$z_p = \frac{1}{2}\left[(\beta_1 + \beta_2) \pm \sqrt{(\beta_1 + \beta_2)^2 - 4\beta_2}\right] \tag{6.68}$$

The poles are located in the z-plane at an angle given by $\phi = \cos^{-1}\left[(\beta_1 + \beta_2)/2\sqrt{\beta_2}\right]$ at a distance $|z| = \sqrt{\beta_2}$ from the origin.

An example of the shaped double-delay canceller design is depicted in Figure 6.26 [6], which illustrates different canceller shapes that result from selecting various values of β_1 ranging from –0.6 to +0.6 and values of β_2 from 0 to 0.5. The values for a double-delay binomial canceller are $\beta_1 = \beta_2 = 0$, which results in an MTI improvement factor of 39.9 dB for the situation depicted. Note that the velocity response of the MTI canceller can be widened or narrowed by selecting the appropriate filter parameters with an accompanying decrease or increase in the MTI improvement factor.

In some cases, we might want the filter of Figure 6.25 to provide two separate and distinct zeros on the unit circle. A configuration that allows this is depicted in Figure 6.27. The transfer function for this configuration is given by

$$H(z) = \frac{z^2 - z(2 - a_1) + 1}{z^2 - z(\beta_1 + \beta_2) + \beta_2} \tag{6.69}$$

where the zeros are located at an angle:

$$\phi_0 = \cos^{-1}\left[\frac{(2 - a_1)}{2}\right] \tag{6.70}$$

on the unit circle and the poles are given by (6.64).

6.3.2.1 Synthesis of Recursive MTI Filters

The use of a recursive MTI filter provides the designer with the flexibility to adjust the MTI velocity response according to changing clutter conditions. One example of the need for this capability occurs in fixed-based air search radar, where the clutter's spectral spread is determined primarily by scanning modulation (see Section 6.2.1). In some radar, the antenna's scan rate is varied to provide more hits on

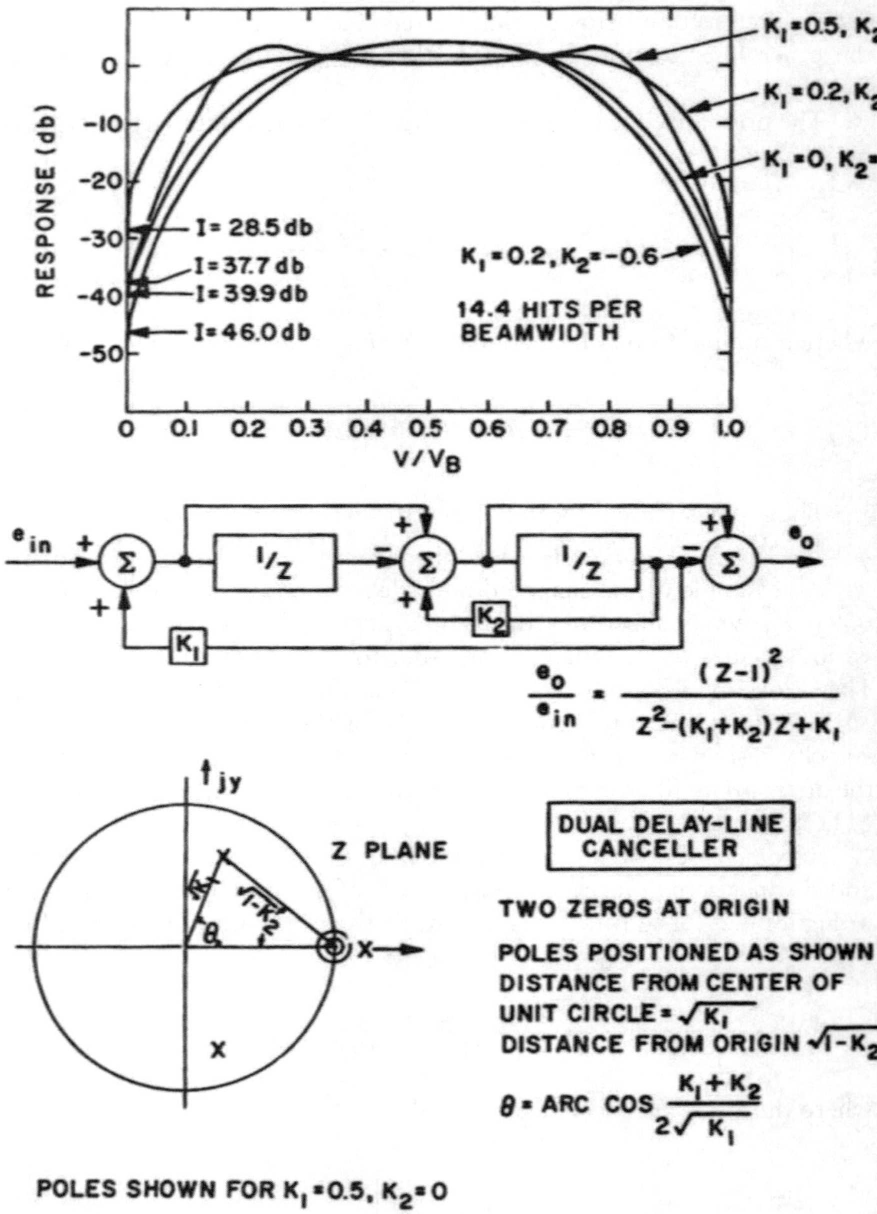

Figure 6.26 Shaped double-delay MTI design example. (*From:* [22, p. 15-25].)

target, which allows detection of weaker targets. Figure 6.28 depicts four velocity responses obtained by using the shaped double-delay canceller configuration of Figure 6.25. The velocity shapes are set to provide a 30 dB MTI improvement factor against ground clutter for the antenna scan rates listed (i.e., 3.3 to 10 rev/min) [4, 23]. This capability is also useful in airborne surveillance radar, where the clutter's spectral spread is determined by the aircraft's velocity and the antenna's azimuth scan position (see Section 6.2.2).

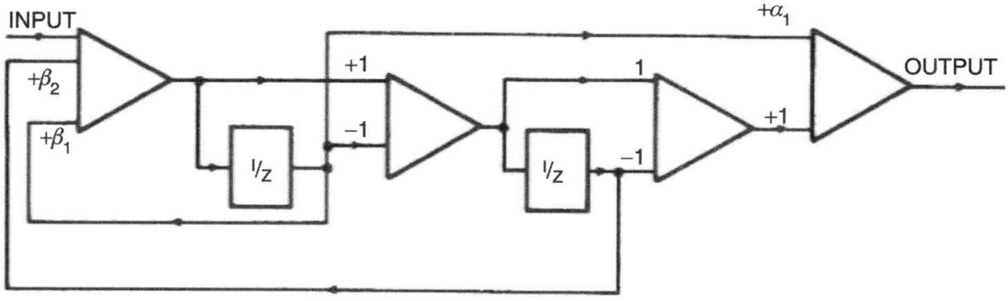

Figure 6.27 Shaped double-delay MTI canceller with adjustable zeros. (*From:* [4, p. 153; 23, p. 196].)

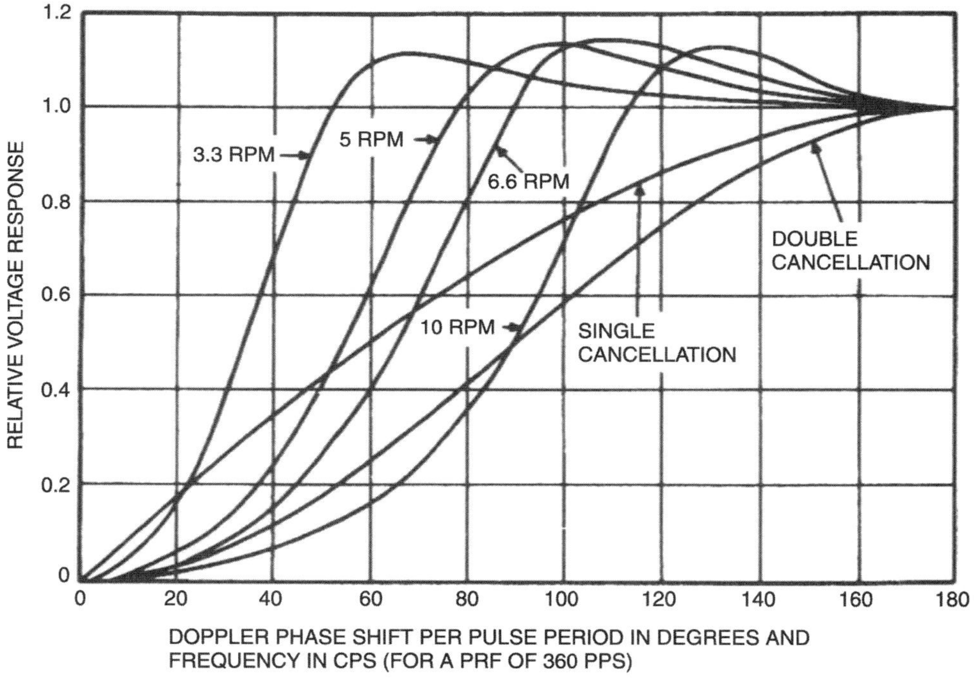

Figure 6.28 Velocity response of shaped double cancellers set for various antenna scan rates. (*From:* [4, p. 156; 23, p. 199].)

The synthesis procedure for designing recursive MTI filters involves selecting a filter shape that provides the maximum velocity coverage consistent with the required MTI improvement factor. Conventional filter synthesis theory can be translated into the sampled data domain by employing a bilinear transformation that translates the Laplace transform domain (p-plane) into the z-transform domain. This allows for translating conventional filter designs that use the Butterworth, Chebyshev, or elliptic transfer functions into a sample-data MTI filter configuration. Once the filter type has been selected (e.g., Butterworth), the synthesis procedure involves a selection of the number of filter poles and the cutoff frequency.

To aid in the selection of filter poles and cutoff frequency, examining the performance of an asymptotic filter is useful, which is defined as

$$H_A(\omega)^2 = \begin{pmatrix} \left(\dfrac{\omega}{\omega_{co}}\right)^{2n}; & \omega \le \omega_{co} \\ 1; & \omega > \omega_{co} \end{pmatrix} \tag{6.71}$$

where n is the number of filter poles and ω_{co} is the radian Doppler frequency dividing the stopband and passband. This filter is essentially flat in the passband and falls off in the stopband according to the number of filter poles. The clutter attenuation for a Gaussian-shaped spectrum is then given by

$$\text{CA} = \frac{\sqrt{\pi\sigma_c^2/2}}{\displaystyle\int_0^{\omega_{co}} e^{-\omega^2/2\sigma_c^2}\left(\frac{\omega}{\omega_{co}}\right)^{2n} d\omega + \int_{\omega_{co}}^{\pi/T\to\infty} e^{-\omega^2/2\sigma_c^2}\, d\omega} \tag{6.72}$$

where σ_c is the clutter's spectrum standard deviation, n is the number of filter poles, and ω_{co} is the filter's cutoff frequency. This equation is plotted in Figure 6.29, where the filter's cutoff frequency is normalized to the clutter's spectral width

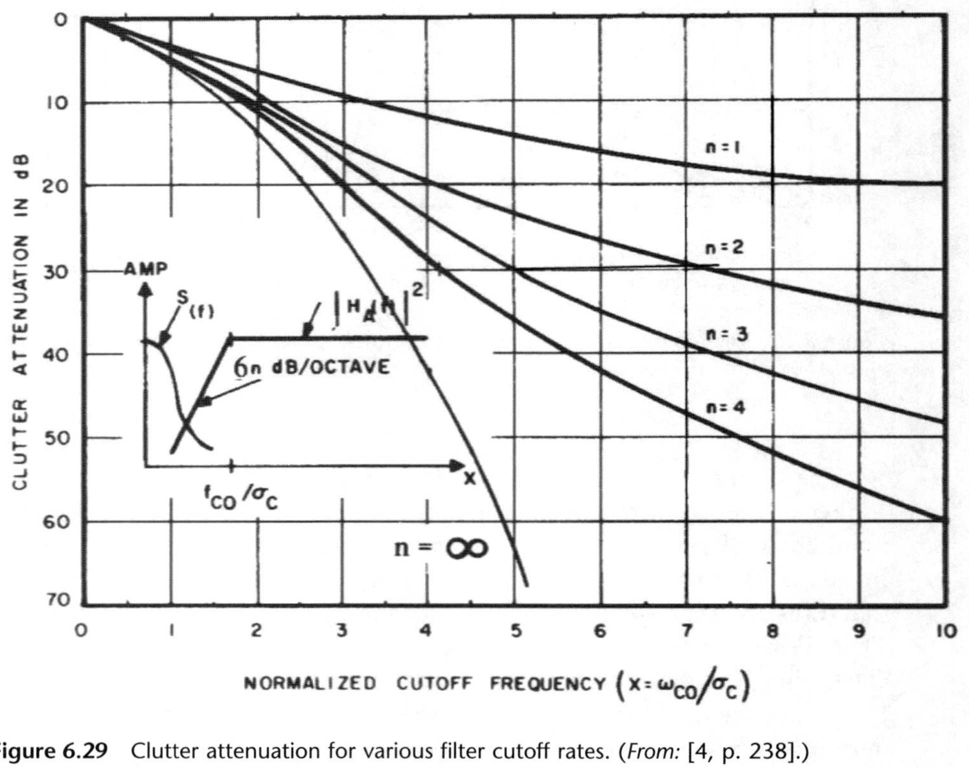

Figure 6.29 Clutter attenuation for various filter cutoff rates. (*From:* [4, p. 238].)

(ω_{co}/σ_c). For example, to achieve 30 dB of clutter attenuation with a double-delay design (i.e., two poles) would require the filter's cutoff frequency to be designed at about seven times the clutter's standard deviation width. The best that can be achieved with an infinite number of delay sections ($n \to \infty$) is to set the filter's cutoff frequency at about three times the clutter's spectral width. Note that the MTI improvement factor is related to the clutter attenuation by the relationship $I = \overline{G} \cdot \text{CA}$, where $\overline{G}$ is the MTI gain. The MTI gain is given for various binomial MTI cancellers in Table 3.1.

The type of filter shape is determined by the desired velocity response characteristic. The Butterworth design provides a monotonic amplitude versus frequency response, where the amplitude is maximally flat about the PRF/2 point in the comb filter response (see Figure 1.6). By allowing a ripple in the passband, the Chebyshev filter, using the same number of poles and zeros, can achieve sharper cutoffs. Elliptic filters yield even sharper cutoffs than the Chebyshev filter for the same network complexity, but allow both passband and stopband ripple. In Figure 6.30, several designs using two-delay Butterworth and Chebyshev responses are depicted for various normalized filter cutoff frequencies ranging from f_{co}/PRF = 0.05 to 0.25.

The bilinear transformation most useful for MTI cancellers takes a conventional lowpass filter design and transforms it into an MTI comb filter response. This transformation is given by

$$z = \frac{p + \Omega}{p - \Omega} \tag{6.73}$$

where p is the pole or zero position in the p-plane, Ω is a scale factor related to the filter's cutoff frequency, and z is the transformed pole or zero position in the

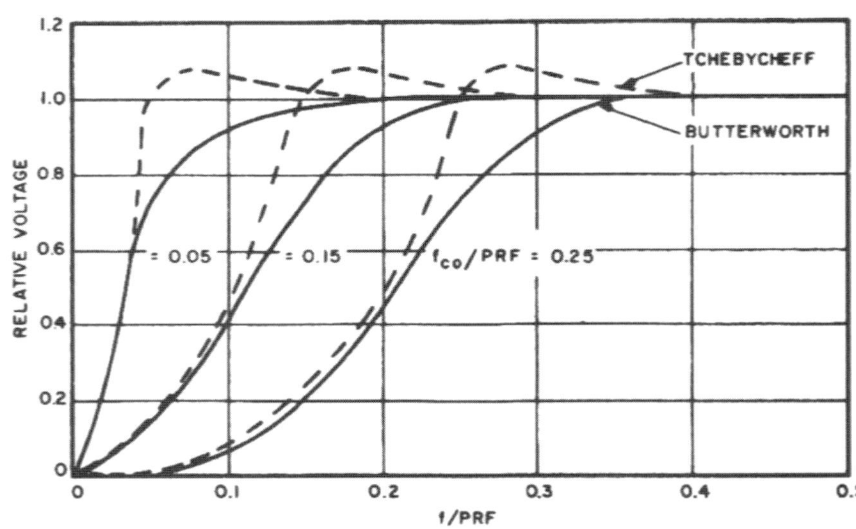

Figure 6.30 Butterworth- and Chebyshev-shaped double-delay MTI filter amplitude responses. (*From:* [4, p. 228].)

z-plane. The converse transformation from the z-plane into the p-plane is given by

$$p = \Omega \frac{z+1}{z-1} \tag{6.74}$$

Figure 6.31 depicts the transformations between the z- and p-planes where the pole-zero pattern of a two-delay Butterworth design is shown in the p-plane. The scale factor is determined by noting that the imaginary point in the p-plane at $+j$ corresponds to the point in the z-plane labeled ①. By using (6.73), we obtain

$$p = +j = \Omega \frac{\cos \phi_0 - j \sin \phi_0 + 1}{\cos \phi_0 - j \sin \phi_0 - 1} \tag{6.75}$$

where $\phi_0 = \omega_{co} T$. Solution of (6.71) determines the scale factor as

$$\Omega = \tan\left(\frac{\phi_0}{2}\right) \tag{6.76}$$

where $\phi_0 = \omega_{co}/\text{PRF}$ represents the fractional angle in the z-plane that corresponds to the cutoff frequency in the p-plane. The scale factor of (6.76) forces the point $j\omega = \pm 1$ in the p-plane to map onto the unit circle in the z-plane at an angle $\pm\phi_0$ that represents a frequency of $f_{co} = \phi_0/2\pi T$.

The poles for a lowpass Butterworth filter are located around a circle in the p-plane at

$$p_k = -\sin \frac{(2k-1)\pi}{2n} + j \cos \frac{(2k-1)\pi}{2n} \tag{6.77}$$

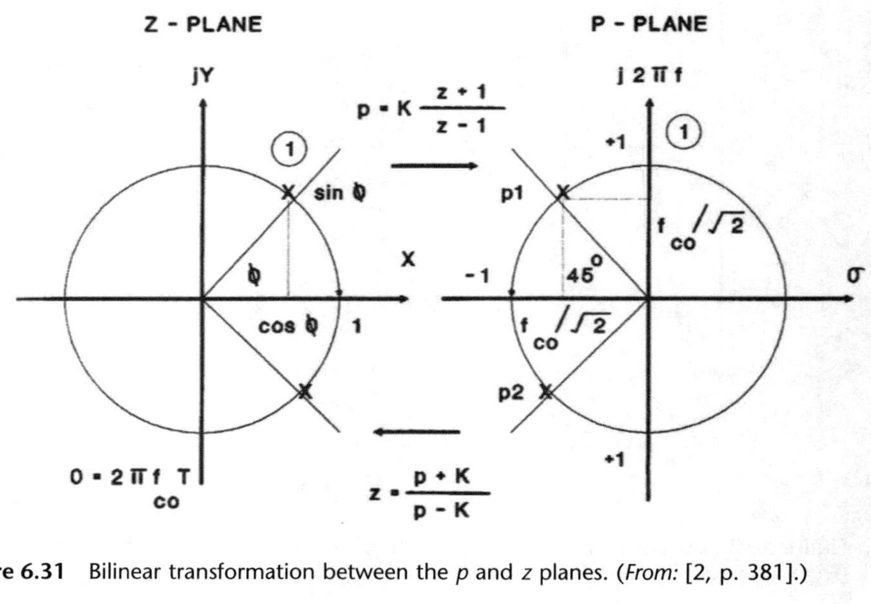

Figure 6.31 Bilinear transformation between the p and z planes. (*From:* [2, p. 381].)

where n is the number of poles (delay sections) and k is from 1 to n. The poles are normalized to the cutoff frequency (ω_{co}) so that the 3 dB point is at $j\omega = 1$. For example, at $n = 2$, the poles are located at $p_{12} = (-1 \pm j)/\sqrt{2}$, whereas for $n = 3$ they are located at $p_{13} = (-1 \pm j\sqrt{3})/2$ and $p_2 = -1$. The Butterworth poles for a shaped double-delay canceller ($n = 2$) would then be located at

$$z_{12} = \frac{1 - \sqrt{2}\tan\dfrac{\phi_0}{2} \mp 1}{\sqrt{2}\tan\dfrac{\phi_0}{2} + 1 \mp j} \tag{6.78}$$

where $\phi_0 = \omega_{co}T$. For example, if the normalized cutoff frequency were f_{co}/PRF $= 0.1$, then the poles would be located at $z_{12} = 0.5715 \pm 0.2936j$ with magnitude $|z| = 0.6425$ at an angle $\phi = 27.2°$. In the shaped double-delay canceller configuration of Figure 6.25, the parameter $\beta_2 = |z|^2 = 0.4128$, whereas the parameter $\beta_1 = 2\ \mathrm{Re}\ z_1 - \beta_2 = 0.7302$. The locus of Butterworth poles for the shaped double-delay canceller is depicted in Figure 6.32.

Alternately, the poles for a lowpass Chebyshev filter are located on an ellipse in the p-plane given by

$$p_k = -\sin\frac{(2k-1)\pi}{2n}\sinh\gamma + j\cos\frac{(2k-1)\pi}{2n}\cosh\gamma \tag{6.79}$$

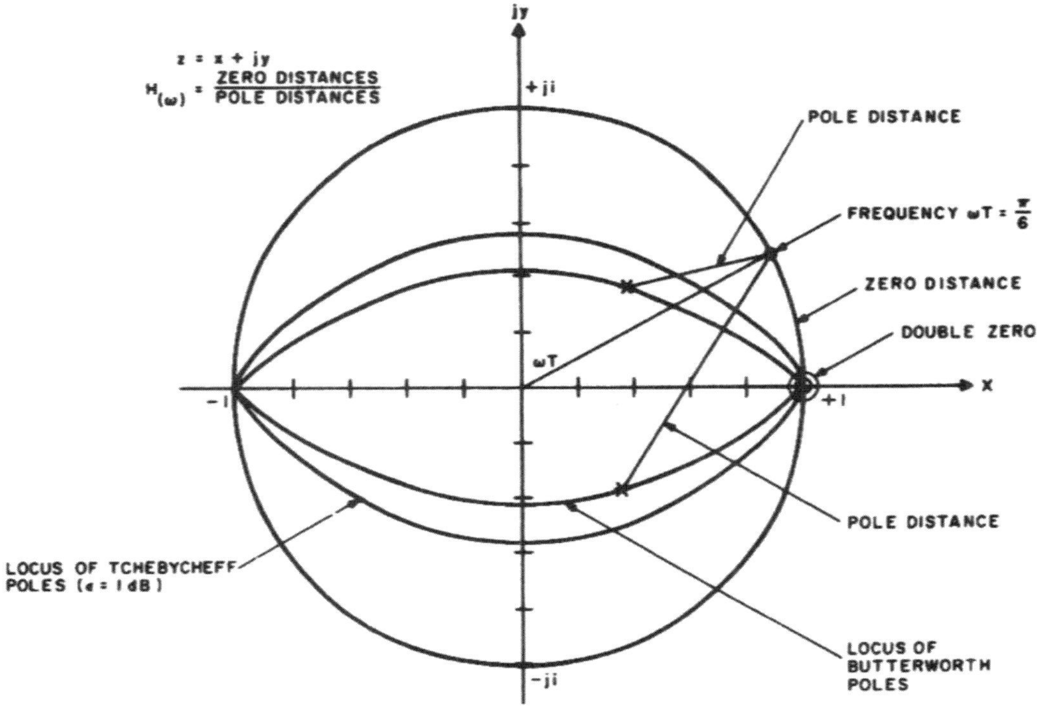

Figure 6.32 Pole-zero patterns of shaped double-delay Butterworth and Chebyshev MTI cancellers. (*From:* [4, p. 225].)

where n is the number of poles and $k = 1, 2, \ldots, n$. The peak-to-valley power ratio of the ripple in the filter passband is equal to $1 + \epsilon^2$ resulting in the ripple amplitude (δ) being equal to

$$\delta = 1 - \frac{1}{\sqrt{1 + \epsilon^2}} \tag{6.80}$$

where ϵ is a filter parameter. The ripple amplitude can be expressed on a decibel scale, resulting in the values of ϵ given in Table 6.1. The values of

$$\sinh\ \gamma = \frac{(\alpha + \beta)^{l/n} - (\alpha - \beta)^{1/n}}{2} \tag{6.81}$$

$$\cosh\ \gamma = \frac{(\alpha + \beta)^{l/n} + (\alpha - \beta)^{1/n}}{2} \tag{6.82}$$

where

$$\alpha = \sqrt{1 + \frac{1}{\epsilon^2}} \tag{6.83}$$

$$\cosh\ \gamma = \frac{(\alpha + \beta)^{l/n} + (\alpha - \beta)^{1/n}}{2} \tag{6.84}$$

where

$$\alpha = \sqrt{1 + \frac{1}{\epsilon^2}} \tag{6.85}$$

$$\beta = \frac{1}{\epsilon} \tag{6.86}$$

can then be found, which allows the value of the Chebyshev filter poles to be determined. The point $j\omega = 1$ is now the point where the response has fallen off by the allowed ripple tolerance rather than by 3 dB as with the Butterworth filter. The locus of Chebyshev poles for the double-delay-shaped canceller is depicted in Figure 6.32.

Table 6.1 Chebyshev Filter Ripple Parameters

$20 \log \delta$	ϵ	ϵ^2
1/2 dB	0.3493	0.1220
1 dB	0.5088	0.2589
2 dB	0.7648	0.5848
3 dB	0.9976	0.9953

As an example of a Chebyshev filter design, consider the problem of designing a three-pole MTI filter with 1 dB passband ripple and a normalized cutoff frequency at f_{co}/PRF = 0.125 [4, 23]. From Table 6.1, the value of $\epsilon = 0.5088$ is obtained for 1 dB ripple, resulting in $\alpha = 2.2051$ and $\beta = 1.9654$. The values of sinh $\gamma = 0.4942$ and cosh $\gamma = 1.1154$ are found, resulting in $\rho_{1,3} = -0.2471 \pm j0.9659$ and $\rho_2 = -0.4942$. Applying the bilinear transformation [see (6.73)] with $\gamma = \tan(45/2)°$ = 0.4142 results in z-plane poles equal to $z_{13} = 0.6000 \pm j0.5838$ with magnitude $|z| = 0.8375$ at an angle of 44.2° and $z_2 = 0.0881$, as depicted in Figure 6.33, which also shows the amplitude response of the lowpass Chebyshev filter.

The complete canceller can be implemented with a cascade formed by using the single-delay configuration given in Figure 6.22 and the shaped double-delay canceller depicted in Figure 6.25. The parameters for the single-delay canceller are $k_B = 0.0881$ and $k_f = 0$. For the shaped double-delay canceller, $\beta_2 = |z_1|^2 = 0.7012$ and $\beta_1 = 2 \text{ Re } z_1 - \beta_1 = 2 \text{ Re } z_1 - \beta_2 = 0.4994$. The double-delay canceller can also be implemented in the canonical form depicted in Figure 6.23. If we set

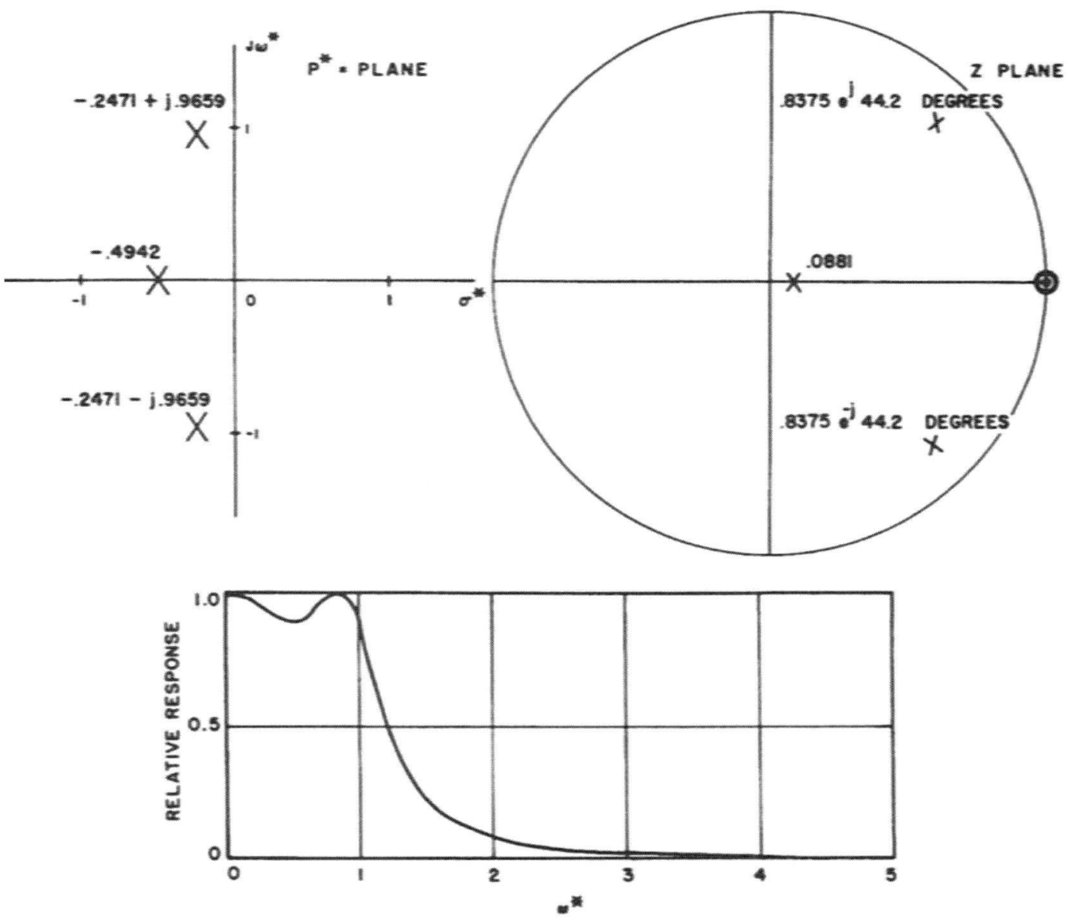

Figure 6.33 Pole-zero plot and frequency response of Chebyshev MTI canceller. (*From:* [4, p. 151; 23, p. 194].)

$k_2 = 1$ and solve (6.63), we have $k_1 = -2$. This selection results in a double zero at $z = 1$. Then, from (6.64), we can determine that $k_3 = 2\ \mathrm{Re}\ z_1 = 1.2001$. Solution of (6.64) then results in $k_4 = -0.7012$. The complete triple-delay canceller is depicted in Figure 6.34 with a transfer function equal to

$$H(z) = \frac{(z-1)^2}{(z-0.0881)(z^2 - 1.2001z + 0.7012)} \tag{6.87}$$

6.3.3 Doppler Filter Banks as Transversal Filters

Doppler filter banks often are employed, either in conjunction with MTI filters (see Section 5.2) or separately to detect targets imbedded in clutter and noise. These filter banks are implemented most efficiently by using digital signal processing techniques that employ the FFT algorithm (see Sections 2.3.3.2 and 2.3.3.3). Most treatments of this subject focus on the computational details of the FFT processing rather than on the basic filtering action of the filter bank. By using a transversal filter representation for the FFT processing, we can gain insight into the properties of such Doppler processing [24].

The kth term of the discrete Fourier transform is given by

$$F(k) = \sum_{i=0}^{n-1} f(i) e^{-j(2\pi/n)ik} \tag{6.88}$$

where $f(i)$ is the signal sample at time points $i = 0, 1, \ldots, n-1$, where n is the number of sample points (i.e., radar pulses processed); and $F(k)$ is the output transform at point $k = 0, 1, \ldots n-1$. If the exponential term of the transform is treated as a complex filter weight given by

$$w_{i,k} = e^{-j(2\pi/n)ik} \tag{6.89}$$

then (6.88) has the form of a transversal filter [see (5.1)]. In this representation, depicted in Figure 6.35, the filter outputs are given by

$$F(i, k) = \sum_{i=0}^{n-1} f(i) w_{i,k} \tag{6.90}$$

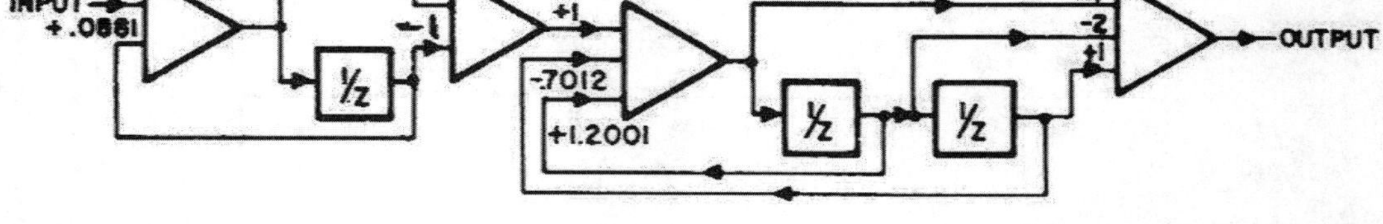

Figure 6.34 Triple-delay MTI canceller with Chebyshev filter response. (*From:* [4, p. 151; 23, p. 194].)

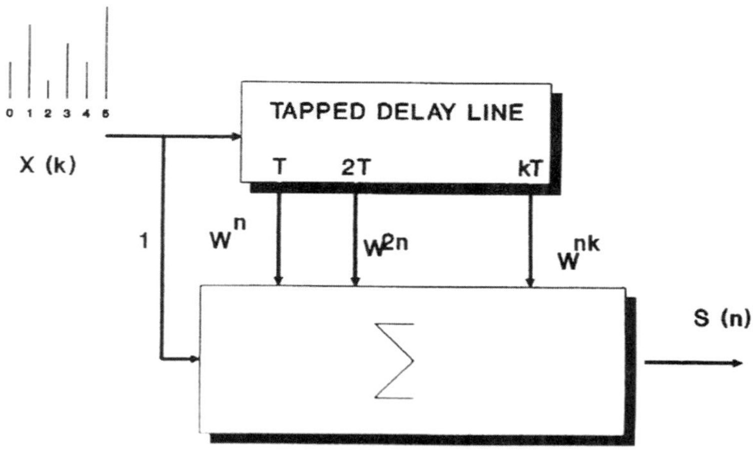

Figure 6.35 Transversal filter representation of an FFT processor. (*From:* [2, p. 385].)

where the index k is the filter number and i is the delay line tap number. Note that n transversal filters are required to complete the representation where $k = 0, 1, \ldots, n - 1$ specifies the particular bandpass filter complex weights into

$$w_{i,k} = e^{-j(2\pi/n)ik} = \alpha_n^{ik} \tag{6.91}$$

where $\alpha_n = \exp[-j(2\pi/n)]$, which allows the z-transform of the transversal filters transfer function to be written as

$$H_k(z) = \left(\frac{z}{\alpha_n^k}\right)^{-(n-1)} \left[\left(\frac{z}{\alpha_n^k}\right)^{n-1} + \left(\frac{z}{\alpha_n^k}\right)^{n-2} + \ldots + \frac{z}{\alpha_n^k} + 1\right] \tag{6.92}$$

where the first term has a magnitude of 1 and represents a time delay. Expansion of the second term by using the relationship $1 + r^2 + \ldots + r^{n-1} = (r^n - 1)/(r - 1)$ results in

$$H_k(z) = \frac{\left(z/\alpha_n^k\right)^n - 1}{\left(z/\alpha_n^k\right) - 1} \tag{6.93}$$

However, $\alpha_n^{-kn} = \exp(j2\pi k) = 1$, allowing the amplitude transfer function of the kth filter of the FFT to be expressed as

$$|H_k(z)| = \frac{z^n - 1}{z - \alpha_n^k} \tag{6.94}$$

The pole-zero pattern then consists of n zeros, equally spaced on the unit circle, which correspond to the nth roots of unity and one pole. The pole occurs at

$$z_p = \alpha_n^k = \exp\left(\frac{-j2\pi k}{n}\right) \tag{6.95}$$

which is located on the unit circle at the same position as one of the zeros. The particular pole location at one of the numerator zeros depends on the filter position (i.e., value of k) selected. For example, if the filter position $k = 0$ is selected, then $z_p = 1$ and the pole zero pattern consists of n zeros equally spaced around the unit circle as depicted in Figure 6.36 with the zero at $z_o = 1$ annihilated by the pole.

The DFT creates each filter passband by selectively deleting the kth zero lying at each of the n complex roots of unity, yielding n passbands. Thus n samples yield n filters, each of order $n - 1$. The stopband in each case is formed by the remaining $n - 1$ zeros evenly spread along the circumference of the unit circle that corresponds to the frequency band between PRF lines.

The transfer function of the kth FFT filter can be evaluated from (6.94) by letting $z = \exp(j\omega T)$, resulting in

$$|H_k(j\omega)| = \frac{\sin(n\omega T/2)}{\sin\left(\frac{\omega T}{2} - \frac{k\pi}{n}\right)} \tag{6.96}$$

where n is the number of FFT points (delay lines) and T is the radar's interpulse period. The transfer function of the FFT filter that provides for coherent integration against fixed targets can be evaluated by setting $k = 0$, resulting in

$$|H_o(j\omega)| = \frac{\sin(n\omega T/2)}{\sin(\omega T/2)} \tag{6.97}$$

Evaluating this function in the limit as $f \to 0$ results in $|H_o(0)| = n\omega T/\omega T = n$, which demonstrates that the coherent integration gain equals the number of pulses (n) being coherently integrated. Figure EX-6.1 in Chapter 2 (also see Figure

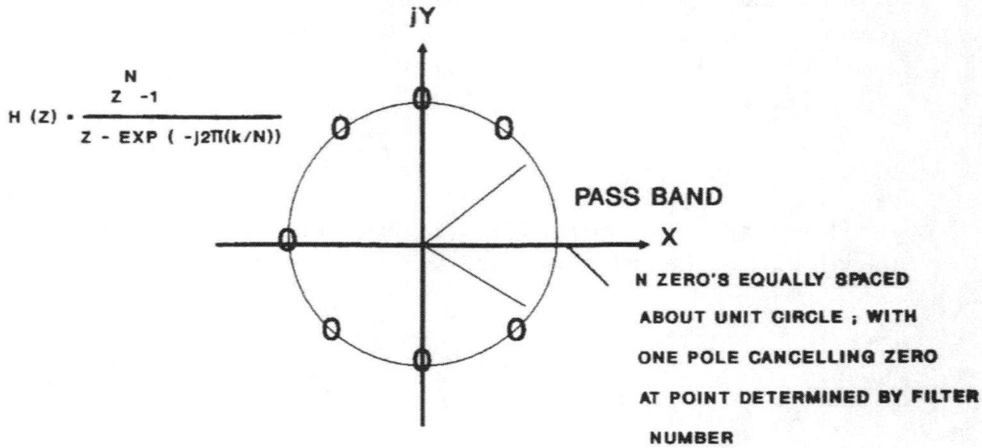

Figure 6.36 Pole-zero pattern of an FFT filter. (*From:* [2, p. 387].)

5.20) depicts the normalized power transfer function ($|H(j\omega)|^2$) for all filters of a coherently integrating FFT filter. Note that the normalized FFT response [see (6.96)] is often approximated by

$$|H_o(j\omega)| = \left| \frac{\sin(n\omega_d T/2)}{(\omega_d T/2)} \right| \tag{6.98}$$

which results in the peak sidelobes being underestimated for far-out sidelobes (see Table 5.2 and Section 5.2).

The sidelobe levels of the filters shown in Figure EX-6.1 result from using a rectangular (uniform) weighting function and can be reduced by using a weighted data window, as depicted in Figure EX-6.2 for Chebyshev weightings. As described in Section 5.2, the weightings can be either applied to the input data or obtained by symmetrical postcombining of adjacent filters on either side of the filter under consideration.

With symmetrical postcombining the transfer function in the z-plane for the FFT transversal filter [see (5.99)] can be written as

$$H_k(z) = (z^n - 1)\left(\frac{1}{z - \alpha_n^k} + \frac{\alpha}{z - \alpha_n^{k+1}} + \frac{\alpha}{z - \alpha_n^{k-1}} \right) \tag{6.99}$$

where $\alpha = 0.5$ for Hann or hanning weighting and $\alpha = 0.426$ for Hamming weighting. For example, when a four-point FFT filter is employed, the pole-zero pattern from Figure 6.36 consists of zeros deployed about the unit circle at $z_0 = \pm 1$ and $z_0 = \pm j$, where one zero is annihilated by a pole determined by the filter location ($k = 0, 1, 2, 3$). Consider the second filter ($k = 1$), which for uniform weighting has zeros located at $z_0 = \pm 1$ and $z_0 = +j$. Evaluation of (6.99) for symmetrical postcombining weighting results in a transfer function ($k = 1$) given by

$$H_1(z) = (z - j) \cdot \left\{ z^2 + \left[\frac{2\alpha j}{(1 + 2\alpha)} \right] z - \left[\frac{1}{(1 + 2\alpha)} \right] \right\} \tag{6.100}$$

where α is the weighting function. For no weighting ($\alpha = 0$), the zeros are located on the unit circle as expected. For hanning weighting ($\alpha = 0.5$), the zeros located on the real axis at $z_0 = \pm 1$ migrate to $z_0 = \left(\pm\sqrt{3} - j\right)/4$.

This analysis indicates that the number of zeros of the z-plane transfer function of the weighted FFT filter depends on the number of samples processed (i.e., number of delay sections). The effect of the weighting is merely to redeploy some of the zeros. In the example for hanning weighting, the zeros on the real axis in the z-plane migrate into the upper half of the z-plane within the unit circle, which has the effect of reducing the filter's sidelobes while broadening the filter's main lobe.

6.4 Staggered PRF

PRF staggering involves the changing of the radar's interpulse period on a pulse-to-pulse basis to provide higher blind speeds and a better low-frequency cutoff of the MTI filter than may be obtained with a constant PRF design by using a binomial canceller. Note from Example 8 in Chapter 3 that staggering also reduces the significant velocity loss associated with an MTI filter that processes targets where there is no *a priori* knowledge of the Doppler shift of the target return. Rather than employing a pulse-to-pulse stagger, we can also use a block-to-block stagger, where the PRF is varied after a number of radar pulses are transmitted at a constant PRF. With a block stagger, the overall MTI velocity response becomes the average of the velocity responses of the individual constant PRF responses.

The advantages of the block stagger (e.g., see the MTD processor; Section 1.6) are that multiple-time-around clutter responses can be cancelled using a coherent MTI system and a system stability (particularly the transmitter) that is easier to achieve than for a system with pulse-to-pulse agility. The composite frequency response of MTI systems using a block stagger with two separate PRFs, in the ratio of 5:4, is depicted in Figure 6.37. Note that the first blind speed of the composite response was increased several times over what it would be for radar operating on only a single PRF. Zero-velocity response occurs only when the blind speeds of each PRF coincide (i.e., at a frequency of $4/T_1 = 5/T_2$ in Figure 6.38). Although the first blind speed may be extended by using more than one PRF, regions of low sensitivity often appear within the composite passband.

The closer the ratio T_1:T_2 approaches unity, the greater the value of the first blind speed and the first null in the vicinity of $f_d = 1/T_1$ becomes deeper. Therefore, the choice of T_1/T_2 is a compromise between the value of the first blind speed and the depth of the nulls within the filter's passband. The depth of the nulls can be reduced and the first blind speed increased by operating with more than two interpulse periods. Four interpulse periods (see Figure 3.27) are generally considered a good design choice for providing a low ripple passband response (i.e., moderate nulls) and a reasonable first blind speed [6].

For many applications, pulse-to-pulse staggering is advantageous because it provides a favorable velocity response compared to the block-to-block stagger approach [6]. For example, consider a double canceller without feedback that has a 36% wide rejection notch and employs a block stagger approach on a scan-to-scan basis. Of the targets, 36% would be lost on each scan, assuming that all targets were uniformly distributed in Doppler over the velocity range of interest, which might well be unacceptable [6]. Nonetheless, with pulse-to-pulse staggering, good response can be obtained on all Dopplers of interest on each radar scan. In addition, a better velocity response can be obtained at some Dopplers of interest than either of the block stagger intervals provide, which results from the modulation, induced on the target by the nonuniform sampling, at a frequency close to the maximum response of the MTI canceller. Pulse-to-pulse staggering may degrade the MTI improvement factor (see Section 3.2.6) attainable in some systems, but in many systems this restriction is not significant or can be eliminated by using time-varying weights in the MTI canceller [6]. An additional advantage of pulse-to-pulse staggering is that it generally removes the need for shaping the MTI canceller response, which considerably simplifies the canceller design and provides superior transient response.

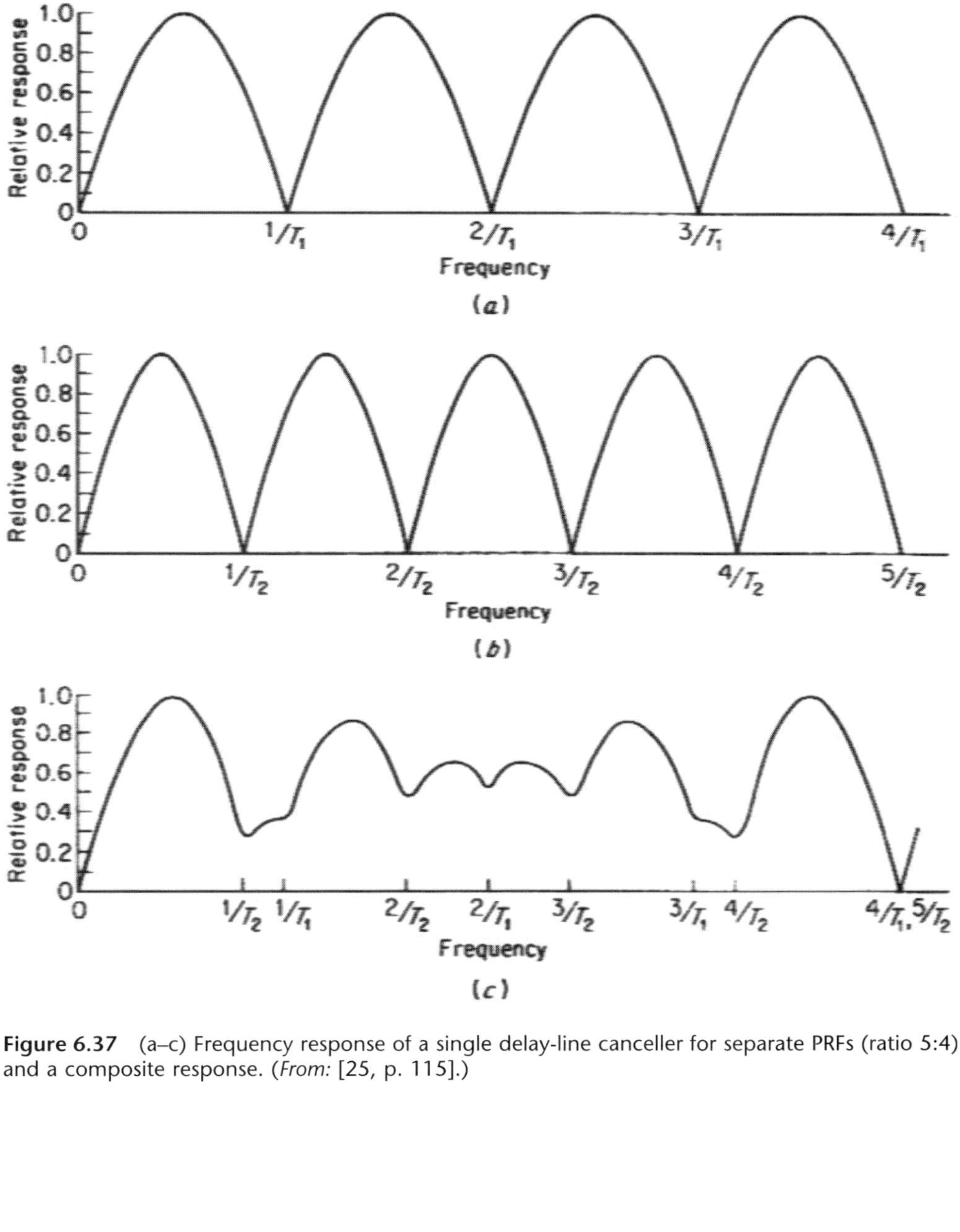

Figure 6.37 (a–c) Frequency response of a single delay-line canceller for separate PRFs (ratio 5:4) and a composite response. (*From:* [25, p. 115].)

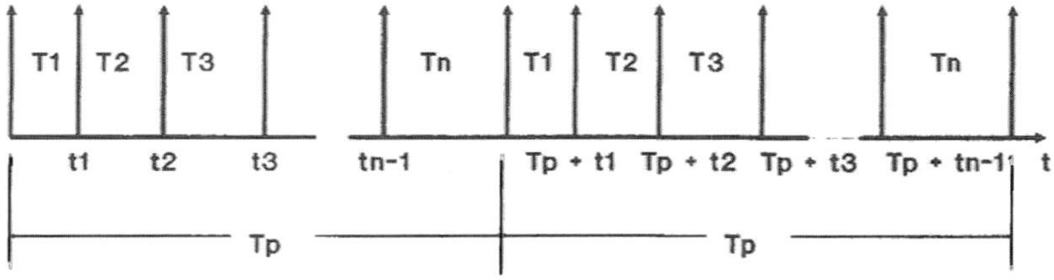

Figure 6.38 General PRF staggering pattern. (*From:* [2, p. 392].)

6.4.1 Blind Speeds with PRF Stagger

When PRF staggering is employed, the radar's interpulse interval is changed on a pulse-to-pulse basis. However, the stagger pattern is periodic in that it repeats after a number of radar pulses are transmitted (see Figure 6.38). The overall period of the n-period stagger is given by

$$T_p = \sum_{i=1}^{n} T_i \tag{6.101}$$

where T_i are the n interpulse periods between the pulses that make up the stagger period. For example, in the two-period ($n = 2$) PRF stagger depicted in Figure 3.24, the two interpulse periods are given by $T_1 = T + T_s$ and $T_2 = T - T_s$ where T_s is the incremental delay forming the stagger sequence, T is the average (nonstagger) period, and $T_p = 2T$ is the overall stagger period. For the three-period stagger ($n = 3$), the interpulse periods are $T_1 = T + T_s$, $T_2 = T$, and $T_3 = T - T_s$ and the overall stagger period is $T_p = 3T$.

The period (T_b) associated with the first blind speed (v_{bs}) of the stagger is given by

$$T_b = \frac{1}{f_{bs}} \tag{6.102}$$

where $f_{bs} = 2v_s/\lambda$ is the Doppler frequency associated with the first null in the velocity response. The interpulse periods associated with the stagger are then given by

$$T_i = n_i T_b \tag{6.103}$$

where n_i are integers defining the stagger ratio. For example, a 5:4 stagger would have interpulse periods equal to $T_1 = 5T_b$ and $T_2 = 4T_b$. The average period of an n-period stagger is then given by

$$T = \frac{n_1 + n_2 + \ldots + n_n}{n} T_b \tag{6.104}$$

If the unstaggered first blind-speed Doppler frequency is given by $f_b = 1/T$, then the resulting stagger blind-speed Doppler frequency from (6.104) is given by

$$f_{bs} = \frac{n_1 + n_2 + \ldots + n_n}{n} f_b \tag{6.105}$$

For example, with a two-period ($n = 2$) 5:4 stagger ($n_1 = 5$, $n_2 = 4$) ratio, the staggered blind speed would be $(5 + 4)/2 = 4.5$ times the unstaggered first blind speed. The three-period stagger depicted in Figure 3.24 with periods $T_1 = T + T_s$, $T_2 = T$, and $T_3 = T - T_s$, which are symmetrical about the center period, have a first blind speed equal to the middle stagger ratio times the unstaggered first blind speed. Hence a 5:4:3 ($n_1 = 5$, $n_2 = 4$, $n_3 = 3$) stagger has a first blind speed equal to $f_{bs} = n_2 f_b = 4f_b$. The four-period stagger (25:30:27:31) velocity response depicted in Figure 3.27 has a first blind speed of 28.25 times the unstaggered first blind speed.

For two- and three-period staggers that use an incremental stagger delay line (see Figure 3.24), it is sometimes convenient to express the stagger using a parameter that expresses the fractional delay represented by the stagger delay, as in this equation:

$$e = \frac{T_s}{T} \tag{6.106}$$

where T_s is the stagger delay period and T is the average interpulse period. The stagger periods for a two-period stagger ($n = 2$) then become $T_1 = T(1 + e)$ and $T_2 = T(1 - e)$ and the stagger ratio (γ) becomes

$$\gamma_2 = \frac{n_1}{n_2} = \frac{1 + e}{1 - e} \tag{6.107}$$

For a three-period stagger, the interpulse periods become $T_1 = T(1 + e)$, $T_2 = T$, and $T_3 = T(1 - e)$, so that the stagger ratio again can be expressed as

$$\gamma_3 = \frac{n_1}{n_3} = \frac{1 + e}{1 - e} \tag{6.108}$$

The fractional stagger period parameter in terms of the stagger ratio can then be obtained from (6.107) or (6.108) as

$$e = \frac{\gamma_{2,3} - 1}{\gamma_{2,3} + 1} \tag{6.109}$$

For example, with a two-period 5:4 stagger, (6.109) can be solved for $e = 1/9$ so that the stagger delay equals $T_s = T/9$. For a three-period 5:4:3 stagger, $\gamma_3 = 5/3$, which when substituted into (6.109) results in $e = 1/4$ so that $T_s = T/4$.

▶ Example 22 PRF Stagger Transfer Functions and Depth of Nulls [22]

The power transfer function for a single canceller and an n-period stagger is given by

$$|H(\omega)|^2 = 2\left[1 - \frac{1}{n}\sum_{i=1}^{n} \cos(\omega T_i)\right]$$

where

$$T_i = \frac{1}{\text{PRF}_i}$$

and the stagger ratio is

$$n_1 : n_2 : n_3 \ldots n_n$$

in descending order. The following relationship describes the stagger:

$$n_1 \text{PRF}_1 = n_2 \text{PRF}_2 = n_3 \text{PRF}_3 = n_n \text{PRF}_4$$

Interactive MATLAB program mtistag1.m calculates the transfer function of this stagger. First the stagger order is entered, then the stagger ratios are entered in descending order, and finally the lowest PRF is entered and the plot is provided. The plot for a third-order stagger with ratios 5:4:3 and a low PRF of 1,000 Hz is shown in Figure EX-21.1. A similar plot with a stagger ratio of 31:30:27:25 is shown in Figure EX-22.2.

The power transfer function for a double canceller and an n-period stagger is given by

$$|H(\omega)|^2 = \frac{1}{n}\left\{[6 - 4\cos(\omega T_1) - 4\cos(\omega T_n) + 2\cos(\omega(T_1 + T_n))] \ldots \right.$$

$$\left. + \sum_{i=2}^{n} [6 - 4\cos(\omega T_i) - 4\cos(\omega(T_{i-1})) + 2\cos(\omega(T_i + T_{i-1}))]\right\}$$

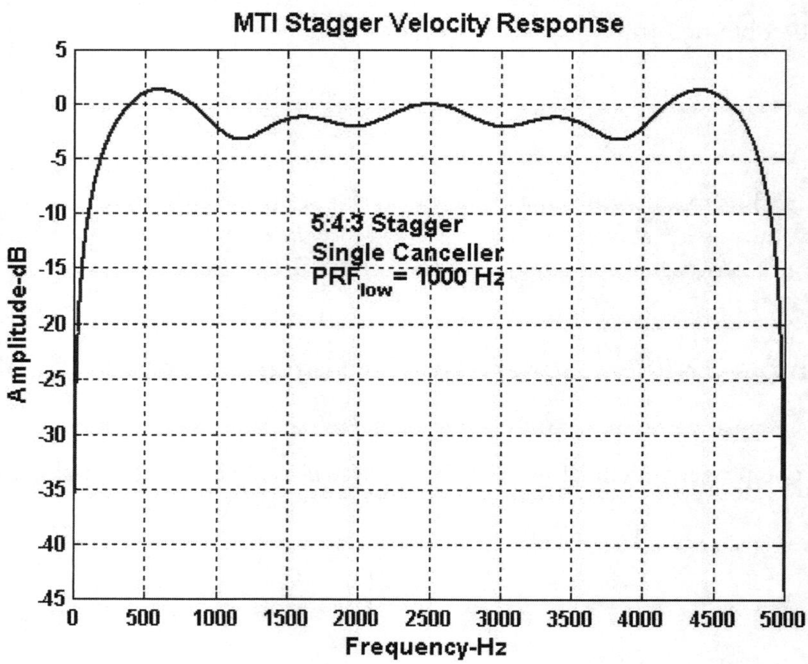

Figure EX-22.1 Transfer function for triple stagger and single canceller.

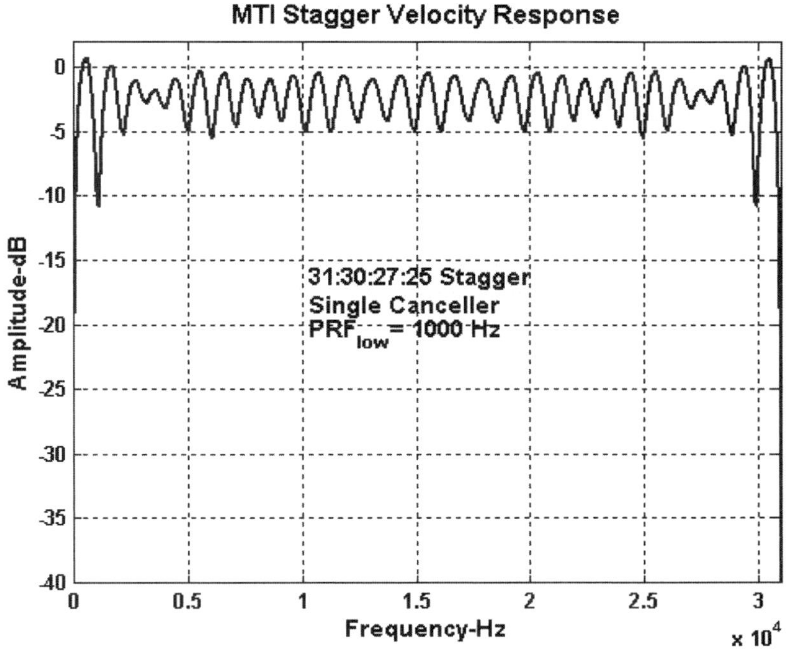

Figure EX-22.2 Transfer function for quadruple stagger and single canceller.

Interactive MATLAB program mtistag2.m calculates the transfer function of an n-period stagger with a double canceller. Figure EX-22.3 plots a 5:4:3 stagger with a double canceller for comparison with that shown in Figure EX-22.1 for a single canceller. Figure EX-22.4 shows the transfer function for a 33:32:29:27 quadruple stagger with a double canceller. This stagger is adopted fom [3] as an example for an analytic method calulating the depth of the first null.

The design of a PRF stagger system invovles a selection of a stagger ratio that eliminates the blind speeds while minimizing the depth of nulls over the Doppler range of interest. Reference [22] suggests that at least a four-period stagger is required to provide an acceptable design. Examination of the above plots indicates that the first null is generally the deepest. Reference [3] provides both an analytic and empirical method for estimating the first null of the stagger. The analytic method [3, 26] calculates a first null of 6 dB for the stagger shown in Figure EX-22.4 that is actually 11 dB. The empirical method [22] utilizes the ratio of the largest period to the smallest period ($T_{\max}/T_{\min}$) as the design parameter.

The plot estimating the first three nulls is shown in Figure EX-22.5. In Figures EX-22.2 and EX-22.4 the first null is 4 dB as compared to 5 dB for the first null estimated from Figure EX-22.5. In Figure EX-22.4 the first null is about 12 dB, whereas the empirical estimate is about 9 dB. The second null is 5 dB, whereas the estimate is about 6 dB.

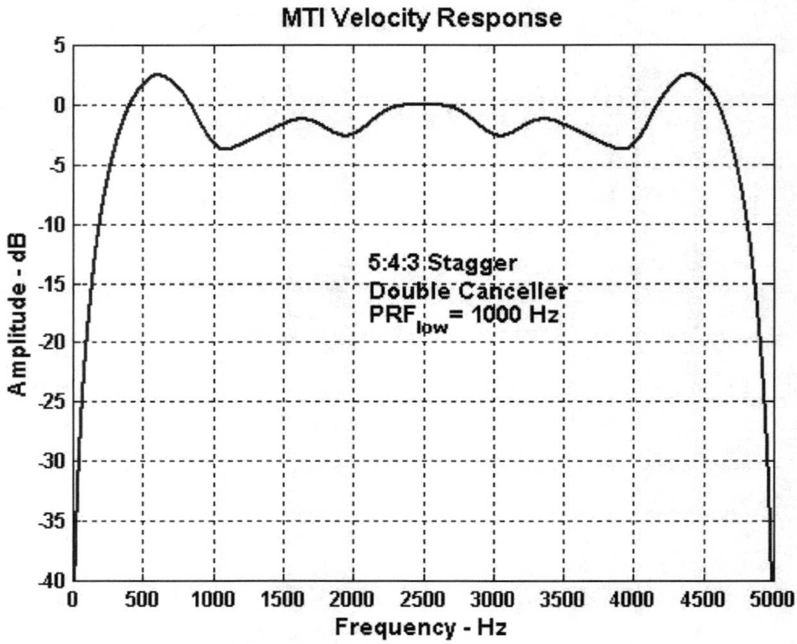

Figure EX-22.3 Transfer function for triple stagger and double canceller.

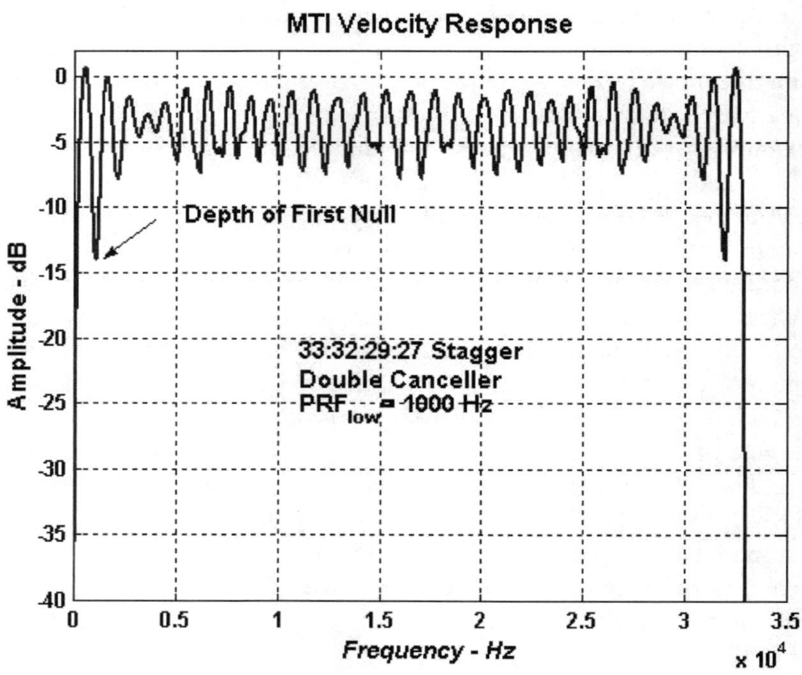

Figure EX-22.4 Transfer function for quadruple stagger and double canceller.

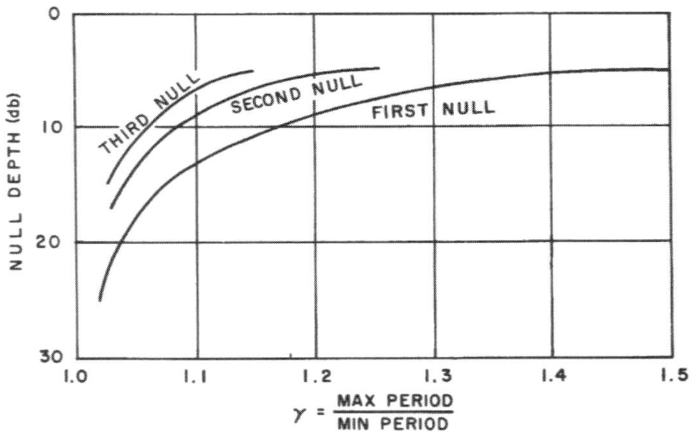

Figure EX-22.5 Estimate of nulls in a PRF stagger. (*From:* [22, p. 15.42].)

▶

▶ Example 23 MTI Staggered-PRF Velocity and Improvement Factor Losses [7]

The MTI velocity loss (see Example 8 in Chapter 3) can be reduced using a pulse-to-pulse PRF stagger. PRF staggering for single and double MTI cancellers is discussed in Section 6.4. The stagger performance is a function of the stagger factor given by

$$e = \frac{T_S}{T}$$

where T_S is the fractional stagger delay period (deviation from mean period) and T is the mean period (average of periods associated with PRFs used in the stagger). The power transfer function for a single canceller with a double stagger is given by

$$|H(\omega)|^2 = 2[1 - \cos(\omega T)\cos(\omega Te)]$$

This can be solved for the first blind speed ($|H(\omega)| = 0$) as

$$f_{BS} = \frac{1}{2Te}$$

The probability of detection for a Swerling 1 target is given by

$$P_d(s) = \exp\left(-\frac{y_b}{1+\bar{s}}\right)$$

where $\bar{s}$ is the average S/N ratio and y_b is the detection threshold. Inserting the stagger transfer function into this equation and simplifying results in

$$P_d = \frac{1}{\pi} \int_0^{\pi} \exp\left(- \frac{y_b}{1 + \frac{\bar{x}}{G} 2 \left[1 - \cos\left(\frac{\theta}{e} \right) \cos(\theta) \right]} \right) d\theta$$

Solving this equation for $\bar{x}$ determines the S/N ratio that provides the desired probability of detection for a single MTI canceller with a double stagger. The expression for a double canceller with a double stagger is given by

$$P_d = \frac{1}{\pi} \int_0^{\pi} \exp\left(- \frac{y_b}{1 + \frac{\bar{x}}{G} 2 \left[3 - 4 \cos\left(\frac{\theta}{e} \right) \cos(\theta) + 2 \cos\left(\frac{2\theta}{e} \right) \right]} \right) d\theta$$

The velocity loss for single- and double-canceller MTI systems with a double stagger is developed in MATLAB program mti_velstagg.m, whereas the associated transfer functions are stored in function programs Homega_2a.m for a single canceller and Homega_2_2a.m for a double canceller. The velocity losses are shown in Figure EX-23.1. When compared to the same system without a stagger developed in Example 8, it is observed that a stagger is necessary to avoid excessive velocity loss. For example, with a P_d of 0.9 a single canceller without a stagger has a loss in excess of 8 dB, whereas the loss is reduced to about 2 dB using a 5:4 stagger.

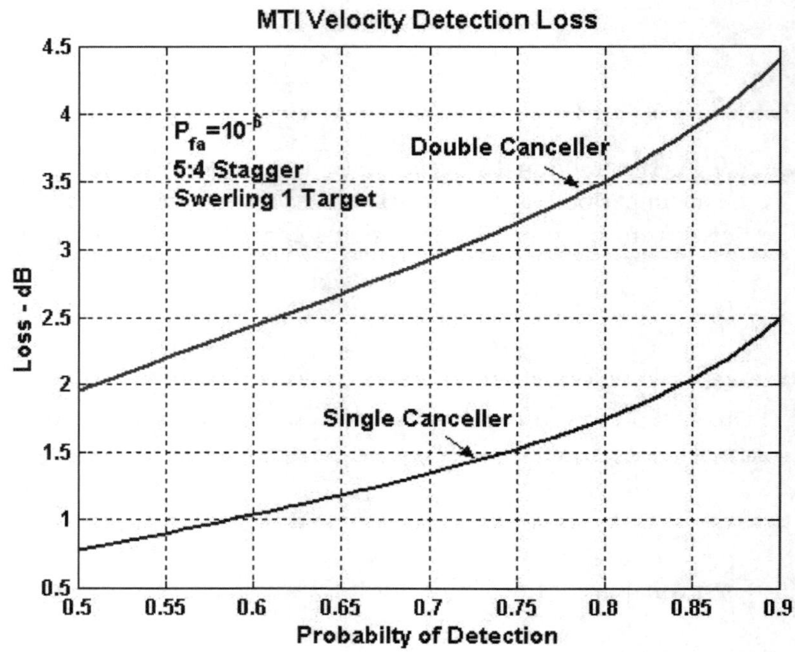

Figure EX-23.1 PRF stagger losses for single and double MTI cancellers.

Note that velocity losses for higher order cancellers can be determined by modifying the function program Homega using the equations found in Section 6.4.

Staggered-PRF MTI systems have a reduced improvement factor compared to that provided by a nonstaggered MTI with the same average period. The improvement factor for a double-canceller MTI can be expressed as

$$I_2 = \frac{1}{1 - \frac{4}{3}\exp\left(\frac{-x}{2}\right) + \frac{1}{3}\exp(-2x)}$$

where

$$x = (\sigma_c T)^2$$

Consider a double-canceller MTI using a two-period stagger with $k = T_2/T_1$ where T_1 and T_2 are the staggered periods. Then

$$T = \frac{T_1(1+k)}{2}$$

and

$$x = \left(\sigma_c T_1 \frac{1+k}{2}\right)^2$$

This can be solved for $\sigma_c T_1$ as

$$\sigma_c T_1 = \sqrt{\frac{4x}{(1+k)^2}}$$

The staggered improvement factor can be expressed as a function of $\sigma_c T_1$ as

$$I_{2S} = \frac{3}{3 - 2F_1 + F_2}$$

where F_1 is the correlation coefficient at the average of T_1 and T_2

$$F_1 = \exp\left(\frac{-(\sigma_c T_1)^2}{2}\right) + \exp\left(\frac{-(k\sigma_c T_1)^2}{2}\right)$$

and F_2 is the correlation coefficient at $T_1 + T_2$

$$F_2 = \exp\left(\frac{-[(k+1)\sigma_c T_1]^2}{2}\right)$$

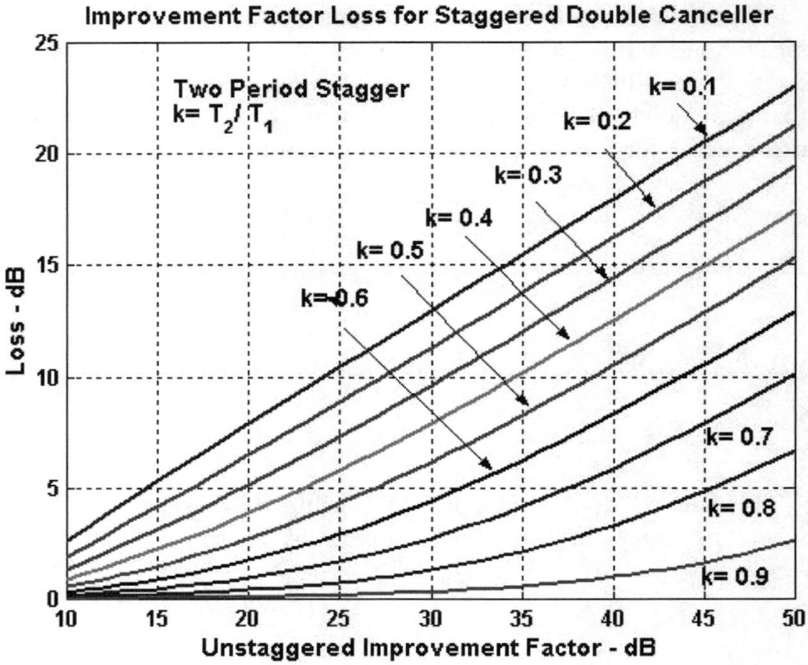

Figure EX-23.2 Improvement factor reduction for staggered double canceller.

The MATLAB program stag_impb.m solves the equations for the improvement factor loss in decibels versus the nonstaggered improvement factor loss in decibels for a double canceller and $k = 0.1, 0.2, 0.3, \ldots, 0.9$. The program first solves the I_2 equation for x using the MATLAB function fzero('f' ,x,options,I2) and then finds $\sigma_c T_1$ and substitutes this into the equation for I_{2S}. The loss is then equal to $I_2 - I_{2S}$ in decibels. The function 'f' is given below. A sample plot is given in Figure EX-23.2.

```
function y=f(x,I2)
% Improvement Factor of Double Canceller
y=I2-(1/ (1-(4/3)*exp(-x/2)+(1/3)*exp(-2*x)));
```

▶

6.4.2 PRF Stagger with a Single-Canceller MTI

The PRF staggering and destaggering block diagram depicted in Figure 3.24 shows that the destaggered signal presented to the MTI canceller occurs at the average rate (i.e., interpulse period equals T) associated with the stagger. Therefore, the destaggered signal can be processed by a single, dual, or triple canceller (with or without shaping) with no modification to the canceller.

In this section, we derive the velocity response when an n-period stagger is applied to a single canceller. The general approach utilized is appropriate to any order nonrecursive or recursive MTI canceller [27, 28], although we discuss only the nonrecursive case represented by binomial and optimal weight cancellers.

The essence of the method is to break the nonperiodic stagger waveform into an equivalent set of n-periodic waveforms that are then applied to the MTI canceller. The response of the canceller to each of the n-periodic waveforms is then combined in the canceller output to form the composite response of the canceller. For example, the two-period stagger waveform depicted in Figure 3.24 can be decomposed into two waveforms each with period $T_p = 2T$, but separated in time by $T_1 = T + T_s = T(1 + e)$. Similarly, the three-pulse stagger can be decomposed into three waveforms, each with period $T_p = 3T$ separated in time by $T_1 = T + T_s = T(1 + e)$ and $T_2 = T$. Finally, the general n-period stagger can be decomposed into n-periodic waveforms each with period $T_p = nT$ separated in time by $T_1, T_2, T_3, \ldots, T_n$. The waveforms for the two- and three-period staggers are depicted in Figure 6.40.

The response of a linear filter to an input waveform can be determined using the convolution integral given by

$$g(t) = \int_{-\infty}^{t} r(\tau)h(t - \tau)\, d\tau \tag{6.110}$$

where r_t is the input waveform and $h(t)$ is the impulse response of the filter. The impulse response of an m-pulse nonrecursive MTI is given by (see Section 2.3.2)

$$h(t) = \sum_{k=0}^{m-1} w_k\, \delta(t - kT_k) \tag{6.111}$$

where w_k are the filter weights (e.g., binomial coefficients with alternating signs for a binomial canceller) and T_k are the delay periods associated with the canceller weightings (see Table 3.1 for binomial weights and Table 5.1 for optimal weights). For example, the impulse response of a single canceller ($m = 2$) is given by $h(t) = \delta(t) - \delta(t - T_k)$, where $\mathbf{w} = (w_0, w_1) = (1, -1)$. Note that the delay period associated with the MTI filter weights would generally be given by $T_k = T$ for a conventional (i.e., nonstaggered MTI) canceller, but must coincide with the actual interpulse periods (i.e., $T_I = n_I T_b$) for a staggered-PRF MTI in order for it to cancel fixed targets. This synchronization is accomplished automatically in the staggered-PRF MTI by the stagger/destaggered action, as depicted in Figure 3.24.

The first step in the derivation of the staggered PRFs' velocity response is to compute the response of the single canceller to the first periodic segment of the decomposed staggered waveform. Let us first consider the two-period stagger shown in Figure 6.39. In accordance with the convolution integral given by (6.110), the MTI filter's output response to a sinusoidal input $r(t) = \exp(j\omega t)$ results from the weighted sum of those past input samples generated by the MTI filter. In the case of a single canceller, two output samples are generated by each input sample, where, due to the time reversal in the convolution integral [i.e., $h(t - \tau)$], the second sample occurs in the past and, hence, contributes to the output. Furthermore, due to the destaggering action of the PRF stagger procedure, the second sample must coincide with the primary sample associated with the second waveform of the decomposed stagger set. The complex frequency response of the first waveform of the decomposed stagger set with a single MTI canceller can then be written as

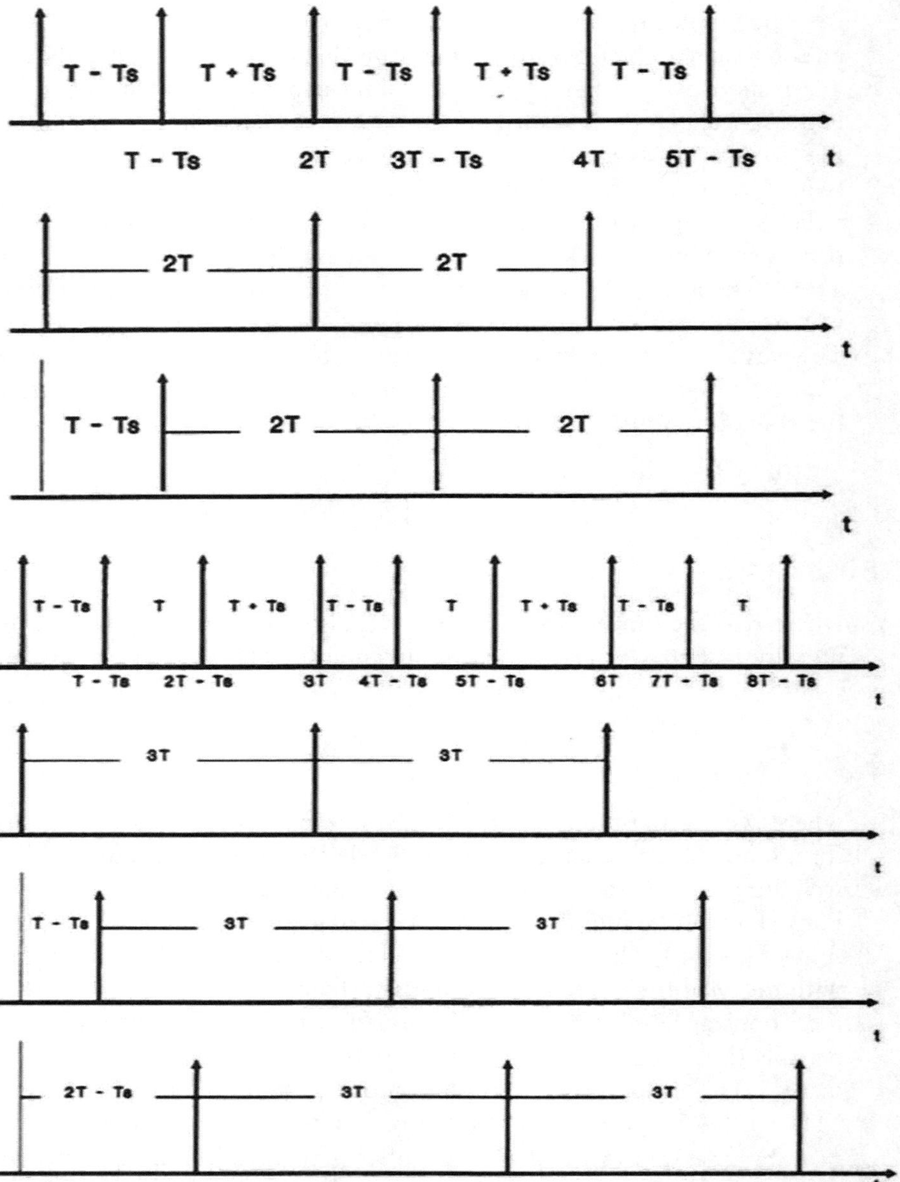

Figure 6.39 Decomposition of two- and three-period stagger waveforms. (*From:* [2, p. 395].)

$$H_1(j\omega) = e^{j\omega kT_p}\left(1 - e^{-j\omega T_2}\right) \tag{6.112}$$

where T_p is the stagger period equal to $2T$ for the two-period stagger, T_2 is the period separating the two decomposed waveforms of the stagger set equal to $T - T_s$, and $k = 0, 1, 2, \ldots$. In a similar manner, the complex frequency response to the second waveform of the decomposed stagger set equals

$$H_2(j\omega) = e^{j\omega(kT_p + T_1)}\left(1 - e^{-j\omega T_1}\right) \tag{6.113}$$

where T_1 is the period that allows the second waveform to be aligned with the first and equals $T + T_s$ for the two-period stagger.

The frequency responses generated by the two decomposed waveforms represented by (6.111) and (6.112) consist of sinusoidal samples with magnitudes given by $|H_I(j\omega)|$, where $I = 1, 2$. The power transfer function can be found by adding the powers from each component as

$$|H(j\omega)|^2 = \frac{1}{2}\sum_{i=1}^{n} |H_i(j\omega)|^2 \tag{6.114}$$

where n is the number of periods in the stagger. Therefore, the power transfer function for the frequency response of a two-period stagger with a single canceller is given by

$$|H(j\omega)|^2 = \frac{1}{2}|1 - e^{-j\omega T_1}|^2 + \frac{1}{2}|1 - e^{-j\omega T_2}|^2 \tag{6.115}$$

This can be written as

$$|H(j\omega)|^2 = 2 - \cos(\omega T_1) - \cos(\omega T_2) \tag{6.116}$$

which further simplifies to

$$|H(j\omega)|^2 = 2[1 - \cos(\omega T)\cos(\omega Te)] \tag{6.117}$$

where T is the average stagger period (i.e., unstaggered period) and $e = T_s/T$ is the fractional stagger delay determined by the stagger ratio [see (6.107)]. Note that, from (6.117), the nonstaggered transfer function ($e = 0$) becomes $|H(j\omega)|^2 = 4\sin^2(\omega T/2)$ as expected for a single canceller.

Equation (6.117) can be extended to find the normalized power transfer function for an n-period PRF stagger followed by a single canceller:

$$|H(j\omega)|^2 = 2\left[1 - \frac{1}{n}\sum_{i=1}^{n}\cos(\omega T_i)\right] \tag{6.118}$$

Specializing to a three-period PRF stagger where $T_1 = T(1 + e)$, $T_2 = T$, and $T_3 = T(1 - e)$ results in

$$|H(j\omega)|^2 = 2\left[1 - \frac{2\cos(\omega Te) + 1}{3}\cos\omega T\right] \tag{6.119}$$

Note that this reduces to the single canceller transfer function for the unstaggered case ($e = 0$). The normalized frequency transfer function for a 5:4:3 PRF stagger and a single canceller are depicted in Figure 6.40, where the response is

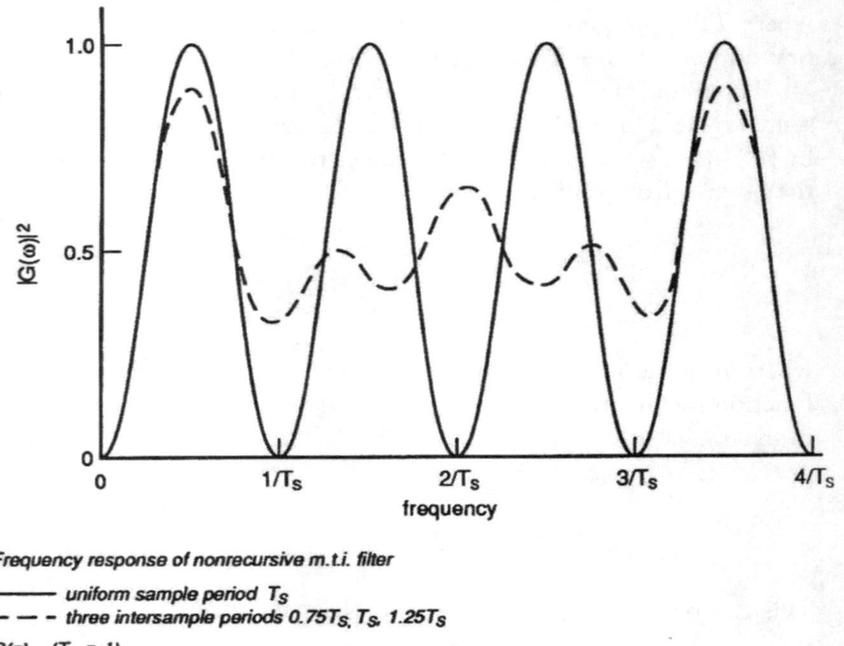

Figure 6.40 Power transfer function of a single-canceller MTI with a 5:4:3 ratios. (*From:* [2, p. 398].)

compared against the unstaggered response using the average period. The variation in the passband is on the order of 3 dB for this configuration where the maximum variation occurs at the band edges.

6.4.3 PRF Stagger with a Dual-Delay Canceller MTI

When a dual-delay MTI canceller is employed with a PRF stagger, three weighted pulses are involved in determining the output response of each segment of the decomposed stagger waveform. The impulse response of a dual-delay line canceller in the stagger configuration is given by

$$h(t) = w_0\,\delta(t) + w_1\,\delta(t - T_k) + w_2\,\delta(t - T_j) \tag{6.120}$$

where the weight set $\mathbf{w} = (w_0, w_1, w_2)$ is equal to $\mathbf{w} = (1, -2, 1)$ for a double canceller and optimal weights are given in Table 5.1.

The power transfer function response for a double-stagger PRF applied to a dual-delay MTI canceller is then given by

$$|H(j\omega)|^2 = \frac{1}{2}\left|w_0 + w_1 e^{-j\omega T_2} + w_2 e^{-j\omega(T_1+T_2)}\right|^2 + \frac{1}{2}\left|w_0 + w_1 e^{-j\omega T_1} + w_2 e^{-j\omega(T_1+T_2)}\right|^2 \tag{6.121}$$

where $T_1 = T + T_s$ and $T_2 = T - T_s$. Specialization to a double canceller ($w_0 = 1$, $w_1 = -2$, $w_2 = 1$) and simplification results in

$$|H(j\omega)|^2 = 6 - 4 \cos \omega(T + T_s) - 4 \cos \omega(T - T_s) + 2 \cos 2\omega T \quad (6.122)$$

which further reduces to

$$|H(j\omega)|^2 = 6 - 8 \cos \omega T \cos \omega Te + 2 \cos 2\omega T \quad (6.123)$$

where $e = T/T_s$ is the fractional stagger delay determined by the stagger ratio [see (6.109)]. Note that for the unstaggered configuration ($e = 0$), (6.123) reduces to $|H(j\omega)|^2 = 16 \sin^4 (\omega T/2)$, which is the conventional double-canceller transfer function. The velocity response for a 63:65 dual-period stagger into a double canceller is depicted in Figure 6.41(a) [6]. Note the deep nulls at the passband edges, which are characteristic of stagger systems using a small number of stagger periods.

The power transfer function response for a triple-stagger PRF applied to a double-canceller MTI is given by

$$\begin{aligned} |H(j\omega)|^2 &= \left|1 - 2e^{-j\omega T_3} + e^{-j\omega(T_2+T_3)}\right|^2 \\ &+ \left|1 - 2e^{-j\omega T_2} + e^{-j\omega(T_1+T_2)}\right|^2 \\ &+ \left|1 - 2e^{-j\omega T_1} + e^{-j\omega(T_1+T_3)}\right|^2 \end{aligned} \quad (6.124)$$

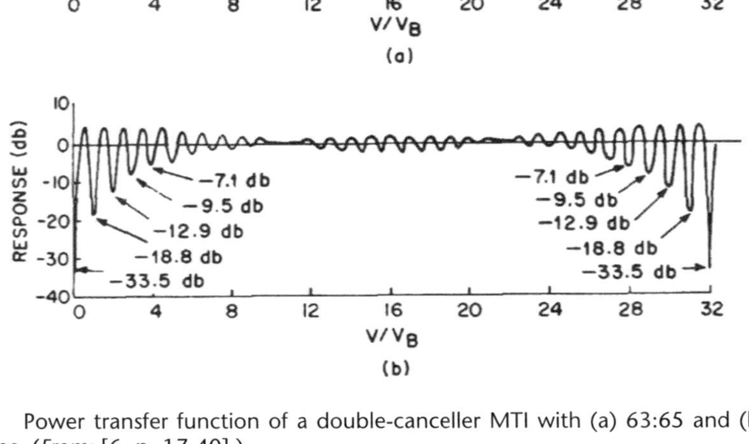

Figure 6.41 Power transfer function of a double-canceller MTI with (a) 63:65 and (b) 31:32:33 stagger ratios. (*From:* [6, p. 17-40].)

which for a symmetrical stagger $T_1 = T + T_s$, $T_2 = T$, and $T_3 = T - T_s$ results in

$$|H(j\omega)|^2 = \frac{1}{3}[18 - 8\cos\omega(T - T_s) - 8\cos\omega(T + T_s) - 8\cos\omega T + 2\cos\omega(2T - T_s) + 2\cos\omega(2T + T_s) + 2\cos 2\omega T] \tag{6.125}$$

This can be further simplified by expressing the periods in terms of the fractional delay stagger $e = T_s/T$ as $T_1 = T(1 + e)$, $T_2 = T$, and $T_3 = T(1 - e)$. The resulting power transfer function for a triple-stagger PRF and a double canceller is given by

$$|H(j\omega)|^2 = \frac{1}{3}[18 - 8(1 + 2\cos\omega Te)\cos\omega T + 2(1 + 2\cos\omega Te)\cos 2\omega T] \tag{6.126}$$

where e can be found in terms of the stagger ratio using equation (6.101). Note that (6.126) reduces to the power transfer function for a double canceller for the unstaggered case ($e = 0$). The velocity response for a 31:32:33 stagger into a double canceller is depicted in Figure 6.41(b) [6]. Comparison with the double-period stagger configuration shown in the figure indicates that the passband ripple is reduced by the additional stagger period.

For the general n-period PRF stagger with a double MTI canceller, the power transfer response can be expressed as

$$|H(j\omega)|^2 = \frac{1}{2}\left|1 - 2e^{-j\omega T_1} + e^{-j\omega(T_1 + T_n)}\right|^2 + \sum_{i=2}^{n}\frac{1}{2}\left|1 - 2e^{-j\omega T_1} + e^{-j\omega(T_i + T_{i-1})}\right|^2 \tag{6.127}$$

This can be simplified to

$$|H(j\omega)|^2 = \frac{1}{n}\left\{[6 - 4\cos\omega T_1 + 2\cos\omega(T_1 + T_n) - 4\cos\omega T_n] + \sum_{i=2}^{n}[6 - 4\cos\omega T_i + 2\cos\omega(T_i + T_{i-1}) - 4\cos\omega T_{i-1}]\right\} \tag{6.128}$$

where T_i, $i = 1, 2, \ldots, n$ are the periods that make up the overall stagger period $\left(\text{i.e., } T_p = \Sigma_{i=1}^{n} T_i\right)$.

The optimum choice of the stagger ratio depends on the velocity range over which there must be no blind speeds and the permissible depth of the first null in the velocity response curve. A four-period stagger ratio is considered a good choice for some applications [6]. It is advantageous to alternate the long and short stagger periods in a sequence to keep the transmitter's duty cycle as constant as possible. The velocity response for a 51:62:53:61:58 PRF stagger ($n = 5$) into a double

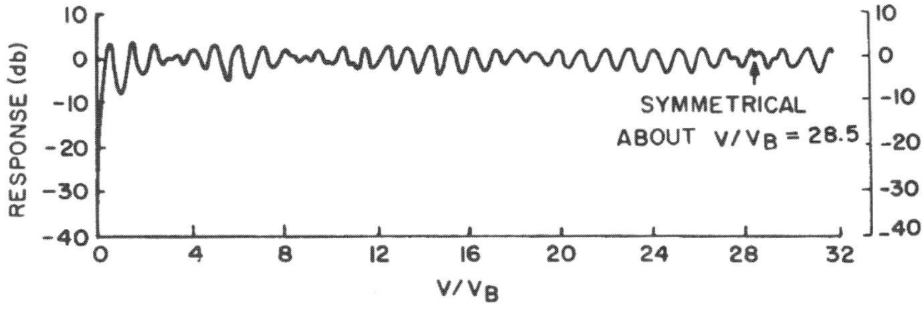

Figure 6.42 Power transfer function of a double-canceller MTI with a 51:62:53:61:58 stagger ratio. (*From:* [22, p. 15.36].)

canceller is depicted in Figure 6.42. Note that the passband ripples are reduced from the two- and three-period staggers depicted in Figure 6.41 and the four-period stagger shown in Figure 3.27.

► Example 24 Improvement Factors for PRF-Staggered MTI

PRF-staggered MTI systems generally have a reduced improvement factor as compared to that provided by a nonstaggered MTI with the same average period. In this example we extend the analysis of Example 23 to derive the improvement factor for a staggered single-canceller MTI and plot the improvement factors for both single- and double-canceller PRF-staggered MTIs. Improvement factors for unstaggered MTI cancellers are also shown for comparison purposes.

Generally, two methods are available to calculate the improvement factors. The first is the correlation coefficient method presented in Example 23. The improvement factor for a single canceller can be expressed as

$$I_1 = \frac{1}{1 - \exp\left(\frac{-x}{2}\right)}$$

where

$$x = (\sigma_c T)^2$$

Consider a single-canceller MTI using a two-period stagger with $k = T_2/T_1$ where T_1 and T_2 are the staggered periods. Then the average

$$T = \frac{T_1(1 + k)}{2}$$

and

$$x = \left(\sigma_c T_1 \frac{1+k}{2} \right)^2$$

This can be solved as

$$\sigma_c T_1 = \sqrt{\frac{4x}{(1+k)^2}}$$

The staggered improvement factor can be expressed as a function of $\sigma_c T_1$ as

$$I_{1S} = \frac{2}{2 - F_1}$$

where F_1 is the correlation coefficient at the average of T_1 and T_2:

$$F_1 = \exp\left(\frac{-(\sigma_c T_1)^2}{2} \right) + \exp\left(\frac{-(k\sigma_c T_1)^2}{2} \right)$$

The result of this analysis is presented in Figure EX-24.1 using MATLAB program impfac_stagb.m. The improvement factors for dual staggered and unstaggered single cancellers are essentially the same (actually there is a loss of 0.05 dB due to staggering), whereas a larger loss is associated with using a double canceller.

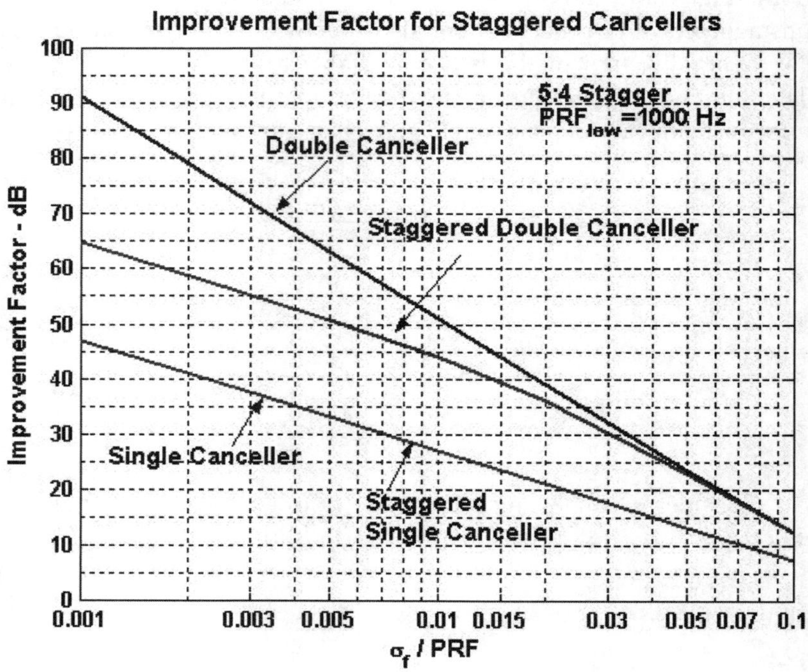

Figure EX-24.1 Improvement factors for dual staggered single and double cancellers.

The second method involves a numerical integration approach to compute the improvement factor as

$$I = G \cdot \text{CA}$$

where G is the MTI's power gain and CA is the clutter attenuation.

For Gaussian-shaped clutter, the input clutter power is given by

$$P_c = \frac{1}{\sqrt{2\pi\sigma_f^2}} \int_0^{f_r/2} \exp\left(-f^2/2\sigma_f^2\right)\, df$$

where f_r = PRF and σ_f is the standard deviation of the clutter spectrum.

The output power from the MTI canceller is given by

$$P_{co} = \frac{1}{\sqrt{2\pi\sigma_f^2}} \int_0^{f_{rbs}/2} |H_{\text{MTI}}(f)|^2 \cdot \exp\left(-f^2/2\sigma_f^2\right)\, df$$

and the power gain is given by

$$G = \frac{\int_0^{f_r} |H_{\text{MTI}}(f)|^2\, df}{f_r}$$

where $|H_{\text{MTI}}(f)|^2$ is the PRF-staggered MTI power transfer function and f_{rbs} is the first blind speed. The improvement factor is then given by

$$I_S = \frac{GP_c}{P_{oc}}$$

The MATLAB program mti_imp_staga.m provides the necessary computer code to calculate the improvement factor of a PRF-staggered MTI using a single or double canceller employing an arbitrary number of stagger periods. The program first calculates the staggered MTI transfer functions using a 5:4:3 stagger. Other stagger periods can be accommodated using the interactive programs mtistag1.m and mtistag2.m described in Example 22. The equations described in this example are then numerically integrated using the MATLAB function trapz(a, b) to compute the improvement factor. For larger first blind speeds, the incremental frequency sample must be adjusted so that there are at least 10 samples within the standard deviation of σ_f of the clutter spectrum. Figure EX-24.2 depicts the improvement factor computed for single and double cancellers for both staggered and unstaggered MTI cancellers.

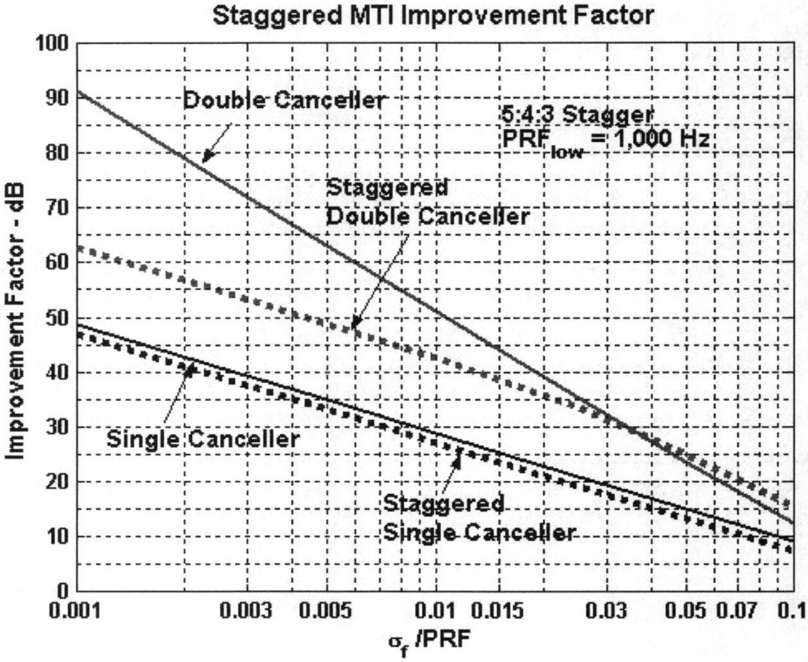

Figure EX-24.2 PRF-staggered MTI improvement factors for triple stagger.

The results are similar to those obtained using the correlation coefficient method showing a large loss when employing a double canceller with large improvement factors and negligible loss using staggered single cancellers.

▶

6.4.4 Synthesis of Staggered MTI Cancellers

The previous sections described the response of simple nonrecursive MTI filters with PRF sampling. Several velocity responses (see Figures 6.41, 6.43, and 3.27) were shown that indicated the effects of varying the stagger order and ratios. It is apparent that different bandpass shapes can be achieved by varying the stagger parameters as well as by eliminating blind speeds within the expected velocity range of the target.

Several procedures have been proposed that allow achieving a specified transfer function by using a staggered PRF MTI canceller [3, 28–31]. However, the procedures generally are complicated and a computer must be used to synthesize a specified staggered MTI design. This is in contrast to the shaped canceller design procedure described in Section 6.3.2.1, where conventional filter theory can be translated into the sampled-data domain by using a simple bilinear transformation. The advantage of the staggered-PRF configuration over the shaped canceller approach is that it provides a superior transient response due to the use of a simple filter (i.e., single or double canceller). The disadvantage of the PRF-staggered design approach is that multiple-time-around clutter is not cancelled.

Some insight into the synthesis of PRF-staggered MTI cancellers can be gained by examining their pole-zero response in the z-plane [28]. First, for the nonuniform sampling that occurs with staggered-PRF designs, the z-transform must be redefined to coincide with the periodic component of the stagger that has period $T_b = 1/f_{bs}$, where f_{bs} is the Doppler frequency of the first blind speed. The z-transform is then defined as

$$z_b = e^{j\omega T_b} \tag{6.129}$$

The first blind speed period is related to the average period (unstaggered period) by

$$T = \frac{n_1 + n_2 + \ldots + n_n}{n} T_b \tag{6.130}$$

where n_i are the stagger numbers and n is the number of stagger periods. Specialization to a symmetrical triple stagger of the average period is

$$T = n_2 T_b \tag{6.131}$$

so that the z_b-transform of a single canceller $[H(z) = 1 - z^{-1}]$ is given by

$$H(z_b) = \left(1 - e^{-j\omega n_2 T_b}\right) = 1 - z_b^{-n_2} \tag{6.132}$$

For example, for a 3:4:5 stagger ratio, $H(z) = 1 - z_b^{-4}$ and four poles and zeros would be distributed around the unit circle, producing null responses at $f_b = 1/T$.

Next, consider the pole-zero pattern of a three-period PRF-staggered design with a single canceller. This can be determined from the extension to (6.115) as

$$|H(z_b)|^2 = |1 - z_b^{-n_1}|^2 + |1 - z_b^{-n_2}|^2 + |1 - z_b^{-n_3}|^2 \tag{6.133}$$

Specialization to a 3:4:5 stagger results in

$$|H(z_b)|^2 = |1 - z_b^{-3}|^2 + |1 - z_b^{-4}|^2 + |1 - z_b^{-5}|^2 \tag{6.134}$$

The zeros of this function can then be found by solving the resulting polynomial:

$$|H(z_b)|^2 = H(z_b)H^*(z_b) = 0 \tag{6.135}$$

When this is done, we find that the zeros move off the unit circle, which provides for extension of the blind speeds, but also causes a deterioration of the clutter attenuation in the vicinity of $z_b = 1$ (see Section 3.2.6) [28].

This discussion indicates that the effect of the PRF stagger is to move the filter zeros away from the unit circle, which is desirable except in the case of the zeros located at $z_b = 1$. This situation can be corrected using time-variable weights for each stagger interpulse period [6, 28]. The effect of this on the staggered-PRF transfer function is given by

$$|H(z_b)|^2 = |w_0 + w_1 z_b^{-n_1}|^2 + |w_0' + w_1' z_b^{-n_2}|^2 + |w_0'' + w_1'' z_b^{-n_3}|^2 \tag{6.136}$$

The synthesis procedure is then to adjust the weights (time variable) for each set of the decomposed stagger (e.g., w_0, w_1; w_0', w_1'; w_0'', w_1''), so that a zero is produced at $z_b = 1$, which provides a deep null for fixed ground clutter [28]. Using the multiple zeros produced by the PRF stagger we also can adjust the weighting coefficients to produce the optimal weights (see Table 5.1) that maximize the MTI improvement factor under the condition that all target velocities are equally likely [28].

References

[1] Barton, D. K., *Modern Radar System Analysis*, Norwood, MA: Artech House, 1988.

[2] Schleher, D. C., *MTI and Pulsed Doppler Radar*, Norwood, MA: Artech House, 1991.

[3] Skolnik, M., *Introduction to Radar Systems*, 3rd ed., New York: McGraw-Hill, 2001.

[4] Schleher, D. C., (ed.), *MTI Radar*, Dedham, MA: Artech House, 1978.

[5] Blackband, W., *Radar Techniques for Detection Tracking and Navigation*, New York: Gordon and Breach, New York: 1966.

[6] Skolnik, M., (ed.), MTI *Radar*, Chap. 17 in *Radar Handbook*, 1st ed., New York: McGraw-Hill, 1970.

[7] Nathanson, F., *Radar Design Principles*, 2nd ed., Mendham, NJ: SciTech, 1999.

[8] Schleher, D. C., *Introduction to Electronic Warfare*, Dedham, MA: Artech House, 1986.

[9] ap Rhys, T., and G. Andrews, "AEW Radar Antennas," AGARD Conf. Preprint No. 139, *Antennas for Avionics*, 1974, pp. 12-1–12-16; also in D. C. Schleher, *MTI Radar*, Dedham, MA: Artech House, 1978.

[10] Andrews, G., *Airborne Motion Compensation Techniques, Evaluation of TACCAR*, Report 7407, Naval Research Laboratory, Washington, D.C., April 12, 1972.

[11] Andrews, G., *Airborne Motion Compensation Techniques, Evaluation of DPCA*, Report 7426, Naval Research Laboratory, Washington, D.C., July 20, 1972.

[12] Hammerle, K., "Cascaded MTI and Coherent Integration Techniques with Motion Compensation," *Proc. IEEE Int. Radar Conference*, Washington, D.C., 1990, pp. 164–168.

[13] Sachs, H., and J. Krstansky, "Controlling RFI Susceptibility in Receivers," *Electronic Industries*, September 1960.

[14] Lemley, L., W. Gleason, and G. Gross, *RF Receivers for Electronic Warfare*, Report 8737, ADB105737, Naval Research Laboratory, Washington, D.C., May 1986.

[15] Andrews, G., *Optimum Radar Doppler Processors*, Report 7727, Naval Research Laboratory, Washington, D.C., May 1974.

[16] Emerson, R., *Some Pulsed Doppler MTI and AMTI Techniques*, Report R-274, Rand Corporation, March 1954, pp. 1–124; also in D. C. Schleher, *MTI Radar*, Dedham, MA: Artech House, 1978.

[17] Capon, J., "Optimum Weighting Functions for the Detection of Sampled Signals in Noise," *IEEE Trans. on Information Theory*, Vol. IT-10, April 1964, pp. 152–159; also in D. C. Schleher, *MTI Radar*, Dedham, MA: Artech House, 1978.

[18] Bellman, R., *Introduction to Matrix Analysis*, New York: McGraw-Hill, 1960.

[19] Schleher, D. C., and D. Schulkind, "Optimization of Nonrecursive MTI," *Proc. IEE Int. Radar Conference*, London, October 1977, pp. 182–185; also in D. C. Schleher, *MTI Radar*, Dedham, MA: Artech House, 1978.

[20] Fowler, C., A. Uzzo, and A. Ruvin, "Signal Processing Techniques for Surveillance Radar Sets," *IEEE Trans. on Military Electronics*, Vol. MIL-5, No. 2, April 1961, pp. 103–108; also in D. C. Schleher, *MTI Radar*, Dedham, MA: Artech House, 1978.

[21] Linden, D., and B. Steinberg, "Synthesis of Delay Line Networks," *IEEE Trans. on Aeronautical and Navigational Electronics*, Vol. ANE-4, No. 1, March 1957, pp. 34–39; also in D. C. Schleher, *MTI Radar*, Dedham, MA: Artech House, 1978.

[22] Skolnik, M., (ed.), "MTI Radar," Chap. 15 in *Radar Handbook*, 2nd ed., New York: McGraw-Hill, 1990.

[23] White, W., and A. Ruvin, "Recent Advances in the Synthesis of Comb Filters," *IRE National Convention Record*, 1957, pp. 186–209; also in D. C. Schleher, *MTI Radar*, Dedham, MA: Artech House, 1978.

[24] Bagley, G., *A Survey of Cancellation versus Integration for Radar Clutter Rejection*, Report 2880, Naval Research Laboratory, Washington, D.C., August 1974.

[25] Skolnik, M., *Introduction to Radar Systems*, 2nd ed., New York: McGraw-Hill, 1980.

[26] Cleetus, G., "Properties of Staggered PRF Radar Spectral Components," *IEEE Trans. on Aerospace and Electronic Systems*, Vol. AES-12, November 1976, pp. 800–803c.

[27] Thomas, H., and N. Lutte, "Z-Transform Analysis of Nonuniformly Sampled Digital Filters," *IEE Proc.*, Vol. 119, No. 11, November 1972, pp. 1559–1567.

[28] Thomas, H., N. Lutte, and M. Jelffs, "Design of MTI Filters with Staggered PRF: A Pole-Zero Approach," *IEE Proc.*, Vol. 121, No. 12, December 1974, pp.1460–1466.

[29] Roy, R., and O. Lowenschuss, "Design of MTI Detection Filters with Nonuniform Interpulse Periods," *IEEE Trans. on Circuit Theory*, Vol. CT-17, No. 4, November 1970, pp. 604–612.

[30] Prinsen, P., "Elimination of Blind Velocities of MTI Radar by Modulating the Interpulse Period," *IEEE Trans. on Aerospace and Electronic Systems*, Vol. AES-9, No. 5, September 1973, pp. 1147–1148; also in D. C. Schleher, *MTI Radar*, Dedham, MA: Artech House, 1978.

[31] Thomas, H., and T. Abram, "Stagger Period Selection for Moving Target Radars," *IEE Proc.*, Vol. 123, March 1976, pp. 195–199.

CHAPTER 7

Pulsed Doppler Systems

By definition, a Doppler radar is one that uses the Doppler effect, which is associated with the radial component of relative radar-target velocity, to select targets having particular radial velocities. A Doppler radar utilizing a coherent pulsed transmitter is referred to as a pulsed Doppler radar. The signals returned from targets illuminated by a PD radar are generally processed in a set of narrowband filters that coherently integrate the signals to which they are tuned. The filters both resolve and enhance signals within a particular velocity band. In search radar applications, the bank of contiguous Doppler filters is used to cover the velocity band of interest to extract specific targets and reject clutter and targets outside that velocity band. Once the target is acquired, only one filter may be needed for tracking.

The PD approach differs from MTI (see Chapter 6) in that the Doppler filter performs coherent integration of the selected signal within a narrow spectral band, whereas the MTI passes targets through a broad comb filter response band. From a theoretical standpoint, the PD radar approximates the optimal processor for detecting a Doppler-shifted return signal in Gaussian clutter and receiver noise when the Doppler frequency of the target is known *a priori* (see Chapter 5). This is in contrast to the MTI approach in which all target velocities are assumed equally likely.

Pulsed Doppler radars can be categorized as low, medium, or high PRF, according to whether Doppler (low PRF), range (high PRF), or range and Doppler (medium PRF) are ambiguous (see Chapter 2). High- and medium-PRF radars have been used primarily in airborne applications, where platform motion clutter spectral spreading causes main lobe ground clutter to extend over tens of knots and where short-range clutter effects are minimized by the aircraft altitude. Use of these range-ambiguous waveforms in surface-based radar is made difficult by the presence of short-range clutter, which competes with long-range targets and has a large power advantage over the targets due to the nth-power increase of clutter power range [i.e., $C/S = k(R_t/R_c)^n$, where $n = 3$ for distributed surface clutter and $n = 4$ for discrete clutter] with respect to the target power. This has caused low-PRF PD radars to assume an increasingly important role in surface-based surveillance because of their superior performance in clutter rejection and the decreasing cost of modern digital processors [1]. An example of low-PRF PD radar is the moving target detector described in Section 1.6.

The most difficult choice among the various PD waveforms involves the airborne case, where high, low, and medium PRFs might be advantageous depending on the specific application. For example, in AEW applications (e.g., the E-2C and AWACS radars) that involve large antenna installations (i.e., on the order of 25 to 30 ft), both low (the E-2C) and high PRFs (the AWACS) have been successfully implemented to detect small RCS airborne targets over both land and water [2]. The high-PRF design has the advantages of rejecting low-speed targets due to ground vehicular traffic and higher angular resolution (through the use of a higher transmitter frequency), which provides an ECCM advantage and better rejection of chaff. A medium-PRF AEW design (Erieye) has been successfully implemented using an active aperture phased-array antenna [3]. A design utilizing fore and aft radome-mounted antennas [2, 4, 5] has been attempted (i.e., the Nimrod radar) for this application, but was apparently unsuccessful due in part to the use of a relatively small azimuth antenna aperture (i.e., on the order of 8 ft) that increased both the clutter power and spectral spreading [see (6.4)] and the computer complexities of handling the large clutter intake while resolving the various range and Doppler ambiguities.

The unambiguous range (R_u in nautical miles) as a function of PRF (in kilohertz) is given by

$$R_u = \frac{80.91}{\text{PRF}_{\text{kHz}}} \tag{7.1}$$

This relationship (plotted in Figure 7.1) illustrates how rapidly the range decreases with an increase in PRF. For example, from an unambiguous range of 400 nmi at a PRF of 200 Hz, the unambiguous range decreases to 10 nmi at a PRF of 8 kHz. The unambiguous velocity (in knots) is also related to the PRF (in kHz) as

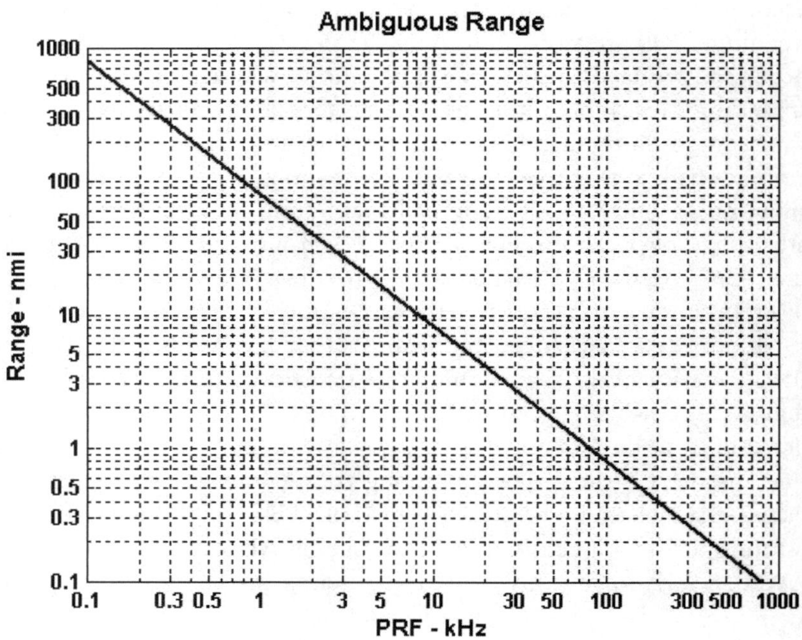

Figure 7.1 Unambiguous range first blind speed Doppler. (*From:* [6, p. 407].)

$$v_u = 971\lambda \text{PRF} \tag{7.2}$$

where λ is the wavelength in meters. By combining (7.1) and (7.2), the result is

$$R_u v_u = \frac{2.3565 \times 10^4}{f_{\text{GHz}}} \tag{7.3}$$

where R_u is in nautical miles, v_u is in knots, and the radar's frequency is in gigahertz. Equation (7.3) can be expressed as an inequality given by $R_u v_u \leq 2.3565 \times 10^4 / f_{\text{GHz}}$, which establishes the unambiguous range and velocity boundary for any waveform.

As an example, consider a low-PRF AEW design that operates at 425 MHz with an azimuth aperture of 25 ft. Assume that we want an unambiguous range of 300 nmi, which from (7.3) results in a first blind speed of 185 knots. The spectral spreading in velocity units at the 3 dB point can be expressed as

$$\Delta v = v_{ac}\,\theta_B \cos\phi_0 \sin\theta_0 \tag{7.4}$$

where v_{ac} is the aircraft's velocity, θ_B is the azimuth beamwidth in radians, ϕ_0 is the depression angle, and θ_0 is the azimuth pointing angle with respect to the ground track. For a 25 ft aperture at UHF, the resulting beamwidth is approximately $\theta_B = \lambda/D = 1/10.8$ rad, resulting in a worst case ($\phi_0 = 0$, $\theta_B = 90°$) velocity spread for a 300 knot aircraft speed of 28 knots. This represents a velocity blockage of $2 \times 28/185$ or 0.3 of the velocity region covered by the radar. On average then, the radar would be capable of only detecting 70% of the targets if they were spread equally in velocity over the entire range of possible target velocities (i.e., 600 knots for aircraft targets). This could be improved by the use of a displaced phase center antenna, which would compensate for the motion-induced spectral spreading (see Sections 3.2.7 and 6.2.2).

As a further example, consider an X-band AI radar with an azimuth aperture of 2.5 ft. Consider that an unambiguous range of 30 nmi is desired in a low-PRF mode. From (7.2), this results in a first blind speed of about 80 knots. The ground clutter spectral spreading ($\psi_0 = 0$, $\theta_B = 30°$) in velocity units for an interceptor speed of 1,500 knots is about 30 knots. This would block 75% of the velocity coverage, which is not acceptable from a design standpoint. To accommodate this situation, we customarily increase the radar's PRF, resulting in the use of medium- or high-PRF waveforms for AI applications.

This example illustrates the problem of employing a low-PRF Doppler radar design for small aperture high-speed AI radar due to the excessive spreading of main lobe ground clutter about the PRF lines, which blocks a significant portion of the velocity region associated with the target. The other significant effect associated with a low-PRF design concerns the Doppler shift induced on the ground clutter that enters the radar through the antenna sidelobes. This sidelobe clutter has velocity components that extend up to the aircraft's ground speed and, in a low-PRF design, generally are aliased throughout the velocity region occupied by the target. The maximum improvement factor of the low-PRF Doppler radar against clutter is thus limited to the integrated two-way antenna main lobe-to-sidelobe ratio (see Section 6.2.2). This is significant in many airborne installations due to the difficulty of achieving a low sidelobe antenna design in this restrictive environment.

The limitation of the airborne low-PRF PD radar led to the development of airborne high-PRF designs whose performance in the case of closing targets does not depend on the level of antenna sidelobes [7]. The ground clutter spectrum returned to a high-PRF design is depicted in Figure 7.2 and extends to Doppler frequencies corresponding to the aircraft's ground speed given by $f_d = \pm 2v_r/\lambda$. The major responses in the spectrum correspond to the altitude return (from beneath the aircraft) and the main lobe clutter response. The clutter spectrum is ambiguous about the PRF lines in accordance with the sampling theorem for pulsed waveforms. From the illustration, we can see that the high-PRF design provides a clutter-free region that extends from the most positive part of the central line of the clutter spectrum (due to the aircraft's forward motion) to the most negative part of the next ambiguous line of the clutter spectrum (due to the antenna's backlobes). In this clutter-free region, target detection is limited only by receiver noise. The minimum PRF is then determined (see Figure 7.2) as the sum of the Doppler frequency of the target most rapidly closing ($f_{dt} = 2\dot{R}_{max}/\lambda$) plus the maximum sidelobe clutter frequency extent ($f_{da} = 2v_r/\lambda$), which is determined by the radar's velocity (v_r). Note, however, that if both closing and receding targets must be resolved, then the minimum PRF must be increased to $PRF_{min} > 2f_{dt} + f_{da}$ because the sidelobes extend both ahead of and behind the aircraft. This ensures that, for example, a closing target response associated with the central line can be distinguished from a receding target response associated with the next ambiguous line of the spectrum. The shorter the wavelength, the higher the Doppler frequencies of the clutter and the target and, hence, the higher the required PRF. Typically, in AI (Airborne Intercept) applications at X-band frequencies, the PRF is on the order of 100 to 300 kHz.

A clutter and target vector diagram that shows the effects of the various types of clutter (i.e., main lobe, altitude line, sidelobe) for different target headings is depicted in Figure 7.3. An aircraft with a scanning antenna is flying in the direction shown. Two targets are indicated, one directly ahead of the aircraft and one at an azimuth angle of 45°. Each target is located at the center of the circles. The three circles, labeled 1, 2, and 3, represent particular ratios of target ground velocity to interceptor ground velocity. A target vector showing velocity ratio and direction is drawn from the center of each target position. The tip of the vector will show whether the target is in the clear or is obscured by clutter. In Figure 7.3, the two

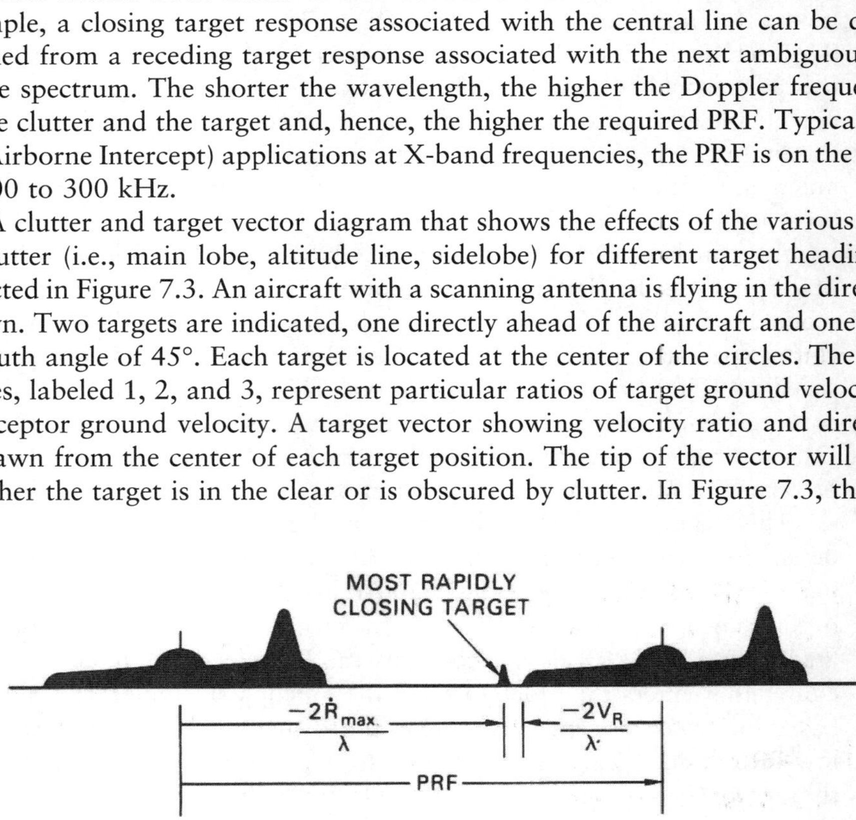

Figure 7.2 Ground clutter spectrum for airborne high-PRF PD radar. (*From:* [8, p. 456].)

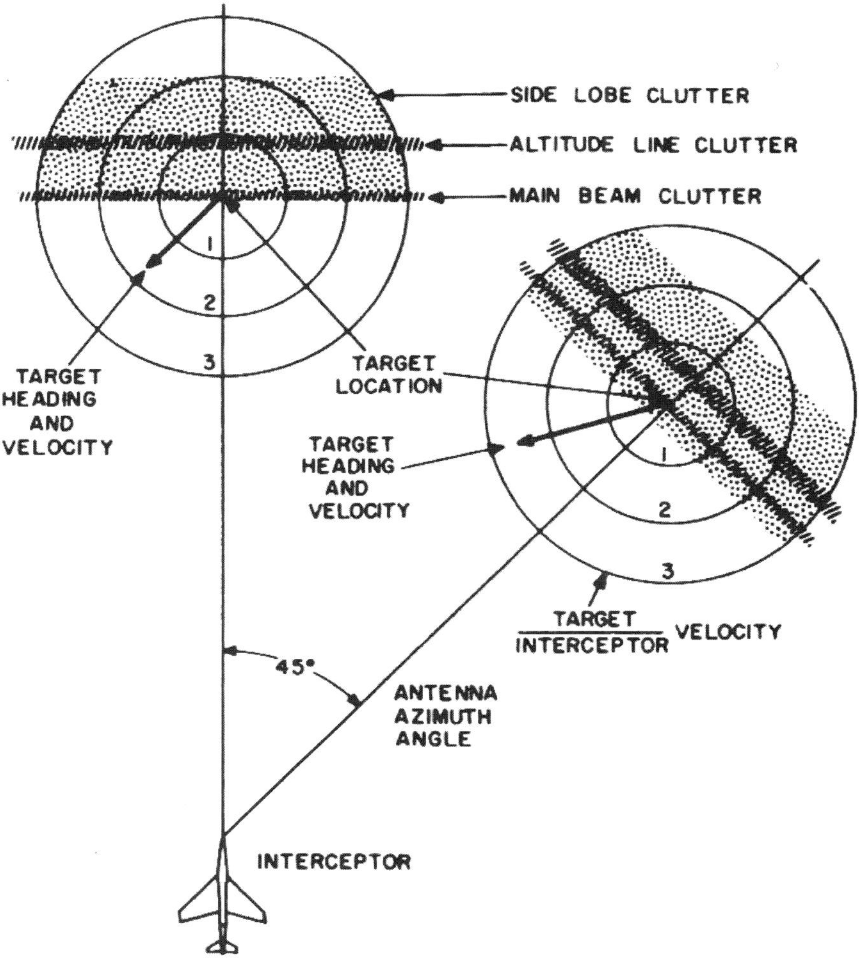

Figure 7.3 Clutter and target vector diagram for a high-PRF Doppler radar. (*From:* [9, p. 118].)

target vectors show the targets to be outside the clutter. If both targets were traveling in the opposite direction, the one straight ahead would be obscured by opening sidelobe clutter, and the target on the 45° bearing, because of its high velocity, would be clear of opening clutter.

Figure 7.3 shows that closing targets and high-speed receding targets (those moving at twice the interceptor speed) can be detected in the clutter-free region. All other velocity targets must compete with clutter to be detected in a high-PRF design. The clutter in the high-PRF case is generally severe due to the large number of ambiguous ranges that results from the waveform design, which causes short-range clutter to fall in the same range cell as a long-range target. The result of this effect is that the high-PRF design is ineffective under tail-chase conditions, which may be a major difficulty in some applications.

The medium-PRF airborne PD radar was developed to get around some of the limitations of low- and high-PRF designs in fighter applications [7, 8]. In particular, the medium-PRF airborne PD radar is designed to work for tail chases as well as closing engagements, thereby providing good all-aspect coverage. For operation at X-band, medium-PRFs range from about 8 to 16 kHz. If the maximum operating range is not exceptionally long, the PRFs can be set high enough to provide adequate separation between the periodic repetitions of the main lobe clutter spectrum without incurring particularly severe range ambiguities. Main lobe clutter can then be isolated from the bulk of the target return on the basis of Doppler frequency, and individual targets can be isolated from the bulk of the sidelobe clutter through a combination of range, Doppler resolution, and low sidelobes. Also, ground-based moving targets, being close to the frequency of the main lobe clutter, can be rejected along with it, without rejecting an unacceptable amount of possible target return (see Figure 7.4).

Due to the ambiguous range and velocity responses associated with a medium-PRF design, the radar must contend with blind zones both in the Doppler spectrum and the range interval being searched. Doppler blind zones are less severe with medium PRFs than with low PRFs because main lobe clutter covers a much smaller portion of the Doppler frequency spectrum. Range blind zones result when target echoes are suppressed by sidelobe clutter simultaneously received from shorter ranges or eclipsed by transmitter pulses. Figure 7.5 depicts how blind zones occur when target echoes at long ranges compete with range ambiguous sidelobe clutter responses from shorter ranges due to the ambiguous range response. Blind zones can be suppressed by switching among a number of different PRFs that have been judiciously selected to provide clear zones in both the Doppler and range domains. Typically, the radar may be cycled through a fixed number of fairly widely spaced PRFs. For example, if eight PRFs are used, the target is in the clear on any three of these, the echoes exceed the detection threshold on all three, the target will be deemed to have been detected, and the range and velocity ambiguities can be resolved. The judicious selection of the various PRFs associated with medium-PRF radar systems is one of the major system decisions associated with the design of this type of radar.

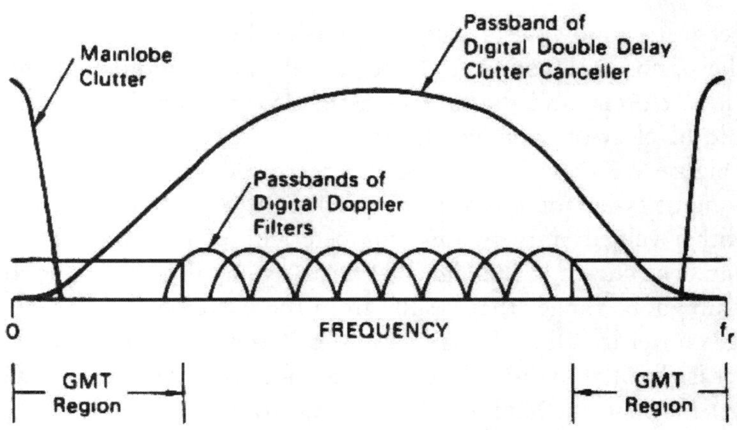

Figure 7.4 Rejection of ground moving targets in medium-PRF PD radar. (*From:* [8, p. 571].)

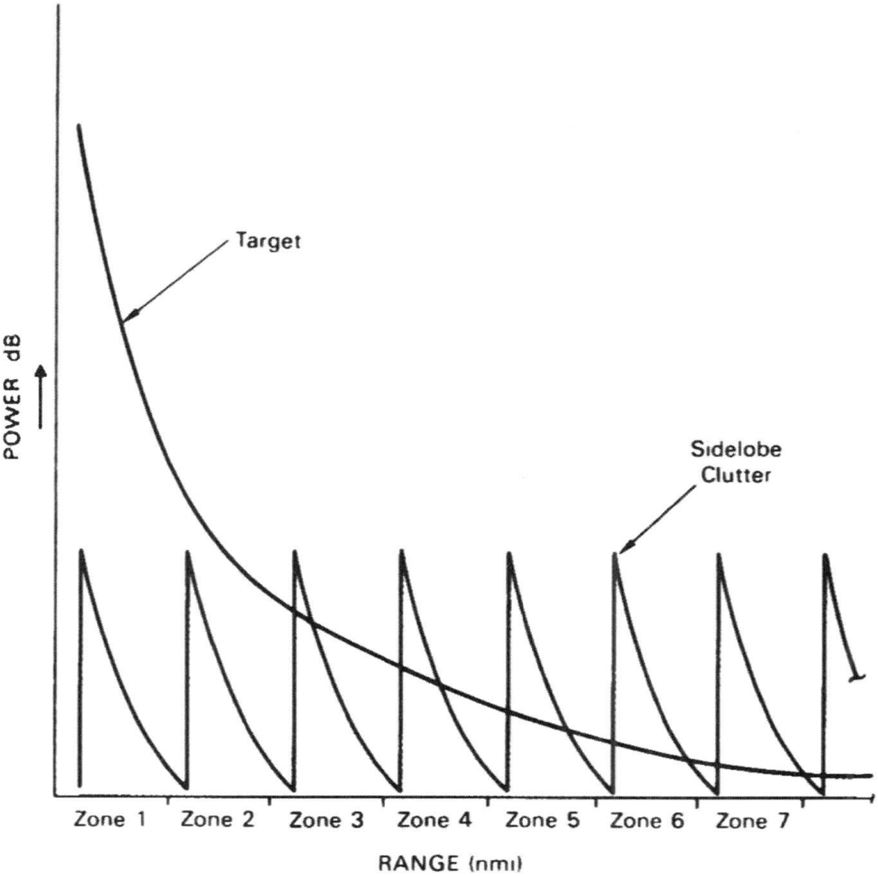

Figure 7.5 Variation of target and sidelobe clutter response in a medium-PRF design. (*From:* [8, p. 447].)

An important class of sidelobe returns that is particularly troublesome to medium-PRF airborne PD radars comes from structures on the ground (i.e., water towers, large buildings) that have an exceptionally large radar cross section (typically from 20 to 60 dBsm). The return from these discrete clutter targets, although reduced by the two-way antenna sidelobes, can be as large as those of desired targets. The strength of these targets is enhanced in medium-PRF designs by ambiguous range, which allows a short-range discrete target to fall in the same range cell as a long-range airborne target. Furthermore, radar platform motion induces a Doppler shift onto these discrete targets, which translates them into the passband of the target Doppler filters where they are detected as if they were aircraft in the radar's main lobe.

One way of dealing with these unwanted targets is to provide the radar with a guard channel. This consists of a separate receiver-processor with its input supplied by a small horn antenna mounted on the radar antenna. The width of the horn's main lobe is sufficient to encompass the entire region illuminated by the radar antenna's principal sidelobes, and the gain of the horn's main lobe is greater than any of the sidelobes. Any detectable target in the radar antenna's sidelobes will therefore produce a stronger output from the guard receiver than from the main radar receiver. When this occurs, the output of the main radar receiver is blanked, thereby eliminating the unwanted target.

7.1 Pulsed Doppler Configurations

There are many different implementations of PD radars depending on the application and whether low-, medium-, or high-PRF waveforms are utilized. Implementations for low-PRF MTI configurations were discussed in Section 6.1, and block diagrams of MOPA and coherent-on-receive types were given in Figures 2.20 and 2.21. In this section, we describe high-PRF configurations that apply to airborne and surface applications; medium-PRF configurations are also discussed.

A block diagram of an airborne high-PRF configuration is depicted in Figure 7.6 [10]. A master frequency control is utilized to control the carrier and PRFs and to provide coherent references for the receiver mixing processes (see Figure 1.10). This process is identical to the MOPA configuration described in Figure 2.20 and Section 6.1 for low-PRF radars. However, the stability requirements are far more severe when used in a high-PRF PD due to the translation of short-range clutter components into target Doppler filters, resulting from the modulation effects of transmitter noise sidebands, which are referred to as *transmitted clutter* (see Section 3.3).

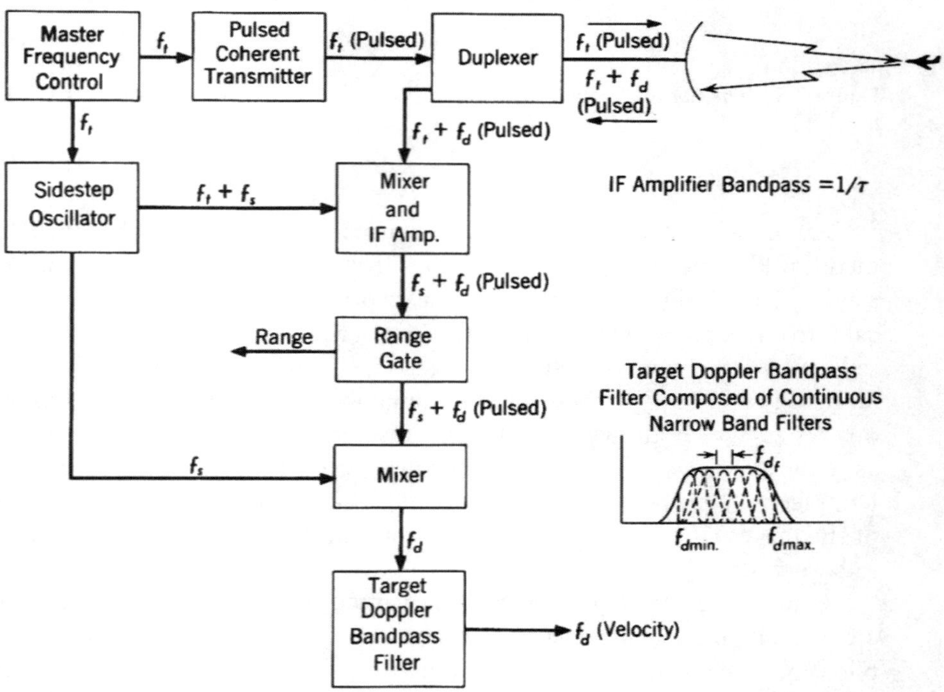

Figure 7.6 Block diagram of a high-PRF PD radar. (*From:* [10, p. 325].)

The master frequency control (STALO) is generally derived from a highly stable crystal oscillator that generates a low fundamental frequency multiplied up to form the STALO operating frequency. The crystal oscillator is also typically employed to generate timing signals (e.g., digital clock) and the sidestep oscillator (COHO) frequency so that all signals in the radar are locked onto a single stable reference. The STALO is then applied to a pulsed coherent transmitter that amplifies this low-powered signal to the power level required for transmission. The power amplifier generally consists of a solenoid-focused gridded traveling wave tube (TWT) or a high-powered klystron amplifier, although current designs use solid-state modular antenna arrays.

The coherent pulsed transmitter is connected to the antenna through a duplexer, which is a high-powered switch that serves the dual function of: (1) passing the high-power transmitted pulses to the antenna while blocking their path to the receiver on transmission, and (2) passing the low-power returns to the receiver while blocking their path to the transmitter on receive. The duplexer is one of the critical components in high-PRF radar because the recovery time of this element determines the maximum duty factor of the high-PRF waveform.

In high-PRF airborne designs, PRFs might range from 100 to 300 kHz, whereas, in medium-PRF airborne radars, the PRF might range from 10 to 30 kHz. The duty factor of the transmitter is given by the ratio of the radar's pulsewidth (τ) to its interpulse period (T) as

$$d_u = \frac{\tau}{T} = \tau \cdot \text{PRF} \tag{7.5}$$

In high-PRF airborne radars, the duty factor might approach 0.5, which is the upper limit set by the necessities that the receiver must be blanked during transmission and the finite-recovery-time duplexer must gate the receiver on during reception. Comparable duty factors for low-PRF designs might range from 0.001 to a maximum of 0.2 if a large pulse compression factor is employed [8]. For medium-PRF radars, the duty factor might range from 0.01 to 0.03, which reflects the moderate pulse compression ratios that are sometimes employed to provide higher range resolution [11].

The average power transmitted by the radar is given by the peak power multiplied by the duty factor:

$$P_A = P_p d_u \tag{7.6}$$

In high-PRF designs, we typically use only the power returned to the radar in the central spectral line of the transmitted spectrum for detection in the receiver. This is equivalent to CW radar that transmits an average power (P_{CW}) of

$$P_{\text{CW}} = P_A d_u = P_p d_u^2 \tag{7.7}$$

Note that there is no dilution of the S/N ratio in this process because the receiver noise power is diminished in exactly the same ratio (d_u) as the signal power when only the central line of the spectrum is processed.

The receiver portion of the airborne high-PRF PD radar shown in Figure 7.6 extracts the Doppler-shifted target return from a background that consists of the received clutter and the inevitable receiver noise. The clutter (i.e., primarily from the ground) can be resolved into components resulting from: (1) main lobe clutter, (2) altitude-line clutter, and (3) sidelobe clutter (see Figure 2.9). Main lobe clutter results from clutter returns (i.e., surface, weather, chaff), which enter the receiver through the main lobe antenna pattern and are large in magnitude (e.g., 60 to 90 dB above receiver noise), but are concentrated in a relatively narrow spectral band, corresponding to the closing velocities of the radar to those scatterers located within the main antenna lobes. Altitude-line clutter results from surface reflections directly beneath the aircraft. It has a large magnitude and its spectrum is centered about zero frequency. Sidelobe clutter is attenuated from main lobe clutter by the ratio of the relative sidelobe-to-main lobe antenna response, but exhibits a wide spectrum extending over a frequency range that corresponds to the aircraft's closing and receding (from backlobes) velocity with respect to the surface. The receiver and associated processor must be linear over the full dynamic range of the main lobe clutter, which may be as large as on the order of 80 to 90 dB above receiver noise to avoid spectral spreading of main lobe clutter (see Sections 3.2.5 and 6.2.3).

The first step in the receiving process is to translate the return signal to IF by mixing with the sidestep oscillator, which is cohered to the master reference oscillator. The signal is then range-gated into range cells that generally correspond to the range resolution of the radar (i.e., the effective pulsewidth of the radar). Range gating serves to reduce the clutter intake into the radar by excluding all returns that are not folded into the range cell occupied by the expected target return (see Figure 1.18). Note that in a high-PRF radar design, the range is highly ambiguous, causing many clutter returns at different ranges than the target to compete with the target return.

Multiple range gates can be distributed throughout the interpulse interval depending on the radar's duty factor. When duty factors approaching 0.5 are utilized, only one range gate is employed, and the return from different ranges cannot be isolated with range gates. In this case, receiver blanking serves the function of a single range gate. However, if the duty factor is much less than 0.5, a significant advantage can be gained by employing multiple range gates in a high-PRF design. Note, however, that range gating must precede Doppler filtering so that each range gate added entails adding a significantly larger number of Doppler filters. This is particularly significant in a high-PRF design (compared to a medium- or low-PRF design) because a great many more Doppler filters are required to provide the same Doppler resolution for high-PRF operation as for medium-PRF operation.

After range gating, the next step in the receiving chain for a high-PRF design is to remove main lobe and altitude clutter returns. One method employs notch filters centered at the main lobe and altitude-line clutter center frequencies [12]. However, we may want to process only the Doppler-clear region, that is, the Doppler region extending from $f_d = 2v_r/\lambda$, particularly if the radar utilizes interlaced medium- and high-PRF waveforms (the medium-PRF design is more effective in the sidelobe clutter region). It is advantageous in this case to employ a bandpass filter that passes the signals having Doppler frequencies that occur in the clear region. A design that uses an analog filter configuration to accomplish this function is depicted in Figure 7.7 [8].

In the analog processor shown in Figure 7.7, the range-gated IF signal is shifted in frequency to place the highest central line sidelobe clutter frequency just below the band of frequencies that the filters are designed to pass. This frequency is equal to $f_d = 2v_r/\lambda$ and can be supplied by using information from the aircraft's inertial navigation system. The first filter (the analog-clear-region filter) eliminates all but the central-line frequencies of the Doppler spectrum. As a result of eliminating all but the central-line return, the output of the analog-clear-region filter is a continuous signal. Because the filter's passband spans only the clear region of the central-line return, this signal consists primarily of target echoes immersed in receiver noise and main lobe clutter residue.

Two analog subband filters further attenuate the clutter and divide the clear region into two contiguous bands. Automatic gain control is applied separately to each subband so that a strong return in one (e.g., clutter residuals) will not desensitize the other. The filter output after AGC is then digitized in an A/D converter for subsequent digital processing. The filters must provide more than 80 dB of attenuation to reject the main lobe clutter. This function is generally performed by using analog processing to reduce the dynamic range required of the A/D converter. To avoid creating Doppler ambiguities, the sampling rate of the A/D converter must be at least equal to the width of the Doppler-clear region in which targets are expected to appear. This width is equal to the Doppler frequency that corresponds to the fastest anticipated target.

Once digitized, I and Q components of the return are supplied to the digital signal processor depicted in Figure 7.8. The returns received during the period the main antenna lobe dwells on the target (T_d) are amplitude weighted and supplied

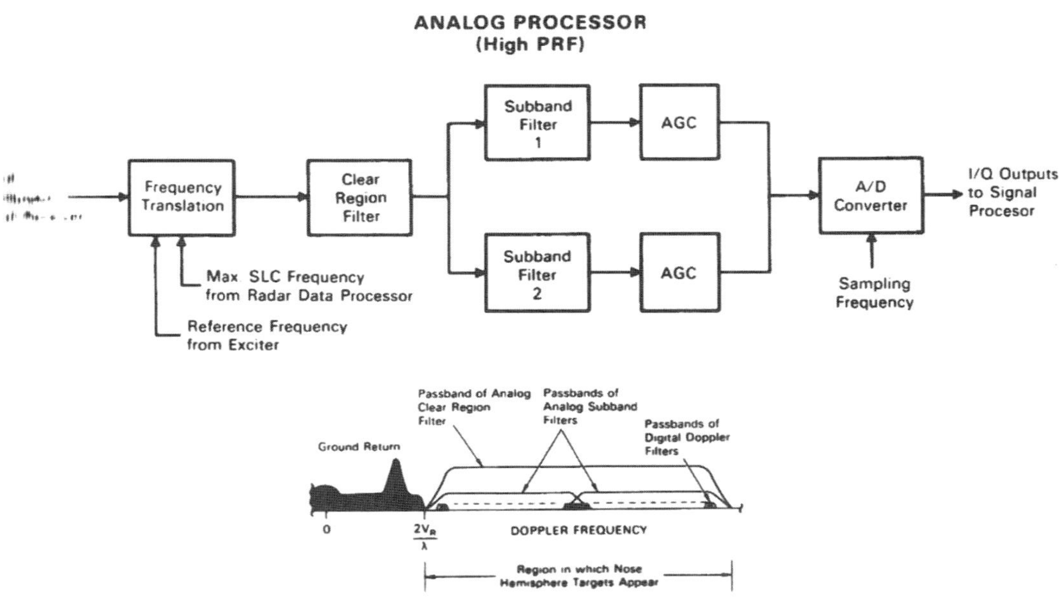

Figure 7.7 High-PRF, clear-region Doppler bandpass filter. (*From:* [8, p. 504].)

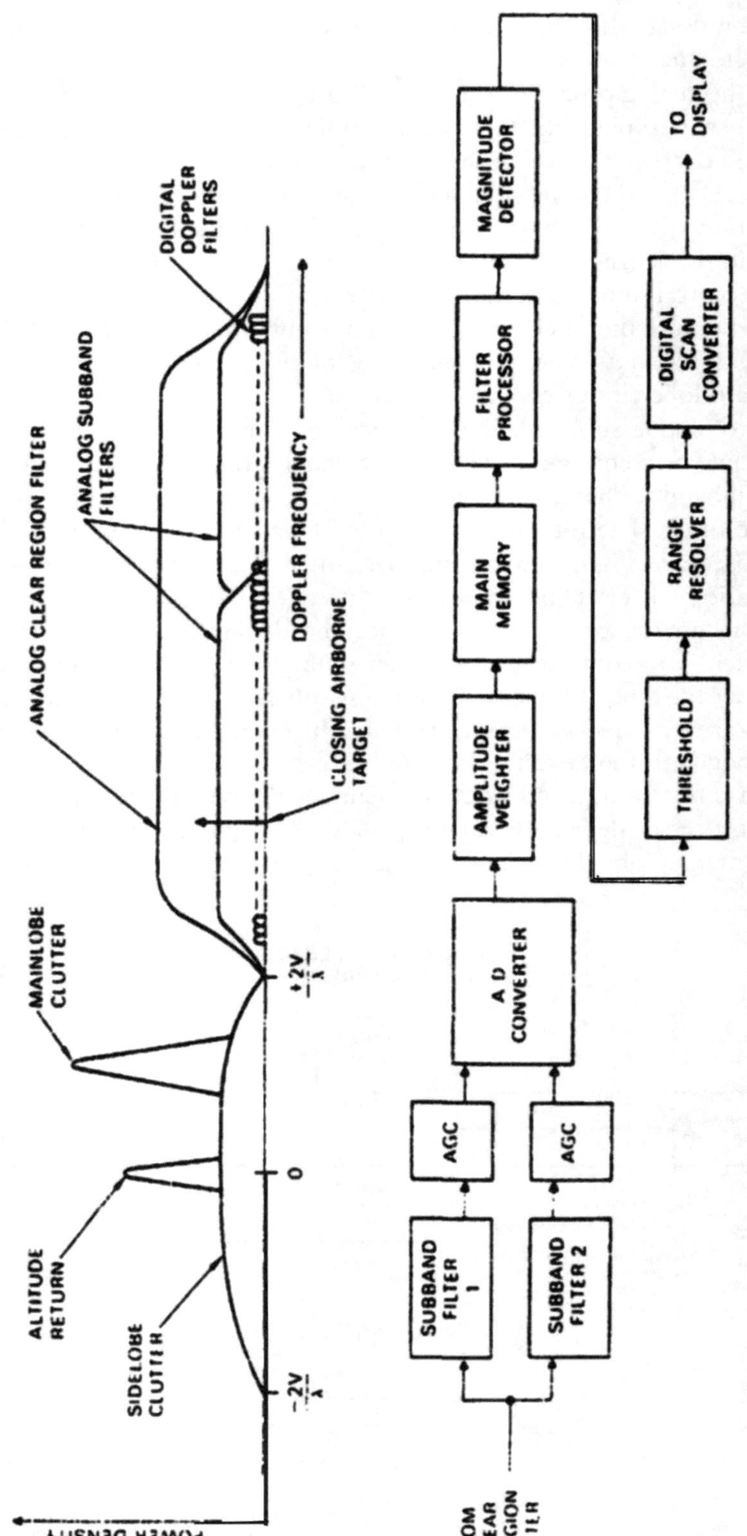

Figure 7.8 High-PRF, Doppler radar digital signal processor. (*From:* [8, p. 570].)

to a bank of Doppler filters. The optimal bandwidth of each Doppler filter is given by $B_D = 1/T_d$ so that the resulting number of Doppler filters in the clear region is equal to

$$n_f = \frac{f_{dt}}{B_D} = \frac{2v_t T_d}{\lambda} \tag{7.8}$$

where v_t is the velocity of the fastest expected target. Note that the coherent dwell often is divided into several *Coherent Processing Intervals* (CPIs). The number of Doppler filters in the clear region is then reduced to $n_f = 2v_t\,\mathrm{CPI}/\lambda$ and postdetection integration of $n' = T_d/\mathrm{CPI}$ samples is generally employed. Weighting reduces the return's spectral sidelobes, thereby providing additional noise and clutter rejection in each Doppler filter and preventing strong targets at one frequency being detected in adjacent filters.

The outputs of the Doppler filters usually formed using the FFT algorithm are then envelope detected, normalized using CFAR processing, and threshold detected to form a target detection. At this point, target detections are highly ambiguous in range, a problem that must be resolved to complete the detection process. In Figure 7.8, an FM ranging procedure is indicated (due to the small number of range gates available in an interpulse interval), which resolves the range ambiguities by changing the frequency of the radar transmitter (see Section 7.3.2). Alternatively, the range ambiguities could be resolved by transmitting multiple PRFs during the period the main lobe of the antenna dwells on the target.

An alternative high-PRF receiver (depicted in Figure 7.9) provides detection of Doppler-shifted targets using all-digital processing. In this configuration, main beam clutter is eliminated by using an MTI canceller (comb) filter, rather than the analog design depicted in Figure 7.7. The A/D converter in this design must handle the full dynamic range of the clutter signals so that on the order of 14 bits plus sign are required. The main beam clutter canceller must also carry the full dynamic range, where the FFT processor need have only enough dynamic range to accommodate sidelobe clutter and the target's dynamic range.

A block diagram of a surface-based PD radar receiver is depicted in Figure 7.10 [14]. This receiver uses range-gated bandpass filters to form the Doppler filters. The filters provide both rejection of clutter returns and detection of Doppler-shifted target returns. In this design, both predetection coherent integration (bandpass filtering) and postdetection integration are performed. The equivalent predetection Doppler bandwidth for this configuration is given by

$$B_e = B_{\mathrm{RF}}^{1-\alpha}(2B_v)^{\alpha} \tag{7.9}$$

where B_{RF} is the IF bandpass filter's bandwidth, B_v is the postdetection bandwidth, and α is a parameter dependent on the ratio $B_{\mathrm{RF}}/2B_v$, which generally ranges from 0.6 to 0.7 for most practical situations [13]. For large values of $B_{\mathrm{RF}}/2B_v$, the parameter α approaches 0.5, which results in an equivalent Doppler bandwidth given by

$$B_e = \sqrt{2B_{\mathrm{RF}}B_v} \tag{7.10}$$

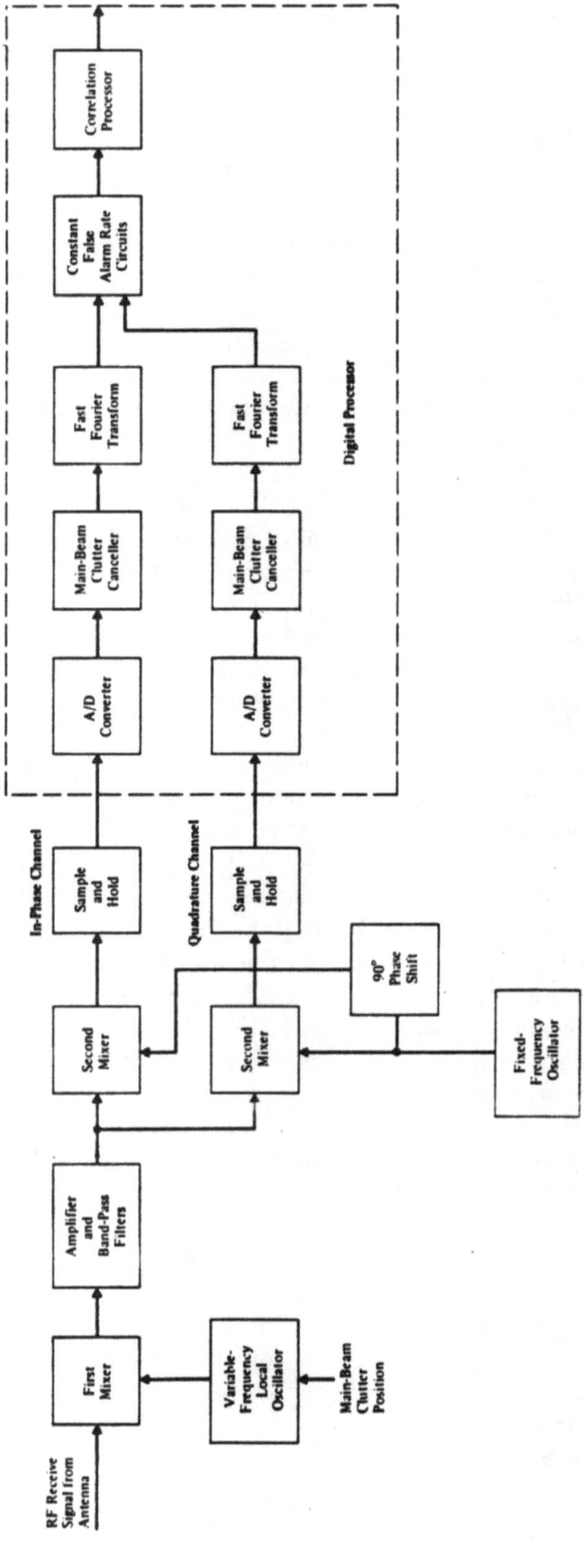

Figure 7.9 High-PRF radar receiver block diagram. (*From:* [13, p. 281].)

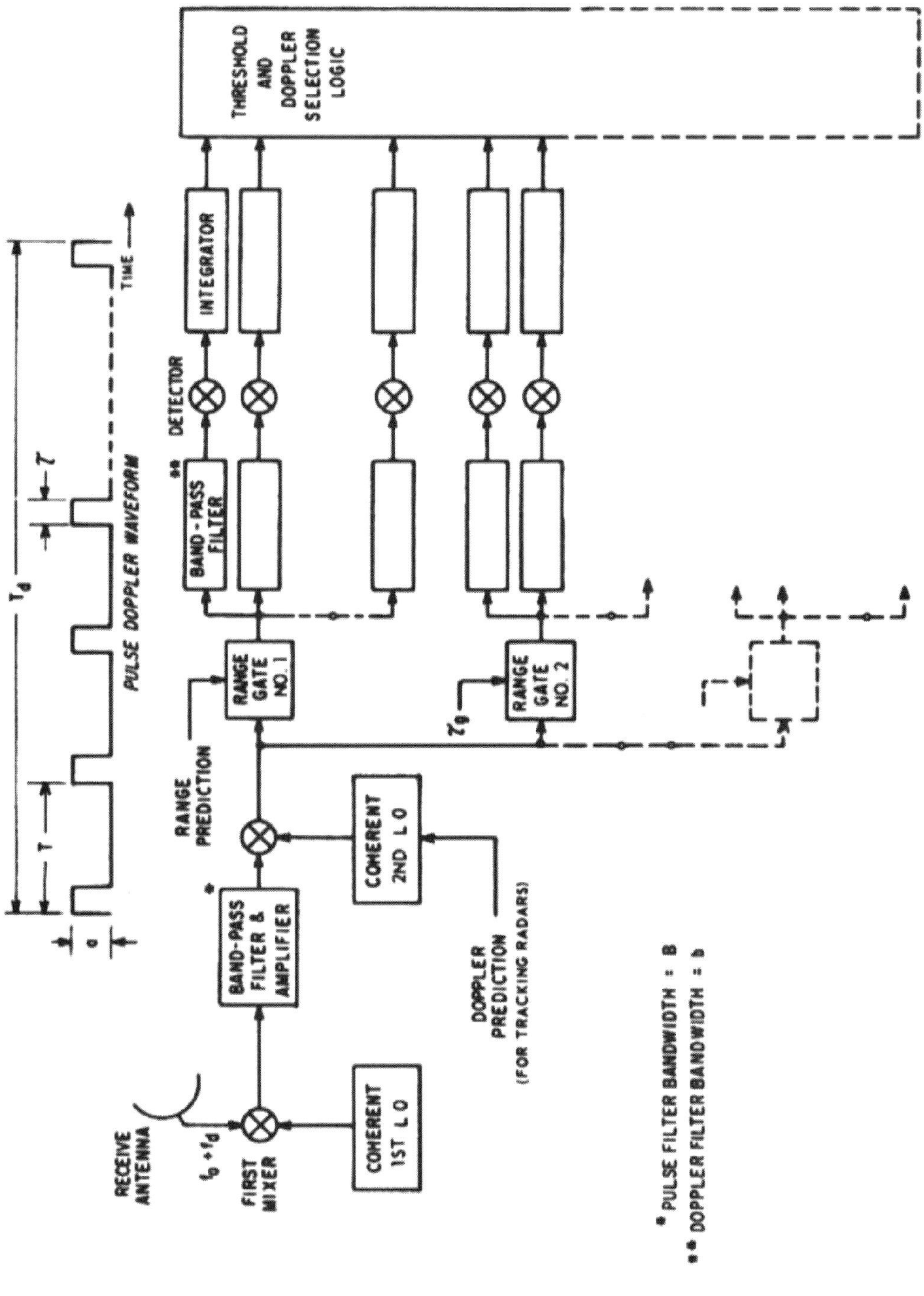

Figure 7.10 PD receiver using range-gated bandpass filters. (*From:* [14, p. 414].)

The bandpass filters utilized in the receiver depicted in Figure 7.10 must provide high attenuation in the stopband while being precisely tuned to a certain Doppler frequency. Generally, crystal filters are needed to satisfy this type of requirement, but they have practical limitations when used in such a design [14]. These limitations have led to the use of range-gated homodyne (zero IF) PD receivers.

A homodyne PD receiver is depicted in Figure 7.11, which is configured for a tracking radar application. A range-gated single-sideband (SSB) mixer translates the microwave RF signal to baseband (zero frequency) in a single step and provides the in-phase and quadrature channels necessary to retain the Doppler information contained in the RF signal. A contiguous Doppler filter bank is synthesized by offsetting the RF reference oscillators into the SSB mixer by the frequency separation desired between the filters. The output of the SSB mixer will be a unipolar video signal when the reference oscillators match the Doppler-shifted RF signal. The low-pass filters (video integrators) that follow the SSB mixer then coherently integrate the Doppler-shifted target echoes. The final output is formed by squaring and adding the outputs of quadrature I and Q channels. The equivalent Doppler bandwidth of the homodyne receiver is thus twice the bandwidth of the lowpass filter ($B_D = 2B_v$).

The homodyne PD receiver provides a flexible method of controlling both the filter separation and bandwidth of the Doppler filter bank. Note that, in this implementation, the video integrator attenuates the clutter returns. Hence, the low-pass filter must have a sufficient number of poles to provide the desired attenuation of the clutter return. In general, this type of filter is implemented using an active low-pass filter (i.e., a filter using an integrated circuit amplifier).

The homodyne receiver can be implemented digitally, which allows synthesizing a large number of Doppler filters. In a digital implementation, the range-gated I and Q returns for each transmitting pulse are applied to an A/D converter, where the digital output is stored in a digital memory. Digital filters can be formed by sequentially rotating the stored return in accordance with the phase pattern expected from a particular Doppler (i.e., the MTD processor) rather than that used in high-PRF processors. A block diagram of a medium-PRF PD radar signal processor is illustrated in Figure 2.13.

Medium-PRF processors generally use a clutter canceller followed by a coherent integrator implemented through the FFT algorithm. Additional gain control is required to prevent saturation of the A/D converter, and narrower Doppler filters are required in the medium-PRF processor to attenuate the higher clutter experienced as compared to a low-PRF processor. Both range and Doppler ambiguities must be resolved in the medium-PRF design, which is generally accomplished by using multiple PRFs.

7.2 Waveform Considerations [8, 11, 15–17]

A basic decision that must be made in the design of PD radar is the selection of the PRF. Low, medium, and high PRFs have been used in various applications (see Chapter 2). Generally, low-PRF (unambiguous range) designs are used in surface-based applications and for airborne radar in look-up or beyond-the-clutter-horizon

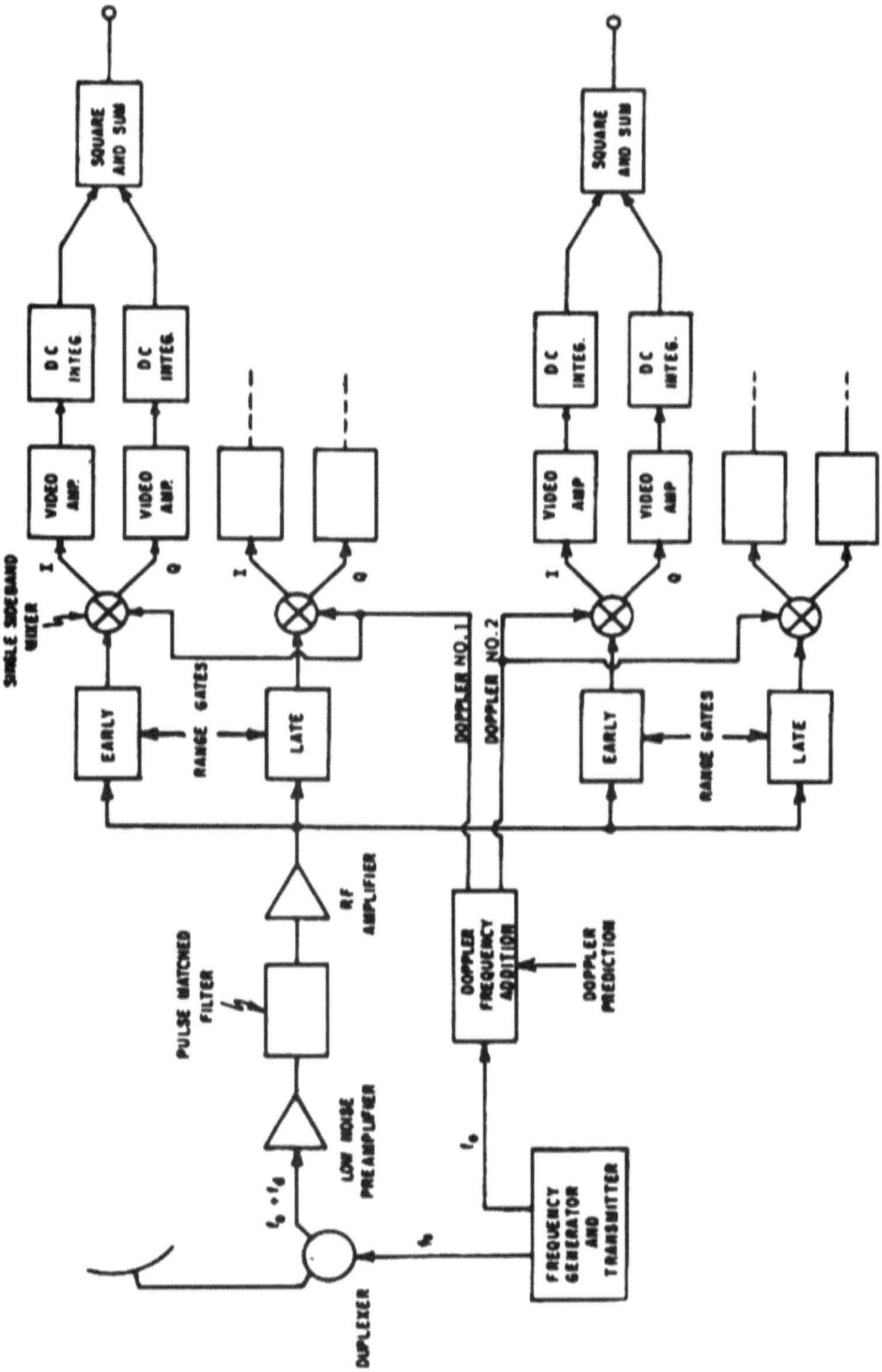

Figure 7.11 PD homodyne receiver. (*From:* [14, p. 419].)

applications. One exception is in airborne applications using antenna motion compensation techniques (see Sections 3.2.7 and 6.2.2) and relatively large antenna apertures (e.g., the E-2C radar). High-PRF (unambiguous Doppler) configurations are used primarily in airborne applications that look down into heavy clutter and are designed to detect closing targets. Medium-PRF (ambiguous range and Doppler) designs are used primarily in airborne applications that require detection of tail-chase targets looking down into heavy clutter with moderate performance against closing targets. Interlaced medium- and high-PRF modes have also been used to provide good all-aspect performance against both closing and chasing targets.

Figure 7.12 depicts the three PRF types in range-Doppler space for an AI application [11]. In general, the interpulse period is divided into range cells (range

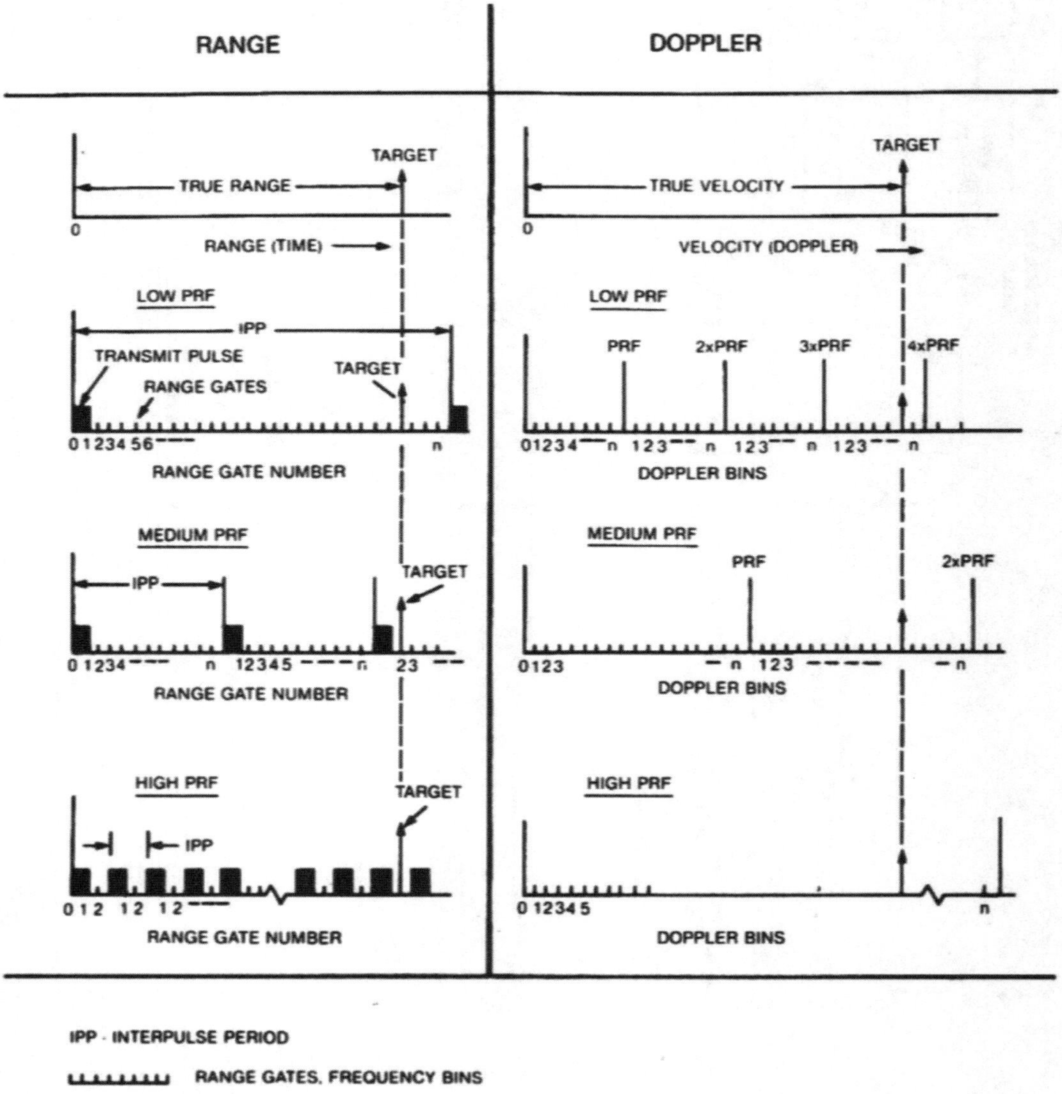

Figure 7.12 PD PRF radars in range-Doppler space. (*From:* [11, p. 303].)

gates) that correspond to the radar's transmitted bandwidth (τ_e = 1/BW), and the Doppler space is divided into Doppler cells (Doppler filters), the bandwidth of which is inversely proportional to the dwell time of the radar antenna's main beam on the target, or the CPI when multiple PRFs are used during the dwell. The number of range gates is greatest for low-PRF radar, whereas only a small number are used for high-PRF radar. Conversely, only a small number of Doppler filters are used in a low-PRF design, whereas a large number are used in a high-PRF design. Although the interpulse period is generally filled with range gates for all types, the low-PRF type is the only design that fills the frequency region between PRF lines with Doppler filters. Generally, the high-PRF design fills only the Doppler-clear region ($2v_r/\lambda < f_d <$ PRF $- 2v_r/\lambda$) with Doppler filters, whereas the medium-PRF design excludes the region occupied by moving vehicular targets from the Doppler coverage (see Figure 7.4).

The principal advantage of a low-PRF design is its ability to restrict clutter intake to just that which is in the same range cell as the target. For example, a rain or chaff clutter region several miles in extent would not interfere with targets not directly in the region, whereas, in a medium- or high-PRF design, this same rain region could clutter the entire unambiguous range interval. Also, sidelobe and main lobe clutter can be restricted by increasing the range resolution of the radar and it is not magnified by the folding of ambiguous clutter returns from short range into the same range cell as the target.

Other features of the low-PRF design include the ability to use a sensitivity-time-control function that allows normalization of the clutter return over the full range of the radar. In addition, range can be directly measured without generating any spurious targets. STC cannot be used in medium- or high-PRF designs because it would attenuate long-range targets of interest as well as near-range clutter. This results in a low-PRF system processor requiring less dynamic range, which allows better control of false alarms. In medium- and high-PRF designs, a range correlation system is required to resolve the resulting ambiguous ranges. When multiple targets are present, they can confuse the range correlator and generate spurious targets called *ghosts*.

On the negative side, in a low-PRF design, unless main lobe clutter is separated in range from the targets the radar encounters, it can be rejected only on the basis of differences in Doppler frequency. The highly ambiguous Doppler response causes a lack of Doppler visibility as depicted in Figure 7.13. Here, the frequency of the main beam ground clutter is translated to zero frequency. The spread of the main lobe clutter spectrum is determined primarily by the aircraft's velocity, azimuth antenna beamwidth, and azimuth look angle. At azimuth angles significantly off the aircraft's ground track (i.e., 30° to 90°), the main lobe spectrum generally occupies a large fraction of the frequency region between PRF lines. The wide Doppler notch blinds the low-PRF PD radar to many targets of interest.

In airborne search radar, the criterion that determines the Doppler rejection notch is primarily that required to reject slow-moving vehicular ground targets rather than main beam clutter rejection because slow-moving target rejection requires a wider notch. In a low-PRF radar design operating over land, ground targets can be confused with low-flying airborne targets due to the aliasing effect of the ambiguous Doppler response. A typical rejection notch width is ±55 knots

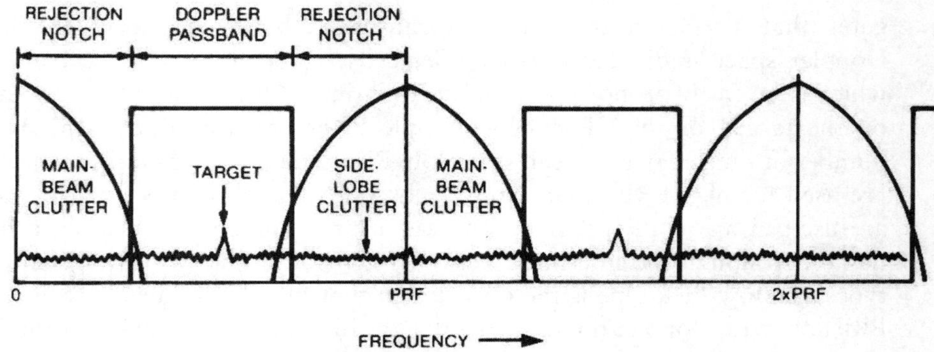

Figure 7.13 Low-PRF PD radar spectrum. (*From:* [11, p. 303].)

with respect to the ground, which translates to a Doppler notch width of ±1,870 Hz at X-band, ±625 Hz at S-band, ±225 Hz at L-band, and ±80 Hz at UHF. This can be compared to main lobe clutter widths by using Figure 7.4. For example, a UHF AEW radar (aperture width equal to 25 ft) moving at 300 knots has a clutter spread of 28 knots, whereas an X-band interceptor radar (aperture width equal to 2.5 ft) moving at 600 knots has a clutter width of 32 knots (see the introduction to Chapter 7). A possible exception, in which the clutter spread exceeds the ground target notch, is the X-band interceptor cited earlier, which moves at supersonic speeds (e.g., 1,500 knots), inducing a clutter spread of 80 knots. Note that the clutter spread due to platform motion can be reduced by using antenna processing techniques (see Sections 3.2.7 and 6.2.2), but the Doppler notch requirement to reject ground targets is difficult to overcome. However, in some applications, ground targets can be resolved by using PRF switching techniques because they generally occur in the first ambiguity zone of the Doppler response (see Figure 7.14).

For the airborne interceptor radar at X-band, using a Doppler notch width of ±1,870 Hz (i.e., ±55 knots), requires that 3,740 Hz of the Doppler spectrum between PRF lines be blind to targets of interest. Because typical low PRFs for airborne radars range from 1,000 to 4,000 Hz, depending on the maximum range of interest, essentially no usable Doppler spectrum remains. In summary, whereas a low-PRF system provides accurate and ghost-free range information, the visible Doppler spectrum available after clutter and rejection of ground-based moving targets is generally too limited for airborne look-down applications.

In an airborne high-PRF design, target Dopplers that appear clear of the sidelobe clutter can be detected in a noise-limited environment, as depicted in Figure 7.15, which shows a simplified downconverted version of a high-PRF airborne radar spectrum. The sidelobe clutter extends to $\pm 2v_r/\lambda$ around the PRF lines, where v_r is the aircraft's ground speed. The main lobe clutter center frequency changes according to the radar's look angle with respect to the aircraft's velocity vector. A closing target is shown having a total Doppler shift that is the target's line-of-sight Doppler plus the main beam clutter Doppler along the LOS. Thus, whenever the target's Doppler along the LOS is greater than the maximum sidelobe clutter

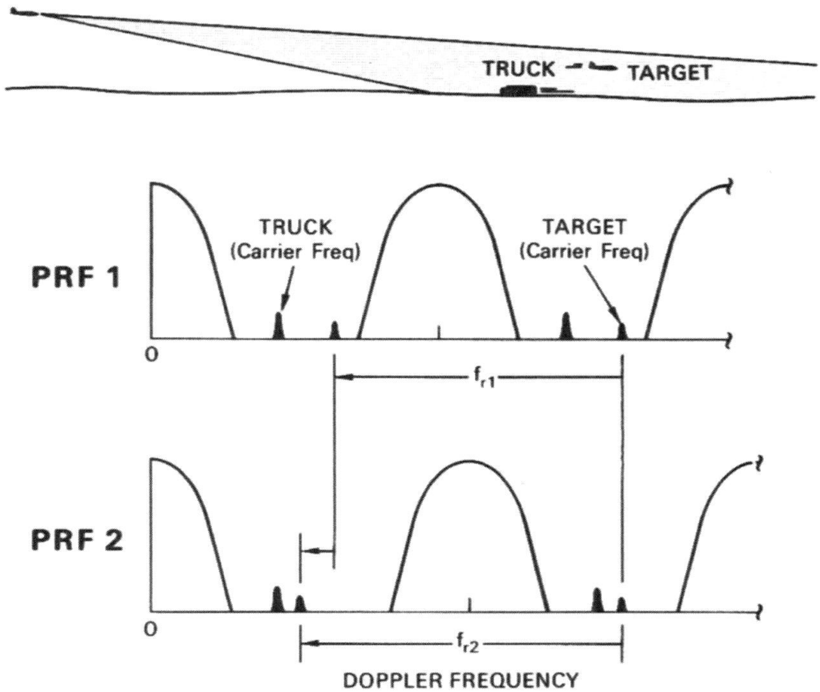

Figure 7.14 Resolving ground targets in a low-PRF Doppler radar using PRF switching. (*From:* [8, p. 435].)

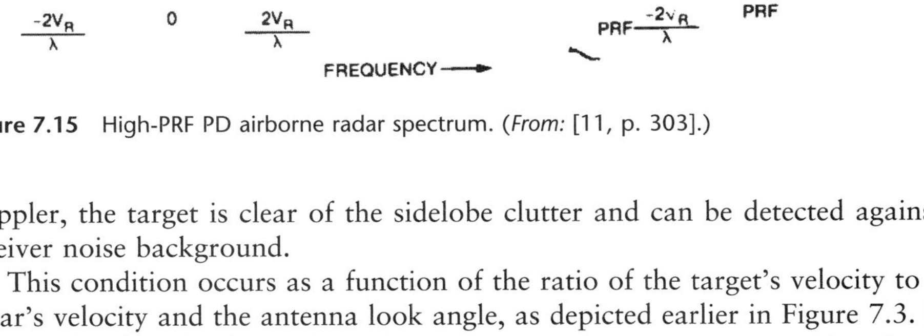

Figure 7.15 High-PRF PD airborne radar spectrum. (*From:* [11, p. 303].)

Doppler, the target is clear of the sidelobe clutter and can be detected against a receiver noise background.

This condition occurs as a function of the ratio of the target's velocity to the radar's velocity and the antenna look angle, as depicted earlier in Figure 7.3. For look angles directly along the radar's velocity vector, any target with a line-of-sight velocity component that is toward the radar becomes clear of sidelobe clutter. As the radar scans to a 90° look angle, the target's velocity component along the LOS must be greater than the radar's ground speed to clear the sidelobe clutter.

A feature of the high-PRF waveform in an airborne application is that slow-moving, ground-based targets can be eliminated because such a wide Doppler spectrum is available for target processing. For example, in the X-band high-speed interceptor example (i.e., ground speed equal to 1,500 knots), the sidelobe clutter spectrum will extend out to 50 kHz. If the expected targets close on the interceptor at a maximum of 600 knots, then an additional 20 kHz of clear Doppler must be provided. If a PRF of 150 kHz, for example, is selected, the lower zone of 80 to 100 knots can be discarded to eliminate ground targets. The result would have only minimal effect in the detection of high-speed closing aircraft targets.

High-PRF airborne radar designs generally operate in a velocity-only search detection mode for initial detection. In this mode range, ambiguities are not resolved and a maximum detection range equivalent to that of a CW radar (i.e., with an average power of $P_A = P_p d_u$) is achieved. When range information is desired, the high-PRF design must employ either frequency modulation of the RF carrier (e.g., linear or sinusoidal frequency modulation) or multiple PRFs during the radar's dwell time on target to resolve the range. This requirement may cause an estimated 20% to 25% degradation in detection range compared to that achieved when range information is not required [11]. A significant limitation of the high-PRF waveform for airborne applications is its limited ability to detect targets moving at velocities that cause their Doppler frequencies to appear in the sidelobe clutter regions. As shown in Figure 7.3, the clutter-free detection regions occur in the forward or closing aspects of the target, whereas for rear or chasing aspects, the target must compete with large sidelobe clutter returns, particularly at low altitudes.

With the high-PRF waveform, range folding causes short-range clutter to compete with long-range targets. In particular, all of the sidelobe clutter folds into a small ambiguous range interval due to the many range ambiguities. For example, with a PRF of 150 kHz, the unambiguous range is only 0.5 nmi and, because sidelobe clutter can be well above receiver noise out to beyond the 10 nmi range, more than 19 ambiguities of clutter will be folded into the PRF interval. Hence, at low altitudes, sidelobe clutter generally will obscure targets located within its velocity domain (i.e., $f_d = \pm 2v_r/\lambda$).

In summary, a high-PRF waveform provides ghost-free target Doppler, the capability to reject slow-moving targets, and a clutter-free target detection region that generally facilitates detection of closing targets. However, the limited all-target-aspect detection, especially at low altitudes in tail chases, generally restricts the use of this waveform to closing targets. In addition, high-PRF designs require complicated methods for resolving range ambiguities and may also suffer from eclipsing problems, in which target returns are lost, either totally or partially, because the receiver is blanked when the target is received due to the high duty factor of the waveform (see Section 7.2.1).

The medium-PRF waveform combines some of the attributes of both high- and low-PRF waveforms, which for airborne applications can provide slow-moving target rejection, all-target-aspect coverage, sidelobe clutter-free regions, and accurate range information. The range and Doppler ambiguities of a medium-PRF design cause the returns from short-range sidelobe clutter to fold into the same Doppler regions occupied by target returns. However, the essence of medium-PRF waveform design is to establish relatively clear regions within the range-Doppler space by the judicious placement of multiple medium PRFs.

The first step in medium-PRF design is to set the Doppler visibility, which is defined as the percentage of the interval between PRF lines not occupied by main lobe clutter (see Figure 2.6). Figure 7.16 shows that the improvement in the velocity-clear region increases from 13% to 70% at an antenna look angle of 30° and an aircraft speed of 400 m/s as the PRF is increased from 2 to 6 kHz. The improvement would be even greater for a 10 kHz PRF, which would provide a clear region of about 83%.

When using a PRF of 6 kHz, the maximum unambiguous range is about 25 km. Search for targets at long range results in multiple-time-around returns, where, for example, a target at 150 km would arrive after 6 pulses have been transmitted. The most important result of the ambiguity is the presence of sidelobe clutter that arises from short ranges, which competes with targets at larger ranges (see Figure 2.12). Because the sidelobe clutter region is most intense at the minimum time delay point, the dominant return is generally caused by the first Doppler surface within the clutter region, as depicted in Figure 7.17 for an airborne application.

In a medium-PRF design, the sidelobe clutter that arises from the transmitted pulse immediately preceding the target echo arrival competes with the target signal. This clutter comes from a ring on the ground at a constant range (R_1) from the aircraft (see Figure 2.11). In general,

$$R_1 = R_t - (n - 1) R_b \tag{7.11}$$

where n is the ambiguous range interval and R_b is the unambiguous range interval determined by the PRF ($R_b = c/2\text{PRF}$). Figure 7.17 illustrates an idealized clutter profile for the case where $n = 2$ and only the clutter from the first ambiguous interval (at R_1) is of concern because all other intervals by comparison return negligible power. The clutter from the first interval decays rapidly as range increases, beginning as an altitude signal (specular return) and changing to a diffused sidelobe

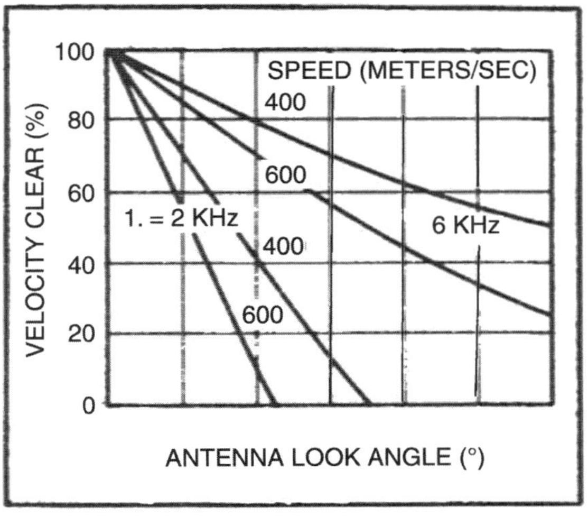

Figure 7.16 Velocity-clear performance with PRF variation for two aircraft speeds. (*From:* [16, p. 133].)

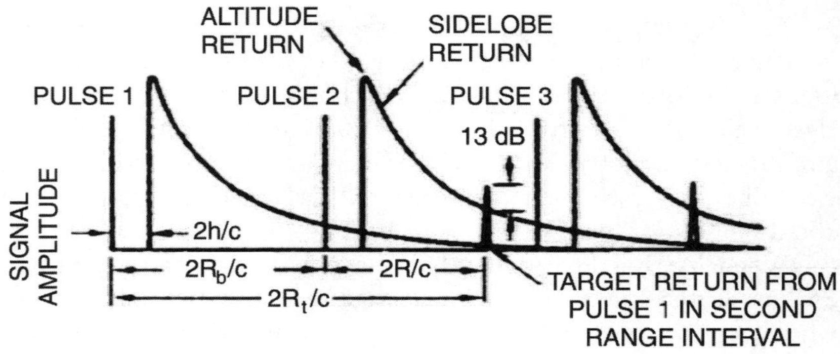

Figure 7.17 Sidelobe return for medium-PRF PD radar. (*From:* [16, p. 303].)

return. When the target appears to be out by several ambiguous range intervals in range, a blind zone will occur in the range dimension where the target signal is not sufficiently larger than the sidelobe clutter to provide detection. This effect is depicted in Figure 7.18, where the target return power from the initial transmitted pulse is superimposed on the ambiguous sidelobe clutter response. As the target moves farther out in range (i.e., becomes highly ambiguous), the range blind zones spread until the target is completely submerged in ground clutter.

In addition to the range blind zones caused by strong sidelobe clutter, other blind zones are caused by eclipsing. While the radar is transmitting (and for the duplexer recovery time thereafter), the receiver is blanked. Consequently, if a target's echoes are received during these periods, the target return is blanked. If the radar transmits narrow pulses, then the receiver blocking may be inconsequential; however, if the pulses are long, as they are in some medium-PRF radars (e.g., with pulse compression), the additional range blanking due to eclipsing can be significant. Figure 7.19 depicts the relative range blanking for representative medium-PRF radar due to sidelobe clutter and eclipsing.

To resolve both range and Doppler blind zones occurring in medium-PRF airborne radars, the radar's PRF is cycled through a fixed number of fairly widely

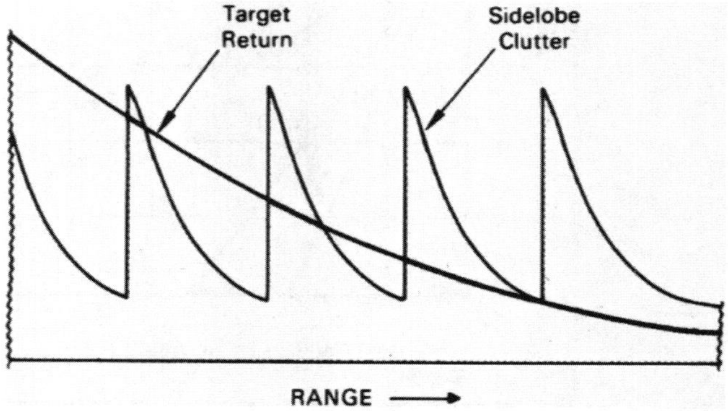

Figure 7.18 Target and sidelobe clutter returns in a medium-PRF airborne radar. (*From:* [8, p. 449].)

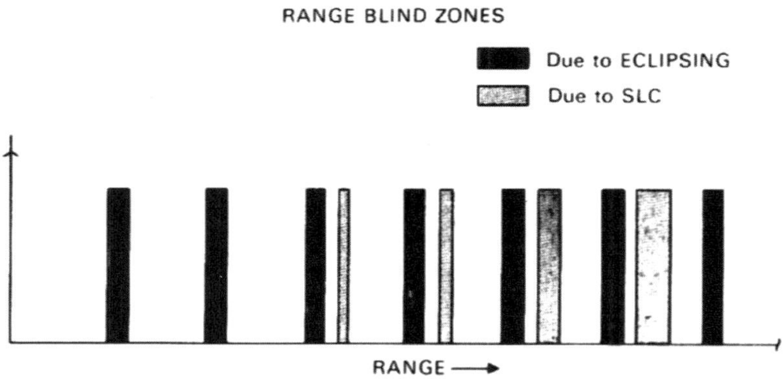

Figure 7.19 Combined range zones due to eclipsing and sidelobe clutter in medium-PRF radar. (*From:* [8, p. 448].)

spaced PRFs while the radar's main lobe dwells on the target. For example, the radar might be cycled through eight different PRFs. The blind Doppler zones for these PRFs due to sidelobe clutter are depicted in Figure 7.20. If a target is in the clear on any three PRFs and its echoes exceed the detection threshold on all three, the target will be deemed detected. The range ambiguities and all spurious target ranges will then be resolved. The optimum PRF code is a function of the operational situation such as radar altitude, clutter levels, and speeds that must be determined for each specific radar.

For a target to be detected during a particular transmitted PRF of a medium-PRF code, it must be simultaneously in both a Doppler-clear region (see Figure 7.20) and in a range-clear zone (see Figure 7.19). If the target's Doppler frequency falls in a Doppler blind zone, its return will not pass through a Doppler filter and be detected, despite the fact that it is in a range-clear zone. If the target is in a

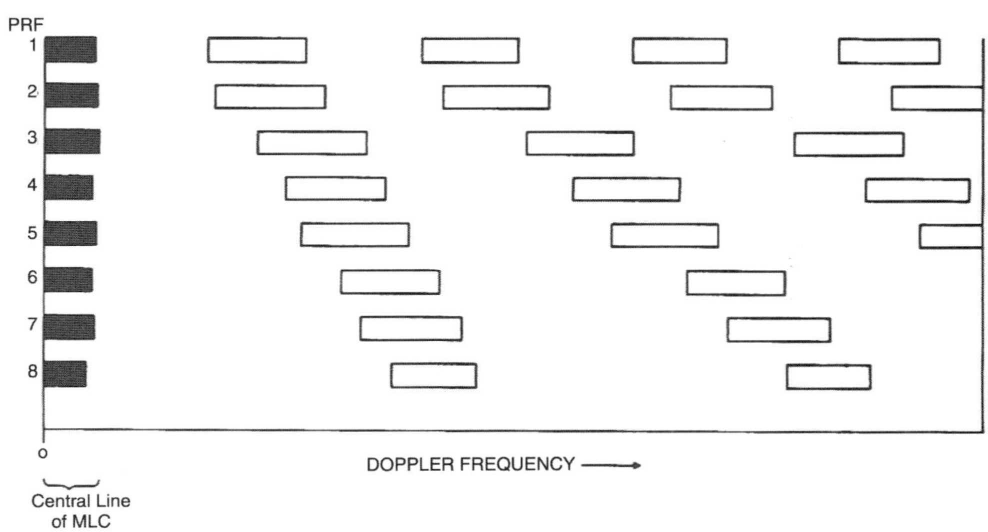

Figure 7.20 Doppler blind zones with multiple PRFs for medium-PRF PD radars. (*From:* [8, p. 446].)

range blind zone, although its return may pass through a Doppler filter, it still will not be detected because the accompanying sidelobe clutter through the filter will cause the detection threshold to be greater than the target response. Figure 7.21 depicts a range-Doppler matrix of representative medium-PRF radar that shows the zones for which the radar is both range clear and Doppler clear for at least three of eight widely spaced PRFs. Note that for tracking radar, the PRF can be adaptively controlled to place the target in clear range and velocity zones, so that only one burst at this PRF need be transmitted, rather than the long, diverse bursts required by search radar.

In a medium-PRF design, the number of PRFs processed during an antenna dwell time (the time the antenna takes to scan over the target) is constrained on the high side by the available processing time. If the time becomes too long, the antenna scan rate drops below an acceptable value. On the low side, the number of PRFs per dwell must be sufficient not only to resolve the range, but also to ensure good Doppler visibility.

Figure 7.22 illustrates the problem of generating spurious reports in a multiple-PRF processing sequence [11]. A three-PRF code is used with a two-out-of-three detection criterion. The PRIs for the three PRFs shown are 7, 8, and 9 range gates. For PRF 1, a target in range gate 4 appears in range gates 4, 11, 18, 25, and so on. A target in range gate 26 appears in range gate 26, modulo-PRI so that for PRF 1, the target appears in range gates 5, 12, 19, 26, and so on. Similarly, for PRF 2, the target in gate 4 appears in range gates 4, 12, 20, 28, and so on. The correlation process for the three PRFs, as shown in Figure 7.22, identifies the

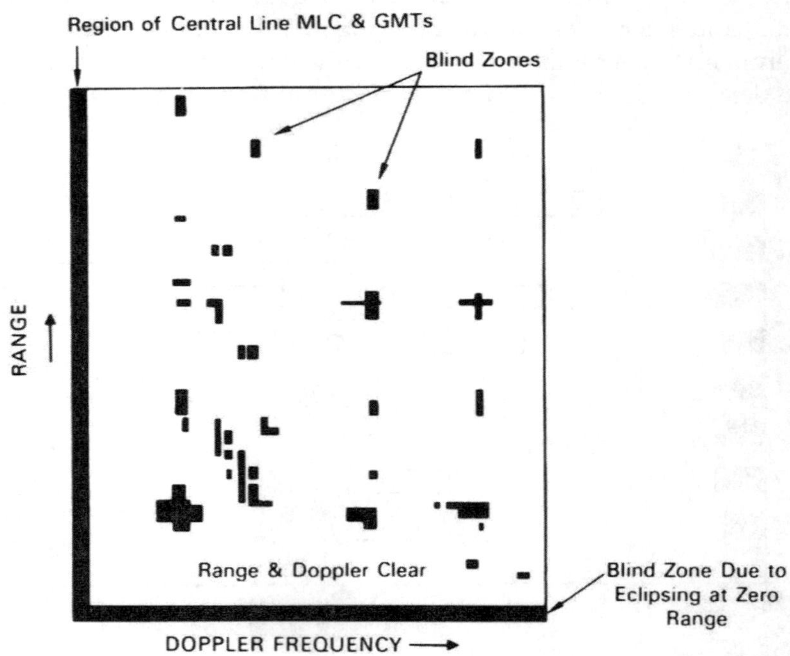

Figure 7.21 Range-Doppler regions in medium-PRF radar that are clear on at least three of eight PRFs. (*From:* [8, p. 448].)

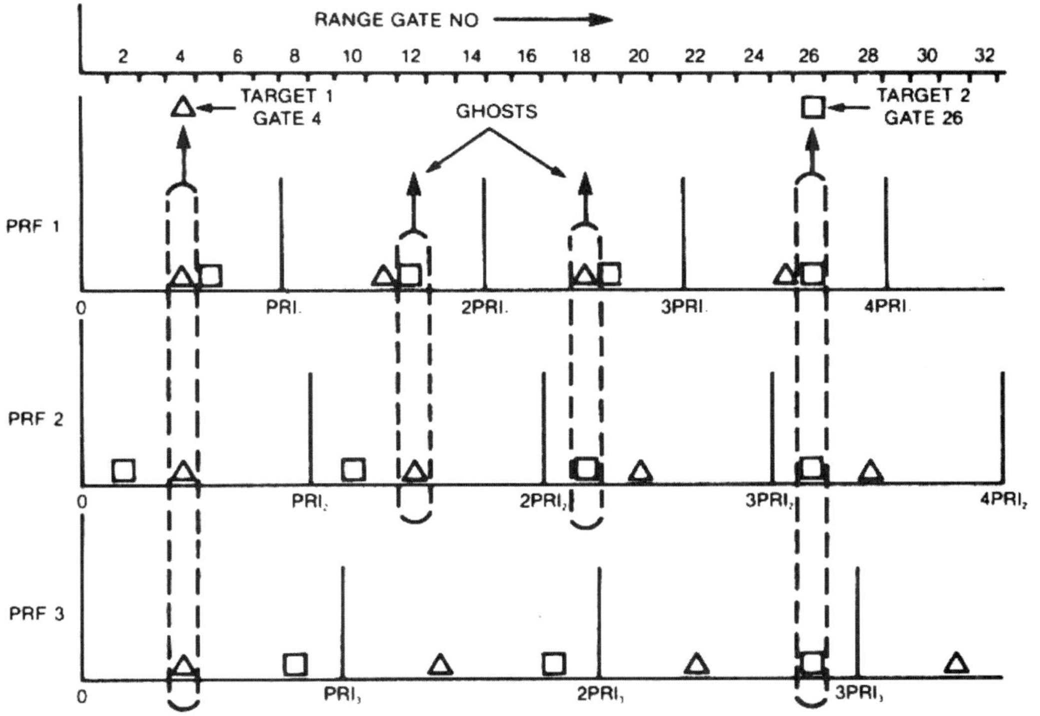

Figure 7.22 Spurious target generation in medium-PRF radar. (*From:* [11, p. 307].)

correct targets in range gates 4 and 26, but generates spurious targets (ghosts) in range gates 12 and 18. In a practical situation, more range gates are used (i.e., higher range resolution), thus tending to minimize ghosts. The PRFs then must be carefully chosen, not only to resolve the range unambiguously, but also to minimize the number of possible cross correlations when dealing with multiple targets.

To analyze the probability of detecting targets in the clear regions of a medium-PRF design, a visibility factor (F_a) in range-Doppler space can be defined as [18]

$$F_a = \left(1 - \frac{\Delta r}{R_u}\right)\left(1 - \frac{\Delta_v}{v_b}\right) \tag{7.12}$$

where Δr is the blind range zone, and Δ_v is the blind velocity zone within a PRF interval, and R_u and v_b are the unambiguous range and velocity responses of the radar (i.e., $R_u = c/2$ PRF, $v_u = \lambda$ PRF/2). During the radar's dwell time on target (T_d), m bursts of different PRFs are transmitted with average duration $T_f = T_d/m$. The detection probability (P_1) per burst will then be $F_a P_d$, where P_d is the detection probability per burst during period T_f, assuming full visibility. In an observation interval $T_d = mT_f$, assuming independent trials, the cumulative probability of at least n successes in m attempts is given by

$$P_c = 1 - \sum_{l=0}^{n-1} \binom{m}{i} P_1 i(1 - P_1)^{m-i} \tag{7.13}$$

For example, if the P_d on each look were 80% and the visibility factor was 50%, the cumulative probability for at least three detections in a burst of eight PRFs would be $P_c = 1 - 0.3 = 0.7$. Equation (7.13) illustrates that, as F_a decreases, the number of PRF bursts (m) must be increased to obtain a high cumulative probability. However, shortening each burst reduces the P_d of each trial and may further reduce the visibility factors in range and Doppler. Optimization of the waveform and the resulting performance ultimately depend on the intensity of the clutter and the stability of the radar system.

The duration of the coherent burst depends on the number of Doppler cells that must be formed between PRF lines and the PRF. Also, coherent processing cannot start until the first echo from the main beam maximum range clutter is received. This restricts the minimum burst period (T_f), particularly for long-range surface radars that might experience chaff or weather clutter at long range and airborne radars when their beam is pointed toward the horizon [18]. If the maximum clutter time delay is $t_{dc} = 2R_c/c$, then the portion of the burst that contributes to the processed output has a duration of

$$T_f' = T_f - t_{dc} \tag{7.14}$$

and the Doppler filter bandwidths and coherent integration gains will depend on T_f' rather than T_f. As a rule of thumb, to use a T_f greater than $4t_{dc}$ is generally desirable [18].

7.2.1 Waveform Duty Factor Considerations [8, 11]

In general, a trade-off must be made in each of the waveform choices (i.e., low, medium, or high PRF) between a high-peak-power and a low-peak-power transmitter. Because the average transmitter power is generally established by the detection requirements of the radar, this choice determines its duty factor [see (7.5)]. In high- and medium-PRF radars, the duty factor affects the range visibility factor (Δr) due to target eclipsing, which ultimately affects the radar's detection probability [see (7.13)]. Where mixed modes are required (e.g., high- and medium-PRF), the choice often is mandated by the requirement for a low-peak-power design in the high-PRF mode because building a coherent transmitter that can simultaneously support both high- and low-power designs is difficult. When this is the case, pulse compression is usually dictated, which allows the average power to be attained in the medium-PRF mode while maintaining the desired range resolution.

An advantage of a high-PRF design is that high average transmitted power can be achieved by simply increasing the PRF. This means that pulse compression is generally not required to attain the desired average power. The resulting duty factor in an airborne design might be as large as between 0.3 and 0.5. When operating at these high duty factors, a considerable fraction of target return is lost due to eclipsing (echoes being received in part or wholly when the radar is transmitting and the receiver is blanked).

Eclipsing sufficiently reduces the effective S/N ratio to leave periodic holes of appreciable size in the radar range coverage. The reduction is determined by the amount of blocking due to the relative timing of the target return and receiver blanking. A target is totally eclipsed only when the period during which the echoes are received coincides exactly with the period during which the receiver is blanked.

The reduction in S/N ratio due to eclipsing for an approaching target is depicted in Figure 7.23. A rapidly approaching target will not remain at an eclipsed range for a long period of time. Also, the decrease in range, which causes the signal strength to increase, fills the gaps in range coverage. In scanning high-PRF radar, a rapidly approaching target may move from one eclipsed region to another between scans, thereby creating a "blind speed" effect [19].

In applications where closing rates may be comparatively low and more continuous detection is required, the length of time that any one range remains eclipsed may be reduced by switching the PRF among different values in a manner analogous to the elimination of blind range zones in a medium-PRF design. In single-target tracking, periodically changing the PRF at appropriate times can generally keep a target in the clear, although the holes in range coverage are not easily eliminated, particularly at short ranges.

Medium-PRF waveforms operate at duty factors that range from 0.01 to 0.03 so that either a high-peak-power or low-peak-power transmitter with pulse compression can be used. In either case, a short processed pulsewidth is required in a medium-PRF waveform to resolve range unambiguously, and also for target range resolution requirements, so that pulse compression is required when using long transmitted pulsewidths. Both pulse compression and long transmitted pulses have deleterious effects on radar performance in a medium-PRF waveform.

Most current medium-PRF radars use binary phase coding to implement pulse compression, although these waveforms suffer from being Doppler sensitive, causing difficulty in achieving low-range sidelobes. The Doppler sensitivity is caused by the shift in phase of individual bits of the phase code by the target Doppler so that, in the extreme, if the last bit is shifted by 180°, the code is no longer matched. A target Doppler that causes this shift is known as the *critical velocity* ($v_c = \lambda/4b\tau$ [m/s]). At this point, large time sidelobes result and the compressed pulse amplitude is distorted. For example, with a 13-bit Barker code using 1 μs bits at X-band,

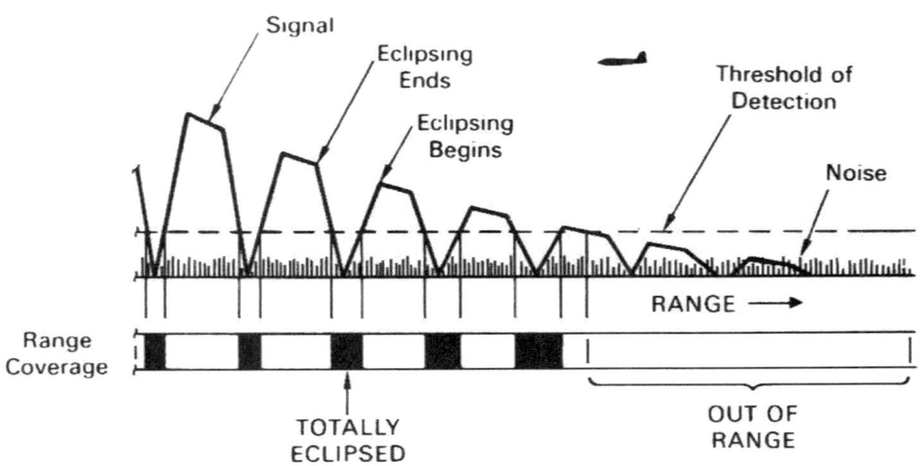

Figure 7.23 Reduction of S/N ratio in a high-PRF radar due to eclipsing. (*From:* [8, p. 465].)

the critical velocity is 1,215 knots. Reference [20] provides MATLAB program pcmissa.m, which demonstrates that the output pulse is distorted beyond recognition for an $f_d/B = 0.07$ where B is the compressed bandwidth.

The time sidelobe requirement on the compressed pulse is typically on the order of 30 dB with a processing loss of 1 dB. However, the sidelobe levels are increased during eclipsing, which generates a source of false targets, thereby causing difficulties in the range correlation portion of the medium-PRF processor. Polyphase coding mitigates both the Doppler sensitivity losses and the higher sidelobe levels, but is difficult to implement.

Additional problems arise in a medium-PRF design using a long transmitted pulse: The eclipsing of short-range targets and those near each ambiguous range and reduction of the range visibility factor (Δr). The short-range target problem can be alleviated by interlacing a short-pulse mode into the waveform. The visibility problem can be solved by increasing the number of bursts in the PRF diversity code, but this lengthens the overall waveform.

In summary, pulse compression can be used to achieve the required processed short received pulsewidths in a medium-PRF waveform, but the problems with Doppler sensitivity, processing losses, time sidelobes, blind ranges, and complexity make this less than an ideal choice. Pulse compression is most compatible with low-PRF designs, where the only fundamental problems are concerned with the detection of short-range targets and filter mismatch due to target Doppler shift.

7.2.2 Antenna Sidelobe Requirements [11, 21, 22]

An important factor limiting the performance of airborne PD radars is the clutter that enters the radar through the antenna's sidelobes. Each possible waveform option responds differently to sidelobe clutter and, hence, the performance requirement on antenna sidelobes is different for low-, medium-, and high-PRF waveforms. The essence of the effect is that sidelobe surface clutter returns are spread throughout the frequency region $f_d = \pm 2v_r/\lambda$ by the Doppler shifts induced onto the clutter scatterer returns, which are illuminated by the omnidirectional antenna sidelobes. In low-PRF designs, sidelobe clutter is aliased into the target Doppler region by the highly ambiguous Doppler associated with this waveform. In high-PRF designs, short-range (R_c) sidelobe clutter is magnified by the factor $(R_t/R_c)^4$ as it competes with long-range target returns (R_t) due to the highly ambiguous range associated with this waveform. In a medium-PRF design, sidelobe clutter is aliased into target Doppler regions and short-range clutter is magnified, but to a lesser degree than in low- or high-PRF designs. The medium-PRF waveform design problem then becomes one of finding PRF choices that provide range and Doppler regions in which the sidelobe clutter is sufficiently benign to allow target detections.

For a low-PRF design, sidelobe clutter appears in the same range cell as the target. The highly ambiguous Doppler response causes the sidelobe clutter PSD to be approximately uniformly distributed between PRF lines (see Figures 2.4 and 2.5). The MTI improvement factor (I_f) is the ratio of the signal-to-clutter ratios at the input and output of the MTI canceller ($I_f = (\mathrm{S/C})_o/(\mathrm{S/C})_{in}$), which can be expressed as $I_f = \overline{G} \cdot \mathrm{CA}$, where $\overline{G}$ is the MTI's power gain (see Table 3.1) and CA is the clutter attenuation. When sidelobe surface clutter is present, the input clutter is the sum of the main lobe clutter (P_{MLC}) and the sidelobe clutter (P_{MLC}/CA), whereas the sidelobe clutter is primarily in the passband of the MTI and passes unattenuated through it. The target power (P_t) is enhanced by the power gain of the MTI, resulting in an output target power equal to $\overline{G}P_t$. The resulting MTI improvement factor is thus given by

$$I = \left[\frac{\overline{G}P_t}{P_{\text{MLC}}/(\text{CA} + P_{\text{SLC}})}\right] \bigg/ \left[\frac{P_t}{(P_{\text{MLC}} + P_{\text{SLC}})}\right] \tag{7.15}$$

which for CA $\to \infty$ and $P_{\text{MLC}}/P_{\text{SLC}} \gg 1$ can be simplified to

$$I = \frac{\overline{G}P_{\text{MLC}}}{P_{\text{SLC}}} \tag{7.16}$$

Further, the ratio $g^2 = P_{\text{MLC}}/P_{\text{SLC}}$ is proportional to the two-way integrated antenna sidelobe power gain of the antenna, which expresses the ratio of the power radiated and received through the antenna's main lobe to that radiated and received through the antenna's sidelobes. Therefore, the maximum MTI improvement factor is given by

$$I_m = \overline{G}g^2 \tag{7.17}$$

where g is the one-way integrated sidelobe (rms) power gain of the antenna with respect to an isotropic antenna (e.g., ordinary, 0 to –5 dBi; low, –5 to –20 dBi; ultralow, –20 dBi; and lower, $g = 1/\overline{G}_{\text{SL}}$).

When a Doppler filter bank is employed in cascade with the MTI canceller in a low-PRF PD design, the improvement factor of the processor is further enhanced because the Doppler filter containing the target excludes a portion of the sidelobe clutter. If the coherent processing interval is T_c seconds long ($T_c \leq T_d$, the target dwell period), $n_f = 1/\text{PRF} \cdot T_c$ Doppler filters can be formed between PRF lines. The resulting maximum improvement factor is thus given by

$$I_m = \frac{\overline{G}g^2}{\text{PRF} \cdot T_c} \tag{7.18}$$

where $\overline{G}$ is the MTI power gain, g is the integrated one-way power antenna sidelobe gain, T_c is the CPI of the radar, and PRF is the radar's pulse repetition frequency.

Note that the improvement factor defines the relative improvement in the S/C ratio due to the processor. The radar's probability of detection can be enhanced, not only by greater processing gain, but also by increasing the S/C power ratio into the processor. This is accomplished in low-PRF PD radars by increasing the transmitter bandwidth that provides finer range resolution, thereby allowing the unambiguous range to be divided into more range cells [see (4.8)].

When using a high-PRF waveform, often only the Doppler spectrum that is clear of sidelobe clutter (see Figure 1.14) is processed, in which case the antenna sidelobe clutter level is of no consequence. In that event, some high-PRF systems use a uniformly weighted antenna aperture to maximize antenna gain and operate with the resulting high sidelobes. If the high-PRF radar operates at relatively high altitudes (e.g., the AEW mission), the sidelobe clutter may be reduced sufficiently to allow detection of targets in this velocity region.

The accurate computation of sidelobe clutter in high-PRF airborne PD radar involves complex geometrical relationships that generally must be handled through a computer program [23–26]. The computational procedure generally involves the integration of sidelobe returns from the ground (range gated by the radar's pulsed response) that occur between isodops (contours of equal Doppler response) corresponding to the frequency extremes of the passband of a Doppler filter (see Figure 1.18 and Section 1.3.4.1).

An idea of the magnitude of the sidelobe clutter for a high-PRF design can be obtained through a rough approximation of the isodop area that is developed as follows [19]. Figure 7.24 depicts the geometry for high-PRF airborne radar by using a flat earth approximation. The area between isodops is approximated as the difference between circular areas (antenna beams projected onto the ground), the diameters of which extend from the intersection of the antenna's boresight with the surface (formed by antenna depression angle ϕ) and two elevation angles ϕ_1 and ϕ_2 that represent the angular difference between two isodops. The clutter area between isodops for the approximation can be shown to equal

$$A_c = \frac{\pi h^2}{4 \sin \phi}\left[\left(\cos \phi - \frac{\sin \phi}{\tan \phi_2}\right)^2 - \left(\cos \phi - \frac{\sin \phi}{\tan \phi_1}\right)^2\right] \qquad (7.19)$$

where h is the radar's altitude, ϕ is the depression angle of the radar to the target, ϕ_1 is the elevation angle to one isodop contour, and ϕ_2 is the elevation angle to a second isodop contour. The average Doppler frequency of the Doppler filter containing the sidelobe clutter represented by A_c is given by

$$f_{\mathrm{SLC}} = \frac{2v_a \cos \phi_a}{\lambda} \qquad (7.20)$$

where ϕ_a is the average elevation angle to the clutter ($\phi_a = \phi_1 + \phi_2/2$).

The Doppler frequency range containing the clutter area A_c is given by

$$\Delta f = f_1 - f_2 = \frac{2v_a}{\lambda}(\cos \phi_1 - \cos \phi_2) \qquad (7.21)$$

which can be solved for the angular difference $\Delta\phi = \phi_2 - \phi_1$ represented by the Doppler filter width $\Delta f = 1/T_c$ as

$$\Delta\phi = \frac{\lambda}{2v_a T_c \sin \phi_a} \qquad (7.22)$$

where v_a is the aircraft's velocity, λ is the radar's wavelength, and T_c is the coherent processing interval.

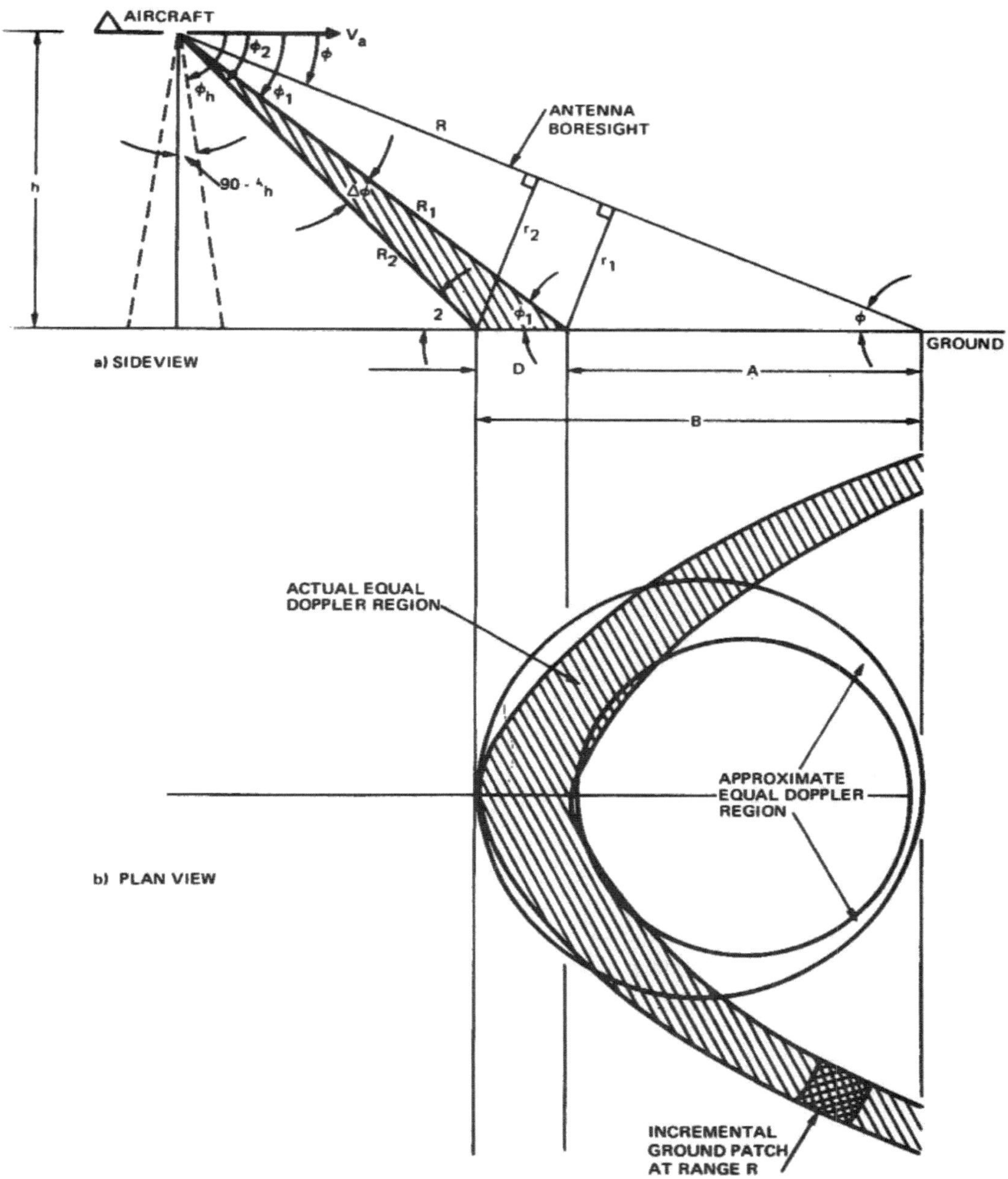

Figure 7.24 Approximation of sidelobe clutter area in a high-PRF PD radar. (*From:* [19, p. 197].)

As an example, consider an AEW radar (f = 1.3 GHz) at an altitude of 25,000 ft with a velocity of 250 knots that is to detect a low-flying 1-m^2 target at 150 nmi in the sidelobe clutter region. Referring to Figure 7.24, the depression angle can be calculated as $\phi = \sin^{-1}(h/R_t + R_t/2r_e) = 2.5°$ (r_e = 4,587 nmi). The Doppler shift of the main beam clutter is given by $f_{MBC} = 2v_a \cos\phi/\lambda$ = 1,113 Hz. Assume that the target Doppler is –500 Hz, which would occur for an aircraft moving away from the AEW radar at approximately 360 knots. The problem is to compute the S/C ratio as a function of antenna sidelobe levels for a Doppler filter set at a –500 Hz center frequency in the sidelobe region.

The first step in the approximate calculation is to compute the elevation angle (ϕ_a) to the isodop from (7.20) as $\phi_a = 63.34°$. Because $\phi_a = \phi_1 + \phi_2/2$, (7.22) can be solved to determine ϕ_1 and ϕ_2 using the further assumption that the coherent processing interval is 30 ms. This results in $\Delta\phi = \phi_2 - \phi_1 = 1.91°$, $\phi_1 = 62.4°$, and $\phi_2 = 64.3°$. Equation (7.19) can then be solved for the clutter area that appears within the 33 Hz ($B_D = 1/T_c$) wide Doppler filter at 500 Hz center frequency as $A_c = 3.7 \times 10^6$ m^2. For farmland terrain ($\sigma^0 = -20$ dB at $\phi = 60°$; see Figure 4.9), the equivalent clutter cross-sectional area is then $\sigma_c = 3.7 \times 10^4$ m^2 modified by the duty factor (d_u) of the PD radar, which is assumed to be $d_u = 0.1$, resulting in $\sigma_c d_u = 3.7 \times 10^3$ m^2. The average range to the sidelobe clutter isodop can be found by averaging $R_1 = h/\sin\phi_1$ and $R_2 = h/\sin\phi_2$ as $\overline{R}_c = R_1 + R_2/2 = 4.6$ nmi. This results in an additional clutter magnification because the returns from short-range clutter ($\overline{R}_c = 4.6$ nmi) and the long-range target ($R_t = 125$ nmi) are both folded into the same Doppler filter. The magnification is given by $\left(R_t/\overline{R}_c\right)^4 = 5.5 \times 10^5$ so that the clutter return in the Doppler filter exceeds the target return by C/S $= \sigma_c/\sigma_t = 2 \times 10^9 = 93$ dB, where both clutter and target are illuminated by an omnidirectional antenna. To define the ratio of the one-way power main lobe antenna gain (G_{ML}) to the one-way power sidelobe antenna gain (G_{SL}) as $\hat{g} = G_{\text{ML}}/G_{\text{SL}}$ would require $\hat{g} = 46.5$ dB sidelobes be deployed. Sidelobes in this region require the use of ultralow sidelobe antennas (i.e., $G_{\text{SL}}/G_{\text{ML}} >$ –40 dB).

In a medium-PRF airborne radar design, sidelobe levels are of great importance and have a direct bearing on system performance. Some insight can be gained by considering the *range extent* of sidelobe clutter for constant antenna sidelobe levels. Figure 7.25 depicts the sidelobe clutter extent for a typical medium-PRF system versus radar altitude, where the range extent is measured in terms of 1 μs range gates [11]. Range extent is defined as the difference in range from the altitude line to the range at which sidelobe clutter equals thermal noise. As is evident if we consider that the average number of 1 μs gates in an interpulse period for an AI radar (PRF = 10 kHz) is about 90, when the sidelobe levels go above 45 dB, a significant portion of the unambiguous range interval is covered by sidelobe clutter. For example, with a 35 dB sidelobe level at 10,000 ft of altitude, 64 gates, or 71% of the unambiguous range interval, have sidelobe clutter greater than thermal noise. This would have a severe effect on the detection performance at those target Dopplers with minimum PRF visibility.

A further consideration in medium-PRF radars involves the detection of large discrete targets in the radar sidelobes. These targets have Doppler frequencies induced by aircraft motion that are aliased into target Doppler regions by the ambiguous Doppler response of the medium-PRF design. In addition, the ambiguous range of the medium-PRF waveform causes short-range sidelobe targets to fold into the same range gates as long-range targets. This situation is further exacerbated in airborne radars by reflection lobes caused by the main radar beam bouncing off the aerodynamically shaped radome surface into the sidelobe region.

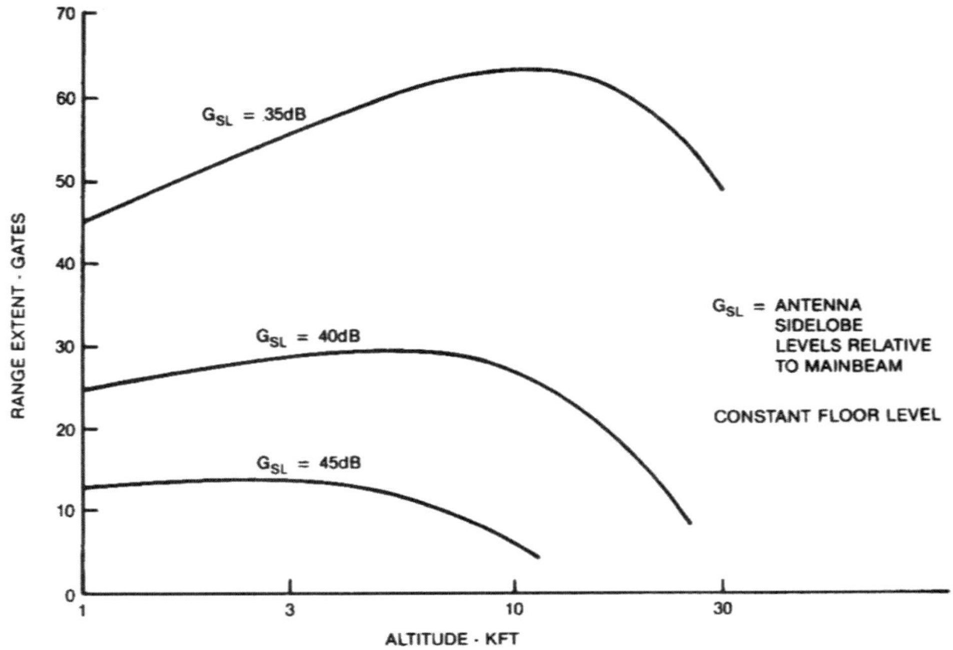

Figure 7.25 Extent of sidelobe clutter in medium-PRF PD radar. (*From:* [11, p. 309].)

For example, if medium-PRF airborne radar can detect a 1 m^2 target at 30 nmi, it also can detect a sidelobe target at 2 nmi of about 200 m^2, assuming a relative sidelobe one-way power gain of –35 dB. One method of dealing with these unwanted targets is to provide the radar with a guard channel, as depicted in Figure 7.26. The guard channel receiver blanks the sidelobe target if its response is greater in the guard band channel (the guard band antenna response exceeds the main lobe antenna's sidelobe response) than in the main channel [8]. Another method of suppressing unwanted sidelobe targets is through an amplitude comparison scheme with a single receiver channel [11]. In this method, the unambiguous range to the sidelobe target is first resolved. A discrete target in the sidelobes can generally be detected only at short ranges due to the two-way attenuation property of the antenna sidelobes. However, bona fide targets in the main lobe generally provide

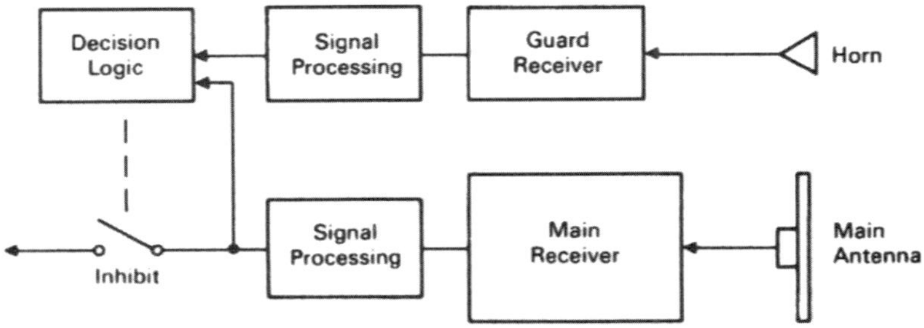

Figure 7.26 Guard channel receiver in medium-PRF PD radar. (*From:* [8, p. 452].)

a large signal at short ranges. Thus, if the detected signal's resolved range is small and its amplitude is smaller than that expected from a short-range minimum size target (e.g., 1 m^2), then the target is assumed to be a spurious sidelobe discrete target.

► Example 25 Isodops

Isodops are lines of constant Doppler shift on the ground observed from airborne radar. The spectrum of the ground clutter can be determined by superimposing range resolution rings on the isodops.

Consider airborne radar moving at velocity v_{ac} directed along the x axis at an altitude h. The velocity vector directed at the point x-y on the ground is given by

$$v_{xy} = v_{ac} \cos \theta_{azo} \cos \phi_{elo}$$

where θ_{azo} is the azimuth angle and ϕ_{elo} is the depression angle relative to the local vertical.

The isodop geometry is depicted in Figure EX-25.1.

Define the following relationships:

$$k = \frac{v_{xy}}{v_{ac}} \qquad \cos(\theta_{azo}) = \frac{x}{xx} \qquad \cos(\phi_{elo}) = \frac{xx}{vx}$$

where

$$xx = \sqrt{x^2 + y^2} \qquad vx = \sqrt{xx^2 + h^2}$$

Then the equations can be solved for y in terms of x, k, and h as follows:

$$y = \sqrt{x^2 \left(\frac{1 - k^2}{k^2}\right) - h^2}$$

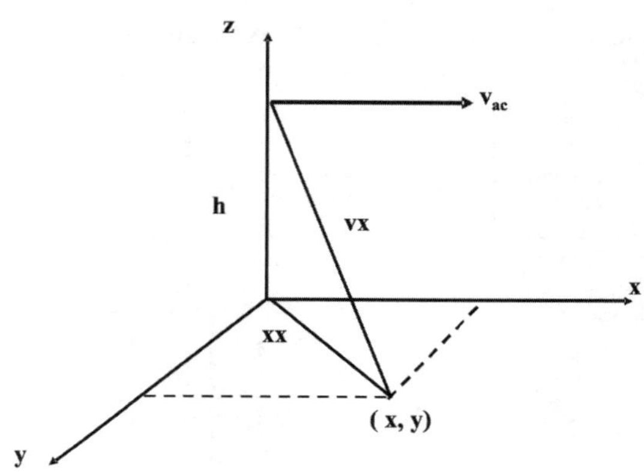

Figure EX-25.1 Isodop geometry.

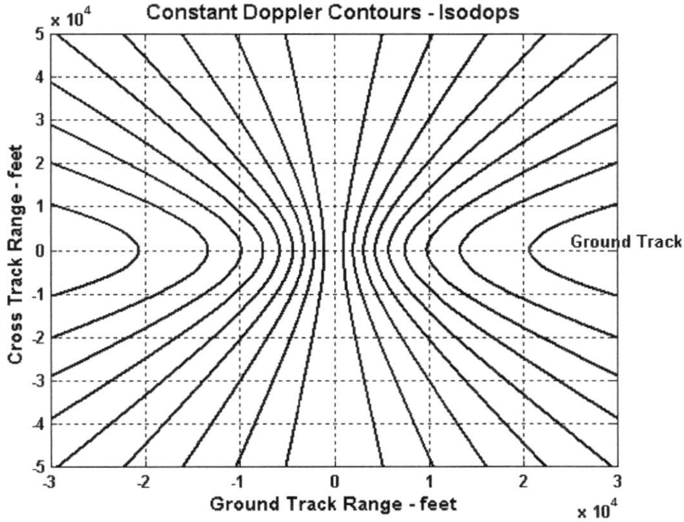

Figure EX-25.2 Constant Doppler isodops from airborne radar.

MATLAB program isodop.m plots the isodops from $x = x_{\min}$ to 30,000 feet where $h = 10{,}000$ feet for $k = 0.1, 0.2, \ldots, 0.9$ and

$$x_{\min} = \frac{h}{\tan(\phi_E)} \qquad \phi_E = \arccos(k)$$

Figure EX-25.2 shows the isodop plots.

▶

7.3 Modern Airborne Interceptor Radars

One of the primary applications of PD radar is in airborne radars that are used in Airborne Intercept (AI) aircraft. These aircraft function to intercept and destroy hostile aircraft and enemy surface assets such as SAM sites. They are generally guided to the vicinity of the threat via a secure data link from the air defense operations center or from the AWACS platform described in Chapter 1. The AI weapon control radar then acquires and locks onto the target and an appropriate missile is launched to destroy the target. Air targets are destroyed using air-to-air missiles that are either active or semiactively radar guided such as the AMRAAM or Sparrow missiles. Alternatively IR missiles such as the Sidewinder can be used for shorter range applications that home in on heat radiated from the adversary aircraft. Ground targets are destroyed using either laser or GPS-guided missiles or bombs. Air survivability dictates that the AI radar must have superior range capability than its adversary in combat engagements. In a head-to-head combat engagement, the aircraft that can fire its missiles first generally is the aircraft that survives. The challenge in AI radar is to maximize its detection range considering the limited *average power-aperture product* dictated by the small diameter available for the radar antenna and limited prime power available from the aircraft's generator while minimizing its own RCS. Other considerations are the small antenna aperture that dictates an operating frequency in X- or K_u-band to provide reasonable angular resolution and the large surface clutter that exists under look-down/shoot-down conditions.

High-PRF PD radar provides several advantages in this application. First, it provides good nose-aspect detection capability against closing targets whereby detection is limited by receiver noise rather than clutter. High average power can be provided without the complication of pulse compression by controlling the PRF to provide a high duty cycle. Main lobe clutter, slow-moving ground targets, chaff, and weather clutter can be rejected without also rejecting air targets. Limitations include degradation of detection range against low closing rate and receding targets due to sidelobe clutter, and that range must be determined using multiple PRFs or frequency modulation of the transmitter frequency, and altitude-line clutter must be rejected by Doppler filtering rather than by range gating. High-PD PRFs generally range from 100 to 300 kHz at X-band frequencies. Eclipsing losses can be significant when high duty cycles are employed.

Some of the limitations of high-PRF PD radars can be mitigated by using a medium PRF that ranges from 8 to 16 kHz at X-band frequencies. The main advantage of medium-PRFs is that they provide the capability of detecting targets in tail-chase situations due to their ability to cope with both main beam and sidelobe clutter. Ground-moving targets can be eliminated so that they are not confused with airborne targets. Also it is possible to determine the range of targets by target delay measurements using the multiple PRFs associated with this technique.

The major limitation is that both low- and high-speed targets must be detected in sidelobe clutter thereby reducing the maximum detection range performance provided by high-PRF PD radar. Both range and Doppler ambiguities must be resolved. Also a special guard channel antenna and receiver must be provided to eliminate strong ground-based discrete clutter. Pulse compression may be necessary to increase the duty factor to that needed to achieve the requisite detection probability.

Generally AI radars use a bar raster scan antenna search pattern where the beam is sequentially swept through the azimuth coverage and then stepped through the elevation coverage in a number of steps (usually eight). To provide all-aspect coverage through the search volume, high- and medium-PRF modes are interlaced from one scan to another. Low PRFs that range from 250 to 4,000 Hz at X-band are employed in real-beam and synthetic aperture ground mapping applications.

Table 7.1 lists a number of operational AI radars and the PRF modes that they employ. Figure 7.27 shows the APG-70 used on the F-15 aircraft. Several antennas are apparent in the figure. A mechanically scanned X-band flat plate planar array antenna transmits and receives the main radar signals. The planar array antenna consists of a contiguous array of waveguide radiators with slots distributed over the surface of the flat plate. The distributed power is adjusted across the array to minimize antenna sidelobes, and its efficiency of more than 90% concentrates most of the radiated power in the antenna's main lobe. Generally antenna radiator elements are spaced $\lambda/2$ apart to minimize the effects of grating and Bragg lobes. When this is done the RCS of the antenna is that of a flat plate, which represents

Table 7.1 Operational Airborne Intercept Radars

Type	*Platform*	*Mission*	*Antenna*	*Freq.*	*Trans*	*PRF Modes*
APG-63/70	F-15	Air combat superiority	Planar array	X-band	TWT	High, medium, low
APG-66/68	F-16	Air combat superiority	Planar	X-band	TWT	High, medium, low
APG-65/73	F-18	Air combat superiority	Planar array	X-band	TWT	High, medium, low
APQ-164	B-1B	Ground target attack/air	PESA	X-band	TWT	High, medium, low
APQ-181	B-2	Ground target attack/air	PESA	K_u-band	TWT	LPI
APG-77	F-22	Air combat superiority	AESA	X-band	SS	LPI
APG-81	F-35	Ground target attack/air	AESA	X-band	SS	LPI

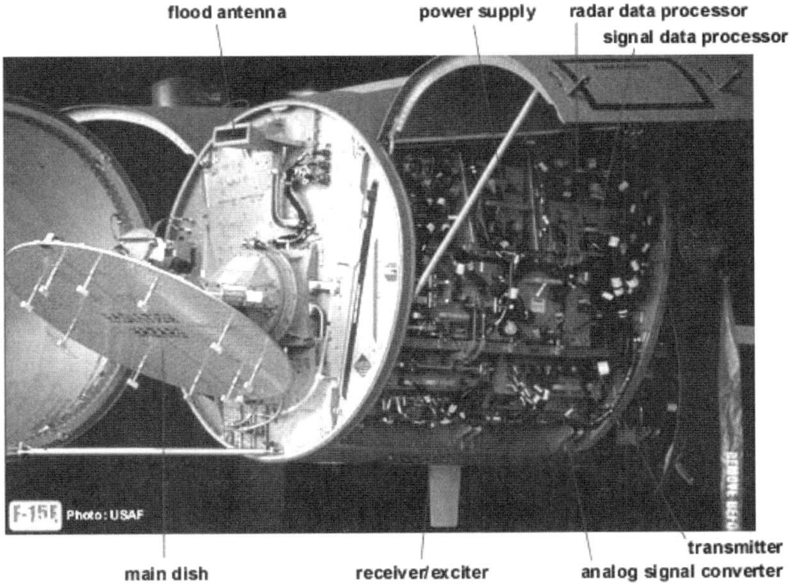

Figure 7.27 APG-70 airborne intercept radar. (*From:* [27, 46].)

a large improvement over previously employed parabolic reflector antennas in certain applications. Their RCS is the same when out-of-band illuminating energy is directed at the boresight of either the flat plate or parabolic reflector. However, in the case of surface-based acquisition radars, the fact that the parabolic reflector refocused the incident energy in the direction of its boresight while the flat-plate antenna reflects the energy so that the angles of incidence and refection are equal causes the RCS of the flat plate to be significantly less than that of the parabolic reflector antenna. The other antennas in the figure are IFF stub elements distributed across the flat plate that are tuned to L-band and the fixed guard band horn antenna, which allows medium-PRF PD radars to suppress strong ground target returns.

The capabilities and configuration of the AI radars employed on the F-16 and F-18 are similar to the architecture and performance of the APG-70. They all use gridded coupled cavity TWT transmitters that provide a peak power between 1.5 and 15 kW. The operational modes provided by the APG-70 are described in Table 7.2.

The APG-70 basic functions are to locate air and ground targets for missile attack. The high-PRF *Velocity Scan* mode locates high-closing-rate targets at maximum radar range. These targets poise the greatest threat to AI aircraft. However, this mode does not provide the range to the target. The interlaced *Range-While-Scan* mode can generally detect and range-on most air targets, but at a reduced range due to ground clutter effects. Up to 10 targets can be tracked simultaneously in the TWS mode. For missile control up to two targets can be tracked with a special 3° scan. A low-resolution ground map can be formed to find targets along the ground track of the aircraft. A high-resolution spotlight SAR can be used outside a ±8° blind sector along the aircraft ground track to locate fixed ground targets. Ground moving target activity can be located using the GMTI mode.

Typical parameters of the APG-63 that operates at X-band includes a peak power of about 3.5 kW, a maximum duty factor of 0.43, PRFs of 0.17 to 308 kHz , pulsewidths of 1 to 21 μs, and an estimated antenna gain of 37 dB. The high-PRF velocity search mode operates with PRFs around 310 kHz with a pulsewidth of 1.4 μs. The medium-PRF, long-range search mode operates with PRFs between 10 and 15 kHz and pulsewidths from 13 to 20 μs augmented with 13-bit pulse compression. The low-PRF ground map mode operates with PRFs of 415 to 830 Hz, and pulsewidths from 13 to 20 μs augmented with 13-bit pulse compression.

Typical parameters of the APG-65 that operates at X-band include a peak power of about 1.7 kW, a maximum duty factor of 0.43, PRFs of 0.3 to 310 kHz,

Table 7.2 Operational Modes of APG-70 AI Radar [27]

Mode	*Coverage*	*PRF Modes*	*Special Techniques*
Range-while-scan	All aspect	High/medium	Interlaced RWSH/RWSM
Velocity scan	Medium-high closure targets	High/medium	No capability for low- and no-closure targets
Range-gated high vector	160 nmi max Low RCS targets	High High	Single PRF/altitude-sensitive half scan rate
Track-while-scan	±15° ±30° ±60°	High/medium	Maximum 10 targets in track
Single target track	Tracks single target	Medium	3° scan/semiactive missile illumination
Dual target track	Tracks two targets	Medium	AMRAAM control
Autoacquisition	Special angle search	High/medium	Acquires and tracks targets automatically
Real-beam ground map	Front search	Low	Low-resolution ground map Laser-guided missile support
High-resolution ground map	Squint SAR	Low	Blind ±8° forward
Ground MTI	Ground-moving targets	Low	Moving target activity

pulsewidths of 0.8 to 50 μs, and an estimated antenna gain of 34 dB. The high-PRF velocity search mode operates with a PRF of about 170 kHz and a pulsewidth of 1.8 μs. The medium-PRF, long-range search mode operates with PRFs between 4 and17 kHz and pulsewidths of 13 μs augmented with 13-bit pulse compression. The low-PRF ground map mode operates with PRFs of 420 to 840 Hz, and pulsewidths from 13 to 20 μs augmented with 13-bit pulse compression. The radar includes a GMTI mode with a low 2,400 Hz PRF and pulsewidth of 1.5 μs augmented with 13-bit pulse compression. The *Doppler Beam Sharpening* (DBS) mode uses a low PRF of 1 to 2 kHz with pulsewidths of 0.12 to 1.35 μs augmented with 13-bit pulse compression.

The modes and capabilities of the AI radars on F-16 and F-18 aircraft are similar to those in the F-15. The APG-73 includes Barker-coded 13:1 pulse compression in the medium-PRF mode and DBS.

The first evolution in technology for this type of radar occurred in the X-band multimode APQ-164 radar employed in the B-1B bomber. This low-level ground attack bomber had a head-on RCS of about 10 m^2 dominated by a few components including the reflection off the radar antenna. By employing a phased-array antenna, it is possible to tilt the planar antenna down 30°, thereby directing the reflections from the antenna away from the illuminating radar. This and treatment of a few other components allowed the head-on RCS to be reduced to about 1 m^2. The 44- by 22-inch space fed phased array (PESA) employs 1,526 phase shifter modules to steer the antenna beam. The antenna is mounted on a roll gimbal that allows it to be locked in forward, broadside, and vertical down positions. Beam positions can be switched in 200 ms, allowing multiple modes to be activated simultaneously. One disadvantage of a phased array is that the antenna gain in azimuth and elevation decreases and the beamwidth broadens as the cos θ of the angle off boresight. This means that the maximum two-way gain loss at an extreme scan position of 60° in azimuth and 30° in elevation would be $L = \cos^2(60°)\cos^2(30°) = 0.1875$ or 7.27 dB. Also the sidelobe level increases. The rest of the radar components employed are similar to those used in the APG-68. Special modes employed in the APQ-164 include automatic terrain following and avoidance and high-resolution SAR.

The K_u-band multimode APQ-181 radar employed in the B-2 stealth bomber uses a similar passive electronic scanned antenna (PESA) technique to that employed in the B-1B. The two low RCS radar antennas located on either side of the bat wing are electronically scanned in two dimensions and feature a monopulse feed design. A strap-down inertial platform is used to measure antenna motion during SAR mode operations. The transmitter employs a liquid-cooled gridded TWT similar to those used in the APG-70 and APG-73 radars and includes a pulse compression mode. During the target search acquisition and identification modes the radar operates in a variable-resolution SAR mode to precisely locate and identify assigned targets. The radar has 21 distinct operating modes including terrain avoidance and following. The important distinct feature of the ALQ-181 is the use of a spread-spectrum LPI waveform designed to minimize the interception of the radar's emissions by ES receivers [28].

The newest innovation in AI radar design is the use of active electronically steered array antennas (AESA) using gallium arsenide (GaAs) monolithic microwave integrated circuitry (MMIC). The AESA antenna is made up of more than 1,000 basic transmit-receive (T/R) modules distributed over the antenna aperture. Each module contains an antenna element, T/R switch, power amplifier, low-noise receiver, and digital phase shifter to scan the beam and control circuitry. Losses in both the transmitting and receiving paths are minimized by placing these elements in the antenna. Also the reliability of the radar is increased because up to 10% of the modules can fail before the system fails. Maintenance is also improved because the high-voltage circuits in a TWT transmitter are eliminated and replaced with low-voltage GaAs elements [44, 45].

Figure 7.28 shows the power output, noise figure, gain, and efficiency of a MMIC module [29]. The power available from a module at X-band is on the order of 10W with an efficiency of 45%, while the noise figure is 1 dB. The reported size of a T/R module used in an X-band AESA phased array is 7 cm long by 1 cm wide by 0.5 cm deep. The low efficiency and concentrated packaging mean that substantial power must be conducted away from the module and that traditional forced-air cooling be replaced with liquid coolant pumped through a cold plate and heat exchanger. An issue with AESA phased array antennas is the high cost of T/R modules that currently cost on the order of $2,500 per module. High volume and cost efficiencies should ultimately reduce this cost, but the original objective of $100 per module appears improbable.

The APG-77 in the F-22 and APG-81 in the F-35 are designed using AESA phased array antennas. Figure 7.29 illustrates the APG-77 AESA antenna, which contains 2,000 elements. The APG-77 is a stealth multimode PD radar with missions of air superiority and precision ground attack. The stealth design features an LPI waveform using spread spectrum techniques, tilted low RCS antenna, and 6 AMRAAM missiles, 2 Sidewinder IR missiles, and a 22 mm cannon all carried internally to preserve the stealth characteristic of the aircraft [30].

Separate T/R modules combined in a circulator are employed to form the T/R pair. The transmitting module contains an RF power amplifier, transmit timing control, data transfer, and voltage regulation capabilities. The receiver module contains an LNA, 6-bit digital phase shifter, post-RF amplification, and receiver protection. The complete T/R is formed using six MMIC chips and five application

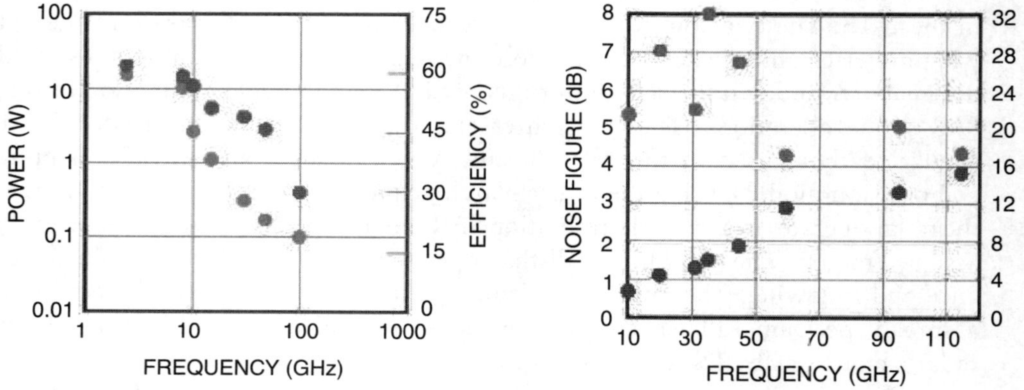

Figure 7.28 Power and noise figure of a MMIC module. (*From:* [29, p. 34].)

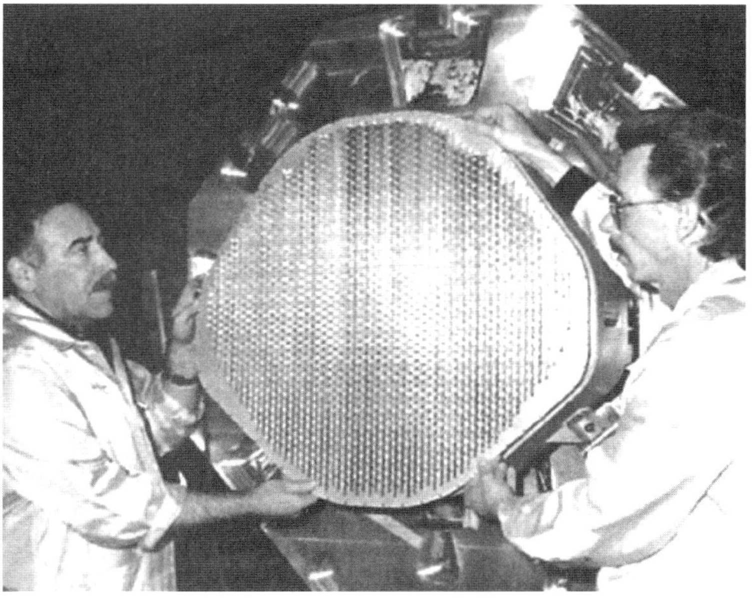

Figure 7.29 APG-77 AESA phased-array antenna. (*From:* [1, p. 550].)

specific integrated circuits (ASICs) and results from a research program that produced a 2 W hybrid module. The various modes illustrated in Figure 7.30 include air and ground search, multiple air target track, weather detection, HPRF velocity search, ISAR target identification, and missile control [30]. The ability of a phased array to move its beam within milliseconds allows multiple modes to be performed simultaneously or interlaced as controlled by the avionics computer. The APG-77 utilizes four guard antennas. Provision for side-panel ASEAs on either side of the aircraft are made to complement the performance of the main array for the large loss experienced when it is scanned to 60° off boresight with a 30° tilt.

The stealth Joint Strike Fighter (JSF) employs the APG-81 multimode PD AESA radar. The air-to-air modes of this radar are similar to those employed in the

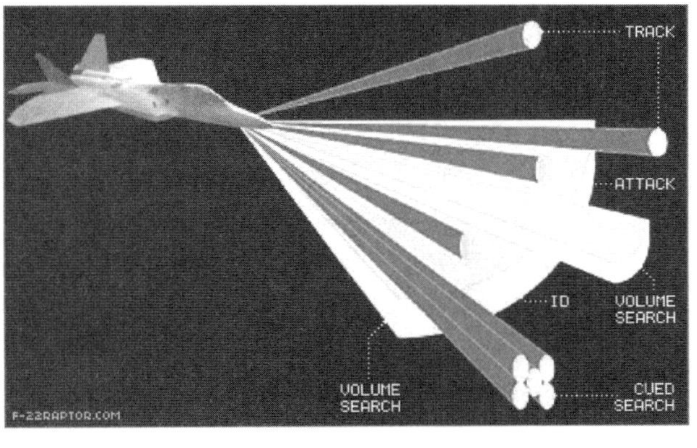

Figure 7.30 APG-77 functional modes. (*From:* [31].)

APG-77, but the air-to-ground mode provides high-resolution mapping, GMTI, and tracking that are similar to that available in the Joint Surveillance and Target Attack Radar System (JSTARS). The circular AESA is depicted in Figure 7.31 mounted on a BAC-111 test bed aircraft. The antenna employs between 1,200 and 1,500 T/R modules of similar design to that of the APG-77. The antenna appears to be tilted up to complement the F-35 stealth design. T/R module antennas are positioned 2.5 cm apart to satisfy grating and Bragg lobe constraints. Figure 7.32 depicts the various air and ground modes provided in the F-35. The AESA radar is complemented with a distributed aperture electro-optical sensor (EOTS) that provides passive detection capability.

AESA radars generally provide greater detection capability than mechanically scanned planar antennas since they minimize RF losses. Also it is possible to modify the aperture distribution function to provide maximum power or low sidelobes. A significant disadvantage of the phased array relative to a mechanically scanned planar array is the substantial scan loss for off-boresight operation. The AESA allows selection of the aperture distribution to accommodate the PRF mode selected. For example, the aperture can be uniformly distributed when a high PRF is selected,

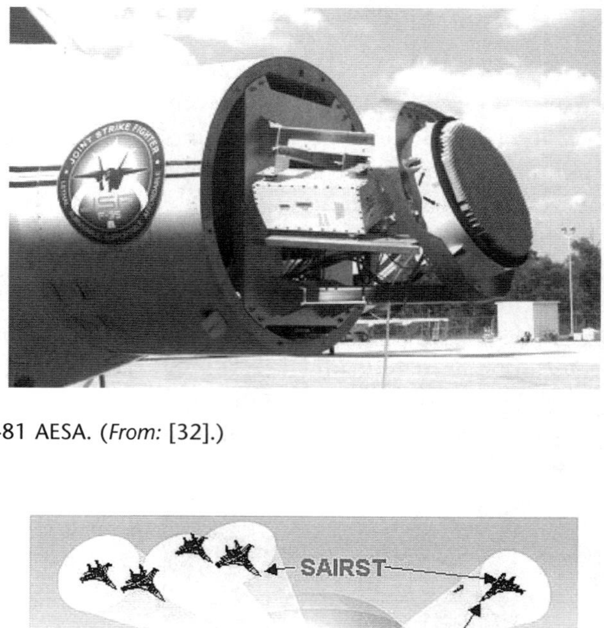

Figure 7.31 APG-81 AESA. (*From:* [32].)

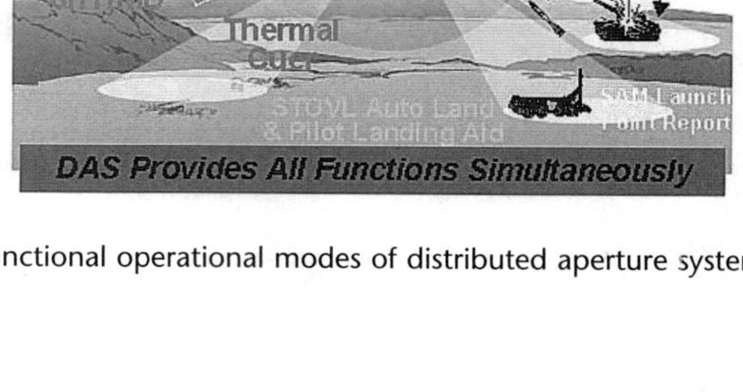

Figure 7.32 F-35 functional operational modes of distributed aperture system. (*From:* [33].)

therefore maximizing the power transmitted, since the processor will eliminate the clutter returns from the high −13 dB sidelobes. On the other hand, medium- and low-PRF operation requires low antenna sidelobes on the order of two-way −50 to −70 dB since the clutter improvement factor is bounded by the integrated two-way sidelobes. The penalty for the low sidelobes required for a medium-PRF design can be assessed using the accompanying figures that apply to a linear array. Figure 7.33 (cheb_bw.m) shows the beamwidth widening generated when a Chebyshev aperture distribution is used to provide low sidelobes. Figure 7.34 (pha_loss.m) depicts the antenna scan loss that results when the antenna beam is scanned off boresight.

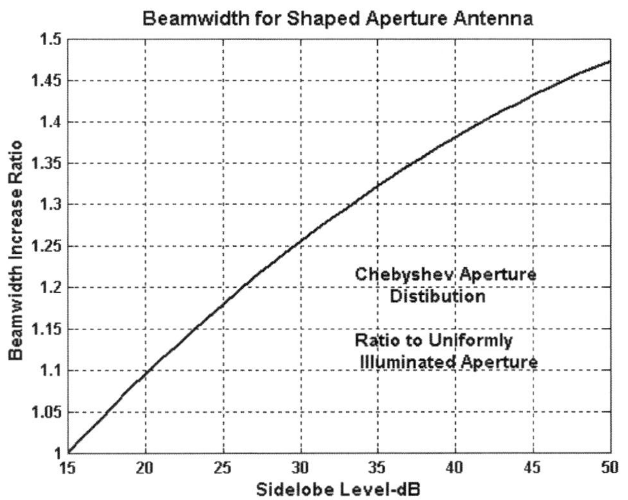

Figure 7.33 Beamwidth widening with Chebyshev aperture shaping.

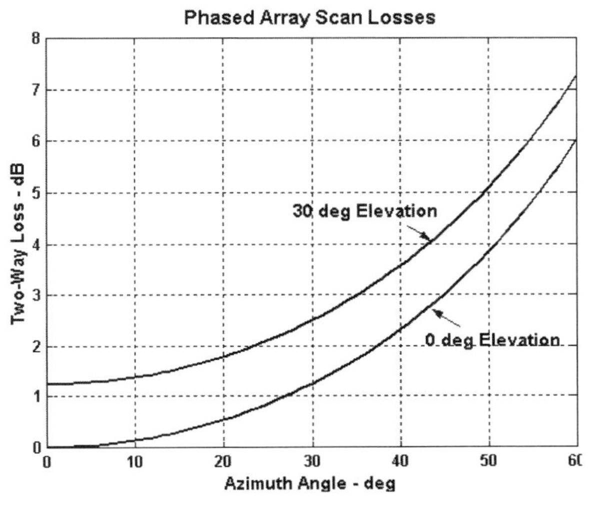

Figure 7.34 Phased-array scan losses (two-way).

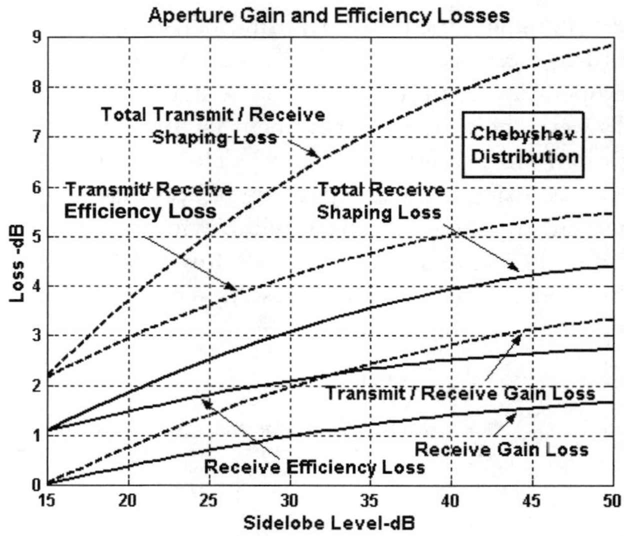

Figure 7.35 Shaped active aperture gain and efficiency losses.

Table 7.3 Characteristics of Operational and Planned AESA Radars [35]

Platform	*Type*	*T/R Modules*	*Max Range (km)*	*Az Track Angles (°)*	*TWS Targets*
F-15C	APG-63(V)3	1,500	144–185	±60	20
F-22	APG-77	1,500	200–230	±60	100
FA/18 E/F	APG-79	1,100	120–130	±60	20
F-16 E/F	APG-80	1,000	100–120	±60	20–50
F-35	APG-81	1,200+	140–160	±60	—
B-2	Dev.	2,000+	—	—	—
Eurofighter	Dev.	1,400	165	±60	—
Global Hawk	—	2,000	—	—	—
JSTARS	—	13,500	—	—	—
Grippen		1,000	100–120	±100	—

Basically, two approaches are used to control the two-way sidelobes. The first, utilized in the Erieye design (see Section 1.1), applies uniform illumination in the transmitting array and shapes the receiving array using a Chebyshev distribution. This has the advantage of transmitting maximum power, but the high transmitter sidelobes are easily detected by an intercept receiver. Also the receiving aperture must provide substantially all of the required sidelobe control. The second approach shapes both transmitting and receiving arrays. This option is most appropriate for LPI designs, but the diminished power in the outer T/R modules reduces the overall transmitted power. Figure 7.35 (ant_beam_loss.m) shows both the antenna gain and illumination efficiency losses for both situations. The F-22 made provision for side panel AESA antennas to compensate for the high losses at the 60° scan angle, but they were not implemented due to the high cost [30].

The use of AESA radars is proliferating because the combat survivability of an AI aircraft depends on the aircraft's performance being comparable to that of their adversaries. The F-15C radar has been upgraded to the APG-63(V) 3/4, the

F/A-18 E/F Super Hornet upgraded to the APG-79, and the F-16 Block 60 upgraded to the APG-80. Developments are in process to convert the B-2 antennas to AESA presumably at X-band rather than the present K_u-band. European developments are proceeding for the Eurofighter and JAS-39 Grippen. Table 7.3 lists estimated parameters for these AESA radars. The maximum range is against a 1 m^2 target [34].

7.4 Resolving Range Ambiguities

A characteristic of medium- and high-PRF PD waveforms is the generation of highly ambiguous range information. When accurate range information is required of these systems, the ambiguous ranges must be resolved. Resolving range ambiguities is not a trivial matter and degradation in the detection range of up to 20% to 25% can result from this process, which generally involves a modulation of the transmitted signal's frequency or PRF [11].

Range ambiguities are generally resolved by using one of two methods. The multiple-PRF method uses a PRF code composed of a number of fixed PRFs. This type of ranging involves sequential measurement of the ambiguous range in each PRF followed by comparison of the measurements to eliminate ambiguities. The second method involves frequency modulating the transmitter by using either a linear or sinusoidal waveform. The time lag between transmission and reception is converted to a frequency difference. By measuring this frequency difference, the time lag and, hence, the range to the target can be determined. Both methods are susceptible to the generation of spurious targets (i.e., ghosts) when multiple targets are present within the same Doppler filter.

The multiple-PRF ranging method generally provides the more accurate range measurement. However, when a high duty factor waveform is employed (e.g., airborne high-PRF design), use of multiple-PRF ranging may not be feasible due to the small number of range gates available to resolve ambiguities. In this case, FM ranging is employed, which is generally less accurate than direct pulse-delay ranging.

Resolution of range ambiguities generally should reduce detection sensitivity from what it would be without ranging. The reduction is primarily due to the need for ranging to divide the total observation time (i.e., the period during which the radar dwells on the target) into a number of segments. For detection, the target generally must be detected independently during each segment of the modulation cycle that resolves the range ambiguities. The time-on-target is thus significantly reduced from the total available observation time. In addition, the probability of detection for each segment must be increased for adequate probability of detecting the target on at least two of the different segments. Both effects cause a significant reduction in the overall probability of detection. A further reduction of observation time may also result from discarding target returns during transient periods associated with changing of the modulation cycle during ranging (e.g., changing of the FM slope during linear FM ranging).

7.4.1 Multiple High-PRF Ranging [19, 36]

In high-PRF airborne PD radars, the use of high PRFs (e.g., 100 to 300 kHz), necessary to prevent clutter from obscuring target Doppler returns, results in highly ambiguous range information. This situation is depicted in Figure 7.36, which shows radar returns for two targets (both ambiguous and unambiguous) that are widely separated in range. Due to the ambiguous range, there is no way of discerning which transmitted pulse caused which target return; hence, true target range is unknown. Another situation that can occur when using a high PRF is the eclipsing of target returns by subsequent transmitted pulses (see Section 7.2).

To overcome these difficulties, multiple PRFs are used sequentially. By observing the target interpulse position in each of the PRFs, the true target range can be deduced. Furthermore, by selecting an appropriate PRF value, the start of an eclipse can be detected and the PRF may be switched to a value that causes the target return to fall within a clear range zone, thereby preventing the eclipse.

The principles involved in a two-PRF ranging system are depicted in Figure 7.37. The PRFs are chosen to have a common submultiple frequency (f_B) with an

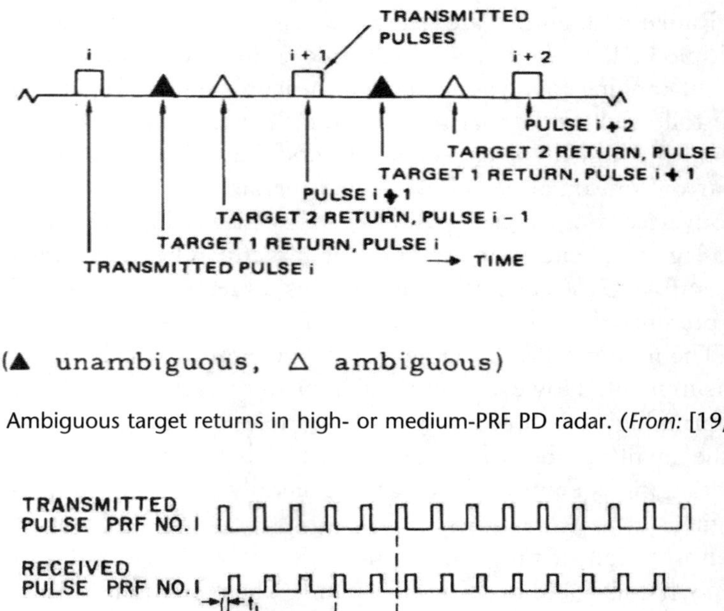

Figure 7.36 Ambiguous target returns in high- or medium-PRF PD radar. (*From:* [19, p. 116].)

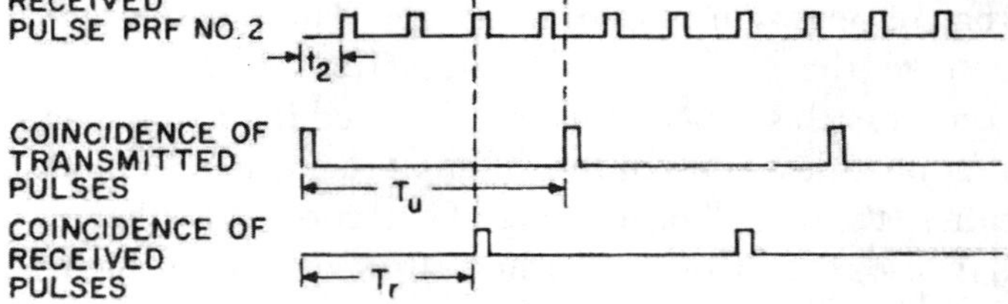

Figure 7.37 Two-PRF ranging system. (*From:* [36, p. 44].)

unambiguous period $T_u = 1/f_B$. The ambiguous range to the target is determined first in PRF_1 and then redetermined for PRF_2. By taking a time comparison of the two ambiguous ranges, made possible by synchronization of the PRFs, we obtain a coincidence pulse that is a measure of the true range.

The common submultiple frequency (f_B) is chosen so that its interpulse period is equal to, or greater than, the expected target range. The minimum value of PRF is chosen to allow a clear-Doppler region free of clutter in which to make a target detection ($\text{PRF} \geq 2f_d + f_t$). The ratio of the PRF to the ranging frequency is set equal to $(N + 1)$ times the ranging frequency. Other values for the second PRF are possible, but this value gives the widest allowed pulsewidth and closest PRF values.

The transmitted pulsewidth has a maximum allowable value, depending on the PRF values and the required resolvable range, depicted in Figure 7.38. From the figure, we can see that

$$(\text{PRF}_1)^{-1} = (\text{PRF}_2)^{-1} + \tau_2 \tag{7.23}$$

where τ_2 is the radar's pulsewidth. The submultiple frequency is given by

$$f_B = \frac{\text{PRF}_1}{N} = \frac{\text{PRF}_2}{N + 1} \tag{7.24}$$

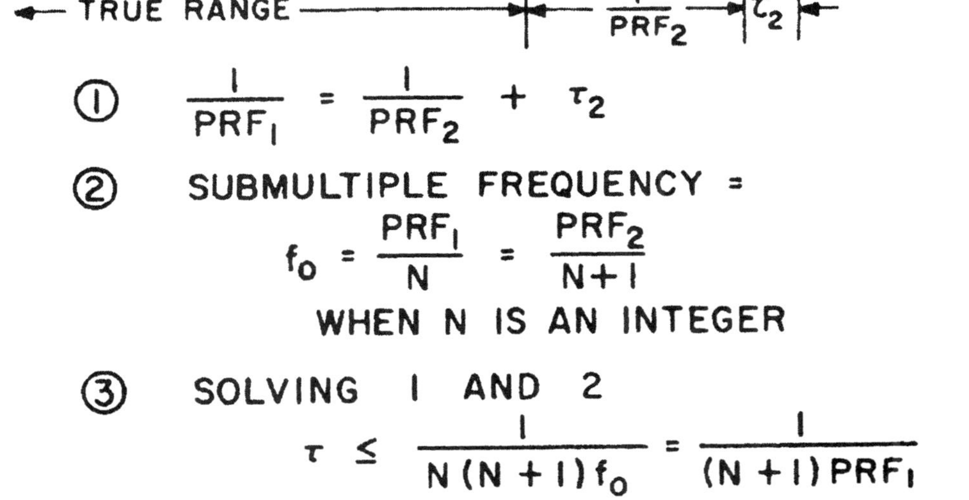

Figure 7.38 Maximum allowable pulsewidth in high-PRF PD radar. (*From:* [36, p. 44].)

where N is an integer. By solving (7.23) and (7.24), the maximum allowable pulsewidth can be determined as

$$\tau_{2\,\max} \le \frac{1}{N(N+1)f_B} = \frac{1}{(N+1)\mathrm{PRF}_1} \tag{7.25}$$

This value for the pulsewidth prevents the pulses adjacent to the true coincidence from overlapping and causing a false coincidence. The resulting maximum pulsewidth is dependent on the range to be resolved and the ratios of the PRFs to the ranging frequency.

As an example of a two-PRF ranging system, consider a medium-PRF radar with a maximum operating range (R_u) of 100 nmi [19]. The common submultiple frequency (f_B) can then be found from

$$f_B = \frac{c}{2R_u} \tag{7.26}$$

where $c = 3 \times 10^5$ km/s and R_u (km) is the maximum desired unambiguous range. For the example, the basic PRF can be determined as approximately $f_B = 810$ Hz. If the ranging parameter is selected as $N = 17$, $\mathrm{PRF}_1 = Nf_B = 13.77$ kHz and $\mathrm{PRF}_2 = (N+1)f_B = 14.58$ kHz [see (7.24)]. Further, the unambiguous range (R_{u_1}) for the first PRF equals

$$R_{u_1} = \frac{c}{2\mathrm{PRF}_1} \tag{7.27}$$

where $c = 3 \times 10^5$ km/s resulting in $R_{u_1} = 10.893$ km and $R_{u_2} = 10.288$ km.

For a target at 60 km, the ambiguous range measured by the first PRF equals $R_{a_1} = 5.535$ km $(60 - 5 \cdot 10.893)$, whereas that measured by the second PRF equals $R_{a_2} = 8.560$ km. If we define the ratio

$$x = \frac{R_{a_1}}{R_{u_1}} \tag{7.28}$$

where x is the fractional range represented by the first ambiguous measurement, the result is $x = 0.508$. Similarly, $y = R_{a_2}/R_{u_2} = 0.832$. Then, from these measurements, the unambiguous target range can be determined by using

$$R_t = (n_1 + x)R_{u_1} \tag{7.29}$$

$$R_t = (n_2 + y)R_{u_2} \tag{7.30}$$

where n_1 and n_2 are integers selected by trial and error so that (7.29) and (7.30) provide the same range (R_t). In the example, $n_1 = n_2 = 5$, resulting in a measured range equal to $R_t = 5.508 \cdot 10.893 = 5.832 \cdot 10.288 = 60$ km.

Alternatively, the *Chinese remainder theorem* can be used to determine the true range from the two ambiguous measurements [12, 37]. From (7.24), the unambiguous ranging period (T_u) is given by

$$T_u = NT_1 = (N+1)T_2 \tag{7.31}$$

where T_1 is the PRI associated with PRF_1 and T_2 is the PRI associated with PRF_2. Dividing by the common multiple $N \cdot (N+1)$ results in

$$\frac{T_u}{N(N+1)} = \frac{T_1}{N+1} = \frac{T_2}{N} \tag{7.32}$$

The choice of $m_1 = N + 1$ and $m_2 = N$ implies that the total unambiguous range is divided into 306 range cells so that

$$m_1 = \frac{T_u}{T_2} = N + 1 \tag{7.33}$$

$$m_2 = \frac{T_u}{T_1} = N \tag{7.34}$$

for the two-PRF ranging code.

In accordance with the Chinese remainder theorem, the unambiguous target range measured in range units ($1 \leq \overline{R}_c \leq m_1 m_2$) is given by

$$\overline{R}_c = C_1 a_1 + C_2 a_2 \ \text{modulo}(m_1 m_2) \tag{7.35}$$

where

$$a_1 = \text{INT}(m_1 x) \tag{7.36}$$

$$a_2 = \text{INT}(m_2 y) \tag{7.37}$$

The constants C_1 and C_2 are determined from

$$C_1 = b_1 m_2 = (1)\text{modulo}(m_1) \tag{7.38}$$

$$C_2 = b_2 m_1 = (1)\text{modulo}(m_2) \tag{7.39}$$

where b_1 is the smallest positive integer that, when multiplied by m_2 and divided by m_1, gives unity as a remainder. The unambiguous range is then given by

$$R_t = \frac{\overline{R}_c R_u}{m_1 m_2} \tag{7.40}$$

For the example $m_1 = N + 1 = 18$ and $m_2 = N = 17$. From (7.38), $b_1 = 17$, and from (7.39), $b_2 = 1$, resulting in $C_1 = 289$ and $C_2 = 18$. The unambiguous target range in range units then becomes

$$\overline{R}_c = 289 \cdot a_1 + 18 \cdot a_2 \text{ modulo}(m_1 m_2) \tag{7.41}$$

which in the example becomes $\overline{R}_c = 99$. Hence, the target range is given by $R_t = (99/306)\ 185.2 = 60$ km.

The maximum range performance of the radar is determined by the average power transmitted. This implies a pulsewidth as wide as possible to minimize the requirement for excessive peak power. As an example, consider a high-PRF airborne PD system that requires PRFs in excess of 172 kHz (at X-band for a 1,400-knot target from an 1,800-knot interceptor) and assume that an unambiguous range of 50 nmi is required [36]. This corresponds to an f_B of 1.62 kHz and an $N = 172/1.62 = 106$. The maximum allowable pulsewidth is then $\tau = 1/(106)\ (107)\ (1{,}620) = 0.05\ \mu$s. This will require a relatively high peak power, which leads to the motivation for using three or more PRFs.

A three-PRF ranging system functions by forming two pairs of PRFs with each pair measuring the range to a different subranging frequency. These two subranging frequencies form, in turn, a two-PRF ranging system that measures the range to the final ranging frequency. A particular choice of the three-PRF ranging parameters [36, 43] that leads to a relatively simple mechanization results in a submultiple ranging frequency given by

$$f_B = \frac{\text{PRF}_1}{N(N+1)} = \frac{\text{PRF}_2}{N(N+2)} = \frac{\text{PRF}_3}{(N+1)(N+2)} \tag{7.42}$$

The unambiguous range is on the order of N^2 times the submultiple frequency as compared to N for the two-PRF ranging system. This results in a value for N with the three-PRF ranging system that is approximately the square root of its comparable value for a two-PRF ranging system.

The Chinese remainder theorem affords a convenient method for decoding the true range from the several ambiguous measurements in the multiple-channel search system [12, 37]. In this method, the period between PRF transmissions is expressed in units of pulsewidth (range cells) by dividing the period into a number of contiguous range cells. Each cell is numbered sequentially as 1, 2, 3, . . . , m_i, where $i = 1, 2, 3$. In the three-PRF ranging system, this results in each PRI being divided into a number of range cell numbers, designated by m_1, m_2, and m_3 ($m_1 > m_2 > m_3$). The three ambiguous range cell numbers, designated by a_1, a_2, a_3, which correspond to the target's measured range for each PRF, are then determined. The target's unambiguous range cell number ($1 \leq \overline{R}_c \leq m_1, m_2, m_3$) is thus given by

$$\overline{R}_c = (C_1 a_1 + C_2 a_2 + C_3 a_3) \text{ modulo}(m_1 m_2 m_3) \tag{7.43}$$

where

$$\begin{aligned} C_1 &= b_1 m_2 m_3 = (1)\text{modulo}(m_1) \\ C_2 &= b_2 m_1 m_3 = (1)\text{modulo}(m_2) \\ C_3 &= b_3 m_1 m_2 = (1)\text{modulo}(m_3) \end{aligned} \tag{7.44}$$

and b_1 is the smallest positive integer that, when multiplied by m_2m_3 and divided by m_1, gives unity as the remainder (b_2 and b_3 are determined in the same manner).

The procedure using the Chinese remainder theorem can be summarized as follows. Once m_1, m_2, and m_3 have been chosen to provide the desired unambiguous range, the constants C_1, C_2, and C_3 are determined [see (7.44)] for this ranging code. The ambiguous range measurements a_1, a_2, and a_3 are determined by measuring the target range in range units for each PRF. Then, the unambiguous target range is determined by multiplying each constant by its ambiguous range [see (7.43)] and dividing by m_1, m_2, m_3 as many times as possible to determine the remainder ($\overline{R}_c$) that equals the unambiguous range in range units. In the three-PRF multiple-ranging system, the maximum allowable pulsewidth is given by [36]

$$\tau_{3\,\max} \le \frac{1}{N(N+1)(N+2)f_B} = \frac{1}{(N+2)\mathrm{PRF}_i} \tag{7.45}$$

where $f_r = N(N+1)(N+2)f_B$ represents the lowest common multiple of all three PRFs. For the previously discussed example of the two-PRF system ($\mathrm{PRF}_1 = Nf_B = 171.72$ kHz, $\mathrm{PRF}_2 = (N+1)f_B = 173.34$ kHz), the value of N now becomes 10 so that $\mathrm{PRF}_1 = (N)(N+1)f_B = 178.2$ kHz, $\mathrm{PRF}_2 = (N)(N+2)f_B = 194.4$ kHz, and $\mathrm{PRF}_3 = (N+1)(N+2)f_B = 213.84$ kHz. However, the principal advantage of the three-PRF ranging system is that the maximum allowable pulsewidth $\tau_{3\,\max} = 0.5\ \mu$s, compared to the value $\tau_{2\,\max} = 0.05\ \mu$s for the two-PRF ranging system. Thus, 10 times the average power may be obtained by using the same peak power with three PRFs instead of two. This advantage is partially offset by the need for three detections of the target, which has a lower probability of success than the two detections required of the two-PRF system. Also, the probability on each trial is reduced because three PRFs must be generated during the time that the radar dwells on the target rather than two PRFs for the two-PRF ranging method.

The block diagram of a three-PRF range measuring system is depicted in Figure 7.39. The common multiple frequency $f_r = (N)(N+1)(N+2)f_B$ is generated by using a crystal controlled oscillator. The three PRFs are then generated by dividing the common multiple frequency (f_r) by N, $N+1$, and $N+2$. This guarantees that all PRFs are phase locked (i.e., synchronized). Any one PRF can be used to key the transmitter by using the PRF selector. All three PRFs are compared in a coincidence circuit, the output of which provides pulses at the ranging frequency (f_B). This is used as a time base for measuring range in a manner analogous to the way that the transmitter pulse is used in low-PRF radar to measure range.

The actual measurement of range is accomplished as follows. The first PRF (PRF_1) is transmitted, and the range gates are searched by means of the range sweep and time modulator. On detection of the target, the range sweep is stopped and the time modulator maintains its delay. The next PRF (PRF_2) is transmitted, and the range gates are searched incrementally in time steps corresponding to the reciprocal of f_r. On target detection, the search stops, leaving the counter at the proper phase to position PRF_2 range gates over the target. The same procedure is then used in PRF_3 to obtain the third ambiguous range measurement. At this time, because all range gates line up with the target, the true range can be measured.

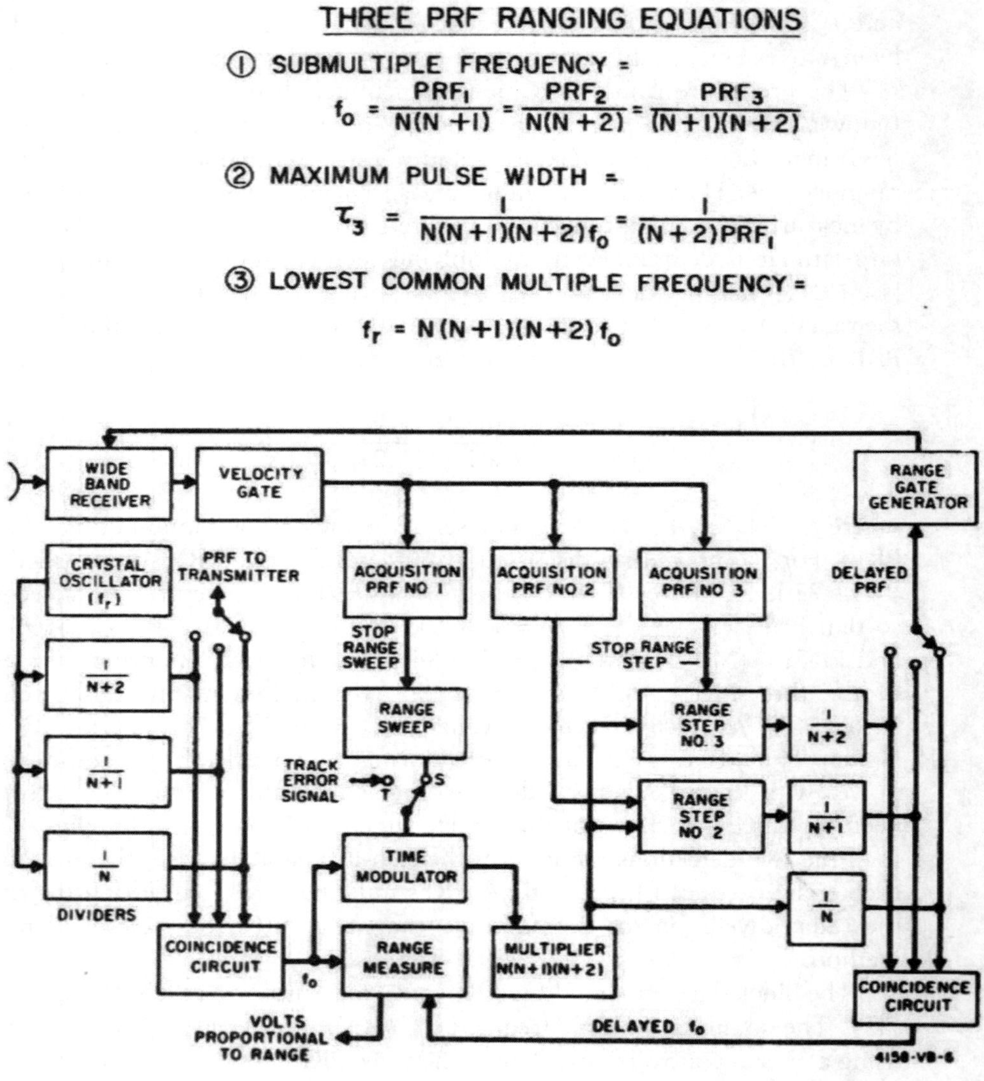

Figure 7.39 Block diagram of a three-PRF range-measuring system. (*From:* [36, p. 45].)

▶ Example 26 The Chinese Remainder Theorem

In medium- and high-PRF PD radars, ambiguous ranges must be resolved to determine the true target range. One method that is used is to employ two- or three-PRF ranging systems in conjunction with the Chinese remainder theorem. In this example, we describe the Chinese remainder theorem and give several examples of its application using interactive MATLAB program chi_rem.m.

To apply this method, multiple PRFs (typically two or three) are sequentially transmitted. Each of the interpulse periods is divided into a different number (m_i) of range gates of equal duration. The numbers must be relatively prime (highest common denominator is 1). By observing the position (a_i) of the target return in each of the interpulse periods it is then possible to deduce the true range using the Chinese remainder theorem.

The desired unambiguous range (R_u) is used to determine the basic ranging frequency given by

$$f_b = \frac{c}{R_u}$$

This establishes the multiple PRFs given for the two- and three-PRF ranging systems as

$$\text{PRF}_1 = m_1 f_b \qquad \text{PRF}_2 = (m_1 + 1) f_b$$

$$\text{PRF}_1 = m_1(m_1 + 1) f_b \qquad \text{PRF}_2 = m_1(m_1 + 2) f_b \qquad \text{PRF}_3 = (m_1 + 1)(m_1 + 2) f_b$$

The resolution for the two cases can be determined to be

$$\Delta R_2 = \frac{R_u}{m_1 m_2} \qquad \Delta R_3 = \frac{R_u}{m_1 m_2 m_3}$$

The unambiguous ranges associated with each PRF for the two cases are

$$R_{u_1} = m_1 \Delta R_2 \qquad R_{u_2} = m_2 \Delta R_2$$

$$R_{u_1} = m_1 \Delta R_3 \qquad R_{u_2} = m_2 \Delta R_3 \qquad R_{u_3} = m_3 \Delta R_3$$

The integer range gate position indicated by the target within each unambiguous PRF is given by

$$a_1 = \text{INT}\left[\frac{R_t \bmod(R_{u_1})}{\Delta R_{2,3}}\right] \qquad a_2 = \text{INT}\left[\frac{R_t \bmod(R_{u_2})}{\Delta R_{2,3}}\right] \qquad a_3 = \text{INT}\left[\frac{R_t \bmod(R_{u_3})}{\Delta R_3}\right]$$

where R_t is the target's range and the R_{u_i} are in compatible units.

The algorithm for the Chinese remainder theorem can then be applied as follows to determine the true target range:

$$M = \prod_{i=1}^{2,3} m_i \qquad t_i = \frac{M}{m_i}$$

$$C_i = t_i b_i = 1 \bmod(m_i) \qquad R_x = \left[\sum_{i=1}^{2,3} C_i b_i\right] \bmod(M)$$

where the b_i values are obtained by trial and error and R_x is the true range in range gate units. The true range is then found using

$$\overline{R_t} = R_x \Delta R_{2,3}$$

As the first example, we calculate the true range for the two-PRF ranging system described in Section 7.3.1. The parameters for this problem are given in Table EX-26.1. The printout for the interactive MATLAB program is given in Table EX-26.2.

The next two examples are taken from [36] and illustrate the need for a three-PRF ranging system with high-PRF PD radar. The requirement is for a high-PRF PD radar with PRFs of about 172 kHz at X-band to detect a 1,400 knot interceptor from an 1,800 knot platform. An unambiguous range of 50 nmi is desired, resulting in a ranging frequency of 1.62 kHz and an $m_1 = 172/1.62 = 106$ for the two-PRF case. The gate width resolution $\Delta R = 50 \cdot 1{,}852/106 \cdot 107 = 8.164$ m. The parameters of this design are shown in Table EX-26.3 and the MATLAB printout is shown in Table EX-26.4.

The problem with this design is that the transmitted pulsewidth is 0.054 μs, resulting in an average duty factor of 0.00986, which provides inadequate transmitted average power. The solution is to use a three-PRF ranging system that allows ranging factors of $m_1 = 9$, $m_2 = 10$, and $m_3 = 11$ [36]. The ranging resolution then becomes $\Delta R = 93.5$m. The parameters for this 3-PRF design are given in Table EX-26.5, and the MATLAB printout is tabulated in Table EX-26.6.

As illustrated in Table EX-26.5, the three-PRF ranging system provides comparable PRFs to that of the two-PRF ranging system, but with a pulsewidth of 0.623 μs, which provides an average duty factor of 0.1, which allows 10.94 greater average power than the alterative design.

Table EX-26.1 Parameters for Two PRF Ranging Problem

m_1	m_2	a_1	a_2	ΔR (m)	PRF_1 (kHz)	PRF_2 (kHz)	R_u (km)	R_x	R_t (km)
17	18	14	9	605.2	14.58	13.77	185.2	99	60

Table EX-26.2 MATLAB Printout for Two-PRF Ranging Problem

```
Number of PRIs used in calculation (2 or 3)? 2
Number of resolution cells for PRI #1? 17
Number of resolution cells for PRI #2? 18
In which resolution cell does the target appear for PRI #1? 14
In which resolution cell does the target appear for PRI #2? 9
What is the length of a resolution cell (in meters)? 605.167
Max. unambiguous range without staggered PRF is for PRI #1 10287.839 m. or 5.555 nmi.
Max. unambiguous range without staggered PRF is for PRI #2 10893.006 m. or 5.8818 nmi.
Max. unambiguous range with staggered PRF is 185181.102 m. or 99.9898 nmi.
Number of range cells is 99 Actual range is 59911.533 m. or 32.3496 nmi.
```

Table EX-26.3 Parameters for High-PRF, Two-PRF Ranging System

m_1	m_2	a_1	a_2	ΔR (m)	PRF_1 (kHz)	PRF_2 (kHz)	R_u (km)	R_x	R_t (km)
106	107	35	73	8.164	171.72	173.34	92.6	7349	60

Table EX-26.4 MATLAB Printout of High-PRF Two-PRF Ranging System

```
Number of PRIs used in calculation (2 or 3)? 2
Number of resolution cells for PRI #1? 106
Number of resolution cells for PRI #2? 107
In which resolution cell does the target appear for PRI #1? 35
In which resolution cell does the target appear for PRI #2? 73
What is the length of a resolution cell (in meter)? 8.1643
Max. unambiguous range without staggered PRF is for PRI #1 865.4158 m. or 0.46729 nmi.
Max. unambiguous range without staggered PRF is for PRI #2 873.5801 m. or 0.4717 nmi.
Max. unambiguous range with staggered PRF is 92599.4906 m. or 49.9997 nmi.
Number of range cells is 7349 Actual range is 59999.4407 m. or 32.3971 nmi.
```

Table EX-26.5 Parameters for High-PRF Three-PRF Ranging System

m_1	m_2	m_3	a_1	a_2	a_3	ΔR (m)	PRF_1 (kHz)	PRF_2 (kHz)	PRF_3 (kHz)	R_u (km)	R_x	R_t (km)
9	10	11	2	1	3	93.5	178.253	160.428	145.843	92.6	641	60

Table EX-26.6 MATLAB Printout for Three-PRF Ranging System

```
Number of PRIs used in calculation (2 or 3)? 3
Number of resolution cells for PRI #1? 9
Number of resolution cells for PRI #2? 10
Number of resolution cells for PRI #3? 11
In which resolution cell does the target appear for PRI #1? 2
In which resolution cell does the target appear for PRI #2? 1
In which resolution cell does the target appear for PRI #3? 3
What is the length of a resolution cell (in meters)? 93.5
Max. unambiguous range without staggered PRF is for PRI #1 841.5 m. or 0.45437 nmi.
Max. unambiguous range without staggered PRF is for PRI #2 935 m. or 0.50486 nmi.
Max. unambiguous range without staggered PRF is for PRI #3 1028.5 m. or 0.55535 nmi.
Max. unambiguous range with staggered PRF is 92565 m. or 49.9811 nmi.
Number of range cells is 641 Actual range is 59933.5 m. or 32.3615 nmi.
```

▶

7.4.2 FM Ranging [8, 38]

As the PRF is increased in high-PRF PD radar, a point eventually is reached where the range is so ambiguous that it cannot be resolved by using the multiple-PRF ranging system. If a range is required, it must be measured indirectly, as in CW radar, by frequency modulating the radar transmitter. Both sinusoidal and linear FM have been employed for this purpose [38]. However, we will primarily describe linear FM ranging, which is the more common of the two types [8].

Linear FM ranging in its simplest form consists of transmitting with a single high-PRF waveform that is periodically ramp modulated in frequency, as depicted in Figure 7.40. The received signal is delayed in time and translated in frequency by the target's Doppler shift as shown in the figure. The linear modulation continues for a period at least several times as long as the round-trip transit time for the most distant target of significance. During the course of this period, the instantaneous

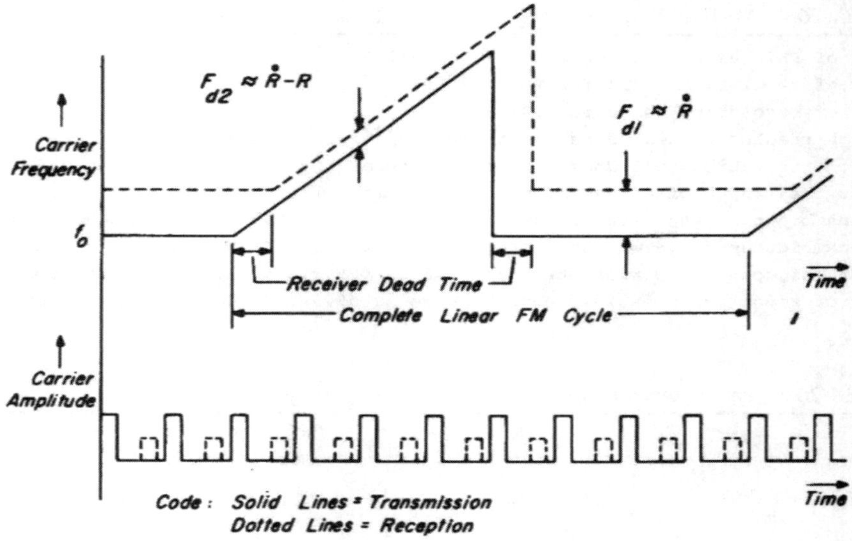

Figure 7.40 Transmitter and receiver frequency in a linear FM ranging system. (*From:* [38, p. 86].)

difference between the frequency of the received echoes and the frequency of the transmitter is measured. During the second half of the ranging cycle, the transmitter frequency remains fixed and, after a short receiver "dead" time, the difference between the transmitted and received signals is again measured.

During the non-FM half of the ranging cycle, the signal frequency is a function of the target velocity, whereas during the FM half, the signal frequency is proportional to the algebraic sum of the target velocity and range. Demodulated returns from targets with high velocities are shifted higher in frequency, and returns from targets with large ranges are shifted lower in frequency during the FM mode. The target range is then obtained as

$$R_{\mathrm{FM}} = \frac{c(f_{d1} - f_{d2})}{2k_{\mathrm{FM}}} \tag{7.46}$$

where f_{d1} (in hertz) is the target's Doppler shift measured during the non-FM part of the ranging cycle, f_{d2} (in hertz) is the measured difference in frequency during the FM part of the ranging cycle, and k_{FM} is the rate of linear FM sweep in hertz per second. A block diagram of high-PRF PD radar using linear-FM ranging is depicted in Figure 7.41.

An alternate two-slope waveform used in linear FM ranging is depicted in Figure 7.42. This waveform uses equal up-slope and down-slope linear ramps, where the magnitude of the slope is equal to that of the rising frequency slope of the previously described waveform (see Figure 7.40). If the target is closing (i.e., it has a positive Doppler), the difference between the frequency of the transmitter and that of the received echoes will be decreased by f_d during the rising frequency segment ($\Delta f_1 = k_{\mathrm{FM}} t_r - f_d$) and increased by f_d during the falling frequency segment ($\Delta f_2 = k_{\mathrm{FM}} t_r + f_d$). Consequently, if the frequency differences for the two segments

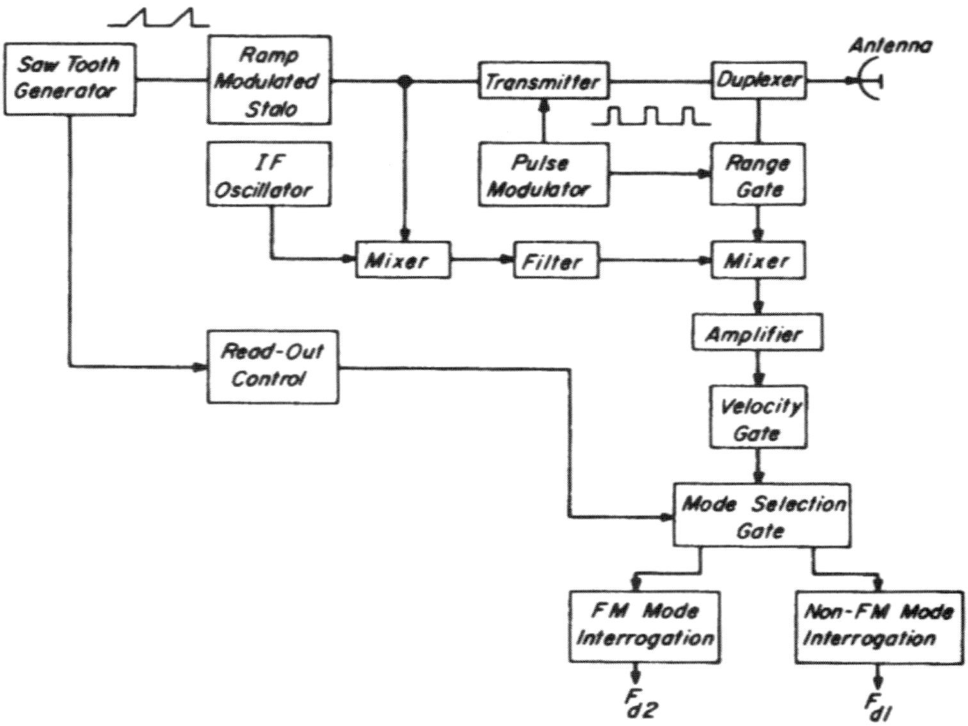

Figure 7.41 Block diagram of linear FM ranging high-PRF radar. (*From:* [38, p. 86].)

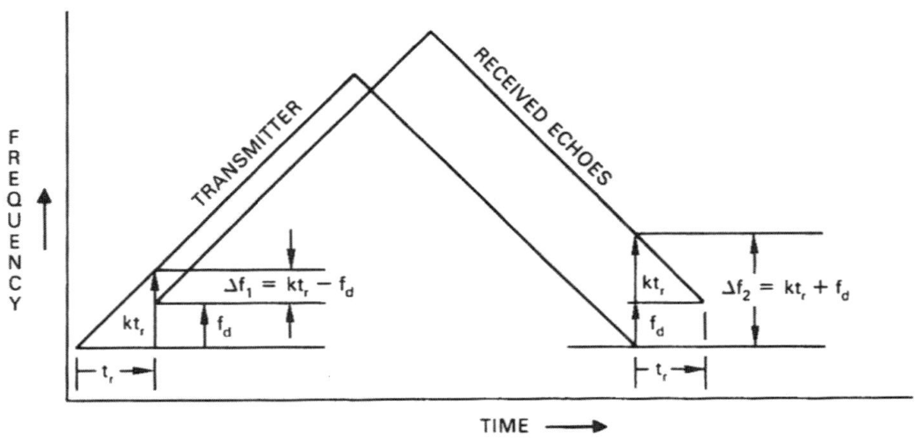

Figure 7.42 A two-slope FM ranging waveform. (*From:* [8, p. 233].)

are added ($\Delta f = \Delta f_1 + \Delta f_2 = 2k_{\mathrm{FM}} t_r$), the Doppler frequency shift is cancelled and only the frequency difference ($2k_{\mathrm{FM}} t_r$) due to the round-trip transit time to the target remains. Thus, the target range can be determined from

$$t_r = \frac{\Delta f_1 + \Delta f_2}{2k_{\mathrm{FM}}} \tag{7.47}$$

where t_r is the round-trip transit time, Δf_1 is the frequency difference between the transmitter and echo frequencies during the rising slope, Δf_2 is the similar frequency difference during the falling slope, and k_{FM} is the rate of change of the transmitter frequency.

For example, assuming no Doppler shift, the measured difference is 10 kHz on each segment, and the transmitter has a slope equal to 10 Hz/μs. From (7.47), the transit time is 1,000 μs, resulting in a range equal to 81 nmi. Now assume that the target has a Doppler shift of 3 kHz. On the up-slope, the measured frequency difference will be 7 kHz, whereas on the down-slope, it will be 13 kHz. Again applying (7.47), the transit time still will be 1,000 μs, illustrating the independence of the waveform on the Doppler shift. Note that, although the example considers a positive Doppler shift, the waveform works equally well on negative Doppler shifts.

If the antenna beam illuminates two targets at the same time, there will be two frequency differences during the first segment of the modulation cycle and two during the second, as depicted in Figure 7.43. Also note that the same situation occurs for the sawtoothed waveform depicted in Figure 7.40. Without further information, determining which pair of difference frequency measurements (i.e., A and x or y) pertains to the same target would be impossible. This ambiguity can be resolved by modifying the ranging waveform to transmit a three-slope ranging waveform, as depicted in Figure 7.44.

The basic three-slope ranging waveform illustrated in Figure 7.44 consists of the two-slope waveform (see Figure 7.42) complemented with a constant frequency segment that allows determination of the target's Doppler frequencies. The target's Doppler frequencies also can be determined by subtracting the two frequency differences measured during the up-slope and down-slope portions of the ranging waveform [i.e., $\Delta f_2 - \Delta f_1 = (k_{FM} t_r + f_d) - (k_{FM} t_r - f_d) = 2f_d$]. Therefore, by subtracting A (or B) from x (or y) and comparing the results with the measured Doppler frequencies, association of the correct pair of Doppler frequency up-slope/down-slope measurements with each other is possible. For example, if $(x - A)$ were twice one of the measured Doppler frequencies, the pairing would be x with A and y with B. Otherwise, y would be paired with A and x with B.

In air-to-air applications, the Doppler frequency shift can be greater than the frequency difference due to the ranging time ($k_{FM} t_r$). This occurs because the rate of change of frequency kFMtr will be only a small fraction of the highest Doppler frequency encountered. This situation is depicted in Figure 7.45, where return echoes are higher than the transmitted frequency for positive Doppler rather than being less than the transmitted frequency, as depicted in Figure 7.42.

The relationships between the measured frequency differences for two or more simultaneously detected targets is interpreted more easily when presented on a separate line for each slope, as depicted in Figure 7.46. The difference frequencies on the rising segment of the FM ranging waveform ($\Delta f_1 = k_{FM} t_r - f_d$; $f_d > k_{FM} t_r$) appear as negative frequencies on the diagram (i.e., A and B), whereas difference frequencies on the falling portion of the waveform (i.e., x and y) and the assumed Doppler frequencies (i.e., f_{d1} and f_{d2}) appear as positive frequencies on the diagram. The differences can be paired by using the Doppler frequencies measured on the

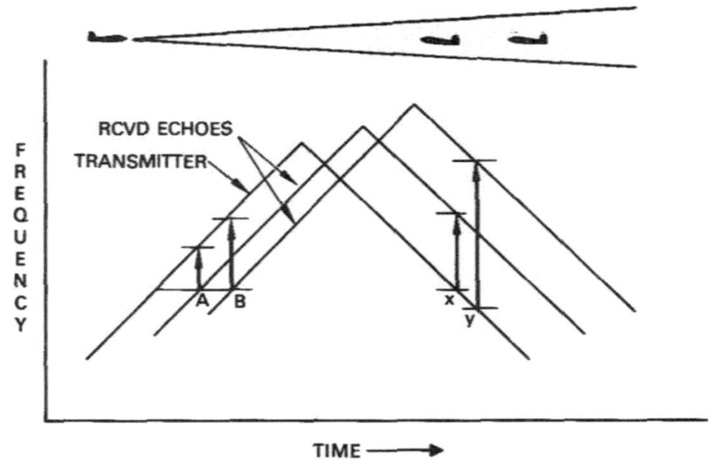

If two targets are detected simultaneously, two frequency differences will be measured during each segment of the modulation cycle.

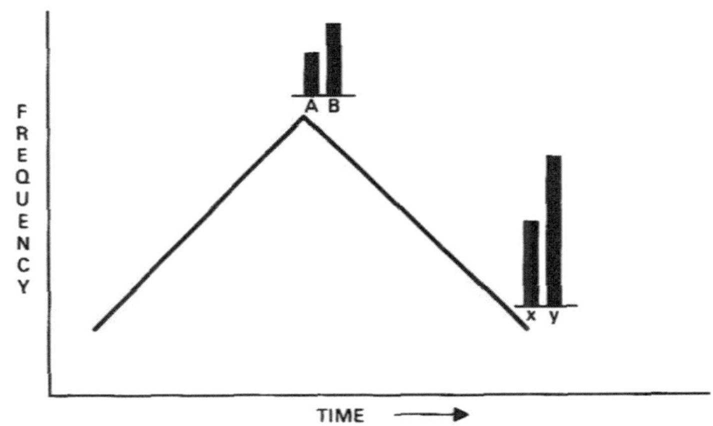

All the radar sees are two frequency differences at the end of each segment. Radar has no way of telling whether A should be paired with x or y.

Figure 7.43 FM ranging with two targets. (*From:* [8, p. 234].)

third segment and connecting those pairs separated by twice the Doppler frequencies f_{d1} and f_{d2} (A and y, B and x on the diagram).

The transit times can then be found from the diagram as $t_{ra} = (f_{d2} - A)/k_{\text{FM}}$ and $t_{rb} = (f_{d1} - B/k_{\text{FM}})$.

When three or more targets are present, we still can pair corresponding targets on the up-slope and down-slope using the procedure described in Figure 7.46;

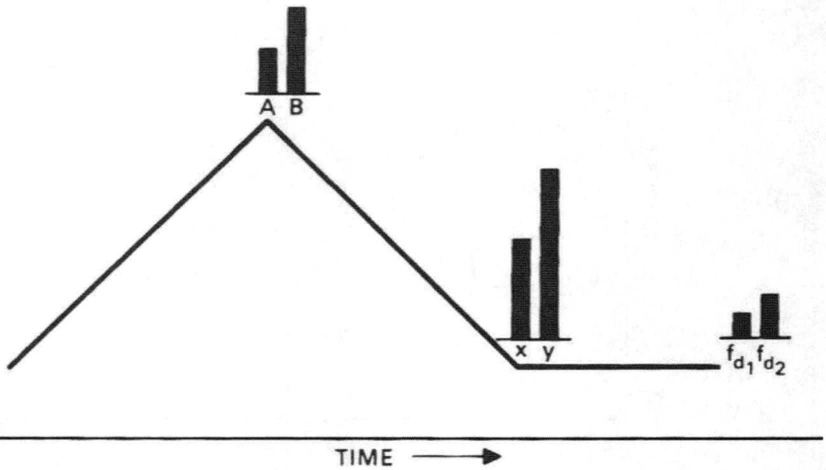

Figure 7.44 Three-slope FM ranging waveform. (*From:* [8, p. 235].)

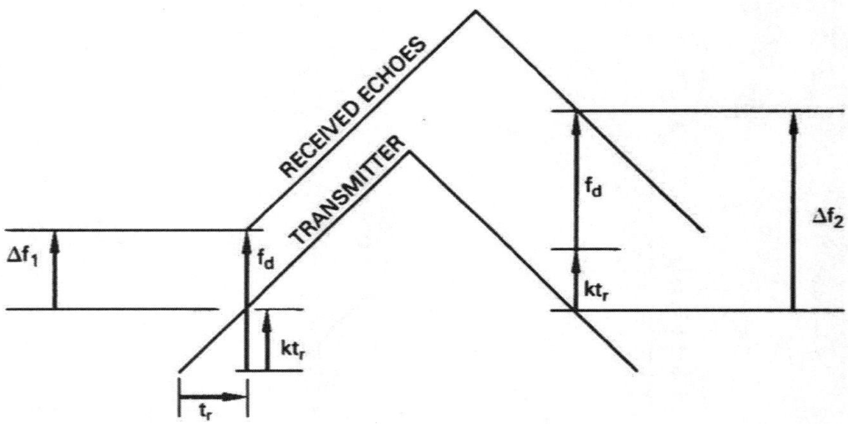

Figure 7.45 Two-slope FM ranging waveforms with Doppler shift greater than ranging frequency difference. (*From:* [8, p. 236].)

however, certain combinations of range and Doppler frequencies may allow multiple pairings (i.e., ghosts). Ghosts can be eliminated by adding more slopes to the waveform code. As with multiple PRFs, if N is the number of slopes, all possible combinations of $N - 1$ simultaneously detected targets can be "deghosted." The problem of ghosts is generally much less severe with FM ranging than with multiple PRF ranging because neither range nor Doppler frequency is ambiguous and, in practice, a three-slope waveform code is generally used.

The accuracy of FM ranging is proportional to the ratio of the frequency resolution of the Doppler filter bank to the rate of change of the transmitter frequency ($k_{FM} = \dot{f}$), as depicted in Figure 7.47. The range accuracy for FM ranging is then given by

$$R_{acc} = \frac{cB_D}{2k_{FM}} \tag{7.48}$$

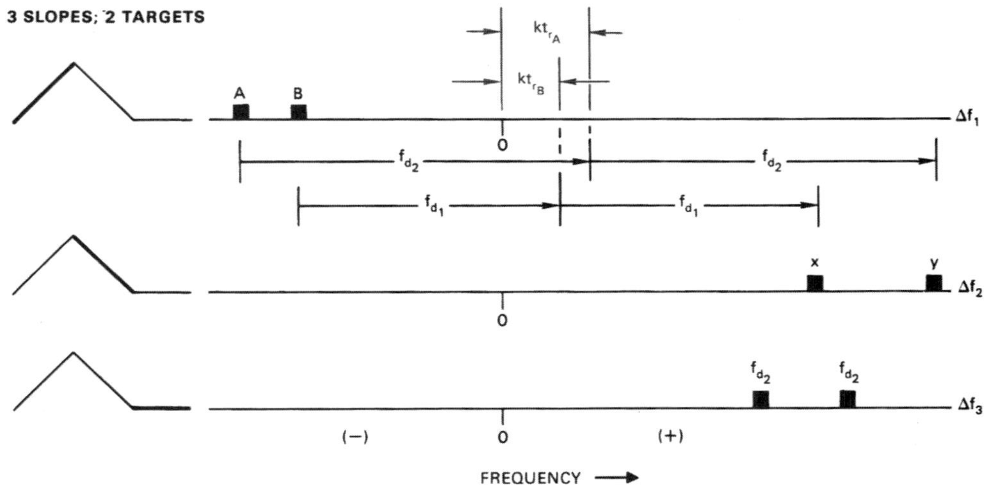

Figure 7.46 Frequency difference plot for FM ranging with two targets. (*From:* [8, p. 236].)

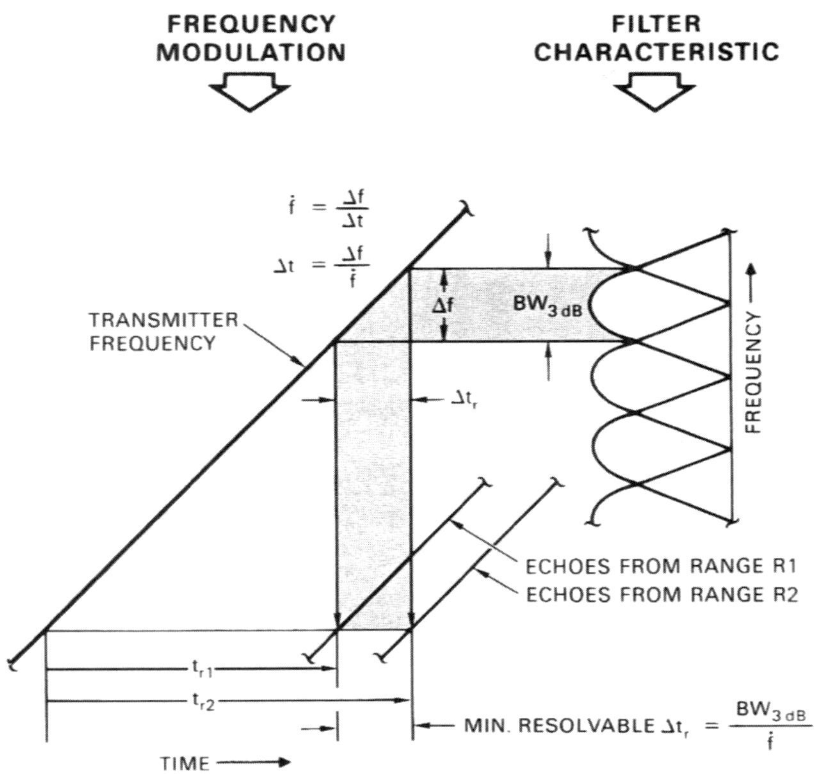

Figure 7.47 Minimum resolvable differences in ranging time. (*From:* [8, p. 462].)

where B_D is the Doppler filter's 3 dB bandwidth, and k_{FM} is the rate of change of the transmitter frequency. For example, if the Doppler filter's bandwidth is 100 Hz and $k_{FM} = 3$ MHz/s, then the range accuracy from (7.48) is 5 km.

From (7.48), we can see that the FM ranging accuracy depends on the Doppler filter bandwidth (B_D) and the rate at which the transmitter frequency is changed (k_{FM}). The greater k_{FM} is, the greater the frequency difference a given transit time will produce and the greater range accuracy that can be achieved. Similarly, the smaller the Doppler filter bandwidth (B_D), the greater the range accuracy. However, the transmitter frequency FM slope (k_{FM}) cannot be made arbitrarily steep nor the Doppler bandwidth arbitrarily small to increase range accuracy.

The minimum Doppler filter bandwidth is determined by the coherent processing interval of the radar, which in a search radar is related to the time the radar antenna illuminates the target (i.e., t_o is the radar observation time to $t_o/3$), which, in turn, limits the minimum Doppler filter bandwidth to $B_D = 3/t_o$ (Hz).

In air-to-air applications, the transmitter's FM rate (k_{FM}) is limited by the smearing of ground clutter returns (which may be received out to ranges greater than 100 nmi) into the Doppler region occupied by target returns. This is the same effect that limits the performance of high-PRF PD radars due to transmitter instability (see Section 3.3). The detection of targets within the Doppler region occupied by the spread clutter spectrum is generally inhibited because target returns are obscured by the clutter. The rate of transmitter frequency change (k_{FM}) must then be held to a small fraction of the maximum target Doppler shift to provide a Doppler-clear region sufficiently wide to allow target detection.

The clutter spectral situation for the detection of long-range targets with various degrees of transmitter FM is depicted in Figure 7.48. Two long-range targets are shown, one with a high Doppler frequency (B) and the other (A) with a Doppler frequency only slightly higher than the highest ground clutter frequency. The first profile shows the normal ground clutter spectrum for high-PRF airborne PD radar with the two targets in the clear-Doppler region. The second profile shows what happens when the frequency shift corresponding to the target's range is comparable to the Doppler shift of the higher closing rate target (B). In this case, main lobe clutter spreads into the normally clutter-clear region and obscures both targets. In the third profile, the frequency shift corresponding to the target's range is a small fraction of the target's Doppler shift. This relationship results in a limited spread of the clutter's spectral width, which has little effect on target detection.

In general, the constraints on transmitter frequency change (k_{FM}) and Doppler filter bandwidth (B_D) are such that, in a typical airborne interceptor application, the range accuracy is on the order of a few miles [8]. This is substantially poorer than can be obtained with pulse-delay ranging. If a maximum detection range is desired in high-PRF PD radar, a sequential detection mode can be employed. Initially, a velocity search mode is employed in which range is not measured. Then, the FM-ranging mode is activated to establish the target's range.

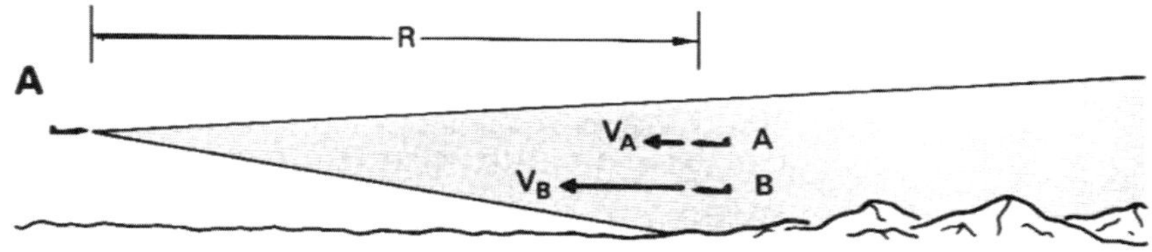

a) Two targets approaching from long range. One has much higher closing rate than the other.

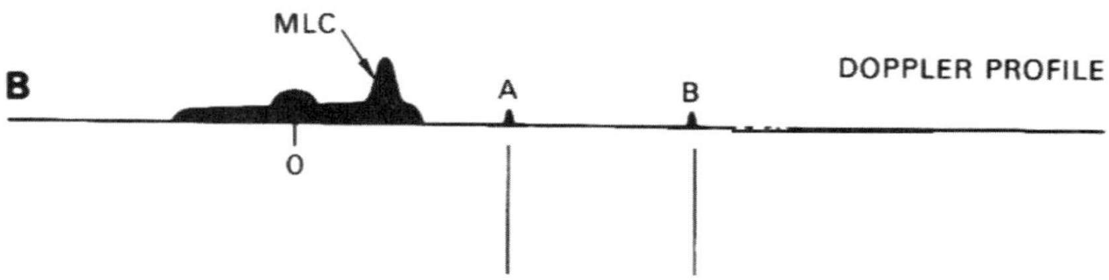

b) Doppler profile, without FM ranging—no clutter spreading.

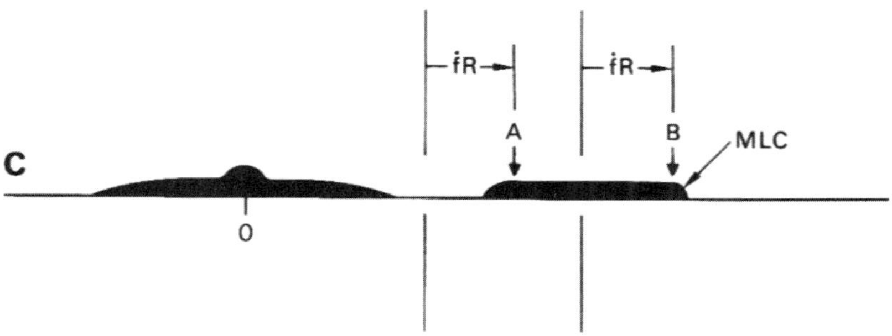

c) Doppler profile, with FM ranging and *large* value of $\dot{f}$. Mainlobe clutter shifts out in frequency and spreads over doppler clear region, obscuring targets.

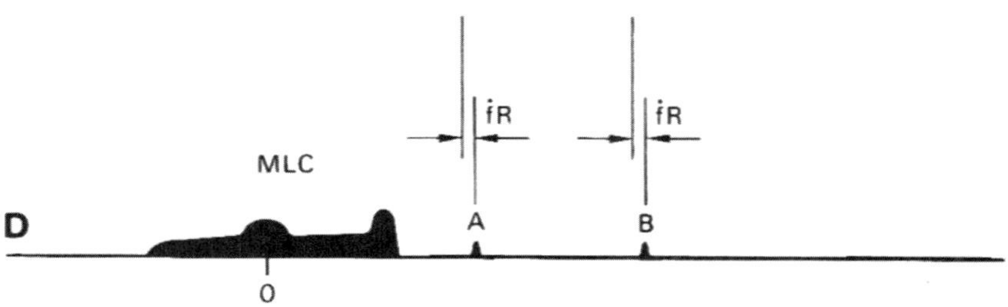

Figure 7.48 (a–d) Ground clutter spectral spreading due to transmitter FM in high-PRF radar. (*From:* [8, p. 463].)

7.5 Main Lobe Clutter

A major factor determining the performance of PD radars is the ability of the radar to attenuate main lobe surface clutter (i.e., ground and sea clutter). Highly range-ambiguous radar waveforms (i.e., high- and medium-PRFs) accentuate main beam clutter, requiring higher clutter attenuation for a given level of performance by these types of waveform design. The basic limits on performance of surface-based high- and medium-PRF designs is generally set by the phase and frequency instabilities of the radar's reference oscillator (STALO) and transmitter, which spread the main lobe clutter into the frequency region occupied by the target's Doppler filters (see Section 3.3). In addition to clutter spectral spreading due to oscillator instability effects, main lobe clutter is also spread by radar platform motion, antenna scanning, and random variations in signal amplitude and phase, as discussed in Section 3.2. These latter effects generally determine the Doppler-clear region available for detection of moving targets in low- and medium-PRF PD designs. In airborne designs, sidelobe clutter can be the limiting factor for low- and medium-PRF designs (see Section 7.2.2), whereas oscillator instability generally determines performance in clutter for a high-PRF design.

Another effect that can limit the performance of PD radars is the spreading of the main lobe clutter spectrum into the Doppler region occupied by targets due to receiver nonlinearity. This problem was covered in Sections 3.2.5 and 6.2.3 for MTI systems. However, in high-PRF PD designs, main lobe clutter can exceed the receiver noise level by 80 to 100 dB due to the folding of range-ambiguous clutter into the target's range cell. Figure 7.49 depicts the spectral spreading of Gaussian-shaped main beam clutter down to the −140 dB level for a soft-limiting type of receiver (see Figure 6.14). This allows the detection performance of high-PRF PD radars to be determined against weak targets in the presence of receiver nonlinearities.

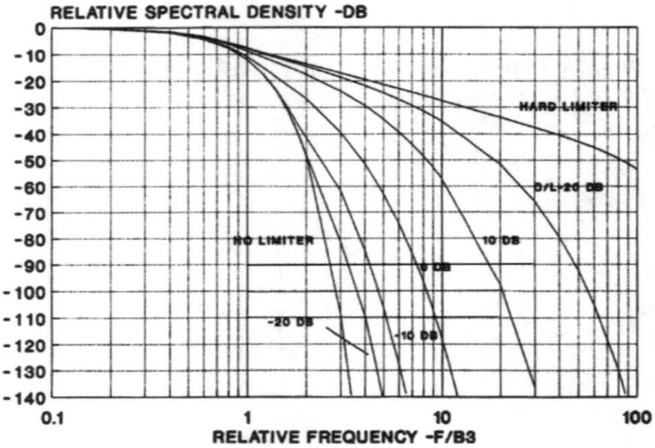

Figure 7.49 Spectral spreading for a nonlinear receiver. (*From:* [6, p. 457].)

▶ Example 27 Altitude Line Return

In airborne radar a strong specular return is reflected from the surface directly below the aircraft. This return is called *altitude line return* because generally its range equals the absolute altitude of the radar. It is transmitted and received through the radar's sidelobes and is generally stronger than the surrounding sidelobe clutter return and may be as strong as or stronger than the main lobe clutter. The geometry of altitude line return is illustrated in Figure EX-27.1.

The return from the ground contained in a pulsewidth is defined by the incidence angle ϕ. The incidence angle is given by

$$\phi = \cos^{-1}\left(\frac{h}{h + R_\tau}\right)$$

where $R_\tau = c\tau/2$ and h is the aircraft's altitude. The radius of the ground return circle can be found as

$$R_G = h \tan \phi$$

resulting in the ground area

$$A_G = \pi R_G^2$$

The maximum altitude line Doppler associated with the edge of the ground circle is given by

$$f_{d\max} = \frac{2v_{ac} \sin \phi}{\lambda}$$

where v_{ac} is the aircraft's velocity. The mean Doppler frequency returned from directly below the aircraft is zero. A plot of the clutter spectrum for a high-PRF PD, including the altitude line, is shown in Figure 3.42. Altitude line clutter is primarily a problem in PD radars that employ high-PRF waveforms because it can be eliminated in low- and medium-PRF PD radars if appropriate range gating is used.

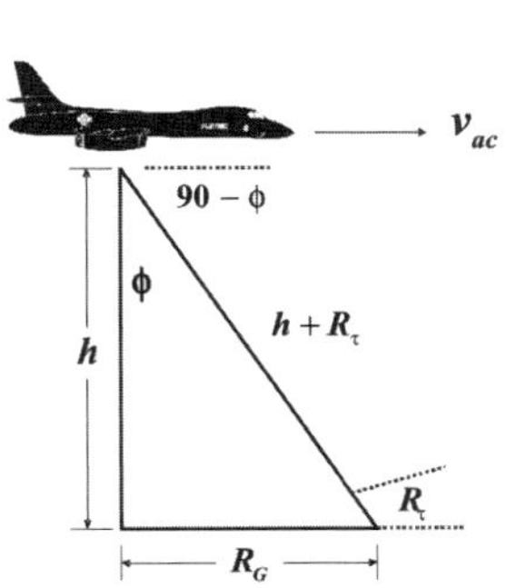

Figure EX-27.1 Altitude line geometry.

The power contained in altitude line return over sea clutter is computed by the MATLAB program alt_lineb.m. The reflectivity of the clutter is a function of the grazing angle ($90 - \phi$) and, hence, varies over the ground circle. Figure EX-27.2 depicts the TSC specular reflectivity (σ^0) model [call MATLAB function tsc(angg, wl, SS)], which is a function of the incidence angle(angg), wavelength, and sea state. Then the ground area is divided into concentric rings each with area ΔA and appropriate incidence angle. The power returned from each ring is then calculated as

$$\Delta P_c = \frac{P_t G_{SL} \lambda^2 \sigma^0 \Delta A}{(4\pi)^3 R^4}$$

The total power is then the sum of the incremental powers and is depicted in Figure EX-27.3 for $P_t = 1{,}500$ W, $\lambda = 0.03$ m, $G_{SL} = -5$ dB, and sea state 0.

The maximum Doppler frequency associated with the altitude line varies with the altitude of the radar. Figure EX-27.4 shows the variation with height for the parameters used in Figure EX-27.3. This is computed in MATLAB program alt_lineb.m.

The spectrum associated with altitude line return is computed by the MATLAB program alt_linec.m. To compute this spectrum, the concentric ground rings segmented into multiple azimuth slices. The Doppler is then computed for each element using

$$\Delta f_d = f_{ac} \sin(\Delta\phi) \cos(\theta_{az})$$

while the power is computed by dividing the power in each concentric ring by the number of azimuth slices. The altitude line spectrum is depicted in Figure EX-27.5.

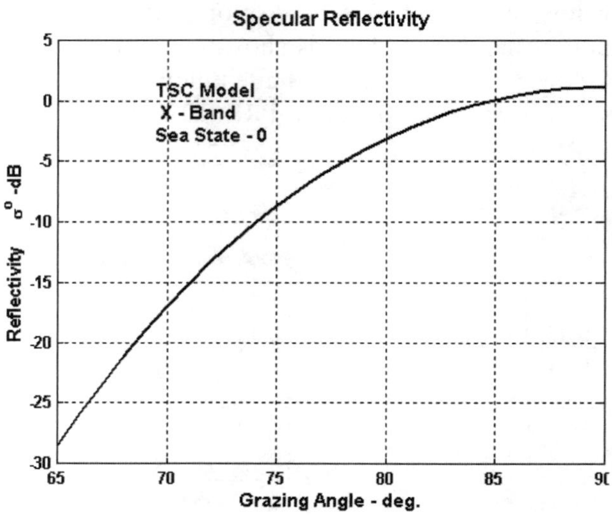

Figure EX-27.2 TSC specular reflectivity sea clutter model.

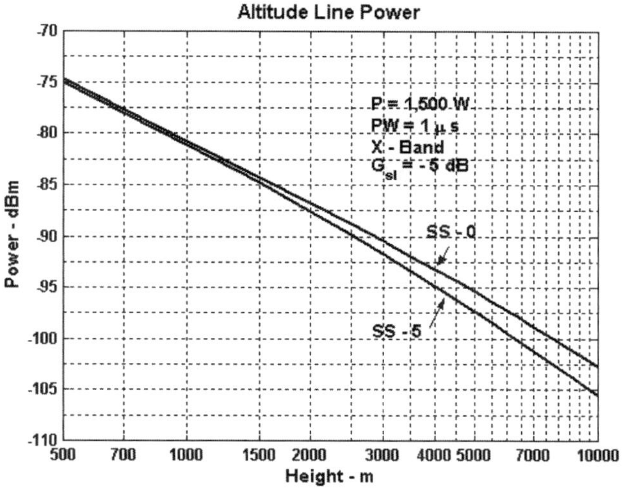

Figure EX-27.3 Altitude line clutter.

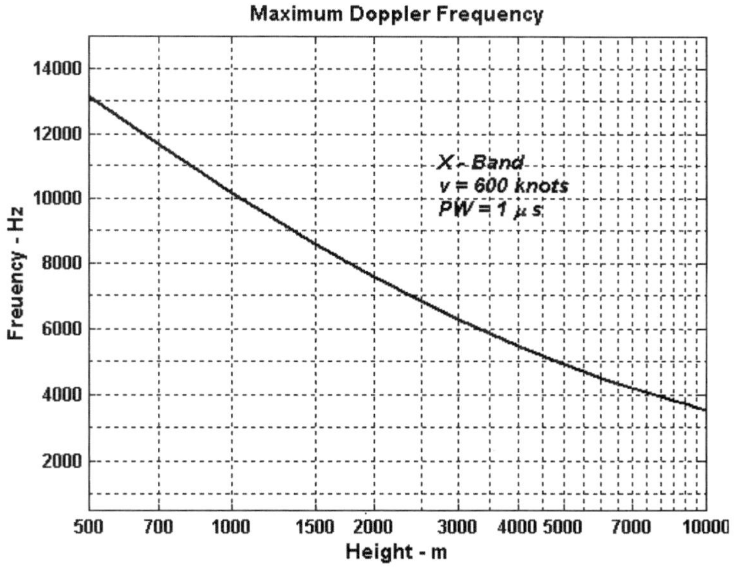

Figure EX-27.4 Maximum Doppler for altitude line clutter.

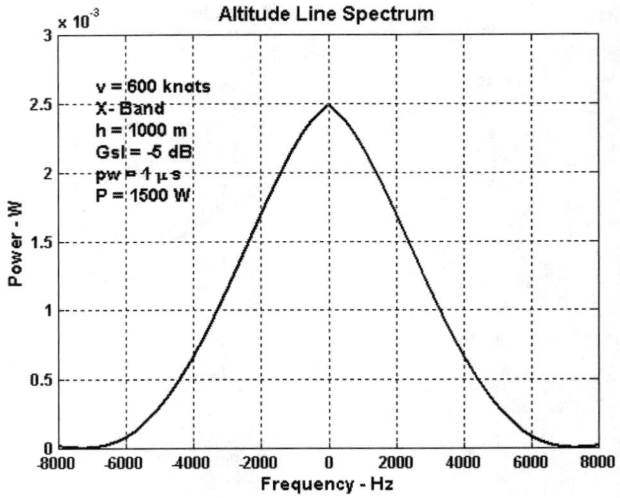

Figure EX-27.5 Altitude line spectrum.

▶

To compute the performance of those designs limited by main lobe clutter, we need to compute the clutter power contained within that portion of the clutter spectrum. The incremental clutter power received from a small patch of ground (see Figure 7.50) on a flat earth can be expressed as

$$dP_c = \frac{P_t G^2(\theta_{az}, \phi_{el}) \lambda^2 \sigma^0 \, ds \, d\theta}{(4\pi)^3 R_c^4 L_c} \tag{7.49}$$

where $G^2(\theta, \phi)$ is the antenna gain in the direction of the clutter cell, θ_{az} is the azimuth angle, ϕ_{el} is the depression angle, σ^0 is the normalized clutter reflectivity

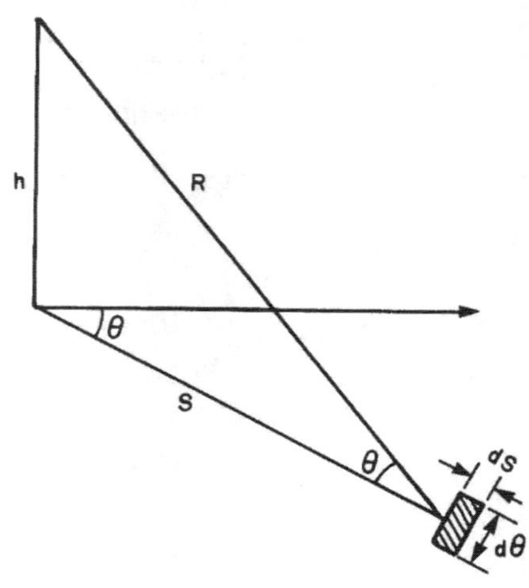

Figure 7.50 Geometry of clutter cell. (*From:* [39, p. 13].)

per unit area, R_c is the range to the clutter patch, and L_c is the radar system clutter losses. For low-PRF radar, the main lobe clutter RCS$(\sigma_{cx}) = \sigma^0 \int_s ds\, d\theta$ can be expressed as

$$\sigma_{cx} = \sigma^0 R_c \theta_{az} \left(\frac{c\tau_e}{2}\right) \sec \psi \tag{7.50}$$

where θ_{az} is the radar's one-way power antenna beamwidth and τ_e is the radar's effective pulsewidth. Substituting this into (7.49) and defining $G^2(\theta_{az}, \phi_{el}) = G_t^2 F_c^4$, where F_c^2 is the clutter pattern propagation factor and G_t is the peak antenna gain and assuming ϕ_{el} is small results in

$$P_{cML} = \frac{P_t G_t^2 \lambda^2 F_c^4 \sigma^0 \theta_{az} \left(\frac{c\tau_e}{2}\right)}{(4\pi)^3 R_c^3 L_c} \tag{7.51}$$

For range-ambiguous clutter, this can be expressed as

$$P_{cML} = \frac{P_t G_t^2 \lambda^2 \theta_{az} R_\tau}{(4\pi)^3 L_c} \sum_{i=1}^{\infty} \frac{\sigma_i^0 F_{ci}^4}{R_{ci}^3} \tag{7.52}$$

where σ_i^0, F_{ci}^4, and R_{ci}^4 are the respective values associated with each of the ambiguous clutter elements. Note that in range-ambiguous radar, the clutter elements closest to the radar contribute the most incremental clutter power. In cases when reflectivity, propagation factor, and attenuation remain relatively constant, the following effective clutter range (R_{ce})

$$R_{ce}^3 = \frac{1}{\sum_{i=1}^{n} \left(\frac{1}{R_{ci}^3}\right)} \tag{7.53}$$

can be substituted into (7.52) [40].

In the summation of (7.52), both the clutter reflectivity (σ_i^0) and the pattern propagation factor generally vary with the range (R_{ci}) to the clutter patch. By using the constant-gamma reflectivity model (i.e., $\sigma^0 = \gamma \sin \psi$), the clutter reflectivity can be removed from the summation. For small depression angles, the relationship $\phi_{el} = h/R$ applies, allowing the clutter power to be evaluated as

$$P_{cML} = \frac{P_t G_t^2 \lambda^2 \theta_{az} \left(\frac{c\tau_e}{2}\right) \gamma h_a}{(4\pi)^3 L_c} \sum_{i=1}^{\infty} \frac{F_{ci}^4}{R_{ci}^4} \tag{7.54}$$

where h_a is the antenna's height above the surface and γ is the constant reflectivity coefficient. For high- and medium-PRF PD airborne radars, the antenna has sufficient height (h_a) that clutter in the first several ambiguous range intervals is not illuminated by the antenna's main lobe. The summation in (7.52), and also in (7.54), if appropriate, will then have a small value of F_{ci} for $i = 1$ and possibly for several successive ambiguities with ranges $R_{ci} = R_{ci} + (i-1)R_a (i > 1)$, where $R_a = c \cdot \text{PRI}/2$ for a constant PRF.

The clutter magnification properties of range-ambiguous radars (i.e., medium- and high-PRF) can be evaluated by computing the S/C power ratio obtained through division of the standard radar equation by (7.52) resulting in

$$\frac{\text{S}}{\text{C}} = \frac{\sigma_t}{\theta_{az} R_t \left(\frac{c\tau_e}{2}\right) L \sum_{i=1}^{\infty} \sigma_i^0 \left(\frac{R_t}{R_{ci}}\right)^3 F_{ci}^4} \tag{7.55}$$

where σ_t is the target's RCS, R_t is the target range, and $L = L_r/L_c$. When the clutter and target range are equal in the peak of the antenna beam ($F_c = 1$), S/C = σ_t/σ_{cx}, where σ_{cx} is the unambiguous clutter RCS given by $\sigma_{cx} = \sigma^0 \theta_{az} R_t (c\tau_e/2)$. For ambiguous clutter ranges (R_{ci}), the clutter reflectivity $\left(\sigma_i^0\right)$ at each ambiguity is then magnified by the factor $(R_t/R_{ci})^3 F_{ci}^4$, which must be summed over all possible clutter ambiguities to obtain the effective clutter RCS.

For high-PRF radars, main lobe clutter power can be estimated by evaluating the clutter power returned to a CW radar (with transmitter power P_t) and multiplying this value by the radar's duty factor ($d_u = \tau/T$). The main lobe clutter power for CW radar can be obtained by replacing the summation in (7.52) with an integral as [40] follows:

$$P_{cML} = \frac{P_t G_t^2 \lambda^2 \theta_{az}}{(4\pi)^3 L_c} \int_0^{\infty} \frac{\sigma^0 F_c^4}{R^3} dR \tag{7.56}$$

where F_c^4 is the two-way (transmitting-receiving) pattern propagation factor that includes the elevation voltage gain pattern of the antenna. For the case of a Gaussian-shaped elevation beam pattern point at the horizon, the pattern propagation factor can be expressed as

$$F_c^4 = \exp\left(-\frac{5.55\phi_{el}^2}{\sigma_B^2}\right) \tag{7.57}$$

where ϕ_B is the antenna's one-way power elevation beamwidth. For small depression angles, $\phi_{el} = h_a/R$ and

$$\int_0^{\infty} \frac{\exp\left(-5.55h_a^2/\phi_{el}^2 R^2\right)}{R^3} = \frac{\phi_{el}^2}{(3.33h_a)} \quad 2 \tag{7.58}$$

The main lobe clutter power for a PD radar (with duty factor d_u) can then be found from (7.56) with a constant σ^0 as

$$P_{cML} = \frac{P_t d_u G_t^2 \lambda^2 \theta_{az} \phi_{el}^2 \sigma^0}{(4\pi)^3 L_c (3.33 h_a)^2} \tag{7.59}$$

where P_t is the radar's peak power, d_u is the duty factor, θ_{az} is the azimuth beamwidth, ϕ_{el} is the elevation beamwidth, h_a is the height of the antenna's phase center above the surface, G_t is the antenna's peak gain, and L_c is the loss associated with clutter detection. By using the antenna's gain-beamwidth relationship $G_t = 4\pi/\theta_{az}\phi_{el} L_n$), where $L_n = 1.2$ is the azimuth antenna pattern loss factor, $\theta_{az} = 1.5\lambda/D_a$, and D_a is the azimuth antenna's aperture (both D_a and L_n have been adjusted for low sidelobes), the result is

$$P_{cML} = \frac{P_t d_u \sigma^0 \lambda D_a}{6\pi L_c L_n^2 (3.33 h_a)^2} = \frac{P_t \sigma^0 \lambda D_a d_u}{400 h_a^2} \tag{7.60}$$

where the second term is evaluated for an azimuth antenna's beam shape loss of $L_c = 1.33$ [40]. For example, consider a surface-based X-band PD radar ($\lambda =$ 3 cm) with a 0.1 duty factor, which uses a 1.2m azimuth aperture with its phase center 2.4m above a surface with a reflectivity coefficient of $\sigma^0 = 0.004$, resulting in a $P_{cML}/P_t = 6.25 \times 10^{-9} = -82$ dB. For a 10 kW transmitter power, the main lobe clutter power is equal to −12 dBm.

Equation (7.56) can also be evaluated by using a constant-gamma reflectivity model ($\sigma^0 = \gamma \sin \phi_{el}$), which, for small depression angles ($\phi_{el} = h/R$), results in the integral:

$$\int_0^\infty \frac{\sigma^0 F_c^4}{R^3}\, dR = \int_0^\infty \frac{\gamma h_a}{R^4} \exp\left(-5.55 h_a^2 / \phi_{el}^2 R^2\right) dR \tag{7.61}$$

$$= \frac{\gamma \phi_{el}^3}{29.4 h_a^2}$$

Substitution of this into (7.56) for a PD radar with duty factor d_u results in

$$P_{cML} = \frac{P_t d_u G_t^2 \lambda^2 \theta_{az} \phi_{el}^3 \gamma}{29.4 (4\pi)^3 L h_a^2} \tag{7.62}$$

This can be further simplified by using $\phi_{el} = 1.5\lambda/D_e$, where D_e is the elevation aperture, and the other simplifications applied in (7.60), resulting in

$$P_{cML} = \frac{P_t d_u \lambda^2 D_a \gamma}{533 D_e h_a^2} \tag{7.63}$$

Note that the main lobe clutter power is inversely proportional to the altitude squared, which results in a less stringent requirement for clutter attenuation in an airborne radar application than to a surface radar. For example, consider an X-band surface-based radar with a transmitter power of 10 kW and a duty factor of 0.1 that uses a circular aperture ($D_a = D_e$) operating at 3m above a terrain with reflectivity 0.1 [18]. The main lobe clutter power from (7.63) is equal to $P_{cML} = 1.9 \times 10^{-5}$ W. Assuming a 1 kHz Doppler filter bandwidth with a system noise temperature of $T_s = 3{,}000$ K results in a noise power for the filter equal to 4×10^{-17} W. The main lobe clutter must then be attenuated by about 120 dB in the target's Doppler filter so that detection is noise limited.

In the airborne case for a circular beam, the clutter RCS can be approximated as

$$\sigma_{cx} = \frac{\sigma^0 \pi (\theta_{3\,\mathrm{dB}} R)^2}{4} \tag{7.64}$$

where θ_{az} is the one-way power antenna beamwidth. Substitution of this result into the radar equation yields

$$P_{cML} = \frac{P_t d_u G_t^2 \theta_{3\,\mathrm{dB}}^2 \sigma^0 \lambda}{\left(\sqrt{2}\, 16\pi\right)^2 R^2 L} \tag{7.65}$$

where d_u is the radar duty factor, $\theta_{3\,\mathrm{dB}}$ is the antenna's one-way power beamwidth, R is the range to the main lobe clutter patch, and L is a system loss term. Substitution of $G = \pi^2 / \theta_{3\,\mathrm{dB}}^2 = (\pi D/\lambda)^2$, where D is the circular antenna aperture diameter results in

$$P_{cML} = \frac{P_t d_u \sigma^0 D^2}{50 R^2 L} \tag{7.66}$$

If we assume the same parameters as for the X-band surface PD radar (i.e., $P_t = 10$ kW, $d_u = 0.1$, $\sigma^0 = 0.1$) with $D = 0.7$ m, $L = 1$, and $R = 10$ km, the result is a main lobe clutter power of $P_{cML} = 10^{-8}$ W. The noise power ($T_s = 3{,}000$ K) in a 1 kHz Doppler filter is 4×10^{-17} W, resulting in a C/N of 2.5×10^8, requiring 84 dB clutter attenuation of main lobe clutter for detection to be noise limited in the Doppler filter.

▶ Example 28 Missile Approach Warning Using Pulsed Doppler Radar

Missile approach warning (MAW) is a necessary function for countering missile attacks on tactical aircraft. MAWs generally warn aircraft that they are under attack by an approaching missile and provide directional and time-to-impact threat information necessary to counter the attack by either dispensing decoys or employing active countermeasures. MAWs are intended to supplement RWRs, which alert the crew to the presence of hostile fire control radar and missiles carrying active radar seekers, but they are important against weapons employing IR/EO passive sensors.

MAWs may be either active or passive. The former use a pulse Doppler radar to detect and track missiles by means of their skin return, whereas the latter respond to either the IR or ultraviolet (UV) signature of the exhaust plume from the missile's rocket engine. The two types of sensors may be combined in a hybrid system, offering the best combined benefits of each [41].

Radar has the advantage of providing all-weather range and range-rate data that enable the calculation of time-to-impact data. Disadvantages involve the difficulty of extracting the small missile RCS target from competing clutter and providing all-aspect accurate AOA data. An additional criticism of active radar is the possibility of its being used as a beacon for an antiradiation-type missile. In addition, LPI waveforms and antenna shielding are necessary when radars are used on stealth aircraft. Discrimination between missiles and rapidly closing aircraft is another difficult problem for active radar MAWs.

A tail attack by a heat seeking missile is the most vulnerable condition faced by a tactical aircraft. Tail warning PD radar is installed on both the B-52 and B-1B aircraft to protect against this situation. Figure EX-28.1 illustrates the operational situation and high-PRF clutter and target spectra for this situation. The missile target is imbedded in sidelobe clutter when the missile's closing velocity on the aircraft is less than the aircraft's forward velocity, dictating a requirement that the backward-looking antenna have low sidelobes in forward-looking directions.

MAW radar has the advantage of providing outputs of target range and range rate, enabling the system to calculate time-to-intercept and thus permitting optimum deployment of countermeasures and time of aircraft evasive maneuvers. Radar also provides the potential to discriminate between approaching missiles and aircraft and to accomplish noncooperative target recognition. A radar-based MAW is effective in all-weather environments at any altitude, but can act like a beacon [41, 42].

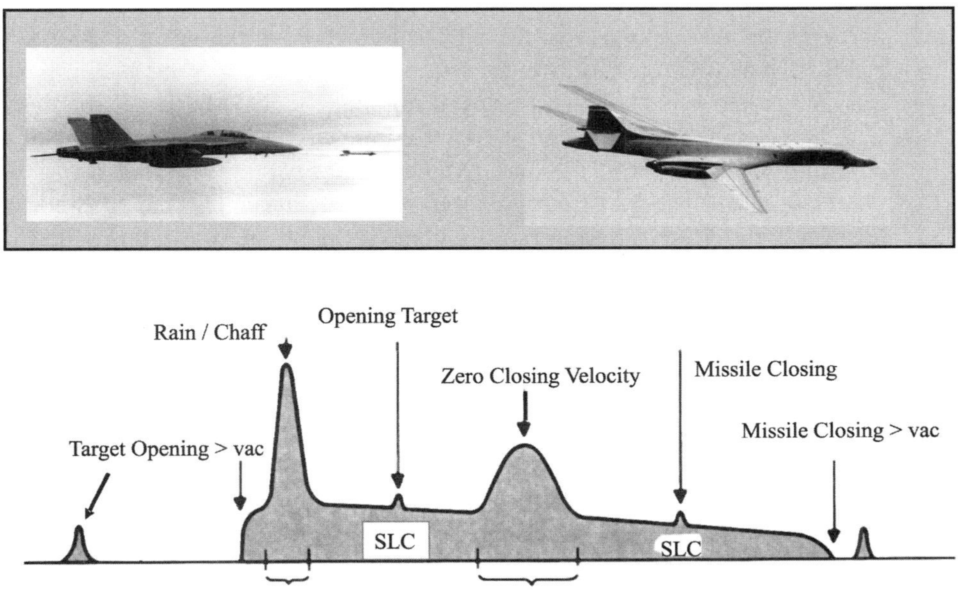

Figure EX-28.1 Operational tail-chase and high-PRF PD clutter spectrum.

The design of pulse Doppler MAW faces two main challenges posed by the small-target RCS (0.01 to 0.1 m^2), the strong ground clutter encountered at low altitudes, and limited space available for antennas in a tactical aircraft. One major task is to achieve a high probability of detection (i.e., 0.9 to 0.95) coupled with a low probability of false alarm at a range that provides adequate time-to-intercept for effective deployment of decoys and possibly for evasive maneuvers. The main impacts of this consideration are on the average transmitter power and the required antenna aperture. The other major challenge is to provide the subclutter visibility (i.e., on the order of 80 to 90 dB) necessary to reject the large ground-clutter signals, particularly at low altitudes that might mask the desired target or cause false alarms.

The basic search radar detection range equation applicable in a noise-limited background is given by [41]

$$R = \left[\frac{P_A A_r t_s \sigma}{4\pi \Psi_s k T_s S_A L_s} \right]^{1/4}$$

where R is the detection range over solid angle Ψ_s, P_A is the average transmitted power in watts, A_r is the effective antenna receiver aperture in square meters, t_s is the time to search solid angle Ψ_s in seconds, σ is the average radar cross section in square meters , Ψ_s is the solid angle search with uniform energy in steradians, $k = 1.38 \cdot 10^{-23}$ W/Hz – K, T_s is the effective system noise temperature in kelvins, S_A is the effective integrated received signal energy-to-noise spectral density ratio E/N_0, and L_s is the total system losses including the pattern-propagation factor.

The search radar equation is primarily independent of frequency except that a number of parameters (i.e., σ, T_s, L_s) may vary with frequency. The range that must be achieved by the MAW is dependent on the warning time required to perform the necessary protective actions, which is on the order of 3 to 5 sec. Table EX-28.1 lists the detection ranges required to provide a 5 sec warning for various missile velocities [41]. The high-speed missile (Mach 5) becomes the design driver with a required detection range of 10.2 km, which is 1.7 times that of the more commonly encountered missiles (Mach 2.5). From (7.9), this translates into a power-aperture product that is of the order of nine times that of the lower speed missile. If the 5.9 km design requirement of the medium-speed missile (Mach 2.5) is adopted, then the warning time is reduced to about 3 sec for the high-speed missile. This warning time may be adequate in many applications.

The RCS of threat missiles varies both as a function of aspect angle and frequency. A good design value for frequencies below 2 GHz is between –10 and

Table EX-28.1 Detection Range Requirements for 5-Sec Warning

		Detection Range Requirements (km)			
Missile Velocity		*Mach 0.95 A/C*		*Mach 0.6 A/C*	
Mach	*m/s*	*Nose*	*Tail*	*Nose*	*Tail*
1.5	516	4.2	1.0	3.6	1.6
2.5	860	5.9	2.7	5.3	3.3
5.0	1720	10.2	7.1	9.6	7.6
(3-s warning)		(6.1)	(4.2)	(5.8)	(4.5)

–15 dBsm. The RCS of small-diameter missiles (e.g., 3 in.) is very frequency sensitive, providing enhancement of 10 dB at 430 MHz and reduction of about 20 dB at 5 GHz. Larger diameter missiles (greater than 8 in.) are less sensitive to frequency and fall within the –10- to –15 dBsm range for aspects to 30 degrees off the missile nose [41].

A typical MAW pulse-Doppler radar design is presented in [42]. The parameters for this design are tabulated in Table EX-28.2. The performance of this system is presented in Table EX-28.3 for a 90% cumulative detection probability and a Mach 0.95 aircraft. Note that the reaction times are generally greater than 3 sec for all attack aspects and greater than 6 sec for the most common missiles. The Doppler filter bandwidth is set at 40 Hz to accommodate missiles with 16-g acceleration.

Radar clutter effects vary as a function of aircraft speed, altitude, and the approach angle of the attacking missile. The most frequently encountered case is a tail-chase missile attacking a high-speed, low-altitude aircraft. Strong ground clutter returns are received from the main lobe, sidelobes, and backlobes of the tail-looking MAW antenna. These signals can be characterized as main lobe, sidelobes, and altitude line clutter, each of which has unique characteristics and affects the design in a different manner.

Main lobe clutter is usually the strongest, owing to the power gain of the antenna's main lobe. For an aft-looking antenna, the main lobe clutter appears to be moving away from the aircraft, producing negative Doppler that can be separated from the missile return (which is moving toward the aircraft and producing positive Doppler) by Doppler filtering. For a nose-mounted antenna, the clutter produces positive Doppler, but the combined closure of the missile and aircraft produces a

Table EX-28.2 Typical MAW Pulsed Doppler Radar

Parameter	*Value*
Transmitter frequency	1.5 GHz
Transmitter average power	500 W
Noise figure	3 dB
Total system losses	10 dB
Antenna effective aperture	40 in.^{2}
Antenna main lobe/backlobe ratio	35 dB
Receiver dynamic range	110 dB
Subclutter visibility	70 dB
Doppler filter bandwidth	40 Hz
Pulse repetition frequency	25 kHz
Pulsewidth	10 μs
Angle coverage	All aspect

Table EX-28.3 MAW Pulsed Doppler Radar Performances

Missile Velocity (Mach)	*Nose Detection (km)*	*Attack Time to Go (sec)*	*Tail Detection (km)*	*Attack Time to Go (sec)*
1.5	7.7	9.2	9.7	51.3
2.5	7.2	6.0	8.4	15.9
5.0	6.2	3.0	7.0	50

much greater positive Doppler, allowing missile detection by Doppler discrimination.

Altitude line clutter for an aircraft in horizontal flight has zero Doppler shift since it is neither approaching nor receding and, thus, can be discriminated from targets of interest. Altitude line clutter is usually received through sidelobes, which have low-relative gain, but can be very strong at low altitudes over water or other surfaces that produce specular-like reflections.

Sidelobe clutter is generally much weaker than either main lobe or altitude line clutter. However, for an aft-looking antenna, backlobe clutter produces positive Doppler that competes with the positive Doppler produced by missiles that are closing on the aircraft at relative speeds less than the aircraft speed. Thus, for this situation, the clutter must be attenuated by providing low antenna backlobes that reduce the clutter magnitude below that of the missile return.

An estimate of the subclutter visibility requirements can be made by estimating the magnitude of the clutter return. The area of the clutter patch at low altitudes is given by $A_c = \theta_{az} \cdot R \cdot R_T$, where θ_{az} is the azimuth antenna beamwidth, R is the range to the patch, and R_T is the range gate length [41]. From Table EX-28.2 at 6 km range, the clutter patch area is $A_c = (20/7 \cdot 2.54) \cdot 6{,}000 \cdot 1{,}500 = 10^7$ m^2. The equivalent RCS of the clutter is then $\sigma_c = \sigma^0 \cdot A_c$, where σ^0 is the backscattering coefficient equal to 0.01 for mountainous and urban return. The signal-to-clutter is given by $S/C = \sigma_T / \sigma_c$, where σ_T is the target RCS of 0.1 m^2. The resultant S/C is then 10^{-5}.

A low false alarm rate of less than one per hour is desired [41]. To search 4π steradians in 1 sec requires the order of $4\pi/(\lambda/D)^2 = 10$ beam positions/s. Because each beam position has three range gates, this results in 30 range gate false alarm positions per second. The Doppler filter bank for each range gate must span a velocity differential of Mach 5 (1,720 m/s) to detect both nose and tail attacking missiles. Each Doppler filter has a velocity range of $v = (\lambda \cdot B_d)/2 = 4$ m/s, resulting in the need for 1,720/4 = 430 Doppler filters per range gate. Hence, in 1 sec, there are 12,900 possibilities for a false alarm; and in 1 hr, there are approximately 10^7 to 10^8 possibilities for a false alarm requiring a $P_{fa} = 10^{-8}$. The signal-to-clutter ratio required for detecting a Swerling case 3 target with $P_d = 0.9$ and $P_{fa} = 10^{-8}$ is 18 dB. Hence, the required subclutter visibility for the MAW described by Table EX-28.2 is 50 + 18 = 68 dB, which can be rounded to 70 dB.

The 70-dB subclutter visibility can be achieved by Doppler filtering in situations where the missile is closing on the aircraft at speeds greater than the missile is closing on the ground. The phase noise from the transmitter (i.e., "transmitter clutter") and from the stable local oscillator in the MAW receiver must be such that random modulations do not spread clutter components into the Doppler filter band [41]. This phase noise must be of the order of –120 dBc [41]. Also, timing of range gates and PWs must have stability in the picosecond region over the Doppler filter processing time.

In tail-chase situations, the backlobes of the antenna induce positive Doppler components on the clutter that range in magnitude up to the forward velocity of the aircraft. When the missile is closing on the aircraft at speeds less than twice the forward velocity of the aircraft, then its Doppler signature must directly compete with the backlobe clutter. This situation requires low antenna backlobes to allow detection of the missile. Because signals are both transmitted and received through the backlobes, a reduction of gain in the backlobes equal to the SCV/2 = 35 dB with respect to main beam gain provides the necessary clutter attenuation for missile detection.

Receiver dynamic range is also important because the strong main lobe and altitude line clutter signals are present in the receiver prior to Doppler filtering. Any nonlinearity in the receiver generates intermods between the various clutter components that spread the clutter spectrum into the Doppler filter passband [41]. A dynamic range on the order of 110 dB is required in the MAW design to allow missile detection at low aircraft altitudes.

The operating frequency of the MAW affects several aspects of its design. First, lower frequencies (i.e., in the UHF band) suffer because antennas with sufficient aperture to provide missile detection cannot provide the main lobe-to-sidelobe gain ratios necessary to meet the subclutter visibility requirements. Also, the resulting large antenna beamwidths at low frequency do not provide the necessary AOA accuracy required to counter the threat. Higher frequencies (i.e., in X- or K_a-band) result in a complex antenna with narrow antenna beams that must be scanned rapidly through the coverage volume to detect the threat. In addition, small-diameter missiles (i.e., on the order of 3 to 5 in.) experience an RCS reduction at high frequencies. Also, weather effects are prevalent at these higher frequencies. The results of a trade-off study considering these issues and others concluded that the L-band (1 to 2 GHz) was the optimum operating frequency for a MAW pulse Doppler radar [41].

Another issue that affects MAW pulse Doppler radar design is that of false alarms generated by aircraft entering the MAW radar's FOV. The false alarms are due to jet engine modulation of the radar return by the aircraft engine that generates Doppler components that correspond to those of missiles. This can generally be controlled by comparing the velocities derived from the Doppler against those derived by differentiations of the range measurements. If they provide the same result, the target is a missile; otherwise, it is an aircraft.

The parameters of a typical MAW pulse Doppler radar are given in Table EX-28.2. Using these parameters and assuming a Swerling 3 type target, a search frame time of 1 sec, a probability of detection of 0.98, and an aperture efficiency of 70%, the detection range provided for antenna diameters from 1 to 12 in. is calculated in MATLAB program maw_ant_pd.m. The result is shown in Figure EX-28.2 for missiles with RCS of 0.1 m^2 and 0.0316 m^2.

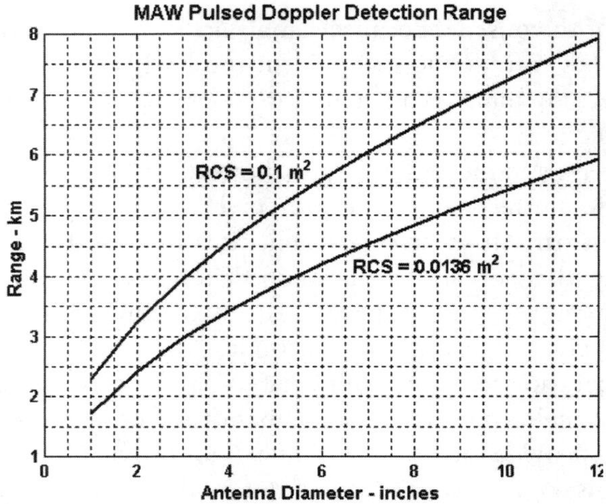

Figure EX-28.2 MAW antenna aperture for various missile detection ranges.

References

[1] Stimson, G., *Introduction to Airborne Radar*, 2nd ed., Mendham, NJ: SciTech Publishing, 1998.

[2] Clarke, J., "Airborne Early Warning Radar," *Proc. IEEE*, Vol. 73, No. 2, February 1985, pp. 312–324; also in M. Skolnik, (ed.), *Radar Applications*, New York: IEEE Press, 1988.

[3] Kronhamn, T., "AEW Performance Improvement with the Erieye Phased Array Radar," *Proc. IEEE National Radar Conference*, Boston, MA, May 1993, pp. 34–39.

[4] Clarke, J., and C. Stewart, "Airborne Radar, Part 1: Air-to-Surface," *Microwave J.*, Vol. 29, No. 1, January 1986, pp. 32–46.

[5] Clarke, J., and C. Stewart, "Airborne Radar, Part 2: Air-to-Surface," *Microwave J.*, Vol. 29, No. 2, February 1986, pp. 44–54.

[6] Schleher, D. C., *MTI and Pulsed Doppler Radar*, Norwood, MA: Artech House, 1991.

[7] Perkins, L., H. Smith, and D. Mooney, "The Development of Airborne Pulse Doppler Radar," *IEEE Trans. on Aerospace and Electronic Systems*, Vol. AES-20, No. 3, May 1984, pp. 292–303.

[8] Stimson, G., *Introduction to Airborne Radar*, Hughes Aircraft Co., El Segundo, CA, 1983.

[9] Goetz, L., and J. Albright, "Airborne Pulsed-Doppler Radar," *IRE Trans. Military Electronics*, Vol. MIL-5, No. 2, April 1961, pp. 116–126; also in D. Barton, (ed.), *CW and Doppler Radar*, Vol. 7 of *Radars*, Dedham, MA: Artech House, 1978.

[10] Povejsil, D., R. Raven, and P. Waterman, *Airborne Radar*, New York: Van Nostrand, 1961.

[11] Long, W., and K. Harriger, "Medium PRF for the AN/APG-66 Radar," *Proc. IEEE*, Vol. 73, No. 2, February 1985, pp. 301–311; also in M. Skolnik, (ed.), *Radar Applications*, New York: IEEE Press, 1988.

[12] Mooney, D., and W. Skillman, "Pulse-Doppler Radar," in M. I. Skolnik, *Radar Handbook*, 1st ed., New York: McGraw-Hill, 1970.

[13] Schleher, D. C., *Introduction to Electronic Warfare*, Dedham, MA: Artech House, 1986.

[14] Nathanson, F., *Radar Design Principles*, New York: McGraw-Hill, 1969.

[15] Aronoff, E., and N. Greenblatt, "Medium PRF Radar Design and Performance," *Proc. 20th Tri-Service Radar Symposium*, 1974, pp. 53–67; also in D. Barton, (ed.), *CW and Doppler Radar*, Vol. 7 of *Radars*, Dedham, MA: Artech House, 1978.

[16] Williams, F., and M. Radant, "Airborne Radar and the Three PRFs," *Microwave J.*, July 1983, pp. 129–133; also in M. Skolnik, (ed.), *Radar Applications*, New York: IEEE Press, 1988.

[17] Hovanessian, S., "Medium PRF Performance Analysis," *IEEE Trans. on Aerospace and Electronic Systems*, Vol. AES-18, No. 3, May 1982, pp. 286–296.

[18] Barton, D. K., *Modern Radar System Analysis*, Norwood, MA: Artech House, 1988.

[19] Hovanessian, S. I., *Radar System Design and Analysis*, Dedham, MA: Artech House, 1984.

[20] Schleher, D. C., *Electronic Warfare in the Information Age*, Norwood, MA: Artech House, 1999.

[21] Morchin, W., "Antenna Sidelobe Requirements for the Three PRFs of an AEW Radar," *Microwave J.*, Vol. 31, No. 9, September 1988, pp. 83–87.

[22] Morris, G., "Antenna Sidelobe Requirements for the Medium PRF Mode of an Airborne Radar," *Microwave J.*, Vol. 32, No. 9, September 1989, pp. 121–134.

[23] Farrell, J., and R. Taylor, "Doppler Radar Clutter," *IEEE Trans. on Aerospace and Navigational Electronics*, Vol. ANE-11, September 1964, pp. 162–172; also in D. Barton, (ed.), *CW and Doppler Radar*, Vol. 7 of *Radars*, Dedham, MA: Artech House, 1978.

[24] Friedlander, A., and L. Greenstein, "A Generalized Clutter Computation Procedure for Airborne Pulse Doppler Radars," *IEEE Trans. on Aerospace and Electronic Systems*, Vol. AES-6, No. 1, January 1970, pp. 51–61; also in D. Barton, (ed.), *CW and Doppler Radar*, Vol. 7 of *Radars*, Dedham, MA: Artech House, 1978.

[25] Ringel, M., "An Advanced Computer Calculation of Ground Clutter in an Airborne Pulse Doppler Radar," *NAECON Record*, 1977, pp. 921–928; also in D. Barton, (ed.), *CW and Doppler Radar*, Vol. 7 of *Radars*, Dedham, MA: Artech House, 1978.

[26] Jao, J., and W. Goggins, "Efficient, Closed-Form Computation of Airborne Pulse-Doppler Radar Clutter," *Proc. IEEE Int. Radar Conference*, Washington, D.C., May 1985, pp. 17–22.

[27] http://web.interware.hu/f15e/techinfo/an_apg-70_eng.htm.

[28] C. Smith, "The B-2 Radar," *Proc. AUTOTESTCON '91*, Anaheim, CA, September 1991, pp. 17–33.

[29] Brookner, E., "Phase Arrays and Radars—Past, Present and Future," *Microwave J.*, January 2006, pp. 24–38.

[30] Malas, J., "F-22 Radar Development," *NAECON 1997 Record*, Dayton, OH, July 1997, pp. 831–839.

[31] http://f-22raptor.com/af_radar.php.

[32] http://kuku.sawf.org/Emerging+Technologies/2667.aspx.

[33] http://www.jsf.mil/f35/f35_technology_das.jpg.

[34] Hommel, H., and H. Feldie, "Current Status of Airborne Active Phased Array (AESA) Radar Systems and Future Trends," *Proc. European Radar Conference 2004*, Amsterdam, 2004, pp. 1517–1520.

[35] "Future DoD High Frequency Radar Needs Resources," *Defense Science Board*, April 2001.

[36] Skillman, W., and D. Mooney, "Multiple High-PRF Ranging," *Proc. IRE 5th National Convention on Military Electronics*, 1961, pp. 37–40; also in D. Barton, (ed.), *CW and Doppler Radar*, Vol. 7 of *Radars*, Dedham, MA: Artech House, 1978.

[37] Clarke, J., "The Chinese Remainder Theorem," Memorandum 3650, DTIC AD-A141 375, Royal Signals and Radar Establishment, February 1984.

[38] Hetrich, G., "Frequency Modulation Techniques as Applied to Pulse Doppler Radar," *IRE Convention Record*, Pt. 5, 1962, pp. 76–86; also in D. Barton, (ed.), *CW and Doppler Radar*, Vol. 7 of *Radars*, Dedham, MA: Artech House, 1978.

[39] Holbourne, P., and A. Kinghorn, "Performance Analysis of Airborne Pulse Doppler Radar," *Proc. IEEE Int. Radar Conference*, 1985, pp. 12–16.

[40] Barton, D. K., (ed.), *CW and Doppler Radar*, Vol. 7 of *Radars*, Dedham, MA: Artech House, 1978.

[41] Black, A., "Pulse Doppler for Missile Approach Warning," *J. Electronic Defense*, August 1991, pp. 44–56.

[42] Russell, W., "AN/ALQ-161 Tail Warning Function," *J. Electronic Defense*, August 1991, pp. 41–43.

[43] Tsui, J., *Digital Techniques for Wideband Receivers,* Norwood, MA: Artech House, 1995.

[44] Skolnik, M., (ed.), *Radar Handbook,* 2nd ed., New York: McGraw-Hill, 1990.

[45] Skolnik, M., *Introduction to Radar Systems,* 3rd ed., New York: McGraw-Hill, 2001.

[46] http://en.wikipedia.org/wki/Image:APG-70.jpg.

CHAPTER 8

Special Topics in Doppler Radar Systems

In this chapter, we discuss a number of special topics related to the use of the Doppler effect in radar systems. These topics generally involve the application of special Doppler signal processing techniques that provide a unique feature within the radar system not achievable otherwise.

Doppler beam sharpening (DBS) is used in airborne radars to provide the capability for real-time surface mapping [1]. The resolution of the surface map is generally sufficient to allow the detection of small fixed targets (e.g., cars, tanks, boats) in the presence of significant background clutter. Typically in this mode the forward-looking antenna scans either side of the aircraft, and with the exception of the region directly ahead of the aircraft, provides a finer azimuth resolution than can be achieved by the real antenna beam. The general principle is that the forward motion of the aircraft imparts different Doppler shifts on the reflectors illuminated by the main antenna beam. Then those reflectors within a Doppler resolution cell (inversely proportional to the antenna's dwell time on the target) can be collected.

Synthetic aperture radar (SAR) provides the capability of synthesizing very high resolution ground maps in an all weather environment from airborne or spaceborne platforms. SAR exploits the forward motion of the platform to produce the equivalent of a sidelooking antenna array that is thousands of feet long. It differs from a real array in that the array elements are collected sequentially as the platform progresses using a relatively small antenna rather than simultaneously in a long antenna. High-resolution SARs employ a focused array that applies an appropriate phase correction to each element of the array to compensate for the phase difference between the signal returned from a straight line path rather than from a circular path. If the array is not focused, it is called an *unfocused array* and provides a reduced resolution as shown later in Example 30.

Maximum resolution is achieved when the SAR is operated in *spotlight mode*, in which the real antenna continually points at the region of interest during the integration time to form the array. SARs require a large amount of processing [2] so that early designs operated by sending received data via a data link to be processed in a ground station. Newer designs such as JSTARS and Global Hawk [3] process the SAR data onboard the platform using advanced digital processing techniques. The long SAR processing time causes moving targets to wash out on the map. Therefore, a GMTI is generally interleaved with the SAR mapping. Markers indicating the moving targets' positions are superimposed on the SAR map and in some cases indicate their range rates [2].

Interferometric SAR (InSAR) [2, 4, 5] can measure the variations in the heights of terrain with a precision approaching the transmission wavelength. Combined with high-resolution SAR, a 3-D topographic map is produced. An InSAR obtains elevation data by determining the elevation angle of the line-of-sight to each resolution cell, the height of the radar, and the slant range to the cell.

Inverse SAR (ISAR) is a technique used operationally to image targets with rotational motion, such as ships, aircraft, and space objects [6]. The technique also has application to instrumentation radar for evaluating the RCS of targets and target models. ISAR is used operationally with the APS-137B (V) 5 radar to classify searchlighted ship targets as to type [7]. The principle employed is that rotation of the ship in pitch, roll, or yaw creates a differential Doppler that, when processed by an FFT, can generate a cross-range image of that aspect of the ship. For example, pitching of the ship causes the top of the masts to have a greater Doppler shift than the bottom of the masts or the superstructure referenced to the center of rotation. Resolving the Doppler profile allows the masts to be imaged. Similarly ISAR images can be created due to yaw or roll motion. Then as the ship pitches, rolls, or yaws ISAR can produce a vertical profile along the major or minor axis of the ship, a horizontal plan view, or a combination of all of these views. However, because the angular rates of the ship's motion are unknown, the ISAR image does not represent true cross-range dimensions as does the range developed from a high-resolution transmitted waveform. ISAR images are far from photographic quality and those who interpret these images are sometimes referred to as "blobologists."

The *Joint Surveillance and Target Attack Radar System* (JSTARS) is a long range, air-to-ground surveillance system employing the sidelooking X-band APY-3 radar, which provides three operational modes: (1) high-resolution SAR imaging for detecting and identifying stationary targets, (2) wide-area GMTI surveillance for situational awareness, and (3) sector GMTI search for battlefield reconnaissance. Targets detected with GMTI are displayed on a digitally stored map or SAR generated maps. The primary innovation in this system is an interferometric GMTI that allows slow-moving ground targets to be detected in the presence of heavy ground clutter. DBS is used to divide the main beam return into Doppler cells whose response is nulled using an interferometer formed by dividing the antenna aperture into three segments. When a moving target is present, its Doppler is displaced from the Doppler response of the ground return in that cell and, hence, not nulled by the interferometer [8, 9]. The limitation to this technique is internal clutter motion, which restricts the minimum ground target speed detection to about 4 knots.

Space Time Adaptive Processing (STAP) is a technique applicable to airborne pulsed Doppler phased-array radars that is a generalization of DPCA processing using the principles of adaptive antennas. Using a phased-array antenna, adaptive weights are applied to each element of the phased array. The weights are adaptively computed to provide the maximum signal-to-interference (clutter-plus-noise-plus-jamming) ratio at the output of the processor. This effectively places antenna nulls over clutter returns and jamming signals while causing the main lobe antenna beam to be focused on the target.

Fully adaptive STAP requires inversion of the large dimension covariance matrix formed in the space-time domain. This is generally beyond the computational capability of currently available processors. Current research in this area is focused on the development of computationally efficient algorithms that allow application of the STAP principle [6, 10]. STAP generally uses the *Sample Matrix Inversion* (SMI) algorithm to develop the adaptive weights [11]. When implementing STAP in a system, a key issue is estimating the covariance matrix used in the SMI algorithm. Errors in estimating the covariance matrix due to the availability of limited samples can cause significant limitations in performance relative to the predicted optimum performance [12].

Generally, MTI radar provides less detection performance than radar that operates in a noise environment. This is due to the many losses associated with detecting targets embedded in multiple-environment clutter. The interactive MATLAB-based MTI Design Tool allows the user to explore the performance of various MTI and MTD processors in an environment consisting of variable intensity combinations of sea clutter, land clutter, precipitation clutter, and chaff. The user enters parameters by overwriting a default MATLAB dialog box. The output is in the form of bar charts.

8.1 Doppler Beam Sharpening

DBS is a method for reducing surface clutter returns in airborne radar using Doppler filtering to narrow the antenna's azimuth beamwidth synthetically. Aircraft motion causes a unique Doppler frequency to be associated with the surface clutter returned from each angular direction within the antenna's main beam pattern (see Section 3.2.3). This causes a spectral spreading of the Doppler clutter return that is a function of the aircraft's velocity, azimuth antenna beamwidth, and the antenna's pointing angle relative to the aircraft's velocity vector. Conversely, the return from a target has a unique Doppler frequency associated with its angular direction. If a narrowband Doppler filter (bandwidth inversely proportional to dwell time) can be placed over the target, only the clutter returned from the angular directions associated with the filter's bandwidth compete with the target. Under favorable conditions, a significant increase in the S/C ratio can be achieved for either stationary or slowly moving targets. The effect is equivalent to reducing the antenna's beamwidth; therefore, the descriptive term *Doppler beam sharpening* [1, 13].

In forward-looking airborne radar that illuminates the earth's surface, all returns from scatterers within a basic resolution cell (see Figure 3.16 and Section 3.2.3) combine to form the radar return. The cell dimensions are proportional to the effective transmitter pulsewidth and the antenna azimuth beamwidth, which is relatively large in airborne applications due to the limited space available for antenna aperture.

The clutter RCS backscattered from the surface for a small beamwidth can be expressed as

$$\sigma_{cx} = \sigma^0 R_c \theta_{az} \left(\frac{c\tau_e}{2} \right) \sec \psi \tag{8.1}$$

where σ^0 is the backscatter coefficient, R_c is the range to the clutter patch, θ_{az} is the azimuth beamwidth, τ_e is the radar's effective pulsewidth, and ψ is the grazing angle from the radar to the clutter cell. The single-pulse S/C ratio for a surface target located in the clutter cell is then given by

$$\frac{S}{C} = \frac{\sigma_t}{\sigma^0 R_c \theta_{az}\left(\frac{c\tau_e}{2}\right) \sec \psi L_c} \tag{8.2}$$

where σ_t is the target's RCS and L_c is a loss term that represents the difference in losses between the clutter and the target. Note that beam shape losses between the clutter and target are generally equal so that their effect should not be included in L_c. For example, assume that a 5 m^2 fixed target (i.e., small car) is to be detected at 10 nmi from a K_u-band airborne radar (λ = 1.85 cm) with a 3 ft horizontal aperture (i.e., $\theta_{az} = 1.5°$) and a 0.2 μs effective pulsewidth. The clutter area for this situation is $A_c = 1.46 \times 10^4$ m^2. For a low grazing angle (i.e., $\psi = 2°$) over farmland terrain, the backscattering coefficient is σ^0 = –30 dB, resulting in S/C = –4.6 dB. MTI will not be effective in this situation because there is no relative motion between the fixed target and the ground clutter. Hence, the target cannot be generally detected with this type of system without the addition of a DBS mode.

The general geometry for DBS is depicted in Figure 8.1. The Doppler shift to the nth point within the range slice is given by

$$f_{dn} = \frac{2v_{ac}}{\lambda} \cos \alpha_n \tag{8.3}$$

where v_{ac} is the aircraft velocity and α_n is the angle between the line from the *n*th point to the radar and the aircraft's velocity vector. As is apparent from (8.3), except for isodops ($\alpha_n = k$), each point on the surface produces a unique Doppler shift. Now assume that the azimuth dimension of the clutter cell is subdivided into a number of smaller cells, each with equally sized angles so that the frequency difference within each cell is given by

$$\Delta f_n = \frac{2v_{ac}}{\lambda} (\cos \alpha_{n1} - \cos \alpha_{n2}) \tag{8.4}$$

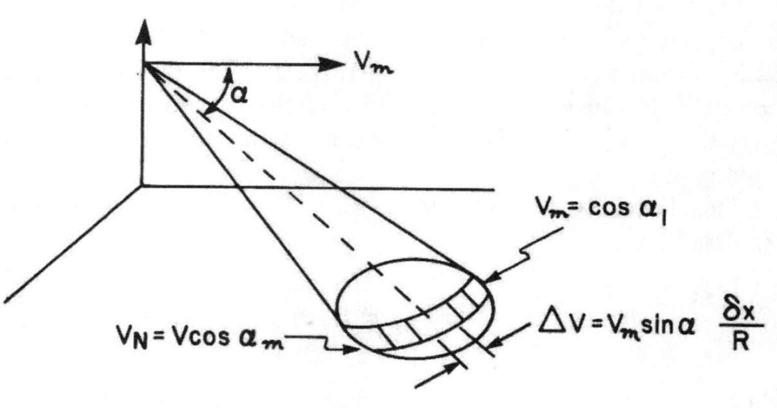

Figure 8.1 DBS geometry. (*From:* [1, p. 470].)

Letting $\Delta\theta$ be the differential angle represented by each subcell such that $\alpha_{n1} = \alpha_m - \Delta\theta/2$ and $\alpha_{n2} = \alpha_m + \Delta\theta/2$, where α_m is the median angle to the center of the clutter cells, results in $\cos\alpha_{n1} - \cos\alpha_{n2} = \Delta\theta \sin\alpha_m$. Setting the Doppler filter bandwidth equal to $B_D = \Delta f_n$ and solving for the Doppler beam-sharpened azimuth angle results in

$$\Delta\theta = \frac{\lambda B_D}{2v_{ac}\sin\alpha_m} \tag{8.5}$$

which can also be expressed in terms of cross-track range resolution as

$$\delta_{az} = \frac{R_c \lambda B_D}{2v_{ac}\sin\alpha_m} \tag{8.6}$$

As is apparent from (8.5) and (8.6), the highest resolutions are achieved with the largest aircraft velocities and when the beam is positioned perpendicular to the aircraft's ground track (i.e., $\alpha_m = 90°$). For the beam positioned along the ground track (i.e., $\alpha_m = 0°$), there is no improvement in resolution due to DBS.

For a scanning surveillance radar, the minimum Doppler resolution is a function of the time that the radar beam dwells on the target ($B_D = 1/T_d$). Hence, the angular resolution (in radians) can be expressed as

$$\Delta\theta = \frac{\lambda}{2v_{ac}T_d\sin\alpha_m} \tag{8.7}$$

where T_d is the target dwell time. The denominator of (8.7) is equivalent to an effective synthetic aperture equal to

$$D_e = 2v_{ac}T_d\sin\alpha_m \tag{8.8}$$

so that the improvement in azimuth resolution and equivalent S/C ratio is given by

$$I = \frac{D_e}{D_{az}} = \frac{2v_{ac}T_d\sin\alpha_m}{D_{az}} \tag{8.9}$$

where D_{az} is the horizontal width of the antenna.

Returning to the previous example, assume that $v_{ac} = 400$ knots and the radar scans at 20°/s [13]. Then, the effective aperture $D_e = 72$ ft at a scan angle of 45° results in an improvement factor of $I = 14.6$ dB and S/C = +10 dB. Alternatively, (8.2) can be modified by using the synthetic aperture beamwidth, resulting in

$$\frac{\mathrm{S}}{\mathrm{C}} = \frac{2\sigma_t v_{ac}T_d\sin\alpha_m}{\sigma^0 R_c\lambda\left(\frac{c\tau_e}{2}\right)\sec\psi L_c} \tag{8.10}$$

The cross-track resolution cell width, from (8.6), is equal to $\delta_{az} = 53$ ft and, hence, additional beam sharpening and higher range resolution can improve the S/C on the fixed target. For example, if the pulsewidth is reduced to 0.1 μs (50 ft), the S/C would be increased to 13 dB.

The spectral spreading of the ground clutter due to platform motion for the situation depicted in Figure 8.2 is given by

$$\Delta f_{3\text{ dB}} = \frac{2v_{ac}\Delta\theta_{3\text{ dB}} \sin \alpha_m}{\lambda} \tag{8.11}$$

For the example given earlier ($v_{ac} = 400$ knots, $\alpha = 45°$, $D_{az} = 3$ ft), the spectral spreading can be evaluated as $\Delta f_{3\text{ dB}} = 400$ Hz and the Doppler filter bandwidth is $B_D = 1/T_d = 13.33$ Hz. To achieve DBS, the range of possible Doppler returns from the surface clutter must be filled by a bank of contiguous Doppler filters. This filter bank is depicted in Figure 8.2 for the example problem and consists of 30 Doppler filters uniformly distributed ±200 Hz about a center Doppler frequency of 15 kHz. The center Doppler frequency is given by

$$\Delta f_{dm} = \frac{2v_{ac} \sin \alpha_m}{\lambda} \tag{8.12}$$

Note that the bandwidths of the Doppler filters depend on the target dwell time and, hence, are constant for a given antenna azimuth scan rate. However, the filters must cover the total spectral spread given by (8.11); hence, a sufficient number of Doppler filters is needed to cover the spectral spread due to the maximum expected aircraft velocity and scan angle.

Note that the analysis generally should include other clutter spectral spreading effects such as antenna scanning modulation, internal clutter motion, and transmitter frequency instability (see Section 3.2). These effects tend to widen further

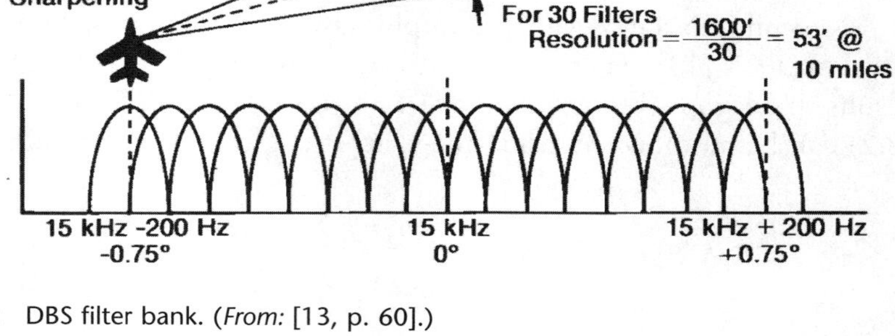

Figure 8.2 DBS filter bank. (*From:* [13, p. 60].)

the clutter spectrum over that caused by platform motion, but they typically are much smaller than platform motion spectral spreading and thus can be neglected. If the target has motion, it will generally translate its Doppler frequency and may cause the target Doppler to fall in a different Doppler filter than its surrounding clutter return.

The DBS resolution provided for the example problem as a function of range and various scan angles is depicted in Figure 8.3. Note that the resolution is better for shorter ranges and larger scan angles. Insight into the Doppler filtering requirements for DBS can be deduced from these curves. For example, at 10 nmi and a scan angle of 45°, 30 Doppler filters (B_D = 13.33 Hz) are required to span the clutter spectrum induced by platform motion. Each filter provides an equivalent azimuth cross-track resolution of 1,600/30 = 53 ft. At 10 nmi and a scan angle of 20.7°, the spectral width is halved, requiring only 15 Doppler filters to cover the surface clutter spectrum. The cross-track azimuth resolution is 1,600/15 = 106 ft for this situation. Therefore, a changing number of Doppler filters must be produced for a constant data time window (T_d), where the number of filters is proportional to the aircraft velocity and the sine of the antenna's azimuth scan angle.

Figure 8.4 shows the DBS image of a steam plant taken from 10,000 ft of altitude at a range of 8 nmi. The azimuth squint angle was 13° with range and cross-range resolution of 100 feet. An aerial photo is included of the scene.

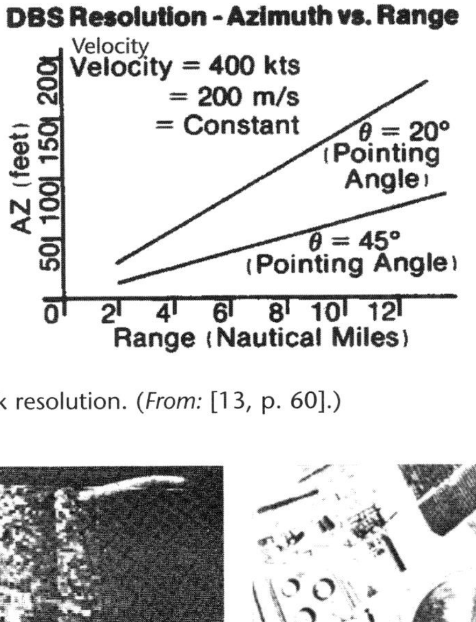

Figure 8.3 DBS cross-track resolution. (*From:* [13, p. 60].)

Figure 8.4 DBS image of steam plant with photo. (*From:* [13, p. 58].)

▶ Example 29 DBS Design

Doppler beam sharpening is a technique used in airborne radar to reduce surface clutter return using Doppler filtering. The overall effect is similar to a decrease in the antenna beamwidth, increasing the cross-range resolution. The aircraft motion associated with the antenna scan causes, for each angular direction, a different Doppler frequency return. As a result, we have a spectral spreading of the Doppler clutter return that is a function of the aircraft velocity, antenna beamwidth, and antenna's pointing angle relative to the aircraft velocity vector.

DBS provides the capability for real-time ground mapping and the potential for increased subclutter visibility against fixed ground targets. The DBS mode can be applied to low- and medium-PRF airborne forward-looking radar and is effective in enhancing air-to-ground weapons delivery operation. DBS is also a principal component in interferometric GMTI as described in Section 8.3.1.1.

In forward-looking airborne radar that illuminates the earth's surface, all returns from scatterers within a basic radar resolution cell combine to form the radar return. The cell dimensions are proportional to the effective transmitter pulsewidth and the antenna azimuth beamwidth, which is relatively large in airborne applications due to the limited space available for the antenna aperture.

A Doppler frequency shift is induced on the return from each scatterer that is proportional to the relative closing velocity of the aircraft to that scatterer. Because scatterers are distributed through the azimuth cell at different angles with respect to the aircraft's velocity vector, their Doppler signatures are different, depending on the relative angular displacement with respect to the center of the cell. This creates a spectrum of Doppler return frequencies distributed about the mean frequency, which is associated with the center of the resolution cell. The spectral spreading is a function of the aircraft's velocity, antenna azimuth beamwidth, and antenna's pointing angle relative to the aircraft's pointing angle.

In the ground-mapping mode, a bank of Doppler filters, each with bandwidth inversely proportional to the scanning antenna's dwell time on the target, is used to process the return. The output of each Doppler filter within the bank then contains only those returns associated with a particular narrow angular segment, which is appreciably smaller than the antenna's azimuth beamwidth. This is the DBS effect. Also, because the returns from targets come from narrow angular regions and hence have distinct Doppler signatures, they tend to fall within a single Doppler filter in the bank. The signal-to-clutter ratio within this filter for both stationary and slowly moving targets is much larger than it would have been if the total clutter return within the azimuth beamwidth were contained within the filter. This is the subclutter visibility improvement effect of the DBS mode.

Consider an aircraft traveling parallel to the ground (constant height relative to the ground) with velocity v_{ac}, operating with forward-looking airborne radar that illuminates the earth's surface. Figure 8.2 shows the general geometry of the Doppler beam sharpening. In this figure, we notice the aircraft traveling with velocity v_{ac}, and its radar antenna pointing downward to the right, making an angle α with the aircraft's velocity vector.

The Doppler shift associated with the ground return can be expressed as

$$f_d = \frac{v_{ac} \cos \alpha}{\lambda}$$

where λ is the radar's operating wavelength and

$$\cos \alpha = \cos \phi \cos \theta$$

with θ the azimuth angle and ϕ the antenna's depression angle. The 3 dB clutter spectrum can then be found as [11, 14]

$$f_{3\,\text{dB}} = f_{ac} \cos \phi_o \left[\theta_B \sin \theta_o + \frac{\cos \theta_o}{2} \left(\theta_B^2 + \frac{R_\tau (\sin \phi_o)^3}{h(\cos \phi_o)^2} \right) \right]$$

where θ_B is the antenna's 3 dB beamwidth , ϕ_o is the mean depression angle, θ_o is the mean azimuth angle, $R_\tau = c\tau/2$, and h is the aircraft's height. The DBS improvement factor is given by

$$I_{\text{dbs}} = f_{3\,\text{dB}} T_d$$

where the dwell time is $T_d = \theta_B / \dot{\theta}$ and $\dot{\theta}$ is the antenna scan rate. If the Doppler filter bandwidth (B_D), equal to $1/T_d$, is greater than $f_{3\,\text{dB}}$ then the improvement factor equals 1. This can be used to find the DBS effective beamwidth as

$$\theta_{\text{dbs}} = \frac{\theta_B}{I_{\text{dbs}}}$$

which can further be used to find the cross-range resolution:

$$\delta x = R \theta_{\text{dbs}}$$

and the effective aperture as

$$D_{\text{eff}} = D\, I_{\text{dbs}}$$

A MATLAB design tool that calculates the DBS parameters can be actuated by typing dbs_proc into the MATLAB Command window. The dialog box shown in Figure EX-29.1 then appears. Then a plot of the 3 dB spectrum width appears, as depicted in Figure EX-29.2. This results in the DBS improvement factor shown in Figure EX-29.3.

Note that the improvement factor is greatest at 90 degrees and one at 0 degrees. This illustrates that DBS is not useful along the aircraft's ground track but performs best in a sidelooking radar such as that used in JSTARS. Next the cross-track resolution plotted in dB/m is illustrated in Figure EX-29.4. As expected the resolution is best at 90 degrees and equals the real beam resolution along the ground track.

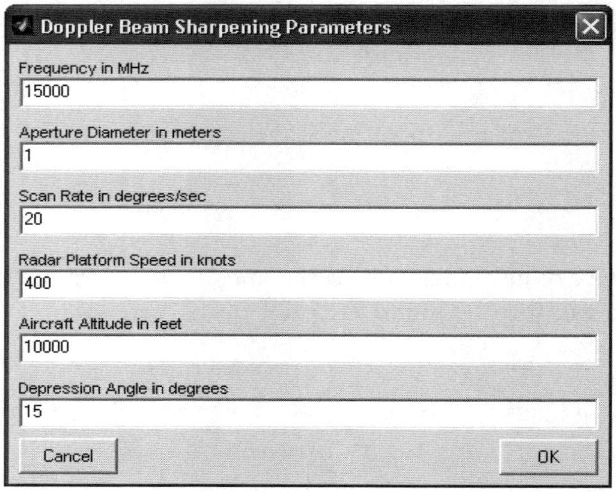

Figure EX-29.1 DBS design parameters.

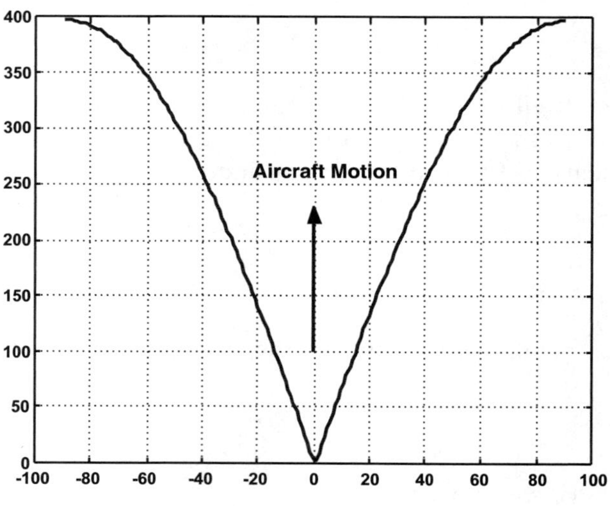

Figure EX-29.2 DBS clutter spectrum bandwidth.

Finally the DBS effective beamwidth is shown in Figure EX-29.5. This is truncated along the ground track since the clutter spectrum does not occupy an FFT filter's beamwidth and becomes significantly less than the real beamwidth at about 20 degrees.

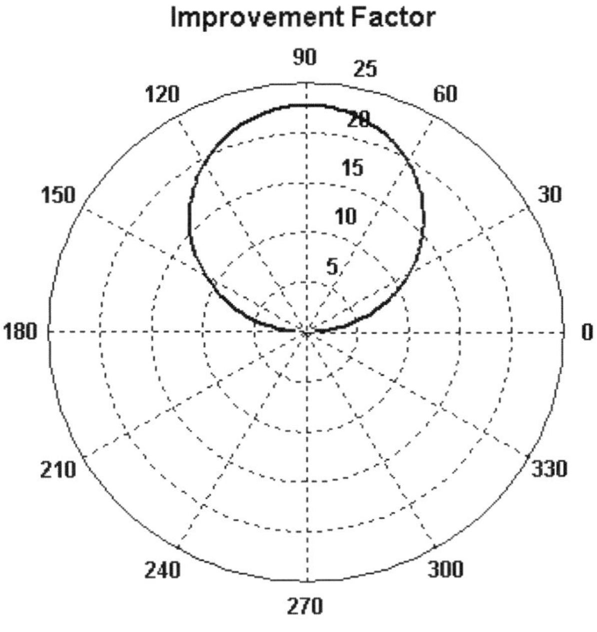

Figure EX-29.3 DBS improvement factor (in decibels) as function of azimuth angle.

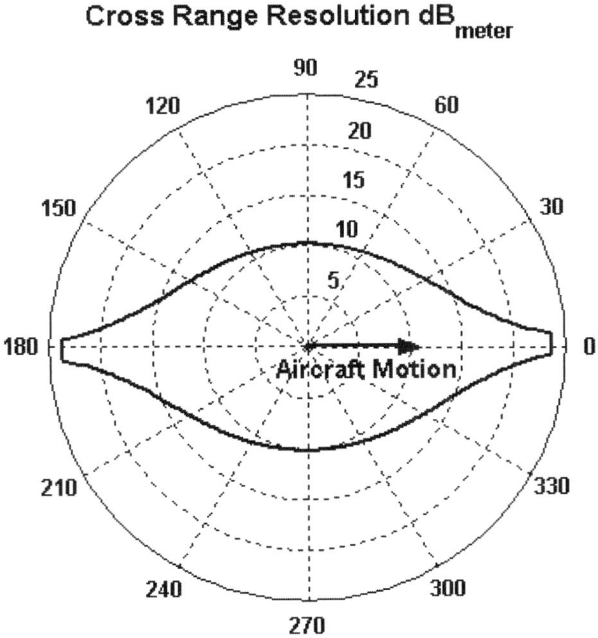

Figure EX-29.4 DBS cross-range resolution as a function of azimuth angle.

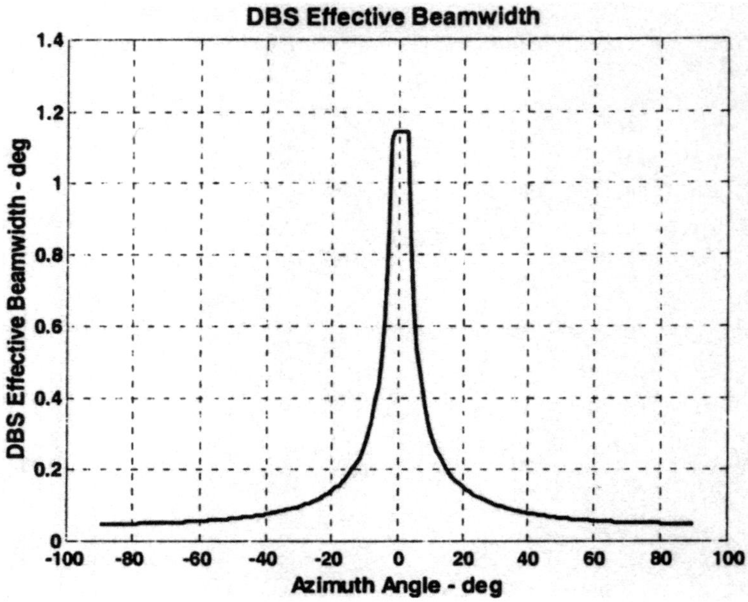

Figure EX-29.5 DBS effective beamwidth.

▶

8.2 Synthetic Aperture Radar

Synthetic aperture radar achieves high resolution in the cross-range dimension by taking advantage of the motion of the vehicle carrying the radar to synthesize the effect of a large antenna aperture. The basic action of the SAR is to produce an azimuth angle resolution higher than can be achieved with a conventional radar. As the term *synthetic aperture* implies, this can be viewed as the result of synthetically creating the equivalent of a large linear array antenna. Alternatively, this can be viewed as dividing the conventional antenna's azimuth cell into a number of Doppler cells, each associated with a particular azimuth segment (i.e., DBS; see Section 8.1). The two models are equivalent, and either can be used to determine the characteristics of SAR.

SAR modes are being incorporated into many airborne types of radar as well as UAVs and other reconnaissance radars. SARs provide photo-like high-resolution imagery of terrain and fixed surface targets at long ranges in all types of weather. In addition they generally provide ground moving target indication (GMTI), allowing them to detect and track slowly moving ground vehicles traveling along roads and in open terrain.

In combat radars the SAR mode assists in mapping, target acquisition, and weapon delivery. Reconnaissance radars provide SAR imagery that allow targets such as mobile missile launchers as well as fixed targets to be located.

The delivery of weapons from an airborne platform against stationary and slowly moving ground targets generally requires a high-resolution image of the ground. This is difficult to achieve with conventional airborne radar in the cross-range direction due to the limited space available for antenna aperture. SAR techniques overcome this limitation through the synthesis of a large aperture by the coherent combination of a number of sequential measurements formed as the antenna moves along the aircraft's flight path. Furthermore the rapid advance of digital signal processing and associated components has made it practical to accomplish the SAR function in an onboard processor. This mode has been incorporated into a number of currently operational airborne radars as described in Section 7.4.

The SAR mode is used primarily in reconnaissance operation to produce a high-resolution strip-map of an area of interest. Practical designs range from noncoherent real beam aperture sidelooking airborne radars (e.g., AN/APS-94) to those using focused synthetic aperture processors (AN/UPD-8). The present trend is toward fully focused SAR using real-time digital processing. Many of the synthetic aperture designs send the unprocessed data to the ground for processing via a wide bandwidth data link.

SAR can also be used in squint and spotlight modes. In a squint mode, the radar antenna looks in the forward direction (15° to 45° with respect to the ground track) of aircraft motion rather than to the side. This mode is similar to DBS.

The spotlight mode causes the antenna to track the target area as the aircraft moves, thereby increasing the potential processing time and, hence, the resolution in a particular region of high interest.

The resolutions achieved by operational strip-map SARs range from 3 m for older systems (i.e., UPD-8) to 1 m for new systems (i.e., Global Hawk) [3, 6]. Resolutions of 0.3 m are available in spotlight mode SAR. This can be contrasted against the feature size necessary for feature resolution of various militarily significant targets listed in Table 8.1. Examination of the table indicates that current SARs allow aircraft and small military vehicles to be identified from ranges on the order of 200 km.

The classic SAR is appropriate for providing high-resolution ground maps, but ill suited for imaging targets such as ships and aircraft that have rotational motion. Unless the differential Doppler shifts, which such motions produce, are accurately predicted and compensated, they tend to defocus the array and blur the image. With slightly different algorithms, these shifts, rather than those due to the radar's

Table 8.1 Feature Recognition Versus Resolution [2]

Feature	*Square Resolution of Cell Size*
Coastline, cities, large discretes Mountain outlines	300 m
Detect major highways, field variations, Large airfields	40–50 m
City street structure, highways, building Shapes, small airfields, road map type Imagery	15–30 m
Vehicles, building, and house details	3–10 m
Aircraft and vehicle types	0.5–3 m

relative velocity, can be used to provide the angular resolution needed for imaging. This is the ISAR technique discussed earlier.

Consider the conventional strip-map SAR geometry depicted in Figure 8.5. The azimuth beamwidth of a uniformly illuminated antenna with real aperture width D_{az} at wavelength λ in radians is given by $\theta_{az} = \lambda/D_{az}$. By direct analogy, the beamwidth of a synthetic array of length L_e is given by the finer beamwidth

$$\theta_s = \frac{\lambda}{2L_e} \tag{8.13}$$

The factor 2 in the denominator is explained as follows [2, pp. 416–417; 15, pp. 21.2–21.3]: In a real array the antenna pattern directivity is determined by progressive phase shifts received at successive array elements during the reception process from points off the antenna boresight. However, in a synthetic array each element both transmits and receives so that its effective antenna pattern is determined by round-trip phase shifts that are double that of a real array. Thus, while the uniformly illuminated real array antenna voltage pattern can be approximated by a sinc(x) function, the corresponding synthetic array results in a sinc$(2x)$ radiation pattern that accounts for a doubling of the phase shifts. The antenna elements in the synthetic array (see Figure 8.5) are formed due to vehicle motion between consecutive radar transmitted pulses and, hence, are separated by the

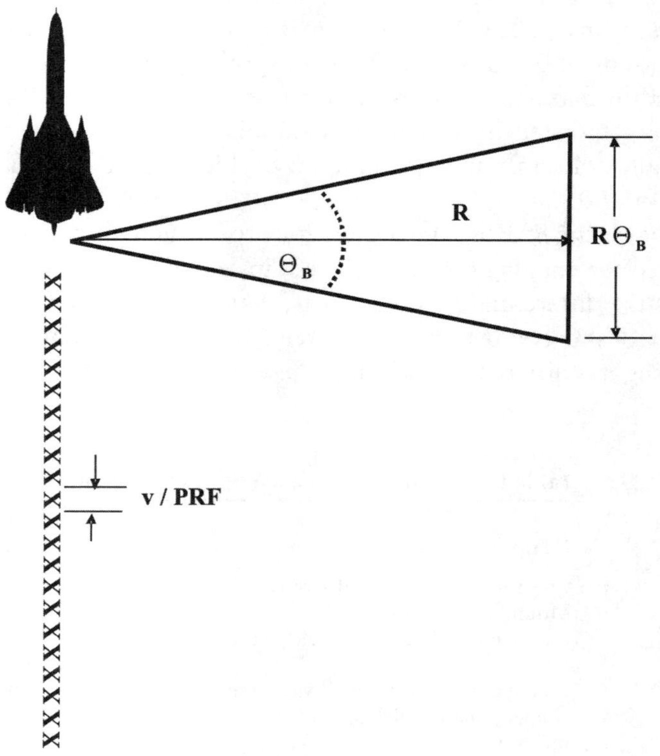

Figure 8.5 Synthetic aperture radar geometry. (*From:* [1, p. 498].)

distance $v_p T$, where v_p is the vehicle velocity and T is the radar's interpulse period. The cross-range resolution of the synthetic array is then given by

$$\delta_x = \theta_s R = \frac{\lambda R}{2L_e} \tag{8.14}$$

The synthetic array contains n elements, each separated by a distance $v_p T$, resulting in a total length $L_e = nv_p T$. The total integration time required to form the synthetic array is given by $t_i = nT$, which can be defined in terms of the Doppler bandwidth as $B_D = 1/nT$, resulting in a cross-range resolution of

$$\delta_x = \frac{\lambda R B_D}{2v_p} \tag{8.15}$$

Note that (8.15) is identical to (8.6) evaluated at a squint angle of $\alpha_m = \pi/2$, demonstrating that the SAR relationships derived from synthetic array and DBS considerations result in identical formulations.

The cross-range resolution of an unfocused array (i.e., quadratic phase variation at each array element are not corrected) can be obtained by defining the synthetic array beamwidth as the beamwidth that results when the difference between the two-way minimum and maximum paths as depicted in Figure 8.6 result in a quarter-wavelength pathlength difference [4]. From this geometry, $R^2 + (L_e/2)^2 = (R + \lambda/8)^2$, which can be solved for the unfocused synthetic aperture length as

$$L_{eu} = \sqrt{R\lambda} \tag{8.16}$$

Substitution of this into (8.15) results in a further expression for the cross-range resolution of an unfocused synthetic aperture array as

$$\delta_{xu} = \frac{\sqrt{\lambda R}}{2} \tag{8.17}$$

For example, if $R = 100$ km with an X-band radar ($\lambda = 3$ cm), the cross-range resolution would be 27 m. Note that the cross-range resolution (unfocused) is a

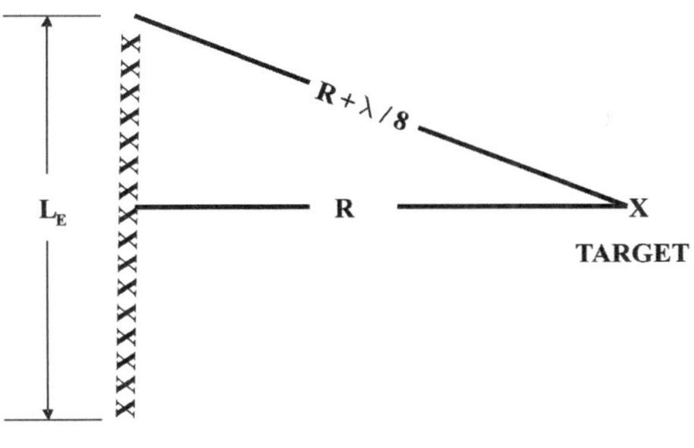

Figure 8.6 Geometry of the unfocused synthetic aperture radar. (*From:* [1, p. 499].)

function of both the range and the radar's frequency, but is independent of the actual antenna's aperture.

The synthetic array can be focused at range R by applying a quadratic phase correction at each position (i.e., for all elements in L_e) of the array equal to $\Delta\phi = 2\pi x^2/\lambda R$, where x is the distance from the center of the synthetic array to the corrected element. A different correction must be applied for each range R and, when focused, all the return echo signals from a scatterer at that range are in phase. The resulting synthetic array is focused over a strip equal to L_E (see Figure 8.5) that must be illuminated by each element of the array, which has a real beamwidth equal to $\theta = \lambda/D$. Substitution of this into (8.30), which applies for the synthetic array, results in $\theta_s = D/2R$ and a focused cross-range resolution of

$$\delta_{xf} = R\theta_s = \frac{D}{2} \tag{8.18}$$

Note that the focused cross-range resolution is independent of both the radar's transmitted frequency and the range at which it is focused, but depends on the real aperture dimensions.

One fundamental limitation of a synthetic array is that the element positions should be spaced a half-wavelength apart to prevent grating lobes. Because the element positions depend on the vehicle velocity (v_p) and the radar's interpulse period ($T = 1/\text{PRF}$), so that $v_p T < \lambda/2$, the resulting PRF should satisfy the inequality

$$\text{PRF} > \frac{2v_p}{\lambda} \tag{8.19}$$

where $f_d = 2v_p/\lambda$ is the Doppler shift of the radar-carrying vehicle. This condition is rarely met and the spacing is generally made to accommodate grating lobes within the array's FOV [2].

The sidelooking antenna's effective length in both the unfocused and focused synthetic aperture arrays may limit the cross-resolution capability of these systems. For example, consider an S-band ($\lambda = 0.1$m) SAR that operates at a 100 km range with an azimuth aperture width of 3.33 m from an aircraft with 300 m/s velocity. The array length for the unfocused case is $L_{eu} = \sqrt{\lambda R} = 100$ m, whereas the cross-range resolution $\delta_x = \sqrt{\lambda R} = 50$ m. Since the vehicle has a velocity of 300 m/s, it takes $t_i = 0.33$ sec to form the array. In the focused case, the array length is $L_e = R\lambda/D = 3$ km, so that the integration time $t_i = 10$ sec. The resolution in the focused case is $\delta_x = D/2 = 1.67$ m. The cross-range resolution is thus inversely proportional to the integration or processing time, as can be deduced by applying (8.6) or (8.15), and can be rewritten as

$$\delta_x = \frac{\lambda R}{2v_p t_i \sin\alpha} \tag{8.20}$$

where t_i is the total integration time and α is the angle between the vehicle's velocity vector and the antenna's boresight ($\alpha = 90°$ in the sidelooking case).

This example illustrates that the cross-range resolution capability of a SAR depends on the integration or processing time required to form the synthetic array, which, in turn, is determined by both the parameters and the geometry of the array and the type of processing utilized (i.e., focused or unfocused). One method for increasing the processing time is to use a spotlight mode SAR in which the real antenna boresight is continually realigned to track a particular target area of interest [16]. The geometry for spotlight SAR is depicted in Figure 8.7. Theoretically, resolutions better than $D/2$ can be obtained with this mode and the only requirements are that the real antenna beamwidth must be large enough to adequately illuminate the desired area to be imaged, and the duration of the illumination must be long enough to obtain sufficient effective rotation of the scene for the desired cross-range resolution [see (8.20)]. Another benefit of this mode is that the scene to be imaged is observed over a range of incidence angles that averages the speckle (specular radar returns associated with a particular viewing aspect) and makes a smoother image.

One interpretation of spotlight SAR is that it is equivalent to the synthetic array radar flying a circular path around the target area, as depicted later in Figure 8.21. This follows from the fact that, once the synthetic array is corrected for the quadratic phase caused by its straight line path, then the phase of the signal returned from a fixed scatterer at the focused range has constant value. A constant phase return would also be returned if the SAR moved on a locus of constant range with respect to the target and hence on a circular path. The equivalent circular arc angle ($\Delta\phi$), which the SAR moves to provide a specified cross-range resolution, can be found by calculating the integration time (t_i) from (8.20) and using this in the relationship

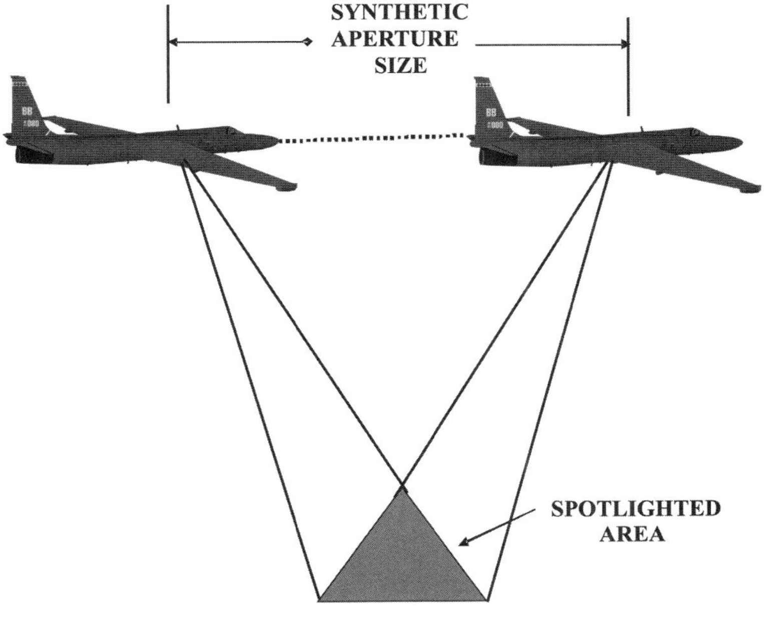

Figure 8.7 Spotlight SAR. (*From:* [17, p. 275].)

$$\Delta\phi = \frac{v_p t_i}{R} \tag{8.21}$$

For this example (v_p = 300 m/s, t_i = 10 sec, R = 100 km), the resulting angle would be $\Delta\phi$ = 3/100 rad = 1.8°.

SAR systems provide a day-or-night, all-weather source of remotely sensed information. They span a wide variety of civil and military applications including observations of areas impenetrable to visible sensing techniques. Military applications include reconnaissance, surveillance, and targeting.

The X-band ASARS-1 radar generates a 10 nmi-wide strip-map 20 to 100 nmi off the SR-71 aircraft's ground track with 3 ft resolution. Alternately a 1 by 1/3-nmi spot mode map with 1-ft resolution can be generated 25 to 85 nmi from the high-altitude SR-71 aircraft. This radar was upgraded into the ASARS-2 and installed into the high-altitude U-2 aircraft. The improvements included a 1-D electronically steered antenna array that provided beam agility in azimuth, allowing rapid beam switching for tracking multiple targets as well as interleaving various SAR modes. In these systems SAR data were transmitted and processed on the ground via high-bandwidth data links because hardware in this time frame was too cumbersome to install aboard the aircraft. The system was further upgraded into the ASARS-2A and used in the unmanned Global Hawk. This system employs a 4.1 by 1.2 ft aperture planar array antenna and transmits 350 W average power. It was initially aimed at SAR image collection and later added limited GMTI using mechanical scanning in both azimuth and elevation that precludes all but the most basic GMTI operation. Further, the low-power aperture product restricts the sensitivity to relatively large RCS targets at short ranges or long revisit times in GMTI search modes. Clutter rejection and the minimum ground target velocity that can be detected are limited by the use of the limited aperture and only two-phase center GMTI processing. Onboard processing is used to reduce the bandwidth of the SAR data, which allows for the long operating range of the Global Hawk and necessitates the use of a satellite data link. The raw SAR data generate 274 Mbps while the processed image is compressed to only 45 Mbps that is compatible with the onboard K_u-band satellite link. As described in Section 7.3, an AESA design is planned for this system that will substantially improve both its SAR and GMTI capabilities.

The K_u-band APY-8 (Lynx) SAR system [18] is employed in the *Predator* unmanned vehicle. It weighs 115 lbs and has a slant range of 30 km in 4 mm/h of rain and up to 60 km in fair weather. It is capable of 0.1 m resolution in the spotlight mode and 0.3 m resolution in the strip-map mode. A 10 m^2 RCS 6 knot or greater radial velocity ground target can be detected at a 25 km slant range. The Predator drone and the APY-8 antenna are shown in Figure 8.8.

In the strip-map mode, resolution is operator adjustable from 0.3 to 3 m while the map can be referenced to the ground or aircraft. Resolution in the spotlight mode is adjustable from 0.1 to 3 m over a range from 3 to 25 km. The GMTI can be scanned over 270° in azimuth with a ground swath of 10 km and a range of 4 to 25 km. Pertinent parameters of the APY-8 are listed in Table 8.2. A Lynx SAR image of a strip-map and 0.1 m spotlight image are depicted in Figure 8.9.

Figure 8.8 Predator UAV and APY-8 antenna. (*From:* [18, p. 5].)

Table 8.2 Characteristics of APY-8 Lynx SAR [18]

Parameter	*Value*
Frequency	15.2–18.2 GHz
Transmitter	TWT
Peak power	320 (700) W
Average power	112 (245) W
Waveform	Linear chirp
Antenna	Reflector/offset feed
Radome	48-cm diameter
Polarization	Vertical
Azimuth BW	3.2°
Elevation BW	7°
Stabilization	3-axis gimbal
Data link	C-band
Weight	115 pounds

Figure 8.9 Lynx APY-8 high-resolution strip and spot images. (*From:* [18, p. 7].)

► Example 30 SAR Design Principles

Modern sidelooking airborne radars employed on both manned and unmanned platforms such as Global Hawk, Predator, JSTARS, U-2, and Astor provide ground reconnaissance, surveillance, targeting, and cueing. These radars operate at X-band

and K_u-band generally using passive antenna technology although AESA designs are currently in development (see Section 7.3). In several of these systems (U-2 and Predator) processing is performed on the ground via a wideband data link that, for example in the U-2 ASARS-2, provides 274 Mbps bandwidth. The U-2 SAR provides 3 m resolution strip maps and 1 ft resolution in its spotlight mode.

Figure EX-30.1 depicts the stripmap geometry utilized in a sidelooking SAR such as that employed in the RQ-4A Global Hawk. The HISAR radar is a derivative of the ASARS-2 radar employed on the U-2. The combined SAR-GMTI stripmap mode provides 6 m resolution over a swath that is 37 km wide (R_{fp}) at ranges from 20 to 100 km. It also contains a SAR spotlight mode that provides 1.8 m resolution over 10 km^2 coverage. The mechanical ±45° azimuth scan antenna has a relatively small aperture of 1.25 m azimuth by 0.37 m elevation, and the pulse compression waveform X-band transmitter provides 3,500 W peak power with a 10% duty cycle. The image processed onboard can be transmitted to a ground station via a 48 in.-diameter satellite antenna data link at a 50 Mbps rate. The Global Hawk can cruise at 650 km/h with an endurance of 36 hours at a maximum operating ceiling of 20 km. Area coverage in the stripmap mode is a maximum of 24,000 km^2/h.

Generally synthetic aperture sidelooking radars employ unfocused, focused, and spotlight modes of operation. Using MATLAB program sar_res.m, the cross-range resolution of real beam, unfocused, and focused sidelooking radar is shown in Figure EX-30.2 for an X-band operating frequency and an antenna with a 10 ft azimuth aperture.

As shown in Section 8.2 the cross-range resolution of an unfocused SAR is

$$\delta_{xu} = \frac{\sqrt{\lambda R}}{2}$$

whereas that of a focused SAR is

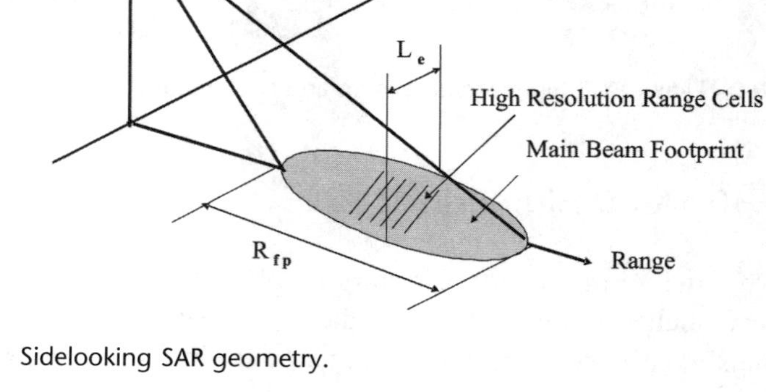

Figure EX-30.1 Sidelooking SAR geometry.

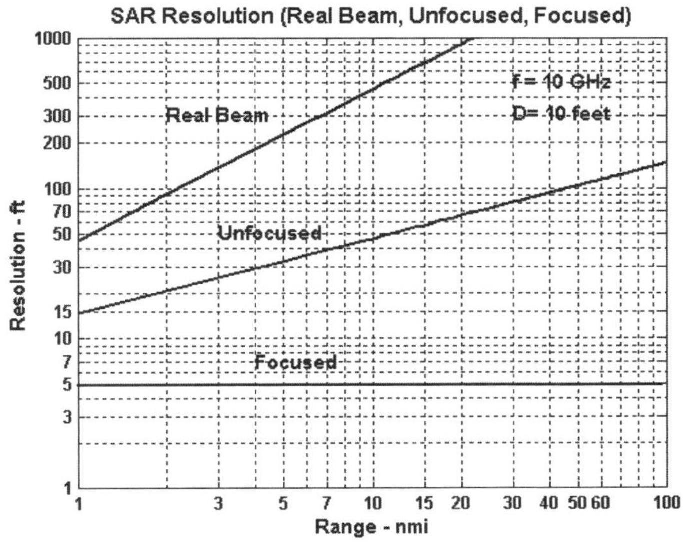

Figure EX-30.2 SAR cross-range resolution.

$$\delta_{xf} = \frac{D_{az}}{2}$$

which is independent of range and depends only on the antenna's aperture width. The limitation in the unfocused case depends on the quadratic phase variation caused by the straight line path of the aircraft rather than a circular path. The phase error in this case is given by

$$\Delta\phi_n = \frac{2\pi d_n^2}{\lambda R}$$

where $d_n = |nv_{ac}T - L_e/2|$ is the distance from element n to the center of the synthesized aperture. This is corrected in the focused array but notice that a correction is needed for each resolution cell, resulting in a large number of computations. In some cases *presumming* of blocks of adjacent elements can be utilized to reduce the number of computations without impairing performance. In the focused array the length L_e of the synthetic array is limited by the footprint of the real antenna beam and is much longer than that of the unfocused array. On an elemental basis a focused array processor must make N^2 phase corrections where $N = L_e\text{PRF}/v_{ac}$.

A critical design parameter is the PRF. Figure EX-30.1 indicates that to avoid ambiguities the minimum PRF is

$$\text{PRF}_{\text{min}} = \frac{c}{2R_{fp}}$$

For example, the Global Hawk's range swath of 37 km results in a minimum PRF of about 4,000 Hz. However, the minimum PRF is generally restricted by the necessity to limit the Doppler spread between points on the ground represented by the main beam's leading and trailing edges given by

$$\text{PRF}_{\text{min}} = \frac{2v_{ac}\,\theta_B \cos\phi_e \sin\theta_{az}}{\lambda}$$

where θ_B is the azimuth beamwidth in radians, ϕ_e is the depression angle, and θ_{az} is the azimuth angle of the antenna boresight with respect to the aircraft's heading. This can be approximated as PRF $= 4v_{ac}/D_{az} = 552$ Hz for the Global Hawk. A more serious limitation on the PRF concerns the generation of grating lobes since the spacing of the array elements generally exceeds $\lambda/2$ due to the forward velocity of the aircraft. Grating lobes are replicas of the main lobe occurring at ever increasing intervals on either side of the main lobe. The limitation here is

$$\text{PRF} > \frac{2v_{ac}}{\lambda}$$

which for the Global Hawk results in a PRF greater than 12 kHz. One solution to this problem is to weight the various returns that form the array so that the PRF is limited by the necessity to remove range ambiguities [2].

Another factor that must be considered is that the near field of the synthetic array is given by

$$R_{nf} = \frac{2L_e^2}{\lambda}$$

For a focused array where $L_e = R\lambda/D_{az}$ the return generally appears in the near field where the beam is not completely formed, resulting in high sidelobes.

The basic concept of SAR is that the aircraft is moving at a constant velocity. Formation of the synthetic array involves an integration that takes between 1 and 10 sec, over which any acceleration must be compensated. This is generally accomplished using an accelerometer mounted on the antenna.

Another problem with SAR is called *speckle*. This results because SAR is a coherent imaging technology measuring both the amplitude and phase of the backscattered return. Each resolution cell of the return contains multiple scatterers that interact, causing destructive and constructive interference and producing a noiselike random signal. This adds a grainy background to the SAR image that must be removed by filtering. Many systems choose to sacrifice the higher resolution possible from the system to reduce the effect of speckle. They do this by averaging the detected powers from neighboring pixels thereby causing the resolution to be degraded [19].

8.3 Interferometric Synthetic Aperture Radar (IFSAR) [2, 5]

IFSAR employs pairs of high-resolution SAR images to generate high-quality terrain elevation maps, called digital elevation maps (DEMs), using phase interferometer techniques. The basic idea is to collect SAR images of the same terrain from very similar positions, retaining phase information, and then to synthesize an image with dimensions equal to the separation between the two measurements.

If, for example, two samples are obtained simultaneously using two antennas on the same aircraft some distance apart in altitude, then any phase difference between the respective returns from the two antennas contains information relative to the angle from which the radar return was received. Combining this with the range information allows one to determine the position in three-dimensions of the image pixel. Thus, from a single pass it is possible to obtain terrain altitude as well as radar reflectivity, thereby creating a DEM. This method was used in the space shuttle program to generate terrain maps of large surface areas to an accuracy of 5 m.

IFSAR obtains the elevation data needed for 3-D mapping by determining the elevation angle of the LOS to the center of each resolution cell mapped. Using the measured elevation angle (θ_e), radar height (H), and slant range (r) to the cell, the cell's height (z) and horizontal distance (y) to the radar are determined. The elevation angle is determined in the same manner that a phase monopulse tracking radar uses to develop its error signal. The general geometry involved in producing IFSAR images is depicted in Figure 8.10 [2]. Radar returns from point (P) in the center of the cell are received by two antennas separated by a cross track baseline (B) connecting the antennas. An angle (θ_L) is defined between the LOS to P and a normal to baseline B. The phases of the coherent radar returns are given by the interferometer output

$$\phi = \frac{2\pi}{\lambda}(r_1 - r_2) \tag{8.22}$$

Using the law of cosines it can be determined that

$$\phi = \frac{2k\pi}{\lambda}\left(r_1 - \left(r_1^2 + B^2 + 2r_1 B \sin\theta_L\right)^{1/2}\right) \tag{8.23}$$

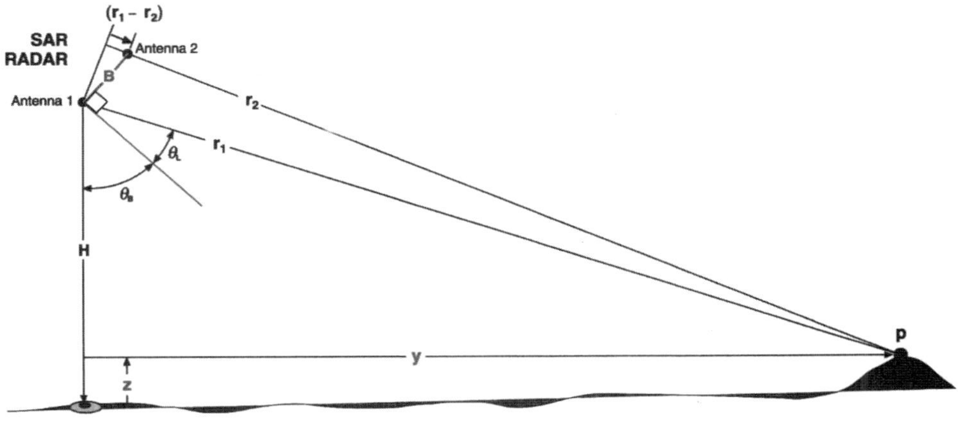

Figure 8.10 IFSAR geometry. (*From:* [2, p. 516].)

where $k = 1$ for one-pass mapping and $k = 2$ for two-pass mapping. This can be solved

$$\theta_L = \sin^{-1}\left(\frac{\lambda^2 \phi^2}{4(k\pi)^2 r_1 B} - \frac{\lambda \phi}{2k\pi B} - \frac{B}{2r_1}\right) \tag{8.24}$$

then

$$\theta_e = \theta_L + \theta_B \tag{8.25}$$

and

$$y = r_1 \sin \theta_e \qquad z = H - r_1 \cos \theta_e \tag{8.26}$$

When collecting returns from successive range increments across the full width of the swath being mapped, the range difference $(r_1 - r_2)$ and, hence, ϕ continuously increases. The short wavelength causes the value ϕ to successively cycle through 2π radians and so is highly ambiguous.

The ambiguities can be resolved by separately making conventional SAR images with the coherent returns received by each antenna. The two images are then coregistered and merged into an interferogram. By adding 2π to the value of ϕ each time a fringe in the interferogram is crossed (a process called *phase unwrapping*), the ambiguities are removed. The horizontal position y and elevation z of each cell are then accurately computed, and the map is topographically reconstructed.

The topographic accuracy critically depends on the accuracy of the phase unwrapping. For high S/N ratios and fringes not too close together, this is a straightforward process. It can become complicated if steep slopes or shadows are encountered. Steep slopes may result in some points at which fringes are crossed being overlaid by others while shadows may cause some points to be missed.

Further, it can be shown that the sensitivity of cell height measurement is given by [5]

$$|\delta z| = \frac{\sin(\theta_L + \theta_B)}{\cos \theta_L}\left(\frac{r_1}{B}\right)|\delta(r_1 - r_2)| \tag{8.27}$$

This equation shows that achieving good height accuracy from significant stand-off ranges requires a large baseline and great precision in measuring r_1 and r_2. Optical imaging systems, with their very fine resolution, can achieve good results with a stereo camera pair on a small baseline in one-pass operation. Conventional SAR-based stereo imaging systems must generally use two-pass operation with significant separation between the two tracks to obtain look-angle differences from the two tracks to the terrain area of interest on the order of 10° to 20° [5]. However, for IFSAR such large baselines are not practical.

An IFSAR image of Washington, D.C., taken from a DC-8 aircraft is shown in Figure 8.11. The different altitudes are colored according to radar-measured terrain elevation [20].

Figure 8.11 IFSAR image of Washington, D.C. (*From:* [20].)

For spaceborne IFSAR systems, the baseline is typically on the order of 100 m, though it can be as large as 1 or 2 km, and the altitude ranges from approximately 250 km (for the space shuttle) to 800 km (for low-earth-orbit satellites) giving an r_1/B on the order of 2,500 to 7,000. For airborne systems, the stand-off range is usually on the order of 10 to 20 km, but the baseline is typically only on the order of a few meters to a foot, so that again r_1/B are on the order of several thousand. Because of this large multiplier, it is necessary to have very small values of $\delta(r_1 - r_2)$ if height errors are to be acceptable. Thus, we need to measure small differences in range from the scatterer to the two receiving apertures.

8.3.1 JSTARS [44]

A sophisticated form of IFSAR is employed in the APY-3 radar that is part of the JSTARS system. As mentioned earlier in the chapter, JSTARS is a long range, air-to-ground surveillance system, designed for battlefield surveillance [2, 4]. It detects, locates, classifies, and tracks hostile moving and stationary targets at a maximum range of 250 km. The radar and sidelooking, slotted planar array antenna are contained in the modified Boeing 707 E-8C aircraft depicted in Figure 8.12. Also illustrated is the ground display obtained during the Desert Storm conflict, showing the retreat of Iraqi forces to Baghdad.

The JSTARS APY-3 radar features several operating modes. The radar's primary mode is wide-area ground surveillance with GMTI (WAS-MTI). It tracks moving ground targets in a 200- by 200-mile area that can be "radar referenced" (moving

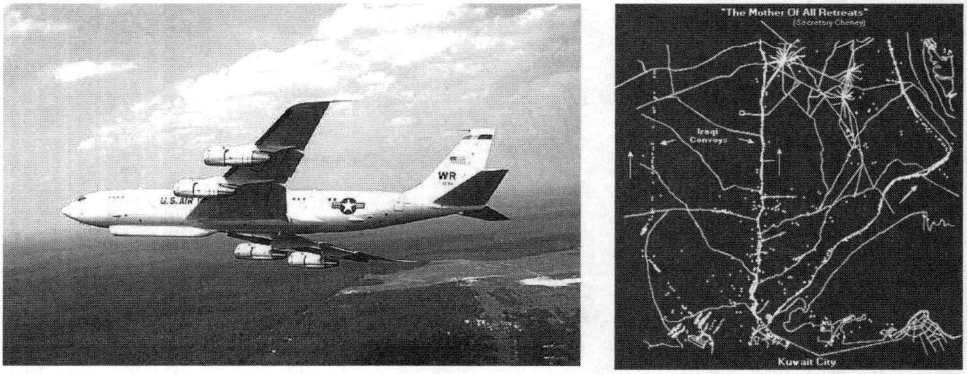

Figure 8.12 JSTARS E-8C aircraft and ground display. (*From:* [21, 22].)

with the aircraft) or "ground referenced," meaning that it remains fixed to a particular set of ground coordinates as long as possible. This mode provides the capability to distinguish wheeled vehicles from tracked vehicles by the fact that the bottom of the track is motionless with respect to the ground, while the top moves at twice the speed of the vehicle. Maritime targets and helicopters can also be detected in this mode.

The sector search mode is essentially a zoom mode that focuses on the detail in a target area of 18 by 18 miles. The high-resolution attack planning mode focuses on moving targets in a 7- by 7-mile area for final target marking.

The synthetic aperture and fixed target indicator (SAR-FTI) mode is used to produce photographic quality maps of the battle area. In SAR-FTI mode, the radar has a range of 110 miles to either side of the aircraft from operating altitude. During an 8-hour mission 50,000 km^2 can be mapped.

The heart of the JSTARS is the APY-3 radar that is the largest airborne radar built in X-band. Its multiport planar-array antenna is contained in a canoe-shaped radome under the forward fuselage of the aircraft. The 24 by 2-ft sidelooking, roll-stabilized PESA antenna can be mechanically oriented to either side of the aircraft and electronically scanned in azimuth through ±60° [45, 46]. The structure of the APY-3 is shown in Figure 8.13 [3]. Its antenna is divided into three equal apertures formed into a multiport antenna structure containing one transmitting port plus four receiving ports. The sum receiving port is used for SAR, while three interferometer ports are used for GMTI.

The antenna aperture consists of a rectilinear grid of 456 by 28 horizontally polarized slots connected to 12 waveguide dual-mode feeds through ferrite phase shifters. A key element of the design is the use of interferometric GMTI, which replaces the classic TACCAR-DPCA–type of motion compensation.

Characteristics of the JSTARS APY-3 radar are given in Table 8.3. Many of the parameters listed are estimated or derived as indicated by an asterisk [23].

8.3.2 Interferometric GMTI [2, 8, 9]

A primary function of JSTARS is to detect slowly moving ground targets in the presence of heavy clutter. One approach, whose MTI improvement factor is limited

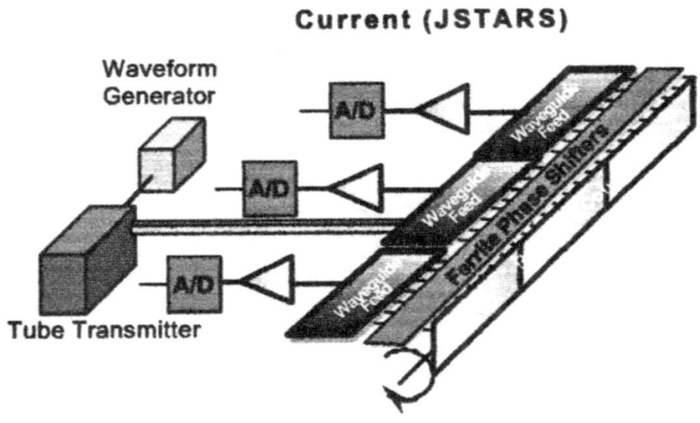

Figure 8.13 JSTARS APY-3 configuration. (*From:* [3].)

Table 8.3 APY-3 Radar Parameters

Parameter	*Value*
Frequency	X-band (9.3–9.8 GHz)*
Transmitter	Coupled cavity TWT
Peak power	50 kW*
Average power	750W*
Waveform	Linear chirp
Bandwidth	42 MHz*
Antenna	Planar array (20 by 2 ft)
Sum gain (transmit/receive)	47 dB
GMTI gain (receiver)	42 dB
Azimuth sum beamwidth	0.3°
Interferometer sub apertures	3
Azimuth GMTI beamwidth	0.9°
EL beamwidth	30°
Polarization	Horizontal
Azimuth sidelobe	–45 dB*
EL sidelobe	–35 dB*
Electronic azimuth scan	±60°
Azimuth scan rate	180°/s
System losses	12–15 dB
Resolution	12–14 ft
Detection range	10–134 nmi
Ground target velocity	1.5–58 knots
SAR image time	30–60 sec

*Derived or estimated.

to the two-way integrated antenna sidelobes, is to employ classic DPCA and TACCAR as described in Sections 3.2.7, 6.2.2, and Example 20 in Chapter 6. If these are executed perfectly, aircraft forward motion is frozen for two pulses and the spread of main beam clutter spectra is due only to clutter internal motion. From Example 3 (see Chapter 1) on internal motion clutter it is seen that a triple MTI canceller would be necessary to provide sufficient MTI improvement factor to allow detection of small RCS slow-moving ground targets in heavy ground clutter. This would deteriorate the DPCA, which cannot compensate for 4 pulses, and any imperfections in the motion compensation would further limit the performance.

In JSTARS an alternative system is employed that has proven successful in practice. This system uses a variation of DBS in combination with phase interferometers to reject main lobe clutter while taking advantage of the small Doppler shifts due to the radial component of the ground target's velocity.

The basic principle of DBS (see Section 8.1 and Example 29) is that the forward motion of the aircraft imparts a unique Doppler shift onto each component of the main beam ground clutter. Then employing Doppler filters it is possible to extract a high-resolution image of the ground illuminated by the antenna's main beam. The 3 dB Doppler bandwidth spread over the main antenna beam is given by

$$\Delta f_{3\text{ dB}} = \frac{2v_{ac}\Delta\theta_{3\text{ dB}} \cos\phi_{elo} \sin\theta_{azo}}{\lambda} \tag{8.28}$$

where f_{ac} is the Doppler frequency due to the aircraft's forward motion, $\Delta\theta_{3\text{ dB}}$ is the antenna's beamwidth, and θ_{azo} and ϕ_{elo} are the antenna's boresight pointing angles. The highest resolution is then achieved for the largest aircraft velocity and when the antenna is pointed perpendicular to the aircraft's velocity vector. The DBS image is then obtained by filling the range of possible Doppler returns by a bank of contiguous Doppler filters. The bandwidth of each Doppler filter is determined by the reciprocal of the time that the antenna dwells on each ground segment of interest (see Figure 8.1).

$$B_{\text{dop}} = \frac{1}{T_d} \tag{8.29}$$

The effect is to subdivide the azimuth beamwidth ($\Delta\theta_{3\text{ dB}}$) into narrow azimuth resolution cells. When a target is moving in a resolution cell, its Doppler will be different than that of the competing ground return in that cell. This translates the moving target into a different Doppler filter that contains ground clutter from a different azimuth angle.This effect is illustrated in Figure 8.14 where a SAR image is shown of the Golden Gate bridge [8], where the images of moving cars are shown on either side of the bridge (due to their moving in opposite directions) so that they appear over the water.

The next step in the process is to eliminate fixed ground returns. This is accomplished by forming an interferometer in each range cell that places a null at every azimuth angle associated with each Doppler filter. The interferometer is implemented as illustrated in Figure 8.15 [2]. As in IFSAR the phases of the coherent radar returns are given by the interferometer output

$$\phi = \frac{2\pi}{\lambda}(\Delta d) = \frac{2\pi}{\lambda} W \sin\theta_n \tag{8.30}$$

where Δd is the difference in distance between the radar return to the two aperture phase centers A and B, and θ_n is the angle off the boresight line. The phase adjustment can be computed as

$$\Delta\phi = \frac{\pi - \phi}{2} \tag{8.31}$$

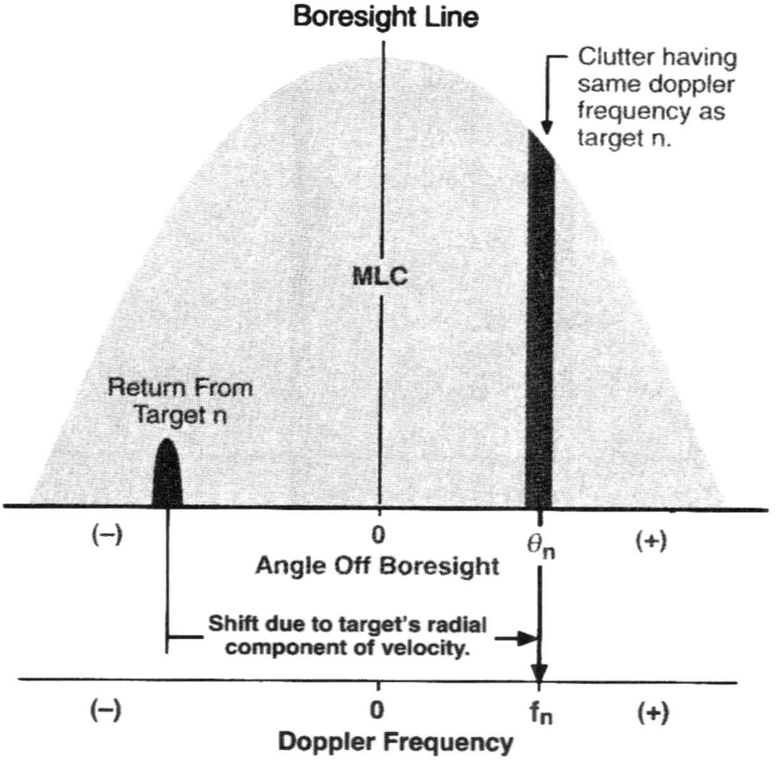

(a)

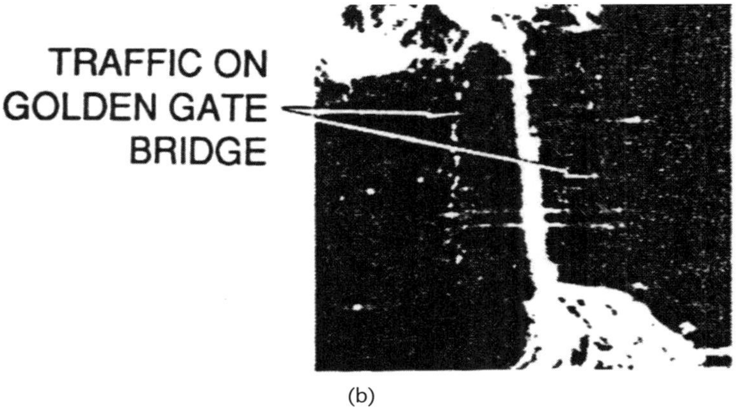

(b)

Figure 8.14 (a, b) DBS image resolution for fixed and moving targets. (*From:* [2, p. 320; 8, p. 446].)

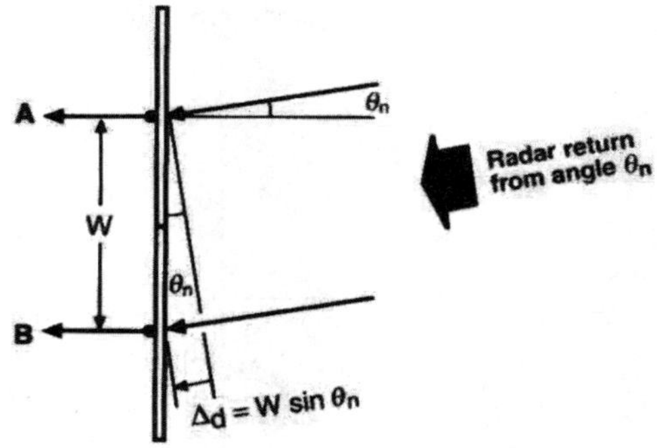

Figure 8.15 Two-segment interferometer to cancel fixed targets. (*From:* [2, p. 320].)

This phase adjustment when made in opposite directions to antenna outputs at apertures A and B will increase the phase difference (ϕ) to π radians as illustrated in Figure 8.16. Then summing the outputs of the two apertures (A' and B') produces a null in the output at angle θ_n. The nulling process must be accomplished after Doppler filter processing to avoid nulling targets along with the ground clutter.

The processor that implements the nulling process is depicted in Figure 8.17. Video outputs of receiving channels A and B are collected in range bins. For each range bin a Doppler filter is formed. Output for each filter (n) is rotated by $+\Delta\phi$ in channel A and $-\Delta\phi$ in channel B. Phase-rotated outputs from apertures A and B are then summed to form the null at angle θ_n. The results of this process are illustrated in Figure 8.18. Note that although the main beam ground return is suppressed at angle θ_n, the target return at this angle is misregistered and appears at an angle that depends on its radial velocity.

The solution to accomplish accurate angle measurements is to add a third aperture by dividing the antenna into three equal horizontal segments as shown in Figure 8.13. Then two interferometers can be formed using the forward-and-center segments and the rear-and-center segments. The ground clutter suppression system described in Figure 8.19 is then applied to both interferometers. Each target return will then appear in corresponding Doppler filters of the two interferometers. Applying phase monopulse processing to each corresponding target's Doppler filter allows an accurate estimate of the targets angle.

From (8.28), $\Delta f_{3\,\text{dB}}$ for JSTARS looking sideways at zero elevation can be computed as 86 Hz for an aircraft velocity of 500 knots. If a DBS factor of 10 is

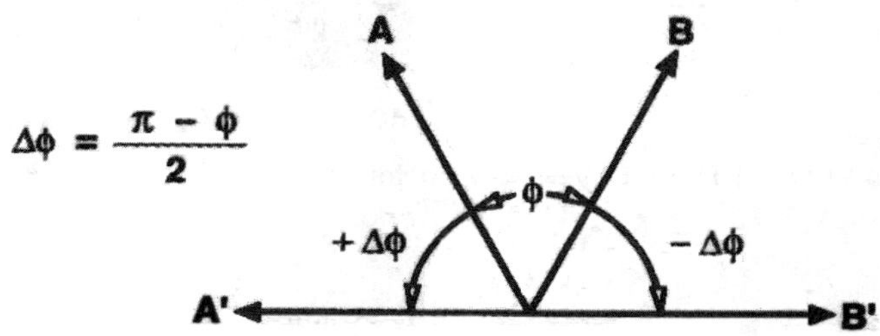

Figure 8.16 Interferometer nulling process. (*From:* [2, p. 320].)

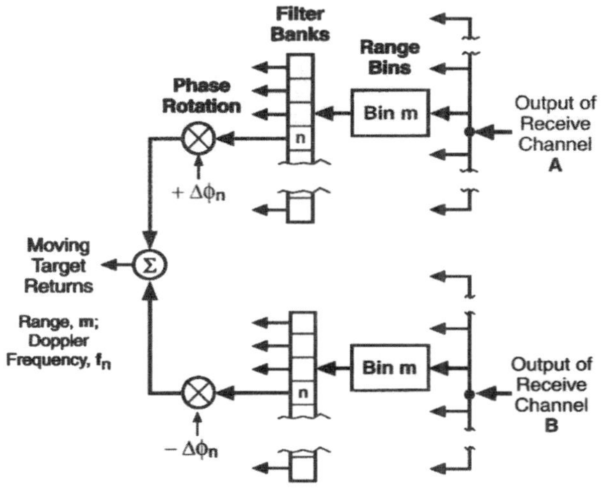

Figure 8.17 Processor to null ground return. (*From:* [2, p. 321].)

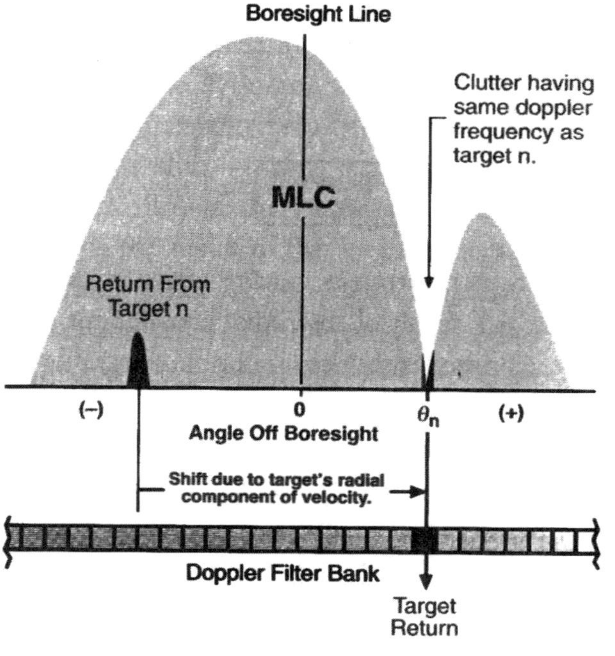

Figure 8.18 Spectrum output of null forming processor. (*From:* [2, p. 320].)

assumed, then a Doppler filter bandwidth of 8 Hz would be reasonable. If a 64-point FFT is applied, then the PRF would be 512 Hz. Therefore, main beam ground clutter (between nulls) would be contained within about a 172 Hz frequency range while the rest of the frequency range would essentially be clutter free. The dwell time from (8.29) is computed as 0.125 sec, resulting in a scan rate of 2.4°/s. For an 8 Hz Doppler filter bandwidth, a radial speed ground target would have

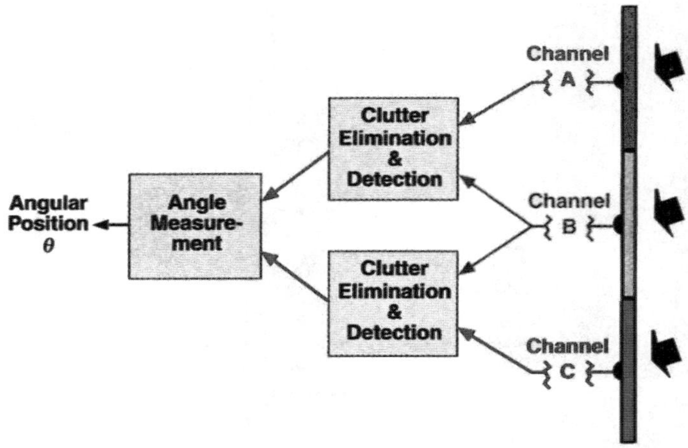

Figure 8.19 Target angle estimation using a three-aperture antenna. (*From:* [2, p. 322].)

to move about 0.3 knot to be detected. However, for those targets within the main beam clutter, internal clutter motion would become the limiting factor. The power spectra for internal motion land clutter is given by

$$P(f) = \frac{\lambda\beta}{4}\exp\left(-\frac{\lambda\beta}{2}|f|\right) \tag{8.32}$$

where β is the exponential shape factor (see Example 3) and λ is the wavelength. MATLAB program land_spec.m plots these spectra, which are shown in Figure 8.20. Depending on wind conditions the two-sided clutter spectra could extend to 60 Hz at the −25 dB point, which would be consistent with Doppler filter sidelobes for this application. The limitation in this case would be 2 knots radial velocity to clear the uncancelled ground clutter.

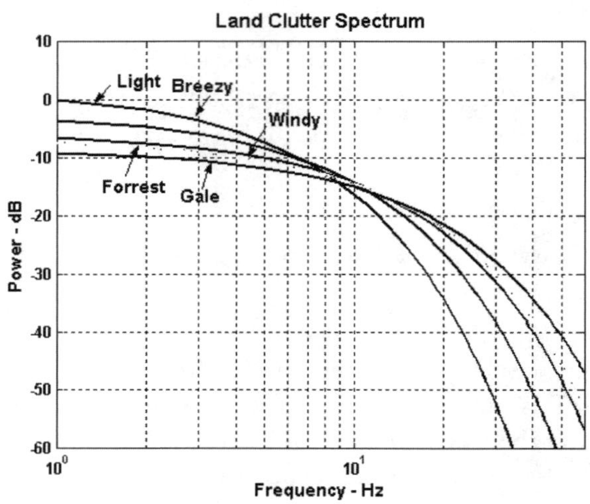

Figure 8.20 Spectra for internal motion land clutter. (*From:* [24].)

8.4 Inverse Synthetic Aperture Radar (ISAR)

The classic SAR has limited ability to image moving targets such as ships, ground vehicles, and aircraft that have rotational motion. Moving targets can be seriously defocused and displaced from their location. *ISAR* is a system that has been developed to take advantage of target rotational motion to provide an image of the target.

Both SAR and ISAR provide a higher cross-range resolution than can be achieved on the basis of the antenna's aperture resolution. In SAR the radar's forward motion allows cross-range points simultaneously received to be separated on the basis of their differential Doppler shifts. In ISAR the principle is the same, but the collecting platform is stabilized and the differential Doppler motion is due to the rotational motion of the spotlighted target about its roll, yaw, and pitch axes. In both systems high resolution is achieved in the range dimension using pulse compression techniques.

The most successful ISAR operation has been in the imaging of ships. In this application pitch motion of the ship causes the mast to have a higher velocity than its superstructure in the length dimension, allowing height information to be derived from the differential Doppler. Roll also allows height information to be determined but in a plane that includes the narrower width of the ship, resulting in less differential Doppler than pitch. Yaw motion provides a plan view of the ship. In general, ISAR generates a perspective view of the ship that can be used to identify the class of ship being imaged. Because a ship's angular rates of pitch, roll, and yaw are not generally known, an ISAR image does not represent true distances in cross-range as contrasted to that in the range dimension.

ISAR is best explained starting with the SAR spotlight mode as depicted in Figure 8.21.

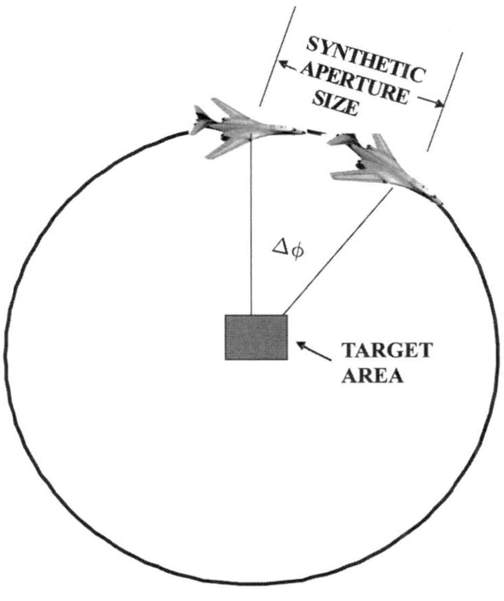

Figure 8.21 Spotlight SAR with a circular flight path. (*From:* [17, p. 275].)

From Figure 8.21, it is apparent that the same data would be collected if the radar were stationary and the target area rotated. When this occurs, the system is called an *Inverse Synthetic Aperture Radar*. ISAR can also function from a moving platform, as depicted in Figure 8.22. As illustrated, with SAR the differences in Doppler frequency that enable us to obtain fine angular resolution are due to the motion of the radar-bearing aircraft. By contrast with ISAR, the differences are due to the angular rotation of the target (e.g., yaw, pitch, and roll) as seen by the radar. In Figure 8.22, the radar's speed is slightly greater than the target's, resulting in point P_1 on the target's tail having a slightly negative rate. Because the target is turning away, point P_2 on the nose has a positive range rate. Thus, the Doppler frequencies of the returns from P_2 and all points between it and P_1 differ in proportion to their distances from P_1.

To image the target, phase corrections are first made to provide motion compensation that removes the variations due to the radar's and target's displacement. The returns from each resolvable range increment are then applied to a separate bank of Doppler filters, and an image is produced from their outputs just as in conventional SAR. The image's cross-range dimension is not necessarily horizontal but rather is perpendicular to the axis about which the target happens to be rotating at the moment. No image can be formed if that axis is colinear with the radar's LOS or if the radar has no rotational motion as viewed from the radar.

A basic 2-D ISAR image can be generated as depicted in Figure 8.23. A series of range profiles is generated by high-range-resolution radar as it observes a rotating target. The range profiles are sampled at a rate that corresponds to the radar's PRF. There are then n range cells, where each range cell has a resolution

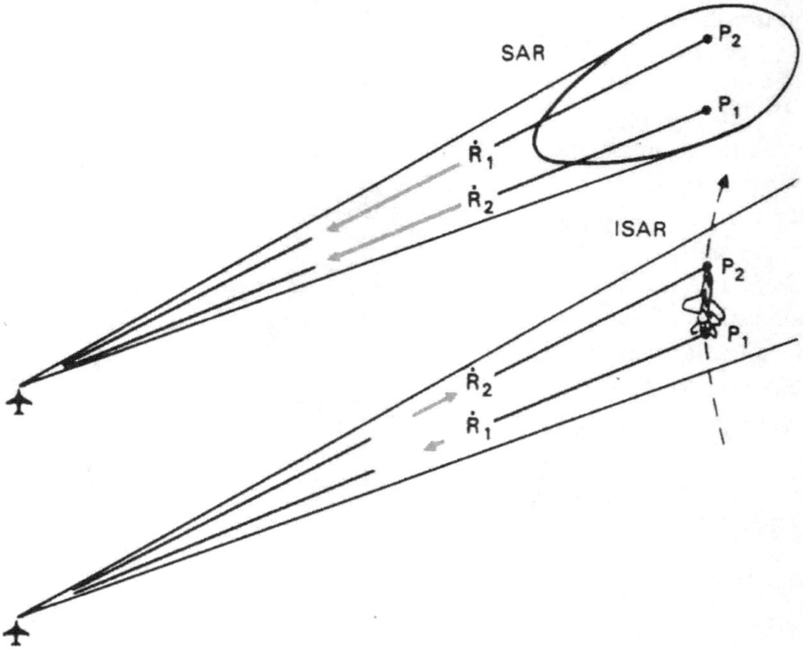

Figure 8.22 Comparison of SAR and ISAR from an airborne platform. (*From:* [2, p. 575].)

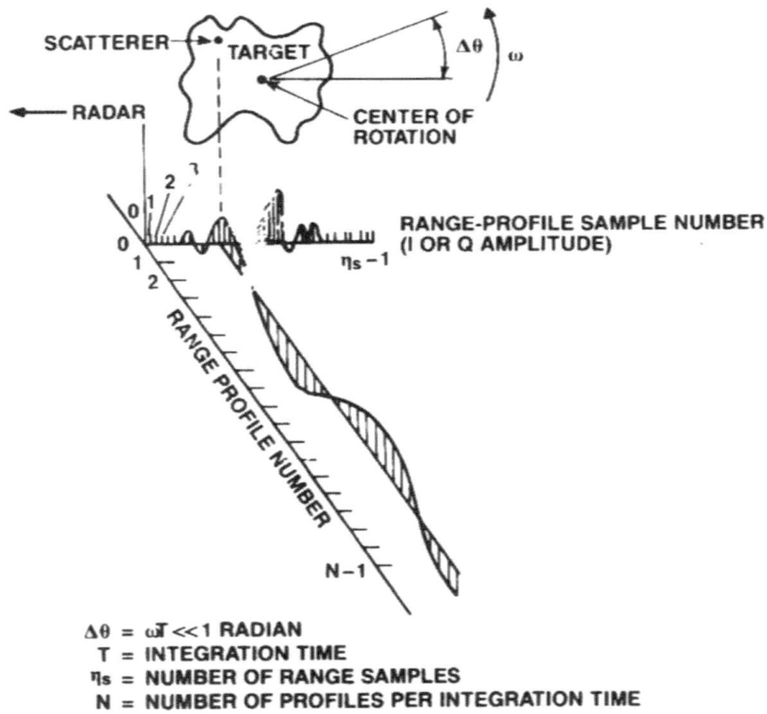

Figure 8.23 ISAR range-Doppler sampling of a target with rotational motion. (*From:* [17, p. 276].)

$\Delta r_s = c\tau_e/2$ ($\tau_e = 1/B_R$, where B_R is the radar's transmitted bandwidth). The n range cells each contain $N = 1/\text{PRF} \cdot t_i$ range profiles, where t_i is the integration time required to produce the desired cross-range resolution as the target rotates through $\Delta\theta$ rad.

The N-range profile samples in each range cell are Doppler filtered to extract cross-range image data. Each target scatterer on the rotating target produces a Doppler signature with a frequency that is proportional to the cross-range distance between the scatterer and the center of target rotation. The magnitude of each scatterer-related Doppler signature is proportional to the reflectivity of that scatterer. The rotating target reflectivity is thus mapped in orthogonal coordinates consisting of slant range and cross range, where the cross-range scale factor is dependent on the target-aspect angular rotation rate. Orientation of the rotation axis relative to the radar's LOS establishes the orientation of the range-Doppler image place.

The Doppler frequency gradient required to obtain high cross-range resolution is generated by the motion of the object relative to the radar. This motion can be generated in a variety of ways, which generally can be related to the simplified case depicted in Figure 8.24 of stationary monostatic radar illuminating a rotating object. The object rotates with uniform angular motion in the x-y plane about the z axis. The object is contained within a coordinate system in range-Doppler space, shown as *isodoppler* and *isorange* lines, where the isorange lines are parallel to the radar-target LOS and the isodoppler lines are perpendicular to the isorange

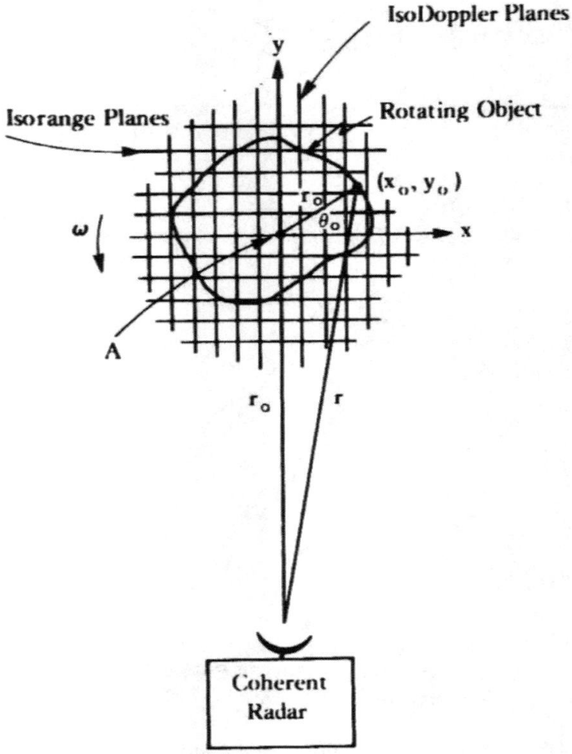

Figure 8.24 ISAR target geometry for range-Doppler imaging. (*From:* [25, p. 364].)

lines. The x axis represents the cross-range dimension, whereas the y axis is the slant-range dimension.

Consider a uniformly rotating scatterer located at the point (x_0, y_0) with angular rate ω and initially positioned ($t = 0$) at angle θ_0 in the coordinate system of Figure 8.24 [25]. The range to this scatterer is given by

$$r = R + v_r t + r_0 \cos(\omega t - \theta_0) \tag{8.33}$$

where R is the range to the center of the object, v_r is its radial velocity, r_0 is the distance of the scatterer to the center of the object, ω is the scatterer's angular rotation rate, and θ_0 is the initial angle of the scatterer with respect to the cross-range axis. The motion-compensated range (i.e., the object's translational velocity Doppler component is removed at the radar) of the scatterer is then given by

$$r = R + x_0 \sin \omega t + y_0 \cos \omega t \tag{8.34}$$

where $x_0 = r_0 \sin \theta$ and $y_0 = r_0 \cos \theta$. The Doppler frequency (f_d) of the returned signal is hence given by

$$f_d = \frac{2}{\lambda}\frac{dr}{dt}\frac{2x_0\,\omega}{\lambda} \cos \omega t - \frac{2y_0\,\omega}{\lambda} \sin \omega t \tag{8.35}$$

If the radar data are processed over a very small time interval about $t = 0$, the target range and Doppler can be approximated as

$$r \approx R + y_0 \tag{8.36}$$

$$f_d \approx \frac{2x_0\,\omega}{\lambda} \tag{8.37}$$

Therefore, by measuring the returned radar signal in terms of slant range ($r_s = R + y_0$) and Doppler frequency (f_d), which allows determination of the cross range ($r_c = x_0 = \lambda f_d/2\omega$), we can estimate the position (x_0, y_0) of the point scatterer.

The result of ISAR processing is then to map or image all scatterers contained within the rotating target in the range-Doppler coordinate system depicted in Figure 8.23. The presence of the object rotation rate (ω) in determining the cross-range component of the scatterer implies that to obtain a properly scaled image of the object the magnitude of ω must be known. Most techniques for estimating the rotation rate depend on *a priori* knowledge or analysis of periodicities in the radar signal level.

Another implicit assumption is that the distance (R) from the radar antenna to the center of the object is constant and known, which implies that any time-varying components are removed by using appropriate motion compensation techniques.

High resolution in slant range can be achieved by a variety of methods involving either very short pulses or coded pulses of long duration that are compressed into short equivalent pulses. The key parameter in this regard is the transmitted radar bandwidth (B_R), which allows a slant range resolution (Δr_s) given by

$$\Delta r_s = \frac{c}{2B_R} \tag{8.38}$$

Being able to achieve a high cross-range resolution (Δr_s) implies a resolution of Doppler frequency (Δf_d) given by

$$\Delta f_d = \frac{2\omega \Delta r_c}{\lambda} \tag{8.39}$$

Because a frequency resolution Δf_d ($\Delta f_d = B_D$) requires a coherent processing time interval, approximately equal to $t_i = 1/\Delta f_d$, this results in a cross-range resolution given by

$$\Delta r_c = \frac{\lambda}{2\omega t_i} = \frac{\lambda}{2\Delta\theta} \tag{8.40}$$

where $\Delta\theta = \omega t_i$ is the angle through which the object rotates during the coherent processing time. This relationship is equivalent to that obtained for spotlight SAR, where the target object remains stationary, but the radar moves with velocity (v_p)

through an angle $\Delta\theta = v_p t_i/R$, which, when substituted into the relationship $\Delta r_c = \delta_x = \lambda R/2v_p t_i$, results identically in $\Delta r_c = \lambda/2\Delta\theta$. This illustrates that the relative rotational motion between the target and radar determines the cross-range resolution in both SAR and ISAR systems.

Any target aspect rotation relative to the radar produces a Doppler gradient across scatterers that are spread out in the cross-range dimension. Aspect rotation is produced by both target translation in any direction other than along the LOS between the radar and target and that due to target angular motion. The rotation produced by target angular motion is generally resolved into pitch, roll, and yaw components. These rotational components for an aircraft target are depicted in Figure 8.25. For aircraft targets seen from ground-based radar, the major source of rotational motion up to X-band frequencies is generally the translational tangential component, whereas for ship targets observed at ranges beyond about 10 nmi, target pitch, roll, and yaw are generally the most significant components [17].

The target geometry for aircraft, where the target has translational velocity v_t in a direction other than the LOS between the radar and the target, is depicted in Figure 8.26. The ISAR image plane (cross range by slant range) is directed in a plane containing the radar LOS vector and an orthogonal vector that contains the tangential component of target velocity. For ship targets, where pitch, roll, or yaw causes rotational motion (see Figure 8.27), the ISAR image plane is directed in a plane containing the radar LOS vector and the rotational motion component. For example, the ISAR image plane for ship roll is depicted in Figure 8.28.

Although all portions of the ship pitch, roll, and yaw at the same angular velocity, their apparent linear velocities sensed by radar fixed in space depend on

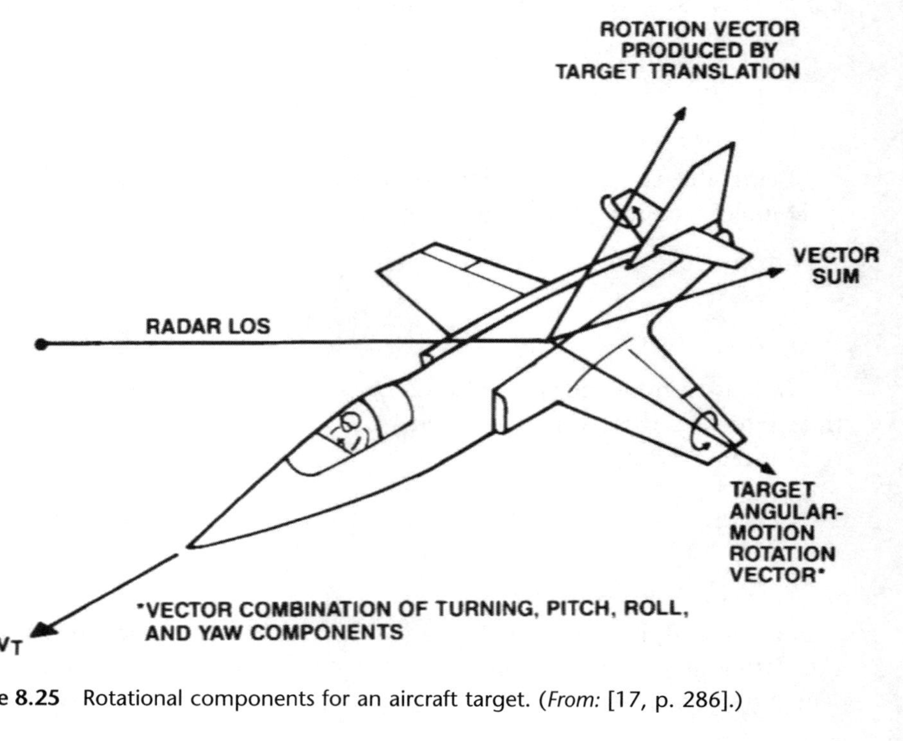

Figure 8.25 Rotational components for an aircraft target. (*From:* [17, p. 286].)

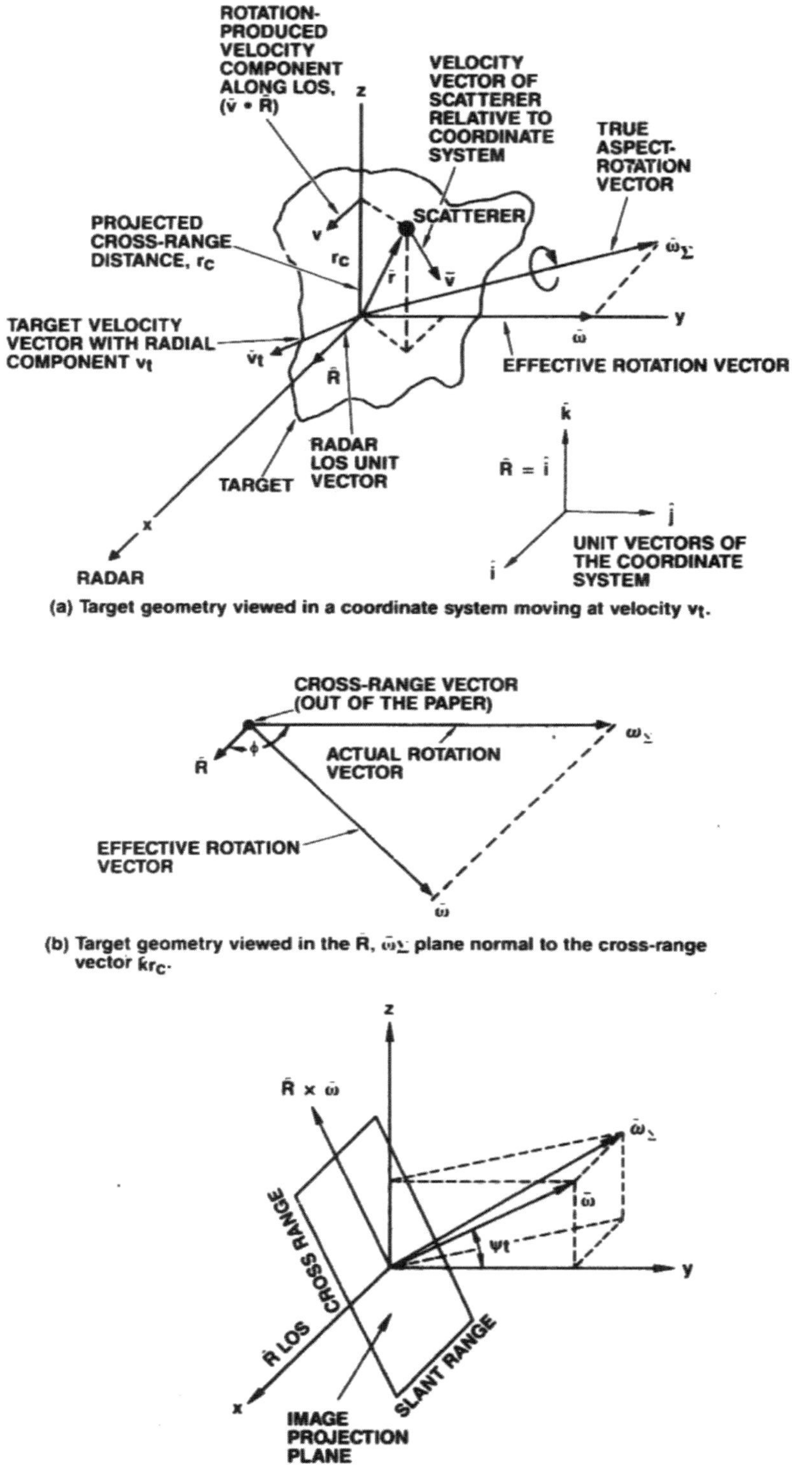

(a) Target geometry viewed in a coordinate system moving at velocity v_t.

(b) Target geometry viewed in the $\hat{R}$, $\bar{\omega}_{\Sigma}$ plane normal to the cross-range vector $\hat{k}r_c$.

(c) ISAR image plane

Figure 8.26 (a–c) Image plane geometry for moving coordinate system. (*From:* [17, p. 288].)

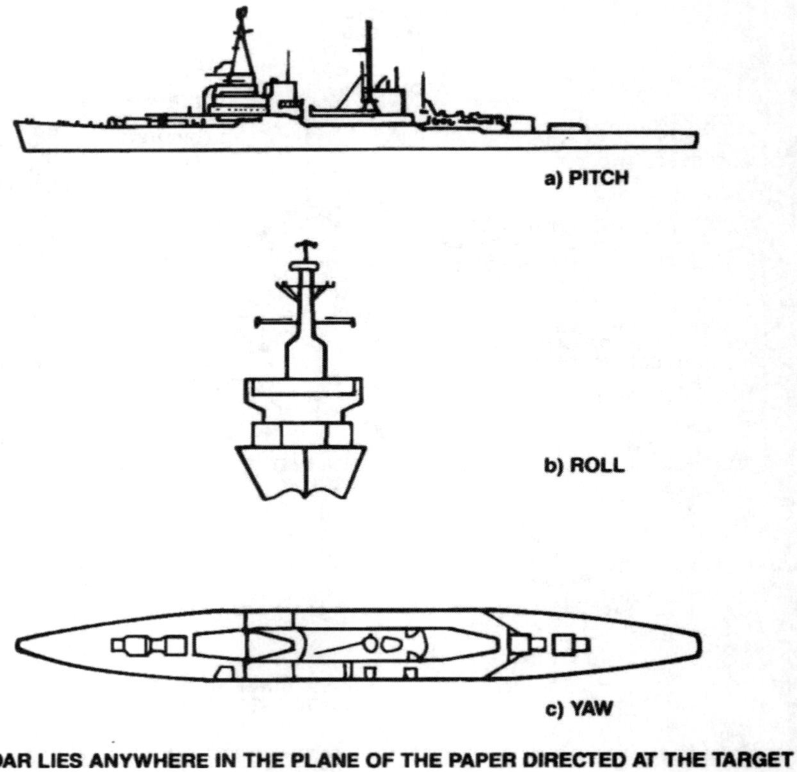

Figure 8.27 (a–c) ISAR image views produced by target pitch, roll, and yaw. (*From:* [17, p. 289].)

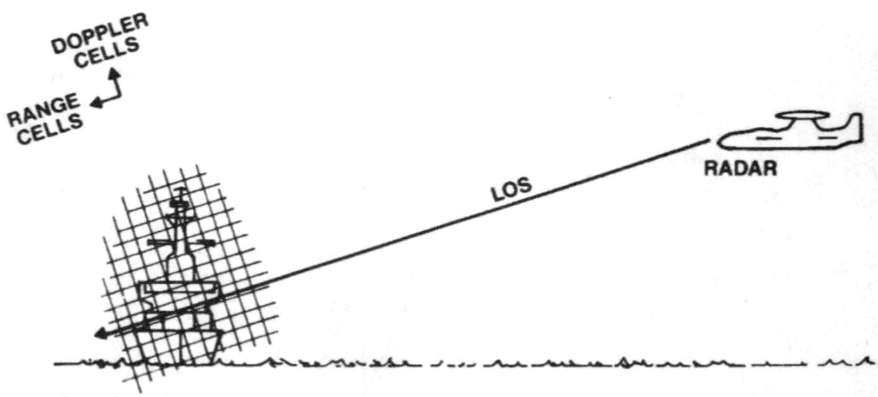

Figure 8.28 ISAR image plane with range and Doppler cells for ship target with roll. (*From:* [17, p. 290].)

their distance from the ship's center of gravity. These different linear velocities create proportionately different Doppler shifts in the echoes returned to a radar fixed in space. As a result of the vessel "pitching," for example, the higher an object is on a ship, such as a superstructure for a radar antenna, the greater will be the Doppler shift of its echo, as depicted in Figure 8.28.

An ISAR image can be generated from a time history of slant-range profiles by performing a FFT on the data points in each slant range cell. This extracts the target Doppler components associated with the cross-range image. The FFT process can be accomplished only if alignment exists along the time history of range profiles. However, the rotational motion of the target generally causes each scatterer to move its position with respect to the radar during image formation (see Figure 8.24). This is also apparent from (8.35) and (8.36), which indicate, respectively, a time variation in slant range and Doppler frequency (proportional to cross range) during the integration time t_i necessary to form an image. The image aberration caused by motion through resolution cells results in a cross-range-dependent astigmatism called *range walk* and a slant-range-dependent cross-range focus error called *variable range rate* [26].

Fine cross-range resolution implies coherent processing over a large rotational angle ($\Delta\theta$), reflected in relatively large variations in slant range [see (8.35)] and Doppler [see (8.36)] for a particular scatterer. Generally, during a processing time interval sufficiently long to give a desired cross-range resolution, points on the rotating object thus may move through several resolution cells. Therefore, the conventional range-delay measurement and Doppler frequency analysis implied by (8.36) and (8.37) will result in degraded imagery for the large processing interval case.

To avoid image degradation caused by target migration through multiple resolution cells when using simple range-Doppler processing, the coherent processing time (t_i) must be limited. Consider the imaging of a body having its cross-range diameter given by D_x and the slant-range diameter of D_y. Hence, points on the body will not move through a slant-range resolution cell (Δr_s) in the processing time (t_i) with angular rotation ($\Delta\theta$) if

$$\Delta r_s > \frac{D_x \Delta\theta}{2} \tag{8.41}$$

Use of the relationship for cross-range resolution [see (8.40)] given by $\Delta r_c = \lambda/2\Delta\theta$ results in

$$\Delta r_c \Delta r_s > \frac{\lambda D_x}{4} \tag{8.42}$$

which imposes a limitation on the product of cross-range and slant-range resolution in terms of the cross-range diameter of the imaged target. Similarly, migration through cross-range resolution cells (Δr_c) is avoided if

$$\Delta r > \frac{D_y \Delta\theta}{2} \tag{8.43}$$

resulting in

$$(\Delta r_c)^2 > \frac{\lambda D_y}{4} \tag{8.44}$$

which imposes a limit on cross-range resolution in terms of the slant-range diameter of the body to be imaged. Note that (8.42) and (8.44) require one to choose a sufficiently small coherent processing time and small object size to avoid migration of points through resolution cells, which causes smearing of ISAR images.

As an example, assume that an X-band radar ($\lambda = 0.03$ m) is used to image a ship target using pitch rotational motion so that the ISAR image plane consists of the ship length for slant range and the ship's height for cross range (see Figure 8.29). Assume that the ship's length is 200 m (D_y) and its height is 30 m (D_x). Hence, the maximum cross-range resolution (Δr_c) for simple processing is given by $\Delta r_c > \sqrt{(0.03)200/4} = 1.5$ m, and the cross range by slant range product is $\Delta r_c \cdot \Delta r_s > (0.03)(30)/4 = 0.225$ m. Further, assuming a pitch angular rate of 1°/s (ω) results in an angular excursion of $\Delta\theta = \lambda/2\Delta r_c = 0.01$ rad or 0.6° to achieve the limit cross-range resolution with a processing time of $t_i = \Delta\theta/\omega = 0.6$ sec. Table 8.4 lists some predicted cross-range resolutions for ship targets by using an S-band ($\lambda = 0.1$ m) radar [17].

Often, finer resolution is desired than can be achieved using the limits imposed by the cell migration constraints given in (8.42) and (8.44). In this case, objects move through range and Doppler resolution cells, and simple range-Doppler imaging will yield blurred images; therefore, the effect of motion through resolution cells must be compensated. Several techniques of compensating for this motion through cells have been developed. These range from linear piecewise approximations to account for the motion to sophisticated polar reformatting techniques [25].

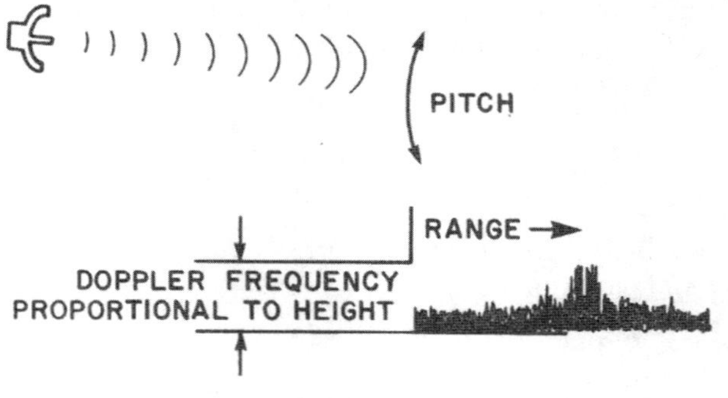

Figure 8.29 ISAR imaging of ship due to pitch rotational motion. (*From:* [27, p. 88].)

Table 8.4 Predicted Cross-Range Resolutions for Ship Targets Using an S-Band Radar in Sea State 5 ($t_i = 1$ sec) [17]

Ship Type	*Motion Type*	*Average Angular Velocity* (ω)	*Average Angular Excursion in 1 sec* ($\Delta\theta$)	*Average Cross-Range Resolution* (Δr_c)
Destroyer	Pitch	1.01°/s	0.0177 rad	2.82 m
	Yaw	0.54°/s	0.0093 rad	5.35 m
	Roll	6.30°/s	0.1100 rad	0.46 m
Carrier	Pitch	0.16°/s	0.0028 rad	17.83 m
	Yaw	0.08°/s	0.0014 rad	35.54 m
	Roll	0.38°/s	0.0066 rad	7.56 m

Polar reformatting is a process applicable to both SAR and ISAR data. It is most practically implemented by using a "stretch" technique for the radar pulse compression processing [25]. With this pulse compression technique, the receiver local oscillator is chirped at a time corresponding to the object's range at the same rate as the transmitted waveform. The returned signal is thus "deramped" so that the return from each scatterer of the rotating object appears as a unique Doppler frequency with respect to the carrier. In the following, we will sketch the fundamental elements of polar reformatting [26].

From Figure 8.24, the complex signal returned from the rotating object can be expressed as

$$f_r(t) = \int_s \sigma(s_0)\, u\left(t - \frac{r}{2c}\right) ds_0 \tag{8.45}$$

where r is the range to the differential scattering surface, $\sigma(s_0)$ is the target reflectivity observed at the receiver over the surface (s_0), and $u(t)$ is the transmitted pulse compression waveform. The Fourier transform of the received waveform is then given by

$$F_r(\omega) = \int_{-\infty}^{\infty} f_r(t) e^{-j\omega t}\, d\omega \tag{8.46}$$

If we substitute for $f_r(t)$ in (8.46) the result is

$$F_r(\omega) = \int_s \sigma(s_0) \int_{-\infty}^{\infty} u\left(t - \frac{r}{c}\right) e^{-j\omega t}\, d\omega\, ds_0 \tag{8.47}$$

which can be simplified to

$$F_r(\omega) = \int_s \sigma(s_0)\, U(\omega) e^{-j(\omega r/2c)}\, ds_0 \tag{8.48}$$

where $U(\omega)$ is the Fourier transform of $u(t)$. The term $e^{-j(\omega r/2c)}$ causes a distortion of the compressed waveform $u(t)$, as r varies with time over the processing interval required to form the ISAR image. The time-varying effects of the range variation r can be removed by multiplying the received signal by a reference signal given by $m(t) = e^{+j(\omega r/2c)}$, which forms the fundamental step in the polar reformatting motion compensation technique.

A block diagram of a simple ISAR system using polar-format processing is depicted in Figure 8.30 [25]. Stretch processing is embodied in the design whereby the pulse compression waveform generator (WFG) is used for both the transmitter and receiver local oscillator. After deramping, the signal bandwidth is limited to those Doppler frequencies represented by the scatterers within the imaged scene

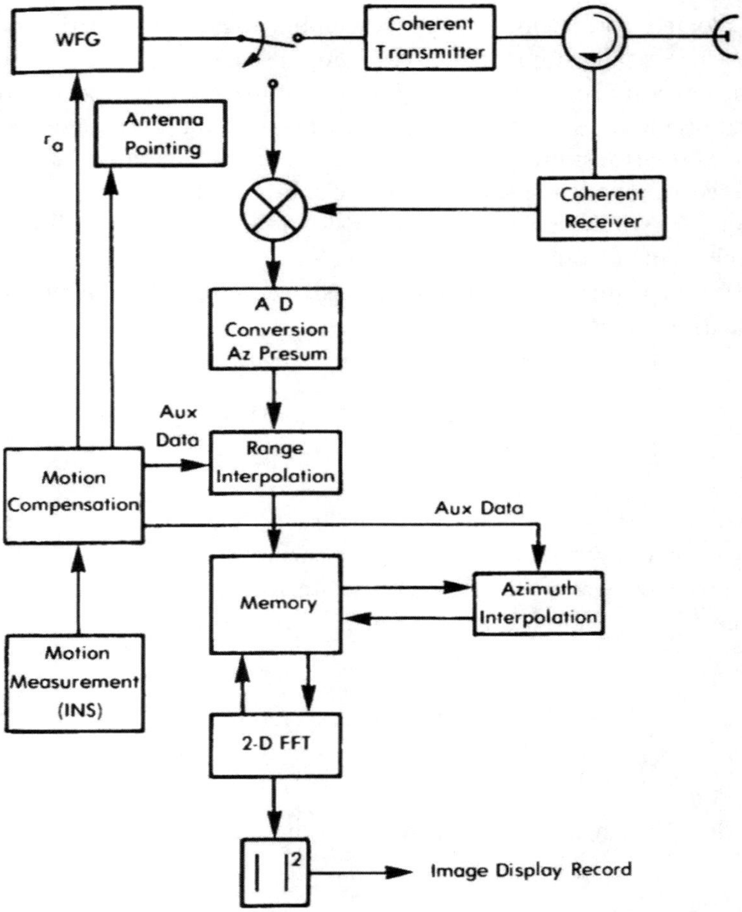

Figure 8.30 Simple ISAR system block diagram with polar-format processing. (*From:* [25, p. 383].)

[see (8.35) and (8.37)]. The use of the linear FM waveform in a deramping scheme on a pulse-to-pulse basis results in signals with frequency and phase determined by the relative range of a scatterer to scene center, thus establishing the conditions required to perform frequency-domain range-Doppler imaging.

Figure 8.31 shows the relative positions of the data samples in polar format collected in the ISAR image place. The samples, represented as black dots, are shown arrayed along radial lines at angle θ_i corresponding to the LOS angles to scene center for each radar transmission, as projected onto a reference plane. The spacing between samples T_r is determined by the original deramped signal projected into the reference plane. The PRF of the radar must be high enough to unambiguously sample the Doppler spectrum, which depends on the range extent of the illuminated scene.

The number of samples along each radial line K is determined by the effective duration of the pulse compression ramp. This duration is directly proportional to the effective RF bandwidth used to obtain slant-range resolution.

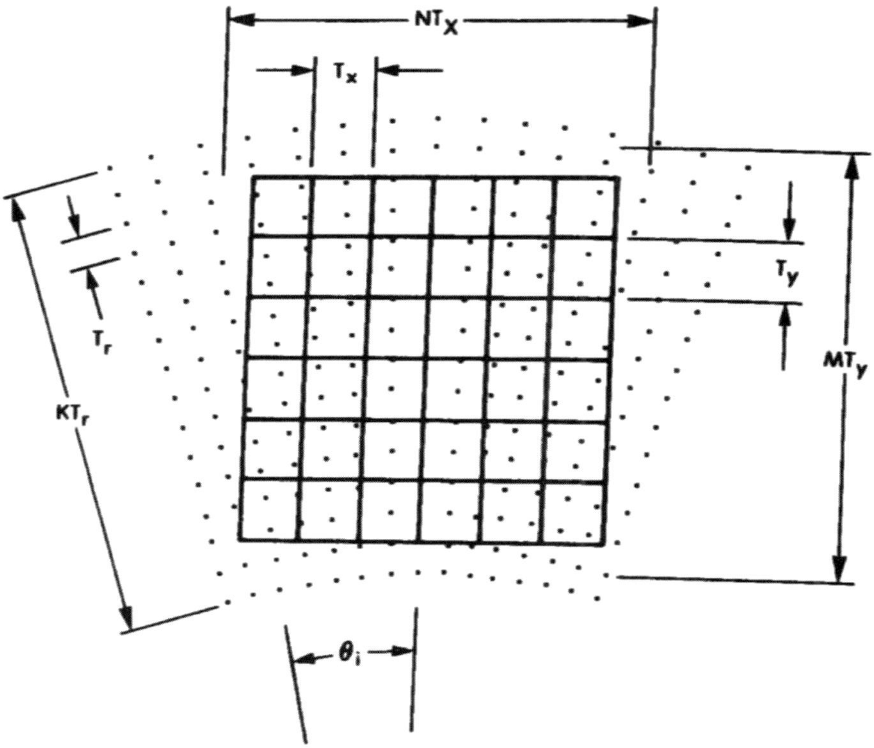

Figure 8.31 Collection digital polar-format geometry. (*From:* [25, p. 386].)

To form an image, the geometric array of samples in Figure 8.31 must be Fourier transformed in two dimensions. To accomplish this, the data must be resampled into a rectilinear grid with spacing T_x and T_y as depicted in the figure. The number of interpolated grid samples in the range and azimuth dimensions determines the output sample spacing in both dimensions of the image.

The rectilinear sample grid can be formed in two separate 1-D interpolation steps. The first step, depicted in Figure 8.32, simply resamples each radar pulse so that the new samples fall on positions along the horizontal lines making up the interpolation grid. This may be accomplished digitally where the input function is a discrete set of uniformly spaced samples, whereas the output samples are computed at a lower sampling rate and with some specified delay with respect to the first sample within each radar pulse.

The second stage of the polar-format interpolation is depicted in Figure 8.33. The azimuth interpolation operates on the samples produced from the range interpolation, only in an orthogonal direction. This may be accomplished digitally with output samples computed at appropriate times for each row of data. However, the input samples to this process cannot be considered equally spaced, and digital processing techniques that take this into account must be used.

After the 2-D interpolation to form the rectilinear signal grid, a complex-valued image is formed by using the 2-D FFT algorithm. Detection of this array produces the ISAR image.

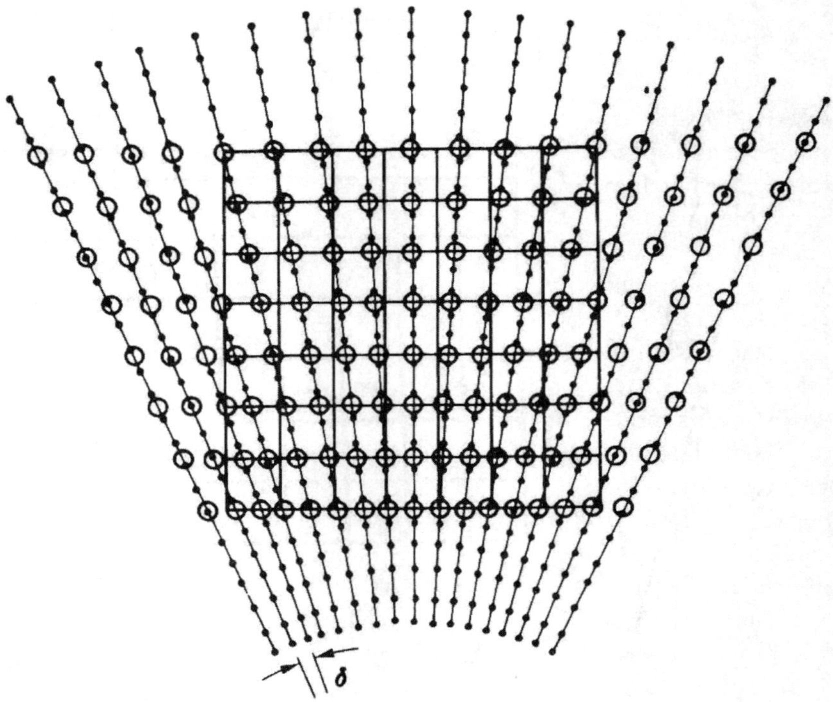

Figure 8.32 Polar-format range interpolations. (*From:* [25, p. 387].)

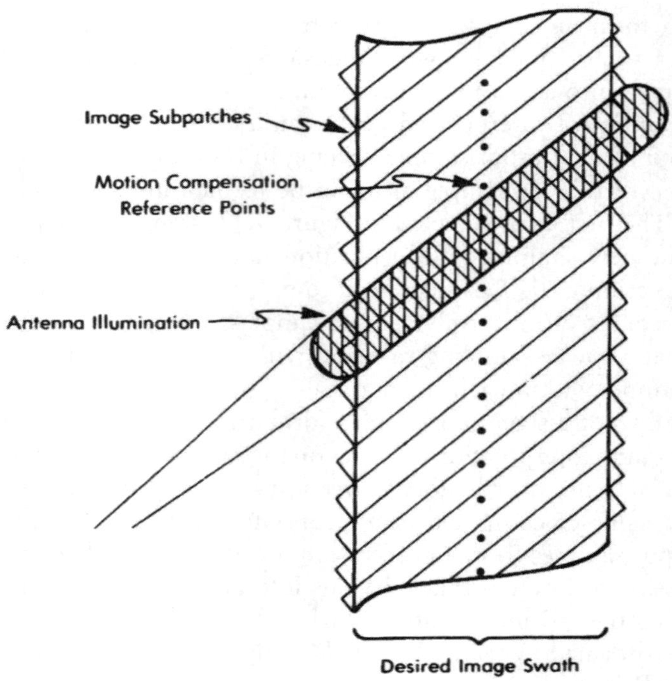

Figure 8.33 Polar-format azimuth interpolation. (*From:* [25, p. 387].)

8.5 Space Time Adaptive Processing for Airborne Radar

STAP [10, 28] is a combined spatial-Doppler filtering technique applicable to airborne and spaceborne low- and medium-PRF PD radars employing phased-array antennas that look down into heavy clutter returns. It provides capabilities for detecting targets in a high-clutter interference background from an airborne platform. Applications include detection of low-velocity ($v > 3$ knots) ground targets from a high-speed moving platform (JSTARS and space-based radar) and detection of targets in jamming and clutter for the AEW mission. The technique effectively compensates for motion-induced Doppler spectrum spreading of the ground clutter (see Section 3.2.3). STAP includes features of Doppler filter processing (MTI and FFT) and adaptive array antenna technology (jammer SLC).

The STAP concept was initially limited by the large processing throughput requirement beyond the capabilities of existing processors. Also STAP requires a separate A/D converter and receiver for each array element only recently practical due to LSI integrated circuit technology. The development of the *Sample Matrix Inversion* (SMI) algorithm provided a fast convergent procedure for adjusting the weights of the processor that makes the method practical [2, 10, 11]. However, the rapid convergence is achieved at the expense of a formidable computational load (approximately N^2 complex multiplications where N is the number of adaptive weights).

The SMI algorithm takes advantage of the fact that coherent returns from the ground can be completely characterized by their complex covariance matrix. Weights for implementing the algorithm are obtained in two steps. Initially, lacking *a priori* knowledge of the clutter and interference situation, a statistical maximum likelihood or equivalent statistical estimate is made for the covariance matrix of the received radiation. The matrix is then inverted, yielding weights for each receiving channel. Thereafter, the matrix is continually updated (adapted) to accurately reflect the varying clutter and interference situation. Each update is based on separate and independent samples of received data from other range increments other than those being processed. The advantage of the algorithm is that only a small number of samples are actually needed for the update, enabling most of the CPI to be devoted to efficiently filtering out the interference. As a result the filter's output contains a high portion of the received signal power providing a high SNR [2].

The clutter return associated with airborne radar is depicted in Figure 8.34. The center value of the spectrum for depression angle ϕ_d is located at

$$f_d = \frac{2v_{ac}}{\lambda} \sin \theta_o \cos \phi_d \tag{8.49}$$

where v_{ac} is the platform velocity, λ is the wavelength, and θ_o is the azimuth angle measured from the side of the aircraft. Note that when the antenna is pointed

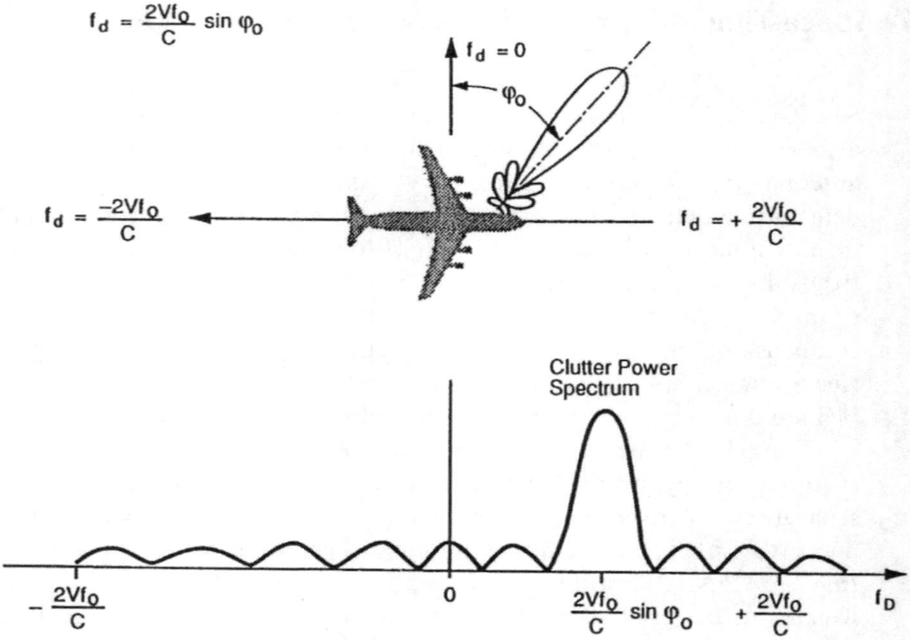

Figure 8.34 Clutter spectrum for airborne radar. (*From:* [29, p. 1016].)

sideways the spectrum is centered at zero frequency and the Doppler frequency ranges over the region $\pm 2 \cdot v_{ac}/\lambda$.

In general, platform motion can be resolved into two components, one parallel and one perpendicular to the antenna aperture. These components have the effect of both spreading the spectrum of the clutter returns and shifting the mean of the spectrum.

With the antenna aperture perpendicular to the velocity vector, the dominant effect is the shifting of the mean of the clutter spectrum. In general, the mean of the Doppler spectrum is determined by the platform velocity, the transmitted frequency, and the angle to the clutter path under consideration. Motion compensation of this effect is generally accomplished using the TACCAR technique, which locks the radar's coherent oscillator to the mean clutter frequency.

The DPCA technique compensates for the velocity component parallel to the antenna's aperture. This is accomplished by either physically or electronically displacing the phase center of the antenna in a direction opposite to the velocity component of the aircraft. A single-canceller MTI can be almost perfectly compensated by DPCA provided that the beam patterns from the two displaced apertures have identical patterns. Compensation of higher order MTI cancellers is difficult, resulting in imperfect compensation for these configurations [6].

When TACCAR and DPCA motion compensation techniques are effectively employed, the overall improvement factor provided by the AMTI is limited to the ratio between the integrated sidelobe level and the main beam gain. This results because the antenna sidelobes essentially cover 360 degrees and the Doppler shifts received through the sidelobes range from $-v_{ac}$ (v_{ac} is the aircraft's velocity) to

$+v_{ac}$. This Doppler spread can cover the entire MTI ambiguous frequency band. As a result, these sidelobe returns are not canceled, and the MTI improvement factor is limited by the ratio of the integrated energy entering the system through the main lobe to that entering through the sidelobes for the two-way antenna pattern [6].

A different way of looking at the motion compensation problem is to plot the clutter return from airborne radar in Doppler-azimuth space using the MATLAB program stfil.m, as illustrated in Figure 8.35. From this illustration it can be seen that all the clutter lies along a single clutter ridge line. Also when the clutter ridge is projected onto the Doppler plane, it is apparent that the sidelobes lie in the MTIs passband limiting the improvement factor performance to the two-way integrated rms sidelobes. However, if the MTI filter's response could be adapted in the spatial (angle) domain as well as the temporal domain, then the null of the MTI filter could be placed on the clutter ridge, thereby providing greatly enhanced performance. This is the essence of *Space-Time* processing that allows detection of slowly moving ground targets and increased performance in clutter and jamming environments [10]. Also note that jamming can be significantly reduced by adapting the nulls in the antenna pattern using procedures developed for sidelobe cancellers (SLC).

The simplest system that combines both spatial and temporal degrees of freedom is DPCA. References [10, 29] show that if all errors are neglected and the element spacing is adjusted so that $\Delta = 2v_{ac}/\text{PRF}$, then the filtering response null is aligned with the clutter ridge. This is illustrated in Figure 8.36, which was obtained from MATLAB program dpcafilc.m. The problem with nonadaptive DPCA is that it is sensitive to antenna errors and temporal and spatial channel matching, and the platform velocity must be known well enough to adjust the interpulse period so that $T = \Delta/2v_{ac}$. These difficulties can be largely overcome by employing an adaptive STAP system. With STAP the radar antenna is tuned to a target at a specific azimuth and Doppler frequency and then adaptively adjusted so that the system maximizes

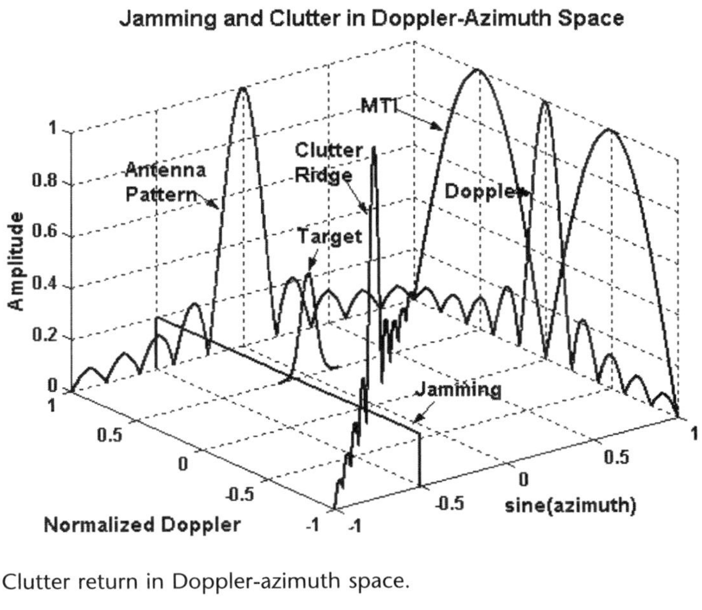

Figure 8.35 Clutter return in Doppler-azimuth space.

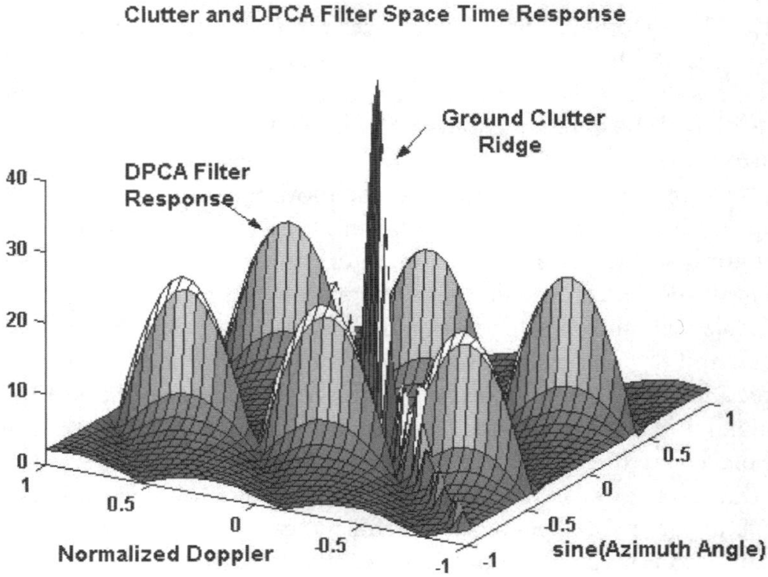

Figure 8.36 DPCA response in spatial-Doppler space with clutter ridge.

the signal-to-interference (clutter-plus-jamming-plus-noise) ratio. However internal clutter motion (see Example 3) and platform crabbing (difference between the platform's heading and ground track velocity due to wind) can cause spectral spreading that moves the clutter off its diagonal ridge, as shown in Figure 8.35.

8.5.1 STAP Processing [10, 28–32, 47]

A block diagram of a generalized STAP processor is shown in Figure 8.37. The transversal filter configuration (see Chapter 5) consists of N-antenna channels each containing M-temporal taps having delay T and weights that are adaptively controlled to maximize the signal-to-interference ratio at the STAP output. The clutter produces N antenna signals given by

$$x_1(t), x_2(t), x_3(t), \ldots, x_N(t) \tag{8.50}$$

at the terminals of antennas 1 through N. The signal at the output of the transversal filter is then given by

$$\begin{aligned} v(t) = {} & w_{11}x_1(t) + w_{12}x_2(t) + \ldots + w_{1N}x_N(t) + \ldots \\ & w_{21}x_1(t-T) + w_{22}x_2(t-T) + \ldots w_{2N}x_N(t-T) + \ldots \\ & \vdots \qquad\qquad \vdots \qquad\qquad \vdots \qquad\qquad \vdots \\ & w_{L1}x_1(t-(M-1)T) + w_{M2}x_2(t-(M-1)T) + \ldots w_{MN}x_N(t-(M-1)T) \end{aligned} \tag{8.51}$$

This can be expressed in matrix form as

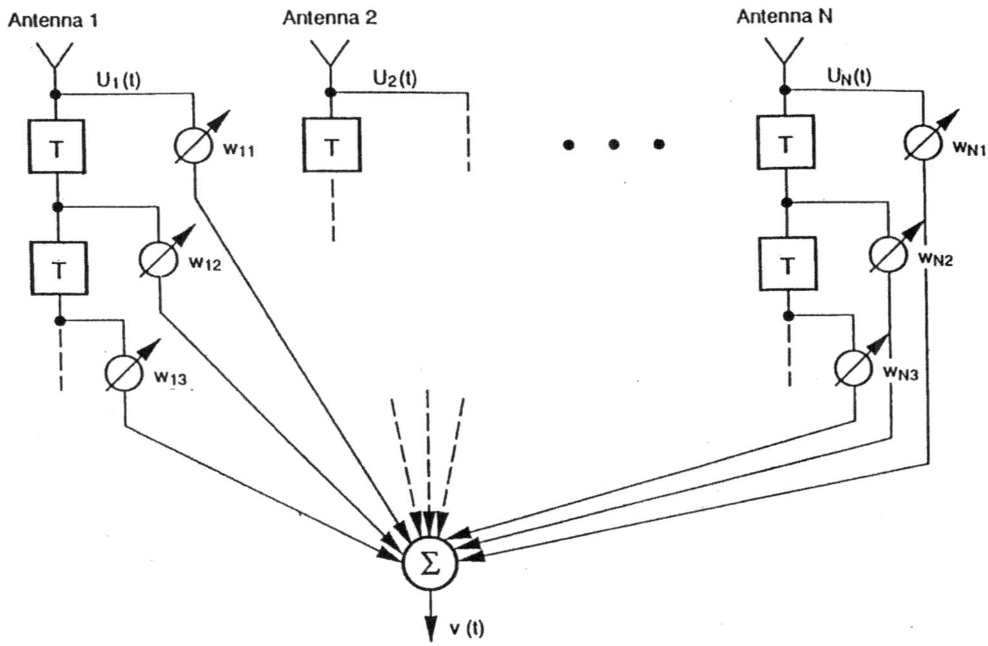

Figure 8.37 Generalized time-space adaptive processor. (*From:* [29, p. 1018].)

$$\mathbf{v} = \mathbf{w}\,\mathbf{x}^T \tag{8.52}$$

where

$$\mathbf{w} = [w_{11}\, w_{12} \ldots w_{1N} \quad w_{21}\, w_{22} \ldots w_{2N} \quad w_{31} \ldots w_{LN}] \tag{8.53}$$

and

$$\mathbf{x} = [x_1(t)\, x_2(t) \ldots x_N(t) \quad x_1(t-T) \ldots x_N(t-(M-1)T)] \tag{8.54}$$

The average clutter power can then be written as

$$C = E\left(|\mathbf{w}\mathbf{x}^T|\right) = \mathbf{w}^* \mathbf{R}_C \mathbf{w}^T \tag{8.55}$$

where the covariance matrix is

$$\mathbf{R}_C = E\left(\mathbf{x}^{T^*}\mathbf{x}\right) \tag{8.56}$$

To include channel noise the covariance matrix is replaced by

$$\mathbf{R}_{C+N} = \mathbf{R}_C + \sigma^2\mathbf{I} \tag{8.57}$$

When a point target is present at a given azimuth and Doppler, it is represented as

$$s = [s_1(t)\, s_2(t)\, s_3(t) \ldots s_N(t-(M-1)T)] \tag{8.58}$$

where s includes any weighting to reduce antenna sidelobes. The corresponding signal power is then given by

$$S = |\mathbf{ws}^T|^2 \tag{8.59}$$

The signal-to-clutter ratio is

$$\frac{\mathrm{S}}{\mathrm{C}} = \frac{|\mathbf{w}\,\mathbf{s}^T|^2}{\mathbf{w}^*\mathbf{R}_{C+N}\,\mathbf{w}^T} \tag{8.60}$$

As derived in [1] the weights that maximize the S/C ratio are

$$\mathbf{w}_{\text{opt}} = \mathbf{R}_{C+N}^{-1}\,\mathbf{s}^{T^*} \tag{8.61}$$

and the maximum signal-to-noise plus clutter ratio is

$$\left(\frac{\mathrm{S}}{\mathrm{N}+\mathrm{C}}\right) = \mathbf{s}\,\mathbf{R}_{C+N}^{-1}\,\mathbf{s}^{T^*} \tag{8.62}$$

The basic STAP data structure can be visualized as a 3-D data cube, as shown in Figure 8.38. These data are collected from L-range samples at the output of N-antenna elements through an M-pulse CPI. Each cell within the data cube is represented by a single complex number. The data cube then consists of $M \cdot N \cdot L$ complex samples that are stored in memory and block processed. Data processors like MATLAB can only process 2-D data so the pertinent data, called a snapshot, is contained in each $M \cdot N$ by 1 column vector. The snapshots are then stacked to complete the data set.

Here we consider only the data received from a single range-gate sample snapshot ($L = 1$). The objective is to adaptively determine the STAP weights in each

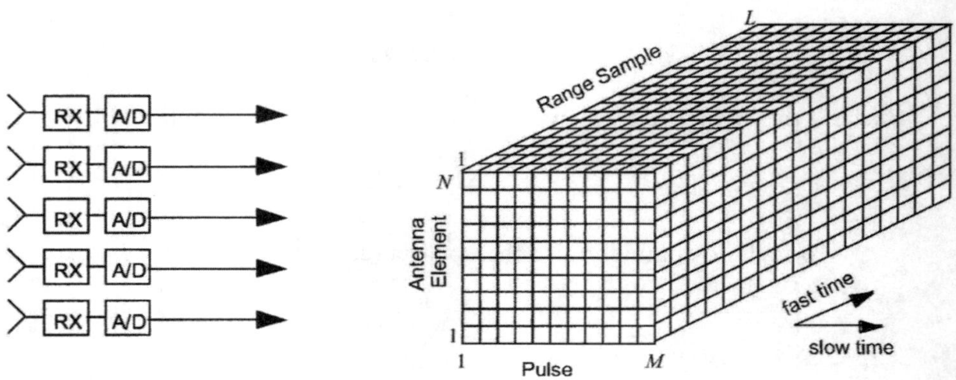

Figure 8.38 STAP data cube. (*From:* [10, p. 10].)

range bin to maximize the signal-to-interference-plus-noise ratio (SINR), thereby determining the presence or absence of targets. Figure 8.39 depicts the scenario for sidelooking radar that illuminates a concentric ring of clutter contained within a range resolution cell. For computational purposes the clutter will be divided into N_o discrete azimuth cells. Note that for each azimuth clutter cell, the Doppler return will be different depending on its azimuth angle in accordance with (8.49). Also the signal from each clutter cell return will generally arrive at each antenna element at a different time depending on its azimuth angle. This situation is depicted in Figure 8.40. The signal phase difference between antenna elements can then be found as $\phi = d \sin \theta / \lambda$ where d is the element spacing and θ is the azimuth angle. Specializing to an element spacing of $\lambda/2$ (to guard against grating lobes within the FOV) results in the spatial steering vector, which is given by

$$\mathbf{a}_N(\theta) = \left[1, e^{-j\pi \sin \theta}, \ldots, e^{-j(N-1)\pi \sin \theta}\right]^T \tag{8.63}$$

The Doppler frequency returned from each clutter patch depicted in Figure 8.39 is also a function of the azimuth angle, resulting in the temporal steering vector given as

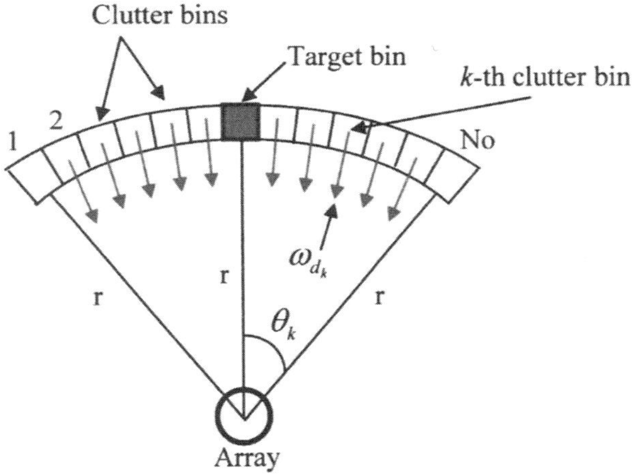

Figure 8.39 Clutter cells for sidelooking radar. (*From:* [31, p. 31].)

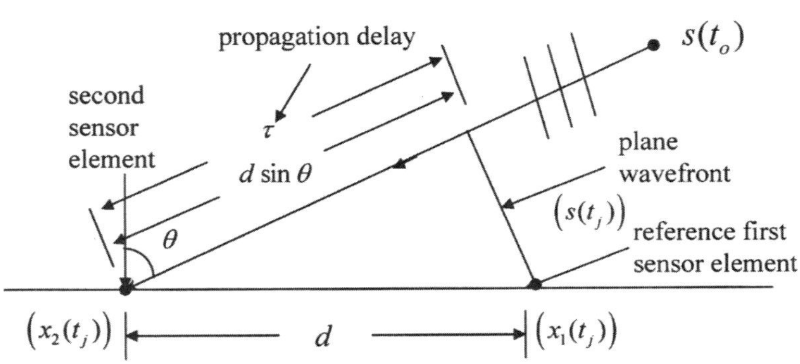

Figure 8.40 Differential time delay between antenna elements. (*From:* [31, p. 20].)

$$\mathbf{b}_M(\theta, \overline{f_d}) = \left[1, e^{-j2\pi\overline{f_d}\sin\theta}, \ldots, e^{-j2\pi(M-1)\overline{f_d}\sin\theta}\right]^T \tag{8.64}$$

where $\overline{f_d}$ is the Doppler frequency return normalized to the PRF given by

$$\overline{f_d} = \frac{2v_{ac}\cos\phi_d T}{\lambda} \tag{8.65}$$

and $v_{ac}\cos\phi_o$ is the platform velocity at the depression angle ϕ_o. The overall $M \cdot N$ by 1 spatial-temporal steering vector can be found as the Kronecker matrix product $\otimes$ [MATLAB function kron(a, b)] given by

$$\mathbf{a}(\theta, f_d) = \mathbf{b}_M(\theta, \overline{f_d}) \otimes \mathbf{a}_N(\theta) \tag{8.66}$$

Application of the SMI method [see (8.61)] requires an estimate of the covariance matrix of the interference (clutter-plus-noise-plus-jamming). This estimate is generally accomplished by excluding the target cell and examining data in range bins surrounding the test cell containing the expected target. This situation is analogous to cell-averaging CFAR systems and suffers from lack of sufficient samples to make an accurate estimate, inclusion of discretes in the estimated data set, unknown interference distributions, and inhomogeneities in the clutter data field. The best situation occurs when the interference is Gaussian distributed while the clutter is homogeneous (the so-called sandpaper earth clutter model) [12]. Then the received voltage signal at the array from the circular clutter range resolution ring at range r_j is represented as

$$\mathbf{x}_c = \sum_{k=1}^{No} \mathbf{c}_k \mathbf{a}(\theta_k, f_{d,k}) + \mathbf{n} \tag{8.67}$$

where c_k is the random amplitude of the kth clutter patch, $\mathbf{a}(\theta_k, f_{d,k})$ is the steering vector that defines the spatial-temporal modulation of the reflected clutter signal, and n is the thermal noise and interference component. If the clutter samples are zero-mean random variables with variance σ_k^2, then the clutter is wide sense stationary and is completely described by the covariance function

$$\mathbf{R}_c = \sum_{k=1}^{No} \sigma_k^2 \mathbf{a}_k \mathbf{a}_k^{T^*} + \sigma_n^2 \mathbf{I} \tag{8.68}$$

where σ_n^2 is the thermal noise power at the sensor input and $\sigma_c^2 = \sum_{k=1}^{No} \sigma_k^2$ is the total clutter power. The clutter-to-noise power is then given by $F = \sigma_c^2 / \sigma_n^2$. The resulting clutter power spectrum that exists along the diagonal clutter ridge (see Figure 8.35) in $\sin(\theta_{az})$-Doppler space can be found as

$$P(\theta, f_d) = \mathbf{a}^{T^*}(\theta, f_d)\mathbf{R}_c\mathbf{a}(\theta, f_d) \tag{8.69}$$

8.5.2 Effect of Internal Clutter Motion on STAP [24, 33]

The diagonal clutter ridge depicted in Figure 8.35 is spread by *internal clutter motion* (ICM) caused by wind-driven vegetation. This can be described using an exponential ground clutter model (see Example 3). The spreading prevents targets whose Doppler places them close to the clutter ridge from being detected. The ratio of dc and ac components is given by

$$r = 10^{[-15.5 \log_{10}(w) - 12.1 \log_{10}(f_c) + 63.2]/10)} \tag{8.70}$$

where w is the wind speed in miles per hour, and f_c is the transmitted frequency in megahertz. The ACF derived in Example 3 is

$$r_c = \frac{r}{r+1} + \frac{1}{r+1} \frac{(\beta\lambda)^2}{(\beta\lambda)^2 + (4\pi T)^2} \tag{8.71}$$

where β is a scale factor determined by wind conditions (see Example 3), λ is the wavelength, and T is the PRI. This can be used to taper the temporal steering vector [see (8.64)] where the taper between the ith and kth elements is given by

$$T_{ik} = r_c((i-k)T) \tag{8.72}$$

The resulting covariance matrix taper (CMT) is given by

$$\mathbf{T}_{ST} = \mathbf{T} \otimes 1_{N\times N} \tag{8.73}$$

where $\otimes$ is the Kronecker matrix product and $1_{N\times N}$ is an N by N matrix of ones. The revised *clutter covariance matrix* is then given by

$$\overline{\mathbf{R}_c} = \mathbf{T}_{ST}\, \mathbf{o} \mathbf{R}_c \tag{8.74}$$

where $\mathbf{o}$ is the Hadamard matrix product (MATLAB function A.*B).

8.5.3 STAP Detection

The covariance matrix when a target is present is given by

$$\mathbf{R} = P_t \mathbf{a}(\theta_t, f_{d,t})\mathbf{a}^{T^*}(\theta_t, f_{d,t}) + \mathbf{R}_c \tag{8.75}$$

where $\mathbf{a}(\theta_t, f_{d,t})$ is the target steering vector and P_t is the signal power. The input signal-to-interference ratio is then $P_t/(\sigma_c^2 + \sigma_n^2)$.

To determine the STAP processor SINR, we must consider the signal and interference signals separately. The received target signal input to the STAP processor in the jth range cell

$$\mathbf{x}_j = \alpha \mathbf{s}_t + \mathbf{n} \tag{8.76}$$

where $\alpha = \sqrt{P_t}$ is a real constant. The signal output of the adaptive STAP processor is

$$v_j(t) = \mathbf{w}^{T^*}\mathbf{x}_j(t) = \alpha\mathbf{w}^{T^*} + \mathbf{w}^{T^*}\mathbf{n} \quad (8.77)$$

where $\mathbf{w}$ are the adaptive weights. The optimum $M \cdot N$ by 1 adaptive weights can be found as

$$\mathbf{w}_{\text{opt}} = \kappa\mathbf{R}_{c+n}^{-1}\mathbf{a}\left(\theta_t, \bar{f}_{d,t}\right) \quad (8.78)$$

where κ is an arbitrary constant and $\mathbf{a}\left(\theta_t, \bar{f}_{d,t}\right)$ is the spatial-temporal target steering vector. From Section 5.1.2 the signal-to-interference power ratio is

$$\text{SINR}_{\text{out}} = \frac{P_t|\mathbf{w}^{T^*}\mathbf{s}_t|^2}{\mathbf{w}^{T^*}\mathbf{R}_{c+n}\mathbf{w}} \quad (8.79)$$

while the S/N ratio is

$$\text{SNR} = \frac{P_t|\mathbf{s}_t^{T^*}\mathbf{s}_t|^2}{\mathbf{s}_t^{T^*}\left(\sigma_n^2\mathbf{I}\right)\mathbf{s}_t} = \frac{P_t NM}{\sigma_n^2} \quad (8.80)$$

A STAP metric that compares the STAP performance relative to the interference-free case is the SINR loss given by

$$\text{SINR}_{\text{LOSS}} = \frac{\text{SINR}}{\text{SNR}} \quad (8.81)$$

The adaptive angle-Doppler pattern given by

$$P_a\left(\theta, \bar{f}_d\right) = |\mathbf{w}^{T^*}\mathbf{a}\left(\theta, \bar{f}_d\right)|^2 \quad (8.82)$$

can be used to evaluate the performance of the adaptive weight vector $\mathbf{w}_{\text{opt}}$ in suppressing the diagonal clutter ridge shown in Figure 8.35. Figure 8.41 displays the capability of STAP to detect targets while suppressing the target clutter ridge.

8.5.4 Sample Matrix Inversion

Operational STAP requires estimation of the interference covariance matrix given by

$$\hat{\mathbf{R}}_c = \frac{1}{K}\sum_{k=1}^{K}\mathbf{x}_i\mathbf{x}_i^{T^*} = \frac{1}{K}\mathbf{Y}_K\mathbf{Y}_K^{T^*} \quad (8.83)$$

where

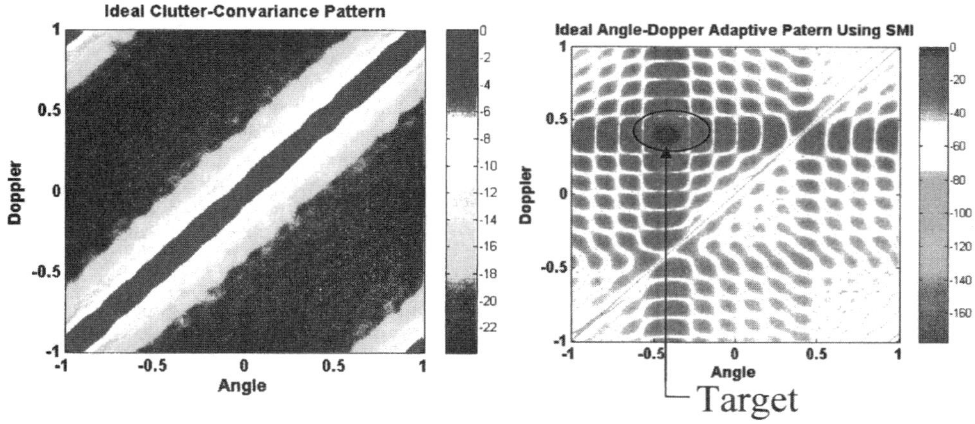

Figure 8.41 STAP target detection. (*From:* [31, pp. 35, 42].)

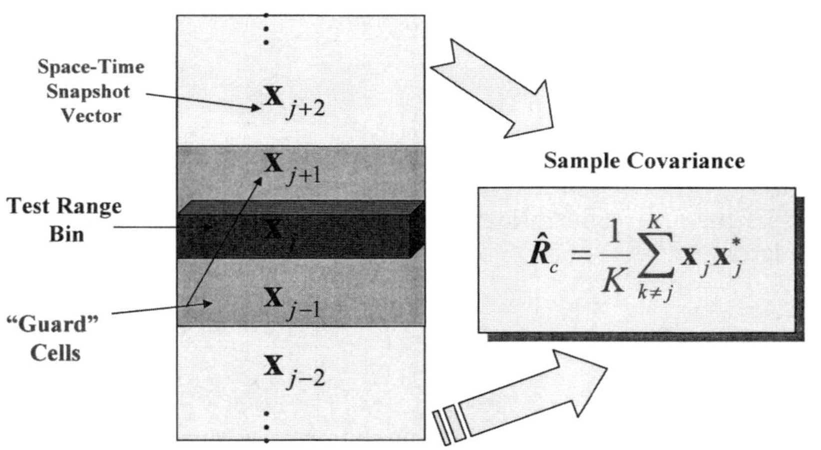

Figure 8.42 Covariance matrix estimation. (*From:* [31, p. 43].)

$$Y_K = [x_1\, x_2\, .\, .\, .\, x_K] \tag{8.84}$$

Samples are drawn from multiple range cells surrounding the target cell as depicted in Figure 8.42. The beamformer power spectrum is given by

$$\hat{P}_B(\theta, f_d) = \mathbf{a}(\theta, f_d)^{T^*}\overline{\mathbf{R}}\mathbf{a}(\theta, f_d) \tag{8.85}$$

while the SMI estimated weights are

$$\mathbf{w}_{\text{SMI}} = \hat{\mathbf{R}}_c\, \mathbf{a}(\theta_t, f_{d,t}) = \left(\frac{1}{K}\,\mathbf{Y}_K\mathbf{Y}_K^{T^*}\right)^{-1} \mathbf{a}(\theta_t, f_{d,t}) \tag{8.86}$$

▶ Example 31 Adaptive Array and Hot Clutter [34, 48]

The linear adaptive array shown in Figure EX-31.1 consists of N antenna elements whose outputs are individually weighted and then summed. The weights are adjusted by an adaptive network so that the array antenna pattern points at the target while jamming sources are minimized.

The output of the ith antenna sensor is given by

$$x_i(t) = u(t)\,\exp[-j(i-1)\pi \sin\phi] + n_i(t)$$

where array spacing $d = \lambda/2$ and $n(t)$ is the noise vector. To tune the array to broadside, adjust the weights

$$\mathbf{w}_q^T = \big(a_1 \quad a_2 \exp(-j\pi \sin\theta_s) \quad a_3 \exp(-j2\pi \sin\theta_s) \quad \ldots \quad a_N \exp(-j(N-1)\pi \sin\theta_{s\theta})\big)$$

where θ_s is the target's angle.

The element S/N ratio is then

$$(\mathrm{SNR})_1 = \frac{E\left[|u(t)|^2\right]}{E\left[|n_i(t)|^2\right]} = \frac{P_S}{\sigma_n^2}$$

while the array S/N ratio for a uniformly illuminated array ($a_i = 1$) is N times greater:

$$(\mathrm{SNR})_N = \frac{N^2 P_S}{N\sigma_n^2} = \frac{NP_S}{\sigma_n^2}$$

The array gain for a uniformly illuminated array is

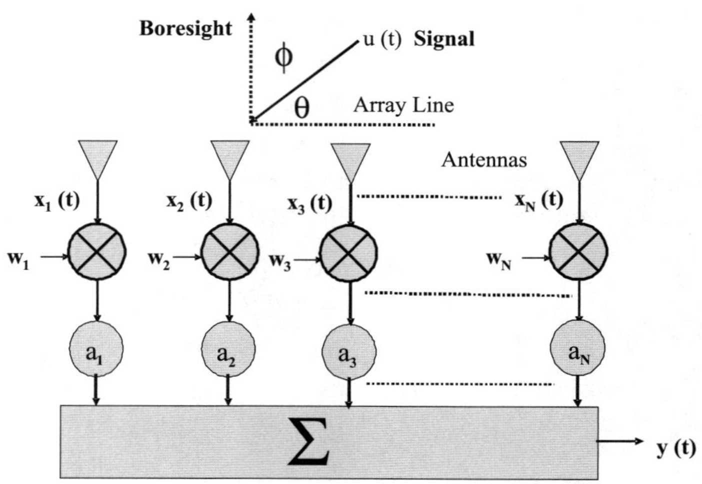

Figure EX-31.1 Adaptive array.

$$G(\phi) = \left[\frac{\sin\left(\frac{N\pi \sin \phi}{2}\right)}{N \sin\left(\frac{\pi \sin \phi}{2}\right)} \right]^2$$

while its null-to-null beamwidth is

$$\phi_0 = 2 \sin^{-1}\left(\frac{2}{N}\right)$$

A windowed array may be employed to reduce the array sidelobes. The amplitude functions $a(i)$ are then adjusted in accordance with, for example, a Chebyshev function that maintains a uniform sidelobe level while providing a sharp transition to the main beam response. The MATLAB function chebwin(n,r) can be used for this purpose where n is the length and r is the specified sidelobe level in decibels.

The antenna pattern for the adaptive array in the presence of high-duty cycle noise jamming can be found using Applebaum's method [34], which requires that the array be optimized when just noise and jamming are present. Basically, the method generates an auxiliary beam that when subtracted from the quiescent beam produces nulls at the jammer angles in the overall beam.

To apply this method, we first find the covariance matrix for noise alone for an N-element array that is given by

$$\mathbf{M}_q = P_n \mathbf{I}_N$$

where $\mathbf{I}_N$ is the identity matrix of order N and P_n is the noise power. Next we generate a diagonal matrix:

$$\mathbf{H} = \begin{pmatrix} 1 & 0 & \cdots & 0 \\ 0 & \exp(j\beta_j) & 0 & 0 \\ \vdots & 0 & \ddots & \vdots \\ 0 & 0 & \cdots & \exp(j(N-1)\beta_j) \end{pmatrix}$$

where

$$\beta_j = \pi \sin \theta_j$$

and θ_j is the angle of the jammer. Then the covariance matrix for the jammer is found as

$$\mathbf{M}_j = \mathbf{P}_j \mathbf{H}^* \mathbf{U} \mathbf{H}$$

where U is an $N \times N$ matrix of ones and P_j is the jammer power. The total covariance matrix is then

$$\mathbf{M} = \mathbf{M}_q + \mathbf{M}_j$$

It can then be shown that the optimum weights [1] that maximize the signal-to-interference ratio are given by

$$\mathbf{w}_{\text{opt}} = \alpha \mathbf{M}^{-1} \mathbf{S}^*$$

where $\mathbf{S}$ is the array steering vector and $\mathbf{M}^{-1}$ is the matrix inverse of the covariance matrix $\mathbf{M}$. The overall array pattern can be found to be

$$G(\theta) = \mathbf{B}^T \mathbf{w}_{\text{opt}}$$

where

$$\mathbf{B}^T = \big(1 \quad \exp(j\pi \sin\theta) \quad \exp(j2\pi \sin\theta) \quad \ldots \quad \exp(j(N-1)\pi \sin\theta)\big)$$

However, it is instructive to separate the pattern into two parts as follows:

$$G(\theta) = G_q(\beta) - \frac{1}{d+N} G_q(\beta_j) C(\beta - \beta_j)$$

where

$$\beta = \pi \sin\theta \qquad d = \frac{P_n}{P_j}$$

and

$$C(x) = \exp\left(\frac{j(N-1)x}{2}\right) \frac{\sin Nx/2}{\sin x/2}$$

The first term is the quiescent pattern; the second term, which is subtracted from the first, is a $\sin(Nx)/\sin(x)$ beam centered on the jammer. This is shown in Figure EX-31.2 and illustrates how the adaptive array cancels jammers in the array sidelobes.

The adaptive array can be shown to cancel a narrowband jammer by the factor P_j/P_n (Figure EX-31.3). For wideband jammers the relationship that applies is given in [35] as

$$\frac{P_0}{P_n}(f) = 1 + \frac{\left\{1 - \cos\left[\pi \sin\theta_j \left(\frac{\Delta f}{f_0}\right)\right]\right\} P_j}{P_n}$$

where θ_j is the jammer's angle off boresight and f_0 is the array's center frequency.

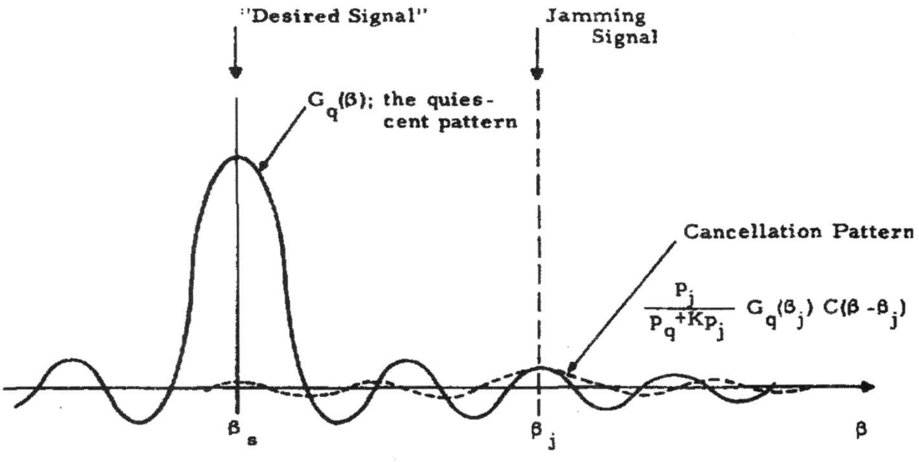

Figure EX-31.2 Adaptive array quiescent and cancellation antenna patterns. (*From:* [34, p. 591].)

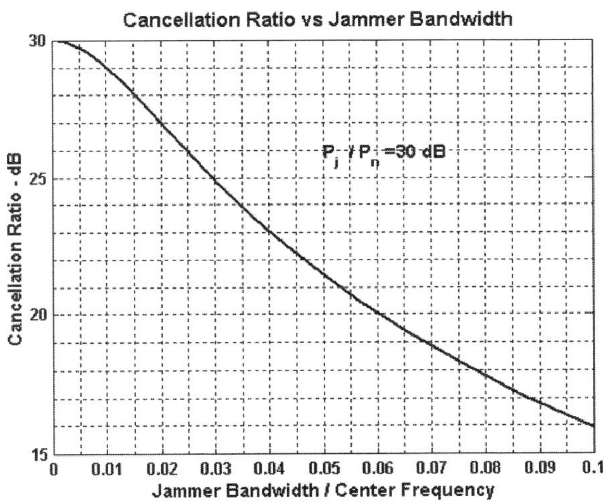

Figure EX-31.3 Degradation in cancellation ratio with jammer bandwidth.

The MATLAB program adap_apple2.m provides a design tool that allows adaptive arrays to be synthesized. The dialog box is shown in Figure EX-31.4.

The results for these inputs are shown in Figures EX-31.5 and EX-31.6 (on a decibel scale). Note that deep nulls are provided at jammer angles of 30°, 45°, and –30°. Error stops are built into the program if the user attempts to place the jammer's angle within the array's main beam or to exceed its degrees of freedom ($N - 1$).

One way to limit the performance of the adaptive array or STAP is to employ *hot clutter*, which can be generated by illuminating either clutter or chaff such that the reflected jammer signals are directed at the array [12, 36]. The spatially dispersed jammer signals then tend to overload the array's degrees of freedom and cause the adaptive array to malfunction. Figure EX-31.7 depicts a simulation using Simulink

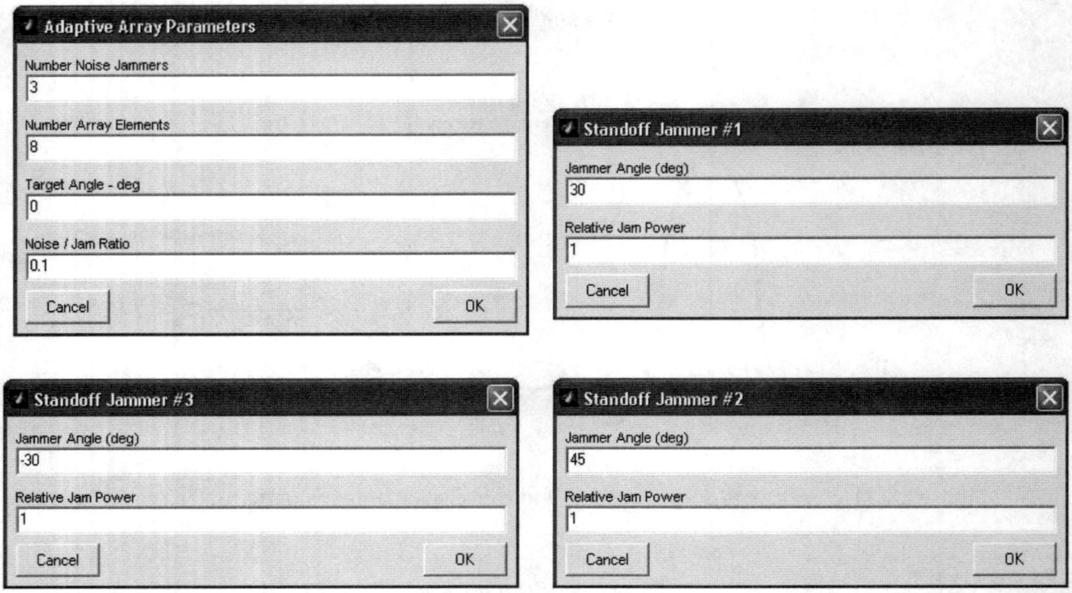

Figure EX-31.4 Dialog boxes for adaptive array design.

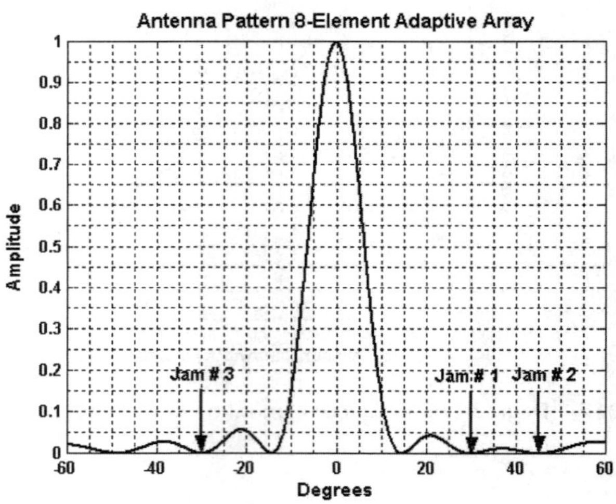

Figure EX-31.5 Pattern for 8-element adaptive array with sidelobe jamming.

which shows that even a small amount of distributed hot clutter has a significant effect on the sidelobe canceller's performance. For example, if the distributed jammer's power is 20 dB below the main jammer's power (JDJR = 20 dB) then the cancellation ratio of the main jammer is reduced by about 20 dB.

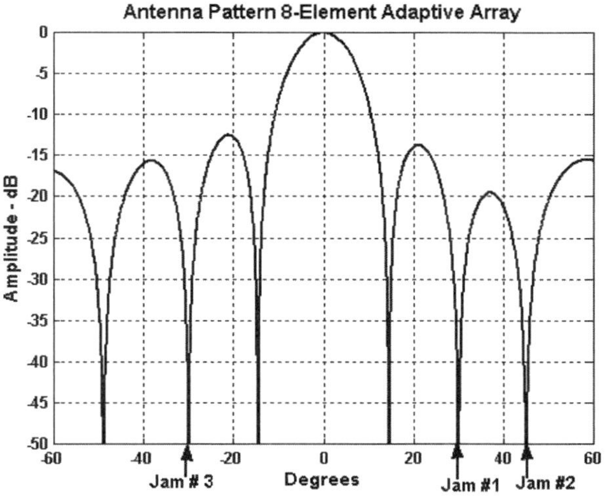

Figure EX-31.6 Pattern for adaptive array on decibel scale with sidelobe jamming.

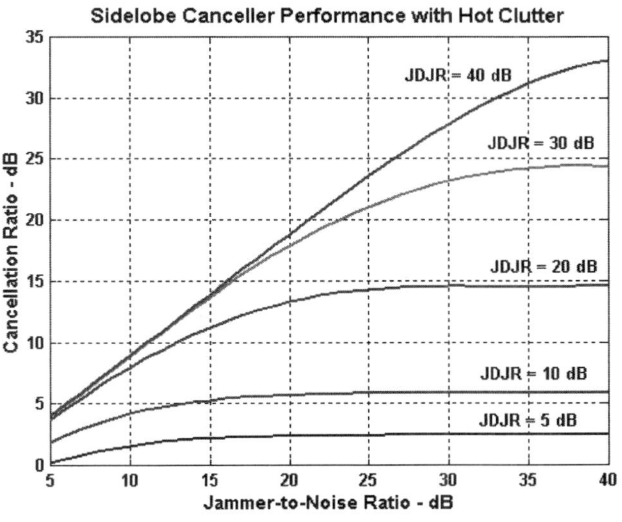

Figure EX-31.7 Hot clutter degradation of SLC cancellation ratio. (*From:* [36, p. 39].)

▶

8.6 MTI Design Tool

In this section we discuss an interactive MATLAB-based design tool that allows the detection performance of surface-based air-search MTI Doppler radar to be determined using different types of processors under a variety of environmental conditions. A number of MATLAB's *Graphical User Interface* (GUI) tools are employed to isolate the user from the task of entering the computer program to modify input data from the default condition that uses the specification for the ASR-9 radar. The design tool uses many of the MATLAB programs developed in the examples provided throughout the text. An advantage of this approach is that if the user is uncertain of a particular parameter, he or she can use or vary the default parameter to determine an appropriate value.

A number of MTI processors are considered, including MTD types. These include single, double, triple, and fourth-order cancellers employing noncoherent integration, cascaded dual-canceller FFT and transversal filters. Either pulse-to-pulse or block-to-block PRF staggers as appropriate are assumed for each of the processors.

The design tool is initiated by entering "mtirangehh.m" into the MATLAB Command Window and pressing enter. A dialog box, called Radar Parameters, opens that contains transmitter, receiver, and antenna parameters as depicted in Figure 8.43. The parameters are either accepted or adjusted appropriately and OK is clicked. Radar Parameters Page 2 appears and again is accepted or adjusted. Note that the parameters FFT or transversal filter (TF) points need be adjusted only if performance of these types of processors is required. A critical parameter is the system stability factor, which contains the combined effect of the cancellation ratio (see Section 3.1.2) and the phase noise (see Section 3.3) [37, 38]. This parameter is often given in the particular radar's specification. The filter sidelobes must be specified for FFT- and TF-based MTD processors to prevent suppression of Doppler filter bands not containing the clutter's primary Doppler response. A Chebyshev-type window is provided for this purpose. If an MTD-type processor is not employed, then the user is further given the alternative of selecting a Uniform

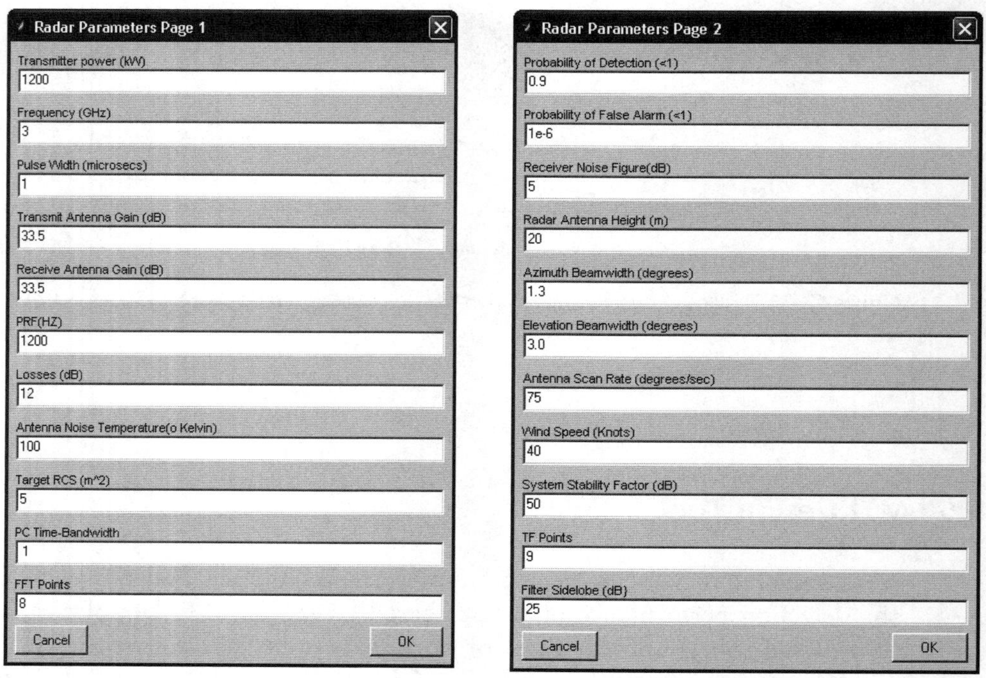

Figure 8.43 Radar Parameters dialog box.

or Chebyshev-type window. Pressing OK on the Radar Parameter Page 2 dialog box opens a menu bar for selecting the Swerling type target as shown in Figure 8.44. The Window Type menu bar then selects the Uniform or Chebyshev window.

The next menu bar then prompts the user to choose one of the four types of clutter to be processed. If Sea Clutter is chosen, then the Sea State menu bar appears with the choice of sea states ranging from 1 to 9. The program assumes that perfect motion compensation is accomplished. Alternatively, if Land Clutter is chosen, then the Terrain Type menu bar is activated with six types of terrain available. For Sea and Land Clutter the default is for clear weather (with atmospheric attenuation), although this can be modified by changing the program. If Precipitation clutter is chosen then a multiclutter environment containing both land and precipitation clutter is activated for MTD-type processors. Otherwise, the MTI cancellers are assumed to be adaptively tuned to the mean Doppler frequency of the precipitation with no land clutter present. After the Terrain Type is chosen, the Precipitation Type menu bar is activated with four choices ranging from light to intense. The program then calculates the detection range, including the attenuation due to the precipitation. When Chaff Type clutter is chosen, the program response is similar to that used for precipitation except that the user is given the choice of Chaff Type ranging from light to heavy in a clear environment.

The program output for the parameters given in Figure 8.43 is presented in a bar chart annotated with a legend detailing the selected parameters. The first example is depicted in Figure 8.45 for wooded hills land clutter with uniform window MTD processing. The first three bars depict the detection range in noise without Doppler processing. The first bar reflects the use of noncoherent integra-

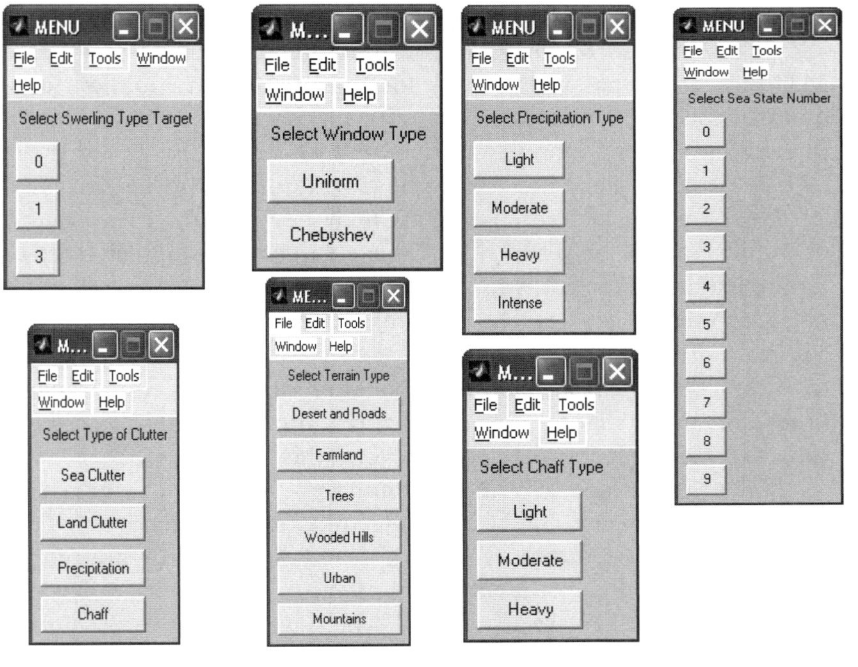

Figure 8.44 Menu bars in MTI design tool.

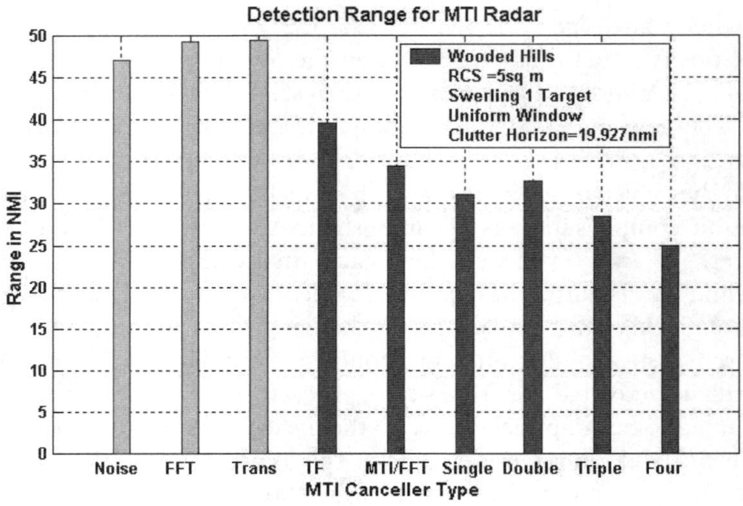

Figure 8.45 Detection range in wooded land clutter with MTD and MTI processors using a uniform window.

tion, whereas the second and third bars include the effect of coherent integration of 8 and 9 pulses. The last four bars depict the detection ranges of single-order through fourth-order MTI cancellers employing noncoherent integration. The double MTI canceller provides the best detection range relative to the other cancellers. The single canceller has a limited improvement factor, and the performance of the higher order cancellers is limited by their increasing velocity loss (see Figure 8.50 later in the chapter) and 50 dB system stability factor, which bounds their improvement factors. The transversal filter MTD provides the best Doppler processor primarily due to its 9-pole clutter attenuation and low velocity loss. The MTI/FFT MTD has the velocity loss and clutter attenuation associated with a double canceller, which causes it to have poorer performance than the transversal filter. Note that all Doppler processors include the use of either a pulse-to-pulse or block stagger since the velocity loss without a stagger is excessive.

Next we consider the use of a Chebyshev window to prevent cross-coupling in the FFT and transversal filter. This enlarges the Doppler filter bandwidth (see Example 6 in Chapter 2) somewhat reducing the crossover loss but allowing more noise into the processor. The resulting detection ranges are shown in Figure 8.46. The noise-limited FFT and transversal filter noise detection ranges are reduced due to the increased noise bandwidth as well as the detection ranges of the MTD processors. Note that the clutter horizon is about 20 nmi so that the noise detection ranges apply for airborne targets beyond this range.

The next example considers Mountain Type land clutter where the response is depicted in Figure 8.47. The detection ranges provided by the MTD-based FFT/MTI and TF processors using a Chebyshev window are reduced from those of wooded hills due to the increased intensity of the clutter. The detection ranges of the MTI cancellers are all reduced. The single canceller does not provide enough improvement factor to be effective. The clutter horizon is extended to 41.5 nmi, requiring Doppler processing over the full operating range of the radar.

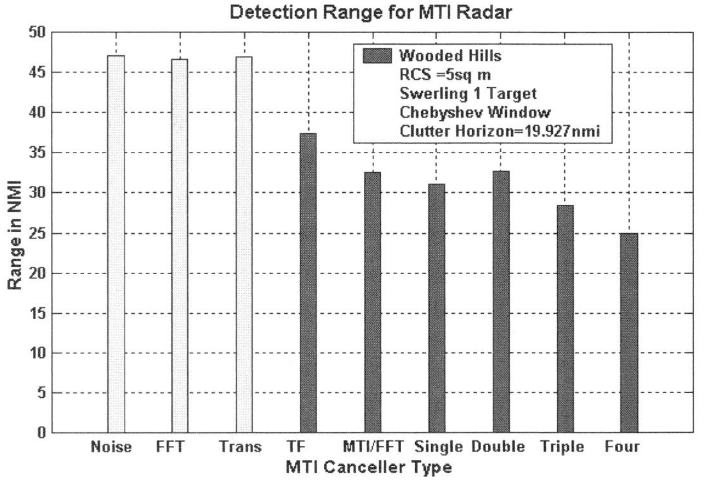

Figure 8.46 Detection range in wooded land clutter with MTD and MTI processors using a Chebyshev window.

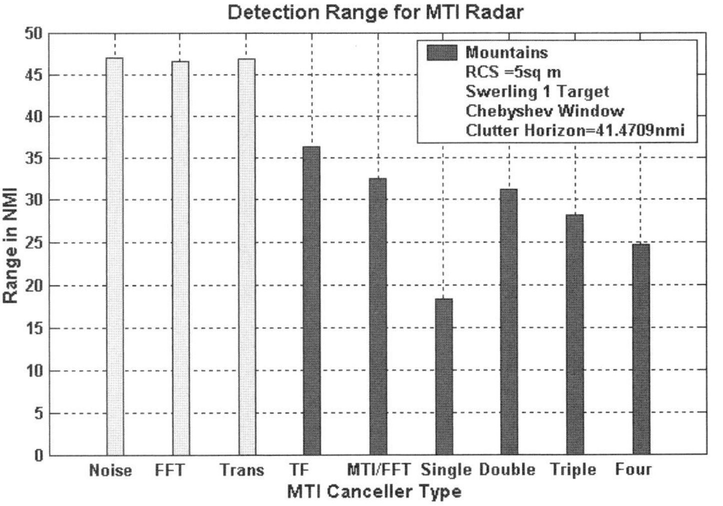

Figure 8.47 Detection range in mountain land clutter with MTD and MTI processors using a Chebyshev window.

We now consider multispectral clutter involving simultaneous land and precipitation clutter. A block stagger and Chebyshev window are employed in the TF/Stagger processor to allow the full power of the MTD design to be realized. The Doppler filters at zero Doppler containing residual land clutter and those containing precipitation clutter are desensitized for each of the two PRFs of the block stagger, thereby causing a relatively large velocity loss. The canceller designs utilize a pulse-to-pulse stagger and are tuned to the mean spectrum of the precipitation while ignoring ground clutter. The response for heavy rain is shown in Figure 8.48 and indicates that only the TF/Stagger processor provides reasonable detection

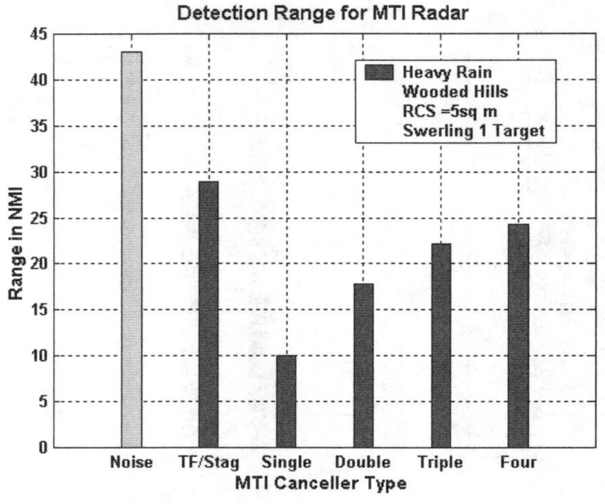

Figure 8.48 Detection range for heavy rain clutter with MTD and MTI processors and a Chebyshev window.

ranges. The problem with the MTI cancellers is that the wind shear component (see Section 4.2.1.2) spreads the clutter spectrum to where a wide canceller notch is required to attenuate the precipitation clutter. However, the wide notches reduce the target response causing a large velocity loss (see Figure 8.50).

Figure 8.49 shows the response of the MTD and MTI processors to moderate chaff. These are similar to that of heavy rain except that the chaff is cut to respond to the radar's frequency while the reflectivity of rain is proportional to the fourth power of frequency. It illustrates that MTIs are relatively ineffective against chaff and a processor employing a Doppler Filter Bank is required.

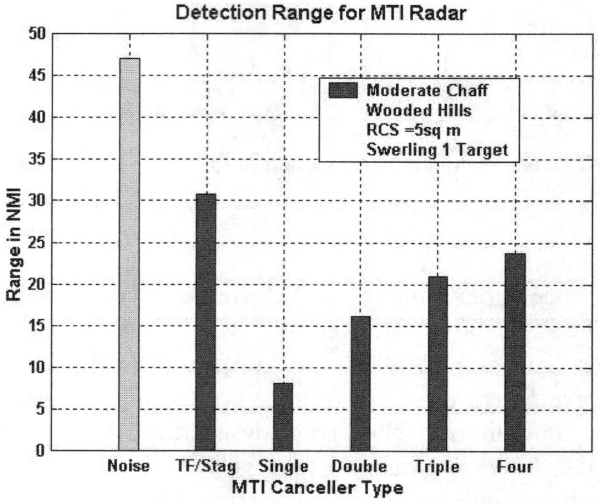

Figure 8.49 Detection range in moderate chaff clutter with MTD and MTI processors and a Chebyshev window.

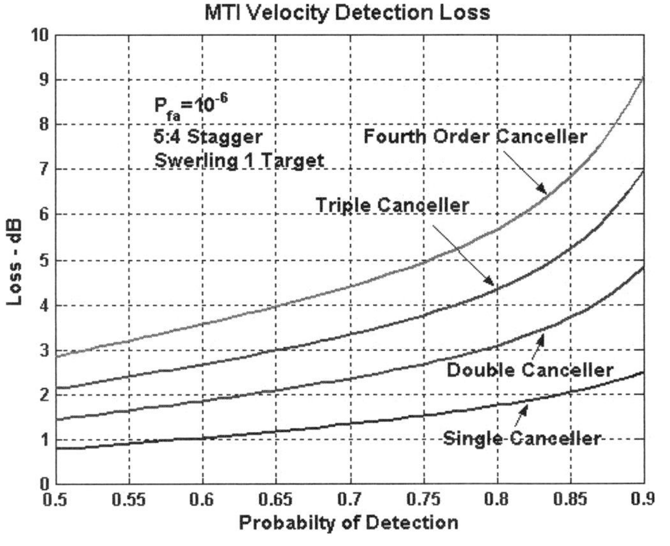

Figure 8.50 Velocity losses for MTI cancellers.

8.6.1 Program Discussion and Notes

The MTI Design Tool allows the detection range to be determined for MTI cancellers of first through fourth order that employ noncoherent integration and MTD-based Doppler processors. One of the important parameters is the system stability factor. Table 8.5 lists these factors and processor type for a number of radars [39]. Note that most of these radars use MTI canceller-type processors. Generally radars in L-band and lower frequencies can suppress precipitation clutter using circular polarization antennas (see Figure EX-13.3 in Chapter 4) rather than using an MTD-type processor, but this is not effective against chaff. The modular nature of the program allows CP processing to be easily incorporated by those skilled in MATLAB programming. In general, solid-state transmitters have the capability to provide much higher stability factors than tube-type transmitters [38].

Table 8.5 Stability Factors and Processor Type for Search Radars [39]

Radar	*Band*	*Stability Factor (dB)*	*Processor Type*
APS-145	UHF	66	3-pulse MTI/FFT(16)
ARSR-4	L	68	8- to 9-pulse MTI
ASR-9	S	50	3-pulse MTI/FFT(8)
SPS-48E	S	41	6-pulse MTI
SPS-48E(U)	S	95	Solid-state transmitter
SPS-49(V)5	L	58	4-pulse FIR MTD
SPS-49A(V)1	L	68	4-pulse FIR MTD
SPS-49A(V)2	L	71	4-pulse FIR MTD
SPS-40E	UHF	54	4-pulse MTI
SPQ-9B	X	90	4-pulse MTI
SPY-1D	S	—	2- to 7-pulse MTI
TPS-59	L	53	3- and 6-pulse MTI
TPS-70/75	S	50	4-pulse MTI

Velocity losses are caused by gaps in the frequency response of Doppler processors due to their filtering action against clutter. In MTI-type processors target returns are generally aliased into the region between zero and the first blind speed. Example 17 in Chapter 5 illustrates that the gaps significantly widen as the order of the MTI canceller increases. Because the aliased frequency of the target return is uncertain, it is generally assumed that the target returns are uniformly distributed over the first PRF region. Example 8 in Chapter 3 demonstrates that large velocity losses result in this situation, whereas Example 23 in Chapter 6 shows that velocity losses can be significantly reduced using PRF staggering, making inclusion of this technique a necessity for any successful design.

Velocity losses for single and double cancellers are given in Example 23 but need to be extended for higher order cancellers included in the MTI Design Tool. The technique is used to first derive the frequency response of the staggered canceller using an extension of (6.112) and then to solve the nonlinear equation for the detection probability in accordance with Example 23. The power transfer function of a triple canceller with a dual stagger is given by

$$|H_3(\omega)|^2 = 10 - 15\cos(\omega T)\cos(\omega Te) + 6\cos(2\omega T) - \cos(3\omega T)\cos(\omega Te) \tag{8.87}$$

while that of a fourth-order canceller with a dual stagger is

$$\begin{aligned} |H_4(\omega)|^2 = 35 &- 56\cos(\omega T)\cos(\omega Te) + 28\cos(2\omega T) \\ &- 8\cos(3\omega T)\cos(\omega Te) + \cos(4\omega T) \end{aligned} \tag{8.88}$$

and the power gain is 20 for the triple canceller and 70 for the fourth-order canceller. The resulting velocity loss is given in Figure 8.50 for cancellers of one to fourth order. This is computed in MATLAB program mti_velstaggx.m. Note that losses become large for higher order cancellers and high probability of detection that points to the use of a double-canceller/FFT design.

The transversal filter stagger also experiences a velocity loss in a clutter environment consisting of ground clutter. An envelope of the TF (see Example 19 in Chapter 5) is used in this calculation. The resulting velocity loss is depicted in Figure 8.51. This is computed in MATLAB program trans_filt_pdc.m. Note that the velocity loss is a function of the clutter-to-noise ratio (F) since the TF null widths about zero frequency and the first blind speed adapt to the intensity of the clutter. When an MTD TF/Stagger design is used in a multiclutter environment of land and precipitation or chaff clutter, then an additional velocity loss is experienced. This is caused by the blanking of the zero frequency and first blind speed filters and those filters containing precipitation or chaff clutter. The velocity loss for the TF/Stagger is calculated online in the program using the MATLAB functions tran_stag.m for land and sea clutter and tran_stagb.m for combined land, precipitation, or chaff clutter. The use of the MATLAB function fzero.m in these programs causes a several-minute delay in the execution of the MTI Design Tool. Unfortunately, the MATLAB *Waitbar* does not function with fzero because it does not contain a "*for loop*."

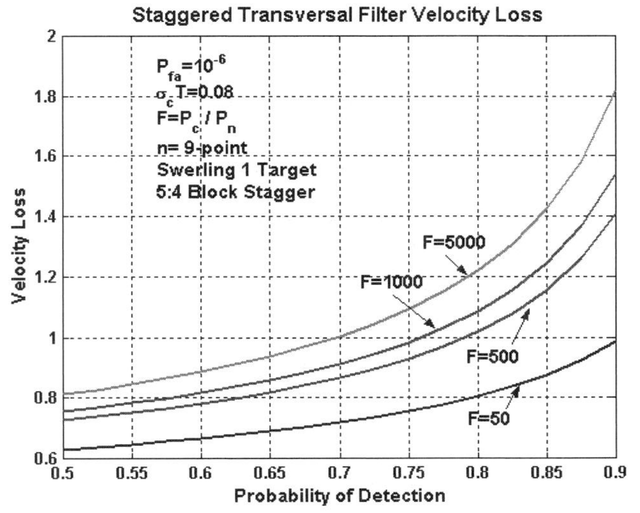

Figure 8.51 Velocity loss of transversal filter in land clutter.

For MTD designs utilizing Doppler filter banks, it is generally necessary to employ a window to reduce the filter sidelobes. The MTI Design Tool employs a Chebyshev window where the user can specify the depth of the sidelobes. Several losses are associated with the use of Doppler filter banks. The first is a crossover loss that accounts for the fact that targets may lie anywhere in the interval between adjacent center frequencies. This is approximated as [40]

$$L_{cx} = 1.25 P_d^{1.25} \left(\frac{1}{bndx} \right)^2 \quad [\text{dB}] \tag{8.89}$$

where *bndx* represents the relative filter 3 dB bandwidth that is 0.89 for a Uniform window and 1.33 for a 50 dB sidelobe Chebyshev window. An interpolation table is included in the program so that the user can specify any sidelobe to 50 dB. The other loss associated with a Chebyshev window is the increase in the Doppler filter's noise bandwidth as the sidelobe level is increased. The window loss is 3.35 dB for 50 dB sidelobes. As with the crossover loss an interpolation table is supplied in the program so that any sidelobe to 50 dB can be specified. Another loss associated with MTI cancellers employing video integration involves the correlation of receiver noise by the MTI filter, which causes a loss in integration gain in a noise environment. The magnitude of this loss is given in Table 5.4 and generally increases as the order of the MTI increases. The overall integration gain of a cascaded MTI canceller and noncoherent video integrator is derived in Example 18 (see Chapter 5) and is a function of both the MTI improvement factor and the clutter-to-noise ratio $\left(F = \sigma_c^2 / \sigma_n^2\right)$. The interference power is given by $I = \sigma_n^2 (1 + F/I_f)$ so that the integration gain is the same as for noise alone when F/I_f is very small and equal to one when F/I_f is large.

The detection factors for the various Swerling-type targets and noise integration gains are calculated using the MATLAB program det_facts (P_D, P_fa, hits).m.

This program uses Albersheim's algorithm (see Example 7 in Chapter 3) to compute the detection factors. The following example, Example 32, compares these factors versus Marcum-Swerling detection factors using algorithms given in [41]. The difference is within 0.3 dB for fixed targets and 0.6 dB for fluctuating targets.

Land clutter reflectivity, over a frequency range of 0.5 to 35 GHz, is calculated using gammas from [42] that are smoothed using a least mean square algorithm. Sea clutter gammas are from [40]. The precipitation detection range is calculated using the MATLAB program rain_range.m, and the Chaff detection range is calculated using chaff_range.m. Both programs use Newton's method. Clear and precipitation attenuation is calculated using MATLAB program range_atten.m, which also employs Newton's method.

► Example 32 Comparison of Marcum-Swerling and Albersheim Detection Factors

In the MTI Design Tool we approximate the Marcum-Swerling (M-S) detection factors using an approximation due to Albersheim [4, 43]. This example compares the two methods for Swerling fixed, case 1, 2, 3, and 4 type targets. Algorithmic programs for calculating the M-S detection factors are given in [41], and Albersheim's procedure is described in Example 7. The approach is to first calculate a threshold Y_b using the MATLAB function threshold (P_{fa}, n). Then MATLAB program mar_swerl_deta.m uses Newton's method to adjust the S/N ratio so that a specified P_d is achieved. In addition, Albersheim's detection factors are calculated using MATLAB function det_factors(P_D, P_fa, hits). Marcum-Swerling detection factors are shown in Table EX-32.1, and Albersheim detection factors in Table EX-32.2.

Figure EX-32.1 compares the Marcum-Swerling and Albersheim's detection factors using MATLAB program mar_swerl_detb.m. In general the error in using Albersheim's approximation is less than 0.6 dB for all types of targets and 0.3 dB for fixed targets. This is more than adequate for the MTI Design Tool considering the accuracy of the clutter backscattering coefficients.

Table EX-32.1 Marcum-Swerling Detection Factors

$P_d = 0.9$
$P_{fa} = 10^{-6}$

n	Yb	Fixed	Swerl_1	Swerl_2	Swerl_3	Swerl_4
1.0000	13.8155	13.1835	21.1436	21.1436	17.2961	17.2961
2.0000	16.6884	10.6539	18.6883	14.8259	14.8261	12.8652
3.0000	19.1292	9.2291	17.3107	12.1372	13.4395	10.7775
4.0000	21.3505	8.2452	16.3617	10.5064	12.4841	9.4489
5.0000	23.4315	7.4980	15.6422	9.3613	11.7596	8.4881
6.0000	25.4126	6.8981	15.0651	8.4897	11.1785	7.7421
7.0000	27.3177	6.3982	14.5847	7.7916	10.6947	7.1355
8.0000	29.1622	5.9707	14.1741	7.2125	10.2812	6.6264
9.0000	30.9571	5.5978	13.8161	6.7196	9.9207	6.1889
10.0000	32.7103	5.2675	13.4992	6.2918	9.6016	5.8063
11.0000	34.4279	4.9714	13.2151	5.9148	9.3155	5.4667
12.0000	36.1144	4.7033	12.9580	5.5783	9.0566	5.1620
13.0000	37.7737	4.4585	12.7232	5.2749	8.8202	4.8858
14.0000	39.4087	4.2334	12.5074	4.9991	8.6028	4.6336
15.0000	41.0221	4.0252	12.3078	4.7463	8.4018	4.4016
16.0000	42.6158	3.8315	12.1221	4.5134	8.2149	4.1870
17.0000	44.1916	3.6506	11.9487	4.2975	8.0403	3.9875
18.0000	45.7512	3.4810	11.7860	4.0964	7.8765	3.8011
19.0000	47.2958	3.3212	11.6328	3.9084	7.7224	3.6264
20.0000	48.8265	3.1704	11.4883	3.7319	7.5768	3.4619

Table Ex-32.2 Albersheim's Detection Factors

$P_d = 0.9$
$P_{fa} = 10^{-6}$

n	Fixed	Swerl_1	Swerl_2	Swerl_3	Swerl_4
1.0000	13.1145	21.1436	21.1436	17.1291	17.1291
2.0000	10.4575	18.4866	14.4720	14.4720	12.4647
3.0000	8.9745	17.0036	11.6509	12.9891	10.3127
4.0000	7.9647	15.9938	9.9719	11.9792	8.9683
5.0000	7.2068	15.2359	8.8126	11.2213	8.0097
6.0000	6.6040	14.6331	7.9421	10.6185	7.2731
7.0000	6.1056	14.1347	7.2526	10.1201	6.6791
8.0000	5.6820	13.7111	6.6857	9.6966	6.1838
9.0000	5.3145	13.3436	6.2066	9.3290	5.7605
10.0000	4.9904	13.0195	5.7933	9.0049	5.3918
11.0000	4.7009	12.7300	5.4308	8.7155	5.0659
12.0000	4.4396	12.4687	5.1087	8.4542	4.7742
13.0000	4.2017	12.2308	4.8193	8.2162	4.5105
14.0000	3.9834	12.0125	4.5569	7.9980	4.2702
15.0000	3.7819	11.8110	4.3172	7.7965	4.0496
16.0000	3.5949	11.6240	4.0967	7.6094	3.8458
17.0000	3.4205	11.4496	3.8928	7.4350	3.6566
18.0000	3.2571	11.2862	3.7031	7.2716	3.4801
19.0000	3.1035	11.1326	3.5261	7.1180	3.3148
20.0000	2.9586	10.9877	3.3600	6.9731	3.1593

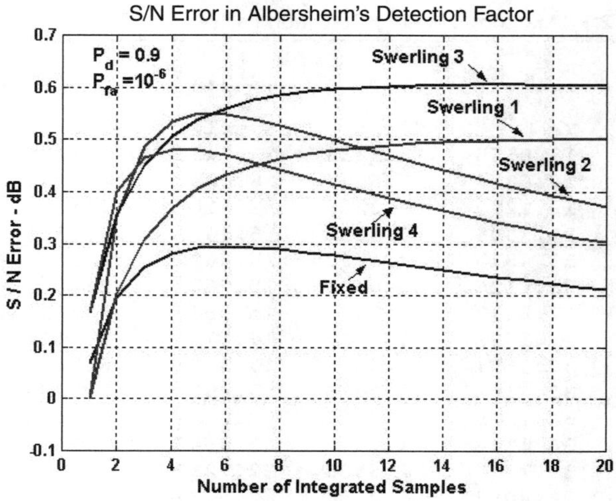

Figure EX-32.1 Comparison of M-S and Albersheim's detection factors.

▶

References

[1] Schleher, D. C., *MTI and Pulsed Doppler Radar*, Norwood, MA: Artech House, 1991.

[2] Stimson, G., *Introduction to Airborne Radar*, 2nd ed., Mendham, NJ: SciTech, 1998.

[3] *Future DoD High Frequency Radar Needs Resources*, AD A391893, Defense Science Board, April 2001.

[4] Skolnik, M., *Introduction to Radar Systems*, 3rd ed., New York: McGraw-Hill, 2001.

[5] M. Richards, M., "A Beginners Guide to Interferometric SAR Concepts," *IEEE AES Systems Magazine*, Vol. 22, No. 9, September 2007, pp. 5–29.

[6] Schleher, D. C., *Electronic Warfare in the Information Age*, Norwood, MA: Artech House, 1999.

[7] Musman, S., D. Kerr, and C. Bachman, "Automatic Recognition of ISAR Ship Images," *IEEE Trans. on Aerospace and Electronic Systems*, Vol. AES-32, No. 4, October 1996, pp. 1392–1404.

[8] Greenspan, M., "Multiport Airborne Radar Techniques for Slow Speed Ground Moving Target Detection and Location," *Proc. 37th Annual Tri-Service Radar Symposium*, Colorado Springs, CO, June 1991, pp. 439–450.

[9] Stockburger, E., and D. Held, *Interferometric Moving Ground Target Imaging*, Proc. IEEE Int. Radar Conference, Washington, D.C., May 1995, pp. 438–443.

[10] Ward, J., *Space-Time Adaptive Processing for Airborne Radar*, Technical Report 1015, Lincoln Laboratories, Lexington, MA, December 1994, pp. 1–203.

[11] Schleher, D. C., (ed.), *MTI Radar*, Dedham, MA: Artech House, 1978.

[12] Griffiths, L., et al., "Space-Time Adaptive Processing in Airborne Radar Systems," *Proc. IEEE Int. Radar Conference*, 2004, pp. 711–716.

[13] Shohat, M., "Doppler Beam Sharpened Attack Radar Uses Magnetron," *Military Electronics/Countermeasures*, October 1979, pp. 58–66.

[14] Povejsil, D., R. Raven, and P. Waterman, *Airborne Radar*, New York: Van Nostrand, 1961.

[15] Skolnik, M., (ed.), "MTI Radar" and "Synthetic Aperture Radar," Chapters 15 and 21 in *Radar Handbook*, 2nd ed., New York: McGraw-Hill, 1990.

[16] Brookner, E., "Synthetic Aperture Radar Spotlight Mapper," in *Radar Technology*, Dedham, MA: Artech House, 1977.

[17] Wehner, D., *High Resolution Radar*, Norwood, MA: Artech House, 1987.

[18] Tsunoda, S., et al., "Lynx: A High Resolution Synthetic Aperture Radar," *Proc. SPIE Aerosense*, Vol. 3704, 1999, pp. 1–8.

[19] Kingsley, S., and S. Quagan, *Understanding Radar Systems*, Mendham, NJ: SciTech, 1999.

[20] FAS, Interferometric Synthetic Aperture Radar (IFSAR), http://www.fas.org/irp/imint/isfar.htm, January 1997.

[21] http://www.gaang.org /Portals/0/JSTARS.jpg.

[22] Muehe, C., and M. Labitt, "Displaced-Phase-Center-Antenna Techniques," *Lincoln Laboratory J.*, Vol. 12, No. 2, 2000, pp. 281–296.

[23] Wang, H., and C. Lee, "Adaptive Array Processing for Real-Time Airborne Radar Detection of Critical Mobile Targets," *Proc. IEEE Adaptive Antenna Symposium*, Melville, NY, November 1992, pp. 95–100.

[24] Billingsley, B., et al., "Impact of Experimentally Measured Doppler Spectrum of Ground Clutter on MTI and STAP," *Proc. Int. Radar Conference*, Edinburgh, Scotland, October 1997, pp. 290–294.

[25] Ausherman, D., et al., "Developments in Radar Imaging," *IEEE Trans. on Aerospace and Electronic Systems*, Vol. AES-20, No. 4, July 1984, pp. 363–400.

[26] Walker, J., "Range-Doppler Imaging of Rotating Objects," *IEEE Trans. on Aerospace and Electronic Systems*, Vol. AES-16, No. 1, January 1980, pp. 23–52.

[27] *Aviation Week*, September 7, 1987, p. 88.

[28] Melvin, W., "A STAP Overview," *IEEE AES Systems Magazine*, Vol. 19, No. 1, January 2004, pp. 19–35.

[29] Barile, E., R. Fante, and J. Torres, "Some Limitations on the Effectiveness of Airborne Adaptive Radar," *IEEE Trans. on Aerospace and Electronic Systems*, Vol. AES-28, No.4, October 1992, pp. 1015–1032.

[30] Richardson, P., "Analysis of the Adaptive Space Time Processing Technique for Airborne Radar," *IEE Proc. on Radar, Sonar, and Navigation*, Vol. 141, No. 4, August 1994, pp. 187–195.

[31] *Space Time Adaptive Processing*, C&P Technology, Closter NJ, http://www.cptnj.com.

[32] Richardson, P., and S. Hayward, "Adaptive Space Time Processing for Forward Looking Radar," *Proc. IEEE Int. Radar Conference*, Washington, D.C., May 1995, pp. 629–634.

[33] Techau, P., J. Bergin, and J. Guerci, "Effects of Internal Clutter Motion on STAP in a Heterogeneous Environment," *Proc. IEEE National Radar Conference*, Atlanta, GA, May 2001, pp. 204–209.

[34] Applebaum, S., "Adaptive Arrays," *IEEE Trans. on Antennas and Propagation*, Vol. AP-24, No. 5, September 1976, pp. 585–597.

[35] Monzingo, R., and T. Miller, *Introduction to Adaptive Arrays*, New York: Wiley, 1980.

[36] Gokton, S., and E. Oruc, "Sidelobe Canceller Jamming Using Hot Clutter," Master's Thesis, Naval Postgraduate School, Monterey, CA , September 2004, pp. 1–111.

[37] Raven, R., "Requirements on Master Oscillators for Coherent Radar," *Proc. IEEE*, Vol. 54, No. 2, 1966, pp. 237–243.

[38] Skolnik, M., (ed.), *Solid State Transmitters*, Chapter 5 in *Radar Handbook*, 2nd ed., New York: McGraw-Hill, 1990.

[39] *Land Clutter Effects on Shipboard Radars*, TAD-PEO, April 11, 1996.

[40] Barton, D. K., and W. Barton, *Modern Radar Systems Analysis Software, Version 2.0*, Norwood, MA: Artech House, 1993.

[41] Schleher, D. C., *Introduction to Electronic Warfare*, Dedham, MA: Artech House, 1986.

[42] Nathanson, F., *Radar Design Principles*, New York: McGraw-Hill, 1969.

[43] Albersheim, W., "A Closed Form Approximation to Robertson's Detection Characteristics," *Proc. IEEE*, Vol. 69, July 1981, p. 819.

[44] Wanstall, B., "Joint Stars Fights to Stay on Target," *Interavia,* November 1988.

[45] Shmitkin, H., "Joint Stars Phased Array Antenna," *IEEE Systems Magazine,* October 1994, pp. 34–40.

[46] Tang, C., "Sensitivity of a Radar as Functions of Its Antenna Characteristics," *IEEE National Radar Conference,* 1993, pp. 250–255.

[47] Guerci, J., *Space-Time Adaptive Processing for Radar,* Norwood, MA: Artech House, 2003.

[48] Skulnik, M., (ed.), *Radar Handbook,* 2nd ed., Chapter 9, "Electronic-Counter-Countermeasures," New York: McGraw-Hill, 1990.

Index

F

G

H

J

K

L

M

Q

R

S

T

U

V

W

Recent Titles in the Artech House Radar Series

David K. Barton, Series Editor

Adaptive Antennas and Phased Arrays for Radar and Communications, Alan J. Fenn

Advanced Techniques for Digital Receivers, Phillip E. Pace

Advances in Direction-of-Arrival Estimation, Sathish Chandran, editor

Airborne Pulsed Doppler Radar, Second Edition, Guy V. Morris and Linda Harkness, editors

Bayesian Multiple Target Tracking, Lawrence D. Stone, Carl A. Barlow, and Thomas L. Corwin

Beyond the Kalman Filter: Particle Filters for Tracking Applications, Branko Ristic, Sanjeev Arulampalam, and Neil Gordon

Cognitive Radar: The Knowledge-Aided Fully Adaptive Approach, Joseph R. Guerci

Computer Simulation of Aerial Target Radar Scattering, Recognition, Detection, and Tracking, Yakov D. Shirman, editor

Design and Analysis of Modern Tracking Systems, Samuel Blackman and Robert Popoli

Detecting and Classifying Low Probability of Intercept Radar, Second Edition, Phillip E. Pace

Digital Techniques for Wideband Receivers, Second Edition, James Tsui

Electronic Intelligence: The Analysis of Radar Signals, Second Edition, Richard G. Wiley

Electronic Warfare in the Information Age, D. Curtis Schleher

ELINT: The Interception and Analysis of Radar Signals, Richard G. Wiley

EW 101: A First Course in Electronic Warfare, David Adamy

EW 102: A Second Course in Electronic Warfare, David L. Adamy

EW 103: Tactical Battlefield Communications Electronic Warfare, David Adamy

Fourier Transforms in Radar and Signal Processing, David Brandwood

Fundamentals of Electronic Warfare, Sergei A. Vakin, Lev N. Shustov, and Robert H. Dunwell

Fundamentals of Short-Range FM Radar, Igor V. Komarov and Sergey M. Smolskiy

Handbook of Computer Simulation in Radio Engineering, Communications, and Radar, Sergey A. Leonov and Alexander I. Leonov

High-Resolution Radar, Second Edition, Donald R. Wehner

Introduction to Electronic Defense Systems, Second Edition, Filippo Neri

Introduction to Electronic Warfare, D. Curtis Schleher

Introduction to Electronic Warfare Modeling and Simulation, David L. Adamy

Introduction to RF Equipment and System Design, Pekka Eskelinen

Microwave Radar: Imaging and Advanced Concepts, Roger J. Sullivan

Millimeter-Wave Radar Targets and Clutter, Gennadiy P. Kulemin

Modern Radar Systems, Second Edition, Hamish Meikle

Modern Radar System Analysis, David K. Barton

Modern Radar System Analysis Software and User's Manual, Version 3.0, David K. Barton

MTI and Pulsed Doppler Radar with MATLAB®, Second Edition, D. Curtis Schleher

Multitarget-Multisensor Tracking: Applications and Advances Volume III, Yaakov Bar-Shalom and William Dale Blair, editors

Principles of High-Resolution Radar, August W. Rihaczek

Principles of Radar and Sonar Signal Processing, François Le Chevalier

Radar Cross Section, Second Edition, Eugene F. Knott et al.

Radar Evaluation Handbook, David K. Barton et al.

Radar Meteorology, Henri Sauvageot

Radar Reflectivity of Land and Sea, Third Edition, Maurice W. Long

Radar Resolution and Complex-Image Analysis, August W. Rihaczek and Stephen J. Hershkowitz

Radar Signal Processing and Adaptive Systems, Ramon Nitzberg

Radar System Analysis, Design, and Simulation, Eyung W. Kang

Radar System Analysis and Modeling, David K. Barton

Radar System Performance Modeling, Second Edition, G. Richard Curry

Radar Technology Encyclopedia, David K. Barton and Sergey A. Leonov, editors

Range-Doppler Radar Imaging and Motion Compensation, Jae Sok Son et al.

Signal Detection and Estimation, Second Edition, Mourad Barkat

Space-Time Adaptive Processing for Radar, J. R. Guerci

Special Design Topics in Digital Wideband Receivers, James Tsui

Theory and Practice of Radar Target Identification, August W. Rihaczek and Stephen J. Hershkowitz

Time-Frequency Transforms for Radar Imaging and Signal Analysis, Victor C. Chen and Hao Ling

For further information on these and other Artech House titles, including previously considered out-of-print books now available through our In-Print-Forever® (IPF®) program, contact:

Artech House	Artech House
685 Canton Street	16 Sussex Street
Norwood, MA 02062	London SW1V HRW UK
Phone: 781-769-9750	Phone: +44 (0)20 7596-8750
Fax: 781-769-6334	Fax: +44 (0)20 7630-0166
e-mail: artech@artechhouse.com	e-mail: artech-uk@artechhouse.com

Find us on the World Wide Web at: www.artechhouse.com